International Farm Animal, Wildlife and Food Safety Law

Gabriela Steier • Kiran K. Patel
Editors

International Farm Animal, Wildlife and Food Safety Law

Springer

FOOD LAW
INTERNATIONAL
LLP
www.foodlawinternational.com

Editors
Gabriela Steier
Food Law International, LLP
Boston, MA
United States

Kiran K. Patel
Food Law International, LLP
Washington, D.C.
United States

ISBN 978-3-319-79243-9 ISBN 978-3-319-18002-1 (eBook)
DOI 10.1007/978-3-319-18002-1

Printed on acid-free paper

This Springer imprint is published by Springer Nature
The registered company is Springer International Publishing AG Switzerland

This Book is Dedicated to
Morrice, Our Love
Liviu, Regina, Fany and Michael
Kanti, Damyanti and Pritesh
And
All Those Beautiful Creatures of Nature
Whom We Wish to give a Voice
and Who Need to be Protected and Treasured

Foreword

In the decades following World War II, we entered into a time called the Cold War. As a small child during this time, I felt a sense of dread that the world I lived in could be destroyed, and I was powerless to eliminate that threat.

Yet, at the very same time, a much more insidious threat was taking hold, and it developed right here on American soil. The practices of industrialism were being applied wholesale to the production of food. Traditional farming, which relied on stewardship of the land, rotation of crops, and raising of animals in pastures, was being unceremoniously eliminated. Driven by technology, commercialism, and greed, agriculture turned into agribusiness. Massive industrial food production factories were built. We were told that this would make our food cheaper and more readily available to a growing human population, but few people considered, or cared about, the built-in downsides. And, as it turns out, the downsides are significant.

Farms that had been in existence for generations were shut down; what used to be the calling of thousands of families throughout America was consolidated into the hands of a few large corporations. The intimate relationship between the farmer, the land, and the food was destroyed.

The raising of farmed animals has changed more rapidly and radically in the years following WWII, than it had in the previous seven thousand years, both in the ways animals are selectively bred and in the ways they are housed and treated. Most of them are forced to live in a state of intensive confinement, in facilities labeled by the federal Environmental Protection Agency as "concentrated animal feeding operations (CAFOs)."

Farmed animals suffer greatly in this completely unnatural housing system; they are held in massive buildings or feedlots, either crowded together or placed into tiny crates, not much bigger than they are. They can't turn around or even lie down comfortably. They don't get to walk, eat grass, breathe fresh air, or do anything that is natural to their species. If you want to get a feel for their living conditions, imagine being strapped into a seat on an airplane with nothing to do, for life.

Today's intensively confined farmed animals are also subjected to painful mutilations, such as debeaking of chickens; dehorning, castration, and tail docking of cattle; and toe removal of turkeys, all done without anesthesia or painkillers.

The breeding practices of agribusiness are intended to maximize the "output" from these animals. Thus, turkeys have been bred to grow a massive breast, which is hard on their hearts, chickens raised for their meat grow so fast that their skeletons can't support this weight, and dairy cows have giant-sized udders to provide massive quantities of milk for human consumption, rather than for their own offspring, who are quickly removed from their mothers.

Quietly and without notice, farmed animals have been reduced to a Cartesian ideal and an animal activist's nightmare: treated as nothing more than unfeeling, unthinking machines. Any emotional connection to industrially raised animals has been severed; any ethical concern for their well-being has been eliminated. They suffer from the denial of all of their natural instincts and needs. Indeed, suffering is their lot for their entire short lives.

And, there are other significant problems. Traditionally, small farms recycled manure from the animals by spreading it on their crops for fertilizer. CAFOs produce so much manure that they overload the system. Most of them collect and store the animal waste in open lagoons or concrete cisterns. CAFO waste is usually not treated to reduce disease-causing pathogens nor to remove chemicals, pharmaceuticals, heavy metals, or other pollutants. Runoff into waterways, such as streams and rivers, can kill fish and contaminate human drinking water supplies.

CAFOs, particularly those raising hogs, also degrade the air quality and are notorious for their odor problems. They produce large amounts of hydrogen sulfide, ammonia, methane, nitrous oxide, and other harmful gases. This pollution comes from the buildings where the animals are housed. The air pollution inside the buildings causes animals to get sick and is potentially deadly to the animals and humans inside if the fans stop operating. Normally, the fans blow the contaminated air to the outside where it may cause health problems for nearby residents.

These impacts on the animals, water, and air quality are but some of the consequences associated with CAFOs. CAFOs are also linked to significant anthropogenic greenhouse gas emissions and climate change, degradation of soils from overfertilization, unsustainable use of water resources, and a significant loss of biodiversity. We have witnessed increased antibiotic resistance in humans due to the nontherapeutic use of antibiotics in the feed given to CAFO animals. Residents in the communities surrounding CAFOs are suffering damage to their health, lifestyles, and home values, and immigrants are exploited as low-paid laborers in terrible working conditions, with no ability to negotiate.

The CAFO approach even has impacts beyond all of this. Overfishing has decimated life in the oceans, and the rate of extinction for wildlife is deeply disturbing. Some, like Wendell Berry, poet, essayist, environmentalist, and traditional Kentucky farmer, began warning of these problems in the early 1970s, but the politicos have not listened. It is now being said that we are the first generation of the Earth's inhabitants who will leave this planet worse off for the future inhabitants than when we arrived.

My life's work has been in the area of litigation to protect animals, and the sad reality is that those of us who seek to protect farmed animals or improve the quality of their lives face a dismal landscape. There is no federal law in the USA that protects animals during the 99 % of the time they are being raised. The majority of state anti-cruelty laws exempt or exclude from their basic protections all farmed animals or the standard, common practices they suffer from. Animals are personal property under the law, and farmed animals are the property of the corporations that cause them intense suffering. These corporations have a stranglehold on legislators at the state and federal level, and litigation to directly impact the animals' well-being is not generally an option.

While environmentalists, food policy experts, and animal protectionists focus on different aspects of the problems caused by CAFOs, most agree that CAFOs represent a damaging and ultimately unsustainable way to raise food. We are finding common cause in a simple acknowledgment that what is bad for farmed animals is bad for the environment and bad for human health. We are questioning the meat industry's narrative that meat is an essential ingredient in our daily diet and asking questions about the ever-burgeoning human population and its impact on the planet. By working together, we hope to build a new way, a way that leads out of industrial food production and back to a reverence for the land, and for food, itself.

International Farm Animal, Wildlife and Food Safety Law brings together many of the leading voices informing us of the problems caused by CAFOs and pointing out solutions. As you will learn, the current food production system is bankrupt. It constitutes a major threat to our American way of life and, indeed, the health of the planet. To quote cartoonist and satirist, Walt Kelly, the creator of Pogo, "We have met the enemy and he is us."[1] As you read these offerings, consider what you can do in your own life, and in your own community, to counteract these problems. Consider what you can do to nurture a movement toward more holistic ways of growing food that is nutritious, wholesome, and ethical.

As Wendell Berry wrote:

> we clasp the hands of those who go before us,
> and the hands of those who come after us;
> we enter the little circle of each other's arms,
> and the larger circle of lovers whose hands are joined in a dance,
> and the larger circle of all creatures,
> passing in and out of life, who move also in a dance,
> to a music so subtle and vast that no ear hears it except in fragments.

I hope we hear it before it is too late.

Joyce Tischler
Founder and General Counsel
Animal Legal Defense Fund
Cotati, CA
United States

---

[1]Kelly included this comment in a cartoon, and it was used on a poster for the first observance of Earth Day in 1970.

Preface

This book is unique. It combines a range of topics but actually zooms in on some of the most pressing aspects of food law, namely, those touching the lives of farm animals and wildlife within the food system. Each part of this book focuses on a distinct set of legal concerns, starting with a basic introduction to farm and food animal law. Cutting-edge insights into industrial animal agriculture are presented in part one and followed by an outstanding set of chapters on marine animals and fishing. Of special importance is the section on wildlife protection, where emerging legal trends in soil conservation and pollinator protection are explained in greater detail.

Topical considerations of this breakthrough book include:

- Animal Welfare and Food Safety Legislation
- Environmental Protection and Clean Energy Overlaps with Animal Law
- Industrial Animal Agriculture Regulations
- Antibiotics Use in Meat Production
- Marine Animal Regulation and (Over-)Fishing Legislation
- Zoologic Diseases and Food Safety Management Legislation
- Pollinator Protection Policies
- Habitat Loss, Agrobiodiversity, and Incidental Wildlife Losses
- Food Policy and Animal Welfare Legislation
- Regulation and Trade Law Issues Focused on Genetically Modified Organisms (GMOs) and Pesticide Use in Agriculture
- Global Issues of Environmental Integrity Within the Law
- Evolving Issues Around the World

The final section provides tools for change, by summarizing, consolidating, and illustrating the statutory frameworks, international treaties, and national laws that practitioners, academics, and anyone inspired by this book will find useful (and necessary) to affect positive change. This final section is a one-of-a-kind consolidation of legal tools, and it is important from an environmental and clean energy perspective because environmental and food lawyers and those implementing laws need to know which tools are currently available to make a difference in food safety, animal welfare, and sustainable agricultural legislation and policy. The editors worked with a group from the Environmental Protection Clinic at the Yale Law School and the Yale School of Forestry and Environmental Studies to assemble the information for this section.

Lawyers can only make an impact in the globalized food production system if they have the proper tools and right priorities to affect change toward a more sustainable, safe, and animal-friendly food production system. With the herein proposed guide to international laws that touch upon the key issues, lawyers, regulators, administrators, and policy-makers around the world will have a reference point for the legal framework that is immediately available to be used to change the field of environmental and food law.

In an effort to make an impact and to help improve food systems, animal welfare protection, and environmental conservation, Food Law International LLP (FLI) started putting this book together. FLI was co-founded by Gabriela Steier and Kiran K. Patel, two attorneys from diverse professional backgrounds, with a passion for international environmental, animal welfare, food and agriculture and climate change law and policy within the context of sustainable food production. FLI's mission is the advancement and development of scholarship and interdisciplinary legal education on all aspects of international food law with the goal to promote environmental sustainability, public health, food safety, animal welfare, and social integrity. Co-founders Steier and Patel have cultivated a global network of authors who are advising and collaborating on FLI's projects and are committed to the same goals as FLI. This book is one of the ways by which FLI seeks to encourage positive change toward a more environmentally sustainable, climate change-resilient, kind, and responsible food system.

We hope this book will inspire conservation of biodiversity and natural resources and protection of animal welfare around the world.

Boston, MA, United States
Washington, D.C., United States

Gabriela Steier
Kiran K. Patel

Acknowledgments

We thank our academic advisors, Prof. William S. Eubanks II, Prof. Randall S. Abate, and Prof. Joyce Tischler; Atty. Zach Corrigan from Food and Water Watch in Washington, DC; our colleagues and friends at the Center for Food Safety in Washington, DC; and all of our accomplished colleagues who contributed to this breakthrough work of scholarship. Special gratitude goes to our mentors who have inspired us to pursue academic excellence and guided our intellectual and professional development: Dean and Prof. Jan M. Levine, Prof. Dr. Kirk W. Junker, Prof. Isabella Perricone, and Prof. Anne de Laire-Mulgrew. The authors who drafted chapters and textboxes for *International Farm Animal, Wildlife and Food Safety Law* are at the forefront of their field and helped us to create the book that we wished we could have had as students embarking upon the field of international food law. Thank you to everyone named and unnamed who supported the development of this book.

Our special thanks go out to our families, who supported and encouraged us during the months of planning, conceptualizing, and editing *International Farm Animal, Wildlife and Food Safety Law*. We especially thank Prof. Dr. Liviu Steier, Dr. Regina Steier, Mrs. Fany Sontag, Dr. Michael J. Nathenson, Mr. Kanti Nema Patel, Mrs. Damyanti Patel, Mr. Pritesh Patel, and, our dear friend, Atty. Alyssa Kaplan. Many heartfelt thanks extend to little Morrice Nathenson Steier for his patience while Mami was working on editing this book.

This book would not have come together as quickly without the help of our editing interns. We thank all of these young professionals and our students for their help and support in creating the book that we all wish we had had for our studies (in alphabetical order): Liza Howard, Christina Papillo, Jessica Rosenblatt, Kimia Shahi, and Daisy Zhang.

Contents

Part V Tools for Change: An Inventory of Global Farm Animal, Wildlife and Food Safety Laws

Editors and Contributors

About the Editors

Prof. Gabriela Steier, B.A., J.D., Esq. Prof. Steier is partner and co-founder of Food Law International (FLI), LLP, and editor in chief of the textbooks *International Food Law and Policy* and *International Farm Animal, Wildlife and Food Safety Law*.

She is an attorney licensed in the USA and focuses on food safety, agriculture law and policy, animal welfare, and GMO issues domestically and in the European Union. Prof. Steier has lectured on these topics and continues her research widely. She holds a B.A. from Tufts University and a J.D. from Duquesne University, and she is pursuing an LLM in Food and Agricultre Law at the Vermont Law School, as well as a doctorate in comparative law at the University of Cologne in Germany.

She worked as an LL.M. fellow in food and agriculture law at the Center for Agriculture and Food Systems at the Vermont Law School. Before joining the Vermont Law School, she completed a fellowship at the Center for Food Safety on Capitol Hill in Washington, DC, a national nonprofit public interest and environmental advocacy organization working to protect human health and the environment by promoting organic and sustainable agriculture. Prof. Steier will soon embark upon the next step in her career as a scholar, advocate and professor of food and environmental law related fields.

She is a distinguished legal scholar and has published widely on international food law, policy, and trade and has earned several awards for her work. Prof. Steier joined the Duquesne University School of Law in Pittsburgh, PA, as an adjunct professor teaching a breakthrough new course in "food law and policy" in 2015 as visiting professor at the University of Perugia, Italy, she also teaches EU-US comparative food law at the Department of Political Sciences and as an experienced editor and with her numerous publications ranging from peer-reviewed articles in international medical journals to law reviews, Prof. Steier has gained widespread interdisciplinary interest. Some of her articles have been on the Top Ten List on SSRN for several months.

Prof. Kiran K. Patel, B.Sc., J.D., LL.M., Esq. is a Partner and Co-Founder of Food Law International (FLI), LLP, and Executive Editor of International Food Law and Policy and International Farm Animal, Wildlife and Food Safety Law. Kiran is also a practicing U.S. patent attorney (licensed to practice in New York State and before the U.S. Patent and Trademark Office) as an Associate Attorney at Kramer & Amado, P.C., in the Washington, DC area. Prior to joining the firm, Kiran worked at boutique intellectual property firms in New York, NY and Washington, D.C. and in the intellectual property practice group of a large, full service law firm. Kiran offers strategic guidance and assistance in the preparation and prosecution of patent applications. He works closely with clients to develop and implement intellectual property strategies that help them to maximize their technological innovations. Kiran' s counsel extends to all stages of patent prosecution, including developing concepts into patentable products; preparing infringement and validity studies; filing for provisional and non-provisional patent applications; assisting in all stages of patent prosecution, and preparing patentability opinions. Kiran is particularly experienced in preparing and prosecuting patent s in various technological fields that span several technologies such as, electrical engineering, computer science and engineering, life science and chemical arts. He also helps clients negotiate licensing agreements and develop strategy for his client's patent portfolios. Kiran's training in biomedical engineering, electrical engineering and computer science gives him an in-depth understanding of the technologies behind his clients' inventions. He has worked with many different types of technologies including semiconductor devices, electronic circuitry, OLEO displays, color sampling algorithms, digital cameras and optical devices. Kiran has also advised clients in the biotechnology, pharmaceutical and medical device industries on gene expression, drug delivery compositions, RNA processing and polymer technologies. Kiran joined Duquesne University School of Law as an Adjunct Professor teaching the intellectual property and environmental law section in the course "Food Law and Policy". As Visiting Professor at the University of Perugia in Perugia, Italy, Kiran also teaches about the connections between food law and U.S. intellectual property law. Kiran has lectured on these topics and continues his research through FLI, LLP. Kiran earned his Bachelor of Science (B.Sc.) in biomedical engineering/ biology with honors from Drexel University in Philadelphia, PA. Subsequently, Kiran earned his Juris Doctor (J.D.) from Duquesne University School of Law in Pittsburgh, PA, on a McDonagh Scholarship and an Academic Award Scholarship. Kiran then earned his Master of Laws (LL.M.) in intellectual property law (specializing in patent law) from Benjamin N. Cardozo School of Law in New York, NY, as a Dean's Merit Scholar. Kiran is currently pursuing his Masters of Science (M.S.) in electrical engineering and computer science at Johns Hopkins University School of Engineering in Baltimore, Maryland. Kiran also serves on the Board of Directors for the Lupus Foundation of America, DMY and several other for-profit and advisory Boards. Kiran is admitted to practice in the Supreme Court of the State of New York (Third Judicial Department of the Appellate Division), the United States Court of Appeals for the Federal Circuit, and the United States District Courts for the Southern District of New York, the Eastern District of New York, the Western District of New York, the Northern District of New York and the U.S. Patent and Trademark Office.

Contributors

Irina Anta Irina Anta graduated from Yale Law School in 2015 and is currently an attorney in the International Department at Miller & Chevalier Chartered, a Washington, DC, law firm. She has always cared about animals and became particularly interested in farm animal issues in law school. During her time at Yale, Irina co-chaired the Student Animal Legal Defense Fund and spent a summer interning at Compassion Over Killing, a nonprofit, which works to fight animal abuse in agriculture.

Amy Atwood Amy Atwood, endangered species legal director and senior attorney, manages and carries out litigation for the center's Endangered Species Program, including efforts to gain protection for species under the Endangered Species Act and to ensure that endangered species are protected and recovered in their native habitats. Before joining the center in 2007, Amy worked as a staff attorney for the Western Environmental Law Center and as an associate attorney for Meyer & Glitzenstein (now Meyer, Glitzenstein & Crystal). She earned her law degree in 2000 from Vermont Law School and received a bachelor's degree in political science from the University of California at Berkeley in 1995.

Desmond Bellamy Desmond Bellamy (B.A. Hons Sydney University, B.Media Hons Southern Cross University) is the special projects coordinator for People for the Ethical Treatment of Animals (PETA) Australia. He is currently researching a Ph.D. dissertation on cannibalism as the process of animalizing humans, preparatory to the more commonplace process of deanimalizing meat.

Lewis Bollard Lewis Bollard is an animal advocate and third-year student at Yale Law School. He recently authored a paper entitled "Ag-Gag: The Unconstitutionality of Laws Restricting Undercover Investigations on Farms." The paper won the Public Justice Hogan/Smoger Access to Justice Essay Contest and was published in the October 2012 issue of the *Environmental Law Reporter*. Lewis has spent two summers with the Humane Society of the United States, most recently in animal protection litigation. He was president of the Yale chapter of the Student Animal Legal Defense Fund and a moot court semifinalist at the National Animal Law Competition. He is originally from Wellington, New Zealand.

Alexander Bruce Alex Bruce is an ordained Buddhist monk in the Tibetan Buddhist tradition and an associate professor with the ANU College of Law where he has taught since 1999. He is a senior fellow of the Higher Education Academy and a fellow of the Oxford Centre for Animal Ethics. Alex worked with the Australian Competition and Consumer Commission from 1992 until 2004 where he was a senior lawyer. In 2003, Alex was attached to the United Nations Conference on Trade and Development where he assisted the UN in capacity building and competition and consumer policies in several African countries. In 2004, Alex created and became a director of Liberation Prison Project (Aust) Ltd. and until 2009 chaplain at Junee Prison where he facilitated classes in philosophy and meditation for groups of prisoners. In 2007, Alex organized and moderated the

"One World – Many Paths to Peace" interreligious symposium with His Holiness the Dalai Lama and hosted by the ANU College of Law. Alex was awarded the 2007 Vice Chancellor's Award for Community Outreach. In 2009, HH Dalai Lama launched Alex's text "One World – Many Paths to Peace" at the Parliament of the World's Religions. In 2010, Alex was awarded the Vice-Chancellor's Award for Outstanding Contribution to Student Learning. His PhD (law) thesis explored the capacity for competition and consumer law and policy to benefit animals. Alex is on the International Advisory Board of the Research Unit for the Study of Law, Society and Religion at the University of Adelaide and is a director of Spiritual Care Australia Limited. Alex is a consultant editor with the *Oxford Journal of Animal Ethics* and a member of the editorial board of the *Australian Animal Protection Law Journal*.

Zach Corrigan Zach Corrigan is the senior staff attorney for Food & Water Watch. He has been an advocate for sustainability and environmental conservation for more than a decade. He began working on seafood safety issues in 2002 when he served as a staff attorney for the US Public Interest Research Group. There, he was a lead advocate for limiting people's exposure to mercury from contaminated fish by fighting for more stringent EPA rules to curb mercury emissions from power plants. In 2004, Zach became a legislative representative for Public Citizen and, shortly thereafter, Food & Water Watch. In May 2006, he became the staff attorney for the organization, working with all of its teams to pursue litigation, regulatory, and legislative solutions for all of the issues on which Food & Water Watch works. Zach graduated from the University of Wisconsin and earned his J.D. from Northeastern University School of Law in Boston. He is a licensed attorney in the District of Columbia and Massachusetts.

Leslie Couvillion Leslie Couvillion is a 2015 graduate of Yale Law School and the Yale School of Forestry and Environmental Studies. At Yale, she served as a teaching fellow for the Environmental Protection Clinic and a student director for the Nonprofit Organizations Clinic and the Visual Law Project (a student-run documentary film production group). Leslie received a B.A. in anthropology from Vassar College in 2008. Prior to law school, she worked at the Orleans Public Defenders office in New Orleans, LA.

Meghan Davis Meghan F. Davis, D.V.M., M.P.H., PhD, is an assistant professor in environmental health sciences at Johns Hopkins Bloomberg School of Public Health. Her research focuses on the intersection of human and animals, with a goal to combat the rise of bacterial antimicrobial resistance on local and global scales. Prior to her current position, she worked in companion- and food-animal general practice as an associate veterinarian. She received her D.V.M. from the University of California at Davis School of Veterinary Medicine in 2000 and her M.P.H. and PhD from Johns Hopkins Bloomberg School of Public Health in 2008 and 2012, respectively.

Thomas Faunce Thomas Faunce completed a PhD on governance of the Human Genome Project at the ANU in 2000 (examiners Prof. Edmund Pellegrino, Prof. George Annas, and Prof. Don Chalmers), and it was awarded the JG Crawford medal (equal best PhD in all fields at the ANU in 2001), named in honor of John Crawford (economist). This has now been published as "Pilgrims in Medicine" by Kluwer Law International (2003) and in which major concepts such as virtue ethics, human rights, and legalism were personified and the text presented in the context of their evolving careers in medicine. He was appointed to positions as a senior lecturer and then associate professor in the ANU College of Law and Medical School from 2004. At the ANU College of Law, he has taught "bioethics, health law, and human rights" in undergraduate and intensive graduate courses, as well as "lawyers justice and ethics" undergraduate course (2004–2006) and the "law reform and justice" course. He designed and has managed since inception the professionalism and leadership (PAL) theme of the ANU Medical School and served on senior committees related to governance of the medical school. His second book *Who Owns Our Health?: Medical Professionalism, Law and Leadership Beyond the Age of the Market State* (UNSW Press and Johns Hopkins University Press 2010) explored governance implications of the hypothesis that the world might one day have a fully privatized health-care system (reviewed by Leeder in MJA) (also reviewed in JAMA). His third book *Nanotechnology for a Sustainable World* (Edward Elgar 2012) (reviewed in World Future Review) made the case that the moral culmination of nanotechnology was global artificial photosynthesis. He was promoted to professor jointly to the ANU College of Law and College of Medicine, Biology and the Environment, by the central administration of the Australian National University in 2012. He was a founding member of the National Biosecurity Centre at the ANU, has served on the board of the ANU Energy Change Institute and the editorial board of the UK *Journal of Medical Humanities*, and edits the Medical Law Reporter for the Australian *Journal of Law and Medicine*. He serves on the ACT Civil and Administrative Appeals Tribunal in medical professional regulatory matters.

Clare Gupta Clare Gupta is a National Science Foundation SEES (Science, Education and Engineering for Sustainability) postdoctoral fellow, Clare Gupta recently completed an interdisciplinary research project that examined sustainability initiatives to "re-localize" agriculture in Hawaii, from a combined political and industrial ecological approach. More specifically, this research examined the different ways in which notions of food self-sufficiency and agricultural sustainability are articulated in Hawaii, the conditions that enable or disable re-localization of food production and the potential social and environmental impacts of re-localization. Previously, her dissertation work examined the implications of wildlife conservation for the livelihoods of rural communities living near protected areas in northern Botswana. This work drew from the fields of human geography and conservation studies to assess human migration patterns, agrarian change, and community-based conservation around Chobe National Park.

Ian Hannam Dr. Ian Hannam is associate professor of international environmental law development and research at the Australian Centre for Agriculture and Law, University of New England, Australia. He is chair of the Specialist Law Group for Sustainable Use of Soil and Desertification of the IUCN World Commission on

Environmental Law and has worked with many countries over many years on environmental law and policy reform.

Dena Jones Dena Jones is the farm animal program manager for the Animal Welfare Institute, a nonprofit organization that has been dedicated since its founding in 1951 to alleviating the suffering inflicted on animals by people. Major goals of the organization include supporting high-welfare farms and achieving humane slaughter and transport for all animals raised for food. Dena has 20 years of experience in animal advocacy. Her areas of expertise include farm animal practices and legal protections, food labeling claims, and public attitudes toward animals and the use of animals for food.

Susan J. Kraham Susan J. Kraham is a senior staff attorney and lecturer in law at Columbia Law School's Environmental Law Clinic. Susan has spent her legal career representing public interest clients with a particular focus on environmental and land use law. Prior to joining the Environmental Law Clinic, Susan served as counsel to the New Jersey Audubon Society. From 1998 until 2005, she was an associate clinical professor in the Environmental Law Clinic at Rutgers Law School, Newark. Susan was a 1992 graduate of Columbia Law School. She also has a masters in urban planning from New York University's Wagner School. After graduation from law school, Susan clerked for the Honorable Justice Gary Stein of the New Jersey Supreme Court. She was a Skadden fellow. Susan was also an Echoing Green fellow where she partnered on a community-based environmental justice project.

Patty Lovera Patty Lovera is the assistant director of Food & Water Watch. She coordinates the food team. Patty has a bachelor's degree in environmental science from Lehigh University and a master's degree in environmental policy from the University of Michigan. Before joining Food & Water Watch, Patty was the deputy director of the energy and environment program at Public Citizen and a researcher at the Center for Health, Environment and Justice.

Kennedy Mwacalimba Kennedy Mwacalimba holds a PhD in public health and policy from the London School of Hygiene and Tropical Medicine (2011). His core expertise is in the areas of policy analysis, epidemiology, public health, and risk analysis. His areas of interest include multi-sectoral risk management policy development, the sociology of risk, zoonoses risk assessment, cost-benefit analysis of zoonotic disease control, food safety, and international livestock and livestock product trade.

He is a former faculty member in public health at the University of Zambia, where he lectured in public health, epidemiology, livestock economics, environmental health, and food safety. He was also the course developer of the emerging and re-emerging diseases course and course leader of the health economics, policy, implementation, and evaluation course under the MSc in One Health Analytical Epidemiology hosted jointly by the Schools of Veterinary Medicine and Medicine of the University of Zambia. He is now an independent policy researcher based in

Indianapolis, Indiana, where he conducts independent medical and pharmaceutical policy, livelihood impact, and One Health research.

Nicole Negowetti She is Policy Director at the Good Food Institute. Previously, she was an assistant professor of law at Valparaiso University Law School. Prior to joining the Valparaiso faculty in 2011, Professor Negowetti practiced commercial and employment litigation at Sheehan Phinney Bass + Green PA in Manchester, NH. While practicing law in New Hampshire, Professor Negowetti taught upper-level writing courses as an adjunct professor at the University of New Hampshire School of Law. After graduating magna cum laude from Franklin Pierce Law Center (University of New Hampshire School of Law), Professor Negowetti clerked for the Honorable Carol Ann Conboy of the New Hampshire Supreme Court.

Kea Ovie Kea Ovie began with studies of law at the Georg-August-University of Göttingen (GER), she successfully completed a bank training. The focus of her studies at the university was public law, particularly the areas of environmental law and renewable energies. Since obtaining the first state exam in 2013, she works as a researcher at the Institute for Agricultural Law (IfLR) of the University of Göttingen. She writes her doctorate in the field of animal welfare legislation and belongs to the interdisciplinary and cross-faculty doctoral program "Animal Welfare in Intensive Livestock Production Systems," promoted by the state of Lower Saxony.

Aurora Moses is a clinical assistant professor of law with the Center for Agriculture and Food Systems at Vermont Law School. Previously she worked at the Center for Food Safety, litigating various sustainable agriculture issues, and was a judicial clerk at the Oregon Court of Appeals. Aurora holds an LL.M. in animal law from Lewis & Clark Law School, where she focused on animal agriculture; a J.D. and certificate in environmental law, also from Lewis & Clark Law School; and a B.A. in linguistics from Reed College. As a law student she was Editor in Chief of *Animal Law Review* and a research scholar in environmental and natural resources law. Aurora grew up in Montana.

Michelle Pawlinger Michelle Pawlinger is farm animal policy associate with the nonprofit Animal Welfare Institute. In this position, she works to improve the lives of animals through the regulatory system and urges companies and animal welfare certification programs to raise their animal care standards. She also lobbies Congress and the USDA. Last year, for example, she helped draft a regulatory petition asking the USDA to better regulate how they approve animal welfare label claims like "humanely raised." Michelle's passion developed from being raised on a horse farm in Miami. "Working and riding on a horse farm helped me understand that animals are not just cute, sweet creatures; they are individuals with unique personalities, like humans," she explains. This awareness inspired her to join a local animal group—Animal Activists of Alachua—while in college at the University of Florida. Through this activism, she learned how farmed animals are slaughtered by the billions each year. In response, she held demonstrations at farming events and vegan-cooking demos on campus and volunteered at farmed animal sanctuaries. "Being an animal lover means more to me than loving animals: it means fighting for them and their freedom." After college, she interned with the Humane Society of the United States in the government affairs department and then lived and worked at a farmed animal sanctuary. But she still wanted to help animals on a larger scale—and that's where law school came in. "Once I discovered Lewis & Clark Law School, I knew there was no other school for me. It had everything: an animal law program, a SALDF chapter, and a community of people dedicating their lives and careers to animals." At Lewis & Clark, Michelle immediately got involved with the Student Animal Legal Defense Fund (SALDF) chapter. She became the SALDF speaker series coordinator and later co-director of her chapter. She helped plan the 20th Annual Animal Law Conference, started a Meatless Monday campaign, and participated in Meat Out, an annual, internationally observed day that encourages people to eat cruelty-free food. She also served as submissions and associate editor for the *Animal Law Review*. In 2012, she was awarded an ALDF Advancement in Animal Law Scholarship. "I was honored to receive the scholarship," she says. "It gave me the freedom to delve into animal advocacy projects that I might not have had the opportunity to explore without the generosity of ALDF." She encourages new animal attorneys to be creative in their plans to help animals. "While in law school, our views often narrow regarding how we can help animals. We start to think that the only way to do so is through the law. We need to create new and exciting ways to help animals," she says. Further, she encourages lawyers to keep sight of the connections between the exploitation of animals and other injustices in the world. "In order to fully help animals we need to connect with other movements to truly make the world a better place for all beings." Michelle lives with her partner Jake in Washington, DC.

Lainie Rutkow Lainie Rutkow, J.D., PhD, M.P.H., is an associate professor of health policy and management at the Johns Hopkins Bloomberg School of Public Health and the assistant director of the Johns Hopkins Center for Law and the Public's Health. Dr. Rutkow uses legal, quantitative, and qualitative research methods to conduct policy analysis and evaluation in areas including chronic disease prevention and emergency preparedness. She is affiliated with the Eastern Region of the Network for Public Health Law. Dr. Rutkow has served as a fellow with the Committee on Government Reform in the US House of Representatives and has worked with the Legal Aid Society of New York's Health Law Unit and the New York Civil Liberties Union. She earned a B.A. from Yale University, a J.D. from New York University School of Law, and an M.P.H. and PhD, in health policy, from the Johns Hopkins Bloomberg School of Public Health.

Anastasia Telesetsky Professor Telesetsky joined the University of Idaho College of Law in 2009, after 8 years of practicing as an attorney in California, Washington, and abroad. Her practice focused on public international law and environmental law. She had the distinction of representing the government of Ethiopia before the Ethiopia-Eritrea Claims Commission at the Permanent Court of Arbitration. In 2003 and 2004, she was a Bosch fellow in Germany where she worked for the

German Foreign Ministry on promoting international food security and assisted in drafting guidelines on implementation for the "right to food." As a Fulbright fellow and a Berkeley Human Rights Center fellow, she collaborated with communities in the Philippines and Papua New Guinea on developing culturally appropriate legal solutions to environmental protection problems. In addition to bachelor's and master's degrees in anthropology, Professor Telesetsky earned her law degree from the University of California-Berkeley (Boalt Hall) and an LL.M. in international law from the University of British Columbia.

Paige Tomaselli Paige Tomaselli is a senior staff attorney at the Center for Food Safety, where she works on law and policy related to genetically engineered crops, organic standards, factory farming, and other food safety issues. Previously, she represented public water suppliers and public agencies in cases involving groundwater contamination and toxic torts at Sher Leff, LLP. She co-wrote a chapter in the *CAFO Reader: The Tragedy of Industrial Animal Factories*, entitled "Changing the Law: The Road to Reform," and has published several other articles on animal welfare and food safety. She frequently speaks at the premier sustainable agriculture and animal law conferences in the USA, and in 2013, she traveled to Japan to speak to the Japanese parliament and ministers of environment and agriculture on the impacts of genetic engineering. In 2011, Paige participated in the Permanent People's Tribunal Session on Agrochemical Transnational Corporations in Bangalore, India, where the attorneys presented dozens of cases illustrating how the sale and use of pesticides undermine internationally recognized rights to health, livelihood, and life to a panel of internationally recognized scholars and scientists. Paige holds a J.D. from Vermont Law School, where she was a member of the Environmental and Natural Resources Law Clinic, published an international comparative animal welfare article through the Animal Legal and Historical Center, and spent time at the University of Siena, Italy, studying international law.

Bernard Vanheusden Bernard Vanheusden specializes in public law and, more specifically, in environmental and energy law. He provides advice to, and represents, authorities, enterprises, and individual clients in legal proceedings.

Bernard Vanheusden is a doctor of law (Catholic University of Leuven, 2007), master of law (Catholic University of Leuven, 2000), and bachelor of philosophy (Catholic University of Leuven, 1996). He wrote his dissertation on brownfields development (Brownfields Redevelopment: naar een duurzame stadsontwikkeling. Rechtsvergelijkende analyse betreffende de sanering van sites—towards sustainable urban development. Comparative law analysis concerning site remediation).

Hillary Vedvig Hillary Vedvig holds a degree from the University of South Carolina in political science and will graduate from Yale Law School in 2017. She was in the eighth generation of her family to grow up on a family farm in Wisconsin. She has worked for the South Carolina State Senate, the Wisconsin State Assembly, and the Department of Justice. She ran a State Assembly campaign in northern Wisconsin and worked at a family law firm in Bozeman, Montana. In law school, she has worked on issues of international human rights, environmental law, and property.

Larissa Walker Larissa Walker is the pollinator campaign director and a policy analyst for the Center for Food Safety. In her role, she integrates national grassroots campaigns with hard-hitting scientific and legal expertise, working with lawmakers on Capitol Hill and regulators at key government agencies to affect positive policy change. Larissa spearheads CFS's pollinators and pesticides campaign, which focuses on protecting bees, butterflies, and other beneficial insects from the harms of pesticides and industrial agriculture. Larissa centered her academic career around environmental policy and theory, with a specific focus on sustainable agriculture and US food policy. She received her master's degree in environmental policy design from Lehigh University and holds bachelor's degrees in political science and philosophy. Larissa is originally from the Hudson Valley region of NY, and now as a resident of DC, she volunteers with FRESHFARM Markets, a regional nonprofit organization promoting local, sustainable food from the Chesapeake Bay watershed.

Heather Wong Heather Wong is a second-year J.D. candidate at Yale Law School, where she became involved with Food Law International's projects through the Environmental Protection Clinic. While in law school, she has also participated in projects to provide legal aid to refugees with the International Refugee Assistance Project and to indigent criminal defendants at New Haven Legal Assistance. She spent the summer after her first year of law school working on prisoners' and disability rights issues at the Texas Civil Rights Project in Austin. Prior to law school, Heather received a B.A. in philosophy and history from the University of Texas at Austin in 2013.

Sylvia Wu Sylvia Wu is a staff attorney at the Center for Food Safety, where she works on law and policy related to genetically engineered crops, factory farming, aquaculture, pesticides, and other food safety issues. Sylvia joined CFS as a full-time legal fellow in 2011, having previously worked at CFS as a law clerk. Sylvia holds a J.D. from UC Berkeley School of Law (Boalt Hall), where she authored a paper on the legal implications of the use of genetically engineered wine yeast in the US wine industry. During law school, Sylvia also worked as a kitchen intern at Corso Trattoria, learning the art of rustic Italian cooking. Sylvia is on the leadership team of Slow Food USA's East Bay chapter, as well as the board of directors for Planting Justice, an Oakland-based nonprofit organization, and is involved in various projects promoting local economy and urban agriculture in the East Bay.

Part I
Introduction to Farm and Food Animal Law

Chapter 1
Environmental Impacts of Industrial Livestock Production

Susan J. Kraham

Abstract Population growth, urbanization, changing economies and food preferences have increased pressure on the agricultural sector and on livestock production and related feed crops in particular. The FAO expects an increase of 70 % in world annual agricultural production from 2005/2007 to 2050 to feed the rising population, which is expected to grow by 40 % over the period (Conforti, Looking ahead in world food and agricultural perspectives to 2050, 2011). Much of the increase in crop (cereal) production is expected to come about as a result of increased demand for feed for livestock (Conforti, Looking ahead in world food and agricultural perspectives to 2050, 2011). To keep up with the demand for animal products, the method of production is changing. In the United States and increasingly around the world, family farms raising small numbers of livestock have given way to industrialized livestock practices often referred to as Concentrated Animal Feeding Operations or CAFOs. Livestock facilities confine ever increasing numbers of animals indoors. Vitamin supplements allow livestock to be confined indoors without sunlight and allow the production of offspring year round, while subtherapeutic use of antibiotics allow livestock to be confined in greater numbers and close quarters, raising the number of livestock that could be produced on a given feedlot or facility (Steinfeld, Livestock in a changing landscape: drivers, consequences, and responses, 2010). Genetics management and nutrition have also allowed animal production operations to intensify, and for the productivity of each animal to increase. For example, in the United States in 1957 it took a broiler chicken 101 days and 17.7 pounds of feed to reach market weight, while in 2001 it took only 32 days and only 5.9 pounds of feed. This has allowed US meat production to skyrocket by over 250 % over the past half-century (Pew Commission, Putting meat on the table: industrial farm animal production in America, 2008). Huge amounts of animal waste are a consequence of industrialized livestock. Inadequate regulation of manure deposition and disposal has resulted in significant air, water, and soil pollution. Animal waste from intensified operations is often disposed of on agricultural land year-round, and in far greater amounts than the land

S.J. Kraham (✉)
Columbia Environmental Law Clinic, Columbia University School of Law, New York, NY, USA
e-mail: skraha@law.columbia.edu

G. Steier, K.K. Patel (eds.), *International Farm Animal, Wildlife and Food Safety Law*, DOI 10.1007/978-3-319-18002-1_1

can absorb. Soils are over-fertilized thus releasing toxic runoff, and leaching contaminants. The runoff can flow into water bodies causing severe ecological harm, and decomposing waste can release dust particles, bacteria, endotoxins, and volatile organic compounds, as well as hydrogen sulfide, ammonia, and other odorous substances into the air (Halden and Schwab, Environmental impact of industrial farm animal production, 2008). Manure often contains many problematic substances including high levels of nitrogen and phosphorous, endocrine disruptors that can interfere with hormonal signaling in animals and humans, antibiotics that can nurture drug-resistant populations in the soil they are reach, resistant forms of bacteria, and arsenic (Halden and Schwab, Environmental impact of industrial farm animal production, 2008). As noted above, the increase in livestock production increases demand for feed crops thus requiring intensification of agricultural land use and resulting in a host of environmental costs on varying levels including increased erosion, lower soil fertility, reduced biodiversity, pollution of ground water, eutrophication of rivers and lakes, and impacts on atmospheric constituents, climate, and ocean waters (Steinfeld, Livestock's long shadow: environmental issues and options, 2006). This chapter will address those impacts. It is organized by medium of impact. Section 1.2 addresses air pollution and climate-change related impacts. Section 1.3 provides background on water consumption and pollution related to industrial livestock. Section 1.4 takes on the range of land-based impacts including habitat, forestry and desertification. The text provides an overview of the impacts but offers specific examples from a number of countries. Many of the impacts addressed are covered in more depth and/or with more specificity in later chapters.

1.1 Introduction

Population growth, urbanization, changing economies and food preferences have increased pressure on the agricultural sector and on livestock production and related feed crops in particular. The FAO expects an increase of 70 % in world annual agricultural production from 2005/2007 to 2050 to feed the rising population, which is expected to grow by 40 % over the period (Conforti 2011). Much of the increase in crop (cereal) production is expected to come about as a result of increased demand for feed for livestock (Conforti 2011).

To keep up with the demand for animal products, the method of production is changing. In the United States and increasingly around the world, family farms raising small numbers of livestock have given way to industrialized livestock practices often referred to as Concentrated Animal Feeding Operations or CAFOs. Livestock facilities confine ever increasing numbers of animals indoors. Vitamin supplements allow livestock to be confined indoors without sunlight and allow the production of offspring year round, while subtherapeutic use of antibiotics allow livestock to be confined in greater numbers and close quarters, raising the number of livestock that could be produced on a given feedlot or facility (Steinfeld et al. 2010).

Genetics management and nutrition have also allowed animal production operations to intensify, and for the productivity of each animal to increase. For example, in the United States in 1957 it took a broiler chicken 101 days and 17.7 pounds of feed to reach market weight, while in 2001 it took only 32 days and only 5.9 pounds of feed. This has allowed US meat production to skyrocket by over 250 % over the past half-century (Pew Commission 2008).

Huge amounts of animal waste are a consequence of industrialized livestock. Inadequate regulation of manure deposition and disposal has resulted in significant air, water, and soil pollution. Animal waste from intensified operations is often disposed of on agricultural land year-round, and in far greater amounts than the land can absorb. Soils are over-fertilized thus releasing toxic runoff, and leaching contaminants. The runoff can flow into water bodies causing severe ecological harm, and decomposing waste can release dust particles, bacteria, endotoxins, and volatile organic compounds, as well as hydrogen sulfide, ammonia, and other odorous substances into the air (Halden and Schwab 2008). Manure often contains many problematic substances including high levels of nitrogen and phosphorous, endocrine disruptors that can interfere with hormonal signaling in animals and humans, antibiotics that can nurture drug-resistant populations in the soil they are reach, resistant forms of bacteria, and arsenic (Halden and Schwab 2008).

As noted above, the increase in livestock production increases demand for feed crops thus requiring intensification of agricultural land use and resulting in a host of environmental costs on varying levels including increased erosion, lower soil fertility, reduced biodiversity, pollution of ground water, eutrophication of rivers and lakes, and impacts on atmospheric constituents, climate, and ocean waters (Steinfeld et al. 2006b).

This chapter will address those impacts. It is organized by medium of impact. Section 1.2 addresses air pollution and climate change related impacts. Section 1.3 provides background on water consumption and pollution related to industrial livestock. Section 1.4 takes on the range of land-based impacts including habitat, forestry and desertification. The text provides an overview of the impacts but offers specific examples from a number of countries. Many of the impacts addressed are covered in more depth and/or with more specificity in later chapters.

1.2 Air Impacts

Animal agriculture produces air pollution and contributes to global climate change. The sources of the pollution are varied. U.S. industrial farms produce more than 400 different types of gases—the odor from swine manure alone contains 331 separate chemical compounds. These emissions include several gases, such as hydrogen sulfide and ammonia, which are hazardous to human health. Moreover, the

400 figure does not include the dust, particulate matter, and endoxins that are released in the course of industrial farming (Sustainable Table 2009).

Perhaps more significant are the ways in which animal agriculture contributes to global climate change through the emission of greenhouse gases. Methane, nitrous oxide, and carbon dioxide are all produced as byproducts of industrial animal agriculture. In total, these gases account for approximately 14.5 % of global greenhouse gas emissions, the equivalent of 7.1 billion tons of CO_2 (Gerber et al. 2013). According to the US Environmental Protection Agency, the agriculture sector was responsible for 7.6 % of US greenhouse gas emissions in 2013 (EPA 2015). The gases come from a myriad of processes, including methane released by animal digestion (enteric fermentation) and the nitrous oxide emitted as a result of nitrogenous fertilizers used on animal feed crops (O'Mara 2011). The decomposition of animal manure also releases methane, nitrous oxide, and carbon dioxide (O'Mara 2011; Sustainable Table 2009). Additionally, animal agriculture contributes CO_2 to the atmosphere through deforestation and fossil fuel use.

1.2.1 Air Pollution

The major issue is the scale of industrial livestock operations today. The huge quantities of animal manure produced by these operations are the largest contributor to factory farm air pollution. According to the USDA, approximately 335 million tons of manure are produced annually on farms in the United States. That manure is stored in tanks and lagoons and then sprayed onto farm fields. As it decomposes the manure releases toxic chemicals into the air. Because the waste sites are often located next to the farms, employees, livestock, and neighboring residents are all exposed to the airborne chemicals. "Hydrogen sulfide, methane, ammonia, and carbon dioxide are the major hazardous gases produced by decomposing manure" (Sustainable Table 2009). Ammonia and hydrogen sulfide are discussed in this section. Since they are green house gases in addition to pollutants, carbon dioxide and methane are discussed in the following section on climate change.

1.2.2 Ammonia: NH_3

Ammonia, or NH_3, is considered a nitrogen (N) fertilizer and a valuable resource. However, NH_3 that escapes into the atmosphere at excessive levels can have negative effects on air quality, ecosystem productivity, and human health (Bittman and Mikkelsen 2009). Livestock production is the largest source of ammonia emission in North America. Livestock manure, specifically, accounts for about 80 % of U.S. ammonia emissions. In North Carolina the hog industry alone produces more than 300 tons of ammonia each day. Most ammonia comes from chicken and hog facilities, and is produced during the decomposition of organic nitrogen compounds in manure (Sustainable Table 2009). Excessively nitrogen-rich feed may not be completely converted into animal byproducts (e.g. eggs, milk, etc.) during the digestive process, with the remainder of nitrogen excreted in urine and manure. The chemical and microbial processes of decomposition then break down the waste, releasing NH_3 into the air. Although barns capture some of the ammonia, that amount is usually released when the manure is applied to the land. Liquid manure has a high initial NH_3 loss after application and, in general, a greater application rate corresponds with a greater rate of loss of N compounds. Additionally, elevated methane from digestion tends to produce a higher pH in manure residue, which affects a higher NH_3 loss. Of the NH_3 lost to the atmosphere, about 40–50 % is from animal housing, 5–15 % from storage, and 40–55 % from land application (Bittman and Mikkelsen 2009).

Atmospheric ammonia has hazardous effects on both animal and human health. NH_3 is an alkaline compound, meaning it is easily absorbed into surfaces. The NH_3 that remains in the atmosphere reacts with acidic compounds (e.g. nitric and sulfuric acids), which forms secondary airborne particulates. These particulates are dangerous because their small size—a diameter of less than 2.5 μm (or PM 2.5) means they are inhaled deeply into the lungs (Bittman and Mikkelsen 2009). Exposure to ammonia can cause irritation of the eyes, skin, and respiratory and cardiovascular tracts and is especially damaging to employees (Sustainable Table 2009).

NH_3 also, notably, has detrimental effects on the environment—both by the formation of fine particulate matter (PM) and by uncontrolled Nitrogen (N) deposition. The PM contributes to atmospheric haze, which occurs when sunlight hits the airborne particles. Although the majority of the damage occurs near livestock production sites, particulate ammonia is also carried by the wind and redeposited in environments that may otherwise have remained pristine; this, in turn, has negative effects on the ecosystem. NH_3 that hits the soil is converted into Nitrate (NO_3^-), a process that releases acidity (H^+). Acidification of soil damages sensitive vegetation, such as lichens and bryophytes. NH_3 that falls on plants, either entering directly through the stomata or as NH_4^+ (via dissolution in water), will be also excrete H^+ through the plant roots. As a fertilizer, NH_3 can increase growth of plants with high-N demand, which has a destabilizing effect on naturally occurring plant species (Bittman and Mikkelsen 2009). In water, increased Nitrogen can promote eutrophication (excessive plant and algae growth), which harms water quality and impacts biodiversity (Chislock et al. 2013).

1.2.3 Hydrogen Sulfide: H_2S

Hydrogen sulfide is primarily a byproduct of hog farming. Like ammonia, this pollutant comes from livestock manure. As animal waste decomposes, sulfur-containing organic matter produces hydrogen sulfide. Hydrogen sulfide is hazardous to heath because it limits the ability of cells to use oxygen. Exposure to high levels of hydrogen sulfide can cause neurologic and cardiac disorders, seizures, comas, and even death. Lesser reactions include skin, eye, and respiratory irritation (Sustainable Table 2009).

The effects of these pollutants on employees are measurable. About 70 % of CAFO employees suffer from acute bronchitis and 25 % from chronic bronchitis. For some employees the effects of dust and gas inhalation are deadly. In one 5-year study, at least 12 employees died from asphyxiation working in manure pits (Sustainable Table 2009).

Residents of communities adjacent to factory farms are also put at risk by airborne pollutants. Hospitalizations increase in people living near farms. One study watched as a hog facility opened near one town in Utah: diarrhea-related hospitalizations increased fourfold and respiratory-related hospitalizations increased threefold over a 5 year period. In Minnesota, the Pollution Control Agency measured hydrogen sulfide concentration in the air of neighborhoods adjacent to industrial hog farms; the amount exceeded the maximum exposure set by the World Health Organization. In Iowa, a 2006 study compared the health of children at elementary schools—one adjacent to a CAFO, and one not. Children who attended the CAFO-adjacent school were significantly more likely to suffer from asthma. The effects of airborne pollutants can be felt in more unexpected ways, as well. Psychological problems, like depression and mood swings, as well as fatigue, are also associated with the airborne pollutants coming off CAFOs (Sustainable Table 2009).

1.2.4 Particulate Matter

As CAFOs expand the scale of their operations, particulate matter (PM) becomes an increasingly threatening pollutant. CAFOs located in arid and semi-arid environments are the most conducive to dust emissions. Emissions in these climates often follow a diurnal pattern known as evening dust peak (EDP)—meaning emissions are at their highest after sunset (Sakirkin et al. 2012).

Particulate matter is measured in terms of its aerodynamic diameter. Most relevant to CAFO pollution is fine particulate matter, which has a diameter of 2.5 μm or less. $PM_{2.5}$, as it is called, is the most threatening to human health, because it is inhaled deeply into the lungs and, as a result, causes respiratory problems (Sakirkin et al. 2012). PM can also cause haze, reduce visibility, and carry bad-smelling odors (Bittman and Mikkelsen 2009).

There are several sources of dust emissions on livestock feedlots. Cattle walking on uncompacted manure, vehicles driving on unpaved roads, hay grinding, grain delivery, and combustion of gases and fuels all emit PM. These types of emissions are considered primary PM, or fugitive dust, which is caused directly by mechanical or chemical processes. When atmospheric conditions are stable these ground-level emissions remain low to the ground (a phenomenon known as "inversion"). In order to limit the amount of primary PM released on feedlots, uncompacted manure should be regularly removed from corral surfaces—the thicker the layer of manure, the higher the potential for dust emissions. Cattle, for example, drag their rear hooves when they walk; the deeper the hoof penetrates into the uncompacted manure, the more PM is released into the air. Alternatively manure kept around a 30 % moisture content also means less PM churned up from hoof contact (Sakirkin et al. 2012).

Secondary PM is formed in the atmosphere. In CAFO operations, secondary PM is most often a result of NH_3 (ammonia) reacting with sulfate, nitrate, and/or chloride ions in the atmosphere. Both primary and secondary PM negatively affect environments beyond their immediate vicinity, and both carry consequences for climate change. In Clean Air Act terminology, primary and secondary PM and fine PM are regulated as one of the six major criteria pollutants with set National Ambient Air Quality Standards (NAAQS). Malodorous PM may also be regulated under the doctrine of nuisance law (Sakirkin et al. 2012).

1.2.5 Climate Change

In 2013, the Food and Agriculture Organization of the United Nations (FAO) reported that green house gas emissions from livestock represent 14.5 % of all human induced emissions, the equivalent of 7.1 billions tons of CO_2 annually (Gerber et al. 2013). Under a business as usual scenario, annual agricultural emissions are projected to grow to 8.2 billion tons of CO_2 equivalent by 2030 (O'Mara 2011). Livestock emissions come in the form of methane, nitrous oxide, and carbon dioxide. The discussion below outlines the major emission pathways for these greenhouse gases.

1.2.6 Methane: CH_4

With a global warming potential 23 times more potent than carbon dioxide, methane is a significant contributor to climate change (Sustainable Table 2009). According to the FOA, methane (CH_4) accounts for about 44 % of the green house gas emissions produced by livestock supply chains, more than any other single source. The amount of methane released is the equivalent of about 3.1 billion tons of CO_2 and accounts for 44 % of all anthropogenic methane emissions. The primary source of CH_4 by far is a digestive process called enteric fermentation—it alone accounts for nearly 40 % of *total* emissions (Gerber et al. 2013). Enteric fermentation is part of the process by which ruminant animals, such as cattle, digest plant materials. In enteric fermentation anaerobic microbes decompose (and ferment) food present in the animal's rumen. This process breaks the food down into simple molecules, allowing ruminants to digest complex carbohydrates that other non-ruminant animals cannot (Pew Center 2009; Gerber et al. 2013). One byproduct of this process is methane.

This situation in exacerbated by the low-quality grain-based feed used by commercial farms today (Sustainable Table 2009). Although such feed fattens their livestock quickly and inexpensively, ruminants are not able to digest it easily, causing the animals to emit more methane per unit of energy ingested (Sustainable Table 2009; Gerber et al. 2013).

Animal manure is also a major source of methane emissions: as organic material in the manure decomposes, some of it is converted into CH_4. According to the FAO, this "occurs mostly when manure is managed in liquid form, such as in deep lagoons or holding tanks" (Gerber et al. 2013). Nonetheless, some methane is also released from the deposition of manure on pastures (O'Mara 2011). In 2013, the FAO estimated that each year manure management contributes enough CH_4 to the atmosphere to account for 4.3 % of all livestock section emissions, as compared with a 39.1 % share contributed by enteric fermentation, making manure a relatively smaller but still significant source of methane pollution (Gerber et al. 2013).

1.2.7 Nitrous Oxide: N_2O

With a global warming potential 310 times greater than carbon dioxide, nitrous oxide (N_2O) accounts for about 29 % of greenhouse gas emissions from animal agriculture. The amount of nitrous oxide released is equivalent to 2 billion tons of CO_2 annually and accounts for 53 % of all anthropogenic N_2O emissions. Like methane, manure is a significant source of nitrous oxide emissions. During manure management—i.e., the storage and processing of manure—N_2O is produced "as part of the N cycle through the nitrification and denitrification of the organic N in livestock dung and urine" (EPA 2015). Manure management is also responsible for indirect N_2O emissions, as when animal waste releases nitrogen into the atmosphere as ammonia (NH_3) that can later transform into N_2O. According to recent FAO estimates, nitrous oxide emissions from manure management account for 5.2 % of all greenhouse gas emissions from livestock supply chains (Gerber et al. 2013).

More significant in terms of nitrous oxide pollution are the emissions related to the production of livestock feed. First, N_2O is released from manure applied to pastures and feed crops. The use of nitrogenous fertilizers on feed crops also significantly increases the amount of mineral nitrogen available in soils, and thus the amount of N_2O produced naturally by the N cycle (EPA 2015). According to one study, the use of synthetic fertilizers was responsible for 68 % of all US nitrous oxide emissions in 2004 (Sustainable Table 2009). Together, nitrous oxide emissions related to feed production accounts for about one quarter of all livestock greenhouse gas emissions (Gerber et al. 2013).

1.2.8 *Carbon Dioxide: CO_2*

The carbon dioxide released in livestock production accounts for about 27 % of sector emissions and 5 % of all anthropogenic CO_2 emissions. The largest source of CO_2 emissions, about 20 %, is from the burning of fossil fuels along all stages of the supply chain. Fossil fuels are used in the manufacture of fertilizers, as well as by the machinery used to manage, harvest, process, and transport feed. Further down the supply chain at the animal production site, fossil fuels are needed to run mechanized operations (e.g., heating and ventilation systems), as well as in the construction of buildings and equipment. When the animals are slaughtered, fossil fuels are needed to process and transport the animal products to retailers (Gerber et al. 2013).

The other major source of CO_2, about 9 % of total emissions from livestock supply chains, is from land use change. According to the FAO, these emissions come mainly from the conversion of natural habitats for pasture and feed crops (chiefly soybeans). The expansion of feed crops and pasture "causes the oxidation of C in soil and vegetation" (Gerber et al. 2013). Deforestation also contributes to climate change because forests play an important role in the carbon cycle by absorbing carbon. One study found that tropical forests absorbed 1.4 billion metric tons of carbon dioxide out of 2.5 billion metric tons absorbed annually (Schimel et al. 2014).

1.3 Water Impacts

Industrial livestock has contributed to the degradation of the world's water resources in two principle ways: consumption and pollution. Consumptive use includes direct animal consumption, feed crop irrigation, and the use of water for processing and other hygiene and safety requirements. Industrial livestock is one of the leading causes of water pollution globally. Animal waste, fertilizer runoff from feed crops, and other sources of wastewater all result in significant adverse impacts to water resources.

1.3.1 *Consumptive Water Use*

The agricultural sector uses more freshwater than domestic and industrial sectors combined (Steinfeld et al. 2006b). Within the agricultural sector, water used for livestock production constitutes nearly one-third of water use (Swanepoel et al. 2010). Most of the water used in livestock production is used for irrigating feed crops, but some water is also used in caring for livestock and processing the animal products. This heavy use of freshwater resources exacerbates water scarcity in many regions—a problem that also is intensified by the effects of climate change (Pimental et al. 1997).

1.3.2 Animals: Care and Processing

Animals must be provided adequate water. A reduction in their water intake can reduce meat, milk, and egg production, and may also lead to health concerns and death. The amount of water used to care for animals will depend greatly on the location and conditions of the facilities. For example, confined animals may consume less water than free-ranging ones because of their lower activity level, yet they may need more water for cooling where the facility's temperatures are high (Steinfeld et al. 2006b).

Water is also a major input when the animal or the animal product moves beyond the farm to the processing facility. The amount of water necessary to process the animal products depends on the products and the methods of production. For example, the processing of poultry generally uses more water than the processing of red meat, in part due to the procedures required to defeather the animals. Local regulations regarding hygiene and quality in food processing activities generally can also increase requirements for water use (Steinfeld et al. 2006b).

1.3.3 Feed Production

The majority of water consumed in livestock production is used to grow feed crops. However, the efficiency with which water is used for feed production depends on the type of livestock and crops and the manner in which the livestock and the crops are produced (Swanepoel et al. 2010). For example, the amount of water consumed in a livestock production system relative to the amount of water available in an area will differ greatly between systems that principally rely on rainfed crops from those which principally rely on irrigated feed crops. Concentrated Animal Feeding Operations rely primarily on irrigated feed crops rather than grazing. Where irrigated feed crops are produced in areas with a shortage of water, it leads to additional water depletion and creates competition with important water uses (Steinfeld et al. 2006b).

The water used for feed crop production contributes to problems with effects beyond increasing water scarcity. Water consumption for feed crop production reduces the amount of water available to a natural ecosystem, contributing to the loss of ecosystem services, the loss of biodiversity, and the degradation of habitats. For example, excessive withdrawals of surface water for irrigation may reduce river flow, jeopardizing wetlands habitats and aquatic species downstream. Additionally, repeated application of freshwater can lead to salinization. Although there is only a small amount of salt in the water used for irrigation, this salt content accumulates affecting the efficiency of plant growth and eventually causing soil infertility. Another concern is waterlogging. When irrigated croplands are not properly drained, excess water can become trapped in the soil. Severe waterlogging displaces oxygen in the soil, killing plant roots and soil microorganisms (Molden 2007).

1.3.4 Case Study: Incentivizing the Efficient Use of Water

Increased demands for water for food production and other uses, combined with increased water stress due to climate change, require the agricultural sector to efficiently use water resources if demand is to be met with the current limited supply. Current water pricing structures, which tend to subsidize or undercharge CAFO's for water, encourage the inefficient use of water (Pimental et al. 1997). In order to promote the efficient use of water, some regions have implemented a pricing structure that incentivizes water conservation.

In the Paraíba do Sul River Basin of southeast Brazil, gradual price increases for water motivated consumers to adopt water efficient technologies and processes and provided an additional source of income that could be invested into the management of the watershed (UNEP 2014). The water pricing reforms that were instituted in Brazil in the late 1990s focused on the emerging practice of pricing water as a resource, also known as "bulk or 'wholesale' water pricing," rather than pricing the service of providing water to consumers, also known as "retail water supply and distribution" (Asad et al. 1999). By setting a bulk price for water, it should encourage an economically efficient allocation of water between agricultural, domestic, and industrial uses, and incentivize also consumers to efficiently use water. Additionally, water pricing systems that internalize the costs of cleaning water encourage the reduction of water pollution. Although water pricing may encourage these benefits, they often are not prioritized over the main goal of generating revenue (Asad et al. 1999). Where there is a focus on revenue generation, it is especially important that water-pricing policies are designed and applied in a way that does not negatively affect the poor by depriving them of meaningful access to this essential resource (Swanepoel et al. 2010).

The relative success in Brazìl at incentivizing the efficient use of water and discouraging non-compliance through water pricing is attributed to four factors. First, the negotiation process with the public rather than a top-down implementation allowed users from all sectors to contribute to the plan. Second, the revenues raised through the water pricing system is required to be reinvested in the river basin. Third, there was an emphasis on social responsibility, including a focus on incentives rather than sanctions to encourage cooperation with the system. Lastly, there was a strong focus on capacity building to implement the project (Formiga-Johnsson et al. 2007).

1.3.5 Water Pollution

Industrial agriculture has led to an increase in water pollution. Most significantly, industrial agriculture produces large quantities of animal waste that cannot be absorbed by the farms that produce them. The various methods used to store or dispose of this waste are often mismanaged in a way that leads directly to water pollution. However, the mismanagement of animal waste is not the only aspect of livestock production that contributes to water pollution. Other elements of the production process that also cause water pollution include the generation of waste-water in the care of live animals or the production of animals products, the increase in soil erosion, and the use of chemical fertilizers and pesticides in feed crop production. In the United States, most of these pollution sources are regulated under the Clean Water Act. However, the effects of these pollutants could also be reduced through improved waste, wastewater, and land management techniques that reduced the amount of pollutants that enter the environment in the first place.

1.3.6 Livestock Waste

The most significant impacts on water pollution from industrial agriculture are caused by the excessive production and mismanagement of livestock waste. Traditionally, livestock and crops were raised in an integrated system, where livestock waste was utilized as a fertilizer resource in crop production and crop wastes were used as livestock feed. However, the increased intensification of livestock production in CAFOs has created a system where the production of livestock and crops has been separated. Although CAFOs continue to apply livestock waste to crops as fertilizer, the amount of waste created on a given farm can no longer be absorbed by the surrounding cropland.

Livestock waste includes a number of contaminants, such as nutrients, heavy metals, pathogens, antibiotics, and hormones, each of which causes its own harmful effects on the environment and human health (Burkholder et al. 2007). Many of these contaminants are intentionally given to the livestock in their feed or as part of a medical treatment to promote growth. However, when the contaminant is not fully absorbed or deteriorated within the animal, then it is excreted in the animal's waste. For example, nutrients such as nitrogen and phosphorous are natural elements that are found in animal feed. However, the amount of nutrients fed to an animal often exceeds the amount that the animal can efficiently absorb. As such, large quantities of these nutrients are still present in the animal's waste. When the waste is not properly disposed of, these contaminants enter surface water and groundwater systems via runoff or leaching due to precipitation after ground application, and leaks or other failures of storage facilities.

Large amounts of nutrients, mainly nitrogen and phosphorus, are introduced into the environment via animal waste each year. Some nutrients created by livestock production can be reused as fertilizer by applying the waste to crop fields, as these same nutrients are necessary for plant growth. However, the amount of nutrients produced usually far exceeds what the operation's land can reabsorb (Vanotti and Szogi 2008). Where there is a surplus of nutrients that cannot be absorbed, it is far more likely that the nutrients will enter the environment via runoff or leaching. When nutrients enter the water, they can lead to eutrophication, which is excessive plant and algae growth. In freshwater and marine ecosystems, eutrophication leads to an overconsumption of oxygen, unappealing flavor and odor of the water, and increased bacterial growth. The overconsumption of oxygen is the most significant of these impacts, as it can alter the balance of plant and animal species in an ecosystem and increase the production of toxins by algae, interfering with human utilization of the water course for recreational or commercial purposes. Phosphorus has not been linked to any direct negative effects on human health. However, high levels of nitrate, a certain form of nitrogen, can pose a risk to human health, poisoning infants and causing abortions and stomach cancers in adults (Steinfeld et al. 2006b).

Heavy metals in livestock production pose many of the same problems as nutrients. Like nutrients, heavy metals are naturally occurring substances that plants and animals require in certain levels for growth; however, the presence of high levels of heavy metals may harm the ecosystem and human health. Heavy metals, such as "copper, zinc, selenium, cobalt, arsenic, iron, and manganese," are commonly added to livestock feed to promote health and growth. However, as explained above, the animals do not absorb the full amount consumed, meaning that most heavy metals are reintroduced to the environment via livestock waste. Since heavy metals are not degradable, they can remain in the ecosystem indefinitely and bioaccumulate through the food chain. Human exposure to high levels of heavy metals has been linked to cancer, anemia, delays in growth, cardiovascular and neurological problems, and many other problems (Vasey et al. 2011).

Livestock waste also contains large amounts, in volume and variety, of bacteria, viruses and other parasites that can affect human health (Burkholder et al. 2007). Some of these pathogens and their effects are well known, such as Camplyobacter, E. Coli, Salmonella, Picornavirus (foot-and-mouth disease), Parvovirus, and giardia lamblia. Each pathogen has its own method of transmission, but some common methods of transmission to humans include transmission via contaminated water, food washed with contaminated water, or food that is improperly prepared. Some of these pathogens may harm livestock as well as humans (Steinfeld et al. 2006b).

The potential for pathogen related illnesses is one of the reasons why livestock production heavily utilizes antimicrobials, including antibiotics. These pharmaceuticals are used for "therapeutic purposes" to treat illnesses, "prophylactically" to prevent illnesses during stressful events, and "routinely . . . to improve growth rates and feed efficiency" (Steinfeld et al. 2006b). However, antimicrobials are not entirely degraded within the animals, and therefore end up in the environment; antimicrobials have been found in groundwater, surface water, and tap water. The use of antibiotics for non-therapeutic purposes has been linked to increased antibiotics resistance of the pathogen species present in the livestock population and the waterways polluted with livestock waste (Burkholder et al. 2007).

Hormones are another pharmaceutical that may be used to increase the efficiency of livestock feed conversion. Like antimicrobials, a significant portion of the hormones used is excreted in livestock waste and can be found in groundwater, surface water, and tap water. Hormones have not been proven to cause negative impacts on human health or the environment, but they have been suggested as an explanation for "developmental, neurologic, and endocrine alterations" in wildlife (Steinfeld et al. 2006b).

1.3.7 Animal Care and Processing

There are many other activities related to livestock production that contribute to water pollution, including the care of animals and the process of animal products. For example, the production of dairy products uses a significant amount of detergents and disinfectants. Also, wastewater from slaughterhouses and meat-processing plants contains high levels of contaminants, such as blood, fat, and solid waste, that could have negative effects on the environment if not properly treated. Furthermore, regulations regarding hygiene and food safety may create additional requirements that increase the amount of wastewater that is produced by these activities (Steinfeld et al. 2006b).

1.3.8 Soil Erosion

Another major cause of water pollution is soil erosion. Livestock production can be linked to soil erosion directly through livestock impacts on grazing lands. The impacts of the animals' hooves can cause compaction of wet soil, loosening of dry soil, the destabilization of stream banks, and the reduction of plant cover. Each of these impacts increases soil erosion. Additionally, livestock production indirectly contributes to soil erosion through land conversion and poor land management practices in feed production areas, which destabilize soil and increase runoff. Sediments transported due to soil erosion are the leading water pollutant in the United States; they obstruct waterways, destroy aquatic ecosystems, disrupt water flow and availability, and contribute to eutrophication (Steinfeld et al. 2006b).

1.3.9 Feed Crop Related Impacts

Additionally, livestock's demand for feed contributes to the use of chemical fertilizers and pesticides, which are both common contributors to water pollution.

Chemical fertilizers and pesticides help combat the effects of decreased soil fertility, increasing production on poorly managed and marginal lands. However, these chemical inputs also migrate into water sources either through runoff into surface waters or by leaching into the groundwater through the soil. Since fertilizers contain high levels of nitrogen and phosphorous and other nutrients, when fertilizers contaminate a water source, they cause many of the same negative effects as nutrients in livestock waste as discussed above. Pesticides can damage the ecosystem by affecting target and non-target species, and they also have adverse impacts on human health when they are present in drinking water and in food (Steinfeld et al. 2006b).

1.3.10 Case Study: CAFO Waste and Water Pollution

Hog farms often are cited as the principle contributors of animal waste to water pollution, but dairy farms, cattle feedlots and poultry farms also use the same lagoon and spray field system of waste management that commonly contributes to water pollution. For example, a lagoon on a dairy farm in Lowville, NY burst in August, 2005, spilling 3 million gallons of cow manure into the Black River, roughly one fourth the volume of the Exxon Valdez oil spill (York 2005). This incident alone polluted over 30 miles of the river. Lethal levels of ammonia and very low levels of dissolved oxygen which led to the death of an estimate of 280,000—370,000 fish over 24 miles of the Black River (New York State Department of Environmental Conservation (DEC) 2014). In response to this incident, health inspectors began testing nearby wells for contamination, and a nearby town that relied on the Black River for part of its public water supply was required to cutoff intake water from the river all together (York 2005). Emergency crews attempted "to dilute the contamination by increasing the water flow to a Black River tributary," (York 2005) but no other cleanup or primary restoration measures were able to be implemented "due to the river conditions at the location of the spill" (DEC 2014). Fortunately, as of 2010 the Black River had shown signs of significant recovery (DEC 2014).

Reports of similar incidents have emerged from various other states as well. In Wisconsin, a sinkhole opened up near a manure spray field, allowing an unknown quantity of manure to leach into the water supply of over a dozen drinking wells of nearby homes. Sixteen people became ill and one was hospitalized due to the effects of this contamination. Contamination of drinking water is not limited to this isolated incident. In one Wisconsin county it is estimated that nearly one-third of private drinking wells are contaminated with high levels of bacteria, and over 20 % of drinking wells tested positive for bacteria in another county. Environmental advocates contribute this contamination to the large farms in the area (Rodewald 2015).

1.4 Land Impacts

Decreasing crop yield growth rates, population growth, urbanization, and a trend towards more meat-intensive diets are resulting in an increased demand for cropland. Land used for agricultural purposes currently makes up around 33 % of the world land area, and cropland specifically makes up 10 %. The United Nations estimates an increase in global agricultural land of between 7 and 31 % until 2050. By 2050 the population will grow to an estimated 9.6 billion, and 70 % of that population will be living in cities (Bringezu et al. 2014). Urbanization and higher incomes are both correlated with demand for meat (Steinfeld et al. 2006a). Trends since the 1990s have already shown a rising consumption of animal based food and leveling of vegetal food. By 2030, global meat consumption is expected to increase by 22 % and milk and dairy consumption by 11 %. Developing countries with rising income levels are driving much of this demand (Bringezu et al. 2014). We are seeing a move to more industrialized production systems to keep up with this demand. Industrialized systems currently account for production of 67 % of poultry meat, 42 % of pig meat, 50 % of eggs, 7 % of beef and veal, and 1 % of sheep and goat meat (Rischkowsky and Pilling 2007). The industrialization of livestock production depends on feed being available at a relatively low cost (Rischkowsky and Pilling 2007). Meat based food requires nearly five times more land per nutrition value than plant based food does (Bringezu et al. 2014) and cropland has thus become a crucial resource.

These changes in demand, coupled with a shift to industrialized agriculture or the Green Revolution, have helped drive a transformation in the global agricultural industry. The industry has shifted from a decentralized system where local farmers grow food for their local communities to a highly centralized, global system of industrialized agriculture (Barker 2007). The move is away from state-centered national systems towards globalized, privatized systems and expanding trade (Bringezu et al. 2014).

The increase in demand for land for agricultural purposes also sees countries making large-scale land acquisitions in an attempt to guarantee their food security (Bringezu et al. 2014). Wealthy countries lacking natural resources have been acquiring farmland in resource-rich developing countries. Around 15–20 million hectares of land were estimated to be the subject of negotiations from 2006 to 2009. Countries engaging in these investments include India, Saudi Arabia, South Korea, the United Arab Emirates and China (Robertson and Pinstrup-Andersen 2010). Abu Dhabi, lacking the water resources to sustain its agricultural needs, for example, is developing almost 30,000 ha of farmland in Sudan to grow alfalfa for animal feed, in addition to maize, beans, and potatoes (Cotula et al. 2009).

China similarly faces water, land, and labor shortages, and only 12 % of its land is arable. This makes it more costly to grow feed grain in China than to have it grown abroad and shipped back to China. China produces and consumes nearly half of all pork, so the demand for feed is pronounced. From 2011 to 2012 nearly 37 % of Brazil's total soy production was exported to China, and this demand is helping drive the conversion of natural ecosystems and pasture to large-scale soy farming. China has also been strategically investing in the international soybean supply chain to strengthen its national security over food. In September 2013 it was reported that China had signed a deal to lease 100,000 ha in Ukraine for 50 years to grow crops and raise pigs. Chinese companies are also at various stages of acquiring large tracts of land in Brazil and Argentina directly (Sharma 2014).

These large-scale acquisition contracts can lead to a number of negative social, economic, and environmental impacts. Intensive agricultural techniques can leave land irreversibly degraded when countries focus on short-term commercial yield and ignore long-term consequences. Local farmers who are displaced will suffer economic losses and turn to wage labor, resulting in a loss of indigenous farming knowledge. The host country also risks food insecurity, since it has given up valuable cropland to produce food for export to the leasing country (Robertson and Pinstrup-Andersen 2010).

1.4.1 Intensification of Agricultural Production

The rising demand for meat is spurring land-use change globally as well as a shift to more industrialized, intensive production methods. The FAO expects about 80 % of projected growth in crop production by 2050 in developing countries to come from intensification in the form of yield increases (71 %) and higher cropping intensities (8 %), though in some regions arable land expansion is expected to account for up to 30 % of crop production growth. In developed countries, total area of arable land has been declining since the mid 1980s, and increases in crop yield have accounted for all production increases and compensated for declining land use during this time (Conforti 2011). New techniques are being used to produce feed crops such as use of high-yield crop varieties, fertilization, irrigation, and pesticide use. This change brings with it a number of environmental impacts. Forests are being cleared, production is being intensified and this is resulting in fertilizer and nutrient pollution, great losses to biodiversity, habitat fragmentation and destruction, and overexploitation of species. These impacts will be discussed in this section.

1.4.2 Intensive Livestock Production

In contrast with the current globalized system where feed is grown in one continent to feed livestock across the globe, livestock production historically relied on local feed inputs. Livestock production took place on pastures, and animals were fed a mix of local food, crop residues, and waste products of human foods. Due to the increasing land scarcity and lack of arable land, however, the industry has relied increasingly on technological advances and new alternative resources to keep up with the demand for increased livestock production. Grazing systems currently predominate and cover 26 % of earth's ice-free surface. For global ruminant production, industrial feedlots make up just a small fraction of systems (as even animals brought to feedlots are raised on pasture first) and are mainly located in North America, and to a lessor degree in Europe and the Near East. However, almost 50 % of pork production and 70 % of poultry production currently occurs in industrialized systems. Over half of that production is happening in developing countries (Steinfeld et al. 2006a).

To meet the increase in demand, meat and milk production is expected to double by 2050 relative to 1999–2001 levels. Developing countries are expected to provide 78 % of the increase in production from 2011 to 2020. This increase will be made possible by greater reliance on industrial farm animal production. These systems are growing six times as fast as grazing systems and twice as fast as traditional mixed farming systems (Humane Society International 2011).

Industrial systems typically involve highly concentrated systems where tens to hundreds of thousands of animals are confined in welfare-depriving conditions (Humane Society International 2011). They generally hold animals of a similar genotype, raised for one purpose, that are fed nutrient-dense industrial feeds and have a high rate of turnover. In the United States these operations are called animal feeding operations (AFOs), and operations holding at least 1000 animal units (where 1 animal unit = 1000 pounds body weight) and where animals are confined or stabled for at least 45 days in any 12 month period are defined by the EPA as concentrated animal feeding operations (CAFOs) (Otte et al. 2007).

The GAO found that the number of large farms in the United States increased from 3600 in 1982 to 12,000 in 2002. The number of animals raised on large farms also increased. By 2002 almost half of livestock and poultry were being raised on large farms GAO (2008). In the United States, 54 % of food animals are concentrated on only 5 % of farms (Halden and Schwab 2008). From 1980 to 2004 offtake per unit of stock of pork, chicken and milk increased by 61 %, 32 % and 21 % respectively. This intensification has resulted from an increase in inputs and technological advances in livestock production techniques including genetics, health, and farm management allowing greater output per animal (Steinfeld et al. 2006a).

1.4.3 Monoculture

Industrialized agriculture is dependent on feed, and due to modern intensive agricultural practices, much of this feed is grown in highly specialized monoculture systems. Monocultures for animal feed make up almost 40 % of global cropland (UNCTD 2013). In these systems farmers produce only a single species of crop, attempting to increase production through the use of high-yield varieties and elimination of weed species. These systems suffer reduced agrobiodiversity. As a result, they lose the benefits of a complex system where a variety of plants and animals provide varying ecosystem services.

Agricultural intensification has led to a dramatic reduction in the types of crops grown worldwide. Humans are able to consume 7000 species of plants, but of these only 150 are commercially valuable. Only 103 species make of 90 % of food crops grown worldwide, and Rice, Wheat, and Maize make up 60 % of the calories and 56 % of the protein people derive from plants. This loss of genetic diversity reduces food stability by making farmers more vulnerable to climactic and other stressors (Thrupp 2000).

Feed crop production's reliance on monoculture systems has numerous ecological impacts. Many species that are attracted to mixed farming systems, where a diverse group of crops and livestock are produced, vanish from monoculture systems (Steinfeld et al. 2006b). Monocultures are more vulnerable to pests and diseases and contain fewer soil organisms and nutrients (Thrupp 2000). They require increased pesticide use to deal with more abundant pests. This can lead to pesticide diffusion along the food chain, which builds pesticide resistance ultimately creating a vicious cycle (Steinfeld et al. 2006b). Heavy pesticide use can also harm beneficial insects and fungi that would otherwise provide useful services (Thrupp 2000). Crop diversity is also a key factor in nutrient use efficiency. Monoculture systems harm soil diversity, which is important for nutrient cycling, pest control, disease control, and superior soil structure (Jackson et al. 2005).

Monoculture systems also jeopardize pollination, an eco-system service wild pollinators, especially wild bees, contribute to crops globally. Pollinators thrive in farms where chemical use is minimized and where there are diverse cropping patterns. Crop rotations, intercropping, and growing different varieties of a single crop are all considered good practices resulting in better crop performance, nutrient availability, pest and disease control, and water management (FAO 2011).

Monocultures lose the benefits agrobiodiversity can provide when different species provide complementary effects. For example, growing mung beans or sweet potatoes with maize can guard against weeds due to the effects from the shade they provide (FAO 2011). An experiment conducted in Yunnan Province, China showed how crop genetic heterogeneity offers greater disease resistance than monoculture. The cool, wet climate of Yunnan Province makes rice grown there particularly susceptible to blast disease. Farmers generally control the disease outbreak through multiple foliar fungicide applications. In 1998 farmers were planting mainly monocultures of a less commercially desirable non-glutinous hybrid rice variety, because the more commercially desirable glutinous variety had lower yields and was highly vulnerable to blast. In the first year of the experiment in 1998, four varieties of rice, including the susceptible variety, were planted in mixtures. Farmers only had to make one foliar fungicide application that year. In the second year, the experiment was expanded and no foliar fungicide applications were required. The experiment was a huge success. In the first year the mixed system blast severity was only 1 % for the susceptible variety compared to 20 % in monoculture systems, and the second year showed similar results. The varieties grown in mixtures additionally had yields that averaged 89 % greater than those grown in monoculture (Youyong 2000).

1.4.4 Soil Degradation and Erosion

Intensive agricultural techniques can result in depletion of natural resources such as soil and water. Organic matter plays an important role in soil by providing substrate for nutrient release and by forming soil structure that increases water-holding capacity and reduces erosion. In intensive cropland in temperate zone agriculture, however, there can be declines in soil organic matter within the first 25 years of cultivation. This decline is typically 50 % of the original Carbon content. In tropical areas losses can occur within just 5 years of conversion to intensive systems. Additionally, as organic matter decomposes it releases large amounts of CO_2 that contribute to climate change (Steinfeld et al. 2006b). Herbicide use and mulching can also have long-term effects on soil microorganisms and processes (Wardle 1999).

Soil erosion rates are influenced by several factors including soil structure, landscape morphology, vegetation cover, rainfall and wind levels, land use, and land management including method, timing, and frequency of cultivation. Water runoff has the greatest effect on erosion, with erosion increasing as infiltration of water into the soil decreases. Croplands under intensive agricultural systems are particularly susceptible to erosion due to the removal of the natural vegetation that would otherwise bind the soil, protect it from the wind, and improve infiltration. Intensive systems also lead to erosion through inappropriate cultivation practices, the mechanical impact of heavy agricultural machines on the land, and depletion of the natural soil fertility (Steinfeld et al. 2006b).

1.4.5 Habitat Deterioration

Intensification of agricultural land use is accompanied by large increases in nitrogen and phosphorous fertilization that can pollute and degrade habitats (Steinfeld et al. 2006b). Increased fertilizer use has accounted for one third to one half of increased yield growth in developing countries since the Green Revolution, and fertilizer use in developing countries has continued to grow at about 3.6 % over the last decade. Globally, however, fertilizer use has stabilized due to a decline in use in developed countries (Conforti 2011).

Crops can only take up a limited amount of the nitrogen and phosphorous that is applied to them, and the rest can become pollution that contaminates habitats. About 40–60 % of nitrogen that is applied to crops is left in the soil or lost to leaching (Steinfeld et al. 2006b). When nitrogen and phosphorous leak into waterways, it can lead to eutrophication of estuaries and coastal seas, loss of biodiversity, and changes in species compositions in terrestrial and aquatic ecosystems. Nitrogen leaching can also lead to groundwater pollution with nitrate and nitrite, increases in the greenhouse gas N_2O, increases in NO_x and resulting tropospheric smog and ozone, and acidification of soils and sensitive freshwaters (Tilman et al. 2001).

About half of the fertilizer nitrogen and phosphorous that is taken up by crops ends up in human and livestock waste streams after consumption, turning livestock waste into a potent pollutant as well. Livestock waste is rarely treated for nitrogen and phosphorous removal, so the nitrogen and phosphorous inputs can eventually end up in surface and groundwater, and nitrogen also volatizes into the atmosphere as ammonia (Tilman et al. 2001). This problem is aggravated by the fact that livestock manure is often poorly managed and unregulated.

In addition to fertilizer pollution, monoculture agriculture offers limited food and shelter for wildlife, and as parcels of land are set aside for intensive agriculture use, wildlife habitats become fragmented (Steinfeld et al. 2006b). Agricultural intensification has been linked to declining farmland bird populations (considered good indicators of overall farmland biodiversity) due to practices such as fertilizer and machinery use that harm habitats and availability of food for birds (Donald et al. 2001).

1.4.6 Ecological Impacts of Reduced Agrobiodiversity

Agrobiodiversity encompasses all biological resources that perform services and functions on which agriculture relies. It includes all crops and livestock, soil organisms, insects, bacteria and fungi, all organisms that act as pollinators, symbionts, pests, parasites, predators, decomposers, and competitors, and the environment in which they exist. It takes into account not just species and genetic resources, but human methods and processes used to exploit these resources (Thrupp 2000). Agricultural production is highly dependent on the eco-system services such as pest control, pollination, soil fertility, and others, provided by both "planned" (aspects of biodiversity controlled by the farmer) and "associated" biodiversity in agricultural eco-systems (Tscharntke et al. 2012).

Increasing demand for livestock production and resulting demand for feed crop has led to four trends that impact agrobiodiversity. First, expansion (extensification) of area used for grazing or crop production, which can harm overall biodiversity by degrading and destroying habitats when land is cleared to allow grazing and farming to expand (Steinfeld et al. 2010). For example, in Latin America growth of large-scale cattle ranching is driving deforestation (Brighter Green 2013). Recently, however, extensification has been giving way to intensification, and there has been increasing pressure on smaller amounts of land to produce greater yields of livestock and crop.

Intensification has led to practices such as monoculture (discussed in-depth in Sect. 1.4.3), and the resulting homogenization of agricultural eco-systems can have a significant impact on the composition and abundance of associated biota such as pests, soil invertebrates and microorganisms, which can in turn affect plant and soil processes (Matson et al. 1997). Since there is a lack of eco-system services being provided, there is a resultant need for increased pesticide and fertilizer use. In systems with reduced agrobiodiversity, not only is there greater exposure to pests and diseases, but monoculture systems can be devastated by an attack. Homogenization on farms also results in fewer soil organisms and nutrients, and the increased reliance on pesticides harms beneficial insects and fungi, which can lead to lowered productivity. In Bangladesh, thousands of farmers were able to eliminate pesticide use and increase yields by 11 % by incorporating fish into their rice paddies, adopting other methods to restore the natural balance between insects and fauna, and planting vegetables on dykes around the edges (Thrupp 2000).

The increasing reliance on greater pesticide use can also be toxic to birds, mammals, amphibians, and fish. Pesticides can reduce the abundance of weeds and insects that are food sources for other species and herbicides can change habitats leading to population decline (Isenring 2010). Pollinators, that can be harmed by pesticide use as well, play a hugely important role and are required for reproduction of almost 90 % of angiosperms, improve production of 70 % of the globally most important crop species, and influence 35 % of global human food supply. There is evidence that having a diverse group of pollinators on site can increase crop yields (Tscharntke et al. 2012).

The third trend is rangeland contraction. This trend positively impacts biodiversity when lands are protected from crop farming or heavy grazing, but can adversely impact biodiversity when it prevents light grazing that has the potential to increase biodiversity (Steinfeld et al. 2010).

Finally, abandonment occurs when farmers abandon grazing areas to allow them to return to their pregrazing vegetation. Abandonment threatens biodiversity in Europe, where 26 out of 196 habitat types considered important due to their high biodiversity value are threatened by abandonment of rural activities (Steinfeld et al. 2010).

Not only has intensification led to a direct erosion of the genetic diversity of livestock, but the methods used to raise and feed livestock also impact biodiversity directly and indirectly. The rest of this section will explore those impacts.

1.4.7 Livestock Genetic Diversity

Agricultural intensification leads to a loss of biodiversity as farmers rely more and more on only a few livestock breeds that are the most productive. Of the 30–40 mammalian and bird species recognized as domesticated, fewer than 14 make up 90 % of the global livestock production (Steinfeld et al. 2010). The FAO classifies around 20 % of total livestock breeds as "at risk." In the 6 years leading up to 2007, 62 breeds went extinct, an average of almost one breed per month. This has been the trend globally as the world moves towards intensive, industrialized production of confined cattle, hogs and poultry. These industrialized systems are devoid of many of the environmental stressors that would otherwise demand a more diverse set of criteria in livestock selection. For example, there is less demand for livestock that are adapted to thrive in local environments. There is similarly less demand for livestock that are naturally resistant to disease because farmers rely instead on veterinary inputs. In these systems feed can make up 60–80 % of production costs so there is greater focus on breeds with high feed conversion ratios. There is also greater demand for species that meet consumer preferences and technical requirements for uniformity of size, fat content, color, flavor, etc. Breeds that are specialized to thrive in these environments have helped lead to genetic erosion of other breeds (Rischkowsky and Pilling 2007).

In contrast, in low to medium external input systems farmers still rely on local breeds that are especially suited to the local conditions, highlighting the importance of maintaining genetic diversity resources. Having diverse genetic resources allows farmers to select from the pool those breeds that satisfy the demands of a particular production system. Further, having access to diverse genetic resources allows selection of particular characteristics that would be well suited to dealing with upcoming environmental challenges—such as increasing demand for livestock products, climate change, and emerging animal diseases (Rischkowsky and Pilling 2007).

There are three broad categories of threats to livestock genetic diversity: trends in the livestock sector, disasters and emergencies, and epidemics and control measures. The security of a breed is linked to its role in livestock systems, and a breed can become threatened if its functions are no longer required. For example, specialized draught breeds are being threatened by the shift to greater mechanization in agriculture. The growing demand for livestock products is leading to a replacement of local breeds by a small selection of high-yielding breeds. Farmers may also cross-breed to produce higher yields, which can lead to genetic erosion. Regulations concerning product uniformity and food hygiene can also reduce the number of marketable livestock products. For example, it has been noted that the current carcass grading system works against small animals and thus disincentives production of certain indigenous cattle breeds. Consumer preferences, like a preference for leaner meat, can also lead to a decline of breeds without those characteristics. Globalization also encourages high specialization in local regions, leading to a decline in diversity (Rischkowsky and Pilling 2007).

Disasters and emergencies, such as natural disasters in addition to war and political instability, can impact genetic diversity through several channels. First, there is the immediate physical impact of the disaster on livestock. Second, social changes that the emergency brings about and interventions that take place to respond to it can have an effect. In particular, "restocking," where external actors provide livestock to a household can influence genetic diversity. Important factors are whether the breeds provided are local or non-local, and whether farmers give them preferential selection for breeding (Rischkowsky and Pilling 2007).

Finally, diseases threaten genetic diversity directly by causing death and indirectly by farmers slaughtering or abandoning certain breeds when disease control measures become too costly or burdensome. Genetic diversity is important to conserve resources that could combat disease. Additionally, research has shown that genetically diverse populations are less susceptible to disease epidemics (Rischkowsky and Pilling 2007).

1.4.8 Grazing Impacts on Biodiversity

Management of livestock grazing can also affect biodiversity of grasslands, with potentially protective and damaging impacts. To some extent livestock grazing can protect rangelands by preventing them from being converted to other uses less suited to maintaining biodiversity, like crop farming and development (Steinfeld et al. 2010). However, grazing also drives deforestation, woody encroachment, and desertification (McAlpine 2009).

Grasses make up about half of the global biomass fed on by livestock and grassland systems cover an estimated 32 % of the world's land area. Grasslands include rangelands, which are fragile ecosystems prone to degradation, loss of biodiversity and water retention capacity, carbon emissions, and reduced productivity, if grazing is not managed properly (Hoffman et al. 2014). Experts disagree on whether livestock grazing can cause changes to ecosystems and biodiversity on rangelands. Factors that may have an impact include how frequently the livestock are moved, the livestock's longevity, and whether the rangelands have a history of grazing. Livestock are more likely to alter biodiversity on rangelands when they are heavily concentrated, graze year-round, or where the rangeland does not have a history of grazing and is thus more sensitive to it.

Livestock grazing can encourage woody encroachment by discouraging the growth of herbaceous plants like grasses and leafy herbs and encouraging the development of woody plants like shrubs and trees in their place. Grazing can also lead to development of non-native short-lived nitrogen-tolerant plants as a result of livestock trampling vegetation and depositing nitrogen, phosphorus and other nutrients near water points. In some instances, grazing can have positive impacts on biodiversity. Light grazing, for example, can remove dominant species that would otherwise control all the natural resources, allowing other species to develop and thrive (Steinfeld et al. 2010).

1.4.9 Other Livestock Related Impacts on Biodiversity

More generally, livestock production threatens biodiversity by generating pollution, spreading diseases, introducing invasive species, and contributing to climate change. Pollution is the greatest cause of loss of biodiversity globally, and livestock are one of the greatest contributors to this pollution. Livestock deposit nitrogen and phosphorous directly through manure and indirectly through fertilizer for feed crops and pastures. This pollution can harm biodiversity directly by killing species or indirectly by affecting habitats (Steinfeld et al. 2010). From 2000 to 2001 the annual total pesticide usage in the United States consisted of 700 million pounds of active ingredient, of which 77 % is used for agriculture, half of which is used for production of grain to feed livestock. Nitrogen present in animal waste applied to field crops for fertilizer can find its way into surface waters (Halden and Schwab

2008), such as the Mississippi where it contributes to eutrophication, and the Gulf of Mexico where it creates dead zones and is responsible for massive fish kills.

Livestock also can impact biodiversity by spreading disease, either through "pathogen pollution" where livestock spread a previously unknown disease to wildlife, or "spillover" where domesticated animals outnumber wild animals, and continually infect wildlife populations with a common disease until the wildlife population goes extinct (Steinfeld et al. 2010). The disease Brucellosis likely was introduced into America through cattle. It now infects elk and bison in Yellowstone National Park and is considered a potential threat that can spill back to cattle that graze at the park boundary (Daszak et al. 2000).

Livestock also can harm biodiversity by introducing nonnative species into foreign environments. Because the nonnative species has no natural predators, may be highly adaptable or do well in human-altered habitats, it can overtake the native species. Livestock itself can be invasive. An example of this is when livestock graze in grasslands that do not have a history of grazing and harm the biodiversity that is present there. Farmers also may introduce nonnative plant species to feed livestock, which outcompete native species and reduce biodiversity. This might occur when farmers introduce certain grasses for pastures and the introduced grasses outspread the natural vegetation (Steinfeld et al. 2010).

As mentioned before, livestock contribute an estimated 18 % of all Greenhouse Gases and climate change already is having an impact on species populations. Feed crop production is a major contributor to these emissions. Livestock production contributes significantly to the three major greenhouse gases: carbon dioxide, methane and nitrous oxide. Livestock account for 9 % of total global anthropogenic emissions of carbon dioxide, 35–40 % of methane emissions and 65 % of nitrous oxide. Climate change is altering species distributions and population sizes and affecting the timing of reproduction and migration as well as the frequency and intensity of pests and disease outbreaks. In Marine ecosystems warming temperatures can kill coral, a species vital for biodiversity because it provides a home to 25 % of marine species (Steinfeld et al. 2006b).

1.4.10 Habitat Change

Livestock production, including growing feed for livestock, changes land-use in a way that can lead to habitat destruction, fragmentation, and degradation. These impacts form a major threat to biodiversity (Steinfeld et al. 2006b). Land used for agricultural purposes currently makes up around 33 % of the world land area, and cropland specifically makes up 10 %. The United Nations estimates an increase in global agricultural land of between 7 and 31 % until 2050. This increase in pasture and cropland comes at the expense of decreases in forests, natural grasslands, and savannahs (Bringezu et al. 2014).

1.4.11 Deforestation and Forest Fragmentation

Deforestation occurs as forest areas are cleared to use the land for livestock or crop production. Forest fragmentation occurs when previously intact forest is broken up and areas are used for livestock or crop production. The remaining forested area becomes a series of isolated forest patches (Steinfeld et al. 2010). This results in habit change and degradation. Deforestation has been occurring at an average rate of about 13 Mega hectares (Mha = ha × 10^6) per year over the last 5 decades, and cropland expansion has been the primary cause (Bringezu et al. 2014). Cattle ranching in Latin America has been the impetus for the conversion of tropical forest. The primary driver there has been clearing land for cattle grazing, but recently conversion of forest to cropland tied to livestock intensification has become a more significant force. Between 2000 and 2005, the Amazon has experienced an estimated 0.6 % rate of deforestation. It has been estimated that 17 % of deforestation in the Brazilian Amazon can be attributed to cropland expansion from 2001 to 2004, primarily for soya to be used for livestock (Steinfeld et al. 2010).

Deforestation can have a particularly destructive impact on species that require large contiguous forests, species that require intact forests to survive, endemic species, and species vulnerable to extinction due to small population sizes (Steinfeld et al. 2010). As land is cleared around forest areas, those edges can no longer support species. This eventually creates islands of forest that are too small to support the populations and leads to extinctions. Forest fragmentation also makes it harder for species to colonize due to the distance between patches (Rudel and Roper 1997). Populations that are particularly vulnerable to forest fragmentation include birds, large predators, primates, butterflies, and solitary wasps. Forest fragmentation also contributes to forest degradation by turning areas with high biodiversity into simplified shrub and grassland with induced flora. Deforestation can interfere with ecological processes like wildlife territory expansion, plant pollination, and seed dispersal (Steinfeld et al. 2010). Thus, the majority of species extinctions are likely due to habitat destruction from tropical deforestation and forest fragmentation (Rudel and Roper 1997).

1.4.12 Desertification and Woody Encroachment

The United Nations broadly defined desertification as "land degradation in arid, semi-arid and dry sub-humid areas resulting from various factors, including climatic variations and human activities" (Horrigan et al. 2002). It often refers to instances where herbaceous cover is replaced with xerophytic shrub cover and bare soil or just bare soil (Steinfeld et al. 2010). Desertification occurs because dryland regions are incredibly vulnerable to over-exploitation and bad land management. Climactic drivers include low soil moisture, changing rainfall patterns, and high evaporation. Human drivers of desertification include over-cultivation, which exhausts the soil, overgrazing land, which removes protective vegetation that guards against erosion, deforestation, which removes trees that bind the soil to land, and poor drainage of irrigation systems, which leads to soil salinization (Hori et al. 2012). Thus poor land management during livestock production contributes to desertification, especially when animals overgraze the land and trees and shrubs are removed subjecting the land to increasing wind and water erosion.

Desertification leads to land degradation that makes the land unsuitable for agriculture (Steinfeld et al. 2010). Land degradation at the biological level manifests a "persistent reduction in biological productivity." The biological productivity reduced depends on the land use. In cropland it might be soil fertility and yield per acre, in rangelands it might be the land's carrying capacity for cattle, and in forests it might be ecosystem services such as water filtration and retention (Welton et al. 2014).

Desertification affects the land's topography, vegetation, and soil. Topsoil is eroded, soil loses fertility, and original vegetation gives way to vegetation of poorer quality Nicholson, (Nicholson et al. 1998). These changes affect the carbon and nutrient cycling of the system. Desertification also results in hydrological changes that make transfer of precipitation to soil less effective, and the net primary productivity per unit of precipitation decreases. There is some suggestion that the resulting increase in bare soil cover can change the ability of the surface to reflect solar energy, resulting in regional and global climate impacts, such as reduced rainfall (Asner et al. 2004). Desertification results in habitat loss affecting migratory bird species, which depend on resources provided by drylands to give refuge during their long flights (UNCCD 2013).

Reversing desertification is very difficult. It can take 500 years to restore just 2.5 cm of soil, which can be lost by erosion in only a few years. Steps that can be taken to combat desertification include restoring soil nutrients, using synthetic fertilizers or natural compost, reducing herd numbers, and giving land time to recover. Diversifying crop and animal production can protect the land by preventing the over-use of any one nutrient. This is because nutrient needs are mixed and the resources removed from land in this case are complementary. Irrigation and reforestation can also help restore land and make it productive again. Installing wind barriers, planting vegetation with roots to fix and protect soil, and a prohibiting grazing may also help recovery (Hori et al. 2012).

Woody encroachment occurs when herbaceous cover is overtaken by woody plants. It differs from desertification in that it involves a smaller loss of herbaceous cover, though herbaceous cover can still decrease (Asner et al. 2004). In the United States the increase in woody cover in non-forest lands ranges from 0.5 to 2 % a year (Anadón et al. 2013). Possible causes of woody encroachment are overgrazing of herbaceous cover (decreasing competition for woody seedlings), fire suppression that enhances woody plant survival, and increases in atmosphere CO_2 and nitrogen pollution that favor woody plant growth. The likelihood of woody encroachment increases if woody plants are already present on the landscape. Woody encroachment decreases quality of land for animal production, but enriches total Carbon and Nitrogen stocks (Asner et al. 2004). Woody encroachment decreases quality of land for animals because the encroaching plants may be of lower nutrient or palpable quality; can be a habitat for pests, parasites, and predators; and can reduce forage production (Steinfeld et al. 2010). Woody encroachment can also affect ecosystem functions like decomposition and nutrient cycling, biomass production, and soil and water conservation. Habitats are affected and savannah like areas in wooded landscapes can vanish due to woody encroachment (Steinfeld et al. 2006b).

1.4.13 Forest Transition and the Conservation of Pastoral Landscapes

Forest transition is defined as the process of returning land used for agricultural purposes to its former forest state. This phenomenon primarily occurs in remote areas with poor soil, and is primarily characterized by pastures that are left to return to forest (Steinfeld et al. 2010). Lands with productive soil in favorable locations by contrast are more likely to remain in production. A study of the Chiguaza region in Ecuador found net increase in forest cover from 1987 to 1997 as abandoned areas reverted to secondary forests (Rudel et al. 2002). Forest transition can have mixed effects on biodiversity. Secondary forests can provide viable habitats for species, which can result in positive effects on biodiversity. However, the impact depends on the species the abandoned agricultural land is replaced with (Meyfroidt and Lambin 2008). Abandoned pastures can sometimes turn into fallow and shrubland with little biological diversity. Thus, in some cases, allowing grasslands with bio diverse resources to be abandoned can result in a loss of biodiversity (Steinfeld et al. 2006b).

1.4.14 Invasive Species

People have been transporting species across habitats for millennia, both accidentally and purposefully, but the fate of these species in their new habitats is difficult to predict. Most do not survive. Of those that survive, only a fraction of species becomes naturalized (forming sustainable populations) (Rejmánek et al. 2005), but some of those naturalized then become invasive (Mack et al. 2000). The IUCN (2000) has defined invasive alien species as "an alien species which becomes established in natural or semi-natural ecosystems or habitat, is an agent of change, and threatens native biological diversity." An alien species is one occurring outside of its natural range or habitat (IUCN 2000). Invasive species form a major threat to biodiversity (McGeoch et al. 2010). Invasive species alter eco-system processes, change community structure, and alter genetic diversity. They also harm native species through predation, competition, hybridization, by introducing pathogens or parasites that can sicken or kill them, and by destroying or degrading their habitat

(Steinfeld et al. 2006b; McGeoch et al. 2010). The adverse impact of invasive species on biodiversity has been increasing in recent decades (McGeoch et al. 2010).

Globally, threatened birds and amphibians are especially vulnerable to invasive species. Thirty percent of threatened birds, 11 % of threatened amphibians, and 8 % of the 760 threatened mammals for which data are available are affected by invasive species. Islands and island species are particularly susceptible to invasive species because of their isolated evolutionary history. Sixty-seven percent of oceanic-island globally threatened birds are affected directly or indirectly by invasive species, compared to 17 % on continental islands, and just 8 % on continents (Baillie et al. 2004).

Livestock production contributes to the introduction of invasive species through habitat change, intentional plant invasion, and animal grazing. In addition, animals may carry invasive species with them across locations. Invasive species can also degrade pastures and have other negative effects on livestock (Steinfeld et al. 2006b).

1.4.15 Deforestation's Impact on Invasive Species

Deforestation in tropical regions to clear land for agricultural use can spread invasive species, including invasive disease species. For example, in South America, approximately 53 million hectares of humid tropical forest in the Brazilian Amazon Basin alone have been converted to pasture, as has some 40 million hectares of native tropical savanna in Colombia, Venezuela, and Brazil. This results in a loss of native vegetation, and in many cases the introduced grasses have spread to and overtaken natural areas. Additionally, deforestation can contribute to the transmission of viruses carrying haemorrhagic fevers that previously circulated benignly in wild animal hosts. The use of irrigation in these areas raises the water table and increases the number of breeding sites available for mosquitos. The problem is exacerbated by agricultural pesticide use that builds pesticide resistance in mosquitos. Infectious disease agents are typically invasive alien species that are devastating to human health and local food and livestock production (Brand 2005).

1.4.16 Introduced Grasses as Invasive Species

Humans have introduced many non-native plant species to new areas in order to feed livestock. Many grasses introduced by humans for pasture are biologically adapted to spread quickly given their abundant and persistent small seeds, an ability to survive under stressful situations, and a tolerance for burning and heavy grazing (Steinfeld et al. 2010).

Temperate grassland in Australia, South America, and western North America has been permanently transformed by human settlement and transplantation of alien plants. While livestock are not solely responsible for the introduction of invasive alien species, they do play a significant role in the process. Two particular characteristics make grasslands especially vulnerable—(1) the lack of large, hooved, congregating mammals during the Holocene period or earlier and (2) dominance by caespitose grasses, which are vulnerable to grazing and trampling. The lack of large, hooved, congregating mammals allowed for the evolution of grasses that are sensitive to grazing animals. In these grasslands the introduction of livestock has resulted in the destruction of native grasses and the dispersal of alien plants through the fur and feces of animals (Mack 1989).

In the United States, domestic livestock grazing and the introduction of weeds have transformed rangelands' plant ecosystems. The majority of weeds have been introduced from other continents, but native species have also spread more rapidly due to management practices such as fire suppression and overgrazing. These native species can spread and reduce overall forage quality or quantity, and can be poisonous to livestock. Before the introduction of annual grasses, the primary native species in the rangelands west of the Rocky Mountains were perennial bunch grasses. However, the native perennial grasses were over-grazed and overcome by introduced winter annual grasses. In some cases the over-grazing has led to increased unpalatable native woody or poisonous species. Suppression of periodic wildfires has also led to an increase in shrub populations, which are typically controlled by burning. Rangeland weeds can interfere with grazing practices; lower yield and quality of forage; increase costs of managing and producing livestock; slow animal weight gain; reduce the quality of meat, milk, wool, and hides; and poison livestock. They can also reduce plant diversity, threaten rare and endangered species, reduce wildlife habitat and forage, alter fire frequency, increase erosion, and deplete soil moisture and nutrient levels (DiTomaso 2000).

1.4.17 Livestock as Invasive Species

Livestock themselves can be considered an invasive species, especially when their impact on native species is not minimized. Livestock (cattle) were domesticated 10,000 years ago from species found in Asia and northern Africa, and taken to other continents that had no previous history of grazing (Steinfeld et al. 2010). Livestock can compete with wildlife for water and food, and threaten the populations of local vegetation as they feed on seedlings. They also play a major role in introducing invasive alien species by dispersing seeds and transmitting disease organisms to populations with no immunity. Many harmful feral populations have also resulted from the introduction of livestock species of economic important to the Americas. The IUCN/SSC Invasive Species Specialist Group (ISSG) classifies feral cattle, goats, sheep, pig, rabbits, and donkeys as invasive alien species (among a total of 22 invasive mammalian species) (Steinfeld et al. 2006b).

Livestock can also play a positive role in managing invasive alien species through prescribed grazing. The goal in this process is to manipulate patterns of defoliation and disturbance to place a target plant at a competitive disadvantage relative to other plants in the community. Achieving the desired outcome requires extensive knowledge about how the herbivore's grazing behavior will affect the eco-system and target plant (U.S. Fish and Wildlife Service 2008).

1.4.18 Feed Crop Related Threats

Many of the impacts from livestock production discussed in this chapter are not only direct livestock impacts but include indirect impacts from production of feed crop for livestock. As discussed before consumption of nonruminant meats has been on the rise and, at the same time, small-scale backyard production of livestock has been decreasing and shifting to more-intensive, large-scale industrial systems. These more intensive systems rely on cereals and processed concentrate feeds for livestock, rather than the traditional household waste food, grass from natural pasture, and other forages low-input systems utilized (Steinfeld et al. 2010). Land devoted to feed crop production has been on the rise and makes up approximately 33 % of total arable land (Hoffman et al. 2014). An increasing amount of cereal production is being used for feed crop. Global use of cereals as feed increased by 0.9 % annually between 1992 and 2002. Maize has been the prevailing feed crop in developing countries, and soybeans have been the fastest growing feed crop with a sixfold increase in total quantity fed to livestock between 1982 and 2002 (Steinfeld et al. 2010).

At the turn of the twenty-first century, approximately 72 % of poultry and 55 % of pigs were raised in global industrialized animal-production systems (Galloway et al. 2007). The feed for animals in such systems often is produced in other regions and thus livestock demand and production in one region can have serious consequences for land-use and crop production in another, distant region (Bringezu et al. 2014). For example, a 7 % increase in crop acreage in Brazil would be required to meet a 10 % increase in export of Brazilian soybeans used to feed chicken and pigs in other regions (Galloway et al. 2007). Demand for feed has been one of the primary drivers of deforestation in Brazil (Bringezu et al. 2014).

Production of feed crop leads to many of the land-degrading impacts discussed in this chapter. The conversion of land for agricultural uses to produce feed crop carries a multitude of impacts associated with intensive crop production such as pesticide and fertilizer pollution, full-scale land conversion, and opportunity cost (Galloway et al. 2007). Agricultural eco-systems face a threat from loss of genetic diversity, soil degradation, nutrient depletion, and the loss of natural pollinators (Steinfeld et al. 2006b). Globally, nitrogen fertilizers applied to feed crops make up 40 % of the total amount manufactured. This leads to emissions of 40 teragrams ($Tg = kg \times 10^9$) CO_2, and vast amounts of nitrogen pollution from nitrogen lost to surrounding air and water, or excreted in the urine and feces of livestock that feed on the crop. Industrialized livestock production relies on immense amounts of water, primarily water used to irrigate the feed crop, as discussed earlier in this chapter (Galloway et al. 2007). These changes in land-use for feed production contribute to loss of wild biodiversity (Hoffman 2011). While ruminants convert feed to meat less efficiently than nonruminants, the majority of ruminant feed is forage from nonarable lands and is made up of elements that humans cannot consume. Nonruminants, however, feed on crops grown on land that could be used to grow foods for direct human consumption, and thus there is also growing competition between feed crops and food crops for land use (Galloway et al. 2007).

1.4.19 Case Study: The Amazon and Cerrado in Brazil

Demand for soybean to feed factory farm raised livestock has led to extensive deforestation in the Brazilian Amazon and the Cerrado, a savannah covering more than one-fifth of Brazil. Only 20 % of the Cerrado is still intact, with agriculture and cattle raising accounting for 50 % of its loss. Between 2002 and 2008 the annual rate of deforestation was at 4 %. Not only is the Cerrado an important carbon sink, but it is home to 5 % of the world's species. The Cerrado is also an important water source for the local community, and rivers generated electricity for 9 out of 10 Brazilians. Chemicals used in agriculture are polluting the rivers and affecting the health of local people, who also fear the soy production and deforestation will lead to water scarcity in the area (Lloyd 2011).

Between 2001 and 2006, 1 million hectares of forest in the Amazon were converted directly for soy production. With pressure from retailers and NGOs, a private sector initiative known as the Soy Moratorium was launched and major soybean traders agreed to not purchase soy grown on lands deforested after July 2006 in the Brazilian Amazon. A recent study found that the Soy Moratorium was effective in reducing deforestation for soy production in the Amazon. In the 2 years before the agreement, almost 30 % of soy expansion occurred through deforestation; after the agreement only 1 % of soy expansion occurred through deforestation. However, soy production still continued to increase by 1.3 Mha over this period. In the Cerrado, additionally, where the agreement did not apply, annual rates of soy expansion through deforestation have ranged from 11 to 23 % from 2007 to 2013 (Gibbs 2015).

The study compared the effectiveness of the Soy Moratorium to the government's official land use policy. The government implemented the Rural Environmental Registry of private properties requiring all rural properties to register by 2016 in order to evaluate compliance with the Forest Code and other government regulations. In Mato Gasso, where 85 % of Amazonian soy is produced, the study found only 115 out of several thousand soy farmers violated the Soy Moratorium since 2006, while over 600 violated the Forest Code over the same period. Thus the study found farmers were more willing to comply with the private sector mechanism rather than government regulations. The Soy Moratorium has the ability to protect the up to 2 Mha of the estimated 14.2 Mha of forest in the Amazon considered suitable for soy production that could be legally cleared under the Forest Code. The study suggested the success of the Soy Moratorium was due to the "(i) a limited number of soy buyers that exert considerable control over soy purchase and finance; (ii) simple requirements for compliance; (iii) streamlined and transparent monitoring and enforcement systems; (iv) simultaneous efforts by the Brazilian government to reduce deforestation; and (v) active participation by NGOs and government agencies." (Gibbs 2015).

1.5 Overexploitation

Overexploitation occurs when renewable natural resources are used at a faster rate than they can be replenished (Millennium Ecosystem Assessment 2005). Species can be overexploited when they are unsustainably harvested for food, medicine, fuel, and material, and for cultural, scientific and leisure activities (Baillie et al. 2004). Examples of overexploitation include overfishing, overlogging, and overgrazing. Overexploitation damages ecosystems and can lead to degradation. Degradation is said to occur when the net supply of ecosystem services is so damaged it is unable to recover on its own within a reasonable period after the damaging action is stopped (Millennium Ecosystem Assessment 2005). Overexploitation is one of the leading factors in biodiversity loss (Steinfeld et al. 2006b). It has been identified as a major threat affecting 30 % of globally threatened birds, 6 % of threatened amphibians, and 33 % of the 760 threatened mammals for which data are available (Baillie et al. 2004). Some of the most commonly overexploited species include marine fish and invertebrates, trees, animals hunted for bushmeat, and plants and animals harvested for medicinal uses and the pet trade. Species that are especially vulnerable tend to be valuable, relatively easy to catch, and to reproduce at relatively slower rates (Millennium Ecosystem Assessment 2005).

Livestock can affect the overexploitation of biodiversity in several ways. First, livestock can compete directly with wildlife. Herders' conflicts with wildlife can lead to the eradication of species as herders quell wild populations threatening livestock through predation or spread of disease. For example, in the early history of domestication herder's feared large carnivores preying on livestock. This led to widespread eradication campaigns that resulted in the local extinction of wolves and bears in Europe. Livestock can also compete with wildlife for natural resources and land access. Second, livestock production can lead to overexploitation of living resources (mainly fish) for use in livestock feed. Finally, biodiversity can be overexploited through the unsustainable focus on fewer, more profitable breeds, leading to erosion of livestock diversity (Steinfeld et al. 2006b).

1.5.1 Livestock's Contribution to Overexploitation of Marine Species

Fish have been exploited for fishmeal used in aquaculture and livestock feed. A report from 2009 placed aquaculture as the largest user of fishmeal accounting for 46 % of fishmeal produced, with pig production using 24 %, poultry using 22 %, and other livestock accounting for the remainder (Hasan and Halwart 2009).

Globally, fish production has been increasing over the last 5 decades. Food fish supply has grown at an average annual rate of 3.2 %, faster than the world human population growth of only 1.6 % FAO (2014). The rising demand for fish protein is being met in part by an increase in aquaculture, which relies on wild-harvested fish products to manufacture feed for captive fish (Millennium Ecosystem Assessment 2005). Food fish aquaculture production has expanded at an average annual rate of 6.2 % from 2000 to 2012. In 2012, more than 86 % (136 million tons) of world fish production went directly to human consumption, with the remaining 14 % designated for non-food uses, 75 % of which was produced into fishmeal and fish oil. Fishmeal and fish oil are important ingredients in most aquaculture feeds to supply necessary nutrients to farmed fish (FAO 2014). In addition to fishmeal and fish oil, low value or "trash" fish are also used as components of feed, or complete feed for farmed fish, crustaceans and molluscan species (Hasan and Halwart 2009). By one estimate, the demand for fishmeal and fish oil are expected to grow along with the expansion of aquaculture and stable global capture fisheries, leading to an 8 % expansion in fishmeal production. However, there has been a trend towards an increasing proportion of fishmeal coming from fish processing by-products, with this proportion increasing from 25 % in 2009 to 36 % in 2010. This use of by-products and waste means fewer whole fish must be used (FAO 2014).

Using wild-caught fish to produce fishmeal and fish oil can have serious implications for food security and aquaculture. As demand for fishmeal grows and as a result prices increase, it can become profitable to shift from small pelagic fish production to fishmeal production. However, in many areas small pelagic fish are a significant, important part of local diets. Since local prices for fish as food cannot compete with international prices for fish as fishmeal, this makes less available a traditional source of cheap protein for the poor. It also incentivizes overfishing stocks (FAO 2014).

Potential solutions to reducing the use of wild fish for feed for aquaculture feeds include substituting terrestrial feed sources, increasing the use of fish-waste (35 % of fishmeal is already produced from fish-processing by-products), greater reliance on extractive species that naturally use available carbon and resources, promoting herbivorous and omnivorous species, and increased investment in innovative technologies (FAO 2014).

References

Anadón JD et al (2013) Effect of woody-plant encroachment on livestock production in North and South America. PNAS 111(35):12948–12953. doi:10.1073/pnas.132058511

Asad M et al (1999) Management of water resources: bulk water pricing in Brazil. World Bank, Washington

Asner GP et al (2004) Grazing systems, ecosystem responses, and global change. Annu Rev Environ Resour 29:261–299. doi:10.1146/annurev.energy.29.062403.102142

Baillie JEM, Hilton-Taylor C, Stuart SN (eds) (2004) 2004 IUCN red list of threatened species: a global species assessment. IUCN, Gland, Switzerland/Cambridge, UK

Barker D (2007) The rise and predictable fall of globalized industrial agriculture. International Forum on Globalization, San Francisco

Bittman S, Mikkelsen R (2009) Ammonia emissions from agricultural operations: livestock. Better Crops 93(1):28–31

Brand K (ed) (2005) South America invaded. GISP Secretariat

Brighter Green (2013) Industrial agriculture, livestock farming and climate change: global social, cultural, ecological, and ethical impacts of an unsustainable industry. Brighter Green and the Global Forest Coalition. http://globalforestcoalition.org/wp-content/uploads/2013/05/MM_Brighter-Green-and-the-Global-Forest-Coalition_WSF_Industrial_Livestock-FINAL.pdf. Accessed 15 Jan 2015

Bringezu S et al (2014) Assessing global land use: balancing consumption with sustainable supply. UNEP

Burkholder J et al (2007) Impacts of waste from concentrated animal feeding operations on water quality. Environ Health Perspect 115:308–312

Chislock MF et al (2013) Eutrophication: causes, consequences, and controls in aquatic ecosystems. Nat Educ Knowl 4(4):10

Conforti P (ed) (2011) Looking ahead in world food and agricultural perspectives to 2050. FAO, Rome

Cotula L et al (2009) Land grab or development opportunity? Agricultural investment and international land deals in Africa. FAO/IIED/IFAD, London/Rome

Daszak P, Cunningham AA, Hyatt AD (2000) Emerging infectious diseases of wildlife--threats to biodiversity and human health. Science 287:443–449. doi:10.1126/science.287.5451.443

DiTomaso JM (2000) Invasive weeds in rangelands: species, impacts, and management. Weed Sci Soc Am 48(2):255–265

Donald PF, Green RE, Heath MF (2001) Agricultural Intensification and the collapse of Europe's farmland bird populations. Proc R Soc Lond 268:25–29. doi:10.1098/rspb.2000.1325

FAO (2011) Biodiversity for food and agriculture: contributing to food security and sustainability in a changing world. FAO and The Platform for Agrobiodiversity Research

FAO (2014) The state of world fisheries and aquaculture: opportunities and challenges. Rome

Formiga-Johnsson RM et al (2007) The politics of bulk water pricing in Brazil: lessons from the Paríba do Sul Basin. Water Policy 9:87–104

Galloway JN et al (2007) International trade in meat: the tip of the pork chop. R Swed Acad Sci 36 (8):622–629

GAO (2008) Report to congressional requesters: concentrated animal feeding operations

Gerber et al. (2013) Tackling climate change through livestock – a global assessment of emissions and migration opportunities. Food and Agriculture Organization of the United Nations (FAO), Rome

Gibbs HK (2015) Brazil's Soy Moratorium. Science 347(6220):377–378

Halden RU, Schwab KJ (2008) Environmental impact of industrial farm animal production. A report of the Pew Commission on Industrial Farm Animal Production. A project of the Pew Charitable Trusts and Johns Hopkins Bloomberg School of Public Health. www.ncifap.org/_images/212-4_EnvImpact_tc_Final.pdf. Accessed 16 Jan 2015

Hasan MR, Halwart M (eds) (2009) Fish as feed inputs for aquaculture: practices, sustainability and implications. FAO, Rome

Hoffman I (2011) Livestock biodiversity and sustainability. Livest Sci 139:69–79. doi:10.1016/j.livsci.2011.03.016

Hoffman I, From T, Boerma D (2014) Ecosystem services provided by livestock species and breeds, with special consideration to the contributions of small-scale livestock keepers and pastoralists. FAO Commission on Genetic Resources for Food and Agriculture

Hori Y, Stuhlberger C, Simonett O (eds) (2012) Desertification: a visual synthesis. UNCCD http://www.unccd.int/en/resources/publication/Pages/default.aspx. Accessed 9 Jan 2014

Horrigan L, Lawrence RS, Walker P (2002) How sustainable agriculture can address the environmental and human health harms of industrial agriculture. Environ Health Perspect 110 (5):445–456

Humane Society International (2011) An HSI report: the impact of industrial farming

Isenring R (2010) Pesticides and the loss of biodiversity. Pesticide Action Network Europe

IUCN (2000) IUCN guidelines for the prevention of biodiversity loss caused by alien invasive species. Fifth meeting of the conference of the parties to the convention on biological diversity, Nairobi, Kenya, 15–26 May 2000

Jackson L et al (2005) Agrobiodiversity: a new science agenda for biodiversity in support of sustainable agroecosystems. Diversitas

Lloyd DJ (2011) Crops for animal feed destroying Brazilian Savannah, WWF Warns. The Guardian. http://www.theguardian.com/environment/2011/apr/11/meat-industry-food. Accessed 26 Jan 2015

Mack RN (1989) Temperate grasslands vulnerable to plant invasions: characteristics and consequences. In: Drake J et al (eds) Biological invasions: a global perspective. Wiley, New York, pp 155–179

Mack RN et al (2000) Biotic invasions: causes, epidemiology, global consequences, and control. Ecol Appl 10(3):689–710

Matson PA et al (1997) Agricultural intensification and ecosystem properties. Science 277:504–509. doi:10.1126/science.277.5325.504

McAlpine CA (2009) Increasing world consumption of beef as a driver of regional and global change: a call for policy action based on evidence from Queensland (Australia), Colombia and Brazil. Glob Environ Chang 19:21–33. doi:10.1016/j.gloenvcha.2008.10.008

McGeoch MA et al (2010) Global indicators of biological invasion: species numbers, biodiversity impact and policy responses. Divers Distrib 16:95–108. doi:10.1111/j.1472-4642.2009.00633.x

Meyfroidt P, Lambin EF (2008) Forest transition in Vietnam and its environmental impacts. Glob Chang Biol 14:1–18. doi:10.1111/j.1365-2486.2008.01575.x

Millennium Ecosystem Assessment (2005) Ecosystems and human well-being: current state and trends, vol 1. Island Press, Washington

Molden D (ed) (2007) Water for food, water for life: a comprehensive assessment of water management in agriculture. Earthscan, London

New York State Department of Environmental Conservation (2014) Restoration and spending plan: marks farm natural resource damages settlement. http://www.dec.ny.gov/docs/fish_marine_pdf/mfarmresplfin.pdf. Accessed 4 Mar 2015

Nicholson SE, Tucker CJ, Ba MB (1998) Desertification, drought, and surface vegetation: an example from the West African Sahel. Bull Am Meteorol Soc 79:815–829

O'Mara FP (2011) The significance of livestock as a contributor to global greenhouse gas emissions today and in the near future. Anim Feed Sci Technol 166–167:7–15

Otte J et al (2007) Industrial livestock production and global health risks. Pro-poor livestock policy initiative

Pew Commission on Industrial Farm Animal Production (2008) Putting meat on the table: industrial farm animal production in America

Pew Center on Global Climate Change (2009) Enteric fermentation mitigation. http://www.c2es.org/docUploads/Enteric-Fermentation-09-09.pdf. Accessed 6 Mar 2015

Pimental D et al (1997) Water resources: agriculture, the environment, and society. Bioscience 47 (2):97–106
Rejmánek M, Richardson DM, Pyšek P (2005) Plant invasions and invasibility of plant communities. In: van der Maarel E (ed) Vegetation ecology. Blackwell, Oxford, pp 332–355
Rischkowsky B, Pilling D (eds) (2007) The state of the world's animal genetic resources for food and agriculture. FAO, Rome
Robertson B, Pinstrup-Andersen P (2010) Global land acquisition: neo-colonialism or development opportunity? Springer Science + Business Media B.V. & International Society for Plant Pathology 2:271–283. doi:10.1007/s12571-010-0068-1
Rodewald M (2015) Manure spills putting water supply at risk. Green Bay Press Gazette. http://www.greenbaypressgazette.com/story/news/investigations/2015/02/06/manure-spills-water-supply/22983669/. Accessed 4 Mar 2015
Rudel T, Roper J (1997) Forest fragmentation in the humid tropics: a cross-national analysis. Singapore J Trop Geogr 18:99–109. doi:10.1111/1467-9493.00007
Rudel TK, Bates D, Machinguiashi R (2002) A tropical forest transition? agricultural change, out-migration, and secondary forests in the Ecuadorian Amazon. Ann Assoc Am Geogr 92 (1):87–102
Sakirkin et al. (2012) Dust emissions from cattle feeding operations. In: Air quality in animal agriculture. Extension. http://www.extension.org/sites/default/files/Dust.pdf. Accessed 6 Mar 2015
Schimel D, Stephens BB, Fisher JB (2014) Effect of increasing CO2 on the terrestrial carbon cycle. PNAS. doi:10.1073/pnas.1407302112
Sharma S (2014) The need for feed: China's demand for industrialized meat and its impacts. IATP
Steinfeld H, Wassenaar T, Jutzi S (2006a) Livestock production systems in developing countries: status, drivers, trends. Rev Sci Tech Off Int Epiz 25(2):505–516
Steinfeld H et al (2006b) Livestock's long shadow: environmental issues and options. FAO, Rome
Steinfeld H et al (eds) (2010) Livestock in a changing landscape: drivers, consequences, and responses, vol 1. Island Press, Washington
Sustainable Table (2009) Air quality. http://www.sustainabletable.org/print/266. Accessed 6 Mar 2015
Swanepoel F et al (eds) (2010) The role of livestock in developing communities: enhancing multifunctionality. University of the Free State and CTA, Cape Town
Thrupp L (2000) Linking agricultural biodiversity and food security: the valuable role of agrobiodiversity for sustainable agriculture. Int Aff 76(2):265–281
Tilman D et al (2001) Forecasting agriculturally driven global environmental change. Science 292:281–284. doi:10.1126/science.1057544
Tscharntke T et al (2012) Global food security, biodiversity conservation and the future of agricultural intensification. Biol Conserv 151:53–59
UNCCD (2013) Migratory species and desertification factsheet. http://www.unccd.int/en/media-center/MediaNews/Pages/highlightdetail.aspx?HighlightID=194. Accessed 9 Jan 2015
UNCTD (2013) Trade and environment review 2014: wake up before it is too late. United Nations
UNEP (2014) Assessing global land use: balancing consumption with sustainable supply. A report of the working group on land and soils of the International Resource Panel
U.S. Fish & Wildlife Service (2008) National wildlife refuge system. Prescribed grazing. Managing invasive plants: concepts, principles, and practices. http://www.fws.gov/invasives/staffTrainingModule/methods/grazing/practice.html. Accessed 15 Jan 2015
U.S. Environmental Protection Agency (2015) Agriculture. In: DRAFT inventory of U.S. greenhouse gas emissions and sinks: 1990–2013. http://www.epa.gov/climatechange/pdfs/usinventoryreport/US-GHG-Inventory-2015-Chapter-5-Agriculture.pdf. Accessed 6 Mar 2015
Vanotti MB, Szogi AA (2008) Water quality improvements of wastewater from confined animal feeding operations after advanced treatment. J Environ Qual 37:S-86–S-96

Vasey DE et al (eds) (2011) Heavy metals. In Berkshire encyclopedia of sustainability: natural resources and sustainability. Berkshire Publishing Group, Great Barrington

Wardle DA (1999) Response of soil microbial biomass dynamics, activity and plant litter decomposition to agricultural intensification over a seven-year period. Soil Biol Biochem 31:1707–1720

Welton S, Biasutti M, Gerrard MB (2014) Legal and scientific integrity in advancing a "Land Degradation Neutral World". Sabin Center for Climate Change Law, Columbia Law School

York M (2005) Workers trying to contain effects of big spill upstate. New York Times. http://www.nytimes.com/2005/08/15/nyregion/15dip.html. Accessed 4 Mar 2015

Youyong Z (2000) Genetic diversity and disease control in rice. Nature 406:718–722

Chapter 2
Globalized Perspectives on Infectious Disease Management and Trade in Africa: A Conceptual Framework for Assessing Risk in Developing Country Settings

Kennedy Mwacalimba

Abstract In the era of globalization, internationalized representations of infectious disease threats have profound implications for understandings of infectious disease problems and their management in developing countries, particularly in Sub-Saharan Africa. By examining the policy implications of the key narratives around public health, animal health and trade, it becomes possible to clarify the relationship between global understandings of infectious disease risk and their impact on the development of local responses to disease problems. We highlight the tensions that resource-constrained countries face in the nexus of animal health-public health and trade, including the perception that resource-constrained countries are both source and victims of potential infectious disease threats. Given this scenario, it is important to think about how developing countries, particularly those in Sub-Saharan Africa, can approach infectious disease risk management as it relates to pandemic scale threats such as avian and pandemic influenza. We outline some of the key considerations in defining and assessing disease risk using avian and pandemic influenza in Zambia as an example. We conclude that the key to the feasibility of the analysis of the risk of multi-sectoral affecting emerging infectious diseases such as zoonotic avian influenza is flexibility in how risk is framed across the public health, animal health and trade systems.

2.1 Introduction

In the era of globalization, internationalized representations of infectious disease threats have profound implications for understandings of infectious disease problems and their management in developing countries, particularly in Sub-Saharan

K. Mwacalimba, BVM, MSc, DLSHTM, PhD (✉)
Zoetis, Florham Park, New Jersey, USA
e-mail: kennedy.mwacalimba@zoetis.com

G. Steier, K.K. Patel (eds.), *International Farm Animal, Wildlife and Food Safety Law*, DOI 10.1007/978-3-319-18002-1_2

Africa. By examining the policy implications of the key narratives around public health, animal health and trade, it becomes possible to clarify the relationship between global understandings of infectious disease risk and their impact on the development of local responses to disease problems. We highlight the tensions that resource-constrained countries face in the nexus of animal health-public health and trade, including the perception that resource-constrained countries are both source and victims of potential infectious disease threats. Given this scenario, it is important to think about how developing countries, particularly those in Sub-Saharan Africa, can approach infectious disease risk management as it relates to pandemic scale threats such as avian and pandemic influenza. We outline some of the key considerations in defining and assessing disease risk using avian and pandemic influenza in Zambia as an example. We conclude that the key to the feasibility of the analysis of the risk of multi-sectoral affecting emerging infectious diseases such as zoonotic avian influenza is flexibility in how risk is framed across the public health, animal health and trade systems.

2.2 Trade, Agriculture and Health

2.2.1 Setting the Stage

It has been known for quite some time now that there are very few human-specific pathogens.[1] Much of the current disease profile in humans owes to either the domestication of animal species or their use during our evolution from hunter-gatherer to agriculturally oriented societies. The human-animal interface is the nexus that permits the cross-species transmission of infectious agents and is represented by a continuum of contacts between humans and animals, either directly or indirectly through their products and their shared environments.[2] 'The human-animal interface'[3] is thus a term that encompasses the wider socio-economic and biological influences of disease transmission and spread, elements which are fundamental to the examination of human-animal infectious disease management. It is the human-animal interface that has arbitrated the transmission of zoonotic diseases and the introduction of novel pathogens into new geographical areas and novel host species. However, while its role in disease transmission is not new, because of globalization, its current ecological dimensions are of a completely different order of magnitude.[4] In essence, modern industrialized society is an

[1]Lloyd-Smith et al. (2009).
[2]Reperant et al. (2013).
[3]Greger (2007).
[4]Reperant et al. (2013).

important source of the expanded ecological pressure at the human-animal interface. The human-animal interface therefore provides an important conceptual framework for the examination of the public and animal health risks of animal-sourced epidemics, and through health policy, their relationship with risk enabling anthropogenic activities. These risk enabling activities include changes in land use, livestock production, chosen routes for economic growth and trade promotion; activities that both foster and enhance zoonosis transmission.[5]

2.2.2 *Globalization and the Relationship Between Trade and Health: The Microbial Perfect Storm*

Globalization plays a central role in shaping the debate around trade and health. This is because it is a comprehensive, multi-faceted phenomenon that is rapidly transforming society.[6] There are different, but important, understandings of what the term globalization means. Lee, Fustukian, and Buse,[7] describe globalization in terms of its spatial, cognitive and temporal dimensions, useful propositions for disaggregating the important aspects of policy that impact on the management of infectious disease risk. However, the key driver of globalization remains the internationalization of commerce; to which, it has been argued, health usually takes a backseat.[8]

According to a National Academies of Science report,[9] the globalization phenomenon has had a snowball effect with regard to infectious disease emergence. It has helped to create the microbial equivalent of 'a perfect storm'. Mann,[10] states this microbial perfect storm will not subside, but will be a recurring event. Changes in land use, livestock production, chosen routes for economic growth and promotion of commerce, climate change etc. are some of the elements that go into the 'perfect microbial storm'. Under livestock intensification, for example, the larger collections of animals provide optimal incubating conditions for the expansion of emerging zoonotic diseases.[11] Globalization is therefore a conduit for infectious disease spread; mainly because of the increased industry, cultural, and microorganism interconnectedness it fosters.[12]

---

[5]Kimball (2006), Greger (2007).

[6]Huynen et al. (2005).

[7]Lee et al. (2002).

[8]Navarro (1998).

[9]*ibid.*

[10]Mann (1990).

[11]Brown (2004).

[12]*Ibid.*

Recent examples of global infectious disease spread, such as Severe Acute Respiratory Syndrome (SARS) and pandemic avian influenza, have negatively impacted on both public health and economies. Such threats have led to a shift to develop policies to respond to these risks, at national, regional and international levels. But because the dynamics, and therefore the risks, of disease emergence differ from location to location, equally important is the integration, within these policy frameworks, of approaches to assessing both the risk's 'local' likelihood and 'impact' to ensure, to the extent possible, the appropriateness of policy responses.[13] This is a challenge, for both developing and developed countries, given the myriad interests that contribute to this 'perfect microbial storm.' It is reasonable to assume, however, that the risk of infectious diseases, and in particular, pandemic scale infectious disease emergence, is unlikely to abate, and as a result, the public health and animal health communities have to think of emerging infectious diseases, their control and the assessment of their risk of occurrence in completely novel ways.

2.2.3 Villain, Accomplice or Innocent Bystander: Trade and Disease Emergence and Spread

Trade and its effects on public health, through disease spread, is a matter of both historical and contemporary policy significance. Historically, disease has spread through traded products and carriage vehicles such as ships, which served as means of introduction of infectious agents into new geographic areas.[14] The link between international trade and the spread of infectious diseases has therefore been recognized for centuries, for example the fourteenth century spread of the 'Black Death' along known international trading routes.[15] It was this recognition that resulted in the International Sanitary Conferences, the first of which was held in France in 1851.[16] At several points in history, trade has been restricted to protect health, with the primary motivation being to minimize interference in trade from health.[17] In recent times, economic interests have taken precedence over health concerns.[18]

[13] Mwacalimba (2012).

[14] Cowen and Morales (2002).

[15] Bettcher et al. (2000).

[16] *Ibid.*, Aginam (2002), Hoffman (2010).

[17] Lee and Koivusalo (2005).

[18] Lang (1999).

Similar arguments have been made concerning global health policy,[19] access to medicines,[20] food safety[21] and infectious disease spread.[22]

Considering the role that trade policy plays in disease spread and control is relevant. Key here is the view of public health proponents that health concerns most of the time plays second fiddle to the interests of global commerce. For instance, Lipson's[23] review of the World Trade Organization's (WTO) health agenda and the study by Shaffer et al.[24] on ethics in public health research both suggest that trade agreements in particular shape national policies on such issues as food safety and health, restricting the capacity of state agencies to regulate these areas. Within this literature are examples that speak to the increasing interconnectedness of infectious disease spread through trade, an anthropogenic activity. They also highlight the importance of the human-animal disease interface.

Admittedly, tensions between trade promotion and health protection have existed in the past, but these tensions are increasing because of globalization.[25] Examples of the international transmission of diseases associated with commerce include the case of Monkeypox in the US in 2003, related to the trade of prairie dogs that had acquired the infection from the African rodents they had been housed with. This led to 71 human cases in six American states.[26] For SARS, bat trade was proposed to be one way in which contact with susceptible amplifying hosts was made at some point in the wildlife supply chain, leading to subsequent market-related human and animal interaction and infection.[27] Live animal markets in Southeast Asia have been implicated in the spread of emerging diseases such as avian influenza, with subsequent human exposure.[28] Even for countries in Africa, African Swine Fever, an animal health problem of trans-boundary animal disease significance, spread rapidly along the Atlantic coast in the dynamic coastal trading networks of West Africa during the late 1990s.[29]

Looking at the trade and health problem from a slightly different perspective, perhaps it is not a simple case of one set of concerns taking pre-eminence over another. The global health governance boundaries are actually being reshaped through the "legally binding" and "soft-law" provisos negotiated and adopted within the respective mandates of multilateral institutions such as the World Health Organization (WHO), the WTO, Food and Agricultural Organization (FAO) and

[19]Lee et al. (2002).

[20]Kerry and Lee (2007).

[21]Rowell (2003).

[22]Kimball (2006).

[23]Lipson (2001).

[24]Shaffer et al. (2005).

[25]Lee and Koivusalo (2005).

[26]Morse (2004), Kahn (2006).

[27]Fevre et al. (2006).

[28]Cowen and Morales (2002), Morse (2004), Karesh et al. (2005).

[29]ALive (2006).

OIE (World Organization for Animal Health).[30] These include international health guidelines, trade agreements and approaches to disease control, all grounded in 'international standards'.

In an attempt to clarify the trade-health relationship and foster greater coherence between the international health and trade communities, a joint WHO/WTO study examining the links between trade and health was published in 2002.[31] This effort did very little to alleviate the concerns of the public health camp and has been described as disappointing by some analysts.[32] However, with the revision of the International Health Regulations (IHRs) in 2005, an important milestone for global public health was reached, enabling the global public health community to attempt to address the more contemporary problems presented by infectious disease threats. Health proponents argue that health compromises continue to be made. Meirianos and Peires,[33] for example, maintain that the revised IHRs made trade-offs between national sovereignty and global health by attempting to guard against global disease spread with minimum interference to trade and travel. So the global health and global trade communities again find themselves at a cross-road insofar as infectious disease control is concerned.

Agriculture, of which animal health is a component, has been pulled into the foray as one of the many interfaces between trade and public health. Perhaps to nudge the animal health camp in particular to align more closely to public health propositions, it has been suggested that the OIE regulations, the animal health counterpart to the IHRs, require a similar revision to better align them with the present-day threats presented by trans-boundary diseases.[34] No attempt has been made to overhaul the OIE regulations, but the international animal health community appears to be moving closer to health by adopting a global perspective on the control of zoonoses.[35] Simultaneously, the international animal health community has taken an active pro-trade stance in their address of issues surrounding trade and health protection. The OIE has been setting international animal health standards for purposes of facilitating safe trade in livestock and livestock products of trade under the Sanitary and Phytosanitary (SPS) Agreement of the WTO in its Terrestrial Animal Health Code.[36] Thus countries that are involved in livestock and livestock product trade are expected to comply with the SPS Agreement in order to reap the full benefits of international commerce.[37] Pushing a free trade agenda,

[30]Aginam (2002).

[31]WHO/WTO (2002).

[32]Howse (2004).

[33]Merianos and Peiris (2005).

[34]*Ibid.*

[35]Blancou et al. (2005).

[36]Bruckner (2009); OIE Terrestrial Animal Health Code 2010, available at http://web.oie.int/eng/normes/mcode/en_index.htm.

[37]Thiermann (2005).

Zepeda et al.[38] uphold (the SPS regulation) that public health measures to ensure food safety and to control plant or animal diseases should be based, as far as is appropriate, on international standards, presumably freeing them from having to justify their animal health policies through analyses of 'risk'. The SPS Agreement espouses the view that measures to protect public health, animal health and plant health should only minimally interfere with trade. It is this 'clause', similarly adopted under the revised IHRs, that has been found to be problematic at national level.[39] Pragmatically, this also demonstrates both the increasing interaction of different areas of international policy in fostering of commercial interests within international health and agriculture.

Similarly a shift in approaches to infectious disease control has occurred, with the entry of internationally important infectious diseases such as SARS and pandemic avian influenza into the world policy arena; that of moving from nation-focused to global-focused control mechanisms.[40] While the merits of a global approach to infectious disease control cannot be disputed, in this shift is an assumed universal acceptance of what infectious diseases should be prioritized on both global and national agendas, the 'risk' they present and how they should be controlled. It is important to understand how developing countries go about responding to these 'global' imperatives, given their unique circumstances. The importance of such research is made especially relevant with the issue of zoonotic risk management.

2.3 The Development Agenda

Global and regional trade present the prospect of involving previously excluded nations in world commerce, thus enabling them to supply more prosperous markets and support and strengthen their economies. This prospect appeals to decision-makers in developing countries, because it promises the positive benefits of trade liberalization such as economic growth and poverty reduction.[41] It has been argued, however, that a liberalized approach to trade, presents novel challenges to public health protection in general and disease prevention and control in particular. Few authors, except as an adjunct, have attempted to include animal health in this discourse, or highlight the combined impacts key policy debates have on development in resource-constrained countries. It is thus clear that an in-depth examination of the public health-animal health-trade nexus as it concerns infectious disease governance in resource-constrained countries is needed to better illuminate important complexities surrounding the development agenda. It is

[38]Zepeda et al. (2005).

[39]Merianos and Peiris (2005).

[40]Fidler (2004a), Lee and Fidler (2007).

[41]Wilkinson and Pickett (2006).

also an important step in highlighting the myriad commonalities and polarities between the developmental needs of the global South and the demands of the global North.

2.3.1 From Global to Local: Developing Countries and Trans-Boundary Infectious Diseases

Trans-boundary animal diseases and their unlikely eradication in the foreseeable future pose a significant problem for developing countries. They almost automatically exclude them from involvement in global trade under WTO regulations.[42] International standards have been used to restrict the direction of trade, on health grounds, from resource-enabled to resource-constrained countries. For instance, Rweyemamu and Astudillo,[43] state that the global distribution of Foot and Mouth Disease (FMD) mirrors the world's economic structure, with industrialized countries generally being free of the disease while developing countries were endemic, which pushes trade in a North–south direction. Furthermore, even with international guidelines and standards provided to facilitate trade, many developing countries have to deal with a range of animal diseases simultaneously; making regulation and technical considerations extremely difficult.[44]

The dominant view is that developing countries pose the greatest risk as sources of infectious diseases.[45] In fact, the FAO's philosophy is to control these diseases at *this* source.[46] This perspective also implies that disease control efforts would focus on the 'global impacting' disease problems from this source, but foster particular methods of control that may not be appropriate for different contexts. Adopting such methods can harm local livelihoods or worse, inadvertently encourage further disease spread.[47] Furthermore, as argued in an analysis of the politics of the securitization[48] of health, a lip service effect may be created, as policy actors in different contexts are pressured to verbalize an infectious disease threat as a priority, but may not treat it as such.[49]

For the world's poorest states, the confluence of interests surrounding global health and global trade therefore presents unique challenges. Global perspectives on infectious disease control and the policies that result have a significant influence

[42] Thomson et al. (2004).

[43] Rweyemamu and Astudillo (2002).

[44] *Ibid*, Upton and Otte (2004).

[45] Hampson (1997), Domenech et al. (2006), Kruk (2008).

[46] Domenech et al. (2006).

[47] Scoones (2010).

[48] Securitization of health is the process through which infectious diseases are viewed as national security threats, particularly with regards to bioterrorism.

[49] Lo Yuk-ping and Thomas (2010).

on development opportunities. Of note is how trade policy such as the SPS Agreements is viewed to exclude developing countries from participating in global trade. More importantly, in these debates, resource-constrained countries are dichotomously perceived as simultaneously needing the most protection and posing the greatest risk. In such contexts, the relationship between public health, animal health and trade is complex and is possibly made more so when issues such as zoonotic risk management and such things as pandemic preparedness are brought into the picture.

2.3.2 Paradigm Shift: Moving Away from the Grown-Up Table

Most developing countries have joined the WTO out of concern that they will be excluded from trade opportunities.[50] Ironically, it appears that by participating in the WTO, less developed countries have been disadvantaged. Therefore, is it cost effective for resource-constrained countries to attempt to meet 'international standards' in trade, or disease control? There are arguments for and against this. Authors like Rweyemamu and Astidullo,[51] for example, have proposed ways in which FMD endemic developing countries could benefit from global trade in livestock and livestock products. On the other hand, others, such as Cumming[52] (citing Jansen et al.[53]) explain how, for instance, the Zimbabwean Government investment in the scaling up of veterinary services and abattoirs to meet European Economic Community (EEC) import standards in the 90s resulted in a net loss to the country because the cost of these renovations exceeded beef export revenues.

To counter the disadvantages faced by developing countries under the current multilateral trading system, nation states have formed alliances with similarly positioned nations. These alliances, and to some extent some civic organizations, are increasingly demanding that the interests of developing countries be better represented at the WTO.[54] Developing countries have also been inward looking, and constituted regional and economic trading blocs, which Roningen and DeRosa[55] contend, put member countries on the path to free trade and its associated benefits, and, politically, are thought to be easier to negotiate because they do not require consensus at the WTO. A plethora of regional and sub-regional committees has emerged on the African continent, forming a complex network of sometimes overlapping trade regions.

---

[50]Shaffer et al. (2005).

[51]Rweyemamu and Astudillo (2002).

[52]Cumming (2010a).

[53]*Ibid.*

[54]Labonte and Sanger (2006).

[55]Roningen and DeRosa (2003).

With this shift to multilateral and regional trade, a growing interest in livestock trade among resource-constrained countries has emerged. An International Livestock Research Institute (ILRI) and the FAO study projected that by 2015, 60 % of meat and 52 % of the world's milk will be produced in developing countries.[56] This study described a "Livestock Revolution" driven by increasing demands for livestock and livestock products in low-income countries as a result of, among other factors, expanding urban populations. These investigators also projected that by 2020, livestock product trade, particularly trade in meat, milk and eggs, would likely be of increasing importance for resource-constrained countries, both in terms of trade between resource-constrained countries and trade with the rest of the world.

Recent evidence suggests that indeed there has been a general increase in the amount of trade in agricultural produce among resource-constrained countries. According to the World Trade Report[57] the share of intra-developing country agricultural exports increased from 31 % in 1990 to 43 % in 2002. It also states that 47.6 % of developing country imports originated from other developing countries. Here again, health commentators assert that the shift to bilateral and multilateral trade agreements is pushing an economic agenda at the cost of health and, it is argued, developing countries are likely to suffer the most.[58] But what are the policy implications of the current shift to regional and bilateral trade agreements and intra-continental trade promotion are for understandings of infectious disease threats?

2.3.3 Unpacking Risk: A Conceptual Framework for Assessing Risk in Developing Country Settings

There is some suggestion in the literature that global policy actors assume infectious disease risk is universally understood, and use this as a platform to drive collaboration in policy responses across sectors at international and national levels. Much of the available literature, understandably, does not fully examine the role that public health, animal health and trade play in multi-sectoral risk management and pandemic preparedness at national level, particularly in resource-constrained settings. While broad themes can be drawn from current knowledge, the discourse on global infectious disease governance and its relationship with global trade is still unfolding.

[56]Delgado et al. (1999); Food and Agriculture Organization of the United Nations, Rome; International Livestock Research Institute, Nairobi. Food, Agriculture, and the Environment Discussion Paper 28.

[57]WTO World Trade Report 2004.

[58]Lee and Koivusalo (2005).

A myriad of events are now perceived to be easily amenable to risk assessment, thanks to the development of scientific approaches to, and the universalization of risk language in, the management of physical, chemical and biological threats. However, when threats such as zoonotic diseases are global rather than local, the vagaries of context, institutions and culture play important roles in the construction of such events as risks, elements that are not exogenous to the technical-scientific processes of risk analysis. This empirical section is based on a policy study that examined the avian and pandemic influenza policy process in Zambia over the 2005–2009 period to suggest a pragmatic way of increasing the efficacy of risk analysis methodology in guiding livestock trade decisions and multi-sectoral disease risk management in resource-constrained contexts. Using the World Organization for Animal Health (OIE) risk analysis framework as an illustration, we demonstrate how the cross-cutting and highly contingent nature of today's infectious disease threats provide learning points for re-conceptualizing the use of risk analysis to inform policy, to better account for the institutional and social phenomena that frame both risk perception and management. While accepting this may be viewed as breaking the conventions of scientific objectivity in the process of risk assessment, we conclude that this approach is necessary for developing economy-friendly multi-sectoral zoonosis risk management strategies in developing countries like Zambia.

2.3.4 *Theoretical Framework: Risk as a Confluence of Probabilistic Science and Social Construction*

A few key theories stand out with respect to understandings of risk within contemporary global society. For instance, Urlich Beck[59] in his seminal book *Risk Society* introduces the theory of reflexive modernization in which the processes of modernity in industrialized societies are posited to be the cause of the emergence of unprecedented and indeterminate risks and hazards, including those presented by infectious diseases. Some of these modernization processes were alluded to earlier, when we discussed the factors that have led to the emergence of infectious diseases. Within this body of work, risk creation, construction and response are intrinsically linked to modernization, and knowledge and science are argued to play a constitutive and sometimes unexamined role in these processes. Beck's views share commonalities with those of another renown sociologist, Anthony Giddens.[60] In Giddens'[61] conception of reflexive modernization, the increasing dependence on

---

[59]Beck (1992).

[60]Giddens (1998), pp. 23–34.

[61]see also Lupton (1999).

society's 'experts' to determine what is and how to respond to 'risk' in societies, has brought with it risk's polar opposite, uncertainty. Risk analysis, for instance, is a process of creating scenarios of risk which are based on contingent scientific knowledge and is therefore subject to change. Uncertainty arises when risk cannot be precisely calculated, e.g. when the probability of occurrence of adverse events are unknown or inestimable.[62] Uncertainty and surprise, in turn, have led to concerns over the validity of purely scientific responses to risk.[63]

Both Beck and Giddens propose a more reflexive approach to risk in which the underpinnings of scientific assertions are drawn out, their situational implications assessed and alternative knowledge bases co-opted, thus taking the risk assessment process out of the 'problematized' purely scientific sphere into a more discursive treatment of 'risk'. An important assertion is made by risk sociologists Douglas and Wildavsky,[64] that although the dangers are real, risk is 'politicized' through several social processes, giving risk a status which is separate from the actual dangers presented by various hazards. Slovic[65] further states that this politicization process makes risk assessment a subjective blend "of science and judgment with important psychological, social, cultural, and political factors".[66] Douglas[67] in particular, presents risk, within a social context, as attributable to an Other. The position of 'Otherness' is subsumed by developing countries, where they are presented as both source and victim of various infectious disease threats. Therefore, a primary focus is to assign blame, first in the global narrative (North to South, or West to East) and then in a regional narrative.

But infectious disease threats are not merely western obsessions misaligned with the needs and subjectivities of developing country contexts. Douglas views risk as 'a socially constructed interpretation and response' to a *real* danger. This is an important consideration in developing risk assessments in resource-constrained settings. Following this train of thought, before a context-relevant and reflexive approach to risk analysis can be proposed, it is important to understand what the framing assumptions of infectious diseases and their impacts are, how they emerge and how they influence the policy process in each context. We will now examine the narratives concerning avian and pandemic influenza, first from an international perspective, and then look at the narratives from the perspective of one developing country, Zambia.

[62]*Ibid.*

[63]see Stirling and Mayer (2000), Millstone (2007), Stirling and Scoones (2009).

[64]Douglas and Wildavsky (1982).

[65]*Ibid.*

[66]see also Horlick-Jones (1998), Pidgeon (1999), Slavic (1999).

[67]Douglas and Wildavsky (1982), Lupton (1999).

2.4 Timeline of the Crisis: The Epidemiology of H5N1 Avian Influenza

The H5N1 problem began at a goose farm in Guangdong Province, southern China in 1996, where it killed around 40 % of the flock.[68] It then spread to three chicken farms in Hong Kong, just adjacent to Guangdong Province, between March and early May of 1997.[69] In May of the same year, a child died of viral pneumonia; the first reported case of zoonotic H5N1 influenza.[70] Following the identification of 17 more human infections that resulted in five deaths between November and December of 1997,[71] H5N1 became recognized as a zoonosis of possible public health concern. As a result, in December 1997, total and rapid depopulation of all poultry in markets and chickens farms in Hong Kong was carried out to control the outbreak, a move that both policy and virology experts believed had averted a human pandemic.[72] In this outbreak, live poultry markets were important in the transmission of the H5N1 virus to other avian species and humans.[73]

However outbreaks had continued to occur in poultry in Hong Kong from 2001 to early 2002, caused by a different H5N1 lineage.[74] In February 2003, during the SARS epidemic, three more human H5N1 infections with two fatalities were identified in China, and according to the WHO, this indicated viral persistence, despite the control measures that had been instituted in 1997.[75] While there is some suggestion that the H5N1 problem had been subdued in 1997,[76] it was in fact, entrenching itself in the poultry systems of Hong Kong, and possibly elsewhere in Southeast Asia, between 1997 and 2003.

Between December 2003 and February 2004, the first wave of an H5N1 panzootic in poultry was reported nearly simultaneously in eight countries in South and Southeast Asia, most of which occurred in commercial poultry establishments. This was followed by a second wave of spread from July 2004.[77] The WHO states that the second wave was associated with more rural settings.[78] The countries initially affected were China, Indonesia, Cambodia, Japan, Laos, Korea, Thailand and Vietnam, with a ninth country, Malaysia, joining the list in August 2004.[79] The pro-poor advocacy NGO, GRAIN, states that the initial outbreaks in

---

[68]Xu et al. (1999), Webster et al. (2002).

[69]Shortridge et al. (1998).

[70]de Jong et al. (1997).

[71]Shortridge et al. (1998).

[72]Fidler (2004b), WHO (2005a), Webster and Hulse (2005).

[73]Shortridge et al. (1998).

[74]Sims et al. (2005).

[75]WHO (2005c).

[76]*Ibid.*

[77]Alexander (2007), Paul et al. (2010).

[78]WHO (2005c).

[79]Sims et al. (2005).

Vietnam, Thailand, Cambodia, Laos and Indonesia all occurred in closed, intensive factory farms.[80] During the first wave, millions of poultry either died or were culled in an effort to control the disease.[81] Human infections were then reported in Hanoi, Vietnam, in January, 2004, a few days prior to a report of large H5N1-related poultry mortalities in two poultry farms in the south of the country.[82] Vietnam had initially experienced an H5N1 outbreak in 2001.[83] In early 2004, during the first wave of the panzootic, the WHO declared the outbreak an unprecedented catastrophe for agriculture in Asia and a "global threat to human health".[84]

Coinciding with the second wave of the panzootic, the period between August and October 2004 saw eight more human fatalities in Thailand and Vietnam.[85] The third wave began in December 2004, involving new poultry outbreaks in Indonesia, Thailand, Vietnam, Cambodia, Malaysia and Laos.[86] Fresh human cases were reported in Vietnam, Thailand and Cambodia.[87] At this point, after reviewing the unfolding situation, a writing committee of the WHO consultation on human influenza established that Vietnam led the human death toll.[88] According to a WHO pandemic threat report,[89] by 2005, H5N1 had 'succeeded' in crossing the species barrier three times; in 1997, 2003, and the period between 2004 and early 2005, which recorded the largest occurrence of human H5N1 cases in the period in question. With the report of migratory birds being affected with H5N1 in Mongolia and China, particularly at Lake Qinghai in China in April 2005, concern grew that this posed a potential risk of southward and westward and therefore global spread of the virus in poultry.[90] Around 6345 birds of different species died in the weeks following the Qinghai outbreak.[91] This is possibly the single most important event linking H5N1 to migratory bird spread. This outbreak singularly raised the profile of the role of migratory birds in the global spread of H5N1.

H5N1 had spread through the diverse market and poultry production systems of Southeast Asia. There is much debate around the primary causes and drivers of the H5N1 problem, revolving around poultry production and marketing practices. An important factor in the Asian panzootic is that ducks appeared to have played a key role in the maintenance of the virus, primarily as silent carriers of H5N1. By 2005, H5N1 had become endemic in the duck population of poultry, providing a reservoir

[80]GRAIN (2007).
[81]WHO (2004).
[82]WHO (2005b).
[83]Sims et al. (2005), Sims and Narrod (2008).
[84]WHO (2004).
[85]WHO (2005c).
[86]Sims et al. (2005).
[87]WHO (2005c).
[88]Beigel et al. (2005).
[89]WHO (2005b).
[90]Chen et al. (2005), Webster and Govorkova (2006), Alexander (2007), Cattoli et al. (2009).
[91]WHO (2005b).

of the virus for other poultry species as asymptomatic shedders of H5N1 influenza.[92] While outbreaks in poultry were still possible, this suggests that in areas where duck production was of less significance, the chances of endemicity could be lower.

2.4.1 The International Narratives in the Global Response to Avian and Pandemic Influenza

Ian Scoones[93] uses 'policy narratives' as framing devices for understanding how disease is understood, identifying which actors are likely to be included or excluded from the policy process, what policy avenues open or close as a result and whose interests are likely to be served. Here we use this approach to outline the dominant global policy narratives in the avian and pandemic influenza crisis that was at its peak in 2005, and then see what national level policy narratives emerged in Zambia in response.

In their research, Scoones and Forster[94] found three primary outbreak narratives driving the global response to avian and pandemic influenza. These were a veterinary narrative, focused on animal health and agricultural livelihoods; a public health narrative focused on human health and disease, and a pandemic preparedness narrative which drove an emergency response. The three outbreak narratives were distilled from a typology of linked debates identifiable in the international policy discourse concerning avian and pandemic influenza. These debates largely revolved around risk and uncertainty, and drove understanding of disease, its implications, and the mitigation responses advocated.

First, the source of the H5N1 threat was Southeast Asia, referred to as an "influenza epicenter".[95] The disease had a visible human health impact, with hundreds of cases logged in three waves by the WHO after the first 18 cases and one death in 1997. However, a lot of uncertainty still existed around both H5N1 evolution as a zoonosis and its effects on public health.[96] Although some understood that public health experts remained uncertain of the likelihood of a human pandemic,[97] the possibility of a pandemic resulted in calls to focus control on the source of this risk; Southeast Asia, and to develop contingencies incase control efforts failed.

Second, because H5N1 was viewed as largely a problem in poultry, the surveillance and control responses championed were considered to be in the veterinary space, with their arsenal of 'tried and tested' methods for disease control. But the

---

[92]Webster and Hulse (2005); Sims et al. (2005); Sims and Narrod (2008).

[93]Scoones (2010).

[94]Scoones and Forster (2010).

[95]see Hampson (1997).

[96]Pitrelli and Sturloni (2007).

[97]Osterholm (2005).

'standardized' approaches adopted worked in some areas and failed in others. For instance, control measures such as culling, disinfection and stamping were successful in controlling H5N1 outbreaks in Europe, but were not as effective in Southeast Asia.[98]

Third involved linkages between poultry production practices, H5N1 epidemiology and disease spread though poultry and poultry product trade and/or migratory bird movement. It was suggested that all parts of the world were at risk of H5N1 incursions as a result of the globalization of trade.[99] Some authors took the view that it was migratory birds that would spread H5N1 across the globe,[100] while others claimed that wild birds were only capable of short range spread.[101]

Third was the potential effect of a human pandemic on the global economy. This concern also drove the 'at source' control initiative. The H5N1 risk mitigation responses largely affected the livelihoods of those in outbreak areas.[102] For example, it was estimated that over 2 billion birds were slaughtered in the effort to control H5N1, and the greatest losses were suffered by the poor.[103] There was also a national level impact as well, where several countries (e.g. Thailand) had their poultry exports prejudiced and rural livelihoods affected by control interventions.[104] This debate thus had links to contentions between business and livelihood interests and controversies over the role of intensive vs. backyard farming in disease spread.[105]

Fourth concerned the development of a multi-sectoral approach response to mitigate the pandemic threat. This included calls to strengthen veterinary control systems in addition to human pandemic preparedness, addressing the pandemic risk at-source but involving human health and other sectors to mitigate the risk.[106] Following outbreaks of H5N1 in Egypt and Nigeria, Africa also popped up on the global public health radar as the next potential reservoir of the H5N1 virus. A WHO Regional Office for Africa risk assessment[107] made sweeping comparisons between Asian and African poultry production systems to justify similarities in risk and provide recommendations for prevention and control. The problem was, however, that the poultry production systems in Africa and Asia are in reality, very different.

The fifth debate involved pharmaceutical interests, covering influenza virus sharing and concerns that genetic sequence information collected from outbreak areas would be used to create vaccines for market that would not be distributed

[98] Yee et al. (2009).
[99] van den Berg (2009).
[100] Normile (2006), Chen et al. (2005).
[101] e.g. Weber and Stilianakis (2007).
[102] Scoones and Forster (2010).
[103] Stirling and Scoones (2009); also Scoones and Forster (2010).
[104] Nicoll (2005).
[105] GRAIN (2006a), GRAIN (2006b), GRAIN (2007).
[106] WHO (2004).
[107] WHO/AFRO (2005).

equitably in case of a pandemic.[108] The policy response was Western countries scrambling to stockpile antiviral drugs and vaccines for 'high level pandemic preparedness efforts', the vaccines of whose production depended on H5N1 virus strains recovered from developing countries.[109] In an attempt to globalize this policy response, there were also calls for affected countries to either develop pharmaceutical capacity or consider non-pharmaceutical interventions.

Linked to this was the sixth debate, involving the 'securitization' framing of the avian and pandemic influenza issue, which, Elbe[110] argued, escalated the controversy over influenza virus sharing. In implementing this 'securitization' approach, Western countries spent massively on pandemic preparedness, with the US and European countries spending approximately US$2.8 billion 'at home' versus US $950 million 'abroad' for disease control 'at-source' by the end of 2008.[111] This forms the background against which resource-constrained countries generated their avian and pandemic influenza intervention policies guided by the WHO global pandemic preparedness plan.[112] These viewpoints can be grouped in four key typologies; risk and uncertainty, effects on food and farming, economy and livelihood impacts and effects on health and extent of disease.

2.4.2 How H5N1 Was Defined in Zambia

Zambia's response to H5N1 was initially motivated by the internationalized outbreak narrative, facilitated by the tripartite alliance of the OIE, FAO and WHO.[113] Furthermore, several national policy actors played a critical role in initially framing H5N1 influenza as an imminent threat, forming a multi-sectoral Task Force on Avian Influenza in 2005. The formation of this committee was facilitated by World Bank funding and FAO and WHO technical expertise. The Task Force comprised representatives from agriculture, health, the poultry industry, academia and local media.[114] There were also representatives from the Ministry of Local Government and Housing, the Zambia Revenue Authority, the Ministry of Home Affairs, the Office of the Vice President, the Ministry of Tourism and Environment, the Ministry of Finance and National Planning, and the United States Agency for International Development (USAID). From this core membership, a technical arm of the Task Force National Avian Influenza Working Group, was constituted,

---

[108]Garrett and Fidler (2007), Fidler (2008).

[109]Elbe (2010).

[110]*Ibid.*

[111]Burgos and Otte (2008).

[112]WHO (2005c), ALive (2006).

[113]Mwacalimba (2012).

[114]Mwacalimba and Green (2015).

comprising an assembly of designated personnel from the ministries of Health and Agriculture.

Zambia initially viewed H5N1 and pandemic influenza as an imminent threat. Perceptions that H5N1 was on Zambia's doorstep were fuelled by unsubstantiated local media reports of bird flu outbreaks in Zambia's poultry, no doubt mirroring the international perspective as seen through the lens of the brewing H5N1 crisis in Southeast Asia. One respondent in the Ministry of Health (Interview 11) explained that the threat to Zambia was being taken seriously at the highest level in the Ministry of Health, with a Cabinet Memo being issued by the then Health Minister, Sylvia Masebo, seeking government input to respond to 'the threat of avian and human influenza that which was coming.'

Although on face value, Zambia seemed to have initially taken a unified stance in dealing with the problem, the array of stakeholders in Zambia's Task Force on Avian Influenza also meant that a number of different understandings of the H5N1 problem were at play. These alternative framing narratives gave impetus to some policy actors, and demotivated others participation in the emergency planning process. They also impacted on the implementation of national policy.

Six unique narratives were identified from interviews with stakeholders, suggesting some different implicit ways of framing the issue in Zambia. These narratives provide insight into the framing assumptions driving the different stakeholders' actions or inactions, in the pandemic preparedness process. More pertinent to this chapter, these narratives also reveal how the threat of H5N1 was constructed as 'risk' and to whom this risk pertained. There were narratives that chimed with the global narratives on avian and pandemic influenza and others that aligned less easily with international viewpoints, in particular, the narratives relating to trade and development.

The first narrative presented H5N1 is an exotic emerging disease. From the onset, there was a strong sense among some veterinary stakeholders that H5N1 was alien to Zambia. An example of this viewpoint is provided in the words of a senior veterinary member of Zambia's Task Force on Avian Influenza, 'We don't have avian influenza as you know. It is an exotic disease to us, but it is a possible emerging disease' (Interview 4). Linked to this was a second narrative framing, which presented H5N1 is an infectious agent of poultry with limited zoonotic potential. This narrative therefore framed H5N1 as a predominantly poultry health concern, described as 'basically ... more of an animal disease which then moves into human beings' (Interview 2). This narrative also justified the need for veterinary leadership in developing any prevention measure to be taken, and reflected the broader international veterinary narrative on H5N1, a problem that required the use of standardized, time-tested technical veterinary approaches to animal disease control.[115]

The problem with this narrative is that in Zambia, at least, the country's veterinary priority lay on controlling cattle diseases. For decision-makers, poultry

[115]*Ibid.*

production was a low priority, 'When you look at our focus, we are more oriented towards cattle. So most of the diseases that affect cattle are given priority. Probably you will look at it and what you get is that birds or bird diseases are not so significant or are not so associated with major economic losses. I think, it's not just because it is avian influenza and it is not there, it's because its poultry and it's not so significant. It's not written but it is implied in the way we do things.' (Interview 4, MACO). Another respondent put it this way, 'I mean you have to remember that they [cattle diseases] already exist in Zambia. We have FMD, it's spreading like fire, ok? So definitely the Government provides funds for FMD, because it's there. The Government will definitely provide funds for CBPP because it's there on the ground, right now.' (Interview 5, MACO).

Although the funding for preparedness planning had largely come from international donor agencies, the resources spent on H5N1 were essentially viewed as wasteful.[116] In the words of a senior veterinary officer, 'Yes you can argue for emergency preparedness but I think over and above, a lot of resources have gone into this (avian influenza) which should have been focused on the more important diseases for the country' (Interview 3, MACO).

The third narrative identifiable from the data was the perception that the greatest risk for H5N1 emergence was rural poultry systems because of their poor biosecurity, low awareness of the H5N1 problem in rural communities and high likelihood of contact with infectious migratory waterfowl. This also resonated with the international narrative presenting the risk of H5N1 as largely emanating from the backyard poultry production systems of Southeast Asia. There were contextual differences between the production systems of Southeast Asia and Zambia in respect to typical farming practices, for instance, the role that rice paddies, duck production and wet markets that went side by side with chicken rearing in Southeast Asia was widely dissimilar to the small flock scavenger chickens reared in rural Zambia.[117] Furthermore, even the concept of backyard production had a different meaning in Zambia. It represented small scale, often commercially oriented stock fed flocks of broiler or layer chickens, raised to supplement household income through chicken and egg sales, using closed structures with some level of restricted access.[118]

Nevertheless, the risk of H5N1 from rural farmers in Zambia was couched in a biosecurity narrative in which smallholder, rural poultry producers were viewed as representing the biggest risk for introducing avian influenza into the country, 'We believe that avian influenza may come from a poor farmer who doesn't believe in biosecurity. Most of these guys lack knowledge. They don't really understand some of these issues. So we think that it is from there, a lack of information and knowledge, that the disease can come,' (Interview 6). Another respondent put it

[116]Mwacalimba (2012).
[117]Mwacalimba (2013).
[118]LSUAC (2008).

this way, 'It is widely accepted that... (breeders and commercial producers), because of their biosecurity levels, chances of them actually getting avian influenza are pretty minor. That's why FAO doesn't want to deal with them. So we are dealing with (sector)[119] three and four; these are the emerging and the traditional farmers,' (Interview 5, MACO).

The fourth narrative, representing the views of stakeholders from the Health sector, centered on H5N1 as a possible zoonotic pandemic threat. This view aligned with the global pandemic preparedness narrative, 'There is an understanding that this disease of birds can now infect human beings. To what extent it affects human beings, you go back to the (WHO) avian and human influenza pandemic phases,' (Interview 11, Ministry of Health). The fifth narrative expressed concern over the potential of a global pandemic to cause widespread social disruption. It specifically focused on concerns over Zambia's capacity to respond to a full blown pandemic. It therefore presented H5N1 as a disease whose treatment in humans was highly technical and resource intensive. This was a practical narrative that weighed Zambia's health system's limited response capacity against a pandemic scale H5N1 outbreak in humans. This narrative thus implicitly favored a preventative response, rather than a preparedness focus. A communication officer put it this way, 'The nature of management of a patient with avian influenza is highly technical and we are not in a position to manage to treat a lot of patients if we had . . . because a lot of them would need to be managed possibly under intensive care kind of management.' (Interview 10, Ministry of Health).

The sixth and final narrative presented H5N1 as a disease that could affect Zambia's trading status. This was downplayed in the national narrative, even if it prioritized H5N1 as a real threat to trade and industry. This was especially pertinent given that perceptions of H5N1 risk, rather than actual incidence, had negatively impacted poultry and poultry product production in the country. This occurred following unfounded media reports of bird flu outbreaks in Zambia, which led to public panic and a scaling down of poultry production due to a reduction in the consumption of poultry and poultry products. The result was an estimated loss of the equivalent of US$7 million over a 3-month period, a significant cost for Zambia's fledgling poultry industry. There was a sense foreboding concerning the impact that an H5N1 outbreak would have on trade, 'It poses a danger to our own exports because once the poultry products . . . from Zambia for example are found

[119]This is based on the FAO poultry classification system in which Sector 1 represent integrated poultry production systems characterized the use of standard operating procedures, high level biosecurity and commercial marketing of birds and their products. Sector 2 production is systems also commercially focused, in which moderate to high biosecurity is practiced. Ideally, poultry are kept indoors continuously, thus preventing contact with other poultry or wildlife. Sector 3 production systems are understood to mean low to minimal biosecurity production with birds and products entering live bird markets. Examples include caged layer farming with birds in open sheds, farms with free ranging poultry or farms producing chickens and waterfowl such as ducks. Sector 4 are systems of production in which there is minimal biosecurity and chickens and their products are consumed locally. FAO Avian Influenza Fact Sheet. Available at http://www.fao.org/docs/eims/upload/224897/factsheet_productionsectors_en.pdf.

to be infected with that avian influenza then we cannot export it' (Interview 17, Ministry of Commerce, Trade and Industry). The Poultry Association of Zambia (PAZ) also embraced this narrative, 'You may wish to know that in the region, it's only this country that has not recorded any major disease outbreak and hence we are considered the cleanest environment in the whole region. And we would want to remain as such' (Interview 16, PAZ).

These narratives tell us a lot about the different perceptions of risk evoked by stakeholders across the animal health, public health and trade sectors. The risk of avian and human influenza was presented in three distinct ways. First, the reality, as understood by decision-makers, was that H5N1 was an 'exotic' disease that was a trade threat. Secondly, there was the understanding of its zoonotic potential and where some of the risks lay, that is, a condition of poultry whose risk of spread is related to poor 'biosecurity'. Third, were public health concerns about the implications for Zambia should an H5N1 incursion occur and become fully zoonotic.

Despite these local understandings, the construction of the policy framework for avian and pandemic influenza preparedness in Zambia was largely driven by the actions of, and financial aid provided by, international agencies.[120] In Zambia's situation, two key points of contagion were identified in the policy process; the first being the traditional sector (as suggested by the FAO), and the second being Zambia's neighbors, with borders consistently described as "porous"[121] (also suggested by the FAO). This prioritization of disease risk mitigation, sidelined the trade and development narratives which spoke to broader public health concerns, including locally important trade and development imperatives, which limited the effectiveness of pandemic preparedness.[122]

The financial pull of the World Bank, FAO and USAID shaped the policy response, reinforcing the animal health framing of the H5N1 problem through several processes.[123] First, they defined the H5N1 problem and its possible sources; Zambia's multiple neighbors, interfaced by porous borders, and its 'high avian influenza risk' traditional poultry production sector. Second, they influenced the nature of intervention programs. Third, the bulk of financing was skewed towards animal health, which, by default, placed a reluctant veterinary department at the helm of policy development. With the agricultural ministry controlling most of the resources, the flow of finances affected the understanding of risk and the politics of the policy process, sustaining an emergency framing from the period between 2005 and 2009 and sending both government veterinary and research institutions alike searching for the elusive H5N1 virus in traditional poultry and wild birds.[124]

---

[120]Mwacalimba (2012).

[121]*Ibid.*

[122]Mwacalimba and Green (2015).

[123]*Ibid.*

[124]*Ibid.*

2.5 Understanding Policy and Risk in the Assessment Risk in Developing Country Settings

In this final section, we will suggest the ways in which understanding policy processes and context could inform risk analysis in such a way as to foster better policy coordination in cases like avian and pandemic influenza prevention and control, paying attention to wider issues such as livestock and livestock product trade. Here we determine (1) the feasibility of conducting an OIE type risk analysis in a manner that informs the development of risk management policies across multiple policy sectors in a resource-constrained country context and (2) present a policy relevant model for risk analysis appropriate for this context.

Thus far, we have reviewed how Zambian policy makers presented their understanding of H5N1 risk in response to the 'global' H5N1 threat. It is important to determine the potential use of these narratives in assessing risk, particularly if assessment outputs are intended to inform the development of context-appropriate policy responses. Because H5N1 is an animal disease with zoonotic potential, we will focus on OIE risk analysis framework, which, by WTO rules, provides the gold standard for the assessment of animal infectious disease risks. Drawing on the preceding discussion, we will begin by highlighting some of the key policy limitations of the current OIE approach to risk analysis, to better define what would aid the risk analysis process in developing country contexts, and what might be lost by conducting a risk analysis in this way. A national level model for an OIE risk analysis within this context will be proposed and its context-specific policy implications assessed, particularly which stakeholders are likely to influence or be influenced by a risk analysis in this context.

2.5.1 Risk Analysis and the Policy Context: Reconceptualising the OIE Risk Analysis Framework

As discussed earlier, separate from the actual dangers presented by various hazards, 'risk' is socially constructed.[125] This has been demonstrated in how different policy actors understood the threat of a zoonotic avian influenza incursion into Zambia. Risk analysis can benefit from social processes like policy making. Perceptions of H5N1 risk were framed differently across various sectors, including the poultry and allied industries, the media, health, agriculture and trade. This implies that different risk perceptions influence policy processes in different but significant ways.

The OIE risk analysis framework,[126] is a science-based method for the assessment of infectious disease risk that is based on the system developed by Covello

[125]Douglas and Wildavsky (1982), Horlick-Jones (1998), Slovic (1998).

[126]Murray et al. (2004).

and Merkhofer.[127] It is a 'Red Book' model[128] heuristic device conceptualized to involve four interacting and iterative stages; hazard identification, risk assessment, risk management and risk communication.[129] This structure makes the framework amenable to a discussion on the policy considerations relevant to its application in different contexts. The OIE risk analysis framework's importance *to* policy is highlighted in the fact that, since the inception of the WTO in 1995, the OIE framework in general has achieved recognition within the WTO SPS agreement as the standard for facilitating trade in animal and animal products.[130] The framework has been, and continues to be, applied to assess animal disease risks for scenarios other than those that are trade-related[131] and has been successfully adapted to a human health setting.[132] It therefore provides a structured approach to risk assessment and is considered to be an iterative and transparent standard for quantifying risk and informing policy.[133]

In the OIE framework, risk assessment is the most technical component of the process and can be a qualitative, semi-quantitative or quantitative assessment of risk on the basis of expert knowledge and/or empirical data.[134] The entire framework, however, is subject to the policy processes of the particular context in which the risk assessment is conducted. This is because the framing of risk, and therefore its assessment, involves the interplay of both contextual and social factors. A weakness of the framework therefore is that it primarily relies on the engagement of expert knowledge, and *their* presentation of the underlying assumptions and the steps followed in the determination of risk.[135] By relying primarily on the knowledge of scientific experts, the framework is blind to social influences, and, in the case of risk management policy development in resource-constrained countries, partial to the narratives on risk voiced by the international agencies holding the purse strings. This is a pertinent potential flaw. As noted with similar technocratic models,[136] funding agencies could select only experts whose viewpoints resonate with their policy agendas, making their assessments of risk highly contestable. Another concern is a lack of robustness in this approach's dealing with the ambiguities of scientific uncertainty and surprise[137] and, a failure to fully engage

[127]Covello and Merkhofer (1993), p. 318.

[128]According to Millstone (2007), the term comes from the red cover of a seminal report produced in 1983 by the National Research Council in the US. This report presented a version of inverted decisionism or technocratic model that is very similar to the OIE risk analysis framework.

[129]Vose (2000), WHO/FAO (2006), OIE Terrestrial Animal Health Code 2010.

[130]Thiermann (2005), OIE Terrestrial Animal Health Code 2010.

[131]MacDiarmid and Pharo (2003).

[132]e.g. Clements et al. (2010).

[133]WHO/FAO (2006), Murray et al. (2004).

[134]Vose (2000), Murray et al. (2004).

[135]Vose et al. (2001), Pfeiffer (2007).

[136]Van Zwanenberg and Millstone (2006), Millstone (2007).

[137]Stirling and Mayer (2000), Stirling and Scoones (2009).

political processes and social choices in addressing risk.[138] One solution is "to develop a more holistic perspective" of risk "that includes explicit consideration of the roles of policy, disease management, and feedbacks between ecosystems and societies."[139]

There are novel approaches that suggest ways of better combining science and policy making such as Millstone's[140] 'co-evolutionary model'. But the preceding critique is not a basis to reject the OIE framework. It is important to suggest *how* it can be made more amenable to social processes. Part of its appeal, as argued by Hueston,[141] is that the framework holds promise for the consideration of policy processes in the assessment of risk. Hueston, however, does not offer any suggestions on how this may be achieved. But the OIE risk analysis framework is useful for assisting decision-makers thinking around particular aspects of risk, which helps inform resource allocation in risk mitigation.[142] We will therefore examine the OIE framework through a policy lens, drawing on insights developed by Millstone[143] wherever they may apply.

2.5.2 Developing a Feasible Risk Analysis for Zambia

The OIE risk analysis framework places the OIE squarely in the centre of the highly political arena of international animal trade. The evolution of the emergency response to avian and pandemic influenza in Zambia, for instance, was a highly political process,[144] a state that cannot be detached from any risk analysis. However, as part of the OIE framework, the OIE's veterinary services evaluation process sets as a benchmark independence from political influence.[145] This separation is impractical. Furthermore, the veterinary profession is potentially limited by this dependence on scientific or authoritative opinion and its exclusion of political and social phenomena.[146] Political and contextual dimensions are just as important as the biological considerations when it comes to the multi-sectoral risk management of emerging, albeit limited zoonosis such as H5N1.[147] As discussed in our case study on Zambia, there were differences in the understanding of risk among sectors at the interface of animal health, public health and trade. The result

---

[138]Millstone (2007).

[139]Cumming (2010b).

[140]Millstone (2007).

[141]Hueston's discussion speaks of the OIE risk analysis framework in very general ways.

[142]MacDiarmid and Pharo (2003).

[143]Millstone (2007).

[144]Mwacalimba (2012).

[145]e.g. Vallat and Pastoret (2009).

[146]Hueston (2003).

[147]Mwacalimba (2013).

was an amorphous understanding of H5N1 risk, stakeholder exclusion in risk management and some inefficient resource considerations.[148]

2.5.3 Policy Considerations for Hazard Identification

There is need to examine the 'what' 'how' 'when' and 'who' interactions of risk as it relate to policy. The first step in conducting an OIE type risk analysis is identifying 'what' the hazard, or source of risk, is. This is hazard identification, defined by the OIE Terrestrial Animal Health Code as "the process of identifying the pathogenic agents which could potentially be introduced in the commodity[149] considered for importation".[150] A hazard is defined as "any pathogenic agent that could produce adverse consequences on the importation of a commodity."[151] This is the technical definition, but there is also a policy equivalent of hazard identification that could be factored into the risk analysis. In a policy sense, hazard identification is the framing of the problem. This is not simply how the agent, for instance, H5N1 avian influenza, is conceptualised, but also how it is *perceived*, as a problem, i.e. its social construction. Perceptions of H5N1 risk were constructed by different international and national policy agendas and evidence in Zambia. Cognisance of these sometimes conflicting interactions in perceptions of risk could potentially enhance the applicability of the OIE risk analysis framework in the context of a country such as Zambia. Tensions such as H5N1's status as a global health concern due to its pandemic potential, a poultry industry or trade concern, or its "exotic" status in the Zambian context, need to be acknowledged before context-specific consensus of this policy problem can be achieved.

As discussed earlier, the H5N1 problem in Zambia was expressed in the three ways conveying how the H5N1 'hazard' was understood by policy makers across animal health, trade and public health. First, the H5N1 'hazard' was seen as "exotic" condition that threatened trade. Secondly, it was a *potential* zoonosis whose risk of incursion lay in traditional poultry flocks with poor "biosecurity", and finally, it carried *plausible* implications for public health if H5N1 became fully zoonotic. These different animal health, trade and economy, and public health framings of the H5N1 problem formed the internal policy response, bringing specific actors to the policy process. The resulting policy framework then addressed four contiguous disease and disease management issues. First, there was the root consideration of H5N1 (or H5N1 *emergency* preparedness) second, there was the consideration of general avian influenzas, third, there was the aspect of human

[148]Mwacalimba (2012).

[149]This definition explicitly mentions commodities intended for importation because the Code's purpose is to facilitate free and safe trade.

[150]OIE Terrestrial Animal Health Code 2010, p. xvii.

[151]Murray et al. (2004), p. v.

seasonal influenzas and fourth, there was the core policy issue of human influenza pandemic *preparedness* and capacity building. These multiple perspectives are the 'framing assumptions' that a risk analyst can use to provide advice in policy making.

In essence, rather than just ask *what* the hazard is, it is also important to ask for *whom* (and *how*) H5N1 avian influenza presents a 'hazard'. To be feasible as a tool to inform policy in a setting such as Zambia, the process of hazard identification should first unpack and properly categorize different policy perceptions into risk statements germane to each policy-relevant stakeholder. This essentially entails that a hazard identification be performed in such a way that it 'maps' how the H5N1 'hazard' relates to general avian influenzas, human influenza and pandemic preparedness *across* sectors. The importance of this is that, as a standalone problem, different levels of priority were accorded to H5N1 in Zambia.[152] It was a high priority in the public health sector and low in the animal health sector, and while other local stakeholders did not know how H5N1 affected them, just the *perception* of H5N1 risk resulted in real consequences in the poultry industry.[153]

A national level risk analysis in a resource-constrained country context would theoretically have a broad audience with divergent conceptions of risk and priorities. The objective in Zambia was to develop a coordinated, multi-sectoral risk management framework. This entails understanding different framings of risk in order to think more adaptively about information gathering for hazard identification. This requires bringing scientific and non-scientific considerations more explicitly in policy processes, thus allowing the appropriate actors participate in the risk analysis process.[154] The process of information gathering may therefore benefit from a stakeholder analysis, beyond international agency considerations, to identify important stakeholders, their viewpoints and information contributions. This is necessary to comprehensively define the hazard and capture information about how the hazard affects, and, more importantly, maybe *affected* by different policy relevant stakeholders. This data gathering process is also important for the risk assessment stage, which is discussed next.

2.5.4 Policy Considerations for Risk Assessment

Risk assessment is "the evaluation of the likelihood and the biological and economic consequences of entry, establishment and spread of a hazard within the territory of an importing country".[155] It has four stages, a release assessment, exposure assessment, consequence assessment and finally, risk estimation. The

[152]Mwacalimba (2012).
[153]*Ibid.*
[154]Slovic (1998).
[155]OIE Terrestrial Animal Health Code 2010. p. xxii.

OIE suggests that the processes of release and exposure assessments require the skills of a veterinary epidemiologist, while the consequence assessment may require an economist's input.[156] However, in addition to being equipped with the framing assumptions of the various interested parties, a risk analyst would be at an added advantage if they had some working knowledge of policy processes beyond a purely 'scientific' viewpoint.

The process of risk assessment begins with a risk question. This defines what can go wrong and how. After hazard identification, the relevant stakeholders formulate the risk questions they intend the risk assessor to help answer, thereby defining the boundaries of the risk assessment. Answering these questions requires a comprehensive process of gathering and collating evidence that describes the risk-relevant epidemiology of the hazard such as host range, vehicles of carriage and transmission, and survival under different environmental conditions.[157] The sources of information considered reliable included libraries, the internet and specialists.[158] While the framework accommodates grey literature,[159] it is partial to 'expert' sources to elucidate, for instance, the virology of H5N1. This is understandable. However, the question of what can go wrong needs to be oriented towards *whom* and *how* each negative outcome is pertinent. The stakeholder analysis at the hazard identification stage and the engagement of these stakeholders at the risk assessment stage could provide important data for an inclusive assessment of risk. Especially in resource-constrained settings where data are scarce, this multi-sectoral data collection process provides a viable data source.

2.5.5 Risk Framing in Zambia and Its Implications for Risk Assessment

In the context of Zambia, although H5N1 had economic impacts and potential health system effects, the risk question actually revolved around H5N1's zoonotic risk. The policy framing and disease mitigation approach focused on preventing an external incursion of H5N1 and less on the local and regional contextual factors that could potentially influence its transmission, establishment and spread. This is typical of approaches to disease control. They emphasize preventing 'contamination' and are sometimes uncritical of 'configuration' or context.[160] In Zambia, decision-makers focused on mitigating disease contamination, as exemplified for instance, by the institution of a partial poultry and poultry product import ban even

[156]MacDiarmid and Pharo (2003).
[157]Pharo (2003).
[158]MacDiarmid and Pharo (2003).
[159]Wooldridge (2000).
[160]Leach et al. (2010).

from countries unaffected by H5N1.[161] But this was a valid concern, given Zambia's prioritization of disease freedom for the purposes of trade.

Zambia's National Response Plan for avian and pandemic influenza prevention and control lists five potential introduction routes for H5N1; live bird imports; poultry product imports; illegal poultry and poultry product trade; returning travellers previously in direct or indirect contact with infected poultry or poultry premises overseas; and aquatic migratory birds. For a risk analyst, these are the modes of 'release' considered pertinent by policy makers in Zambia.

There are deeper issues to consider. For example, In Zambia there were tensions between the preoccupation with the temporal concern of H5N1 risk (when will this happen?), and externally defined evidence on the spatial concerns of risk (how could this happen?): "According to Food and Agriculture Organisation (FAO), although Zambia is currently free of the virus, the country is at high risk because of many neighboring countries, which has led to increase in human traffic and trade in poultry and poultry products".[162] What policy makers really lacked was a clear mapping ('configuration') of how an H5N1 incursion and outbreak might occur in the Zambian context. A risk assessment, guided by stakeholder-relevant risk questions, would assist policy-makers and stakeholders focus more deeply on 'how' an incursion and outbreak might occur. This would help to better define resource allocation in risk management.

In Zambia, commercial breeders were perceived to have high biosecurity. However, it was argued by the independent NGO GRAIN[163] that many of the H5N1 outbreaks in Southeast Asia occurred in large commercial institutions with poor biosecurity. If such perspectives can, at the very least, be considered, then, other than illegal cross border trade, human travelers and migratory birds, Zambian breeders provided an important link to the global poultry industry. Potentially, 'big poultry' in Zambia (Sector 1 and 2) was also at risk of acquiring H5N1. Another important factor is that the poultry industry in Zambia had orientated itself towards poultry exports, implying that should Zambia have an outbreak, it could be a potential source of H5N1 for its trading partners. In terms of risk assessment, it is thus very important to consider the 'configuration' of the risk system to better inform disease management.

For Zambia, three interlinked risk systems would have to be considered in the weighing of H5N1 risk release in this context. These three risk systems are the biological risk, the ecological risk and the policy risk. These are essentially the 'map' that a risk assessor should develop to determine the risk of release, exposure and consequence(s) of an H5N1 incursion. The biological risk system would draw on virology and epidemiology, as this is a technical exercise. The ecological risk system is the poultry production system at play in the Zambian context, encompassing production characteristics and the nature and extent of interaction

[161]Mwacalimba 2013.

[162]Zambia's National Response Plan on avian influenza, 2008 version, p. 6.

[163]GRAIN (2007).

among production systems, processing systems and market distribution systems. In the case of Zambia, the production systems include the traditional backyard production systems, semi-commercial housed production systems, emergent production systems, commercial production systems and commercial breeding systems. Together, the biological and ecological risk systems determine the likely points at which first, the production systems interface each other (and hence the routes by which H5N1 could spread from system to system), and second, the human exposure to H5N1 may occur (defining the human-animal interface for Zambia). The policy system includes, but is not limited to; the identification of the institutions, resources, stakeholders and policies available for risk management. These are important in identifying the type and feasibility of interventions that already exist to mitigate this risk.

A release assessment would begin by determining the current disease status of countries with which Zambia has trade dealings (existing trade agreements, known trading partners etc.). The next step would then focus on verifying the claim that Zambia's poultry breeders, of which only six hatcheries supplying the entire commercial poultry industry (including emergent and small scale production systems),[164] in 2009 actually had the levels of biosecurity and surveillance systems in place to support the claim that they were at low risk of an H5N1 incursion.[165]

A conceptual scenario diagram for the assessment of multi-sectorial zoonotic risk in Zambia should represent the routes for introducing (contamination) zoonotic H5N1 into the population of interest and potential routes of spread (configuration) (Fig. 2.1). In determining the risk of H5N1 release, the product is diseased poultry or their products and the possibility of biological carriage via human travel or via aquatic migratory birds. For poultry and poultry products, a risk analyst can trace the movement of these commodities through the entire production system, by conceptualizing physical pathways through the supply chain from hatcheries, producers, small scale producers, finally to markets (formal and informal), overlaid by a biological pathway defining host-pathogen interaction and an examination of biosecurity measures throughout the supply chain. Additional details would include pathways for the biological carriage of H5N1 via human travel and aquatic migratory birds.

Interestingly Fig. 2.1 presents many of the key issues around risk identified by a policy analysis. It is a conceptual example of how a risk assessment might present H5N1, incorporating trade, public health and animal health. These sectors provide possible policy mitigation points, trade "surveillance" (through border and import controls, including poultry and poultry products in transit), veterinary surveillance (domestic commercial and traditional poultry, food safety, poultry markets and wild poultry) and human surveillance (port health, hospital and health centre

[164]According to Zambia's National Response Plan for Avian and Human Influenza (2008), the country has four poultry production systems. These are commercial sector, emerging sector, small scale (also called backyard production) sector and the Village/free range sector.
[165]Mwacalimba (2013).

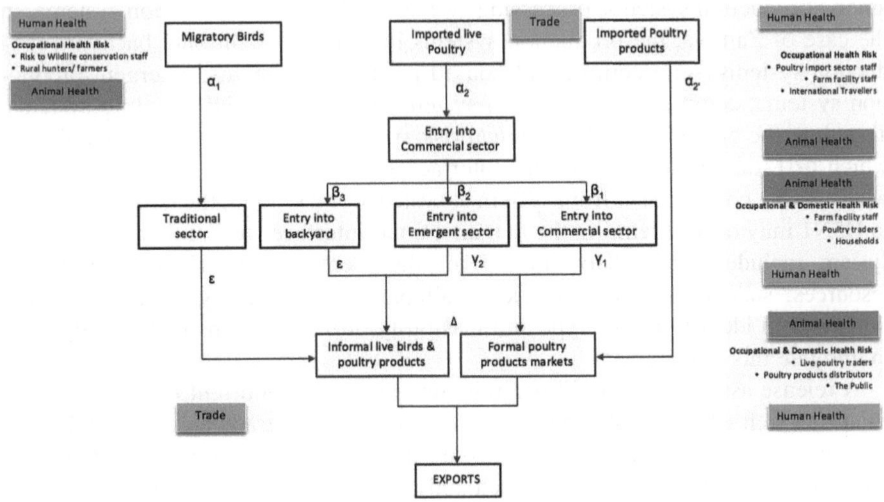

Fig. 2.1 A conceptual risk model for the assessment of the risk of H5N1 introduction in Zambia. A different version of this risk scenario has been presented elsewhere (Mwacalimba 2013). This representation however, does not explicitly differentiate distributive and occupational exposure risks for H5N1. It highlights instead the ecological foci for Human Health, Animal Health and Trade. The *arrows* show the direction of poultry and poultry products flow within the supply chain (physical pathways) as well as the routes of H5N1 within Zambia (biological pathways). *Greek symbols* represent data inputs for this risk scenario (presented in Table 2.1)

surveillance and food safety). This is a 'policy pathway', or more accurately, a 'risk management policy pathway', since the movement of poultry and poultry products is defined by complex socio-economic and policy interactions. The physical pathways in Fig. 2.1 could consider trade agreements and SPS protocols, poultry production and marketing, avian influenza surveillance in humans and poultry and food safety.

We should also bear in mind that resource and managerial aspects of risk should not be separated from the assessment process.[166] The scenario diagram here attempts to include these critical aspects of risk management. For a risk assessment to be policy relevant, it must relate avian and pandemic influenza control to trade policy activities to zoonotic H5N1 risk. This includes an assessment of each sector's roles and actions in each risk pathway, including some consideration of existing legal and policy frameworks, mandates and provisos. This implies the 'practical' data inputs for a risk assessment to inform avian and pandemic influenza control policy. In Fig. 2.2, we aggregate the biological, ecological and policy risk systems, with the primary focus being on how public health is affected, which was the root concern for the development of the avian and pandemic influenza control policy. The Greek letters represent the parameters of the risk areas that could

[166]Horlick-Jones (1998).

Table 2.1 Conceptual parameters explaining symbols used in the conceptual risk model shown in Fig. 2.1

Risk pathway parameter	Interpretation	Data input
α_1	• Seasonal migration of wild aquatic birds • Contact rates with local aquatic ducks • Contact rates with traditional poultry from communities living near large water bodies with the most migratory bird activity	• Ornithological data for Zambia • Husbandry practices in traditional flocks • Identification of areas where contact between traditional and aquatic birds is most likely
α_2 $\alpha_{2'}$	• Border entry protocols for live poultry • Inspection protocols (SPS) • Source verification • Transit vehicle inspection protocols • Quarantine procedures • Personnel at checkpoints • Border entry inspection for poultry products • Inspection protocols (SPS) • Source verification • Transit vehicle inspection protocols • Quarantine procedures • Food safety protocols	• SPS protocols at ports of entry • Import data • Import permits • Quarantine procedures for imported breeding stock • avian influenza surveillance and control
β_1 $\beta_{1'}$ β_2 β_3	• Monitoring of biosecurity measures and husbandry in the commercial producer sector • Monitoring of biosecurity for poultry from the emerging sector coming into the producer sector • Monitoring of biosecurity and husbandry in the emerging sector • Monitoring of biosecurity and husbandry in the small scale sector	• Poultry sector description data • Sectoral activities around avian influenza surveillance and control • Human influenza surveillance • Food safety
γ_1 γ_2	• Monitoring of poultry product food safety for the commercial sector prior to marketing • Monitoring of poultry product food safety protocols for the emerging sector prior to marketing	
Δ	• Informal product markets • Surveillance of poultry products from the formal (emerging) sector • Surveillance of poultry products from the informal (small scale) sector	
E	• Informal live animal markets • Monitoring of poultry health in informal markets • Monitoring of health and mixing in informal markets	

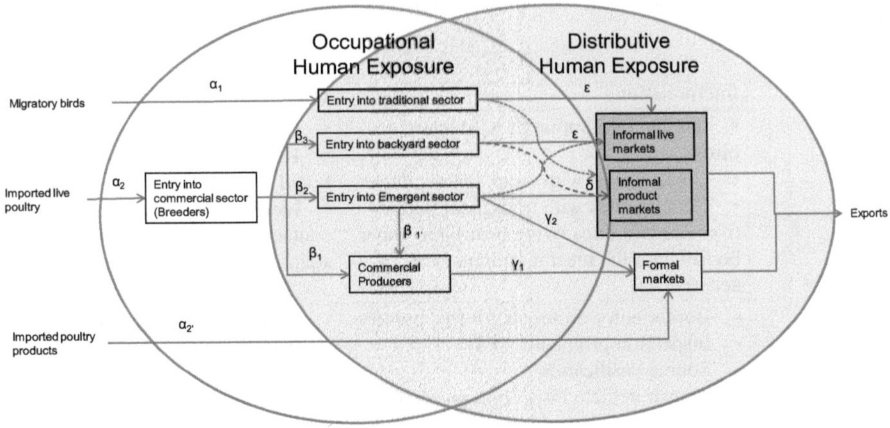

Fig. 2.2 A conceptual risk model for the assessment of the risk of H5N1 introduction in Zambia

potentially be assessed in relation to H5N1 epidemiology, ecology and policy that would need to be instituted to reduce the risk of trade-related H5N1 introduction. These parameters are explained in the summary provided in Table 2.1. In addition to expert opinion and the literature on H5N1 epidemiology, this could potentially form the basis for the analytical framework for the risk release, exposure and consequence.

Such a model would examine (1) Release assessment; involving a consideration of the trade-related, human travel related and aquatic migratory bird related pathways through which viable H5N1 could be introduced into Zambia from affected trading partners and regions including border inspection and SPS protocols; (2) Exposure assessment; involving a consideration of the pathways through which Zambian poultry and poultry products and high risk humans (occupational exposure) and consumers (distributive exposure) could be exposed to H5N1 following trade-related, human travel-related and aquatic migratory bird-related introduction. This would examine biosecurity, husbandry, wild bird and domestic bird contact rates and food safety protocols (3) Consequence assessment; involving a consideration the ways in which an H5N1 outbreak(s) would affect public health, the economy, or be spread further to Zambia's export markets. This would examine losses to the poultry and related industries, health system effects and the wider economic consequences (4) Risk estimation; involving a summary of the previous estimates.

Conceptualized like this, such a risk model would examine zoonotic H5N1 entry into poultry production systems, release *into* these poultry production systems, possible routes of exposure of other poultry flocks and the risk of human beings contacting potentially infected poultry and poultry products. The aspect of exposure of human beings to zoonotic H5N1 is fundamental, since in Zambia, for policy-makers at least, the zoonotic risk was more important than the effect on poultry industry productivity.[167] Broadly speaking, the human populations at most risk

[167]Mwacalimba and Green (2015).

would probably be those working very closely with poultry (occupational exposure) and those working in poultry trade (distributive exposure). An important aspect for the risk assessment to address is actually 'when' human exposure is likely to occur following detection of outbreaks in poultry. The literature shows that in other contexts, human cases occurred between a month (Egypt) to a year (Nigeria) after poultry outbreaks were detected.[168] This provides the possible timeframe in which the health system would have to respond in order to reduce the risk of further spread.

2.6 Roles in Multi-Sectorial Risk Management in Zambia

In the OIE code, risk management is defined as "the process of identifying, selecting and implementing measures that can be applied to reduce the level of risk".[169] The stages of risk management are risk evaluation, option evaluation, implementation and monitoring and review.[170] Risk evaluation involves determining whether or not the risk calculated by a risk assessor requires intervention. Option evaluation seeks to define the nature of this intervention and identify the various approaches available to manage risk.

An important policy question that contextualizes a modified version of the OIE risk analysis framework is 'who' manages risk? In the scientific risk literature, it has been stated, rather definitively, that risk management, is undertaken by *risk managers* knowledgeable in policy and in possession of the appropriate level of technical know-how to communicate efficiently with persons assessing risk.[171] The same body of work states that risk analysis is intended to assist *decision-makers* weigh the risks posed by particular courses of action.[172] It is important to unpack the terms 'decision-maker' and 'risk-manager'; in developing country contexts, they could refer to different sets of policy actors or the same set of people. In the development of multisectoral zoonosis risk management policy, for instance, the term 'risk manager' is actually fluid, applicable to a multitude of stakeholders. In Zambia's case, for example, there was a 20-person Task Force on Avian and Human Influenza, the Department of Veterinary and Livestock Development, the Ministry of Health, the FAO, the WHO, USAID, the National Agricultural Information Service (NAIS) and the Ministry of Health's Health Education Unit, all working to 'manage' the same risk. In addition, there were stakeholders such as the Poultry Association of Zambia, who acted to manage media-generated *perceptions* of risk: 'We realized that avian influenza gained a lot of prominence in the press and the prominence was full of fake things and people were just downloading the

[168]WHO (2010).

[169]OIE Terrestrial Animal Health Code 2010, p. xxii.

[170]Murray et al. (2004).

[171]Vose et al. (2001).

[172]MacDiarmid and Pharo (2003).

Turkey situation or the China situation and making it appear as if it was a Zambian situation. In the 3 months of the AI prominence in the media . . .So we woke up from slumber and took a leading role in the sensitization of our members and also the general public. . .' (Interview 16 PAZ).

Another consideration for conducting risk analyses in this setting is the importance of the *risk analyst* understanding policy processes, and being comfortable with *explaining* the technical aspects of the risk assessment in accessible ways to different stakeholders. In the context of Zambia and H5N1, at least, there were various technical and non-technical strands to the preparedness effort as it related to the management of H5N1 risk; the public health response, based on the Integrated Disease Surveillance and Response (IDSR) framework; the animal health response, guided by a National Response Plan and also the omitted, but potentially important, trade response, based on their capacity to translate, support or implement trade agreements for the purpose of animal health, plant health or human health protection.[173] It would be unrealistic to expect that 'risk managers' in these different policy communities would have "the appropriate level of technical background to communicate effectively with risk assessors."[174]

The risk analysis process should assist effective risk management by highlighting response system vulnerabilities *across* sectors in the process of characterizing the dynamics of zoonotic risk, and explain these vulnerabilities to the appropriate audience. Risk analysis' role in informing decision making would be enhanced if risk is considered across the entire policy spectrum and not just from one viewpoint. In this endeavor, a risk analyst should understand that in multi-sectoral settings, there are differences in priorities, norms and policy frameworks which can impact upon the development of a risk management policy. Such challenges include properly linking the risk management policy response to livestock trade and holistically addressing the various conceivable modes of disease introduction and routes for human exposure. The purpose of this exercise is to bring different stakeholders to view risk management in mutually inclusive ways. The risk management process should therefore draw both on multiple framings of risk and the resources of a wide pool of policy relevant stakeholders, to aid the process of assigning roles and resources more appropriately across sectors.

2.7 Policy Considerations for Risk Communication

As long as a policy issue is on the agenda, risk communication is not just about a unidirectional communication to stakeholders about risk, but an evolving process of continuous dialogue across sometimes different epistemic communities. The OIE Code defines risk communication as "the interactive transmission and exchange of

[173]Mwacalimba (2013).

[174]Vose et al. (2001), p. 814.

information and opinions throughout the risk analysis process concerning risk, risk-related factors and risk perceptions among risk assessors, risk managers, risk communicators, the general public and other interested parties".[175] For the OIE risk analysis to be useful, there has to be a consideration of who the policy relevant *risk communicators* are for a given risk problem in a given context. There is need to also think about how risk is communicated to, and by, different policy stakeholders, including the public, the media, farmers, medics, veterinarians, decision-makers across different sectors and even the donor community. Equally important is that in the process of risk communication, a context appropriate *forum* is used, allowing for as much dialogue and feedback in the risk analysis process as possible. Such a risk analysis process would derive most benefit if presented both on a forum capable of reaching the largest number of policy relevant stakeholders and in ways that engages the interests of each policy relevant stakeholder. This entails structuring risk in flexible and stakeholder inclusive ways across public health, animal health, trade and more widely. Such a forum and approach would provide an audit of the multi-sectoral zoonosis risk management policy, including the identification of the institutions, resources and policies available for risk management. More importantly, it would be able to mobilize the necessary resources and institutions to enforce and evaluate the risk management response.

2.8 Conclusion

This chapter analyzed narratives on public health, animal health and trade, to better understand the relationship between global understandings of infectious disease risk, and how they impact on the development of local responses to disease problems. We discussed the tensions faced by resource-constrained countries in animal health-public health and trade, with particular emphasis on these countries being potential sources of infectious disease threats. This raised the question of how developing countries, should approach infectious disease risk management as it relates to pandemic scale threats such as avian and pandemic influenza.

Framing assumptions have significant, but sometimes unacknowledged influence on the policy process. Millstone[176] states framing assumptions influence the questions posed, the type of evidence used or excluded, and even how this evidence is interpreted. Therefore, the key to the feasibility of the analysis of the risk of multi-sectoral affecting emerging infectious diseases such as avian and pandemic influenza is flexibility in how risk is framed across the public health, animal health and trade systems. This requires taking explicit notice of multiple risk framings from a diverse cross-section of stakeholders, to better negotiate risk analysis and risk management. Based on the understanding that risk is socially constructed, we

[175]OIE Terrestrial Animal Health Code 2010, p. xxii.

[176]Millstone (2007).

argued that in the process of hazard identification, it is the recognition of the various framing assumptions that construct the risk in each context that will help foster wider stakeholder inclusion. This, in turn, will take into account the multiple perspectives that exist in resource-constrained countries.

By basing the hazard identification on framing assumptions, the OIE risk analysis framework can be made amenable to more open and inclusive evidence gathering and interpretation, thus treating 'risk' and its assessment in a more discursive manner. For instance, rather than seeking to answer one, externally influenced, risk question as the current conception of the OIE risk analysis framework would probably do,[177] this approach uses framing assumptions to develop *multiple* risk questions that speak to the interests of multiple policy relevant stakeholders. Furthermore, by acknowledging these framing assumptions throughout the various stages of the risk analysis process, a better map of the local risk management context can be developed that examines both the scientific aspects of local configuration and the politics of policy processes.

Adopting this approach would help reshape the face of the current multi-sectorial risk management response in developing countries, in which exist uncoordinated, narrow and fragmented framings, overly influenced by international agency funding, evidence and advice. The risk analysis would need to be applied in such a way that it assists stakeholders align resource and institutional priorities to the prevention and management of an infectious disease incursion. By analysing the feasibility of the applicability of the OIE risk analysis framework through a policy lens, this chapter attempted to demonstrate that given the interactions between local context, risk assessment and risk management policy, the relationship between policy and risk cannot be viewed as linear. Therefore, in the context of multi-sectoral risk management, risk assessors should consider taking into account political and social phenomena in the process of risk assessment.

References

Aginam O (2002) International law and communicable diseases. Bull World Health Organ 80:946–951

Alexander DJ (2007) An overview of the epidemiology of avian influenza. Vaccine 25:5637–5644

ALive (2006) Avian influenza prevention and control and human influenza pandemic preparedness in Africa: assessment of financial needs and gaps. In: Fourth international conference on avian influenza, Bamako, Mali, December 2006

Beck U (1992) Risk society: towards a new modernity. Sage publications, London

Beigel JH, Farrar J, Han AM, Hayden FG, Hyer R, de Jong MD, Lochindarat S, Nguyen TK, Nguyen TH, Tran TH, Nicoll A, Touch S, Yuen KY (2005) Avian influenza A (H5N1) infection in humans. N Engl J Med 353:1374–1385

Bettcher DW, Yach D, Guindon GE (2000) Global trade and health: key linkages and future challenges. Bull World Health Organ 78:521–534

[177] see MacDiarmid and Pharo (2003).

Blancou J, Chomel BB, Belotto A, Meslin FX (2005) Emerging or re-emerging bacterial zoonoses: factors of emergence, surveillance and control. Vet Res 36:507–522
Brown C (2004) Emerging zoonoses and pathogens of public health significance - an overview. Rev Sci Tech 23:435–442
Bruckner GK (2009) The role of the World Organisation for Animal Health (OIE) to facilitate the international trade in animals and animal products. Onderstepoort J Vet Res 76:141–146
Burgos S, Otte J (2008) Animal health in the 21st century: challenges and opportunities: Pro-poor livestock policy initiative, Research report 09–06
Cattoli G, Monne I, Fusaro A, Joannis TM, Lombin LH, Aly MM, Arafa AS, Sturm-Ramirez KM, Couacy-Hymann E, Awuni JA, Batawui KB, Awoume KA, Aplogan GL, Sow A, Ngangnou AC, El Nasri Hamza IM, Gamatie D, Dauphin G, Domenech JM, Capua I (2009) Highly pathogenic avian influenza virus subtype H5N1 in Africa: a comprehensive phylogenetic analysis and molecular characterization of isolates. PLoS One 4, e4842
Chen H, Smith GJ, Zhang SY, Qin K, Wang J, Li KS, Webster RG, Peiris JS, Guan Y (2005) Avian flu: H5N1 virus outbreak in migratory waterfowl. Nature 436:191–192
Clements AC, Magalhaes RJ, Tatem AJ, Paterson DL, Riley TV (2010) Clostridium difficile PCR ribotype 027: assessing the risks of further worldwide spread. Lancet Infect Dis 10:395–404
Covello VT, Merkhofer MW (1993) Risk assessment methods: approaches for assessing health and environmental risks. Plenum Press, New York, p 318
Cowen P, Morales R (2002) Economic and trade implications of zoonotic diseases. In: Burroughs T, Knobler S, Lederberg J (eds) The emergence of zoonotic diseases: understanding the impact on animal and human health. National Academies Press, Washington
Cumming DHM (2010a) Responses and reactions to Scoones et al. 'Foot-and-mouth disease and market access: challenges for the beef industry in southern Africa'. Pastoralism 1:13
Cumming GS (2010b) Risk mapping for avian influenza: a social–ecological problem. Ecol Soc 15:32
de Jong JC, Claas EC, Osterhaus AD, Webster RG, Lim WL (1997) A pandemic warning? Nature 389:554
Delgado CM, Rosegrant M, Steinfeld H, Ehui S, Courbois C (1999) Livestock to 2020: the next food revolution. International Food Policy Research Institute, Washington
Domenech J, Lubroth J, Eddi C, Martin V, Roger F (2006) Regional and international approaches on prevention and control of animal transboundary and emerging diseases. Ann N Y Acad Sci 1081:90–107
Douglas M, Wildavsky A (1982) Risk and culture. University of California Press, Berkeley
Elbe S (2010) Haggling over viruses: the downside risks of securitizing infectious disease. Health Policy Plan 25:476–485
Fevre EM, Bronsvoort BM, Hamilton KA, Cleaveland S (2006) Animal movements and the spread of infectious diseases. Trends Microbiol 14:125–131
Fidler DP (2004a) Germs, governance, and global public health in the wake of SARS. J Clin Invest 113:799–804
Fidler DP (2004b) Global outbreak of avian influenza A (H5N1) and international law
Fidler DP (2008) Influenza virus samples, international law, and global health diplomacy. Emerg Infect Dis 14:88–94
Garrett L, Fidler DP (2007) Sharing H5N1 viruses to stop a global influenza pandemic. PLoS Med 4, e330
Giddens A (1998) Risk society: the context of British politics. In: Franklin J (ed) The politics of risk society. Polity Press, Cambridge, pp 23–34
GRAIN (2006a) Bird flu crisis: small farms are the solution not the problem: seedling. July 2006. http://www.GRAIN.org/seedling_files/seed-06-07-11.pdf=12
GRAIN (2006b) The top-down global response to bird-flu: against the GRAIN. http://www.grain.org/articles/?id
GRAIN (2007) Bird flu: a bonanza for 'Big Chicken.' Against the Grain. http://www.grain.org/articles/?id=22

Greger M (2007) The human/animal interface: emergence and resurgence of zoonotic infectious diseases. Crit Rev Microbiol 33:243–299
Hampson AW (1997) Surveillance for pandemic influenza. J Infect Dis 176(Suppl 1):S8–S13
Hoffman SJ (2010) The evolution, etiology and eventualities of the global health security regime. Health Policy Plan 25:510–522
Horlick-Jones T (1998) Meaning and contextualizing in risk assessment. Reliab Eng Syst Saf 59:79–80
Howse R (2004) The WHO/WTO study on trade and public health: a critical assessment. Risk Anal 24:501–507
Hueston WD (2003) Science, politics and animal health policy: epidemiology in action. Prev Vet Med 60:3–12
Huynen MM, Martens P, Hilderink HB (2005) The health impacts of globalization: a conceptual framework. Glob Health 1:14
Kahn LH (2006) Confronting zoonoses, linking human and veterinary medicine. Emerg Infect Dis 12:556–561
Karesh WB, Cook RA, Bennett EL, Newcomb J (2005) Wildlife trade and global disease emergence. Emerg Infect Dis 11:1000–1002
Kerry VB, Lee K (2007) TRIPS, the Doha declaration and paragraph 6 decision: what are the remaining steps for protecting access to medicines? Glob Health 3:3
Kimball AM (2006) Risky trade: infectious diseases in the era of global trade. Ashgate Publishing, London
Kruk ME (2008) Emergency preparedness and public health systems lessons for developing countries. Am J Prev Med 34:529–534
Labonte R, Sanger M (2006) Glossary of the World Trade Organisation and public health: part 1. J Epidemiol Community Health 60:655–661
Lang T (1999) Diet, health and globalization: five key questions. Proc Nutr Soc 58:335–343
Leach M, Scoones I, Stirling A (2010) Dynamic sustainabilities: technology, environment, social justice. ISBN 978-1-84971-093-0
Lee K, Fidler D (2007) Avian and pandemic influenza: progress and problems with global health governance. Glob Public Health 2:215–234
Lee K, Koivusalo M (2005) Trade and health: is the health community ready for action? PLoS Med 2, e8
Lee K, Fustukian S, Buse K (2002) An introduction to global health policy. In: Lee K, Buse K, Fustukian S (eds) Health policy in a globalizing world. Cambridge University Press, Cambridge
Lipson DJ (2001) The World Trade Organization's health agenda. BMJ 323:1139–1140
Lloyd-Smith JO, George D, Pepin KM, Pitzer VE, Pulliam JRC, Dobson AP, Hudson PJ, Grenfell BT (2009) Epidemic dynamics at the human-animal interface. Science 326(5958):1362–1367. doi:10.1126/science.1177345
Lo Yuk-ping C, Thomas N (2010) How is health a security issue? Politics, responses and issues. Health Policy Plan 25:447–453
LSUAC (2008) Partnerships for food industry development – public-private partnerships for HPAI prevention and mitigation in Africa. Louisiana State University Agriculture Centre. A U.S./Togolese/Zambian Partnership. First & Second Quarterly Report (October 1, 2007 – March 31, 2008). Funded by USAID
Lupton D (1999) Risk. Routledge, London
MacDiarmid SC, Pharo HJ (2003) Risk analysis: assessment, management and communication. Rev Sci Tech 22:397–408
Mann J (1990) The global lesson of AIDS. New Scientist, June 30, 1990
Merianos A, Peiris M (2005) International health regulations. Lancet 366:1249–1251. doi:10.1016/S0140-6736(05)67508-3)
Millstone E (2007) Can food safety policy-making be both scientifically and democratically legitimated? If so, how? J Agric Environ Ethics 20:483–508

Morse SS (2004) Factors and determinants of disease emergence. Rev Sci Tech 23:443–451
Murray N, McDiarmid SC, Wooldridge M, Gummow B, Morley RS, Weber SE, Giovannini A, Wilson D (2004) Introduction and qualitative risk analysis. In: Handbook on import risk analysis for animals and animal products, vol 1. World organisation for animal health (OIE), Paris
Mwacalimba KK (2012) Globalised disease control and response distort ion: a case study of avian influenza pandemic preparedness in Zambia. Crit Public Health. doi:10.1080/09581596.2012.710739
Mwacalimba KK (2013) Pandemic preparedness and multi-sectoral zoonosis risk management: implications for risk assessment of avian influenza in Zambian trade, health and agriculture. LAP LAMBERT Academic Publishing. ISBN 978-3-659-34609-5
Mwacalimba KK, Green J (2015) 'One health' and development priorities in resource-constrained countries: policy lessons from avian and pandemic influenza preparedness in Zambia. Health Policy Plan 30(2):215–222. doi:10.1093/heapol/czu001
Navarro V (1998) Comment: whose globalization? Am J Public Health 88:742–743
Nicoll A (2005) Avian and pandemic influenza--five questions for 2006. Euro Surveill 10:210–211
Normile D (2006) Avian influenza. Evidence points to migratory birds in H5N1 spread. Science 311:1225
Osterholm MT (2005) Preparing for the next pandemic. N Engl J Med 352:1839–1842
Paul M, Tavornpanich S, Abrial D, Gasqui P, Charras-Garrido M, Thanapongtharm W, Xiao X, Gilbert M, Roger F, Ducrot C (2010) Anthropogenic factors and the risk of highly pathogenic avian influenza H5N1: prospects from a spatial-based model. Vet Res 41:28
Pfeiffer DU (2007) Assessment of H5N1 HPAI risk and the importance of wild birds. J Wildl Dis 43:S47–S50
Pharo HJ (2003) The impact of new epidemiological information on a risk analysis for the introduction of avian influenza viruses in imported poultry meat. Avian Dis 47:988–995
Pidgeon NF (1999) Social amplification of risk: models, mechanisms and tools for policy. Risk Decis Policy 4(2):145–159
Pitrelli N, Sturloni G (2007) Infectious diseases and governance of global risks through public communication and participation. Ann Ist Super Sanita 43:336–343
Reperant LA, Cornaglia G, Osterhaus AD (2013) The importance of understanding the human-animal interface: from early hominins to global citizens. Curr Top Microbiol Immunol 365:49–81. doi:10.1007/82_2012_269
Roningen VO, DeRosa DA (2003) Zambia in regional and extra-regional free trade agreements: estimates of the trade and welfare impacts. Vorsim/Potomac Associates, Arlington
Rowell A (2003) Don't worry, it's safe to eat: the true story of GM food, BSE, and foot and mouth. Earthscan, London
Rweyemamu MM, Astudillo VM (2002) Global perspective for foot and mouth disease control. Rev Sci Tech 21:765–773
Scoones I (2010) The international response to avian influenza: science, policy and politics. In: Scoones I (ed) Avian influenza: science, policy and politics, Pathways to sustainability series. Earthscan, London
Scoones I, Forster P (2010) Unpacking the international response to avian influenza: actors, networks and narratives. In: Scoones I (ed) Avian influenza: science, policy and politics, Pathways to sustainability series. Earthscan, London
Shaffer ER, Waitzkin H, Brenner J, Jasso-Aguilar R (2005) Global trade and public health. Am J Public Health 95:23–34
Shortridge KF, Zhou NN, Guan Y, Gao P, Ito T, Kawaoka Y, Kodihalli S, Krauss S, Markwell D, Murti KG, Norwood M, Senne D, Sims L, Takada A, Webster RG (1998) Characterization of avian H5N1 influenza viruses from poultry in Hong Kong. Virology 252:331–342
Sims L, Narrod C (2008) Understanding avian influenza – a review of the emergence, spread, control, prevention and effects of Asian-lineage H5N1 highly pathogenic viruses. United Nations Food and Agriculture Organization (FAO). www.fao.org/avianflu

Sims LD, Domenech J, Benigno C, Kahn S, Kamata A, Lubroth J, Martin V, Roeder P (2005) Origin and evolution of highly pathogenic H5N1 avian influenza in Asia. Vet Rec 157:159–164
Slavic P (1999) Trust, emotion, sex, politics, and science: surveying the risk-assessment battlefield. Risk Anal 19(4):689–701
Slovic P (1998) The risk game. Reliab Eng Syst Saf 59(1):73–77
Stirling A, Mayer S (2000) A precautionary approach to technology appraisal? – A multi-criteria mapping of genetic modification in UK agriculture. TA-datenbank-Nachrichten 3:39–51
Stirling AC, Scoones I (2009) From risk assessment to knowledge mapping: science, precaution and participation in disease ecology. Ecol Soc 14:14
OIE Terrestrial Animal Health Code 2010. Available at http://web.oie.int/eng/normes/mcode/en_index.htm
Thiermann AB (2005) Globalization, international trade and animal health: the new roles of OIE. Prev Vet Med 67:101–108
Thomson GR, Tambi EN, Hargreaves SK, Leyland TJ, Catley AP, van 't Klooster GG, Penrith ML (2004) International trade in livestock and livestock products: the need for a commodity-based approach. Vet Rec 155:429–433
Upton M, Otte J (2004) The impact of trade agreements on livestock producers pro-poor livestock policy initiative. Research Report. RR Nr. 04–01
Vallat B, Pastoret P-P (2009) The role and mandate of the World Organisation for Animal Health in veterinary education. Rev Sci Tech 8(2):503–510
van den Berg T (2009) The role of the legal and illegal trade of live birds and avian products in the spread of avian influenza. Rev Sci Tech 28:93–111
Van Zwanenberg P, Millstone E (2006) Risk communication strategies in public policy-making. In: Dora C (ed) Health, hazards and public debate: lessons from risk communication from the BSE/CJD saga. WHO
Vose D (2000) Qualitative versus quantitative risk analysis and modelling. In: Rodgers CJ (ed) Risk analysis in aquatic animal health. Proceedings of the OIE international conference, 8–10 February 2000
Vose D, Acar J, Anthony F, Franklin A, Gupta R, Nicholls T, Tamura Y, Thompson S, Threlfall EJ, van Vuuren M, White DG, Wegener HC, Costarrica ML (2001) Antimicrobial resistance: risk analysis methodology for the potential impact on public health of antimicrobial resistant bacteria of animal origin. Revue Scientifique et Technique (International Office of Epizootics) 20(3):811–827
Weber TP, Stilianakis NI (2007) Ecologic immunology of avian influenza (H5N1) in migratory birds. Emerg Infect Dis 13:1139–1143
Webster RG, Govorkova EA (2006) H5N1 influenza — continuing evolution and spread. N Engl J Med 355:2174–2177
Webster R, Hulse D (2005) Controlling avian flu at the source. Nature 435:415–416
Webster RG, Guan Y, Peiris M, Walker D, Krauss S, Zhou NN, Govorkova EA, Ellis TM, Dyrting KC, Sit T, Perez DR, Shortridge KF (2002) Characterization of H5N1 influenza viruses that continue to circulate in geese in southeastern China. J Virol 76:118–126
WHO (2004) Press Release: Unprecedented spread of avian influenza requires broad collaboration-FAO/OIE/WHO call for international assistance. 27 January 2004
WHO (2005a) Avian influenza: assessing the pandemic threat. http://www.who.int/csr/disease/influenza/H5N1-9reduit.pdf
WHO (2005b) H5N1 avian influenza: timeline 28 October 2005. World Health Organization. http://www.who.int/csr/disease/avian_influenza/Timeline_28_10a.pdf
WHO (2005c) WHO Global influenza preparedness plan: the role of WHO and recommendations for national measures before and during pandemics: World Health Organization, WHO/CDS/CSR/GIP/2005.5
WHO (2010) Cumulative number of confirmed human cases of avian influenza A/(H5N1) reported to WHO, WHO. Available at: http://www.who.int/csr/disease/avian_influenza/country/cases_table_2010_11_19/en/index.html

WHO/AFRO (2005) Influenza pandemic risk assessment and preparedness in Africa. World Health Organization Regional Office for Africa, Brazzaville

WHO/FAO (2006) Food safety risk analysis: a guide for national food safety authorities. WHO/FAO, Rome. Available at: ftp://ftp.fao.org/docrep/fao/009/a0822e/a0822e00.pdf

WHO/WTO (2002) WTO agreements and public health – a joint study by the WHO and the WTO secretariat

Wilkinson RG, Pickett KE (2006) Income inequality and population health: a review and explanation of the evidence. Soc Sci Med 62:1768–1784

Wooldridge M (2000) Risk analysis methodology: principles, concepts and how to use results In: Rodgers CJ (ed) Risk analysis in aquatic animal health. Proceedings of the OIE international conference, 8–10 February 2000

Xu X, Subbarao K, Cox NJ, Guo Y (1999) Genetic characterization of the pathogenic influenza A/Goose/Guangdong/1/96 (H5N1) virus: similarity of its hemagglutinin gene to those of H5N1 viruses from the 1997 outbreaks in Hong Kong. Virology 261:15–19

Yee KS, Carpenter TE, Cardona CJ (2009) Epidemiology of H5N1 avian influenza. Comp Immunol Microbiol Infect Dis 32:325–340

Zepeda C, Salman M, Thiermann A, Kellar J, Rojas H, Willeberg P (2005) The role of veterinary epidemiology and veterinary services in complying with the World Trade Organization SPS agreement. Prev Vet Med 67:125–140

Chapter 3
Global Approaches to Regulating Farm Animal Welfare

Lewis Bollard

Abstract Industrialized animal agriculture is rapidly spreading around the globe. But animal welfare protections have failed to keep up, resulting in billions of farm animals suffering. This article chronicles the slow rise of global laws and institutions intended to mitigate that suffering. In particular, it focuses on the evolving animal welfare policies of the European Union, World Animal Health Organization, World Bank, and Food and Agriculture Organization. It also addresses how farm animal welfare is becoming integrated into World Trade Organization case law and bilateral free trade agreements. I argue that international laws and institutions, though currently failing to protect farm animal welfare, provide a promising framework for more extensive protections in future. And I provide some recommendations on how to implement those future protections.

3.1 Introduction

In the last half-century, as global laws and institutions have proliferated, so too that has industrialized animal agriculture. The two share common causes: globalization, the growing world population, and growing prosperity. But until recently the two were not closely linked. Industrialized animal agriculture has largely escaped even domestic regulation, allowing large animal welfare problems to fester. It is no surprise, then, that global laws and institutions—driven by domestic actors who were largely ignoring farm animal welfare—chose to ignore the issue too.

L. Bollard, BA, JD (✉)
Open Philanthropy Project, Washington, DC, USA
e-mail: lewis.bollard@gmail.com

G. Steier, K.K. Patel (eds.), *International Farm Animal, Wildlife and Food Safety Law*, DOI 10.1007/978-3-319-18002-1_3

3.1.1 The Globalization of "Factory Farming"

In the last half-century, animal agriculture has expanded dramatically around the world, driven by the huge growth in poultry consumption. In 1961, when the Food and Agriculture Organization (FAO) first began collecting data, it estimated that the world's farms held 400 million pigs, 900 million cattle, and 4 billion chickens, ducks, and turkeys.[1] By 2013, the FAO estimated that the world's farm animal populations had grown to a billion pigs, 1.3 billion ducks, 1.5 billion cattle, and an astonishing 24 billion chickens, ducks, and turkeys.[2] During the same time period, the world's human population grew from 3 billion to 7 billion people. So, while the pig population grew roughly proportionally to the human population (and the cattle population grew more slowly), the population of chickens, ducks, and turkeys grew three times faster than the human population.

At the same time, animal agriculture has industrialized. Until the mid-twentieth century, the majority of the world's farm animals were raised in low-density pastoral systems, in which a family cared for a small herd of animals, often spanning several different species.[3] But in the early twentieth century, farmers in the developed world discovered that they could increase efficiency by specializing in just one species, and more intensively rearing those animals. These farmers were aided in their intensification efforts by the invention of antibiotics and medicated feeds, which allowed them to keep animals alive in unnatural, crowded environments.[4] They were also aided by the invention of artificial insemination in the early twentieth century, which allowed them to select for animals with greater productivity and a heightened ability to withstand intensive confinement. At the beginning of the twentieth century, no farm animals were artificially inseminated; by the century's end, almost all were.[5]

The Green Revolution in the 1950s and 1960s brought industrialized agriculture to the developing world. The Green Revolution focused on innovations in irrigation, chemical fertilizers, and crop genetic selection, which significantly improved crop efficiency, more than doubling crop yields after 1960. These efficiencies in turn increased the value of agricultural land, raising the opportunity cost of grazing livestock outside on potential crop land. These efficiencies also drove down the cost of grains, reducing the cost of running a "factory farm," where animals confined indoors and fed grains rather than grazing on pasture outside. Indeed, the industrialization of animal agriculture has largely focused on pigs and poultry—and not cattle and sheep—because pigs and poultry thrive on grain-based diets, whereas ruminants require roughage.[6]

---

[1]FAO (2014a).

[2]FAO (2014a).

[3]Cochrane (1993), pp. 122–150.

[4]Cochrane (1993), pp. 122–150.

[5]Wilmot (2007).

[6]Sherman (2002), p. 79.

China in particular has driven a huge portion of the growth in factory farming since the early 1980s. Traditionally, the Chinese consumed very little meat, and animal agriculture was small-scale, with most households owning a pig or two.[7] But after Deng Xiaoping began opening China's economy in 1978, the nation's economy boomed, fueling a surge in meat consumption. Since then, per capita consumption of meat has quadrupled, driven by huge increases in pork and chicken consumption.[8] Much of that increased consumption is of industrially-raised meat, which is often favored by consumers who view it as more strictly regulated and thus safer than backyard-raised meat.[9] Since 2006, when China experienced an outbreak of porcine reproductive and respiratory syndrome virus, the government has encouraged the consolidation and industrialization of the nation's pork industry to improve food safety and reduce supply shocks.[10] That effort has succeeded: in 2005, most Chinese pigs were raised on backyard farms with fewer than 50 pigs; by 2015, less than a third of Chinese pigs were.[11] Today, half of the world's pork is raised and consumed in China, and Chinese pork consumption is still growing rapidly.[12]

The globalization of industrial animal agriculture has been a boon for multinational agribusinesses, which now control the global supply chains in pork, beef, and poultry production. Brazilian-based JBS has become the world's largest meat company through acquisitions in Australia, Argentina, and the United States, including the purchase in 2007 of Swift & Company, then America's third largest beef and pork processor. JBS now has estimated annual revenues of $50 billion[13]—greater than the Gross Domestic Product of Ethiopia or Panama. In 2013, Chinese-based Shuanghui/WH Group purchased America's largest pork producer, Smithfield Foods, to become the world's largest pork producer. And Tyson Foods and Cargill—with annual revenues of $34 billion and $20 billion respectively—produce much of the rest of the world's beef, pork, and poultry.[14] Tyson Foods alone has greater annual revenues than Phillip Morris International, the world's largest private tobacco company.[15] Even many distinctive meat brands—like Pilgrim's Pride chicken, Jimmy Dean's sausages, Swift beef and pork, Hillshire Farm pork, Wright bacon, Shady Brook turkey, and Armour pork—are in fact owned by these four huge multinational meat companies.

[7] Schneider and Sharma (2014), p. 12.

[8] FAO (2014a).

[9] Schneider and Sharma (2014).

[10] Schneider and Sharma (2014), p. 13.

[11] Schneider and Sharma (2014), p. 19.

[12] Schneider and Sharma (2014), p. 14.

[13] Samora (2014).

[14] Clyma (2014).

[15] Fortune (2014).

3.1.2 The Animal Welfare Effects of the Globalization of "Factory Farming"

The spread of industrial animal agriculture has likely created more suffering to animals than any other event in human history. The vast majority of the tens of billions of animals slaughtered around the world every year are raised in indoor confinement systems, where they are deprived of sunshine, clean air, or enough space to engage in natural behaviors. They are routinely physically mutilated, separated from their mothers at a premature age, and slaughtered in inhumane manners.

In 1964, the British government commissioned a technical committee to investigate the animal welfare effects of intensive farming methods then coming into widespread use. The resulting "Brambell Report," named after Professor Rogers Brambell, who led the committee, concluded that modern factory farm systems severely undermine animal welfare.[16] In particular, the committee recommended that all farm animals should at least have enough room to stretch, lie down, turn around, and groom themselves; that mutilations, such as de-beaking hens and castrating pigs and cows, should be done only when necessary and only with pain relief; and that flooring, lighting, and ventilation in factory farms should be optimized for animal welfare—not just for efficiency. But the Report was largely ignored by the Government. Eight years later, a member of the committee wrote in *New Scientist* that the government had missed a unique opportunity to stem the rise of factory farming practices.[17]

Almost a half century later, the Pew Commission on Industrial Farm Animal Production reached similar conclusions to the Brambell committee.[18] The Commission, made up of 15 experts—including a former U.S. Secretary of Agriculture and several leading animal scientists—concluded that the spread of factory farming had harmed animal welfare. It recommended the phase out of the most intensive confinement systems—including veal crates, gestation and farrowing crates for breeding pigs, and battery cages for hens.[19] It also recommended a host of other reforms reminiscent of the Brambell Report's recommendations, including an end to painful mutilations and improvements to flooring and housing systems.[20] The Commission noted that, "Recently, the [factory farming] model has begun to spread to all corners of the world, especially the developing world."[21]

To this day, most animals on factory farms do not enjoy the basic animal welfare standards called for by the Brambell Report and the Pew Commission. The poultry

---

[16] Brambell (1965).

[17] Ewer (1973).

[18] Pew Commission (2008).

[19] Pew Commission (2008).

[20] Pew Commission (2008).

[21] Pew Commission (2008).

industry confines billions of "broiler" chickens in giant sheds, where they are deprived of sunshine, fresh air, or the space to perform natural behaviors.[22] The industry has selectively bred chickens to grow so fast—over three times faster than they grew 50 years ago—that many can no longer walk naturally, and likely experience chronic pain.[23] At slaughter time, these chickens are caught in huge groups—with catchers handling 1000–1500 animals per hour—resulting in rough handling, bruising, and broken limbs.[24] And chickens are typically slaughtered in unreliable electric stun baths and cutting lines—a process that the U.S. Department of Agriculture reports results in almost a million birds boiled alive every year in the U.S. alone.[25]

Chickens in the egg industry have it even worse. On egg factory farms, most hens are confined in battery cages, so named because they are stacked in "batteries" up to six levels high.[26] In these wire-mesh cages, which hold from three to nine birds, each bird typically receives just 48 to 72 square inches of space—about the size of a shoe box—and no space to nest, scratch, or perch.[27] Each battery shed can house tens of thousands of hens—the American median shed stocking density is 30,000–70,000 hens—while several million hens can be confined in one facility.[28] Most male chicks are killed at birth, often by being ground up alive, and female chicks are "de-beaked"—a painful mutilation done to stop hens from cannibalizing each other in extreme confinement conditions.[29]

Indeed, most farm animals of all species are now subjected to intensive confinement and acts that inflict acute pain. Almost all pigs have their tails and genitals mutilated at birth without the use of painkillers, and most breeding pigs live almost their entire lives completely immobilized as they alternate between gestation crates (for pregnancy) and farrowing crates (for giving birth).[30] A fast rising portion of fish are now raised in aquaculture operations that are essentially water-based factory farms, with the fish intensively confined in unclean water, where parasites may feed on their faces.[31] Dairy cows are forced to carry excess milk, causing pain to their udders, and their unwanted male calves are often sold into the veal industry, where some producers continue to use "veal crates"—two by six foot metal enclosures that prevent calves from turning around or socializing with other calves.[32] And although beef cattle likely have it the best of any

[22] HSUS (2014a).
[23] HSUS (2014a).
[24] HSUS (2014a).
[25] Kindy (2013).
[26] HSUS (2008).
[27] HSUS (2008).
[28] HSUS (2008).
[29] HSUS (2008).
[30] HSUS (2010).
[31] Yue (2013).
[32] HSUS (2014b).

farm animal, with most raised on open pasture ranches for the majority of their lives, most will still endure a period of feedlot confinement, and painful castration and branding.[33]

3.1.3 The Failure of Domestic Laws at Addressing Globalized Factory Farming

Domestic laws have failed to keep up with the globalization of factory farming practices. Developed nations barely regulate factory farming within their borders, let alone globally. In the United States there is no federal law preventing cruelty to animals on the farm, and most state animal cruelty laws specifically exempt routine agricultural practices.[34] The result is that acts that would be considered torture if done to a dog or a cat—like castration without painkillers—are perfectly legal when done to a pig on a factory farm.[35] The European Union has better protections for farm animals (see Sect. 3.2 below), but these protections largely only apply to farm animals within their borders. They do not apply, for example, to a European agribusiness that expands its operations into China and then re-imports the meat into Europe. Indeed, even New Zealand, known for its strong animal welfare laws, has been powerless to stop the nation's largest dairy cooperative, Fonterra, from establishing factory farms with essentially no animal welfare oversight in China.[36]

Developing nations have widely varying levels of protections for animals. Most African nations still lack even basic animal protection laws, according to the non-profit group World Animal Protection.[37] Russia has anti-cruelty provisions in its general penal code, which should cover farm animals, but it has no standards for farm animals specifically.[38] Brazil has adopted more comprehensive regulations protecting farm animals during raising, transport, and slaughter—likely to maintain access to the higher welfare European market—but the protections are still minimal.[39] Mexico has adopted farm animal welfare provisions through its Federal Animal Health Act 2007, but enforcement of these provisions appears to be inconsistent.[40]

The lack of strong standards in developing nations matters because they are increasingly the world's dominant producers of animal products. Nine of the world's ten largest poultry producers are developing nations—China, Brazil,

[33]HSUS (2014c).
[34]Wolfson and Sullivan (2004).
[35]Wolfson and Sullivan (2004).
[36]Radio New Zealand (2010).
[37]Animal Protection Index (2014).
[38]Animal Protection Index (2014).
[39]Animal Protection Index (2014).
[40]Animal Protection Index (2014).

Indonesia, Russia, India, Iran, Mexico, Columbia, and South Africa (the one exception is the United States).[41] Eight of the world's ten largest egg producers are developing nations,[42] and five of the world's ten largest pork producers are developing nations.[43] This trend is only likely to increase as multinational meat producers expand their operations in developing nations, to take advantage of lower costs, and laxer labor, environmental, and animal welfare regulations. It will also be spurred by rising domestic demand for animal protein as developing nations get richer.

The rise of factory farming in China, in particular, has the potential to undermine much of the progress on farm animal welfare in developed nations. China now has more pork-producing pigs—almost 700 million as of 2012—than the other nine largest pork producing nations combined.[44] China produces more eggs—almost 500 billion as of 2012—than the other nine largest egg producing nations combined.[45] And although the poultry production market is more diversified—the United States has almost as many chickens as China, and Brazil is not far behind—China slaughtered almost 10 billion meat-producing chickens in 2012.[46] Major American chicken producers have also began investing in China, drawn by demand from fast food chains—in China, a new fast food outlet opens every 18 hours.[47] Both Cargill and Tyson are investing in fully integrated chicken production facilities in China that promise to produce safe chicken at low prices.[48] China has also become a major importer of American pork or chicken, especially of pieces that Americans won't eat, like pig ears, intestines and feet.[49] Most of China's growing number of farm animals are confined in factory farms.

China's legal framework has been slow to respond to this situation. China historically has had no meaningful animal welfare law, let alone specific protections for farm animals. A 2011 survey found that two thirds of Chinese have never even heard of the concept of "animal welfare."[50] In 2005, the Tenth People's Congress adopted the "Animal Husbandry Law of the People's Republic of China," but it is focused on increasing agricultural productivity and protecting food safety, not on improving animal welfare.[51] Indeed, the Law's 73 Articles never mention protecting the welfare of animals, while they promote practices that

---

[41]HSI (2012a).
[42]HSI (2012b).
[43]HSI (2012c).
[44]HSI (2012c).
[45]HSI (2012b).
[46]HSI (2012a).
[47]Everett (2014).
[48]Everett (2014).
[49]Everett (2014).
[50]You (2014).
[51]Chinese National People's Congress (2005).

are likely to be detrimental to animals' welfare—for example, increased selective breeding for growth and intensification of production systems.

But there are finally signs in China of progress on farm animal welfare. The 2011 survey found that, although Chinese are more concerned with food safety and food quality than animal welfare, roughly two thirds would support a national law to protect farm animals.[52] And a majority disapproved of routine factory farming practices, including killing chickens in front of other chickens and housing pigs on concrete floors (after the respondents were told by surveyors that pigs preferred to live on dirt).[53] In August, 2008, the Chinese Commerce Ministry added humane slaughter requirements to its regulations on the slaughter of pigs, although it remains unclear how widely the regulation is obeyed.[54] And in September, 2009, a panel of scholars proposed China's first comprehensive animal protection law, which would cover the majority of animals, including farm animals—although the National People's Congress had not yet acted on the law at the time of this article's publication.[55]

3.2 Europe's Farm Animal Welfare Reforms

Europe pioneered the first transnational approach to farm animal welfare, and is now spreading that approach globally. Because of Europe's history of progressively improving its animal welfare standards, its standards are today more stringent than those of any other government or international institution (though some companies have adopted more stringent standards—for example Whole Foods Market). As such, Europe's standards could provide a model for higher global welfare standards in the future.

3.2.1 The Council of Europe's Standards

Europe has long led the world in transnational farm animal welfare standards, although these began as largely symbolic principles. In 1976, before some European nations had even developed domestic farm animal welfare standards, the Council of Europe began drafting pan-European standards. The Council of Europe was founded after World War II to promote unity between European nations, although by 1976 it had expanded to regulate economic and other issues. Much of this regulation would later be subsumed under the European Union, when

[52]You (2014).
[53]You (2014).
[54]Xinhua (2009).
[55]International Fund for Animal Welfare (2009).

it was founded in 1993. But the Council of Europe continues to this day with a broader membership than the EU (47 member states as of mid 2016). And even when the Council was adopting animal welfare standards in 1976, it already included members that would never join the EU, like Turkey and Switzerland.

The Council of Europe's (1976) "Convention for the Protection of Animals Kept for Farming Purposes" bound the Council's member states to follow a set of principles for animal welfare.[56] In particular, it provided that animals should be housed and cared for in a manner "appropriate to their physiological and ethological needs."[57] And, in stark contrast to intensive confinement systems becoming common at the time, the Convention specifically provided that "[t]he freedom of movement appropriate to an animal . . . shall not be restricted in such a manner as to cause it unnecessary suffering or injury."[58] But these standards lacked any penalty provisions for non-compliance, leading a European Court to conclude that "[i]t is clear from the very wording of [the Convention's] provisions that they are indicative only" and not enforceable against the member states.[59] Indeed, for several decades after the adoption of the Convention, European nations that had ratified it continued to use production systems like battery cages and gestation crates that clearly denied animals their "freedom of movement."

3.2.2 The European Union's Standards

In the 1990s, a separate body, the Council of the European Communities (later to become the Council of the European Union; here collectively referred to as the "Council"), developed these general principles into specific directives. These directives only applied to the members of the European Economic Community, and later the European Union—both of which had more limited memberships than the Council of Europe (the EU had 28 member states as of 2016). But they still created the world's first enforceable transnational farm animal welfare protections. In 1991, the Council adopted a directive providing minimal standards for the housing of veal calves in new facilities.[60] Although this directive allowed the continued use of veal crates and tethering, it established a strong enforcement system, which would ensure welfare improvements when the European Commission modified the directive to abolish veal crates and tethering in 1997.[61] In particular, the directive provided that member states must adopt the standards and

[56]Council of Europe (1976).

[57]Council of Europe (1976).

[58]Council of Europe (1976).

[59]Judgment of the Court of 19 March 1998—The Queen v Minister of Agriculture, Fisheries and Food, ex parte Compassion in World Farming Ltd.

[60]Council of Europe (1991).

[61]European Commission (1997).

inspect their producers for compliance with them, and report to the Commission on the result of those inspections.[62]

The directives on egg and pork production had even more pronounced effects on animal welfare, creating the first transnational bans on the use of battery cages and gestation crates. In 1999, the Council adopted standards for the protection of laying hens, which noted that the "the welfare conditions of hens kept in current battery cages and in other systems of rearing are inadequate."[63] The directive required the phase out of barren battery cages by 2012 (most producers have complied by transitioning to roomier colony cages, although some nations are still not in compliance with the directive). In 2001, the Council adopted pig welfare standards which largely prohibited the use of gestation crates—although it did not prohibit farrowing crates, where sows are kept after giving birth, or the use of gestation crates during the first four weeks and final week of pregnancy (about a third of the sow's 16.5 week pregnancy).[64] The Commission adopted parallel pig welfare standards that set basic standards for flooring, noise, lighting, and other housing factors.[65]

These directives acted as both a sword and a shield: forcing European nations to raise their animal welfare standards, but also shielding them from animal welfare based trade restrictions from higher welfare nations. This became apparent when the European Court of Justice ruled on a case about British veal calf imports in 1998. Britain had long imposed higher animal welfare standards for veal calves than the rest of Europe, prompting British dairy producers to export their calves to European nations with more lenient standards. In 1995, the animal advocacy group Compassion in World Farming sued in British court to try to stop this trade in veal calves across the English Channel. The British court referred the case to the European Court of Justice, which found that not only was England not required to stop the export of veal calves to lower welfare nations, but it could not do so.[66] The Court reasoned that the European free trade area required states to harmonize their export and import standards, barring states from "adopt[ing] stricter measures for the protection of calves other than provisions applying within their own territory."[67]

Europe's reforms were driven by supportive public opinion and a pro-regulatory legislative system. A 2001 survey found that the European public reported high levels of concern for animal welfare, with a consensus that consumers would benefit from improved animal welfare through the improved quality, safety, and taste of the

[62]Council of Europe (1991).

[63]Council of Europe (1999).

[64]Council of Europe (2001).

[65]European Commission (2001).

[66]Judgment of the Court of 19 March 1998—The Queen v Minister of Agriculture, Fisheries and Food, ex parte Compassion in World Farming Ltd.

[67]Judgment of the Court of 19 March 1998—The Queen v Minister of Agriculture, Fisheries and Food, ex parte Compassion in World Farming Ltd.

products coming from better treated animals.[68] Matheny and Leahy (2007) argue that Europe's reforms can be attributed to a long tradition of animal protection, a vibrant field of academic animal welfare science, reliance on an independent Scientific Committee on Animal Health and Welfare, a less powerful agricultural lobby than in the United States, and a smaller agricultural export market.[69] These factors are still largely lacking in the rest of the world, but European advocacy is increasingly providing the needed boost to improve farm animal welfare.

3.2.3 European Advocacy for Global Farm Animal Welfare Regulation

Europe is now pushing for other nations to harmonize their farm animal welfare standards upward, through free trade agreements. Traditionally, free trade agreements (FTAs) have ignored animal welfare. For example, the North American FTA (NAFTA) signed in 2008 contained separate agreements on environmental and labor cooperation, but no similar provisions on animal welfare. Thus these agreements have typically treated animal welfare restrictions on trade as presumptively invalid, reducing nations' ability to regulate animal welfare.

But the EU has begun adding explicit animal welfare provisions to bilateral FTAs, forcing their trade partners to improve their standards. The EU has recognized both that its domestic producers risk being undercut by imports from countries with lower animal welfare standards, and that its citizens' desire for more humane treatment of animals is undermined when imports are not held to their own standards.

The EU first asked for the inclusion of animal welfare language in an FTA in 2003, in negotiations over a bilateral agreement with Chile. Although the resulting EU-Chile FTA did not include any enforceable standards, it recognized that "[t]his Agreement aims at reaching a common understanding between the Parties concerning animal welfare standards."[70] The FTA also established an aim of harmonizing Europe and Chile's animal welfare standards over the long-term. Moreover, the countries seem to be seeking to harmonize both of their standards upwards—i.e. not by lowering Europe's standards to Chile's. Cabanne (2013) argues that the EU-Chile FTA has "played a positive role in the institutionalization of animal welfare in Chile, in particular for livestock production."[71] In the years since entering the FTA, Chile has passed an animal welfare law that includes farm

[68]Harper and Henson (2001).

[69]Matheny and Leahy (2007).

[70]EU-Chile agreement on sanitary and phytosanitary measures applicable to trade in animals and animal products, plants, plant products and other goods and animal welfare (2003), Article 1 of the Annex.

[71]Cabanne (2013).

animals and created a sub-department to cover animal welfare issues. It has also joined with Uruguay to establish a Collaborating Centre on Animal Welfare with the support of the OIE.

The EU has since included animal welfare provisions in new bilateral FTAs. In 2011, the EU included animal welfare language in an FTA with South Korea.[72] In 2013, Andrea Gavinelli, the head of the Animal Welfare Unit of the European Commission, announced that "[w]e are sitting in trade negotiations wishing to have animal welfare included. It doesn't matter which country we are talking about, this is the mandate we have got and this is what we are doing."[73] In EU-Japan trade talks that were ongoing when this article went to publication, animal welfare was again on the agenda.[74] And in ongoing EU-US negotiations over the Trans-Atlantic FTA, EU negotiators have reputedly attempted to insert animal welfare provisions, only to be rebuffed by US trade negotiators.[75]

3.3 Other Global Approaches to Farm Animal Welfare

Beyond the EU, other global institutions have been slow to adopt animal welfare policies. Partly this is a question of mandate: these institutions were founded to promote international harmony and spur global economic growth, not to regulate the treatment of animals. But partly it is a refusal to see animal welfare as a legitimate issue: even as these institutions have tackled issues beyond their harmony and economy mandates—whether human rights, corruption, or environmental protection—they have still been slow to take up animal welfare. Finally, though, this appears to be changing. There seems to be a growing recognition amongst global institutions that they need to be proactive on animal welfare, and especially on farm animal welfare, to avoid contributing to the problem and to keep up with broader global trends.

3.3.1 The World Bank

International financial institutions have not only failed to address the globalization of factory farming—they have often unintentionally abetted it. The World Bank, a United Nations agency that provides loans to developing nations to spur growth and reduce poverty, typically took no position on animal welfare or factory farming. But the World Bank's International Finance Corporation (IFC), which finances private

[72]HSI (2011).

[73]Farming UK (2013).

[74]European Commission (2014).

[75]Kopperud (2014).

sector projects in developing nations that are meant to boost economic growth, repeatedly invested in huge agribusiness projects. These projects typically spread American-style factory farms with no regard for the welfare of the animals confined inside of them. Indeed, a 2013 report found that the IFC had invested millions of dollars in factory farms in developing nations that use battery cages and gestation crates—two of the most controversial factory farm practices, both now largely illegal in the EU.[76]

The IFC has belatedly recognized the need to consider the animal welfare implications of these investments. In 2006, the IFC published a "Good Practice Note on Animal Welfare in Livestock Operations," a set of guidelines for agribusiness projects that the IFC invests in.[77] The Note recommended that farm housing systems should "allow all animals space to stand, turn around, stretch, sit, and/or lie down comfortably at the same time."[78] But the Note did not actually prohibit the use of battery cages, gestation crates, or other confinement systems that deny animals the ability to perform these basic behaviors, and the IFC has apparently since invested in projects involving these extreme confinement systems. [79] The Note also referenced at length the minimal animal welfare audit standards of the U.S. National Council of Chain Restaurants and the Food Marketing Institute, apparently endorsing these standards.[80] And the IFC's guidelines were voluntary, with the IFC not making clear whether disobeying the guidelines would actually hurt a project's chance of receiving financing.[81]

In 2014, the IFC updated its "Good Practice Note on Animal Welfare in Livestock Operations," but largely failed to strengthen the language. The new guidelines replaced the reference to the U.S. chain restaurants' animal welfare standards with a reference to the significantly stronger standards of the Global Animal Partnership, a nonprofit group that sets the standards for meat sold in Whole Foods Market supermarkets.[82] And the new guidelines acknowledged an "international trend to move from the use of sow stalls to group housing systems."[83] But the 2014 Note also maintained the voluntary nature of the guidelines, encouraged nations to adopt the weak OIE animal welfare standards, and even referenced the abysmal standards of the U.S. Chicken Council, a trade group for largely industrial chicken producers.[84] And the 2014 revisions removed the recommendation from the 2006 Note that animals should be given enough space to turn around (although it

[76]HSI, Compassion in World Farming, Four Paws (2013).

[77]International Finance Corporation (2006).

[78]International Finance Corporation (2006).

[79]HSI, Compassion in World Farming, Four Paws (2013).

[80]International Finance Corporation (2006).

[81]International Finance Corporation (2006).

[82]Compare International Finance Corporation (2006) with International Finance Corporation (2014).

[83]International Finance Corporation (2014).

[84]International Finance Corporation (2014).

was not immediately clear what the effect of this change would be, since the IFC had seemingly ignored this guidance since 2006).[85]

3.3.2 The World Organisation for Animal Health (OIE)

The World Organisation for Animal Health (OIE) has developed the first internationally applicable farm animal welfare standards. The OIE (the acronym refers to the organization's historic French name) is the intergovernmental organization tasked with improving animal health worldwide. Although the OIE operates on a voluntary basis, its standards have significant sway across its 180 member states. The OIE identified animal welfare as a priority in its 2001–2005 strategic plan—the first international institution outside of the EU to do so. In 2002, the OIE established an Animal Welfare Working Group, whose recommendations the OIE adopted a year later.[86] In 2004, the OIE integrated animal welfare guiding principles into its Terrestrial Animal Health Code, its set of guidelines for the transportation and handling of animals, which until then had largely concerned animal health and zoonotic diseases. Since then, OIE member nations have adopted 12 global animal welfare standards, covering more than 100 pages of specific guidelines, which have been integrated into the Terrestrial Animal Health Code.[87]

The OIE standards are minimal, largely enshrining current industry practices. For example, the "broiler chicken production system" standards only provide vague standards on preventing overcrowding—rather than a specific space requirement per bird—and fail to address over-breeding practices that cause birds to suffer.[88] The poultry slaughter standards—while significantly stronger than standard practices in the American poultry industry, which is not bound by any humane slaughter standards—fail to provide measurable goals for preventable cruelties, like live birds immersed in the scalding tank.[89] And the pork and poultry standards say nothing about intense confinement in gestation crates and battery cages. This is perhaps to be expected since new standards must receive the consensus approval of all 180 member states of the OIE.

But the OIE standards nonetheless likely provide the most promising route to improving animal welfare standards globally, for three reasons. First, the OIE standards are the only standards with buy-in from the vast majority of the world's nations. The OIE's 180 member states include all of the largest meat producing nations and many developing nations that are yet to adopt any domestic animal welfare laws. The OIE standards not only require these states to commit to

[85]International Finance Corporation (2014).

[86]OIE (2014a).

[87]OIE (2014b), Chapter 7: Animal Welfare.

[88]OIE (2014b), Article 7.10.4.

[89]OIE (2014b), Article 7.5.7.3(b).

implementing the OIE animal welfare standards—though it is unclear how many have actually done so—but also involve them in preparing the standards through the OIE's consensus-based approach.[90] The OIE also holds ongoing workshops and conferences to build the capacity of national regulators to adopt and enforce the OIE standards.[91] And the OIE has developed animal welfare strategic plans to implement the standards in each of its five regions, while appointing an implementation liaison in each of its 180 member states.[92] The OIE has thus developed a global animal welfare infrastructure that no one other organization can match.

Second, under World Trade Organization (WTO) rules, the OIE standards are a presumptively valid basis for regulating trade. The WTO Agreement on the Application of Sanitary and Phytosanitary Measures (SPS Agreement) formally recognizes the OIE standards as the international reference for animal health.[93] That means that high-welfare nations could likely refuse to allow the import of animal products from other nations unless the animals were treated in compliance with the OIE standards. The flipside, of course, is that this makes it hard for high-welfare nations to impose standards on the sale of animal products that surpass the OIE standards. That likely explains why the European Union—which has the world's highest farm animal welfare standards—has taken such an active role in improving the OIE standards and ensuring their worldwide adoption.

Third, the standards acknowledge pro-animal welfare principles, which could form the basis for stricter standards in the future. In particular, the "Guiding Principles" for the standards include the "five freedoms" first laid out in the Brambell Report (freedom from hunger, thirst, and malnutrition; freedom from fear and distress; freedom from physical and thermal discomfort; freedom from pain, injury, and disease; and freedom to express normal patterns of behavior).[94] If applied consistently, the five freedoms would require OIE member states to do away with battery cages, gestation crates, and other routine factory farm practices that prevent farm animals from expressing their normal patterns of behavior. The OIE standards also acknowledge "[t]hat the use of animals carries with it an ethical responsibility to ensure the welfare of such animals to the greatest extent practicable"—an aspirational principle that provides a basis for standards evolving upwards as nations develop.[95] (For more on how these standards could be improved, see Sect. 3.5.2).

[90]OIE (2013).

[91]*See*, e.g., OIE (2012).

[92]IFC (2014).

[93]*See* Agreement on Sanitary and Phytosanitary Measures: Preamble, Article 3, and Annex A, paragraph 3(b).

[94]OIE (2014b).

[95]OIE (2014b).

3.3.3 Other International Institutions

A number of other international institutions play more tangential roles in farm animal welfare. Amongst them two stand out for their proactive approaches to the topic. The European Bank of Development and Reconstruction, a multinational development bank owned by 64 nations and two EU institutions, had historically invested in private agribusiness projects without concern to animal welfare. Indeed, since its founding in 1991, it had invested almost $10 billion in agribusiness projects in largely developing nations—all without requiring any assurances on animal welfare.[96] But after a campaign led by non-profit animal protection groups and members of the European Parliament, the Bank announced in May, 2014, that it will only fund agribusiness projects in future that meet or exceed the EU farm animal welfare standards.[97] And in 2014, the Bank cosponsored a report comparing animal welfare standards in the beef, pork, and poultry industries between European nations and major developing nations.[98] The Bank's new proactive approach to animal welfare has the potential to significantly improve conditions at new factory farms in the developing world.

The Food and Agriculture Organization of the United Nations has also increasingly made farm animal welfare a priority. The FAO, with 192 member nations, is the UN agency tasked with combating hunger and promoting agriculture development. Until recently, despite considering environmental sustainability and other factors, the FAO had ignored animal welfare. But in 2008, the FAO hosted a meeting of experts on farm animal welfare, although their resulting report lacked specific recommendations; instead it included vague suggestions like "[a]nimal welfare should not be treated as a stand-alone issue," and nations should "search for improvements that will be practical in the given situation."[99] The FAO also now runs an online "Gateway to Farm Animal Welfare," which publishes news, resources, and events relating to farm animal welfare.[100] And FAO legal experts help nations draft new animal welfare legislation and build their capacities to monitor farm animal welfare. In October 2014, the UN Committee on World Food Security even included a fleeting mention of farm animal welfare in its "Principles for Responsible Agricultural Investments."[101] The FAO certainly has the potential to be a powerful force for improving farm animal welfare in the future.

[96]HSI, Compassion in World Farming, Four Paws (2013).

[97]Humane Society International, EBRD Sets Milestone for Animal Welfare: Adopts Rules to Stop Financing Extreme Confinement of Farm Animals.

[98]FAO (2014b).

[99]FAO (2008).

[100]*See* http://www.fao.org/ag/againfo/themes/animal-welfare/aw-awhome/en/?no_cache=1.

[101]UN Committee on World Food Security (2014), p. 16.

3.4 International Trade Law and Farm Animal Welfare

Traditionally free trade agreements, and particularly the World Trade Organization (WTO), have impeded farm animal welfare regulations by treating them as a barrier to trade. But that may be changing, as the EU pushes for inclusion of animal welfare provisions in new trade agreements (see Sect. 3.2.2 above) and as the WTO appears to loosen its opposition to animal welfare-based trade restrictions. In particular, a recent WTO dispute resolution opinion, in *EC-Seals*, suggests that the WTO framework is becoming more open to animal welfare regulations.

3.4.1 Trade and Farm Animal Welfare

International trade regulations are crucial to progress on farm animal welfare. As of 2007, roughly 9 % of the world's meat was traded across national borders—a percentage that is likely to increase as the world further globalizes.[102] The European Commission has explained the risk that this poses to domestic animal welfare standards: "[A]nimal welfare standards, notably those concerning farm animal welfare, could be undermined if there is no way of ensuring that agricultural and food products produced to domestic animal welfare standards are not simply replaced by imports produced to lower standards."[103] For example, after Britain adopted a ban on the tethering and crating of mother pigs in 1999, European nations with lower-welfare standards began sending lower-welfare and lower-cost pork to Britain, undercutting higher-welfare British producers.[104] These imports undermined the effectiveness of England's animal welfare standards by putting higher-welfare British producers at a comparative disadvantage, and reducing the number of animals covered by Britain's standards. By 2005, half of all pork sold in British supermarkets was imported, and the majority of this imported pork was produced using confinement systems illegal in Britain.[105]

The EU partly mastered this problem by imposing animal welfare standards on a transnational basis. International trade rules do not stop nations from mutually agreeing to shared animal welfare regulations, as European nations effectively have done. But even the EU's animal welfare standards are threatened by imports from countries not bound by the EU's standards. Thus European nations—and other nations committed to advancing farm animal welfare—are seeking ways to regulate the animal welfare standards of products imported from foreign countries. The problem is that trade law has traditionally looked on such animal welfare import regulations as unfair barriers to trade.

[102]Matheny and Leahy (2007).

[103]European Commission (2000).

[104]Matheny and Leahy (2007).

[105]Matheny and Leahy (2007).

3.4.2 WTO Regulation of Animal Welfare-Based Trade Restrictions

The WTO, which governs and promotes international trade between its 160 member nations, has historically been hostile to animal welfare regulations. The General Agreement on Tariffs and Trade (GATT), which governs most WTO disputes, was designed to eliminate trade barriers, and as such makes no mention of animal welfare. Sometimes, eliminating trade barriers has positive animal welfare effects—for example, by reducing subsidies for grains fed to factory farmed animals and ending subsidies for the export of live farm animals in Europe. But the adoption of WTO standards has also encouraged the spread of factory farming. Schneider and Sharma (2014) argue that China's decision to cut tariffs on soybean imports to ensure its accession to the WTO was integral to the surge in pig factory farming since 2006 because it vastly reduced the cost of animal feeds.[106]

Furthermore, two principles of the GATT appear to explicitly preclude nations from imposing animal welfare standards on the import or sale of foreign goods. Article I of the GATT establishes the "most favored nation status" principle, which requires members to treat all trade from all other WTO members on an equal, non-discriminatory basis.[107] This principle appears to prohibit nations from favoring the imports of animal products from nations with higher animal welfare standards. Article III of the GATT establishes the "national treatment" principle, which requires nations to treat imported products no less favorably than "like products" of domestic origin.[108] This principle appears to prohibit nations from imposing domestic animal welfare standards on imported goods because low and high welfare animal products are both "like products," in that they both have the same physical appearance and use.

The WTO's Agreement on Technical Barriers to Trade (TBT) further entrenched the traditional view that animal welfare standards are not a legitimate basis to discriminate against foreign trade.[109] The TBT Agreement distinguishes between "process and production methods" that affect a product's final characteristics (e.g. a method of making safer cars) and those that do not (e.g. a cleaner method of making cars).[110] The former are a permissible basis for discrimination in trade; the latter are not. The TBT Agreement classifies animal production methods in the latter category—as not affecting a product's final characteristics—and thus bars nations from discriminating between animal products based on how they were produced.[111] As

[106]Schneider and Sharma (2014), p. 13.

[107]*See* General Agreement on Tariffs and Trade (1994) 1867 U.N.T.S. 187.

[108]*See* General Agreement on Tariffs and Trade (1994) 1867 U.N.T.S. 187.

[109]Thiermann and Babcock (2005).

[110]Agreement on Technical Barriers to Trade (1994) 1868 U.N.T.S. 120.

[111]Agreement on Technical Barriers to Trade (1994) 1868 U.N.T.S. 120.

such, the TBT Agreement appears to largely forbid nations from imposing animal welfare standards on imported products.

But three potential exceptions exist in the WTO framework to the seeming prohibition on sales or import restrictions based on animal welfare. First, animal welfare measures that can be justified as "animal life" or "animal health" measures may be permissible under the TBT Agreement.[112] That Agreement permits states to adopt technical regulations if they pursue a "legitimate objective," are non-discriminatory toward "like products," and do not function as a disguised restriction on trade.[113] One of the "legitimate objectives" recognized by the Agreement is the protection of "animal or plant life or health."[114] But the TBT agreement only applies to measures that are considered "technical" regulations—and these measures still have to survive review under the GATT.

Second, animal welfare measures may be permissible under the WTO Agreement on the Application of Sanitary and Phytosanitary Measures (SPS) if a nation can show that they are necessary to stop the threat of animal disease. The Agreement gives nations broad authority to impose regulations to protect animal life and health. But the SPS Agreement only applies to measures enacted with the objective of preventing the spread and risk of disease in a territory.[115] And nations must be able to show that the measures are (a) scientifically justifiable, (b) necessary to stop the disease, and (c) are no more trade-restrictive than necessary.[116] So far, no WTO member nation has tried to make that difficult showing for an animal welfare regulation.

Third, animal welfare measures may be permissible under Article XX of the GATT. Article XX provides for ten exceptions to the general prohibitions on trade restricting regulations outlined in Articles I and III of the GATT.[117] Two of these exceptions allow measures "necessary to protect public morals" and necessary to protect "animal life or health."[118] But these exceptions impose a tough "necessity" standard: a measure is not necessary under Article XX if "an alternative measure which [the member nation] could reasonably be expected to employ and which is not inconsistent with other GATT provisions is available to it."[119] And Article XX has a jurisdictional limit: the measure must be enacted to protect public morals, or animal life or health, in the enacting country—i.e. Article XX cannot be used to justify measures designed to protect the health of animals in a foreign nation.[120]

---

[112] Agreement on Technical Barriers to Trade (1994) 1868 U.N.T.S. 120.

[113] Agreement on Technical Barriers to Trade (1994) 1868 U.N.T.S. 120.

[114] Agreement on Technical Barriers to Trade (1994) 1868 U.N.T.S. 120.

[115] Thiermann and Babcock (2005).

[116] Thiermann and Babcock (2005).

[117] *See* General Agreement on Tariffs and Trade (1994) 1867 U.N.T.S. 187.

[118] General Agreement on Tariffs and Trade (1994) 1867 U.N.T.S. 187.

[119] General Agreement on Tariffs and Trade (1994) 1867 U.N.T.S. 187.

[120] General Agreement on Tariffs and Trade (1994) 1867 U.N.T.S. 187.

This makes it hard for a nation to invoke Article XX to justify a restriction on the import of products based on how animals were treated in a foreign nation.

3.4.3 The WTO EC-Seal Products Decision

The conventional wisdom was that the three exceptions outlined above to the WTO's prohibition on animal welfare based trade barriers were not enough to justify most animal welfare regulations. Thiermann and Babcock (2005) reflected this common view when they argued that "attempting to resolve animal welfare concerns under the WTO may not be the best place to address these issues."[121]

But that conventional wisdom was called into question by the recent *EC-Seal Products* decision by the WTO dispute settlement body. In that case, Canada and Norway challenged the EU's 2009 ban on the import and marketing of seal products. They argued that the European ban violated GATT Articles I and III and the TBT Agreement because it imposed an unnecessary restriction on trade and discriminated against Canadian and Norwegian products. In particular, Canada and Norway argued that the EU had discriminated against them by banning their seal products while continuing to allow the import of seal products from subsistence hunting by indigenous communities and from marine resource management culls.

On November 25, 2013, the WTO Dispute Settlement Panel largely rejected Canada and Norway's arguments, in a decision that could have precedential effects for farm animal welfare trade restrictions.[122] The Panel determined that the EU had a legitimate basis to restrict trade after finding that seals hunts are likely to cause inhumane treatment to seals. The Panel ruled, however, that the exception for seal products from subsistence indigenous hunting was not "designed and applied in an even-handed manner," and that the exception for seal products from marine resource management culls lacked a legitimate basis.[123] Thus the Panel rejected the ban under the TBT Agreement and the Article XX "chapeau," which requires that justified trade restrictions not operate in a way that amounts to arbitrary or unjustifiable discrimination.[124] On May 22, 2014, the WTO Appellate Body largely affirmed this opinion, although it determined that the TBT Agreement did not apply at all, since the European ban was not a technical regulation.[125]

The Panel and Appellate Body rulings likely represent a victory for animal welfare, despite their immediate effect of striking down Europe's ban. The faults that both bodies found with the EU seal products import ban may be fixable, since both bodies upheld the core purpose of the EU's ban. (Although the EU will likely

[121]Thiermann and Babcock (2005).

[122]WTO Panel Reports (2013).

[123]WTO Panel Reports (2013).

[124]WTO Panel Reports (2013).

[125]WTO Appellate Body Report (2014).

face stiff resistance from Denmark, which controls Greenland, when it tries to restrict the indigenous hunting exception to the seal products ban to comply with the WTO decision.) Crucially, the Panel rejected Canada and Norway's proposed alternative to the EU Seal Products ban of conditioning market access for seal products on labeling and certified compliance with animal welfare standards, finding that this alternative would not achieve the same benefits for seal welfare as the ban.[126] And the Appellate Body rejected Canada and Norway's new argument on appeal that Europe could only ban seal hunts if it imposed similar bans on slaughterhouses and all other forms of animal abuse.[127] (While farm animal advocates might support such bans, this argument was clearly intended as a *reductio ad absurdum*, since the EU was never going to ban slaughterhouses just to keep its ban on the sale trade in place.).

Most importantly, the WTO Panel and Appellate Body concluded that animal welfare could be a legitimate basis for restricting trade under Article XX of the GATT. The Panel ruled that the core of the EU ban was permissible under Article XX's exception for regulations "necessary to protect public morals," because it addressed the moral concerns of the European public about seal welfare.[128] The Appellate Body upheld that ruling, noting that animal welfare is an important moral value in Europe, and that the ban was clearly intended to address the immorality of Europeans consuming seal products from inhumane commercial hunts.[129] (This leaves open the question of whether Article XX could be used to justify regulations designed to improve the morality of animal welfare practices in another country. The Appellate Body focused on the moral concerns of European citizens, but the cruelty of seal clubbing was occurring largely in Canada.).

3.5 Recommendations for Future Global Progress on Farm Animal Welfare

As this article has illustrated, the current framework of global rules and institutions fails to adequately protect farm animals' welfare. Some scholars have called for a new global framework to regulate animal welfare, for example through a UN Declaration on Animal Welfare. While such a declaration would surely improve animal welfare, its passage currently appears unlikely. In the meantime, though, the international community could achieve a lot of incremental progress on farm animal welfare through the existing institutions and rules. In particular, the international community should do three things: (1) push for a more liberal interpretation of the GATT's Article XX and the SPS agreement; (2) develop stronger

[126]WTO Panel Reports (2013).

[127]WTO Appellate Body Report (2014).

[128]WTO Panel Reports (2013).

[129]WTO Appellate Body Report (2014).

voluntary best practices at the OIE and FAO; (3) implement binding standards through free trade agreements.

3.5.1 Interpret WTO Agreements Liberally

The WTO should treat farm animal welfare as a legitimate basis for restraints on trade. Although WTO members are unlikely to succeed in altering the relevant WTO governing agreements—whether the GATT or TBT or SPS agreements—they could push for more liberal interpretations of the existing agreements by the WTO dispute resolution bodies. In particular, the WTO dispute resolution bodies could build on their recent decisions in *EC-Seals* and make clear that both the SPS Agreement and Article XX of the GATT justify trade restrictions based on farm animal welfare practices. Since the WTO dispute resolution bodies do not issue advisory opinions, this would require one WTO member to challenge the animal welfare based trade restrictions of another, as Canada and Norway did in *EC-Seals*.

A more liberal WTO interpretation of the SPS Agreement and Article XX would allow the world's nations to preserve and extend their farm animal welfare standards. It would make the EU's animal welfare standards more effective by allowing the EU to ban the import of lower-welfare animal products from non-EU nations. It would encourage other nations to adopt similar standards to gain access to the EU's 500 million consumers (and the consumers of other nations, like India, that are increasingly regulating farm animal welfare). And it would reduce the political opposition to raising animal welfare standards from farmers, who fear that their more expensive high welfare products will be undercut by low welfare imports. This could transform the current "race to the bottom" for animal welfare standards into a "race to the top."

3.5.2 Develop Best Practices at International Institutions

The OIE should strengthen its voluntary animal welfare standards and do more to ensure their adoption. The current OIE standards are a good start: they recognize the "five freedoms" of farm animal welfare and the need to comprehensively address the welfare of all farm animals from birth until slaughter. But despite these lofty principles, the current standards largely enshrine current, inhumane industry practices. That's likely the result of the consensus-based process, where animal industries have a significant say in the standards through their national governments. But the OIE could achieve real change if it announced that all standards must conform to the "five freedoms." That condition alone would require the OIE to transform its standards, since its current slaughter standards do not ensure animals freedom from pain and its current confinement standards do not ensure laying hens and pigs freedom to express natural behavior. It would also set a clear vision for developing nations as they adopt new animal welfare standards.

The FAO should also strengthen its work on farm animal welfare by declaring its support for higher welfare systems and its opposition to low-welfare practices. The FAO has the expertise and the international reach to impact animal welfare in developing nations. The FAO also has the unique ability to bring together industry groups, national governments, and animal protection organizations—as it is already doing. But while the FAO has been active in bringing these parties to the table, and providing a forum to discuss animal welfare, it has appeared largely agonistic on animal welfare outcomes. The FAO should adopt a clearer position, in favor of constant improvement and opposed to practices like gestation crate confinement that harm animals.

3.5.3 *Implement Binding Standards Through Free Trade Agreements*

Nations with high farm animal welfare standards should nudge other nations to adopt higher standards through binding standards in FTAs. The EU has led the way with its agreements with Chile and South Korea, and its insistence that future FTAs include animal welfare provisions. These provisions both ensure a level playing for higher-welfare European producers and encourage improvements in other nations. They are also unlikely to receive WTO scrutiny, since the WTO only hears cases from complaining nations, and a nation that has signed the animal welfare provisions of an FTA is effectively agreeing to not complain.

The EU should continue demanding the inclusion of these provisions in FTAs, and other nations should follow suit. The United States, in particular, should move from its dogged opposition to animal welfare reforms to a more progressive approach of promoting the issue in trade negotiations. Other nations that have championed animal welfare—from India to Israel and New Zealand—should demand that other nations sign animal welfare provisions in future FTAs.

Ultimately a series of bilateral FTAs containing animal welfare provisions could form the basis for the issue's inclusion in future multilateral trade negotiations—whether through the WTO, or through a regional body like the Asia Pacific Economic Cooperation (APEC) organization. The presence of a number of nations that have already adopted bilateral FTAs with animal welfare provisions will make the issue easier to address multilaterally. Then the world will finally have a form of global farm animal welfare regulation.

3.6 Conclusion

Global institutions need to adopt stronger farm animal welfare policies, and nations need to establish the basis for future international farm animal welfare laws. The failure of global institutions and laws to keep pace with the rise of factory farm in the twentieth century has allowed severe animal abuse to occur. But global institutions and laws are finally seeking to redress this gap. They now need to speed up the pace of farm animal welfare reform. The moral credibility of the global order depends on it.

References

Animal Protection Index (2014) World animal protection. Http://api.worldanimalprotection.org/indicators. Accessed 3 Jan 2015

Brambell FWR (1965) Report of the technical committee to enquire into the welfare of animals kept under intensive livestock husbandry systems. Her Majesty's Stationery Office, London

Cabanne C (2013) The EU-Chile free trade agreement—a boost for animal welfare. Eurogroup for Animals, Brussels

Chinese National People's Congress (2005) Animal husbandry law of the People's Republic of China. Http://www.npc.gov.cn/englishnpc/Law/2007-12/13/content_1384134.htm. Accessed 3 Jan 2015

Clyma K (2014) Slideshow: ranking the meat and poultry industry, Meat + Poultry, April 4, 2014. http://www.meatpoultry.com/articles/news_home/Business/2014/04/Slideshow_Ranking_the_meat_and.aspx?ID={4EB1C430-7CF8-41B3-8457-9FB76625F9F8}. Accessed 24 Jan 2015

Cochrane W (1993) The development of American agriculture: a historical analysis. University of Minnesota Press, Minneapolis

Council of Europe (1976) European convention for the protection of animals kept for farming purposes, Strasbourg, 10.III.1976. Http://conventions.coe.int/Treaty/en/Treaties/html/087.htm. Accessed 3 Jan 2015

Council of Europe (1991) Council Directive of 19 November 1991 laying down minimum standards for the protection of calves (91/629/EEC). Http://ec.europa.eu/food/fs/aw/aw_legislation/calves/91-629-eec_en.pdf. Accessed 3 Jan 2015

Council of Europe (1999) Council Directive 1999/74/EC of 19 July 1999 laying down minimum standards for the protection of laying hens. Http://eur-lex.europa.eu/LexUriServ/LexUriServ.do?uri=OJ:L:1999:203:0053:0057:EN:PDF. Accessed 3 Jan 2015

Council of Europe (2001) Council Directive 2001/88/EC of 23 October 2001 amending Directive 91/630/EEC laying down minimum standards for the protection of pigs. Http://eur-lex.europa.eu/legal-content/EN/TXT/PDF/?uri=CELEX:32001L0088&from=EN. Accessed 3 Jan 2015

European Commission (1997) Commission Decision of 24 February 1997 amending the Annex to Directive 91/629/EEC laying down minimum standards for the protection of calves (97/182/EC). Http://ec.europa.eu/food/fs/aw/aw_legislation/calves/97-182-ec_en.pdf. Accessed 3 Jan 2015

European Commission (2000) European Communities proposal: animal welfare and trade in agriculture. Http://www.wto.org/english/tratop_e/agric_e/ngw19_e.doc. Accessed 3 Jan 2015

European Commission (2001) Commission Directive 2001/93/EC of 9 November 2001 amending Directive 91/630/EEC laying down minimum standards for the protection of pigs. Http://eur-lex.europa.eu/legal-content/EN/TXT/PDF/?uri=CELEX:32001L0093&from=en. Accessed 3 Jan 2015

European Commission (2014) EU-Japan trade talks continue, Europa, January 31, 2014, http://trade.ec.europa.eu/doclib/press/index.cfm?id=1024. Accessed 3 Jan 2015

Everett (2014) China's rising meat consumption draws U.S. companies, Midwest Center for Investigative Reporting, May 15, 2014, http://investigatemidwest.org/2014/05/15/chinas-increasing-chicken-consumption-draws-u-s-companies/

Ewer T (1973) Farm animals in the law, New Scientist, October 18, 1973

FAO (2008) Capacity building to implement good animal welfare practices. http://www.fao.org/fileadmin/user_upload/animalwelfare/AW_Exp-meeting_EN_1.pdf. Accessed 24 Jan 2015

FAO (2014a) International livestock database. Http://faostat.fao.org/site/573/DesktopDefault.aspx?PageID=573#ancor. Accessed 3 Jan 2015

FAO (2014b) Review of animal welfare legislation in the beef, pork, and poultry industries. http://www.fao.org/3/a-i4002e.pdf. Accessed 24 Jan 2015

Farming UK (2013) EU presses for animal welfare standards in trade agreements, Farming UK, May 25, 2013. Http://www.farminguk.com/news/EU-presses-for-animal-welfare-standards-in-trade-agreements_25596.html

Fortune (2014) Fortune 500 list. Http://fortune.com/fortune500/supervalu-inc-94/. Accessed 24 Jan 2015

Harper G, Henson S (2001) Consumer concerns about animal welfare and the impact on food choice, EU FAIR CT98-3678. Http://ec.europa.eu/food/animal/welfare/eu_fair_project_en.pdf. Accessed 3 Jan 2015

HSI (2011) HSI applauds animal protection provisions in EU-South Korea free trade agreement. Http://www.hsi.org/world/europe/news/releases/2011/06/eu_south_korea_fta_063011.html

HSI (2012a) An HSI fact sheet: top 10 global producers: chickens & chicken meat. http://www.hsi.org/assets/pdfs/hsi-fa-white-papers/top_10_chicken_meat_producers.pdf. Accessed 24 Jan 2015

HSI (2012b) An HSI fact sheet: top 10 global producers: hen eggs. http://www.hsi.org/assets/pdfs/hsi-fa-white-papers/top_10_producers_of_eggs.pdf. Accessed 24 Jan 2015

HSI (2012c) An HSI fact sheet: top 10 global producers: pigs and pig meat. http://www.hsi.org/assets/pdfs/hsi-fa-white-papers/top_10_producers_of_pigs.pdf. Accessed 24 Jan 2015

HSI, Compassion in World Farming, Four Paws (2013) International finance institutions, export credit agencies and farm animal welfare. http://www.hsi.org/assets/pdfs/hsi_ifi_report_june_2013.pdf. Accessed 24 Jan 2015

HSUS (2008) An HSUS report: the welfare of animals in the egg industry. Http://animalstudiesrepository.org/cgi/viewcontent.cgi?article=1016&context=hsus_reps_impacts_on_animals. Accessed 3 Jan 2015

HSUS (2010) An HSUS report: the welfare of animals in the pig industry. Http://animalstudiesrepository.org/hsus_reps_impacts_on_animals/28/. Accessed 3 Jan 2015

HSUS (2014a) An HSUS report: the welfare of animals in the chicken industry. Http://files.givewell.org/files/shallow/industrial-agriculture/HSUS%20Chickens.pdf. Accessed 3 Jan 2015

HSUS (2014b) An HSUS report: the welfare of cows in the dairy industry. Http://www.humanesociety.org/assets/pdfs/farm/hsus-the-welfare-of-cows-in-the-dairy-industry.pdf. Accessed 3 Jan 2015

HSUS (2014c) An HSUS report: the welfare of animals in the meat, egg, and dairy industries. Https://www.petfinder.com/docs/library-articles/animal-welfare-overview.pdf. Accessed 3 Jan 2015

International Finance Corporation (2006) Good practice note on animal welfare in livestock operations. Http://www.ifc.org/wps/wcm/connect/7ce6d2804885589a80bcd26a6515bb18/AnimalWelfare_GPN.pdf?MOD=AJPERES. Accessed 3 Jan 2015

International Finance Corporation (2014) Good practice note: improving animal welfare in livestock operations. Http://www.ifc.org/wps/wcm/connect/c60a668043904 1ae9a25ba869243d457/IFC-GPN-AnimalWelfare_2006vs2014.pdf?MOD=AJPERES. Accessed 3 Jan 2015

International Fund for Animal Welfare (2009) China drafts nation's first animal protection law, September 28, 2009. Http://www.prnewswire.com/news-releases/china-drafts-nations-first-animal-protection-law-62358622.html
Judgment of the Court of 19 March 1998 - The Queen v Minister of Agriculture, Fisheries and Food, ex parte Compassion in World Farming Ltd. Http://eur-lex.europa.eu/legal-content/EN/TXT/?uri=CELEX:61996CJ0001. Accessed 3 Jan 2015
Kindy K (2013) USDA plan to speed up poultry-processing lines could increase risk of bird abuse, Washington Post, October 29, 2013. Http://www.washingtonpost.com/politics/usda-plan-to-speed-up-poultry-processing-lines-could-increase-risk-of-bird-abuse/2013/10/29/aeeffe1e-3b2e-11e3-b6a9-da62c264f40e_story.html
Kopperud S (2014) 'Global animal welfare' — who's on first? Brownfield Ag News, October 3, 2014. Http://brownfieldagnews.com/2014/10/03/global-animal-welfare-whos-first/
Matheny G, Leahy C (2007) Farm-animal welfare, legislation, and trade. Law Contemp Probl 70:325–358
OIE (2012) Third global conference on animal welfare: implementing the OIE standards – addressing regional expectations. Http://www.oie.int/eng/AW2012/intro.htm. Accessed 3 Jan 2015
OIE (2013) International trade: rights and obligations of OIE member countries. Http://www.oie.int/fileadmin/Home/eng/Internationa_Standard_Setting/docs/pdf/Legal_rights_and_obligations/A_Rights_and_obligations_April_2013.pdf. Accessed 3 Jan 2015
OIE (2014a) OIE's achievements in animal welfare. Http://www.oie.int/animal-welfare/animal-welfare-key-themes/. Accessed 3 Jan 2015
OIE (2014b) Terrestrial animal health code. Http://www.oie.int/index.php?id=169&L=0&htmfile=titre_1.7.htm. Accessed 3 Jan 2015
Pew Commission on Industrial Farm Animal Production (2008) Putting meat on the table: industrial farm animal production in America. Pew Charitable Trusts, Washington
Radio New Zealand (2010) Fonterra intensive dairy farming in China criticized, February 3, 2010. Http://www.radionz.co.nz/news/national/2116/fonterra-intensive-dairy-farming-in-china-criticised
Samora R (2014) UPDATE 1-CEO of Brazil's JBS meat company sees 2014 revenues up 22 pct, Reuters, November 13, 2014. Http://www.reuters.com/article/2014/11/13/brazil-jbs-china-idUSL2N0T319B20141113
Schneider M and Sharma S (2014) China's pork miracle? Agribusiness and development in China's pork industry. Institute for Agriculture and Trade Policy. Http://www.iatp.org/files/2014_03_26_PorkReport_f_web.pdf. Accessed 24 Jan 2015
Sherman D (2002) Tending animals in the global village: a guide to international veterinary medicine. Lippincott Williams & Wilkins, Philadelphia
SPS Agreement: Preamble, Article 3, and Annex A, paragraph 3(b). Http://www.wto.org/english/tratop_e/sps_e/spsagr_e.htm. Accessed 24 Jan 2015
Thiermann AB, Babcock S (2005) Animal welfare and international trade. Rev Sci Tech Off Int Epiz 24(2):747–755
UN Committee on World Food Security (2014) Principles for responsible agricultural investments. Http://www.fao.org/fileadmin/templates/cfs/Docs1314/rai/CFS_Principles_Oct_2014_EN.pdf. Accessed 24 Jan 2015
Wilmot S (2007) From 'public service' to artificial insemination: animal breeding science and reproductive research in early twentieth-century Britain. Stud Hist Philos Biol Biomed Sci 38 (2):411–441
Wolfson D, Sullivan M (2004) Foxes in the hen house: animals, agribusiness, and the law: a modern American fable. In: Sunstein C, Nussbaum M (eds) Animal rights: current debates and new directions. Oxford University Press, Oxford
Xinhua (2009) Humane slaughtering spearheads China's drive to promote animal welfare, September 5, 2009. Http://china.org.cn/environment/2009-09/05/content_18519951.htm

You X (2014) A survey of Chinese citizens' perceptions on farm animal welfare. PLoS ONE 9 (10):1371. Http://journals.plos.org/plosone/article?id=10.1371/journal.pone.0109177. Accessed 24 Jan 2015

Yue S (2013) An HSUS report: fish and pain perception. Http://www.humanesociety.org/assets/pdfs/farm/hsus-fish-and-pain-perception.pdf. Accessed 3 Jan 2015

WTO Panel Reports (2013) European Communities – measures prohibiting the importation and marketing of seal products, WT/DS400/R and WT/DS401/R

WTO Appellate Body Report (2014) European Communities – measures prohibiting the importation and marketing of seal products, WT/DS400/AB/R and WT/DS401/AB/R

Chapter 4
Voluntary Standards and Their Impact on National Laws and International Initiatives

Dena Jones and Michelle Pawlinger

Abstract Numerous private entities—both national and international in scope—have developed or are in the process of developing nonregulatory standards to assure consumers that animals and natural resources used in agricultural production are properly treated. This chapter describes the differing approaches of three countries: one that uses voluntary standards to supplement legal standards (United Kingdom), one that uses voluntary standards as a substitute for legal standards (United States), and a third that uses voluntary standards to assist in interpreting and enforcing legal standards (Canada). The impact of these voluntary standards on international animal welfare initiatives is also discussed.

4.1 Introduction

Worldwide, 70 billion land animals are slaughtered for food each year.[1] Nine billion of these animals are killed in the United States, 650 million in Canada, and 1 billion in the United Kingdom[2] (together approximately 15 % of the worldwide total). A growing number of animals in agriculture are covered under voluntary animal welfare and environmental stewardship standards in an attempt to assure consumers that the animals and the environment are not abused in the production process.

---

[1]Food and Agriculture Organization of the United Nations (FAO) Statistics Division (2015). This number is an approximation and does not include animals killed for indigenous meats.

[2]See US Department of Agriculture (USDA), National Agricultural Statistics Service (NASS) (2015a), p. 5; USDA, NASS (2015c), p. 5; See Agriculture and Agri-Food Canada (2014); see Royal Society for the Prevention of Cruelty to Animals (2008). The slaughter numbers are approximations calculated using these slaughter reports and statistics.

D. Jones (✉) • M. Pawlinger
Animal Welfare Institute, Washington, DC, USA
e-mail: Dena@awionline.org

G. Steier, K.K. Patel (eds.), *International Farm Animal, Wildlife and Food Safety Law*, DOI 10.1007/978-3-319-18002-1_4

The International Organization for Standardization (ISO), the world's largest developer of voluntary international standards, defines "standards" as documents that provide "requirements, specifications, guidelines or characteristics that can be used consistently to ensure that materials, products, processes and services are fit for their purpose."[3] Voluntary animal welfare standards apply to the processes used to raise, transport, and slaughter animals, while voluntary environmental stewardship standards relate to the resources used in the production of agricultural products.[4]

Voluntary animal welfare and environmental stewardship standards are most often associated with private entities. Various private entities have established standards, and continue to develop these standards, as consumer demand increases for higher-quality products. Private standards are set by individual companies, industry trade associations, and independent third-party certifiers. Animal welfare and environmental stewardship standards have expanded in scope and in depth in all three sectors. At present, the global animal agriculture industry is impacted by a variety of differing and in some cases contradictory standards, and this trend is expected to continue.

Consequently, efforts are underway through international initiatives to harmonize individual sets of standards. The World Organization for Animal Health (commonly referred to by its French initialism "OIE"), for example, is an intergovernmental organization aiming to advance global animal health.[5] It maintains animal welfare standards that aspire to be a global solution to individualized standards.[6] In addition to the OIE, the ISO has several sets of environmental standards and is in the process of developing animal welfare standards similar to those of the OIE.[7] GLOBALG.A.P., another international organization seeking to harmonize agriculture standards worldwide, maintains voluntary animal welfare standards.[8] This chapter will discuss how individual private standards and those of the OIE, ISO, and GLOBALG.A.P. compare and how they impact one another.

While private voluntary standards are more prevalent, voluntary standards are sometimes established by governments. Organic production standards, for instance, are often regulated by governmental entities. Governments often require farmers to meet higher-than-industry animal welfare and environmental stewardship standards

[3]International Organization for Standardization (ISO) (no date), Standards.

[4]This chapter will focus on animal welfare, but will provide a review of environmental stewardship voluntary standards.

[5]World Organization for Animal Health ("OIE") (2015a), About us.

[6]OIE (2015b), OIE's achievements in animal welfare.

[7]See ISO (no date), ISO 14000-environmental management.

[8]GLOBALG.A.P. (no date), GLOBALG.A.P. animal welfare add-on.

for organic production, and rely on third-parties to verify compliance on organic farms.[9] Section 4.4 of this chapter focuses on third-party certification, and voluntary organic standards are included in that discussion.

The main focus of the chapter, however, is the effect of private voluntary standards on national laws and international initiatives. The chapter concentrates on voluntary standards in the United States, Canada, and the United Kingdom in order to demonstrate three distinct approaches to voluntary standards and their impacts on national and international policy initiatives. Section 4.2 details corporate approaches to voluntary standards, including insight into the origins of corporate interest in animal welfare and environmental stewardship, and the impact of corporate national and international standards. Section 4.3 reviews the role of trade associations in animal welfare and environmental stewardship, and explains how their impact on animal welfare differs in the reviewed nations. Section 4.4 discusses independent, third-party certification organizations and their impact on animals and the environment. Section 4.5 provides case studies of how voluntary standards impact national laws and international initiatives.

4.2 Self-Regulation Through Corporate Policies

4.2.1 Overview of the Origin and Role of Farm Animal Welfare and Environmental Stewardship Policies in Corporate Entities

Social responsibility has been a longstanding issue within corporate development, but companies have often separated philanthropic endeavors from business practices.[10] As the world economy globalized, companies seeking a competitive advantage began incorporating socially responsible practices into their business strategies.[11] These socially responsible practices—often referred to as "Corporate Social Responsibility," or CSR—typically include plans to improve worker safety, human health, the environment, and the community in which a corporation is involved.[12]

---

[9]Organic production could also be discussed under trade associations, as in certain countries organic production is certified by organic industry associations, such as in the UK where the leading organic certifier is the United Kingdom's Soil Association.

[10]Pastore (2013), p. 109.

[11]Pastore (2013), p. 109.

[12]According to the World Business Council for Sustainable Development, Corporate Social Responsibility is "the continuing commitment by business to behave ethically and contribute to economic development while improving the quality of life of the workforce and their families as well as of the local community and society at large." Watts and Holme (1999), p. 3.

According to the Food and Agriculture Organization of the United Nations (FAO), "CSR has become an important part of business. It is an explicit set of business principles, developed and adopted by the companies themselves to suit their specific procurement, manufacturing, logistics, marketing, and other business circumstances."[13] Companies embrace CSR policies because they can positively impact communities, better retain employees, and set examples for respective industries.[14] Perhaps the most obvious reason for implementing CSR policies is economic advantage; CSR policies can improve a company's image and thus help set it apart from others.[15]

Companies tend to focus their CSR policies on areas that are already regulated by governments or on issues upon which society places great value.[16] For instance, the natural environment, which is highly regulated and of high societal interest, is often a key element of corporate social responsibility.[17] Animal welfare is a newer concept in corporate policies, but it is gaining traction.[18] A 2014 international report on corporate animal welfare policies, titled *The Business Benchmark on Farm Animal Welfare* (BBFAW),[19] found that only 41 % of companies publish farm animal welfare-related objectives and targets.[20] The BBFAW report observes, however, that there are tangible signs of an increase in animal welfare policies.[21]

The following section will review corporate animal welfare and environmental stewardship policies in the United States, Canada, and the United Kingdom.

4.2.2 *Corporate Standards for Farm Animal Welfare and Environmental Stewardship*

4.2.2.1 United States

In the 1980s, corporations increasingly added environmental stewardship to their corporate social responsibility policies.[22] Another decade would pass before corporate entities began to develop animal welfare policies, as well.[23] Prior to this,

[13]Pastore (2013), p. 109.

[14]Thorpe (2013).

[15]Portney (2005), p. 112.

[16]Davis et al. (2006), p. 8.

[17]See Portney (2006), p. 108 (maintaining that environmental responsibility is part of CSR).

[18]Amos and Sullivan (2014), p. 4.

[19]Amos and Sullivan (2014). This report assessed businesses in three core areas: (1) management commitment and policy, (2) governance and management, and (3) leadership and innovation.

[20]Amos and Sullivan (2014), p. 7.

[21]Amos and Sullivan (2014), p. 8.

[22]Rondinelli and Berry (2000), p. 1.

[23]See Singer (1998), Ch. 5 (explaining how McDonald's became the first fast food company to make small improvements in its supply chain).

a few companies had general statements that mimicked industry standards and did little to ensure compliance.[24] Numerous companies continue to use vague statements for both environmental and animal welfare policies; however, businesses are making strides to improve in both areas.

Animal welfare policies were developed, in part, because pressure from animal advocacy groups—including graphic exposés at slaughterhouses—drove companies into action.[25] Fast food companies were among the first corporations to form specific policies regarding animal welfare.[26] McDonald's implemented animal welfare auditing at its suppliers' slaughterhouses; Wendy's and Burger King quickly followed suit, and over time slaughterhouses made modest improvements to their practices.[27]

Along with fast food restaurants like McDonald's and Wendy's, food producers, hotels, high-end restaurants, and supermarket chains began devising animal welfare policies and statements. While fast food restaurants originally developed only slaughterhouse policies, they pushed corporations across the country to open dialogue on animal-raising practices as well. For instance, Aramark, one of the largest private companies in the United States, has announced plans to (1) purchase eggs that are produced only from cage-free hens by 2020, (2) purchase pork only from crate-free sows and veal only from crate-free calves by 2017, and (3) "address animal welfare issues associated with fast growth of broiler chickens and turkeys."[28]

There are two main trends among corporate animal welfare policies in the United States: companies tend to make incremental versus sweeping changes, and they grant suppliers lengthy phase-in periods for compliance. Incremental changes allow companies to improve in areas that are demonstrably important to consumers, while acknowledging that changes can be financially burdensome for producers. Long phase-in periods allow producers time to gradually pay for improvements and for expensive equipment to depreciate. For example, in 2007, Smithfield Foods, the country's largest pork producer, announced it would phase out the use of sow gestation crates from company-owned facilities in 10 years, and in 2014 the company expanded this requirement to contract-farmers, requiring compliance in 8 years.[29]

---

[24]See Singer (1998), Ch. 5 (showing McDonald's general statement on animal welfare before it implemented basic, but specific standards).

[25]See People for the Ethical Treatment of Animals (no date).

[26]See Singer (1998), Ch. 5.

[27]Grandin (2005), pp. 370–373 (explaining that slaughterhouses able to properly stun animals 95 % of the time increased from 3 out of 10 to 9 out of 10).

[28]Aramark (2015).

[29]Smithfield (2007); Smithfield (no date), Housing of pregnant sows. As another example of the main trends in corporate policies, in 1999, McDonald's created slaughterhouse standards. In 2012, it announced that by 2017 it would source pork from producers *committed* to going crate free, and by 2022 would only source pork from supply chains free of gestation crates. The company also

While there has been significant progress in terms of the development of corporate social responsibility policies that address the raising and slaughtering of animals, the actual impact on animal welfare is unclear. This is because (1) numerous companies still use general statements as opposed to specific animal welfare requirements,[30] (2) company auditing and reporting of animal welfare indicators is relatively new, and (3) many phase-in periods have yet to come to fruition.[31] Kroger, the largest grocery chain in the United States for example, does not obligate suppliers, but merely encourages them, to eliminate the use of gestation crates.[32] Additionally, the company does not publically report the percentage of animals in confinement versus group housing in its supply chain. Starbucks, the largest coffeehouse company worldwide, states that it incentivizes suppliers to phase out gestation crates for pregnant pigs and battery cages for egg-laying hens, but does not provide public progress reports.[33] However, companies are beginning to require auditing and/or reporting from their suppliers. In 2014, for example, several companies, including Smithfield, Jack in the Box, and Wendy's, announced that they will start requiring progress reports from suppliers.[34]

Along with animal welfare policies, corporations also create environmental stewardship policies. Corporations have been refining their environmental policies over the last 60 years—evolving from managing crises as they occurred to a more proactive approach.[35] Companies have moved in this direction because of public demand for environmental protections, new technologies making advancements easier, and long-term cost savings.[36] Corporations often focus on water quality and use, energy efficient operations, and waste management. For example, Starbucks had a goal to reduce its water usage 25 % by 2015.[37] Additionally, in 2014, 98 % of new Starbucks stores were LEED-certified [38] when they opened.

---

plans to develop a "verification system to assess compliance" with its standards. McDonald's (2014), p. 21.

[30]For example, while a number of US food companies have prohibited their suppliers from using gestation crates, they have not provided alternative welfare requirements, such as the type of space, flooring, bedding, and feeding systems. As a result, it is unclear to what degree new policies will improve animal welfare for sows.

[31]Amos and Sullivan (2014), p. 9 (concluding that "reporting on farm animal welfare remain [s] underdeveloped across. . .Management Commitment and Policy, Governance and Management, Innovation, and Performance.").

[32]Kroger (2012).

[33]See Starbucks Coffee (no date), Animal welfare-friendly practices statement.

[34]Smithfield (no date), Housing of pregnant sows; Wendy's (2014); Jack in the Box (2014).

[35]Berry and Rondinelli (1998), p. 39.

[36]Berry and Rondinelli (1998), pp. 39–40.

[37]Starbucks (2014), p. 7. At the end of 2014 Starbucks had reached 23 %.

[38]LEED stands for Leadership in Energy and Environmental Design.

While countless corporations have established environmental policies, it is important to remember that there is a distinction between creating policies and implementing them.[39] Companies may tout a commitment to protecting the environment, but this does not always translate into an improved environmental record. One study notes that service companies, which include companies in the food service industry, "appear to be less likely to implement...environmental policies than those in the manufacturing, chemical manufacturing, and oil and gas industry sectors."[40] The study concludes that this is due to the fact that environmental upgrades improve business performance to a greater degree for the other three industry sectors than for the service industry.[41]

The impact of corporate environmental and animal welfare policies are not fully known, but large strides have been made in the years since their introduction. New corporations will continue to join the growing list of companies creating voluntary policies, and companies with animal welfare practices will continue to make new commitments as long as (1) consumers continue to show that animal welfare and the environment are important to them, (2) new technologies help incentivize strong corporate policies, and (3) advocacy organizations campaign for improvements. For animal welfare, the extent of the impact will become more evident as phase-in periods end and reporting requirements expand.

4.2.2.2 Canada

Relatively few international food corporations, which include retailers, food service companies, producers and processors, are headquartered in Canada. Some international corporations with operations within Canada are committed to global corporate policies for farm animals and environmental stewardship, while others follow country-specific policies. For example, as of 2015, Burger King's transition to cage-free eggs and crate-free pork in the United States does not apply to its operations in Canada.[42] Conversely, Wendy's announcement in 2014 that it is phasing out use of gestation crates for sows applies to its suppliers in both the United States and Canada.[43] McDonald's 2012 announcement that, as of 2022, it will no longer buy from suppliers using gestation crates, is specific to stores in the United States but appears to impact Canadian stores, as well, since they source their pork products from US suppliers.[44]

---

[39]See Ramus and Montiel (2005).

[40]Ramus and Montiel (2005), p. 394.

[41]Ramus and Montiel (2005), p. 394.

[42]Canadian Federation of Humane Societies ("CFHS") (no date), Putting an end to gestation stalls in Canada.

[43]CFHS (no date), Putting an end to gestation stalls in Canada.

[44]CFHS (no date), Putting an end to gestation stalls in Canada.

Several Canadian retailers have committed to eliminating the use of gestation crates, including supermarket chain Loblaws[45] and coffee shop chain Tim Hortons.[46] In addition, in 2013, the Retail Council of Canada (composed of Co-op Atlantic, Canada Safeway, Costco Wholesale Canada, Federated Cooperatives Limited, Loblaw Companies Limited, Metro Inc., Sobeys Inc., and Walmart Canada) committed to sourcing pork products from sows raised under alternative housing as defined by the Canadian revised code of practice for the raising of pigs.[47]

While Canada has had far fewer producers offering to change their production practices than the United States, a few have announced commitments. Maple Leaf, Canada's largest pork producer, intends to phase out the use of gestation crates by 2017,[48] and Olymel, another Canadian pork producer based in Quebec, estimates that all pigs slaughtered by its company in Canada will come from crate-free farms by 2022.[49] In terms of trade associations, the Manitoba Pork Council has pledged to encourage producers to eliminate gestation crates by 2025,[50] and Manitoba Egg Farmers is requiring that egg farmers building new facilities or renovating existing ones utilize enriched caging or a noncage housing system, beginning in 2015.[51]

4.2.2.3 United Kingdom

Several major multinational food corporations are headquartered in the United Kingdom, including Compass, the world's largest contract food service company, with operations in over 50 countries, and Tesco, the world's third-largest food retailer. UK-based companies have attained the highest average scores in the 2014 BBFAW report, although the authors caution that small sample sizes have the potential to skew the results.[52] The survey included 18 UK companies, or 22 % of the total companies reviewed. Those 18 companies were evenly split between the production, retail/wholesale, and restaurant/bar sectors.[53] The UK-based companies in the retail/wholesale sector scored significantly better than those in the other two sectors, with all six UK retail/wholesale companies scoring in the top three (of six) tiers, and all six UK bar/restaurant companies landing in the bottom three tiers of the survey results.[54]

---

[45] Loblaw (no date).

[46] Tim Hortons (no date).

[47] Post (2013).

[48] Maple Leaf (no date).

[49] Olymel (no date).

[50] Manitoba Pork Council (2011), p. 39.

[51] The Poultry Site (2013).

[52] Amos and Sullivan (2014), p. 30. Average overall score for UK-based was 47 %, compared with 25 % for European companies, 30 % for US companies, and 30 % for all 80 companies included in the survey.

[53] Amos and Sullivan (2014), pp. 85–86.

[54] Amos and Sullivan (2014), p. 9.

Several individual UK-incorporated companies have made serious commitments to improving farm animal welfare in their supply chains. For example, Cranswick, a UK food producer, allows no genetic modifications or growth hormones, requires pre-stunning of all animals at slaughter, and is moving toward no castration of pigs and no routine beak trimming of chickens.[55] It also is sourcing some meats from organic or free-range systems.[56] Marks and Spencer Group, which operates stores in 40 international locations and was one of only three companies to score in the top tier of the BBFAW report, also allows no routine mutilations[57]; in addition, it has set limits on the length of time animals can be transported,[58] and the company has created a payment scheme that rewards dairy farmers who implement high welfare standards.[59]

UK-based businesses are also setting the bar for sustainability commitments. Marks and Spencer has made 100 individual commitments toward becoming a sustainable business.[60] In 2014, the company became the first retailer to obtain three Carbon Trust Standard Certifications for carbon, water, and waste. Also in 2014, Marks and Spencer achieved carbon neutrality for all of its own operations and joint ventures across the world, by reducing emissions, sourcing renewable electricity, and buying and retiring carbon offsets. In 2014, the company's carbon dioxide emissions were down 24 % from 2006/2007 baseline levels. It achieved this by using electricity more efficiently, reducing gas leaks from refrigeration, and employing better waste recycling methods.[61] Similarly, the Compass Group plans to reduce both carbon dioxide emissions and water use by 2020 to 20 % below 2014 levels.[62] Also by 2020, the company intends to serve wild caught and aquaculture seafood only from environmentally responsible sources, or from sources that are on a clear path toward sustainability.[63]

4.2.2.4 International

Animal welfare is a growing concern for businesses around the world. In the BBFAW report, 84 % of participating companies acknowledged animal welfare as a "business issue."[64] According to the report, this is the first step toward

[55]Amos and Sullivan (2014), pp. 39–45.

[56]Amos and Sullivan (2014), p. 37.

[57]Amos and Sullivan (2014), p. 43.

[58]Amos and Sullivan (2014), p. 46.

[59]Amos and Sullivan (2014), p. 50.

[60]Marks and Spencer (no date), p. 2.

[61]Marks and Spencer no date, p. 6.

[62]Compass Group (no date), p. 8.

[63]Compass Group (no date), p. 6.

[64]Amos and Sullivan (2014), pp. 7, 33. This is a notable increase from the 71 % in the 2012 and 2013 Benchmark reports.

implementing animal welfare policies. Reasons for the growing international concern with animal welfare noted by the report include "the 2013 European horsemeat scandal, tightening regulatory requirements on animal welfare and on food safety and quality, investor concerns about how food companies are managing animal welfare and other risks in their supply chains, and consumer interest in issues around food quality, safety, provenance, and traceability."[65]

While the report's authors are encouraged by the continued interest in animal welfare, the report concludes that the food industry is not effectively managing or reporting animal welfare.[66] The report identified seven animal welfare issues of interest to businesses: close confinement, the use of genetically modified or cloned animals, the use of growth promoting substances, the use of antibiotics for prophylactic purposes, routine mutilations, pre-slaughter stunning, and long-distance live transportation.[67] Only 10 % of companies had specific policies on long distance travel; 14 % had policies on pre-slaughter stunning, and 23 % had policies on routine mutilations.[68] Additionally, the number of policies on close confinement and genetic modification actually decreased from 2013 to 2014.[69]

International corporations' animal welfare policies often vary by geographic location. For example, the international corporation Sodexo, one of the largest food services companies in the world, provides all suppliers with general animal welfare policies, but sets specific animal welfare guidelines for different countries.[70] In North America, the 39 million shelled eggs purchased each year by Sodexo must come from cage-free hens, and animal welfare audits are required for "fully integrated" suppliers.[71] In the Netherlands, the company supports welfare initiatives but is not as specific as it is with its North American guidelines.[72] In the BBFAW report, only 1 % of companies made a universal commitment to avoid extreme confinement practices; 8 % committed to prohibiting growth-promoting substances, and 3 % committed to avoiding routine mutilations.[73] Changes in standards based on geographical location are dependent on societal pressure and expectations in the region, supply availability, varying legal obligations, and differing welfare benchmarks.

The need for harmonization of food standards across regions of the globe has been addressed by an organization known as GLOBALG.A.P. (formerly named EUREPGAP), which claims to be the "world's leading farm assurance program."[74]

[65]Amos and Sullivan (2014), p. 14.
[66]Amos and Sullivan (2014), p. 48.
[67]Amos and Sullivan (2014), p. 35.
[68]Amos and Sullivan (2014), p. 35.
[69]Amos and Sullivan (2014), p. 36.
[70]Sodexo (2013), pp. 1–5.
[71]Sodexo (2013), p. 4.
[72]Sodexo (2013), p. 4.
[73]Amos and Sullivan (2014), p. 37.
[74]GLOBALG.A.P. (no date), History.

It was established in 1997 as an initiative by European retailers to provide their supermarket consumers with assurance of producer compliance with Good Agricultural Practices. The organization has developed standards for product safety, environmental impact, and the health, safety and welfare of workers and animals. In the area of animal welfare, GLOBALG.A.P. offers certification in transport and in husbandry for meat chickens and finishing pigs.[75] Although GLOBALG.A.P. is active in more than 100 countries, as of 2015, a majority of participating producers and retailers were headquartered in Europe.[76] The program, whose standards exceed legal requirements of the European Union, provides one option for harmonizing environmental, food safety, and animal care standards at the international level.

4.3 Self-Regulation Through Industry Trade Associations

4.3.1 Overview of the Origin and Role of Industry Trade Associations

Prior to World War II, most chicken meat in the United States came from a surplus in egg-production flocks and Americans scarcely consumed it.[77] In the late 1940s and early 1950s, vertical integration and new technologies—evisceration machinery, modified chicken breeds, and confinement practices—allowed for the commercial expansion of the chicken meat industry.[78]

As a result of increased chicken production, the growing industry created the National Broiler Council (NBC), now called the National Chicken Council (NCC). This industry trade association represents producers, processors, hatcheries, and all other segments within the market.[79] The NBC played a key role in increasing consumer demand for chicken products: aggressive advertising campaigns such as "chicken, the high-protein, low-calorie meat" and "chicken, the food of the future," along with increased federal lobbying, helped boost annual chicken consumption from 20 pounds per capita in the 1940s to over 80 pounds per capita in 2010.[80]

Trade associations such as the NCC are created to standardize and promote industries. Globally, companies from an industry market or sector align to form trade associations and use their combined efforts to influence the public, lobby

[75]GLOBALG.A.P. (no date), Animal welfare add-on.

[76]GLOBALG.A.P. (no date), Membership.

[77]Perry et al. (2012), p. 3.

[78]Reimund et al. (1981), p. 4; Also see Perry et al. (2012).

[79]See National Chicken Council (NCC) (2012e).

[80]Macdonald (2014), p. 7; Perry et al. (2012), p. 3. See NCC (2012c).

governing bodies, provide a unified voice, and create (not always successfully) an overall positive and respectable image of the industry.[81]

A key role of industry associations is to set unified guidelines for those within the association, and industry as a whole, to follow. Trade associations do this in order to forestall government intervention, develop industry practices, and promote good public relations.[82] Industry standardization can properly police the industry when stringent standards are set, companies have an incentive to follow the standards, and an enforcement mechanism is in place to ensure compliance. On the other hand, standard-setting can be used as a tool by trade associations to appease consumers and governments, without establishing competitive and auditable standards.[83]

The following section will review industry trade associations' animal welfare and environmental stewardship standards. This will include a discussion of how standards are administered, and the level to which producers conform to industry standards in the United States, Canada, and the United Kingdom. (For the purpose of this chapter, farm animal welfare assurance schemes created by or for the conventional meat, dairy, and egg industries will be discussed as industry trade association programs in this section, while national organic programs and farm animal certification programs administered by animal welfare organizations will be covered in the following section on third-party standards.)

4.3.2 *Trade Association Standards for Farm Animal Welfare and Environmental Stewardship*

4.3.2.1 United States

In the United States, trade associations represent all sectors of animal agriculture, including eggs, dairy, and meat. United Egg Producers (UEP) and the NCC respectively represent 95 % of all egg and chicken production.[84] The North American Meat Institute (NAMI), formed in 2015 from the merger of the American Meat Institute (AMI) and the North American Meat Association, represents 95 % of red meat and 70 % of turkey processors and suppliers in the United States.[85] NAMI overlaps with the National Cattlemen's Beef Association (NCBA), representing "America's one million cattle farmers and ranchers," and with the National Turkey

[81]See LaBarbera (1983), pp. 58–59.

[82]See LaBarbera (1983), pp. 58–59.

[83]See LaBarbera (1983), p. 58.

[84]NCC (2012d); United Egg Producers (UEP) (2004a). In 2014 the egg industry produced approximately 99 billion eggs, while the chicken industry slaughtered 9 billion chickens for consumption. USDA, NASS (2015b), pp. 7, 12.

[85]NAMI (2015a). In 2014 the cattle industry slaughtered approximately 31 million cattle and the turkey industry slaughter 236 million turkeys. USDA, NASS (2015b), p. 5; USDA, NASS (2015a), p. 5.

Federation (NTF), representing 95 % of the turkey industry. The National Pork Producers Council represents 43 state pork associations and is closely connected to the quasi-governmental body, the National Pork Board (NPB).[86]

Quasi-governmental bodies have characteristics of both governmental bodies and private entities.[87] A subset of quasi-governmental bodies are agency-related nonprofit corporations, which can significantly differ from one another, but always have a legal association with a federal department or agency.[88] The United States Department of Agriculture (USDA) uses several of these nonprofit corporations for commodity research and promotion, mirroring the role of trade associations.[89] The NPB, established under the Pork Promotion, Research, and Consumer Information Act of 1985,[90] is one of these entities and maintains voluntary pig welfare standards for the industry.[91] The Cattlemen's Beef Promotion and Research Board, another quasi-governmental body, has created animal welfare standards with the NCBA through their Beef Quality Assurance Coalition.[92]

Animal welfare standards, whether created through quasi-governmental bodies or traditional trade associations, cover similar categories of on-farm husbandry: physical alterations, space allowance, air quality, lighting, euthanasia, and handling procedures. For instance, UEP guidelines allow debeaking of chickens up to 10 day of age, 67–86 square inches of space per hen, and ammonia levels up to 25 parts per million (ppm).[93] The NPB, through its Quality Assurance Standards and supplemental reference manuals, states that ammonia should not exceed 25 ppm, pigs must have space to lie down and stand up, males can only be castrated within 7 days of birth, and immobile pigs with a body condition score[94] of 1 should be euthanized (while all other sick or injured animals should receive "timely euthanasia").

Along with on-farm guidelines, trade associations often cover animal care standards during transport and at slaughter. NCC transport recommendations endorse a stocking density that allows birds to sit in a single layer, and state that

---

[86]NAMI (2015a); National Pork Producers Council (no date). In 2014 the pork industry slaughtered 106 million pigs. USDA, NASS (2015a), p. 5.

[87]See Kosar (2011).

[88]Kosar (2011), p. 12.

[89]Kosar (2011), p. 13; See the Pork Promotion, Research and Consumer Information Act (7 U.S.C. § 4801) as an example of quasi-governmental bodies mirroring trade associations. The purpose of the law is to create "an effective and coordinated program of promotion, research, and consumer information designed to strengthen the position of the pork industry in the marketplace; and maintain, develop, and expand markets for pork and pork products."

[90]7 U.S.C. §§4801–4819.

[91]See generally National Pork Board (no date), Pork quality assurance plus, site assessment guide 2.0; See also National Pork Board (NPB) (2003).

[92]Dunn (2006), p. 2 (showing the relationship between the National Cattlemen's Beef Promotion and Research Board and the National Cattlemen's Beef Association (NCBA)). See generally, Beef Quality Assurance (no date).

[93]UEP (2014), pp. 9, 21–22, 32.

[94]National Pork Board (no date), Pork quality assurance plus, pp. 24, 28–29; NPB (2003), p. 10.

corrective action must be taken if "dead on arrival" (DOA) [95] rates exceed 0.5 %.[96] Additionally, the cattle industry maintains a manual dedicated to transportation through the Beef Quality Assurance program; the NTF provides its industry with a slaughter manual, and NAMI provides core criteria recommendations for the transport and slaughter of mammals.[97]

Guidelines published by trade associations are generally inadequate to properly address animal welfare. They are often performance-based—qualitative descriptions without quantitative standards. For instance, the NPB's space allowance recommendation, that pigs have enough space to lie down and stand up, allows for significant variation in raising practices. Even when trade industry guidelines are output-based, such as UEP's space requirements, they do not represent a high standard of care for animals.[98] Furthermore, trade association guidelines are missing several essential welfare components—environmental enrichment, access to the outdoors, and pain control for physical alterations are scarcely addressed.

Not only are these guidelines insufficient, they are not consistently implemented or audited to incentivize conformity, and all trade association guidelines are explicitly voluntary. The NCC and the NTF provide auditing guidelines for producers and recommend third-party audits to limit bias; however, producers are not penalized if found to be nonconforming.[99] Similarly, pork producers can use NPB guidance on their own or they can become Pork Quality Assurance Plus Certified, which entails a training session and test of standards, but does not require compliance with the NPB's *Swine Care Handbook*.[100] In contrast, UEP allows producers to use the logo "United Egg Producer Certified" on packaging only if they comply with UEP guidelines.[101] Producers opting in to this program are audited by an independent company or the Agricultural Marketing Service (AMS), an agency within the USDA responsible for facilitating "fair marketing" of agricultural products.[102] Producers are audited yearly and must file quarterly compliance reports with UEP.[103]

In addition to animal welfare recommendations, trade associations often develop environmental impact policies. All animal agriculture trade associations in the United States advertise responsible environmental practices, but commitment to

[95] "Dead on arrival" is a term used by the industry and the USDA to describe birds that have died prior to arrival at the place of slaughter.

[96] NCC (2014), pp. 11–12.

[97] See generally Beef Quality Assurance (no date), Master cattle transport guide; See also National Turkey Federation ("NTF") (2012b), Animal care best management practices for the production of turkeys; and AMI (2013).

[98] See Sect. 4.4.2 below for examples of high-welfare animal care standards.

[99] NCC (2012a); NTF (2012a), p. 3.

[100] The *Swine Care Handbook* is used as a reference.

[101] UEP (2004b).

[102] UEP (2004b); USDA, Agricultural Marketing Service (AMS) (2014).

[103] UEP (2014), pp. 6, 8.

responsible policies varies from association to association. The NCC, for example, provides a public statement describing the importance of environmental policies, but does not provide producers with guidance on how to properly make positive changes for the environment.[104] On the other hand, NAMI encourages producers to use "Environmental Management Systems,"[105] even providing tools and incentives to get started. Several associations, including NAMI, also have environmental stewardship awards to incentivize progress.[106]

Trade associations' animal welfare and environmental policies are often used to forestall government intervention. Currently, there is no federal law regulating on-farm treatment of animals, and 98 % of animals raised for food are exempt from the few existing federal laws that address animal welfare. Additionally, animal agriculture is frequently exempt from environmental regulations. For example, under the Clean Air Act agriculture facilities often do not meet the threshold requirements for regulation, and under the Clean Water Act "agriculture stormwater discharges" are expressly exempt from the "point source of pollution" definition.[107]

Overall, animal agriculture trade association guidelines help to ensure a minimal level of care for animals and the environment. However, guidelines are implemented mainly to forestall government intervention, reassure consumers when producers' practices are brought into question, and ultimately to promote the industry.

4.3.2.2 Canada

Trade association standards have more influence in Canada than in the United States, and they are arrived at through a far more transparent and deliberative process. In Canada, animal agriculture's efforts to address farm animal care are coordinated by the National Farm Animal Care Council (NFACC).[108] It describes itself as "the only organization in the world that brings together animal welfare groups, enforcement, government and farmers under a collective decision-making model for advancing farm animal welfare."[109] Council members include commodity associations; processor associations; animal welfare associations; retail, restaurant, and food service associations; veterinary associations; and provincial farm

[104]See NCC (2012b).

[105]According to the Environmental Protection Agency, an "Environmental Management System is a set of processes and practices that enable an organization to reduce its environmental impacts and increase its operating efficiency." Environmental Protection Agency (2015).

[106]NAMI (2015b); NCBA (2015).

[107]33 U.S.C. § 1362.

[108]National Farm Animal Care Council ("NFACC") (no date), About NFACC.

[109]NFACC (no date), About NFACC.

animal care councils, with nonvoting members from the federal government and the research community.[110] The Canadian Federation of Humane Societies (CFHS) is a founding member of the NFACC and, in 2015, represented the only animal welfare organization on the council.[111]

The NFACC was established in 2005 and it facilitated consultations that created a code development process in 2006.[112] The process was pilot-tested in 2007 and 2008 to revise the Code of Practice for the Care and Handling of Dairy Cattle, which was released in 2009. In addition to dairy cattle, codes of practice have been developed for beef cattle, equines, farmed deer, goats, sheep, pigs, and for transport of farm animals, although not all of these were created under the new code development process. As of 2015, the NFACC was updating its codes for bison, chickens and turkeys, egg-laying hens, rabbits, and veal calves.[113] The codes address such animal care issues as housing systems and space provisions for animals; painful practices such as castration, dehorning, and tail docking; care and treatment of sick and injured animals; use of electric prods; and other handling and euthanasia methods.[114]

The CFHS views the codes as a compromise between regulations for on-farm care of animals, which it supports, and having no recognized standards for farm animal husbandry. It identifies the main advantage of the voluntary approach as the ability to develop and revise the codes more quickly and more cost-effectively, and notes that the codes were "established with the expectation that they would be reviewed every 5 years and revised according to new scientific knowledge and technological advances."[115] The council operates on a consensus model. As with the rulemaking process in some countries, the process brings together diverse stakeholders, and input is accepted from the general public.[116] Financial support for the code-development process comes from the federal government through the Department of Agriculture and Agri-Food Canada.[117]

The NFACC acknowledges that the codes of practice alone are insufficient to ensure farm animal care; a mechanism is required to assess producer compliance with the standards.[118] Accordingly, the council has developed an animal care assessment process to complement its codes. An initial animal care assessment model was developed using the dairy code of practice, which was pilot-tested during 2012 and 2013 by the Dairy Farmers of Canada. Based on the pilot, an Animal Care Assessment Framework was revised and finalized by the NFACC in

---

[110] NFACC (no date), About NFACC.

[111] CFHS (no date), Codes of practice and the National Farm Animal Care Council.

[112] NFACC (no date), About NFACC. See also NFACC (2015), NFACC code of practice development process.

[113] NFACC (no date), Codes of practice for the care and handling of farm animals.

[114] CFHS (no date), Codes of practice and the National Farm Animal Care Council.

[115] CFHS (no date), Codes of practice and the National Farm Animal Care Council.

[116] NFACC (no date), About NFACC.

[117] CFHS (no date), Codes of practice and the National Farm Animal Care Council.

[118] NFACC (no date), Animal care assessment framework.

2014.[119] In addition to this process, individual industry trade groups have developed their own auditing programs.[120]

4.3.2.3 United Kingdom

Most farm animals in the United Kingdom are raised under the standards of a farm assurance program or "scheme."[121] The programs vary greatly in their requirements for the housing and handling of animals raised for food, yet all claim to ensure high levels of animal welfare.[122] Included are programs created and/or administered by the food industry, the national government, and independent, nonprofit organizations.

Assured Food Standards (also known as "Red Tractor") is a UK food assurance scheme that covers over 78,000 participating farm enterprises that sell their food to one of 350 packers licensed to use the Red Tractor logo on their packaging.[123] Launched in 2000 by UK farmers, food producers, and retailers, current program standards cover five kinds of farm animals (chickens, pigs, dairy cattle, beef cattle, and sheep). Additionally, Red Tractor recognizes as equivalent similar farm animal assurance programs operating in Ireland, Scotland, and Wales. The program is owned by the UK food industry but is independently operated. Red Tractor actively manages the certification bodies that police their standards, and its independent inspectors conduct over 60,000 inspections a year.[124] The British Lion Quality program for eggs addresses environmental protection and animal welfare for hens, with an emphasis on *Salmonella* surveillance and traceability.[125] Turkeys are covered by the Quality British Turkey program of the British Poultry Council.[126] Quality Meat Scotland, which is associated with the UK's Assured Food Standards program, operates an assurance scheme that covers more than 90 % of the beef, lamb, and pork produced in Scotland.[127]

These schemes represent the conventional industry standard for farm animal welfare in the United Kingdom and ensure little more than compliance with minimum legislative requirements.[128] This is illustrated by the results of a

---

[119]NFACC (no date), Animal care assessment framework.

[120]CFHS (no date), Codes of practice and the National Farm Animal Care Council. Industry groups that have developed auditable animal care programs include Chicken Farmers of Canada, Egg Farmers of Canada and the Canadian Turkey Marketing Association.

[121]Compassion in World Farming & OneKind (2012), p. 3.

[122]Compassion in World Farming & OneKind (2012), p. 3.

[123]Assured Food Standards (Red Tractor) (no date).

[124]Assured Food Standards (Red Tractor) (no date).

[125]Lion Egg Farms (no date).

[126]Quality British Turkey (no date).

[127]Quality Meat Scotland (no date).

[128]Compassion in World Farming & OneKind (2012), p. 58. While the UK farm assurance standards, such as those of the "Red Tractor" program, may only reflect minimum legal standards, they are considerably higher than US trade association guidelines.

comprehensive review of UK farm animal welfare assurance schemes conducted in 2012 by Compassion in World Farming and OneKind, two UK-based animal protection charities. The groups scored six UK farm assurance schemes on their performance on various animal welfare criteria grouped into five core areas: the animals' environment; husbandry; handling, transport, and slaughter; genetics and breeding; and auditing.[129] The Red Tractor, Quality Meat Scotland, and British Lion Quality program for eggs consistently scored significantly lower on animal welfare than the British and Scottish organic programs and the food certification program administered by the Royal Society for the Prevention of Cruelty to Animals (RSPCA).[130]

4.4 Independent Regulation Through Third-Party Certification

4.4.1 Overview of Third-Party Certification Programs

Third-party certification provides companies with the opportunity to have their practices and procedures evaluated by an independent auditor. The auditor certifies that the company is compliant with standards developed by the third party. This provides companies a means to assure consumers that the products they purchase meet a certain quality. In the context of this chapter, independent auditors assess compliance with animal welfare and environmental stewardship standards set by third-party certification organizations. The standards are generally above industry guidelines, and certification is used to show consumers that farmers did not mistreat animals or abuse the environment. Once compliance is demonstrated, companies can use a certifier's logo on product packaging (which increases its value), or to gain access to specific markets.[131]

Certification programs are helpful in reducing consumer confusion in the often-puzzling marketplace.[132] Third-party certifiers publish their certification requirements, which makes it easier for interested consumers to decipher animal-raising and environmental stewardship practices. Additionally, most third-party certifiers require that auditors go onto farms to ensure compliance with standards, and are generally transparent in their auditing procedures.[133] If a farm is not compliant with

---

[129]Compassion in World Farming & OneKind (2012), pp. 6–7.

[130]Compassion in World Farming & OneKind (2012), p. 58.

[131]Santacoloma (2013), p. 11. For instance in order to sell certain meat products at Whole Foods one must be certified by Global Animal Partnership (GAP) (discussed below).

[132]Anders et al. (2007), pp. 650–651.

[133]For example GAP and Animal Welfare Approved (AWA) (discussed below) maintain manuals explaining the procedures necessary to become certified. GAP (2014); AWA (no date), Animal welfare policy manual.

a program's standards it may be obligated to remove the certification logo on products until the nonconformance is fixed.

While third-party certification programs are helpful for consumers, and generally provide higher than industry standards, they have several drawbacks. For example, auditing standards range in breadth and do not always require audits for each farm used by a single producer.[134] There is also a wide range of animal welfare standards among different third-party programs, and as standards get higher, the number of producers and farmers opting to participate tends to decrease because the standards are harder to meet.[135] Consequently, few animals are typically covered under the programs with the highest standards. Additionally, consumers are commonly not aware of the wide spectrum of animal care associated with third-party certification programs and often believe they are purchasing products from animals raised to their own perception of "humane," even though this may not be the case.[136] Arguably, the most significant shortcoming of third-party certification programs is that their contribution to animal welfare is currently unproven. The few studies that have compared the health and welfare of animals raised on conventional farms with animals raised on organic or higher-welfare farms have failed to demonstrate a clear difference.[137] Additional research is needed to better understand the impacts of these programs on the animals themselves.

The following section will provide an overview of third-party certification programs found in the United States, Canada, and the United Kingdom. Third-party standards within these nations are generally higher than industry guidelines, and more comprehensive than corporate policies.

[134]See and compare GAP (2014), AWA (no date), Animal welfare policy manual, and Humane Farm Animal Care (HFAC) (2014c).

[135]For example American Humane Certified (AHC) (no date) (a program of American Humane Association (AHA)) has lower standards than AWA, but covers over 1 billion animals, while AWA covers significantly fewer.

[136]For example in a survey commissioned by Consumer Reports 66 % of participants thought the claim "humane" meant that animals had access to the outdoors, and nearly 80 % believed the claim should mean animals have outdoor access. Consumer Reports (2014), p. 10.

[137]See Bergman et al. (2014), Main et al. (2003), Norwegian Scientific Committee for Food Safety (2014), Napolitano et al. (2009), and Ruegg (2009).

4.4.2 *Farm Animal Welfare and Environmental Stewardship Certification Programs*

4.4.2.1 United States

In the United States, as of 2015, four third-party certification programs focused solely on animal welfare: American Humane Certified (AHC), Animal Welfare Approved (AWA), Certified Humane,[138] and Global Animal Partnership (GAP).[139] In addition, three third-party certification programs certified farms for environmental stewardship: USDA Organic,[140] Food Alliance Certified (FAC), and Certified Naturally Grown (CNG).[141] Of the countries reviewed in this chapter, the United States has the most third-party certification programs, and they cover the greatest number of animals.

Animal welfare third-party certification programs developed in the United States after farmers and consumers came to recognize the problems associated with intensive farming.[142] Farmers wanted to provide an alternative to industry practices and showcase this to consumers. In 1989, the Animal Welfare Institute (AWI), a nonprofit animal welfare organization, developed high-welfare animal care standards for raising pigs.[143] These standards led to the first USDA-approved animal-raising label ("Pastureland Farms") on meat packaging.[144] In 2006, after several years updating and expanding its standards, AWI developed the AWA program.[145] Meanwhile in 2000, American Humane Association founded the first third-party certification program in the United States, called "Free Farmed" (later renamed "American Humane Certified").[146] Certified Humane developed standards a few years later, and GAP launched its standards in 2010.[147]

---

138 HFAC administers the Certified Humane program.

139 AHA (no date); AWA (2013); Certified Humane (2015a); GAP (2015b) About (GAP is a step-level program with unique standards ranging from Step 1 to Step 5+).

140 As of the writing of this chapter USDA organic regulations mention animal welfare, but do not provide detailed standards. See 7 C.F.R. § 205.239.

141 7 U.S.C. §§ 6501–6523 (organic enabling statute); Food Alliance (no date), About Food Alliance; Certified Naturally Grown (CNG) (2015a). Both Food Alliance and CNG have animal welfare requirements incorporated into their standards, but they are not the main focus of either program.

142 Sullivan (2013), p. 391.

143 Animal Welfare Institute (2015).

144 Animal Welfare Institute (2015).

145 In 2014, AWA became a program of the Trust for Conservation Innovation's "A Greener World" project. The Animal Welfare Institute remains associated with the AWA, but no longer administers the program.

146 AHA (2013b).

147 GAP (2015c), History (explaining the program started in 2008, but the organization did not set out standards until a few years later).

Together, the four programs cover over 10 % of animals raised for food in the United States.[148] AHC, the largest of the certification programs, covers more than 1 billion animals,[149] whereas, 290 million and 96.7 million animals are raised under GAP and Certified Humane standards, respectively.[150] The number of animals covered under AWA, considered the most stringent of the four, is much lower. All four programs maintain standards for the care of beef cattle, bison, meat chickens, pigs, and turkeys.[151] Three of the programs (GAP excluded) have standards for additional species such as dairy cattle and egg-laying hens.[152]

Each program has its own unique standards, which range from just above industry guidelines to high-welfare pasture-based programs. Certified Humane, GAP Steps 1 and 2, and AHC allow feedlots and do not require outdoor access for all animals.[153] GAP Steps 3 to 5+ and AWA require compliance with higher welfare standards and provide animals with opportunities to perform more natural behaviors: GAP Step 3 requires continuous outdoor access; GAP Steps 4 to 5+ and AWA have pasture-based standards.[154]

Environmental third-party certification programs also play a large role in animal production in the United States. USDA Certified Organic was created after inconsistent state and regional organic standards complicated interstate marketing of organic products.[155] Organic certification is now a voluntary program run by the AMS; the agency accredits third-party certifiers to audit producer compliance with regulatory requirements.[156] CNG developed as an alternative to organic certification for small farmers who sell most products intrastate.[157] FAC originated from a collaborative project between universities in Oregon and Washington aiming to create incentives for sustainable agriculture practices.[158] USDA Certified Organic is by far the biggest of the three programs. In 2011, Certified Organic covered over 37 million animals, and this number continues to grow.[159] FAC certifies approximately 330 small to mid-size farms, and CNG certifies over 700.[160]

---

[148]There are 9 billion animals slaughtered for food in the United States each year. See USDA, NASS (2015a), p. 5; USDA, NASS (2015c), p. 5.

[149]AHA (2015).

[150]GAP (2015d); HFAC (2014a), Annual report, p. 1.

[151]See Certified Humane (2015b); GAP (2015a), 5-step standards (GAP is in the process of writing standards for several additional species); AWA (no date), Standards; AHA (no date), Science-based standards.

[152]See Certified Humane (2015b); GAP (2015a), 5-step standards (GAP is in the process of writing standards for several additional species); AWA (no date), Standards; AHA (no date), Science-based standards.

[153]HFAC (2014b), p. 10; AHA (2013a), p. 19; GAP (2009), p. 15.

[154]GAP (2015d); AWA (2013).

[155]Sustainable Agriculture Research & Education (2012).

[156]7 C.F.R. §§205.500- 205.510; USDA (2014).

[157]CNG (2015b).

[158]Food Alliance (no date), History of Food Alliance.

[159]USDA, Economic Research Service (2013).

[160]Food Alliance (no date), About Food Alliance; See CNG (2015c).

All three environmental certification programs have both livestock and crop production standards.[161] Organic certification prohibits the use of specific substances, requires use of tillage and cultivation practices that minimize soil erosion, and requires soil fertility to be managed through cover crops, rotations, and application of plant and animal materials.[162] CNG standards are based on USDA organic standards, but have several distinctions—mostly related to cost, paperwork requirements, and animal treatment.[163] FAC's livestock standards are more comprehensive and cover, *inter alia*, soil and water conservation, pest and disease management, and wildlife habitat conservation.[164]

While there is a large spectrum of care provided to animals and the environment through third-party certification programs, all have some standards higher than the industry baseline. The number of animals and the amount of land impacted by certification programs continues to grow, and consumers are becoming increasingly concerned with animal welfare—making them more inclined to seek out and purchase "humane" and "environmentally-friendly" products.[165] This does not mean, however, that third-party certification programs will be able to solve all problems of industrial agriculture, but they likely will continue to help raise standards for farm animals and the environment in the United States.

4.4.2.2 Canada

In Canada, as of 2015, two independent food certification programs addressed farm animal welfare, and two other certification programs addressed both environmental stewardship and farm animal welfare. Overall, Canadian third-party programs certify fewer producers and cover far fewer animals than their US counterparts. Products from welfare certification programs based in the United States, including AWA, Certified Humane, and GAP, are also available in grocery stores in limited areas of Canada.

The British Columbia Society for the Prevention of Cruelty to Animals (BC SPCA) launched its "SPCA Certified" program in 2002. As with other animal welfare certification programs, SPCA Certified is based on the principle of

---

[161]They also maintain animal welfare standards, but these are not as comprehensive as those of certification programs dedicated solely to animal welfare; FAC goes even further and maintains worker safety standards.

[162]See generally §§ 7 C.F.R. 205.1- 205.690; 7 C.F.R. § 205.203.

[163]CNG (2015d).

[164]Food Alliance (no date), Sustainability standards for livestock operations.

[165]See Grimshaw et al. (2014), pp. 443–444 (demonstrating that nearly 70 % of participants in a survey conducted by Texas A&M University believed that animal welfare is important). The number of animals covered by AHC soared over 1000 % in 4 years, and organic production increased over 11 % from 2013 to 2014.

"Five Freedoms" for farm animals,[166] and their standards exceed those of the Canadian animal agricultural industry's codes of practice.[167] As of 2015, the program certified more than 20 producers of eggs, dairy, chicken, pork, and beef products in three of Canada's provinces (British Columbia, Alberta, and Saskatchewan).[168] Since BC SPCA's launch, 1.8 million animals have been raised under the program's standards.[169] The Winnipeg Humane Society also launched a certification program in 2002, this one based on organic standards.[170] Significantly smaller in scope than SPCA Certified, the program offers certified humane meat in a few Winnipeg-area markets.[171]

Local Food Plus, introduced in 2006, certifies farmers according to standards that represent seven key tenets of sustainable agriculture: (1) reduce or eliminate synthetic pesticides and fertilizers, (2) avoid the use of hormones, antibiotics, and genetic engineering, (3) conserve soil and water, (4) ensure safe and fair working conditions, (5) provide healthy and humane care for livestock, (6) protect and enhance wildlife habitat and biodiversity, and (7) reduce on-farm energy consumption and greenhouse gas emissions.[172] By 2015, the Land Food People Foundation–administered program had certified over 200 producers and processors, and partnered with nearly 100 retailers and other food service companies.[173]

The largest certification program operating in Canada that impacts farm animal welfare and environmental stewardship is Canada Organic. With higher national regulatory standards than those of USDA Certified Organic,[174] the program sets

---

[166]The concept of Five Freedoms originated in the United Kingdom in 1965 with issuance of the Report of the Technical Committee to Enquire into the Welfare of Animals kept under Intensive Livestock Husbandry Systems, also referred to as "the Brambell Report." This stated that farm animals should have freedom "to stand up, lie down, turn around, groom themselves and stretch their limbs." As a result of the Brambell Report, the Farm Animal Welfare Advisory Committee was established, which disbanded when the Farm Animal Welfare Council (FAWC) was formed in 1979. FAWC eventually developed what is currently known as the list of Five Freedoms for farm animals. They are: (1) freedom from hunger and thirst, (2) freedom from discomfort, (3) freedom from pain, injury or disease, (4) freedom to express normal behavior, and (5) freedom from fear and distress. See Farm Animal Welfare Council (FAWC) (no date).

[167]CFHS (no date), Farm animal welfare certification in Canada.

[168]British Columbia Society for the Prevention of Cruelty to Animals ("BC SPCA") (no date).

[169]CFHS (no date), Farm animal welfare certification in Canada.

[170]CFHS (no date), Farm animal welfare certification in Canada.

[171]Winnipeg Humane Society (no date).

[172]Land Food People Foundation (no date), About.

[173]Land Food People Foundation (no date), Projects.

[174]In June 2009 the governments of Canada and the United States entered into an agreement on the trade of organic products. The two countries' systems were recognized as equivalent with four exceptions, one of which is space allowances for animals. Products from animals raised in the United States may not be sold as organic in Canada unless the stocking densities set out in Canadian organic regulations are met. See Canadian Food Inspection Agency (no date).

minimum space requirements and requires access to the outdoors for all animals.[175] Unfortunately, Canada does not conduct frequent surveys for all types of products in the organic sector (i.e., dairy, eggs, meats); however, in 2008, more than 2 million animals were raised under Canada Organic, a large majority of whom were meat chickens and turkeys.[176]

4.4.2.3 United Kingdom

The "Freedom Food" program (renamed "RSPCA Assured" in 2015) was launched in 1994 with standards for laying hens, pigs, beef cattle, dairy cattle, and sheep based on the Five Freedoms concept for farm animals.[177] By 2010, 1000 labeled product lines were available, which doubled to 2000 product lines in 2013.[178] By 2015, the United Kingdom's only farm assurance scheme dedicated solely to farm animal welfare had more than 3500 participating businesses, and was covering 43 million land animals and 140 million salmon. Since its creation, 600 million animals, representing 10 animal species, have been raised under the program's standards.[179] As of 2013, 54 % of all ducks, 31 % of pigs, and 70 % of salmon raised in the United Kingdom were covered under the program.[180] That year, McDonald's UK switched to 100 % Freedom Food—produced pork.[181]

The RSPCA program likely has been the inspiration for all humane food certification programs currently operating worldwide; in fact, the standards of the AHC ("Free Farmed") and Humane Farm Animal Care ("Certified Humane") programs at the time of their launch in the United States mirrored the RSPCA's standards almost exactly.[182] In the Compassion in World Farming and OneKind (2012) survey of farm assurance programs, Freedom Food consistently scored well above UK industry-backed programs (described in Sect. 4.3.2.3). In fact, for some species (dairy cattle, turkeys, and egg-laying hens), Freedom Food scored higher than the Scottish organic program.[183]

---

[175]See Organic production systems and general principles and management standards, CAN/CGSB-32.310-2006. See also Canadian Federation of Humane Societies (no date), Farm animal welfare certification in Canada.

[176]Agriculture and Agri-Food Canada (no date).

[177]Freedom Food (no date), Our history.

[178]Freedom Food (no date), Our history.

[179]Freedom Food (no date), Facts and figures.

[180]Freedom Food (no date), Impact report 2013.

[181]Freedom Food (no date), Our history.

[182]Since the launch of the welfare certification programs in the US in the early 2000s, RSPCA has continually revised and upgraded its standards, while the standards of the Certified Humane program have remained relatively unchanged, and the American Humane standards have been lowered significantly (examples include a shorter weaning period for pigs, less light and space for meat chickens, and allowing confinement to cages for egg-laying hens).

[183]Compassion in World Farming & OneKind (2012), pp. 16–53.

The oldest organic certification program in existence is administered by the United Kingdom's Soil Association. It was launched in 1946 by a group of farmers, scientists, and nutritionists who saw a connection between farming practices and the health of plants, animals, people, and the planet.[184] The Soil Association program and its high animal welfare and environmental standards have influenced pasture-based farming systems in many countries, including the AWA program in the United States. The program is the leading organic organization in the United Kingdom, certifying over 70 % of all organic products sold in that country,[185] and scored the highest of any UK farm assurance program in the 2012 Compassion in World Farming and OneKind survey. The Soil Association standards were deemed superior to the standards of the industry-backed programs, the RSPCA's Freedom Food program, and the Scottish organic program for all eight species of farm animals covered by the survey.[186] Soil Association standards are also higher than the EU organic minimum standards in several areas, including animal welfare.[187]

4.5 Impact of Voluntary Standards on National Laws and International Initiatives

4.5.1 International Initiatives on Animal Welfare

As described in previous sections, individual multinational food corporations, along with international food assurance programs (such as GLOBALG.A.P.), have attempted to address the treatment of farm animals on a global level. The leading international initiative impacting animal welfare, to date, has been the development and adoption of guidelines by the OIE. The OIE was established in 1924 to address animal diseases at the global level and is accepted worldwide as the intergovernmental organization responsible for improving animal health.[188] It is recognized by the World Trade Organization, and as of 2015, had a total of 180 member countries and territories.[189] Originally formed to address animal health, in recent years the scope of the organization's mission has been expanded to include animal welfare.

[184]Soil Association (no date), Who we are.

[185]Soil Association (no date), Our work 2014.

[186]Compassion in World Farming & OneKind (2012), pp. 16–53. For example, for pigs, the Soil Association scored 81 out of 100 possible points, compared with 71 for the Scottish organic program, 58 for Freedom Food, 29 for Quality Meat Scotland, and 27 for Assured Food Standards ("Red Tractor").

[187]Soil Association (no date), Organic standards. The United States and the European Union have signed an organic equivalency agreement despite the two sets of respective standards differing significantly, particularly in the area of animal welfare.

[188]OIE (2015a), About us.

[189]OIE (no date), The 180 member countries.

In 2005, the OIE adopted the first international farm animal welfare standards, for the transport of animals by sea, land, and air; the slaughter of animals; and the killing of animals for disease control purposes. Subsequently, the OIE developed and adopted standards for the raising of beef cattle, dairy cattle, meat chickens, and farmed fished, and for the transport, slaughter, and killing for disease control purposes of farmed fish.[190]

In 2008, the International Committee of the OIE raised the issue of private animal health and welfare standards, which it characterized as a "problem" due to the fact that these standards are established unilaterally by private entities without direct involvement of governments. The committee noted that individual OIE members were concerned regarding the potential for private standards to conflict with the official standards established by the OIE. In 2009 an ad hoc group on private standards was convened to examine the possible risks and rewards presented by private standards for food safety and animal welfare in regard to international trade. The group distributed a questionnaire to all OIE members, the results of which demonstrated a significant difference of opinion, particularly between developed and developing countries.[191]

In February 2010, the OIE convened a meeting with global private standard-setting organizations, including GLOBALG.A.P. and the Global Food Safety Initiative.[192] While it was agreed that the basis for private standards on food safety should be the existing standards of the OIE and Codex Alimentarius[193] (and any relevant national and regional legislation), no definitive approach for animal welfare was identified.[194] In May 2010, the World Assembly of Delegates of the OIE adopted a resolution on public and private standards in animal health and welfare. This recommends "the implementation of the OIE animal welfare standards as reference standards that apply globally."[195]

In order to promote the implementation of its animal welfare standards, the OIE is supporting an initiative by the ISO to develop a technical specification on animal welfare management for organizations in the food supply chain.[196] The stated purpose of the technical specification, which is only in the drafting stage as of 2015, is "to improve the living conditions of animals raised for food production

---

[190]OIE (2015b), OIE achievements in animal welfare.

[191]OIE (no date), Implications of private standards in international trade of animals and animal products. An executive summary of a report on the questionnaire's findings may be accessed at: http://www.oie.int/fileadmin/Home/eng/Internationa_Standard_Setting/docs/pdf/en_executive_20summary.pdf.

[192]Global Food Safety Initiative (no date), What is GFSI? The scope of GFSI is limited to food safety and does not extend to issues related to animal welfare, the environment, or ethical sourcing.

[193]Codex Alimentarius is a collection of international food standards set out by the Codex Alimentarius Commissions, which the FAO established in 1961. Codex Alimentaruis (2015).

[194]OIE (no date), Implications of private standards in international trade of animals and animal products.

[195]OIE (2010).

[196]ISO (no date), ISO/WD TS 34700.

around the world." The ISO aims to do this by (1) providing a management tool that facilitates implementation of the OIE animal welfare guidelines, (2) providing guidance for the "integration and mutual recognition of additional provisions from public or private standards and relevant legislation, on condition that they meet at least the OIE TAHC [Terrestrial Animal Health Code]," and (3) facilitating the integration of animal welfare principles in business relations between suppliers and their customers.[197]

4.5.2 Impact of Voluntary Standards

Voluntary standards are particularly relevant in the United States and Canada due to a general lack of federal laws addressing the treatment of farm animals. The Humane Methods of Slaughter Act[198] and the Twenty-Eight Hour Law[199] (which addresses transportation) are the only laws expressly addressing farm animal welfare in the United States.[200] Neither law protects birds (which account for 98 % of all animals killed for food in the United States), and there have been few, if any, prosecutions of truck drivers under the Twenty-Eight Hour Law. Similarly, federal protections for farm animals in Canada are limited to laws governing transport and slaughter, with additional limited coverage for animal cruelty under the criminal code. However, voluntary standards have impacted the lives of animals raised for food and influenced government regulation of food animal production in these two countries. Some areas where voluntary standards have influenced government regulation are described below.

4.5.2.1 Case Study: Farm Animal Cruelty

In Canada, as in the United States, there are no national laws protecting the welfare of animals raised for food while they are on the farm.[201] Furthermore, the criminal code that prohibits willful cruelty to animals in Canada does not apply to meat chickens and other birds.[202] Instead, the treatment of farm animals is generally addressed at the provincial level, which results in a lack of consistency in how abuse or neglect of farm animals is managed in the country. While compliance with

---

[197]ISO (2014).

[198]7 U.S.C. §§ 1901–1907.

[199]49 U.S.C. 80502.

[200]The Twenty-eight Hour Law does not specifically exempt birds from its purview; however the USDA has interpreted the law to exclude birds. 7 U.S.C. §1902.

[201]Farm Animal Council Network (2013). See also CFHS (no date), Realities of farming in Canada.

[202]Government of Canada, Criminal Code, Section 446—Cruelty to Animals.

the codes of practice for the care and handling of farm animals is voluntary, four provinces—Manitoba,[203] Newfoundland and Labrador,[204] Prince Edward Island,[205] and Saskatchewan[206]—reference the codes of practice in their animal protection laws and/or regulations. Consequently, police officers and SPCA inspectors in those provinces may cite the codes of practice as representing acceptable husbandry practices in court proceedings. The codes of practice can be offered as a credible standard even in provinces that lack a specific reference to them in their cruelty law.[207]

4.5.2.2 Case Study: Crates for Veal Calves

In 2007, two prominent American veal producers—Strauss Veal and Marcho Farms—pledged to stop using veal crates within 10 years.[208] These announcements followed two decades of campaigning by American animal protection advocates on the issue. Soon after the corporate announcements, the American Veal Association—the trade association for the industry in the United States—resolved to encourage all producers of veal to make the same commitment.[209] Since the Strauss Veal and Marcho Farm's announcements, eight American states have limited or banned the use of veal crates.[210] This example demonstrates that, as large companies improve their animal welfare policies, they are likely to influence their respective trade associations. Furthermore, as individual corporate practices become established in trade association guidelines, laws will more easily change.

[203]Government of Manitoba, Animal Care Regulation 126/98 of the Animal Care Act (C.C.S.M. c. A84).

[204]Government of Newfoundland and Labrador, Regulation 36/12, Animal Protection Standards Regulation under the Animal Health and Protection Act, 2012.

[205]Government of Prince Edward Island, Animal Health and Protection Act Chapter A-11.1, Animal Protection Regulations, PEI Reg EC 71/90.

[206]Government of Saskatchewan, The Animal Protection Act, 2000, Chapter A-21.1 Reg 1, as amended by Saskatchewan Regulations 32/2015.

[207]See Farm Animal Council Network (2013).

[208]Humane Society of the United States (2012), p. 2.

[209]Bakke and American Veal (2007).

[210]Ariz. Rev. Stat. Ann. § 13-2910.07; Cal. Health and Safety Code §25990; Colo. Rev. Stat. Ann. §35-50.5-102; Me. Rev. Stat. tit.7 § 4020; Mich. Comp. Laws Ann. § 287.746; Ohio Admin. Code § 901:12-4, 901:12-5-03; R.I. Gen. Laws Ann. §4-1.1-3; 302 KAR 21:030. Most state's veal confinement laws have vague language, allowing farmers to use crates so long as they provide space to turn around, lie down, and stand up.

4.5.2.3 Case Study: Gestation Crates for Sows

In 2007, Smithfield Foods, the largest pork producer in the United States, committed to phasing out gestation crates from its supply chain.[211] Shortly after, Canada's largest pork producer, Maple Leaf, announced its intention to phase out the use of gestation crates.[212] Other major American and Canadian pork producers followed the lead of Smithfield and Maple Leaf. Since the time Smithfield made its original commitment, six American states have banned or limited the use of gestation crates.[213] Moreover, in 2014, the NFACC released its revised Code of Practice for the Care and Handling of Pigs, which includes a prohibition on the use of sow crates for all newly built or rebuilt facilities in Canada after July 1, 2014.[214] It seems unlikely that the NFACC would have taken this step without prior action from at least one major pork producer or trade industry group. As of 2015, the OIE had yet to adopt standards for the welfare of animals in pork-production systems; however, the organization was expected to address pig welfare in the near future. Whether the OIE standards ultimately prohibit intensive confinement may well be decided by the adoption of voluntary positions against confinement by private food corporations and industry trade associations, particularly those operating on an international level.

4.5.2.4 Case Study: Battery Cages for Hens

As described earlier in the chapter, as of 2015, many major American and Canadian food corporations had adopted positions in regard to the manner in which egg-laying hens are housed. Although no major American egg producers have eliminated the use of conventional battery cages as of 2015, enriched cages have been adopted by Manitoba Egg Farmers.[215] This move may lead to a revision of Canada's NFACC standard for hen housing, which is in the process of being updated as of 2015. In the United States, third-party certification standards for egg-laying hens have significantly impacted hen-housing laws in at least two states. The Oregon Legislature passed a law in 2011 regulating cage requirements for egg-laying hens. According to the law, by 2026, producers selling eggs in Oregon must meet "standards equivalent to the requirements for certification of enriched colony facility systems established in the American Humane Association's farm

---

[211] Smithfield (2013).

[212] Maple Leaf (no date).

[213] Cal. Health and Safety Code § 25990; Colo. Rev. Stat. Ann. § 35-50.5-102; Me. Rev. Stat. tit. 7, § 4020; Mich. Comp. Laws Ann. § 287.746; Ohio Admin. Code § 901:12-8; Or. Rev. Stat. § 600.150; R.I. Gen. Laws Ann. §4-1.1-3. Arizona (Ariz. Rev. Stat. Ann. § 13-2910.07) and Florida (Fla. Const. art. X, § 21) did however outlaw crates before Smithfield made its decision.

[214] NFACC (2014), p. 11.

[215] The Poultry Site (2013).

animal welfare certification program."[216] Washington also codified AHC's enriched colony cage standards in 2011.[217]

4.5.2.5 Case Study: Tail Docking of Cattle

Following the adoption by the OIE of standards for the welfare of animals in beef cattle production, trade associations representing beef cattle producers in both the United States and Canada made related revisions to their animal care guidelines. For example, the OIE standards recommend that producers not dock the tails of cattle, noting that research shows that increased space per animal and proper bedding are effective in preventing a condition known as "tail tip necrosis," the reason commonly given for routine tail docking in beef cattle.[218] In 2013, the NFACC revised its beef cattle code of practice to include a prohibition on docking the tails of beef cattle except on the advice of a veterinarian, and a recommendation that stocking densities in slatted-floor facilities be lowered to reduce tail injuries.[219] Similarly, in 2014 the NCBA published supplemental animal care guidelines consistent with the OIE beef cattle standards. The supplemental guidelines cover castration, dehorning, branding, and tail docking in a manner very similar to the OIE, and in some cases the language is taken verbatim from the OIE standard.[220] The previous NCBA guidelines contained no prohibition against the practice of routine tail docking, and the standard for castration was that it be performed prior to 120 days; this was lowered in the 2014 guidelines to 3 months, consistent with the OIE.[221]

---

[216]ORS 632.840. There are several incremental changes that producers must meet before 2026 to comply with the Oregon law. AHC has separate standards for enriched cages and cage-free production.

[217]Wash. Rev. Code §69.25.065, §69.25.107.

[218]OIE (2014).

[219]NFACC (2013), p. 25.

[220]For example, the tail docking section states: "Tail docking has been performed in beef cattle to prevent tail tip necrosis in confinement operations. Research shows that increasing space per animal and proper bedding are effective means in preventing tail tip necrosis. Therefore it is not recommended for producers to dock the tails of beef cattle." Beef Quality Assurance ("BQA") (2014), p. 2.

[221]See BQA (no date), The cattle industry's guidelines for the care and handling of cattle, p. 7. See also OIE.

4.5.2.6 Case Study: Organic Production

In 1990, the United States Congress passed the Organic Food Production Act (OFPA) in order to "establish national standards governing the marketing ... [of] organically produced products."[222] In order to affix the claim "organic" to a product, a producer must be in full compliance with the OFPA. The law gives the USDA authority to write animal care regulations for organic production.[223] The regulations, which the USDA first promulgated in 2001, state that all animals must have year-round access to the outdoors with direct sunlight, fresh air, and exercise areas.[224] Despite taking over 10 years to finalize, the organic regulations are overly general in terms of animal care, allowing for significant variation in practices. For instance, 38 % of organic egg farms give birds less than 2 square feet of space, while 24 % provide from 2 to less than 3 square feet, 25 % provide from 3 to less than 5 square feet, and 13 % provide 5 or more square feet. Additionally, outdoor access varies greatly in organic egg production; some birds are raised on pasture while others are confined to giant barns with small enclosed porches that certifiers qualify as outdoor access.[225]

The National Organic Standards Board (NOSB), USDA's organic advisory board, spent roughly a decade drafting more reliable welfare regulations for the USDA to promulgate.[226] As part of its deliberative process, the NOSB reviewed standards of the AHC, Certified Humane, GAP, and AWA programs. Third-party certification standards influenced the NOSB's recommended regulations in several ways, including how birds should be handled before slaughter and how long animals may be transported before rest and feed are provided. As of the writing of this chapter, the USDA has drafted but not yet finalized new welfare regulations based on the NOSB recommendations.

4.5.2.7 Case Study: Livestock and Poultry Slaughter

In 2008, an undercover investigation documenting animal cruelty at a California slaughterhouse led to the largest meat recall in United States history. A significant portion of the meat produced at the slaughter facility had been destined for the federal school lunch program,[227] which is overseen by the AMS. The agency spends approximately $1.5 billion each year procuring products for this program.[228] After the recall incident, the AMS implemented slaughter and handling standards

---

[222] 7 U.S.C. § 6501.
[223] 7 U.S.C. § 6509(2).
[224] 7 CFR §205.239.
[225] USDA, Animal and Plant Health Inspection Service (2013), p. 11.
[226] See National Organic Standards Board (2001).
[227] Martin (2008).
[228] See USDA, AMS (2015a).

for livestock commodity purchasing. In order for the AMS to consider purchasing meat, the supplier must meet certain animal welfare standards.[229] These standards are based on the 2013 edition of the AMI's *Recommended Animal Handling Guidelines & Audit Guide*.[230] Bidding suppliers must, *inter alia,* ensure that all animals are rendered unconscious, and that no more than 1 % are compromised (injured) when arriving at the slaughterhouse.[231] The AMI guidelines only apply to mammals; consequently, as of 2015, there are no animal welfare purchasing requirements for poultry suppliers.

Poultry slaughter in the United States is regulated by the Poultry Products Inspection Act (PPIA),[232] which aims to ensure that poultry products are properly labeled and unadulterated.[233] The PPIA gives the USDA authority to institute regulations when necessary to fulfill its purpose.[234] One such regulation states that birds must be slaughtered "in accordance with good commercial practices."[235] However, the United States has no legal definition for the term "good commercial practices." Instead the USDA's Food Safety and Inspection Service (FSIS) utilizes the NCC's animal welfare guidelines when auditing slaughterhouses for "good commercial practices."[236] Inspectors issue "Memorandums of Interview"[237] when slaughterhouses are noncompliant with NCC standards.[238] The FSIS's sister agency, the AMS, also uses NCC guidelines in its "Process Verified Program."[239] The program allows producers to use a label claim, such as "humanely raised," on packaging that also states "USDA Process Verified" after the AMS audits the producer for compliance with its own animal care standards, which are typically based on minimal industry guidelines.[240]

[229]AMS (2015b).

[230]AMS (2015b), pp. 2–3.

[231]AMI (2013), pp. 45–53; USDA, AMS (2015b).

[232]21 U.S.C. §§ 451–472.

[233]21 U.S.C. § 452.

[234]21 U.S.C. § 463.

[235]9 C.F.R. § 381.65.

[236]70 Fed. Reg. 56624–56626.

[237]Memorandums of Interview record non-regulatory deficiencies at slaughter plants, while Noncompliance Records document a failure to meet a regulatory requirement. In a recent FSIS notice the agency prohibited inspectors from quoting NCC in Noncompliance Records.

[238]Food Safety and Inspection Services (2015).

[239]AMS (no date), Official listing of approved USDA process verified programs, p. 20.

[240]AMS (2015c).

4.6 Conclusion

Voluntary standards have the potential to significantly impact the manner in which farm animals are raised, transported, and slaughtered, particularly in countries such as Canada and the United States where few national legal standards exist. One country may have a dozen or more sets of private standards for the raising of a particular animal species. The variety of differing—and sometimes contradictory—private standards is fueling a call for international harmonization. While harmonizing standards may facilitate their adoption by countries in certain regions of the world without animal welfare standards, harmonization may also constrain the development and acceptance of higher standards. The actual impact on animals of voluntary standards—and regulatory standards for that matter—is largely unknown. Appropriate species-specific animal welfare indicators must be identified and then routinely measured to determine to what extent various husbandry standards impact the lives of animals raised for food.

References

7 C.F.R. §§ 205.1-205.690
9 C.F.R. § 381.65
7 U.S.C. §§ 1901–1907
7 U.S.C. § 4801–4819
7 U.S.C. §§ 6501–6523
21 U.S.C. §§ 451–472
33 U.S.C. § 1362
49 U.S.C. § 80502
70 Fed. Reg. 56624–56626
302 Kentucky Administrative Regulations 21:030
Agriculture and Agri-Food Canada (no date) Certified organic production in Canada 2008. http://www.agr.gc.ca/eng/industry-markets-and-trade/statistics-and-market-information/by-product-sector/organic-products/organic-production-canadian-industry/certified-organic-production-in-canada-2008/?id=1276033208187. Accessed 18 May 2015
Agriculture and Agri-Food Canada (2014) Slaughter reports. http://www.agr.gc.ca/eng/industry-markets-and-trade/statistics-and-market-information/by-product-sector/red-meat-and-live stock/red-meat-market-information-canadian-industry/slaughter/?id=1415860000003. Accessed 15 June 2015
American Humane Association (no date) About American Humane Association. http://www.humaneheartland.org/about-us. Accessed 15 June 2015
American Humane Association (no date) Science-based standards. http://www.humaneheartland.org/our-standards. Accessed 15 June 2015
American Humane Association (2013a) Animal welfare standards for beef cattle
American Humane Association (2013b) Humane Euthanasia. http://humaneheartland.org/our-farm-programs/american-humane-certified/2-uncategorised. Accessed 15 June 2015
American Humane Association (2015) Celebrate those who grow our food, and do it right. http://www.humaneheartland.org/the-certified-blog/item/celebrate-those-who-grow-our-food-and-do-it-right. Accessed 15 June 2015

American Meat Institute (2013) Recommended animal handling guidelines & audit guide: a systemic approach to animal welfare. http://animalhandling.org/ht/a/GetDocumentAction/i/93003. Accessed 15 June 2015

Amos N, Sullivan R (2014) The business benchmark on farm animal welfare 2014 report. http://www.bbfaw.com/wp-content/uploads/2015/02/BBFAW_2014_Report.pdf. Accessed 15 June 2015

Anders S, Souza Monteiro DM, Rouviere E (2007) Objectiveness in the market for third-party certification: does market structure matter? Paper presented at the 105th EAAE seminar, Bologna, Italy 8–10 March 2007

Animal Welfare Approved (no date) Animal Welfare Approved policy manual. http://animalwelfareapproved.org/wp-content/uploads/2015/04/Policy-and-Guidelines-v14.pdf. Accessed 15 June 2015

Animal Welfare Approved (no date) Standards. http://animalwelfareapproved.org/standards/. Accessed 15 June 2015

Animal Welfare Approved (2013) About. http://animalwelfareapproved.org/about/. Accessed 15 June 2015

Animal Welfare Institute (2015) History of AWI's leadership on establishing and upholding farm animal standards. https://awionline.org/content/history-awis-leadership-establishing-and-upholding-farm-animal-standards. Accessed 15 June 2015

Aramark (2015) Aramark establishes animal welfare policy. http://www.aramark.com/about-us/news/aramark-general/aramark-establishes-animal-welfare-policy. Accessed 15 June 2015

Arizona Revised Statute Annotated § 13–2910.07

Assured Food Standards (Red Tractor) (no date) FAQs. http://www.redtractor.org.uk/faqs. Accessed 21 May 2015

Bakke, American Veal Association (2007) Resolution. http://www.americanveal.com/wp-content/uploads/2011/10/GRP_HOUSING_RESOL1-0507.pdf. Accessed 15 June 2015

Beef Quality Assurance (no date) Master cattle transport guide

Beef Quality Assurance (no date) National manual for cattle care & handling guidelines. http://www.bqa.org/CMDocs/bqa/CCHG2015_Final.pdf. Accessed 15 June 2015

Beef Quality Assurance (no date) The cattle industry's guidelines for the care and handling of cattle. http://www.bqa.org/Cmdocs/BQA/GuidelinesfortheCareandHandlingofCattle.pdf. Accessed 3 June 2015

Beef Quality Assurance (2014) Supplemental guidelines. http://www.bqa.org/CMDocs/bqa/Supplemental%20Guidelines%202014_.pdf. Accessed 3 June 2015

Bergman MA, Richert RM, Cicconi-Hogan KM et al (2014) Comparison of selected animal observations and management practices used to assess welfare of calves and adult dairy cows on organic and conventional dairy farms. J Dairy Sci 97:4269–4280

Berry MA, Rondinelli DA (1998) Proactive corporate environmental management: new industrial revolution. Acad Manag Exec 12:38–50

British Columbia Society for the Prevention of Cruelty to Animals (no date) SPCA Certified farms. http://www.spca.bc.ca/welfare/farm-animal-welfare/spca-certified/spca-certified-farms/. Accessed 18 May 2015

California Health and Safety Code §25990

Canadian Federation of Humane Societies (no date) Codes of practice and the National Farm Animal Care Council. http://www.cfhs.ca/farm/codes_of_practice. Accessed 15 May 2015

Canadian Federation of Humane Societies (no date) Farm animal welfare certification in Canada. http://www.cfhs.ca/farm/humane_labelling_in_canada. Accessed 18 May 2015

Canadian Federation of Humane Societies (no date) Putting an end to gestation stalls in Canada

Canadian Federation of Humane Societies (no date) Realities of farming in Canada. http://www.cfhs.ca/farm/farming_in_canada. Accessed 18 May 2015

Canadian Food Inspection Agency (no date) Canada – US organic equivalence arrangement – overview. http://www.inspection.gc.ca/food/organic-products/equivalence-arrangements/us-overview/eng/1328068925158/1328069012553. Accessed 18 May 2015

Certified Humane (2015a) Our mission. http://certifiedhumane.org/. Accessed 15 June 2015
Certified Humane (2015b) Standards. http://certifiedhumane.org/how-we-work/our-standards/. Accessed 15 June 2015
Certified Naturally Grown (2015a) About CNG. https://www.naturallygrown.org/about-cng. Accessed 15 June 2015
Certified Naturally Grown (2015b) Brief history of certified naturally grown. https://www.naturallygrown.org/about-cng/brief-history-of-certified-naturally-grown. Accessed 15 June 2015
Certified Naturally Grown (2015c) List of producers. https://www.naturallygrown.org/producers. Accessed 15 June 2015
Certified Naturally Grown (2015d) Livestock standards. https://www.naturallygrown.org/programs/livestockstandards. Accessed 15 June 2015
Codex Alimentaruis (2015) Codex timeline from 1945 to the present. http://www.codexalimentarius.org/about-codex/codex-timeline/en/. Accessed 15 June 2015
Colorado Revised Statutes Annotated §35-50.5-102
Compass Group (no date) Sustainability platform: a vision for 2020. http://compass-usa.com/Pages/Sustainabilityfor2020.aspx. Accessed 21 May 2015
Compassion in World Farming and OneKind (2012) Farm assurance schemes and animal welfare: how the standards compare. http://www.onekind.org/uploads/publications/120323-farm-assurance.pdf. Accessed 21 May 2015
Consumer Reports (2014) Food labels survey. http://www.greenerchoices.org/pdf/consumerreportsfoodlabelingsurveyjune2014.pdf. Accessed 15 June 2015
Davis GF, Whitman MN, Zald MN (2006) The responsibility paradox: multinational firms and global corporate social responsibility. Ross School of Business Paper No. 1031
Dunn BH (2006) Beef Quality Assurance: present and future in a review of the beef quality assurance program for the joint evaluation advisory committee, Reno, Nevada July 10, 2006
Environmental Protection Agency (2015) Environmental Management Systems (EMS). http://www.epa.gov/ems/. Accessed 15 June 2015
Farm Animal Council Network (2013) A summary report on farm animal welfare law in Canada
Farm Animal Welfare Council (no date) Five freedoms. http://webarchive.nationalarchives.gov.uk/20121007104210/http:/www.fawc.org.uk/freedoms.htm. Accessed 4 June 2015
Florida Constitution article X, § 21
Food Alliance (no date) About food alliance. http://foodalliance.org/about. Accessed 15 June 2015
Food Alliance (no date) History of food alliance. http://foodalliance.org/about/history. Accessed 15 June 2015
Food Alliance (no date) Sustainability standards for livestock operations. http://foodalliance.org/livestock/livestockops. Accessed 15 June 2015
Food and Agriculture Organization of the United Nations Statistics Division (2015) Download data. http://faostat3.fao.org/download/Q/QL/E. Accessed 15 June 2015
Freedom Food Limited (no date) Facts and figures. http://www.freedomfood.co.uk/aboutus/facts-and-figures. Accessed 22 May 2015
Freedom Food Limited (no date) Impact report 2013. http://www.freedomfood.co.uk/aboutus. Accessed 22 May 2015
Freedom Food Limited (no date) Our history. http://www.freedomfood.co.uk/aboutus/history. Accessed 22 May 2015
Global Animal Partnership (2009) Global animal partnership 5-step animal welfare rating standards for beef cattle. http://glblanimalpartnership.blob.core.windows.net/standards/Beef%20Cattle%20Welfare%20Standards.pdf. Accessed 15 June 2015
Global Animal Partnership (2014) Pilot GAP policy manual. https://glblanimalpartnership.blob.core.windows.net/other/GAP%20Policy%20Manual.pdf. Accessed 15 June 2015
Global Animal Partnership (2015a) 5-step standards. http://www.globalanimalpartnership.org/5-step-program/standards. Accessed 15 June 2015

Global Animal Partnership (2015b) About. http://www.globalanimalpartnership.org/about. Accessed 15 June 2015
Global Animal Partnership (2015c) History. http://www.globalanimalpartnership.org/about/history. Accessed 15 June 2015
Global Animal Partnership (2015d) Welcome to global animal partnership. http://www.globalanimalpartnership.org/. Accessed 15 June 2015
Global Food Safety Initiative (no date) What is GFSI? http://www.mygfsi.com/about-us/about-gfsi/what-is-gfsi.html. Accessed 8 June 2015
GLOBALG.A.P. (no date) Animal welfare add-on. http://www.globalgap.org/uk_en/what-we-do/globalg.a.p.-certification/globalg.a.p.-00001/Animal-Welfare/. Accessed 3 June 2015
GLOBALG.A.P. (no date) History. http://www.globalgap.org/uk_en/who-we-are/about-us/history/. Accessed 3 June 2015
GLOBALG.A.P. (no date) Membership. http://www.globalgap.org/uk_en/who-we-are/members/. Accessed 3 June 2015
Government of Canada Criminal Code, Section 446 – cruelty to animals
Government of Manitoba, Animal Care Regulation 126/98 of the Animal Care Act (C.C.S.M. c. A84)
Government of Newfoundland and Labrador, Regulation 36/12, Animal Protection Standards Regulation under the Animal Health and Protection Act, 2012
Government of Prince Edward Island, Animal Health and Protection Act Chapter A-11.1, Animal Protection Regulations, PEI Reg EC 71/90
Government of Saskatchewan, The Animal Protection Act, 2000, Chapter A-21.1 Reg 1, as amended by Saskatchewan Regulations 32/2015
Grandin T (2005) Special report: maintenance of good animal welfare standards in beef slaughter plants by use of auditing programs. J Am Vet med Assoc 226:370–373
Grimshaw K et al (2014) Consumer perception of beef, pork, lamb, chicken, and fish. Meat Sci 96:443–444
Humane Farm Animal Care (2014a) Annual report. http://certifiedhumane.org/wp-content/uploads/2015/03/Annual_Report_2014-Final.pdf. Accessed 15 June 2015
Humane Farm Animal Care (2014b) Beef cattle. http://certifiedhumane.org/wp-content/uploads/2014/02/Std14.BeefCattle.1J.pdf/. Accessed 15 June 2015
Humane Farm Animal Care (2014c) Program/policy manual. http://certifiedhumane.org/wp-content/uploads/2014/07/Pol14.2J.pdf. Accessed 15 June 2015
Humane Society of the United States (2012) An HSUS report: the welfare of animals in the veal industry. http://animalstudiesrepository.org/cgi/viewcontent.cgi?article=1028&context=hsus_reps_impacts_on_animals. Accessed 15 June 2015
International Organization for Standardization (no date) ISO 14000-environmental management. http://www.iso.org/iso/home/standards/management-standards/iso14000.htm. Accessed 15 June 2015
International Organization for Standardization (no date) ISO/WD TS 34700. http://www.iso.org/iso/catalogue_detail.htm?csnumber=64749. Accessed 8 June 2015
International Organization for Standardization (no date) Standards. http://www.iso.org/iso/home/standards.htm. Accessed 15 June 2015
International Organization for Standardization (2014) WG 16 consultation on the outline of the TS on animal welfare proposed by the drafting group
Jack in the Box (2014) Animal welfare at Jack in the Box Inc. http://www.jackintheboxinc.com/assets/animal-welfare-report.pdf. Accessed 15 June 2015
Kosar RK (2011) The quasi government: hybrid organizations with both government and private sector legal characteristic, congressional research service. https://www.fas.org/sgp/crs/misc/RL30533.pdf. Accessed 15 June 2015
Kroger (2012) Kroger provides update on animal welfare. http://ir.kroger.com/Cache/1001185186.PDF?Y=&O=PDF&D=&FID=1001185186&T=&IID=4004136. Accessed 15 June 2015

LaBarbera PA (1983) The diffusion of trade association advertising self-regulation. J Mark 47 (1):58–67
Land Food People Foundation (no date) About. http://landfoodpeople.ca/about. Accessed 18 May 2015
Land Food People Foundation (no date) Projects. http://landfoodpeople.ca/projects. Accessed 18 May 2015
Lion Egg Farms (no date) British Lion quality code of practice. http://www.lioneggfarms.co.uk/information/british-lion-quality-code-of-practice/. Accessed 21 May 2015
Loblaw (no date) 2012 Corporate social responsibility report. http://www.loblaw-reports.ca/responsibility/2012/targets-and-achievements/source-with-integrity/. Accessed 15 May 2015
MacDonald J (2014) Technology, organization, and financial performance in U.S. broiler production, Information Bulletin No. 126, p. 7. http://www.ers.usda.gov/media/1487788/eib126.pdf. Accessed 15 June 2015
Main DC, Whay HR, Green LE et al (2003) Effect of the RSPCA Freedom Food scheme on the welfare of dairy cattle. Vet Rec 153:227–231
Maine Revised Statute title 7 § 4020
Manitoba Pork Council (2011) Embracing a sustainable future. http://manitobapork.com/wp-content/uploads/2012/12/Embracing-a-Sustainable-Future-Full-Version-FINAL.pdf. Accessed 15 May 2015
Maple Leaf (no date) Sustainability report. http://www.mapleleafsustainability.ca/social.html. Accessed 15 May 2015
Marks and Spencer (no date) Your M&S plan: a report 2014. http://planareport.marksandspencer.com/. Accessed 21 May 2015
Martin A (2008) Largest recall of ground beef is ordered. New York Times (18 February 2008). http://www.nytimes.com/2008/02/18/business/18recall.html?_r=0. Accessed 15 June 2015
McDonald's (2014) 2014 McDonald's sustainability update. https://www.aboutmcdonalds.com/content/dam/AboutMcDonalds/2.0/pdfs/2014_sustainability_report.pdf. Accessed 15 June 2015
Michigan Compiled Laws Annotated § 287.746
Napolitano F, De Rosa G, Ferrante V et al (2009) Monitoring the welfare of sheep in organic and conventional farms using an ANI 35 L derived method. Small Rumin Res 83:49–57
National Cattlemen's Beef Association (2015) Environmental stewardship award. http://www.beefusa.org/environmentalstewardshipaward.aspx. Accessed 15 June 2015
National Chicken Council (2012a) Animal welfare for broiler chickens. http://www.nationalchickencouncil.org/industry-issues/animal-welfare-for-broiler-chickens/. Accessed 15 June 2015
National Chicken Council (2012b) Environment. http://www.nationalchickencouncil.org/industry-issues/environment/. Accessed 15 June 2015
National Chicken Council (2012c) History of the National Chicken Council. http://www.nationalchickencouncil.org/about-ncc/history/. Accessed 15 June 2015
National Chicken Council (2012d) Overview. http://www.nationalchickencouncil.org/about-ncc/overview/. Accessed 15 June 2015
National Chicken Council (2012e) U.S. chicken industry history. http://www.nationalchickencouncil.org/about-the-industry/history/. Accessed 15 June 2015
National Chicken Council (2014) Animal welfare guidelines and audit checklist. http://www.nationalchickencouncil.org/wp-content/uploads/2014/04/NCC-Guidelines-Broilers-April2014.pdf. Accessed 15 June 2015
National Farm Animal Care Council (no date) About NFACC. http://www.nfacc.ca/about-nfacc. Accessed 15 May 2015
National Farm Animal Care Council (no date) Animal care assessment framework. http://www.nfacc.ca/animal-care-assessment. Accessed 15 May 2015
National Farm Animal Care Council (no date) Codes of practice for the care and handling of farm animals. http://www.nfacc.ca/codes-of-practice. Accessed 15 May 2015

National Farm Animal Care Council (2013) Code of practice for the care and handling of beef cattle. https://www.nfacc.ca/pdfs/codes/beef_code_of_practice.pdf. Accessed 3 June 2015
National Farm Animal Care Council (2014) Code of practice for the care and handling of pigs. https://www.nfacc.ca/pdfs/codes/pig_code_of_practice.pdf. Accessed 3 June 2015
National Farm Animal Care Council (2015) NFACC code of practice development process
National Pork Board (no date) Pork Quality Assurance Plus site assessment guide 2.0. http://porkcdn.s3.amazonaws.com/sites/all/files/documents/PQAPlus/V2.0/SiteAssessment/SiteAssessmentGuideV2.0.pdf. Accessed 15 June 2015
National Pork Board (2003) Swine care handbook. http://porkcdn.s3.amazonaws.com/sites/all/files/documents/PQAPlus/V2.0/TrainingAdults/SwinecareHandbookv2.0.pdf. Accessed 15 June 2015
National Pork Producers Council (no date) About us, mission and governance. http://www.nppc.org/about-us/. Accessed 15 June 2015
National Turkey Federation (2012a) Animal care best management practices. http://www.eatturkey.com/sites/default/files/NTF%20Production%20Welfare%20-%202012%20FINAL.pdf. Accessed 15 June 2015
National Turkey Federation (2012b) Animal care best management practices for the production of turkeys. http://www.eatturkey.com/sites/default/files/NTF%20Slaughter%20Welfare%202012%20-%20FINAL.pdf. Accessed 15 June 2015
National Organic Standards Board (2001) Pasture livestock committee recommendation. http://www.ams.usda.gov/AMSv1.0/getfile?dDocName=STELPRDC5057231. Accessed 15 June 2015
North American Meat Institute (2015a) About. https://www.meatinstitute.org/index.php?ht=d/sp/i/204/pid/204. Accessed 15 June 2015
North American Meat Institute (2015b) Environmental achievement awards program. https://www.meatinstitute.org/ht/display/ShowPage/id/11491/pid/11491. Accessed 15 June 2015
Norwegian Scientific Committee for Food Safety (2014) Comparison of organic and conventional food and food production, Part II: animal health and welfare in Norway. Doc. No.: 11-007-2
Ohio Administrative Code § 901:12–4, 12–8, 12-5-03
Olymel (no date) Olymel and animal welfare. http://www.olymel.ca/en/commitments/animal-welfare/. Accessed 15 May 2015
Oregon Revised Statute § 600.150
Oregon Revised Statute §632.840
Organic production systems general principles and management standards. CAN/CGSB-32.310-2006, amended Oct. 2008, Dec. 2009 and June 2011
Pastore A (2013) FAO's strategic vision to engage with the private sector. In Maybeck A, Redfern S (eds) Voluntary standards for sustainable food systems: challenges and opportunities. Rome, p 109
People for the Ethical Treatment of Animals (no date) Peta's milestones for animals. http://www.peta.org/about-peta/milestones/. Accessed 15 June 2015
Perry J et al (2012) Poultry production in the United States. In: Broiler farms' organization, management and performance agriculture, Information Bulletin No. 748, p 3. http://www.ers.usda.gov/media/256034/aib748b_1_.pdf. Accessed 15 June 2015
Portney P (2005) Corporate social responsibility an economic and public policy perspective. In: Hay BL, Stavins RN, Vietor RHK (eds) Environmental protection and the social responsibility of firms perspectives from law, economics and business. RFF Press, Washington
Post A (2013) Grocers agree to eliminate pig gestation crates. Winnipeg Free Press. http://www.winnipegfreepress.com/business/agriculture/Pig-gestation-crates-to-be-eliminated-at-Manitoba-farm--205400701.html. Accessed 15 May 2015
Quality Meat Scotland (no date) QMS. http://www.qmscotland.co.uk/qms. Accessed 15 May 2015
Ramus CA, Montiel I (2005) When are corporate environmental policies a form of greenwashing? Bus Soc 44:377–414
Reimund D et al (1981) Structural change in agriculture: the experience for broilers, fed cattle, and processing vegetables. U.S. Department of Agriculture, Economics and Statistics Service, p 4
Revised Code of Washington §69.25.150

Rhode Island General Laws Annotated §4-1.1-3
Rondinelli DA, Berry MA (2000) Corporate environmental management and public policy bridging the gap. Am Behav Sci 44:168–187
Royal Society for the Prevention of Cruelty to Animals (2008) The slaughter of food animals
Ruegg PL (2009) Management of mastitis on organic and conventional dairy farms. J Anim Sci 87:43–55
Santacoloma P (2013) Nexus between public and private food standards: main issues and perspectives. In: Maybeck A, Redfern S (eds) Voluntary standards for sustainable food systems: challenges and opportunities. Rome
Singer P (1998) Ethics into action. Rowan & Littlefield Publishers, Inc, Lanham
Smithfield (no date) Housing of pregnant sows. http://www.smithfieldcommitments.com/core-reporting-areas/animal-care/on-our-farms/housing-of-pregnant-sows/. Accessed 15 June 2015
Smithfield (2007) Smithfield foods makes landmark decision regarding animal management. http://files.shareholder.com/downloads/SFD/264465172x0x173866/A793AB68-6806-4E6E-AF82-CABDF36FC791/SFD_News_2007_1_25_General.pdf. Accessed 15 June 2015
Smithfield (2013) Smithfield foods global hog production operations moving towards complete conversion to group housing. http://investors.smithfieldfoods.com/releasedetail.cfm?releaseid=731294. Accessed 15 June 2015
Sodexo Group (2013) Sodexo Group position paper: animal welfare. http://sodexousa.com/usen/Images/Sodexo-Group-Animal-Welfare-Position-Paper-Final-december2013342-797767337-843001.pdf. Accessed 15 June 2015
Soil Association (no date) Organic standards. http://www.soilassociation.org/Whatisorganic/Organicstandards. Accessed 3 June 2015
Soil Association (no date) Our work 2014. http://www.soilassociation.org/annualreview. Accessed 22 May 2015
Soil Association (no date) Who we are. http://www.soilassociation.org/aboutus/whoweare. Accessed 22 May 2015
Starbucks Coffee (no date) Animal welfare-friendly practices statement. http://globalassets.starbucks.com/assets/a228b865c3aa45938b7508cb82a17cf1.pdf. Accessed 15 June 2015
Starbucks (2014) Global responsibility report. http://globalassets.starbucks.com/assets/ea2441eb7cf647bb8ce8bb40f75e267e.pdf. Accessed 15 June 2015
Sullivan S (2013) Empowering market regulation of agricultural animal welfare through product labeling. Anim Law 19:391–422
Sustainable Agriculture Research & Education (2012) History of organic farming in the United States. http://www.sare.org/Learning-Center/Bulletins/Transitioning-to-Organic-Production/Text-Version/History-of-Organic-Farming-in-the-United-States. Accessed 15 June 2015
The Poultry Site (2013) Manitoba egg producers to phase out conventional cages. http://www.thepoultrysite.com/poultrynews/30239/manitoba-egg-producers-to-phase-out-conventional-cages/. Accessed 15 May 2015
Thorpe D (2013) Why CSR? The benefits of corporate social responsibility will move you to act. Forbes. http://www.forbes.com/sites/devinthorpe/2013/05/18/why-csr-the-benefits-of-corporate-social-responsibility-will-move-you-to-act/. Accessed 15 June 2015
Tim Hortons (no date) Animal welfare at Tim Hortons. http://www.timhortons.com/us/en/social/animal-welfare.php. Accessed 15 May 2015
United Egg Producers (2004a) About Us. http://www.unitedegg.org/default.cfm. Accessed 15 June 2015
United Egg Producers (2004b) Animal welfare http://www.unitedegg.org/AnimalWelfare/. Accessed 15 June 2015
United Egg Producers (2014) Animal husbandry guidelines for U.S. egg laying flocks.
US Department of Agriculture, AMS (no date) Official listing of approved USDA process verified programs. http://www.ams.usda.gov/AMSv1.0/getfile?dDocName=STELPRD3320450. Accessed 15 June 2015

US Department of Agriculture, AMS (2013) National organic program. http://www.ams.usda.gov/AMSv1.0/nop. Accessed 15 June 2015
US Department of Agriculture, Agriculture Marketing Service (AMS) (2014) About AMS. http://www.ams.usda.gov/AMSv1.0/ams.fetchTemplateData.do?template=TemplateD&navID=AboutAMS&topNav=AboutAMS&page=AboutAMS&acct=AMSPW. Accessed 15 June 2015
US Department of Agriculture, AMS (2015a) AMS annual purchasing summary. http://www.ams.usda.gov/AMSv1.0/getfile?dDocName=STELPRDC5099583. Accessed 15 June 2015
US Department of Agriculture, AMS (2015b) Federal purchase program specifications for animal handling and welfare. http://www.ams.usda.gov/AMSv1.0/getfile?dDocName=STELPRDC5084518. Accessed 15 June 2015
US Department of Agriculture, AMS (2015c) LPS process verified program. http://www.ams.usda.gov/AMSv1.0/processverified. Accessed 15 June 2015
US Department of Agriculture, Animal and Plant Health Inspection Service (2013) Layers 2013 Part IV: reference of organic egg production in the United States. http://www.aphis.usda.gov/animal_health/nahms/poultry/downloads/layers2013/Layers2013_dr_PartIV.pdf. Accessed 15 June 2015
US Department of Agriculture, Economic Research Service (2013) Table 3: certified organic and total U.S. acreage, selected crop and livestock, 1995–2011.
US Department of Agriculture, Food Safety and Inspection Services (2015), FOIA response 2014–0092, Good commercial practices Aug 2012-Nov 2013 final response (on file with the Animal Welfare Institute)
US Department of Agriculture, National Agricultural Statistics Service (NASS) (2015a), Livestock slaughter. http://usda.mannlib.cornell.edu/usda/nass/LiveSlau//2010s/2015/LiveSlau-01-22-2015.pdf. Accessed 15 June 2015
US Department of Agriculture, NASS (2015b) Poultry-production and value 2014 summary. http://usda.mannlib.cornell.edu/usda/nass/PoulProdVa//2010s/2015/PoulProdVa-04-30-2015.pdf. Accessed 15 June 2015
US Department of Agriculture, NASS (2015c) Poultry slaughter 2014 summary. http://usda.mannlib.cornell.edu/usda/current/PoulSlauSu/PoulSlauSu-02-25-2015.pdf. Accessed 15 June 2015
Washington Revised Code §69.25.065, §69.25.107
Watts P, Holme L (1999) Corporate social responsibility. World Business Council for Sustainable Development, Geneva
Wendy's (2014) Wendy's animal welfare program. https://www.wendys.com/en-us/about-wendys/animal-welfare-program. Accessed 15 June 2015
Winnipeg Humane Society (no date) Purchase humane food. http://www.winnipeghumanesociety.ca/humane-certified-meat-and-eggs. Accessed 18 May 2015
World Organization for Animal Health (no date) About us. http://www.oie.int/about-us/. Accessed 8 June 2015
World Organization for Animal Health (no date) Implications of private standards in international trade of animals and animal products. http://www.oie.int/international-standard-setting/implications-of-private-standards/. Accessed 8 June 2015
World Organization for Animal Health (no date) The 180 member countries. http://www.oie.int/about-us/our-members/member-countries/. Accessed 8 June 2015
World Organization for Animal Health (2010) Resolution no. 26: roles of public and private standards in animal health and animal welfare. http://rpawe.oie.int/fileadmin/doc/eng/Resolutions/78_GS_2010_Resolution_26.pdf. Accessed 8 June 2015
World Organization for Animal Health (2014) Animal welfare and beef cattle production systems. Terrestrial Animal Health Code, Chapter 7.9
World Organization for Animal Health (2015a) About us. http://www.oie.int/about-us/. Accessed 15 June 2015
World Organization for Animal Health (2015b) OIE achievements in animal welfare. http://www.oie.int/animal-welfare/animal-welfare-key-themes/. Accessed 8 June 2015

Chapter 5
Treatment of Unwanted Baby Animals

Desmond Bellamy

Abstract There is something profoundly primal in the sight and the sound of a baby that draws an emotional and empathetic response from adults, even those of a different species. The removal and slaughter of baby farmed animals must, therefore, be carried out as invisibly as possible, as most viewers would find the spectacle intolerable. Consensual selective blindness is an essential ingredient of all animal agriculture, but particularly the treatment of unwanted babies, considered "waste products". Case studies of male dairy calves and male cockerels illustrate the difficulty of promoting compassionate ethical positions in cases where animals have no commercial value.

> Nothing more strongly arouses our disgust than cannibalism. . .. And yet we ourselves make much the same appearance in the eyes of the Buddhist and the vegetarian. We consume the carcasses of creatures of like appetites, passions, and organs with ourselves; we feed on babes, though not our own; and the slaughter-house resounds daily with screams of pain and fear.—Robert Louis Stevenson[1]

5.1 Introduction: Worthless Commodities?

5.1.1 Baby Animals

There is something profoundly primal in the sight and the sound of a baby that draws an emotional and empathetic response from adults, even those of a different species. A mother deer, for example, will run protectively toward sounds of distress from baby seals, marmots and even baby humans.[2] Children will rush to see lambs,

---

[1]Stevenson (2012), Chapter XI.

[2]Holmes (2014).

D. Bellamy (✉)
PETA Australia, Byron Bay, NSW, Australia
e-mail: coloursofbyron@gmail.com

G. Steier, K.K. Patel (eds.), *International Farm Animal, Wildlife and Food Safety Law*, DOI 10.1007/978-3-319-18002-1_5

kids, calves and chicks in petting zoos. Adults will often, at the sight of a baby of any species, feel a surge of oxytocin, a peptide that stimulates both protective urges and lactation, but can also promote intergroup bias and conflict.[3] This unconscious chemical reaction is paradoxical as it incites empathy with the baby, while also promoting a primitive tribalism, which potentially separates us from the "other" people or species of which that baby may be a member.

The ultimate forms of intergroup bias interspecially are racism, sexism, etc., while intraspecially it manifests as anthropocentrism—the conviction that only humans have significant moral worth or standing. This anthropocentrism allows the objectification of non-human animals, and human ones, to the extent that they are justifiably bought, sold, exploited and killed as commodities, or exterminated as vermin or waste products. This paradox is clearest in our treatment of the most vulnerable of creatures: baby farmed animals. The removal and slaughter of these infants must be carried out as invisibly as possible, as most viewers would find the spectacle intolerable. Consensual selective blindness is an essential ingredient of all animal agriculture, but particularly the treatment of unwanted babies.

Most of the animals whose flesh is eaten by humans are little more than babies when slaughtered. "Broiler" chickens (grown for flesh rather than eggs), who make up over 90 % of the land animals killed for human consumption, are only 7 weeks old when killed. That is just 1 % of their natural lifespan of 15–20 years.[4] By then, selective breeding has created "baby birds with huge overblown bodies".[5] Even if the retail packaging states "raised cage free," or "humanely raised", their bellies will often be a mass of raw, angry, red sores from sitting in excrement, their legs being too weak to carry their huge frames.[6] In the US, up to 80 % of birds go to slaughter with bruises or fractures.[7] Pigs and lambs are typically killed at 5–7 months and veal calves stumble from their crates to the abattoir truck at 4 months.[8] Animals destined for slightly longer lives regularly undergo various agonising treatments when still babies. Boars are routinely castrated shortly after birth to avoid an unpleasant odour known as boar taint.[9] Cattle industry websites recommend that dehorning or disbudding take place before 6 weeks and anaesthetics and anti-inflammatory drugs are "highly recommended".[10] Castration is routinely done at a young age without anaesthetic.[11] This is despite the fact that anaesthetic drugs for animals are extremely cheap, "less than a dime per hundredweight".[12]

---

[3]De Dreu et al. (2011).
[4]Patterson (2002), p. 116.
[5]Davis (1998).
[6]Kristof (2014).
[7]Rollin (2008), p. 252.
[8]Patterson (2002), p. 116.
[9]The Pig Site (2012).
[10]Hanson (2014).
[11]Gandhi (2013).
[12]Maday (2014).

The abuse of the very young is even more evident where animals are brought into the agricultural process as by-products of a production process. The case studies in this paper are of cockerels (baby roosters) who are slaughtered simply because they cannot lay eggs and are therefore worthless, and baby bull calves of dairy cows (bobby calves), that are born simply to cause their mothers to lactate, and have minimal commercial value.

Carol Adams has analysed the process by which non-human animals, and many humans, particularly women, are objectified, fragmented and consumed.[13] Animals lose their identities under the technology of the abattoir, but also from their definition as "food-producing units", which neutralises any moral claim the animals may have had to consideration and justifies what Derrida called "non-criminal putting to death".[14] Once disassembled, the body parts are also redefined: as "beef", "pork" or stripped of the ability to possess their parts, so a lamb's leg will instead be called a "leg of lamb".[15] Cooking and seasoning further disguises the body parts, until finally they can be consumed as anonymous commodities. The absence of the real animal is often reinforced by the substitution of anthropomorphised cartoon animals on the packet for whom the consumer can identify "appropriate" affective feelings.[16]

5.1.2 *The Business Case for Animal Welfare*

The industry counters criticisms of legal shortcomings in animal welfare regulation by arguing that existing statutes are working fine because welfare makes good business sense.[17] Agriculture industry spokespersons regularly claim that commodification of non-human animals works in favour of welfare, in that business people will avoid damaging any part of their valuable stock commodities. Modern slaughterhouses try to maximise profits by utilising as many parts of the animal as possible. Upton Sinclair observed in his exposé of the Chicago stockyards over a century ago, "No tiniest particle or organic matter was wasted".[18] Any abuse would, by this argument, devalue the meat of the victim's body, as injured or even frightened animals lose bodily tone, taste different, or may even become unfit for consumption. For example, the industry group representing chicken meat producers argues that "growers care about their flocks", that "[h]igh standards of bird welfare make good economic sense, and ensuring that chickens are well-fed, healthy and comfortable is in the best interests of birds and growers" and that "[o]ur customers -

[13] Adams (2010), pp. 73–74.

[14] Derrida (1974), p. 278.

[15] Adams (2010), p. 74.

[16] Stewart and Cole (2009), p. 462.

[17] Ellis (2013), pp. 358–9.

[18] Sinclair (1965), p. 50.

quick service restaurants and supermarkets most prominently - have exacting standards like us and audit bird welfare systems".[19]

Ellis[20] and many others have comprehensively discredited this argument, with Goodfellow and Radan arguing that "the economic literature shows animal welfare and productivity are in conflict".[21] For the purpose of this paper, it is sufficient to note that, even if the argument of economic interest aligns with welfare was valid, it would not benefit animals who are born as waste by-products since being fragmented for consumption is not the purpose for which they were bred. Their intact bodies, therefore, have not even the potential commercial value of the animals heading for the knives.

5.1.3 The Industrialisation, Proximity and Visibility of Production

There is no longer any such thing in the world of industrial agriculture as a "chicken" or a "cow". Like modern tools, technology and monoculture plants, animals have been specifically bred or genetically modified to fill a market niche and to grow as big as fast as possible. Chickens who lay eggs are slow growing, smaller and do not grow enough muscle to cost-justify slaughtering them for human consumption, while those who are bred for eating are too young and often incapable of laying eggs. Beef cattle are bred for size and marbling of their muscles, while dairy cattle are specialised to produce abnormally large quantities of milk. Milk, eggs and meat have a dollar value; live farmed animals do not, except as potential commodities.

Siobhan O'Sullivan points out that Western animal protection movements began at a time when animals were far more visible than today. The visibility of a particular species of animal determines how much protection that animal can expect.[22] To most city dwellers, agricultural animals have become largely invisible.[23] Any abuse is emotionally challenging and so is deliberately and routinely hidden from the public gaze. Certainly, most consumers who happily buy milk or eggs or lamb chops would be horrified to see baby animals killed in front of them, as happens every minute behind high walls. This applies even more so to baby animals, who are kept from public gaze for most or all of their brief lives.

[19] Australian Chicken Meat Federation Inc. (2011).

[20] Ellis (2013), p. 363.

[21] Goodfellow and Radan (2013).

[22] O'Sullivan (2011), p. 4.

[23] Ibid., p. 60.

5.1.4 Feminised Protein Production

A major segment of the animal agriculture market comes from the exploitation of the female reproductive system. Adams suggests milk and eggs be called "feminized protein" and demonstrates that female animals are doubly exploited—for the products of their reproductive system when alive and, when spent, for their meat.[24] This is related to the subject of this chapter, baby animals, but is of course a major field of study in itself.

The point that needs to be made briefly here is that the production of dairy products and eggs does not just affect the female animals from whom they are taken, but millions of their babies who are waste products of these industries. Therefore, those who choose to be vegetarian, a term regularly substituted for the more accurate ovo-lacto vegetarian, are often causing more suffering than the occasional meat-eater, in that they replace meat in their diet with large quantities of eggs and dairy. The abolitionist philosopher Gary Francione states, "there is probably more suffering in a glass of milk than in a pound of steak".[25]

5.2 Legal and Financial Frameworks

5.2.1 The Limitations of Animal Law

Legally, non-human animals are primarily considered property[26] and penalties, therefore, tend to protect owners rather than victims. Anticruelty legislation is increasingly common, but there is still none in Iran, Lebanon, Iraq, Somalia or Ethiopia.[27] China has minimal wildlife and property protection;[28] anti-cruelty legislation was introduced in 2009 but is yet to be enacted.[29] China is the world's largest farming nation and is increasingly taking on Western factory farming practices such as gestation crates, battery cages and beak-trimming which are being phased out elsewhere.[30] While moves are underway to remedy this, at publication time the legislation introduced in 2009[31] has still not been enacted due to industry opposition.[32] In fact, animal welfare has only just been incorporated

---

[24]Adams (2010), p. 21.

[25]Francione (2008a).

[26]Wagman and Liebman (2011), p. 26.

[27]Ibid., p. 28.

[28]Ibid., p. 30.

[29]Wedderburn (2009).

[30]Tobias (2012).

[31]Wedderburn (2009).

[32]Personal communication from Hailey Chang, PETA Asia.

into the curriculum for veterinary training in China.[33] In the US, the Humane Society observed in a recent report on cruelty that many states "specifically exclude livestock or any 'common' agricultural practices from their cruelty laws, and even when good laws exist, it can sometimes be difficult to convince law enforcement to make an arrest".[34]

In many other agricultural countries, application of laws intended to protect animals from cruelty are often delegated to police or charitable organisations, who are limited both by lack of resources and legal restrictions on entering private property.[35] Even where abuse can be proven, anticruelty laws are widely overwritten by exceptions for animals used for food, experimentation or some forms of entertainment such as hunting.[36]

Laws combating cruelty were first mooted in the UK in 1796 and, 200 years later, it was estimated that there were some 3500 provisions relating to animals in UK legislation.[37] However, animal protection legislation, where it exists, tends to be inconsistent and full of loopholes. In the Australian state of New South Wales, for example, the Prevention of Cruelty to Animals Act (1979) states that (9.1) confined animals must be provided adequate exercise. However, the next section (9.1A) states that this does not apply to a "stock animal other than a horse" or "an animal of a species which is usually kept in captivity by means of a cage", thus excluding almost all confined animals: those used by the agricultural industry.[38] Similarly, several defences are available (Section 24) including allowances for branding, castration, dehorning, tailing or mulesing (the removal of layers of skin from the backside of lambs to cause scarring against flystrike), as well as hunting or preparing animals for human consumption, as long as "no unnecessary pain" is inflicted.

5.2.2 *Codes of Conduct*

Since the overwhelming majority of the animals that interact with humans are farm animals,[39] the many and varied exceptions built into legislation or presented in codes of practice that overwrite legislation make laws regarding animal welfare largely ineffective. In Australia, for example, Model Codes of Practice and Welfare Standards and Guidelines vary between states but, in most jurisdictions, complying with a code is accepted as a defence against prosecution under the animal welfare

[33]Wang (2014).

[34]Humane Society US (2014).

[35]Sunstein and Nussbaum (2004), p. 210.

[36]Ibid., p. 6.

[37]Radford (2001), pp. 3–4.

[38]NSW *Prevention of Cruelty to Animals Act* (1979).

[39]Norwood and Lusk (2011), p. 4.

legislation.[40] As the codes are written with significant input from the animal industries, it is not surprising that the codes are far less rigorous than even the vague anticruelty legislation passed by parliaments. The codes are usually voluntary and are widely ignored.

5.2.3 Is Welfare Reform Through the Law Even Possible?

Sharman claims in her paper on farm animal welfare in Australia[41] that no major farm animal reforms have yet been achieved. This raises the question of how farm animal reforms can be contested and won through legal reform alone. If welfare reforms are to be successful, they can only be driven by supply or demand. Concentrating on supply requires legislation to manage the industry's activities through compliance (voluntary guidelines) or deterrence (punitive regulations). In most jurisdictions, welfare regulation reflects both approaches, but tends overwhelmingly to the compliance model.[42] The demand method attempts to persuade consumers to insist on welfare improvements or boycott products seen as produced by cruel methods. At one end of this spectrum is the Australian RSPCA's "paw of approval" seal, which attempts to identify 'humanely' processed animal products; "If you see the RSPCA logo on a carton of eggs, packet of pork, chicken or turkey, you can be assured that animals involved in the production of these products were raised under high animal welfare standards."[43]

At the other end of the demand spectrum are abolitionists like Gary Francione, who reject all reforms as counter-productive tokenism and advocate boycotts of all animal exploitation, and a vegan world.[44] Between those positions are the major animal activist organisations, such as the Humane Society, Animals Australia (AA), Animal Liberation and PETA which lobby for more humane farming methods while simultaneously encouraging supporters to forego all animal products. This paper looks at both aspects of the campaign for welfare reform through the lens of two issues: the slaughter of about 700,000 male dairy calves annually in Australia as 'waste products', the commercially worthless by-products of keeping milking heifers pregnant in order to maintain high levels of milk production[45] and the debate over the slaughter of baby cockerels in the European Union.

---

[40] Dale and White (2013), pp. 155–156.

[41] Sharman (2013), p. 81.

[42] Goodfellow (2013), p. 186.

[43] RSPCA.

[44] Francione (2008a).

[45] Donovan (2008), p. 250.

5.2.4 Animal Law Versus Agricultural Profits

Improving the lot of animals, whether by liberating them or resizing their pens, requires significant effort or sacrifice by humans, rewarded only by clear consciences (since the animals cannot offer any reciprocal material trade-off). Ellis[46] states that improving animal welfare makes humans more comfortable with continuing exploitation by offering a pretence of protection. Powerful forces back this pretence—the cultural attachment to animal products (meat, dairy and eggs) and the huge profits generated from developing and fulfilling these desires.

Voiceless, the Australian "animal protection institute", points out that the movement promoting animal law reform is growing rapidly, as evidenced by nine Australian universities now offering animal law courses.[47] Their patron, JM Coetzee, adds that industry has a huge advantage in resourcing and access to government but "it is impossible to believe that, in the end, justice and compassion will not triumph."[48]

Others are less sanguine. Animal regulations are overwhelmingly updated and overseen by vested interests, and the regulations are often incoherent, fragmented in authority and rife with conflicts of interest.[49]

5.3 Global Statistics of Farmed Animals

The statistics below are taken from the United Nations Food and Agriculture Organisation (FAO) and industry sources. The vast numbers of animals killed are probably conservative estimates in that they ignore small abattoirs and home slaughter, and in any case include only land animals, ignoring the billions of sea creatures who die of suffocation or are killed in other ways each year.

There were 1.5 billion cattle on the planet in 2013.[50] Brazil was the biggest holder of cattle overall with 217 million, just topping India at 214 million. China held 113 million and the US, 89 million. Australia had 29 million, a number well ahead of its human population.

FAO figures indicate a world population of 260 million dairy cows in 2011.[51] The country with the largest population was India with almost 45 million, followed by Brazil with 23 million, China with 12 million and Ethiopia, Pakistan and the US close behind with around 10 million each.

[46]Ellis (2013), p. 345.

[47]Voiceless (2009).

[48]Ibid.

[49]Ellis (2013), p. 353.

[50]FAOSTAT (2013).

[51]Dairyco (2014).

The world population of pigs in 2013 was 977 million,[52] with 482 million of them in China. Next was the US with 65 million and Brazil with 39 million.

The planet held 1.17 billion sheep in 2013,[53] with 185 million in China, 75 million in Australia, 75 million in India and 52 million in Sudan.

There are almost 22 billion chickens in the world[54] at any one time: 5.5 billion in China, 1.9 billion in the US, 1.8 billion in Indonesia and 1.3 billion in Brazil. As mentioned, broiler chickens are slaughtered at about 7 weeks, so while there are "only" 22 billion alive at any one time, at least 50 billion are slaughtered each year.[55] Egg laying hens live until their egg production declines and make up some 7 billion of the total population.[56]

5.4 Worthless Animals: Case Studies

Most animals, as observed above, are still babies or infants when killed. Waste product animals, those who are simply by-products of feminised protein production and have minimal or no financial value in themselves, are robbed of even that brief life. Male and many female dairy calves are typically removed from their mother within 12 h of birth and shipped to slaughter or to veal production facilities at around 5 days old. The case has been argued in Australia that a calf should be at least 8 days old before the arduous transport to the slaughterhouse but these claims have been rejected by industry on cost grounds. Baby chicks of layer hens are separated into male and female and the males ground up, gassed or suffocated within hours of hatching.

5.4.1 The Myth of the Happy Cow

Cows are represented in children's literature as creatures that naturally give milk. In fact, dairy cows are similar to humans: mammals, with a comparable gestation period, who only lactate after giving birth. Most cows never see a bull. Semen is collected artificially and sold to dairy farmers for artificial insemination.[57] Selectively bred to produce much larger quantities of milk than their wild ancestors, cows lactate for up to a year, during which time they will be artificially impregnated so that they are almost always lactating from the previous birth and pregnant for the

[52]FAOSTAT (2013).

[53]Ibid.

[54]Ibid.

[55]United Poultry Concerns.

[56]The Poultry Site (2015).

[57]Galloway (2014).

next. The side effect of this cycle is, of course, a calf, that half the time is male and becomes a rival for his mother's milk—the farmer's commercial product. Bull calves are a financial drain to the dairy industry, requiring feeding and transport to the slaughterhouse. The cuts of beef from their bodies, if they are allowed to reach maturity, are lower yield and worth less than those of cattle bred for beef. The cost of rearing, transport and slaughter are often less than the meat's value, particularly when the market is depressed.[58]

Bull calves are removed from the mother at about 12 h old and the first milk, which contains colostrum, is taken from the mother and manually delivered by the farmer, who may choose to give most of it to the females whom he wishes to keep for future herd replacement.[59] Since farmers are squeamish about killing babies on the farm,[60] they are fed watered milk or substitutes for a few days, and then either castrated for use as beef cattle, fattened and processed into veal or shipped to the abattoir at about 5 days old to be processed into dog food, pharmaceuticals and calf leather.[61] The cost of transport is often more than the value of the new-borns, and in Australia and New Zealand they were often sold for one shilling or 'a bob'[62]; thus the name "bobby calves". The stress on newly born and weaned animals is severe—they are placed in a truck, often handled roughly by men facing deadlines, sick from the motion of the vehicle and desperately hungry and thirsty after lengthy transportation.[63] Pollan, despite promoting omnivorism, states that weaning is the most stressful time for farmers and animals: "the cows will mope and bellow for days and the calves, stressed by change of circumstance and diet, are prone to get sick."[64] Claims are often made that activists indulge in sentimental anthropomorphism, but studies comparing calves separated abruptly to those who can see their mothers show significant physiological and behavioural differences.[65] Unless we are Cartesians, the distress of the cow and calf are self-evident.

None of this is illegal. The Australian industry body, Dairy Australia, posts interviews with farmers, transporters and functionaries of abattoirs stating that all welfare regulations are followed.[66] The process of removing a baby from his or her mother on the day of birth is claimed to "minimise risk of disease transfer … and lower the stress for both cow and calf".[67] Norwood and Lusk point out that if babies are to be separated from their mothers, doing it once is probably less traumatic than

---

[58]The Herald Scotland (2014).

[59]Norwood and Lusk (2011), pp. 142–143.

[60]Donovan (2008), p. 250; Humphreys (2013).

[61]Humphreys (2013).

[62]Bardsley (2012).

[63]Singer (2009), p. 17.

[64]Pollan (2006), p. 71.

[65]Price et al. (2003), p. 116.

[66]Dairy Australia, *Care of bobby calves on-farm.*

[67]Dairy Australia, *Care of bobby calves on-farm.*

doing it twice a day before milking.[68] Farmers claim that this is part of giving their animals the best of care because they "wouldn't make any money if they mistreated them".[69]

Animal advocates disagree, pointing to the inexorable trend of increasing productivity at the expense of welfare, more so in the dairy industry than in ungulate meat production. A cow could live to the age of twenty,[70] during which she would normally interact with her calf for 9–12 months. Instead, she is kept lactating and pregnant almost continuously, with the aid of what even farmers call a 'rape rack' and is 'spent' and sent for slaughter at 5–7 years.[71] Added to that is the distress of having each calf taken from her within hours of birth. Cows "will bellow for days, pace the spot where they gave birth, and stop eating. Then they'll produce a season's worth of milk and be led straight back to the rape rack."[72] The calf, meanwhile, will be raised as a new milk machine or, if male, transported for butchering.

The dairy industry maintains that it is committed to ensuring that "our calves are provided with a safe, healthy environment for the whole of their lives."[73] As the "whole of their lives" will be under a week for at least half of the calves born,[74] the focus of campaigns that have gone toward improving this 'safe, healthy environment' are very narrow compared to other campaigns that may follow several stages of a particular animal's life and treatment.

5.4.2 International Context

Although this case study specifically examines Australian law and codes, the situation is not radically different in other developed nations. In the USA, federal laws regarding cattle are mostly concerned with transportation and slaughter. Individual states tend to allow the industry to set its own standards, and do not deal with animal cruelty except in cases that fall outside "accepted industry practice."[75] A typical case is the Maryland Criminal Law, which states that cruelty laws do not apply to "customary and normal veterinary and agricultural husbandry practices including dehorning, castration, tail docking, and limit feeding."[76] Use of anaesthesia for such clearly painful practices is uncommon due to the cost.[77] Male

68Norwood and Lusk (2011), p. 145.
69Dairy Mom blogspot.
70Mendelson (2012), p. 131.
71McWilliams (2013).
72Ibid.
73Dairy Australia, *Australian dairy farmers are committed to caring for their calves.*
74Humphreys (2013).
75Turk (2007).
76Ibid.
77Ibid.

dairy calves in the US are mostly harvested for beef, with about 30 % killed at a young age for veal.[78] Although the individual tethering stalls that gave veal such a bad name among consumers have mostly been replaced with larger individual stalls, these are still too small to allow the calf to turn around.[79]

In the UK, calves are considered unfit for transport if they have an unhealed navel, under the welfare of animals (transport) order 1997.[80] Prior to 1997, the UK exported over 400,000 calves to the Continent each year for veal. This market collapsed with the confirmation of BSE (Mad cow disease).[81] Subsequent market recoveries were short-lived due to foot and mouth outbreaks in 2001 and bovine tuberculosis in 2005. The result is that thousands of male calves are now regularly shot at birth.[82] In Scotland, lethal injection is apparently preferred,[83] indicating that the meagre flesh on the animals is not worth leaving uncontaminated.

In the EU, about 6 million calves are slaughtered each year for veal, about half in France and The Netherlands.[84] The EU "slaughter premium system" and differing veal prices within the EU often result in young animals having to undergo lengthy transport to abattoirs.[85] The EU legislated to ban veal crates in 1997 but standards relating to feeding still need improvement.[86]

Canada has recently banned veal crates following a widely publicised investigation of routine cruelty by animal activist organisation Mercy for Animals. In a rare case of animal protection laws being used in a farm situation, an ex-employee of a major veal supplier, Delimax, was charged with two counts of cruelty following release of a video in which workers were seen kicking, punching, and beating baby calves chained inside crates so narrow they couldn't turn around or even lie down comfortably.[87]

While New Zealand is far smaller in geography and population than Australia, the dairy industry is a very large component of its GDP and therefore, any discussion of animal welfare is always assessed against economic barriers. While Australia slaughters 700,000 bobby calves per annum, New Zealand more than doubles that number, peaking at over 2 million in the year to September 2014.[88] As the value of these newborns is less than the cost of transporting them to slaughter, many are killed on the farm, leading the government's National Animal Welfare

---

[78]Norwood and Lusk (2011), p. 141.

[79]Ibid., pp. 141–142.

[80]DEFRA.

[81]The Dairy Site (2010).

[82]Ibid.

[83]The Herald Scotland (2014).

[84]Eurogroup for animals (2010), p. 31.

[85]Ibid.

[86]Ibid.

[87]Mercy For Animals (2014).

[88]Fox (2014).

Advisory Committee to recommend laws be amended to prevent beating the babies to death.[89]

5.5 The Industry in Australia and Codes of Practice

Australian Bureau of Statistics figures indicate that 689,500 calves were slaughtered in the 12 months to September 2014.[90] Milk production is centred in the state of Victoria, which produces about 66 % of the national total of 9.2 billion litres, up 40 % from 5.4 billion in 1979–80.[91] Of the 1.69 million dairy cows in Australia in 2013–14, 1.11 million were in Victoria. Interestingly, the number of dairy farms has fallen from 22,000 in 1979 to 6,300 today while the total herd has fallen by about 10 %, indicating a trend to large-scale industrial farming of high-yielding dairy cattle, replacing the family farmer and "contented cows" that the industry likes to feature in its advertising. [92]

Legal reform in Australia is complicated by the fact that most relevant legislation is determined at State level.[93] State and Territory laws are general-purpose criminal laws that apply to all animals, but then exclude farmed animals by classifying them as 'stock'.[94] This disambiguation effectively makes farmed animals 'disappear from the law'.[95] In the 1970s, animal welfare reform overseas, particularly in Europe, gave rise to fears that Australia would be perceived as outdated in its treatment of farmed animals, which was only perceived as a problem in that it might affect trade.[96] Codes of practice were developed at this time to provide guidance to farmers. As these were developed, they were, nominally at least, based on the Farm Animal Welfare Council's "Five Freedoms." These were freedom:

1. From hunger and thirst
2. From discomfort
3. From pain, injury and disease
4. To express normal behaviour
5. From fear and distress[97]

[89]Cronshaw (2014).

[90]Australian Bureau of Statistics (2014).

[91]Dairy Australia, *Latest production and sales statistics.*

[92]Dairy Australia, *Cows and farms.*

[93]E.g. NSW Prevention of Cruelty to Animals Act (1979).

[94]Sharman (2013), pp. 75–6.

[95]Wolfson and Sullivan (2004), p. 206.

[96]Dale and White (2013), pp. 152.

[97]Ibid.

Most of the States and Territories adopted Codes of Practices in the 1980s based on the Model Codes developed by the Primary Industry Ministerial Council (PIMC).[98] This left the decision on which Codes to implement to individual States and Territories, which failed to institute consistent and uniform adoption of even minimum standards. Accordingly, the Federal government from 2005 started developing the Australian Animal Welfare Strategy (AAWS), which aimed to create "a more consistent and effective animal welfare system"[99] which would add enforceable standards as well as voluntary guidelines.

5.6 The Issue: Time-Off-Feed

The first standard developed was "The Australian Standards and Guidelines for the Welfare of Animals: Land Transport of Livestock"[100] in 2008, endorsed by the PIMC in 2009. The draft standards proposed that bobby calves aged 5–30 days and travelling without their mothers should be transported "in less than 18 hours from last feed with no more than 12 hours spent on transports."[101] As delivery was often followed by an overnight stay and then the commencement of slaughter, the sticking point of the proposal became the acceptability of leaving these new-borns hungry for 30 h from last feed to slaughter. This seems to fly in the face of all of the "Five Freedoms." Public submissions were invited by May 2008[102] and the 16 submissions that mentioned bobby calves[103] resulted in a number of "unresolved issues" which mostly revolved around welfare versus cost.[104] A study of transport mortality from 1998 to 2000 in Victoria showed that 1430 calves (0.64 %) died on 1376 consignments, out of a total 220,519 sent to abattoirs.[105] Extensive studies of bobby calf transportation indicate that their welfare "may be seriously compromised."[106] Both the advocates of animal welfare and the meat processing industries called for the minimum age for transport to be raised from 5 to 8 days, but producers argued that this would cause "significant extra cost". The 18 h time-off-feed (TOF) was argued back and forth, and the difficulties in assessing calf welfare as well as proving their ages were considered. The decision was to make no

[98]Ibid, p. 153.

[99]Australian animal welfare strategy.

[100]Ellis (2010), p. 32.

[101]Primary Industries Ministerial Council (2011).

[102]Australian Animal Welfare Standards and Guidelines, *Land Transport Consultative Process* (2013).

[103]ENVision Environment Consulting (2011).

[104]Primary Industries Ministerial Council (2011).

[105]Cave et al. (2005).

[106]Hemsworth et al. (1995), p. 167.

decision, but to examine ways to improve calf welfare and revise the standards "over the next two years."[107]

Animal Health Australia (AHA) was charged with preparing a "science-based standard" for maximum allowable TOF. Public consultation was invited by AHA on their preferred option: "a maximum of 30 hours without a liquid feed from the time of last feeding to the next feed or slaughter of the calf."[108] Submissions were invited from 4 January to 3 February 2012, and resulted in 6000 email submissions plus 33 detailed written submissions from industry and welfare organisations and government departments.[109] The flood of emails came largely through a sustained campaign by Animals Australia (AA) and the RSPCA. AA placed quarter-page ads in newspapers around Australia headed "Do you want to know a secret?"[110] Much publicity was engendered, such as an article in *The Australian* that called the issue Dairy's "dark secret".[111]

The AHA reported that the bobby calf issue was emotive (implying unreasonable or impractical) and that, while the bulk of email submissions called for shorter TOF or questioned the need for transport at all, there was not unanimous support for a shorter TOF option whereas there was "good support for a 30 hours TOF limit from some government and all industry respondents."[112] As a result of this consultation, and despite the 6000 email submissions, AHA concluded "the 30 hours TOF option [be] recommended for government endorsement." The rationale was largely that this standard would set a mandatory maximum TOF, whereas previously there had been a divergent set of model codes, which were applied at law "at best as guidance or a defence to a prosecution."[113] Dale and White bluntly state that this was an example of science being "commissioned to support a pre-determined standard."[114] The report admitted that the science used had several shortcomings in terms of the climatic period chosen, the methods used to test stress and the fact that the report was commissioned by the Dairy Industry.[115] Although a consensus was as far away as ever, the AWS website confirmed that industry had agreed to implement the 30 h TOF, with a possible further review mooted for 2014.[116] Ellis observes that this is "unsurprising" as this was the industry's preferred position.[117] Dale and White point out that, since industry controls the funding for research

---

[107] de Witte (2009), p. 155.

[108] Australian Animal Welfare Standards and Guidelines, *Appendix 3* (2011).

[109] Ibid.

[110] Animals Australia, *Do you want to know a secret?*

[111] Neales (2011).

[112] Australian Animal Welfare Standards and Guidelines (2011), n 40, 2.

[113] Ibid, p. 5.

[114] Dale and White (2013), p. 175.

[115] Australian Animal Welfare Standards and Guidelines, *Appendix 3* (2011), pp. 9–10.

[116] Australian Animal Welfare Standards and Guidelines, *Bobby Calf Time Off Feed Standard.*

[117] Ellis (2013), p. 347.

(including the matched government funding), it is quite likely that researchers will often conclude that the status quo does not damage animal welfare.[118]

The proposed 2014 review did not happen. The Federal government elected in September 2013 scrapped the AAWS Advisory Committee, which oversaw the development of the strategy, in November,[119] then cut funding from the Strategy itself in its Mid-Year Economic and Fiscal Outlook statement in December 2013.[120] This effectively cements the limbo status quo of divergent and largely unenforceable State and Territory standards and guidelines. The decision by the Federal government to dissolve the AAWS and cut funding to Animals Australia, which had been recompensed for its participation in the Committee, has delighted supporters of the animal industries.[121]

5.7 Case Study: Baby Cockerels in the European Union

There are more than 5 billion egg-laying hens around the globe, at least 3 billion of which are confined to battery cages.[122] Conditions in such cages are dire: stretching a wing is impossible, and part of their beaks is sliced off to stop the aggressive behaviour caused by such overcrowding. Unable to build a nest or even scratch in dirt, each hen will lay about 250 eggs each year for up to 2 years, after which, despite a potential life span of 20 years,[123] the pressure on her body will leave her "spent". She will then be jammed into a crate and taken to slaughter, the rough treatment probably causing several of her weakened bones to break. Shackled upside down, she will be run past an automated knife that will cut her throat unless she jerks out of its way and is plunged, fully conscious, into the scalding de-feathering tanks.[124]

Most existing legal or regulatory changes around the world concentrate on the fate of the hen, the size of the cage or the health implications for humans of various modes of production, but few consider the newly hatched and economically valueless baby chick. For example, the UK has its own laws which cover poultry and require registration of any premises with over 50 birds, and an environmental permit for premises which carry over 40,000 birds. However, the Eggs and Chicks (England) Regulations 2009 only cover the EU marketing standards, mainly related to human health concerns from Salmonella and other infections.[125] New Zealand

[118]Dale and White (2013), p. 176.
[119]Vidot (2013a).
[120]Vidot (2013b).
[121]Bettles (2013).
[122]Wagman and Liebman (2011), p. 68.
[123]Back Yard Chickens (2008).
[124]DeGrazia (2012), p. 219.
[125]Gov. UK (2012).

has instituted Minimum standard no. 17—humane destruction that calls for "very rapid and complete fragmentation of the egg or day-old chick into small particles" for unhatched eggs or chicks up to a day old.[126]

In the US, agricultural policy is set by the Department of Agriculture, an organisation set up in 1862 to support farmers. The USDA also dispenses nutritional advice to the public about the health benefits of animal protein despite clear conflicts of interest in that its committees are made up largely of representatives of the meat, dairy and egg industries.[127] About 95 % of eggs sold in the US still come from caged hens.[128] Since the US Humane Slaughter Act does not apply to poultry, most hens will be fully conscious throughout slaughter.[129] In September 2014, there were an estimated 301 million laying hens in the US. Laying rate was 77.7 per 100 hens, so they laid almost 234 million eggs every day. 63 companies control over 1 million hens, and these companies represent 87 % of total production. 99 % of hens exist under the control of 172 companies with flocks larger than 75,000 hens. 16 of these companies have flocks of over 5 million birds.[130] One company, Cal-Maine Foods, has 29 million hens in production. Seven of the ten biggest egg-producing companies in the world are in the US.[131]

In 2008, California voters passed Proposition 2 by a large majority, which banned battery cages from 2015. Effectively, this will mean that most hens in California will no longer be cramped in cages allowing 67 square inches per bird, which is 30 % smaller than a sheet of letter paper, and instead will enjoy a sumptuous 116 square inch "colony cage" that allows each hen 25 % more room than that same sheet of paper. Other farmers will move to barn laid eggs.[132] As well as voting, consumers have made their opinions felt to food manufacturers, many of which are moving to use only cage-free eggs in their products.[133] Abolitionists such as Francione opposed Proposition 2 at the time, arguing that the benefits to animals were minimal, and such reforms were easily by-passed and just served to make consumers feel better about continuing exploitation. Francione's recommendation was that those organisations supporting Proposition 2 and similar legal and political campaigns should instead use their funds to promote veganism.[134]

Some 250 million male chicks are euthanized each year in the US, along with defective or slow-hatching females. They may be ground up alive, gassed, electrocuted or simply dumped in rubbish bags to suffocate or be crushed.[135] There are no

[126]National Animal Welfare Advisory Committee (2012), p. 31.
[127]Marcus (2012), p. 346.
[128]Friedrich (2013).
[129]DeGrazia (2012), p. 219.
[130]American Egg Board (2014).
[131]World Poultry (2012).
[132]LA Times (2014).
[133]Ibid.
[134]Francione (2008b).
[135]United Poultry Concerns.

federal laws guaranteeing humane euthanasia in hatcheries,[136] only guidelines from the American Veterinary Medical Association, which states that maceration and gassing with high concentration of CO_2 is acceptable, but smothering in bags and containers is not.[137]

5.8 Media Exposés

In 2009, animal rights group Mercy For Animals (MFA) posted graphic footage of chicks at the Hy-Line Hatchery in Iowa being carried by conveyor belt: the males to a macerator or giant mincing machine which ground them up alive and the females to a de-beaking machine.[138] MFA called on Walmart and the 50 largest grocery chains to require all eggs sold in their stores to have a label reading "Warning: Male chicks are ground-up alive by the egg industry."[139]

The video was widely covered by media, including CBC News and Associated Press. The industry body United Egg Producers called the MFA demand "almost a joke", noting that MFA was calling for the end of egg consumption altogether,[140] suggesting that therefore their motivation somehow called the evidence into question. The company involved, Hy-Line, announced an investigation into the video, particularly the issue of chicks falling off the belt and dying on the floor, but pointed out that "instant euthanasia" is standard procedure supported by the veterinary and scientific community.[141] A spokesperson for United Egg Producers made the situation very clear when he said, "There is, unfortunately, no way to breed eggs that only produce female hens. If someone has a need for 200 million male chicks, we're happy to provide them to anyone who wants them. But we can find no market, no need."[142]

An MFA spokesperson stated that "most people would be shocked to learn that 200 million chicks are killed a year"[143] just because they have no commercial value. From anecdotal evidence, this appears to be quite accurate: people whom I have told about the fate of male chicks are almost without exception shocked and surprised, although logically it is obvious that this must be the case. Like the unexamined belief that cows somehow produce milk with no real causal connection to a baby, most people assume chickens lay eggs, without considering the situation of those who can't. Similar exposes have appeared all over the world, for example,

[136]CBC News (2009).
[137]AVMA (2013), pp. 100–102.
[138]Mercy For Animals (2009).
[139]Ibid.
[140]CBC News (2009).
[141]Ibid.
[142]Ibid.
[143]Ibid.

in the UK[144] and India, where male baby chicks are often killed by suffocation in plastic bags.[145]

5.9 The Egg Industry in Germany and the European Union

Europe prides itself on having a leading role in animal welfare reform, based on the principle of accepting the sentience of animals. In 2009, the Treaty of Lisbon amended the EU Treaty into the Treaty on the Functioning of the European Union (TFEU). It added an Article 13 to an impressive list of other articles, which covered human rights, discrimination and sustainable development. This stated:

> In formulating and implementing the Union's agriculture, fisheries, transport, internal market, research and technological development and space policies, the Union and the Member States shall, since animals are sentient beings, pay full regard to the welfare requirements of animals, while respecting the legislative or administrative provisions and customs of the Member States relating in particular to religious rites, cultural traditions and regional heritage.[146]

The Regulations spelling out the permitted procedures were enacted in 2009 to go into force January 1, 2013.[147] These regulations have led to an interesting test case in Germany where the slaughter of baby chicks has been banned in the state of North Rhine-Westphalia, based on an interpretation of Article 13 and the Regulations. If successful, this would likely lead to the necessity to raise the chicks to an age where they could be slaughtered for meat production.

Overall, an estimated 330 million chicks are killed within 3 days of hatching in Europe.[148] EU regulations (chapter 2 part 2) call for the "instantaneous maceration and immediate death of the animals"[149] as a minimum standard, although it does not rule out other methods of killing such as cervical dislocation[150] where no other methods are available for stunning. However, section 20 of the regulations states that "many killing methods are painful for animals. Stunning is therefore necessary to induce a lack of consciousness and sensibility before, or at the same time as, the animals are killed."[151] Section 21 defines stunning in terms of loss of consciousness, which relates to the animal's "ability to feel emotions and control its voluntary

[144]Poulter (2010).

[145]Saraswathy (2014).

[146]European Union. *The EU and animal welfare: policy objectives.*

[147]European Union. Council regulation (EC) No 1099/2009.

[148]Buhl (2013), p. 3.

[149]European Union. Council regulation (EC) No 1099/2009, p. L303/23.

[150]Ibid.

[151]European Union. Council regulation (EC) No 1099/2009, p. L303/3.

mobility."[152] It is clear from the videos cited above that the chicks being dropped into the maceration machine are fully conscious and mobile. Stunning is defined in the regulations as "any intentionally induced process which causes loss of consciousness and sensibility without pain, including any process resulting in instantaneous death"[153] and maceration is actually listed as a "stunning method" in Table 1[154] which describes it as "the immediate crushing of the entire animal" and states that it applies to chicks under 72 h old and egg embryos.

However, Article 4 "stunning methods" clearly states that, "animals shall only be killed after stunning.... The loss of consciousness and sensibility shall be maintained until the death of the animal."[155] Clearly, the chicks are very much conscious at maceration, and it could therefore be argued that mandatory stunning has not taken place before slaughter. This is the understanding of the Internal Directive of the Ministry of Climate Protection, Environment, Agriculture, Nature and Consumer Protection of the state of North Rhine-Westphalia in Germany, which banned the shredding of male chicks under the as yet unpublished Directive VI-5 4201-722 (viewed and analyzed by German lawyer Amelie Buhl).[156] This is a directive to subordinate authorities to issue ordinances banning the practice of shredding chicks within their jurisdictions. The Directive quotes the German Animal Welfare Act section 17, which states that there is a penalty of up to 3 years' imprisonment for "killing of a vertebrate without good reason."[157] Buhl follows the line of reasoning under which the Münster public prosecutor's office determined that, although the long-term practice of maceration was exculpatory in a current prosecution, the previous tolerance of the act did not justify its continuance into the future.[158] Although expecting the ruling to be challenged, the Environment Ministry expects that farmers will cease maceration and instead feed the chicks until they are big enough for slaughter for human consumption.[159] This will take approximately 17 weeks instead of the 5 weeks for specially bred 'broiler' chickens[160] but the interpretation of the Animal Welfare Act is that killing simply for economic reasons is not a "good reason". In terms of European Union law, member states are expected to abide by the EU regulations, but under Section 26, they may legislate "more extensive protection of animals at the time of killing".[161] On that basis, there is no reason why a German state or a nation cannot extend the protection of animals under the umbrella of EU provisions, particularly if that is to

[152]Ibid.

[153]European Union. Council regulation (EC) No 1099/2009, p. L303/8.

[154]European Union. Council regulation (EC) No 1099/2009, p. L303/19.

[155]European Union. Council regulation (EC) No 1099/2009, p. L303/9.

[156]Buhl (2013), p. 1.

[157]Animal Welfare Act (2006), section 17.

[158]Buhl (2013), p. 2.

[159]Sewell (2013).

[160]Buhl (2013), p. 7.

[161]European Union. Council regulation (EC) No 1099/2009, p. L303/17.

make it consistent with national legislation. Buhl points out that the actions of the state of North Rhine-Westphalia will attract attention nationally and internationally, and hopes that the federal government will take the opportunity to issue similar regulations on a national basis.[162]

5.10 Innovations

Several new procedures have been discussed and trialled in the animal agriculture industry that may reduce some of the worst aspects of baby animal welfare abuses. While some are mere window dressing and others seem impractical, it is possible that some scientific or political breakthroughs may reduce the number of unprofitable animals born, thereby somewhat ameliorate the situation in which these newborns find themselves.

5.10.1 Free Range

As stated earlier, most animals whose flesh is used for human consumption are barely more than babies when slaughtered. Relevant, although outside the scope of this paper, is the general issue of improving the welfare of those animals who are destined to reach the preferred age of slaughter. This requires at minimum some workable regulations that ensure control over the mechanisation of birth, feeding, transportation, death, and preferably a break from the mechanisation of factory farming and the return to some sort of "free range" farming where concerns other than profit determine treatment. Even Michael Pollan, who is far from an animal rights or vegan advocate, states that:

> All it would take to clarify our feelings about eating meat, and... begin to redeem animal agriculture, would be to simply pass a law requiring all the sheet-metal walls of all the CAFOs, and even the concrete walls of the slaughterhouses, to be replaced with glass... we would not long continue to raise, kill and eat animals the way we do.[163]

5.10.2 Sexed Semen

In the dairy and egg industries, males are unfortunate by-products, waste materials that must be disposed of as quickly as possible. The holy grail of the dairy industry, where the cost of producing a heifer calf is far greater than that of producing a

[162]Buhl (2013), p. 7.

[163]Pollan (2006), pp. 332–333.

chick, is the ability to sort bull semen so that cows can be inseminated with only female sperm to produce female (heifer) calves. Sexed semen has been under development for some time and is readily available but still considered "unprofitable."[164] Fluorescence activated cell sorting is widely practiced, particularly in the US and Canada, but the time involved in sorting and the reduced conception rates from the resulting "straws"[165] all result in higher costs.

A 2010 survey of farmers in Victoria, Australia, showed that the cost per heifer calf born was more than double when using sexed semen rather than "conventional" artificial insemination: $160 versus $70.[166] The semen straws were $60 compared to $20 for conventional straws. Only 14 % of farmers surveyed were using sexed semen, quoting cost, low conception rates and doubts about the genetic merits of the bulls.

There is much speculation in the dairy and beef industries about the potential for sexed semen. If only heifers are born, the market may become saturated and prices collapse, leading farmers to restrict sexed breeding to their best milk producers and to inseminate the other cows with beef bull semen, to produce calves that grow bigger. This will still result in a calf and therefore lactation, but the calf will be larger and more suitable for rearing for meat.[167] The other side of this equation is that larger calves may result in a harder birth and may lead to long recovery time or even culling of the mother.[168]

Although widespread adoption of sexed semen would reduce the number of bobby calves born and therefore sent for slaughter or to veal facilities, it appears that economic considerations will keep it of limited use in the near future at least.

5.10.3 Termination Before Hatching

New technology may be on its way that would enable male chick embryos to be destroyed while still in the shell. This would eliminate the maceration or gassing of newly hatched chicks. Unilever, the world's third largest consumer products company, has announced it is shifting its production to the use of cage-free eggs and egg substitutes, and is now looking at commercially viable technology to determine the sex of embryos in the shell.[169] This process will apparently use "gel-based" technology to achieve ovo-gender identification (sexing) of eggs, the males of which will then be macerated before hatching.[170]

---

[164]De Vries.

[165]Weigel (2003), p. E120.

[166]McMillan (2010).

[167]The Herald Scotland (2014).

[168]Ibid.

[169]Humane Society International India (2014).

[170]Saraswathy (2014).

5.10.4 More "Humane" Slaughter

As mentioned above, New Zealand is legislating to stop farmers killing calves with blunt force trauma on the farm. Whether this is actually worse than a trip to the abattoir, terrified, hungry and thirsty, is something that is almost impossible to know.

Scientists in Israel, where chicks are routinely killed by suffocation and crushing in plastic bags,[171] are developing an electrocution method that is claimed to be much quicker than current methods (particularly the plastic bag). This electrocution method is still being tested.[172]

5.10.5 No-Kill Farms

Campaigns against killing of baby animals sometimes find results in individual farms. An example is Elgaar in Tasmania, where "we do not separate calves from their mums and no Elgaar calves are sent to the abattoir".[173] The website maintains that calves are allowed to suckle for 2–4 months, and males are then sent to another farm to be raised for organic beef or kept as bulls for the herd.[174] The females are milked for up to a decade and then put out to graze for the rest of their lives—one lived until the age of 38, compared to the life-span of 4 years that Elgaar claims is the average in the industry.[175] Similarly, Barambah Organics in Queensland states "the calves that are born on our property stay within our care. Our calves are not considered by us to be waste products" and generally stay with the farm until about 4 years of age.[176] B.-d. Farm Paris Creek in South Australia goes further—it is run by vegetarians—heifers are kept with their mother for about 3 months, while bobby calves are kept with their mother for at least a week and then often "adopted" by cows on other farms who have lost babies or have too much milk for their own calf. After that they may be raised as vealers for 4 months or organic beef (about a year) and sent to an abattoir, which will "humanly euthanize them"[177] (sic).

Even the marketing of veal, long seen as one of the most ethically distasteful products, is being somewhat rehabilitated with the marketing of rosé veal as "high welfare veal."[178] Kilkenny Rosé Veal in Ireland was awarded the Compassion in World Farming "Good Dairy Award" in 2011 and does not use the crates and the

---

171 Shamir et al. (2013).

172 Ibid.

173 Elgaar.

174 Ibid.

175 Shop Ethical! (2010).

176 Barambah Organics.

177 B—d. Farms Paris Creek.

178 Kilkenny People (2014).

low-iron diet which results in white veal and bad press. They buy male calves from dairy farms especially for this purpose and feed them a nutritious, high-fibre diet.[179]

5.11 Public Awareness: The "Face" of the Animal

In public debate, emotion is often far more effective than rational argument. Animals are de-realised, hidden from view, so that we can continue to eat, wear and experiment on them without having a face-to-face relationship.[180] The big brown eyes of bewildered calves and images of fluffy chicks falling into macerators arguably played much greater roles in the bobby calf and chick shredding debates than all the discussions of economics and welfare. A picture really does tell a thousand words when it comes to vulnerable babies, as discussed above. Stories and videos about "the hidden horror that the egg industry does not want you to see"[181] and "Dairy's dark secret"[182] are routinely accompanied by scenes of endearing baby animals facing horrendous deaths.

The French philosopher Emmanuel Levinas saw human ethics in the face of the "Other," which "manifests itself by the absolute resistance of its defenceless eyes.... The infinite in the face...brings into question my freedom, which is discovered to be murderous and usurpatory."[183] The face of the other, Levinas concludes, "is what forbids us to kill."[184] Levinas made it clear that he was speaking about the human face, although he added, "without considering animals as human beings, the ethical extends to all living beings. We do not want to make an animal suffer needlessly..."[185] The American philosopher Calarco argues that Levinas' attempt to draw a clear distinction between humans and other animals in the questions of ethics is "not just bad biology – it is also bad philosophy, inasmuch as it critically reinforces... metaphysical anthropocentrism..."[186] He concludes that "today philosophy finds itself faced by animals... what philosophy is now encountering, and what Levinas's philosophy tries desperately but unsuccessfully to block or dissimulate, is the simple fact that we know neither what animals can do nor what they might become."[187]

One of the most promising strategies for animal campaigns, it seems to me, is to have humans look into the face, the "defenceless eyes", of vulnerable baby animals,

[179] Ibid.

[180] Taylor (2008), p. 64.

[181] Poulter (2010).

[182] Neales (2011).

[183] Levinas (1990), p. 294.

[184] Levinas (1985), p. 86.

[185] Levinas (1988), p. 172.

[186] Calarco (2013), p. 62.

[187] Ibid., p. 63.

then choose to discard our "murderous" freedom and connect with an ethics of avoiding needless suffering.

5.12 Conclusion

Is reform possible beyond the productivity improvements that may benefit animals as a side-effect of improving profitability? How can we assign value to a 'waste product'?

Steven Wise asks the "core question" of morality and law: "are things or beings or ideas valuable because we value them or because they are inherently valuable?"[188] Animals, in law, are almost universally considered 'property,' which Adams calls a "device used to deny moral culpability."[189] Supporters of animal agriculture contend that property status protects animal welfare; for example Posner states (as do many industry websites) that "people tend to protect what they own."[190] Pollan goes further, stating that domestication is not a form of slavery but rather a "symbiosis" that ensures the survival of the species, if not the welfare of the individual animals.[191] However, baby animals in the dairy and egg industries are in the invidious position of being commodities without value. If, therefore, such commodities lack Wise's 'inherent value', a determination that they are without personal value to humans must make any treatment of them acceptable and any legal reforms either superficial or subject to blocking by vested interests. Meanwhile, a significant section of the animal rights movement will settle for nothing less than "the purest philosophical position",[192] total abolition of animals' property status.

The bobby calf campaign is a good example of this clash of paradigms. The failed attempts at reform over 8 years seem to evidence both Sharman's statement[193] that no major reform has been secured, and Francione's assertion that any welfare reforms that do happen are only side effects of innovations that are motivated by higher productivity and profits. However, despite or perhaps because of its legislative failure, this campaign crystallised the issue around the 30-hour time off feed question and led to widespread discussion and outcry which has dramatically raised public awareness of the calves' plight. Further progress will not come from 'supply-side' legislative impositions of standards, but rather from industry response to consumer 'demand'—the clamour against bobby calving or the wide-scale boycott of dairy products. Some smaller farms are already responding

188Wise (2000), p. 66.
189Adams (2009), p. 30.
190Posner (2004), p. 59.
191Pollan (2006), p. 320.
192Favre (2004), p. 236.
193Sharman (2013), p. 81.

that they do not bobby calf, instead keeping calves with their mothers for extended periods before incorporating them into the milking herd or sending them for slaughter. Dairy Australia has recognised community concern with several programmes to improve supply-chain handling of calves.[194] Bizarrely, the RSPCA[195] has encouraged Australians to eat more veal, to motivate farmers to postpone calves' slaughter. The movement to ovo-lacto vegetarianism (as opposed to veganism), which sees no ethical dilemma with eating eggs and dairy products, has marginally affected the meat industry but been a big boost for the dairy and egg industries.

The banning of maceration of chicks in the German state of North Rhine-Westphalia will be worth watching for legal developments. The complex legal arguments from general anti-cruelty legislation to the conclusion that farmers must keep baby cockerels alive until ready for slaughter for consumption (whatever the financial implications), rather than shredding them at hatching, could perhaps start a trend for more welfare-friendly interpretations of existing legislation around the world. Insofar as most legislation and codes of conduct are drafted with the close cooperation of the animal industries, however, it is more likely that politicians will simply take more care in future amendments that unprofitable welfare reforms cannot be interpreted from legislative changes.

Is there hope that the ethics of compassion could be applied to baby animals of no commercial value? Posner, while rejecting 'ethical' arguments for animal rights, believes that people are willing to recognise the inherent value of non-human animals, regardless of their commercial value, if they are made aware of their needs (and can see minimal personal costs).[196] However, it is unrealistic to expect that the millions of baby bulls and billions of baby cockerels will be saved from slaughter if there is no possibility of financial recompense, so the only long-term strategy for the reduction or abolition of bobby calving and cockerel maceration or gassing is reducing demand for eggs and dairy products through the development of public empathy. Surveys show that consumers overwhelmingly want to buy "humanely raised" animal products.[197] Children are an obvious target for persuasion as they tend to observe moral issues with far less social mediation.[198] At the same time, the burgeoning capital and labour costs of producing eggs and milk, together with the plummeting farm gate price, may in fact be more effective than any campaigns activists can devise.[199]

As for activism, it is apparent that people will generally avoid campaigns that make them feel uncomfortable or appear to have personal costs.[200] Future

[194]Dairy Australia 2010, "Calf management across the supply chain".

[195]Humphreys (2013).

[196]Posner (2004), p. 66.

[197]Kristof (2014).

[198]Anderson (2006).

[199]Mendelson (2012), p. 137.

[200]Cooney (2011), p. 35.

campaigns, while aiming to convert people to plant-based diets, will most effectively start with the empathetic image of the vulnerable baby animal. Getting people to see the face of the baby cockerels, to look into the eyes of a bobby calf, as the AA campaign[201] did, is far more effective than handing out pamphlets of vegan recipes.

References

Adams C (2010) The sexual politics of meat: a feminist-vegetarian critical theory. Continuum, New York

Adams W (2009) Human subjects and animal objects: animals as other in law. J Anim Law Ethics 3(1):29–51

American Egg Board (2014) Check out the latest facts on the egg business. http://www.aeb.org/farmers-and-marketers/industry-overview. Accessed 17 Nov 2014

Anderson J (2006) Why do young children choose to become vegetarians? (8 August). Cited in http://www.gse.harvard.edu/news/10/09/morality-meatlessness-why-children-choose-vegetarianism. Accessed 16 Aug 2014

Animal Welfare Act (2006) German Federal Law Gazette [BGBl.] Part I pp. 1206 and 1313. Amended 9 December 2010. http://www.cgerli.org/fileadmin/user_upload/interne_Dokumente/Legislation/TierSchG2011.pdf. Accessed 12 Dec 2014

Animals Australia. Do you want to know a secret? http://www.animalsaustralia.org/features/newspaper-ads-speak-up-for-bobby-calves.php. Accessed 24 Sept 2014

Animals Australia. What you never knew about dairy. http://www.animalsaustralia.org/issues/dairy.php. Accessed 24 Sept 2014

Australian Animal Welfare Standards and Guidelines. Appendix 3 - bobby calf time off feed - public consultation report (6/4/2011). http://www.animalwelfarestandards.net.au/files/2011/05/Bobby-Calf-ToF-Consultation-Report-Final.pdf. Accessed 17 Dec 2013

Australian Animal Welfare Standards and Guidelines. Bobby calf time off feed standard. http://www.animalwelfarestandards.net.au/land-transport/bobby-calf-time-off-feed-standard/. Accessed 14 Dec 2013

Australian Animal Welfare Standards and Guidelines (2013) Land transport consultative process (10 December) http://www.animalwelfarestandards.net.au/land-transport/consultative-process/. Accessed 17 Jan 2014

Australian animal welfare strategy. http://www.australiananimalwelfare.com.au/content/about-aaws. Accessed 12 Dec 2013

Australian Bureau of Statistics (2014) 7215.0 - Livestock products, Australia, September. http://www.abs.gov.au/AUSSTATS/abs@.nsf/DetailsPage/7215.0Sep%202014?OpenDocument. Accessed 17 Dec 2014

Australian Chicken Meat Federation Inc. (2011). Chicken welfare matters: healthy chickens mean a healthy industry. http://www.chickenwelfare.com.au/. Accessed 12 June 2014

AVMA (American Veterinary Medical Association) (2013) AVMA guidelines for the euthanasia of animals: 2013 Edition. Illinois. https://www.avma.org/KB/Policies/Documents/euthanasia.pdf. Accessed 3 Nov 2014

B—d. Farms Paris Creek. http://www.bdfarmpariscreek.com.au/index02.php?id=11#bobbycalfs. Accessed 12 Dec 2014

[201]Animals Australia, *What you never knew about dairy.*

Back Yard Chickens (2008) http://www.backyardchickens.com/t/35213/how-long-do-chickens-live. Accessed 6 Oct 2014

Barambah Organics, About the Barambah Farm. http://www.barambahorganics.com.au/barambah-difference/the-farm.aspx. Accessed 12 Dec 2014

Bardsley D (2012) Rural language - The evolution of rural language. Te Ara - the Encyclopedia of New Zealand, updated 14 November. http://www.TeAra.govt.nz/en/photograph/18586/the-bobby-truck. Accessed 7 Aug 2014

Bettles C (2013) Activists stripped of govt funds. The Land 19 December. http://www.theland.com.au/news/agriculture/agribusiness/general-news/activists-stripped-of-govt-funds/2682172.aspx?storypage=0. Accessed 4 Jan 2014

Buhl A (2013) Legal aspects of the prohibition on chick shredding in the German State of North Rhine-Westphalia. Global Journal of Animal Law. Issue 2. http://www.gjal.abo.fi/gjal-content/2013-02/article1/Article_Buhl_FINAL.pdf. Accessed 7 Nov 2014

Calarco M (2013) Zoographies: the question of the animal from Heidegger to Derrida. Columbia University Press, New York

Cave J, Callinan A, Woonton W (2005) Mortalities in bobby calves associated with long distance transport. Aus Vet J 83(1–2):82–84

CBC News (2009) Male chicks ground up alive at egg hatcheries. September 1, http://www.cbc.ca/news/male-chicks-ground-up-alive-at-egg-hatcheries-1.823644. Accessed 12 Oct 2014

Cooney N (2011) Change of heart: what psychology can teach us about spreading social change. Lantern Books, Brooklyn

Cronshaw T (2014) Govt reviewing bobby calves killing rules. NZ Farmer 27 Feb. http://www.stuff.co.nz/business/farming/agribusiness/9771288/Govt-reviewing-bobby-calves-killing-rules. Accessed 6 June 2014

Dairy Australia, Care of bobby calves on-farm. http://www.dairyaustralia.com.au/Animal-management/Animal-welfare/Bobby-calves/Care-of-bobby-calves-on-farm.aspx. Accessed 4 Aug 2014

Dairy Australia (2010) Calf management across the supply chain. December, http://www.dairyaustralia.com.au/~/media/Documents/Animal%20management/Animal%20welfare/Bobby%20calves/2010-12%20-%20Calf-management-across-the-supply-chain.PDF. Accessed 15 Aug 2014

Dairy Australia, Cows and farms. http://www.dairyaustralia.com.au/Markets-and-statistics/Farm-facts/Cows-and-Farms.aspx. Accessed 4 Sept 2014

Dairy Australia, Latest production and sales statistics. http://www.dairyaustralia.com.au/Markets-and-statistics/Production-and-sales/Latest-statistics.aspx. Accessed 12 Sept 2014

Dairy Mom blogspot, Dairy farmers care for their cows like you care for your pets. http://thedairymom.blogspot.com.au/2010/11/dairy-farmers-care-for-their-cows-like.html. Accessed 13 Aug 2014

DairyCo Market Information (2014) 14 January. http://www.dairyco.org.uk/market-information/farming-data/cow-numbers/world-cow-numbers/#.VIFIJ9KhxQA. Accessed 12 Nov 2014

Dale A, White S (2013) Codifying animal welfare standards: foundations for better animal protection or merely a facade? In: Sankoff PJ, White SW, Black C (eds) Animal law in Australasia: continuing the dialogue, 2nd edn. The Federation Press, Sydney

Davis K (1998) UPC's realtor files lawsuit to stop Perdue. Poultry Press (Fall/winter), http://www.upc-online.org/fall98/perdue_lawsuit.html. Accessed 22 Oct 2014

De Dreu C, Greer L, Van Kleef G, Shalvi S, Handgraaf M (2011) Oxytocin promotes human ethnocentrism. Proc Natl Acad Sci 108(4):1262–1266

DEFRA (Department for Environment, Food and Rural Affairs). Moving and selling calves. http://adlib.everysite.co.uk/adlib/defra/content.aspx?id=000IL3890W.182WH9KMQQWZJ5. Accessed 2 Nov 2014

DeGrazia D (2012) Meat-eating. In: Imhoff D (ed) The CAFO reader: the tragedy of industrial animal factories. Watershed Media, pp 219–224

De Vries A, The economics of sexed semen in dairy heifers and cows. Animal Sciences Department, Florida Cooperative Extension Service, Institute of Food and Agricultural Sciences, University of Florida. http://edis.ifas.ufl.edu/an214. Accessed 21 Aug 2014

de Witte K (2009) Development of the Australian animal welfare standards and guidelines for the land transport of livestock: process and philosophical considerations. J Vet Behav: Clin Appl Res 4(4):148–156
Derrida J (1974) 'Eating well,' or the calculation of the subject. Points... Interviews, 1974–1994:255–87
Donovan N (2008) Challenging the art of disinformation in Australian animal law. Aust Law Libr 16:243, 245
Elgaar Farm. We deeply care for our cows and calves, and bulls. http://www.elgaarfarm.com.au/our-cows-calves-and-bulls.html. Accessed 12 Dec 2014
Ellis E (2010) Hot topics: legal issues in plain language: animal law. Legal Information Access Centre
Ellis E (2013) The animal welfare trade-off or trading off animal welfare? In: Sankoff P, White S, Black C (eds) Animal law in Australasia: continuing the dialogue, 2nd edn. The Federation Press, Sydney
ENVision Environment Consulting (2011) Results of the public consultation for the Draft Land Transport Standards & the Regulatory Impact Statement. http://www.animalwelfarestandards.net.au/files/2011/02/Summary-of-the-consultation-response-results.pdf. Accessed 12 July 2014
Eurogroup for animals (2010) Areas of concern: analysis of animal welfare issues in the European Union. In http://www.animalwelfareintergroup.eu/wp-content/uploads/2011/10/EurogroupForAnimals-AreasOfConcern2010.pdf. Accessed 18 Oct 2014
European Union (2009). Council regulation (EC) No 1099/2009. 2 On the protection of animals at the time of killing. 4 September. http://eur-lex.europa.eu/LexUriServ/LexUriServ.do?uri=OJ:L:2009:303:0001:0030:EN:PDF. Accessed 18 Oct 2014
European Union. The EU and animal welfare: policy objectives. http://ec.europa.eu/food/animal/welfare/policy/index_en.htm Accessed 18 Oct 2014
FAOSTAT (2013) Food and Agriculture Organisation of the United Nations, Statistics Division. http://faostat3.fao.org/browse/Q/QA/E. Accessed 19 Sept 2014
Favre D (2004) A new property status for animals: equitable self-ownership. In: Sunstein CR, Nussbaum MC (eds) Animal rights: current debates and new directions. Oxford University Press, New York
Fox A (2014) Big jump in bobby calf kill. NZ Farmer 25 November. http://www.stuff.co.nz/business/farming/beef/63489037/big-jump-in-bobby-calf-kill. Accessed 18 Dec 2014
Francione G (2008a) Interview with Gary Francione, author of Animals as persons, essays on the abolition of animal exploitation, Columbia University Press Blog. http://www.cupblog.org/?p=283. Accessed 11 July 2014
Francione G (2008b) What to do on proposition 2? The abolitionist approach. http://www.abolitionistapproach.com/what-to-do-on-proposition-2/#.VKCmasHCgI. Accessed 11 July 2014
Friedrich B (2013) The cruelest of all factory farm products: eggs from caged hens. The Huffington Post, January 14. http://www.huffingtonpost.com/bruce-friedrich/eggs-from-caged-hens_b_2458525.html. Accessed 7 Oct 2014
Galloway J (2014) It's strictly semen business. NZ Farmer. 25 December. http://www.stuff.co.nz/business/farming/dairy/64412341/its-strictly-semen-business. Accessed 27 Dec 2014
Gandhi M (2013) Cruelty meted out to cattle. Mathrubhumi, 11 June. http://english.mathrubhumi.com/news/columns/faunaforum/cruelty-meted-out-to-cattle-1.10632. Accessed 17 Aug 2014
Goodfellow J (2013) Animal welfare law enforcement: to punish or persuade? In: Sankoff P, White S, Black C (eds) Animal law in Australasia: continuing the dialogue. The Federation Press, Sydney
Goodfellow J, Radan P (2013) Why market forces don't protect animal welfare. The Conversation 1 July. https://theconversation.com/why-market-forces-dont-protect-animal-welfare-15501?utm_medium=email&utm_campaign=Latest+from+The+Conversation+for+2+July+2013&utm_content=Latest+from+The+Conversation+for+2+July+2013+CID_77ac2eaec7f6534226b650ee84419667&utm_source=campaign_monitor&utm_term=Why%20market%20forces%20dont%20protect%20animal%20welfare. Accessed 16 Dec 2014

Gov.UK (2012) Poultry farms: general regulations. Department for Environment, Food & Rural Affairs. https://www.gov.uk/poultry-farms-general-regulations. Accessed 22 Nov 2014
Hanson M (2014) Disbudding/dehorning resource. Dairy Herd Management. December 3. http://www.dairyherd.com/advice-and-tips/disbuddingdehorning-resource. Accessed 20 Dec 2014
Hemsworth PH et al (1995) The welfare of extensively managed dairy cattle: a review. Appl Anim Behav Sci 42(3):161–182
The Herald Scotland (2014) The value of beef and milk production is all a matter of considered breeding. 24 November. http://www.heraldscotland.com/business/farming/the-value-of-beef-and-milk-production-is-all-a-matter-of-considered-breeding.1416826403. Accessed 2 Dec 2014
Holmes B (2014) Primal pull of a baby crying reaches across species. New Scientist (2987). http://www.newscientist.com/article/mg22329873.100-primal-pull-of-a-baby-crying-reaches-across-species.html#.VGlJcTShxQA. Accessed 29 Sept 2014
Humane Society International India (2014). Unilever: killing of male chicks in egg industry must end. 3 September. http://www.hsi.org/world/india/news/releases/2014/09/unilever-baby-chicks-090314.html. Accessed 11 Oct 2014
Humane Society U.S. (2014) Animal cruelty facts and statistics. October 29. http://www.humanesociety.org/issues/abuse_neglect/facts/animal_cruelty_facts_statistics.html?credit=web_id98058733#Livestock. Accessed 11 Nov 2014
Humphreys H (2013) Call for better life for dairy's rejects. The Age (Melbourne), 13 October. http://m.theage.com.au/victoria/call-for-better-life-for-dairys-rejects-20131012-2vff7.html. Accessed 1 Aug 2014
Kilkenny People (2014) Madigans of windgap toast of Rosé veal market. 28 November. http://www.kilkennypeople.ie/news/farming-news/madigans-of-windgap-toast-of-rose-veal-market-1-6443089. Accessed 1 Dec 2014
Kristof N (2014) Abusing chickens we eat. The New York Times. http://www.nytimes.com/2014/12/04/opinion/nicholas-kristof-abusing-chickens-we-eat.html?_r=0. Accessed 11 Dec 2014
LA Times (2014) California's egg-laying hens to get their breathing room. December 26. http://www.latimes.com/opinion/editorials/la-ed-hens-eggs-california-proposition2-ab1437-20141226-story.html. Accessed 28 Dec 2014
Levinas E (1990) Difficult freedom: essays on Judaism. The Athlone Press, Maryland
Levinas E (1985) Ethics and infinity. Duquesne University Press, Pittsburgh
Levinas E (1988) The paradox of morality, an interview conducted by T. Wright, P. Hughes and A. Ainley. In: The provocation of Levinas: rethinking the other, trans. Benjamin and Wright. Routledge, London/New York, pp 168–80
McMillan D (2010) Victorian dairy farmer perceptions of sexed semen. Dairy Australia Conference. http://www.dairyaustralia.com.au/~/media/Documents/Industry%20overview/Dairy-community-and-networks/Industry-forums/Au%20Dairy%20Conference/ADC%202010/Victorian%20dairy%20farmer%20perceptions%20of%20sexed%20semen.pdf. Accessed 1 Dec 2014
McWilliams J (2013) Milk Of human kindness denied to dairy cows. Forbes. http://www.forbes.com/sites/jamesmcwilliams/2013/10/25/milk-of-human-kindness-denied-to-dairy-cows/. Accessed 17 July 2014
Maday J (2014) Pain mitigation for dehorning cows. Bovine Veterinarian, December 16. http://www.bovinevetonline.com/advice-and-tips/practice-tips/pain-mitigation-dehorning-calves. Accessed 21 Dec 2014
Marcus E (2012) Dismantlement: a movement to topple industrial animal agriculture. In Imhoff D (ed) The CAFO reader: the tragedy of industrial animal factories. Watershed Media
Mendelson A (2012) The milk of human unkindness. In: Imhoff D (ed) The CAFO reader: the tragedy of industrial animal factories. Watershed Media
Mercy For Animals (2009). Hatchery horrors: the egg industry's tiniest victims. http://www.mercyforanimals.org/hatchery/. Accessed 4 Dec 2014
Mercy For Animals (2014) Breaking: cruelty charges filed following MFA investigation at Canada veal factory farm. MFA Blog at http://www.mfablog.org/breaking-cruelty-charges-filed-following. Accessed 22 Dec 2014

National Animal Welfare Advisory Committee (2012) Animal welfare (layer hens) code of welfare 2012. NZ Ministry for Primary Industry. http://www.biosecurity.govt.nz/files/regs/animal-welfare/2012-layer-hens-code-web.pdf. Accessed 13 July 2014

Neales S (2011) Killing of young calves is dairy industry's 'dark secret' The Australian November 24. http://www.theaustralian.com.au/news/nation/killing-of-young-calves-is-dairy-industrys-dark-secret/story-e6frg6nf-1226204115002. Accessed 29 July 2014

Norwood F, Lusk J (2011) Compassion by the pound: how economics can inform the farm animal welfare debate. Oxford University Press, Oxford

NSW Prevention of Cruelty to Animals Act (1979) at http://www.austlii.edu.au/au/legis/nsw/consol_act/poctaa1979360/. Accessed 1 July 2014

O'Sullivan S (2011) Animals, equality and democracy, The Palgrave Macmillan animal ethics series. Palgrave Macmillan, New York

Patterson C (2002) Eternal Treblinka: our treatment of animals and the Holocaust. Lantern Books, New York

Pollan M (2006) The omnivore's dilemma: a natural history of four meals. Penguin Press, New York

Posner R (2004) Animal rights: legal, philosophical and pragmatic perspectives. In: Sunstein C, Nussbaum M (eds) Animal rights: current debates and new directions. Oxford University Press, New York

Poulter S (2010) The disturbing conveyor belt of death where male chicks are picked off and killed so you can have fresh eggs. Daily Mail, 4 November. http://www.dailymail.co.uk/news/article-1326168/Secret-footage-shows-millions-British-chicks-killed-year.html#ixzz3NFECmzjo. Accessed 3 Oct 2014

Price E et al (2003) Fenceline contact of beef calves with their dams at weaning reduces the negative effects of separation on behavior and growth rate. J Anim Sci 81(1):116–21

Primary Industries Ministerial Council (2011) Australian standards and guidelines for the welfare of animals land transport of livestock regulatory impact statement – abridged. http://www.animalwelfarestandards.net.au/files/2011/02/Abridged-Public-Consultation-Version-of-the-Regulation-Impact-Statement-March-2008.pdf. Accessed 28 Aug 2014

Radford M (2001) Animal welfare law in Britain: regulation and responsibility. Oxford University Press, Oxford

Rollin B (2008) The ethical imperative to control pain and suffering in farm animals. In: Armstrong SJ, Botzler RG (eds) The animal ethics reader. Routledge, London, pp 248–259

RSPCA, Shop Humane. http://www.rspca.org.au/shophumane/. Accessed 2 July 2014

Saraswathy M (2014) Crushing the chicks right after they are hatched. Business Standard 22 September. http://www.business-standard.com/article/companies/crushing-the-chicks-right-after-they-are-hatched-114092200497_1.html. Accessed 29 Oct 2014

Sewell A (2013) Day-old male chicks given reprieve from shredder in German state. Digital Journal. September 27. http://www.digitaljournal.com/article/359160. Accessed 5 Aug 2014

Shamir N, Barlev E, Klein E, Gross F, Pinkas A (2013) Humane killing of male chicks at the laying branch. Agricultural Research Organization, Institute of Agricultural Engineering. 25 August. https://archive.today/T20gw#selection-233.1-233.75. Accessed 2 Oct 2014

Sharman K (2013) Farm animals and welfare law: an unhappy union. In: Sankoff P, White S, Black C (eds) Animal law in Australasia: continuing the dialogue. The Federation Press, Sydney

Shop Ethical! (2010) Elgaar Farms responds on bobby calves. 23 January. http://www.ethical.org.au/blog/elgaar-farms-responds-on-bobby-calves/. Accessed 3 Aug 2014

Sinclair U (1965) The Jungle. Penguin Modern Classics, UK

Singer P (2009) Animal liberation: a new ethics for our treatment of animals. Harper Perennial

Stevenson RL (2012) In the south seas. USA: Pennsylvania State University. http://www.gutenberg.org/files/464/464-h/464-h.htm. Accessed 19 Nov 2014

Stewart K, Cole M (2009) The conceptual separation of food and animals in childhood. Food 12(4):457–476

Sunstein C, Nussbaum M (2004) Animal rights: current debates and new directions. Oxford University Press, New York

Taylor C (2008) The precarious lives of animals: Butler, Coetzee, and animal ethics. Philos Today 52(1):60–72

The Dairy Site (2010) Veal production in the UK. http://www.thedairysite.com/articles/2360/veal-production-in-the-uk. Accessed 12 Dec 2014

The Pig Site (2012) Myth–busting boar taint. May 30. http://www.thepigsite.com/articles/3950/mythbusting-boar-taint. Accessed 1 Sept 2014

The Poultry Site (2015) Global Poultry Trends 2014: Rapid Growth in Asia's Egg Output. http://www.thepoultrysite.com/articles/3446/global-poultry-trends-2014-rapid-growth-in-asias-egg-output/. Accessed 23 Jul 2016

Tobias M (2012) Animal Rights in China. Forbes. http://www.forbes.com/sites/michaeltobias/2012/11/02/animal-rights-in-china/. Accessed 19 Oct 2014

Turk D (2007) Detailed discussion of cattle laws. Animal Legal and Historical Center/Michigan State University

United Poultry Concerns. Chickens. http://www.upc-online.org/chickens/chickensbro.html. Accessed 21 Aug 2014

Vidot A (2013a) Federal Government scraps welfare advisory group. ABC. http://www.abc.net.au/news/2013-11-08/animal-welfare-committee-scrapped/5079284. Accessed 28 Aug 2014

Vidot A (2013b) Few surprises for agriculture in economic update. ABC. http://www.abc.net.au/news/2013-12-17/rural-regional-myefo/5161978. Accessed 28 Aug 2014

Voiceless (2009) The animal law toolkit (December) http://www.voiceless.org.au/sites/default/files/VoicelessFinalToolkit_010210small_1.pdf. Accessed 20 May 2014

Wagman B, Liebman M (2011) A worldview of animal law. Carolina Academic Press, Durham

Wang Q (2014) Book emphasizes animal welfare. China Daily USA. 28 October. http://usa.chinadaily.com.cn/china/2014-10/28/content_18813534.htm. Accessed 27 Nov 2014

Wedderburn P (2009) China unveils first ever animal cruelty legislation. The Telegraph 18 September 2009. http://blogs.telegraph.co.uk/news/peterwedderburn/100010449/china-unveils-first-ever-animal-cruelty-legislation/. Accessed 9 Nov 2014

Weigel K (2003) Exploring the role of sexed semen in dairy production systems. J Dairy Sci 87 (Supplement 0):E120–E130. doi:10.3168/jds.S0022-0302(04)70067-3. Accessed 28 Aug 2014

Wise SM (2000) Rattling the cage: toward legal rights for animals. Perseus Books, Cambridge

Wolfson DJ, Sullivan M (2004) Foxes in the hen house: animals, agribusiness and the law. In: Sunstein CR, Nussbaum MC (eds) Animal rights: current debates and new directions. Oxford University Press, New York

World Poultry (2012) Ranking the world's major egg producers. http://www.worldpoultry.net/Home/General/2012/1/Ranking-the-worlds-major-egg-producers-WP009929W/. Accessed 2 Aug 2014

Part II
Industrial Animal Agriculture

Chapter 6
Industrial Animal Agriculture in the United States: Concentrated Animal Feeding Operations (CAFOs)

Aurora Moses and Paige Tomaselli

Abstract In the United States, industrial animal factories called "CAFOs" (concentrated animal feeding operations) raise most land-based food animals, reducing their own production costs by intensively confining farm animals. However, they do so at the expense of the animals, who suffer horrific institutionalized abuses through intensive confinement, as well as the public, which endures public health endangerment and environmental degradation from CAFOs' air and water pollution. Federal environmental laws potentially govern the industry's pollution, but these laws have been largely ineffective at reining in CAFO environmental harms. State and federal laws have also failed to address CAFO animal abuses. Further, the CAFO industry has successfully promoted state laws that limit the public's ability to document and communicate CAFO threats to public health, the environment, and animal welfare. However, some hope remains: citizen groups diligently and creatively use legal challenges and legislative advocacy to address the worst CAFO practices, and the American public is increasingly alarmed by CAFOs' lax oversight and supportive of reforms in regulating this industry.

6.1 Introduction

> We have given up the understanding . . . that our land passes in and out of our bodies just as our bodies pass in and out of our land; that as we and our land are part of one another, so all who are living as neighbors here, human and plant and animal, are part of one another, and so cannot possibly flourish alone . . . .—Wendell Berry[1]

---

[1]Berry (1977), p. 22.

A. Moses (✉)
Center for Agriculture and Food Systems at Vermont Law School, South Royalton, VT, USA
e-mail: aurora.paulsen@gmail.com

P. Tomaselli
Center for Food Safety, Washington, DC, USA
e-mail: PTomaselli@CenterforFoodSafety.org

G. Steier, K.K. Patel (eds.), *International Farm Animal, Wildlife and Food Safety Law*, DOI 10.1007/978-3-319-18002-1_6

For thousands of years, farm animals generally had access to the outdoors and the freedom to move around,[2] and large farms were limited to dozens or hundreds of animals.[3] Not now, though. Beginning in the middle of the twentieth century, farm animal producers increasingly looked to the profitable new mass production of consumer goods, adapting that model by intensively confining extraordinary numbers of farm animals and mechanizing their oversight.[4] In fact, each year, approximately 9 billion land-based farm animals are raised and killed in the United States, mostly in industrial operations.[5]

While these animal factories—also called concentrated animal feeding operations (CAFOs)—increase profits for large-scale producers, they do so at the expense of the public, which shoulders the hidden costs of CAFOs through federal subsidies, environmental degradation, and public health impacts, as well as at the expense of farm animals' welfare.[6] Such abuses of the public interest are made possible by largely business-friendly laws and regulation. Nevertheless, advocates for farm animals, human health, and environmental protection—which are all, of course, intimately related—have made significant progress in recent decades in addressing CAFO harms through creative and diligent use of legal tools.

6.1.1 Defining CAFOs

CAFOs are industrial operations that confine extraordinary numbers of farm animals. The federal government characterizes 450,000 industrial U.S. animal factories as "animal feeding operations" (AFOs), which it defines as operations that "congregate animals, feed, manure and urine, dead animals, and production operations on a small land area" and import feed rather than allowing animals to graze or otherwise range to seek food.[7] CAFOs are a subset of AFOs, operating on a larger scale; of AFOs, approximately 15 % are considered large enough to qualify as CAFOs.[8]

These large CAFOs confine tens of thousands or even millions of animals. For example, a single CAFO might confine as many as 800,000 pigs or 2 million chickens.[9] The animals in CAFOs are densely housed in cages and pens, fed

---

[2]Pew Charitable Trusts (2008), p. 31.

[3]*See generally* Rollin (2010), pp. 6–14 (discussing the rise of industrialized animal agriculture).

[4]*Id.*

[5]Humane Soc'y U.S. (2014).

[6]Gurian-Sherman (2008), pp. 3–5; Pew Charitable Trusts (2008), p. 6.

[7]U.S. Envtl. Prot. Agency (2014); U.S. Envtl. Prot. Agency (2015); *see* 40 C.F.R. § 122.23(B)(2) (defining "animal feeding operation").

[8]U.S. Envtl. Prot. Agency (2015).

[9]U.S. Gov't Accountability Office (2008), p. 1.

industrial feed quite unlike the diets for which they evolved, and restricted to the extent that they are unable to engage in almost all of their natural behaviors, such as foraging for food, grooming, and establishing social communities.[10] Unsurprisingly, then, due to poor physical and mental health, CAFO animals are highly susceptible to a wide range of diseases and ailments, many of which can, in turn, jeopardize consumers' health.[11]

To compensate for the ill health of intensively farmed animals and promote faster growth, CAFOs typically add drugs such as antibiotics and growth hormones to animals' feed and water, directly inject them into animals, or administer them via ear implants or tags.[12] Farm animals do not completely metabolize most antibiotics, but instead excrete a significant portion of them—up to 80 or 90 %—in manure.[13] The vast majority of antibiotics are administered non-therapeutically, and this non-therapeutic use of antibiotics promotes dangerous antibiotic-resistant strains of bacteria.[14] Like antibiotics, other feed additives, synthetic growth hormones, and various drugs are excreted in animals' manure, often eventually entering waterways.

The federal government reports that AFOs generate about 500 million tons of animal waste each year, which is more than three times the sanitary waste that American humans generate.[15] Of that waste, approximately 300 million pounds per year come from CAFOs.[16] In small-scale animal agriculture, animal manure fertilizes crops and restores nutrients to soil.[17] CAFOs, however, produce prodigious quantities of animal manure and wastewater that far exceed the carrying capacity of the land around them.[18] CAFOs thus typically store the manure and wastewater they produce and periodically dispose of an estimated 90 % of it by spreading it on nearby fields, purportedly as fertilizer.[19] However, rather than benefitting crops, the concentrated and drug-laden waste from CAFOs degrades both air and water quality.

---

[10]Pew Charitable Trusts (2008), p. 33.

[11]*Id.*, p. 11.

[12]National Pollutant Discharge Elimination System (NPDES) Concentrated Animal Feeding Operation (CAFO) Reporting Rule, 76 Fed. Reg. 65,431, 65,434 (Oct. 21, 2011).

[13]*Id.*

[14]Pew Charitable Trusts (2008), p. 6.

[15]National Pollutant Discharge Elimination System Permit Regulations and Effluent Limitation Guidelines and Standards for Concentrated Animal Feeding Operations (CAFOs), 68 Fed. Reg. 7176, 7180 (Feb. 12, 2003).

[16]*Id.*

[17]Gurian-Sherman (2008), p. 1; U.S. Envtl. Prot. Agency (2002), p. 13.

[18]68 Fed. Reg. at 7180.

[19]U.S. Envtl. Prot. Agency (2002), p. 13; U.S. Gov't Accountability Office (2008), pp. 1–2.

6.1.2 CAFO Harms to Health and the Environment

Every year, CAFOs produce hundreds of millions of tons of animal manure and wastewater, which they commonly dispose of by applying it to nearby fields or shipping it offsite.[20] According to the federal government, CAFO waste contains various toxic pollutants, including nutrients like nitrogen and phosphorus; solid manure and materials mixed with manure, such as bedding and litter, spilled feed, hair, feathers, and animal corpses; pathogens; potentially toxic trace elements like arsenic; odorous/volatile compounds like carbon dioxide, methane, and ammonia; antibiotics; and drugs, pesticides, and hormones.[21]

CAFOs commonly apply manure to land far in excess of what the land can absorb,[22] so excess waste runs off into waterways, polluting the water and causing algal blooms that harm aquatic plans, kill fish, and ultimately contribute to "dead zones" that are largely uninhabitable for aquatic organisms and affect an estimated 173,000 miles of U.S. waterways.[23] CAFO-generated pollutants also enter the environment through overflows from waste storage, leaching into soil and ground water, and volatilization of hazardous compounds.[24] Through these routes and others, agriculture is the leading contributor of pollutants to identified water quality impairments in American rivers and streams.[25]

CAFOs also pollute the air, emitting significant quantities of particulate matter, greenhouse gases, including methane, and toxic compounds, such as hydrogen sulfide and ammonia.[26] Ammonia and hydrogen sulfide are gases produced by decomposing animal manure or other organic matter.[27] Both gases can cause human respiratory illnesses, lung inflammation, and vulnerability to diseases such as asthma, and exposure to hydrogen sulfide can even be lethal for CAFO workers.[28] These gases also sicken the farm animals in CAFOs, leading to decreased activity, weakened immune systems, and breathing disorders, and leaving burns on confined birds.[29] While ammonia, hydrogen sulfide, and particulate matter are of greatest concern for human and farm animal health, these pollutants,

---

[20] 76 Fed. Reg. at 65,433–34.

[21] National Pollutant Discharge Elimination System Permit Regulations and Effluent Limitation Guidelines and Standards for Concentrated Animal Feeding Operations, 66 Fed. Reg. 2960, 2976–79 (Jan. 12, 2001).

[22] Pew Charitable Trusts (2008), p. 23.

[23] 68 Fed. Reg. at 7181; Pew Charitable Trusts (2008), p. 25.

[24] 68 Fed. Reg. at 7181.

[25] *Id.*

[26] Pew Charitable Trusts (2008), p. 27.

[27] CERCLA/EPCRA Administrative Reporting Exemption for Air Releases of Hazardous Substances From Animal Waste at Farms, 73 Fed. Reg. 76,948, 76,950 (Dec. 18, 2008).

[28] Pew Charitable Trusts (2008), p. 16; U.S. Envtl. Prot. Agency, Animal Waste: What's the Problem?

[29] Humane Soc'y U.S. (2010), p. 5; Pew Charitable Trusts (2008), p. 86.

as well as volatile organic compounds and greenhouse gases such as methane, also cause substantial ecological damage.

Once airborne, ammonia, for example, can travel over 300 miles before being deposited on the ground or in water.[30] The U.S. Environmental Protection Agency (EPA) estimates that 80 % of ammonia emissions in the United States originate from farm animal waste.[31] These CAFO air emissions contribute to soil acidification, as well as water eutrophication, which is an excessive nutrient level that causes dense plant growth and, subsequently, the death of aquatic animals due to a lack of oxygen.[32] CAFOs also release greenhouse gases, including nearly a third of the United States' emissions of methane,[33] and CAFOs' greenhouse gas emissions have increased by more than 50 % over the last 2 decades.[34]

6.1.3 CAFO Regulation

Due to the lobbying power of the agribusiness industry, CAFOs have mostly escaped regulation of their harms to farm animals, human health, and the environment. For example, although slaughter is subject to regulation for certain species, no federal law governs the treatment of farm animals in CAFOs,[35] and most state anticruelty laws exempt the treatment of farm animals.[36] Further, the only federal environmental law that expressly governs CAFOs, the Clean Water Act, has been undermined by limitations on agencies' abilities to prevent water pollution. And CAFOs have been successful in many states at maintaining the veil around their practices by advocating state "ag gag" and "veggie libel" laws that, respectively, prohibit the public from documenting or discussing CAFO harms.

Nevertheless, in recent decades, public interest advocates have found creative ways to harness the law to address CAFO harms. As discussed below, these advocates have worked to improve CAFO reporting to federal agencies—which otherwise are often unaware even of CAFOs' locations, numbers of animals, and air and water emissions—and strengthen regulatory prevention of CAFO air and water pollution, and they have brought suits under federal laws governing groundwater contamination and the reporting of release of hazardous pollutants. Advocates have also brought constitutional challenges to the state ag gag laws that stifle the public's ability to identify and communicate CAFO harms. Finally, to bypass the existing

[30]Nat'l Res. Def. Council, Facts about Pollution from Livestock Farms.

[31]Doorn et al. (2002), p. 1.

[32]Pew Charitable Trusts (2008), p. 25.

[33]Copeland (2010), pp. 21–22.

[34]U.S. Envtl. Prot. Agency (2012), pp. 2–13.

[35]Wolfson and Sullivan (2004), p. 207.

[36]*Id.*

laxness of CAFO regulation, citizen groups have helped pass state anti-confinement laws that establish limitations on CAFO conditions.

6.2 Overview of U.S. Laws Regulating CAFOs

Various federal and state laws potentially affect CAFO management. For example, federal laws mandate limitations on air and water pollution as well as the reporting of releases of hazardous pollutants, and they could reduce harms to public health and the environment. However, agencies have been reluctant to apply those environmental laws to CAFOs. Concerning farm animal welfare, both states and the federal government have laws that govern animal treatment. But those laws largely exempt farm animals, and the CAFO industry has helped pass "ag gag" and "veggie libel" statutes that further limit public advocacy for animals. Those limitations notwithstanding, citizen groups have made great progress for animal protection with state-level anti-confinement laws. Finally, federal antimonopoly laws offer promise in restricting CAFO consolidation, and thus corporate control of this industry, but only if enforcement improves.

6.2.1 Environmental Laws

Several federal environmental laws could, in theory, stem CAFO air and water pollution. But they have so far provided little in the way of actual environmental or public health protection. For example, although the Clean Water Act expressly applies to CAFOs, courts have construed this law to preclude EPA from proactively applying permitting requirements to CAFOs that are likely to pollute the nation's waters, and the agency has recently declined even to use its water pollution reporting authority to request information about CAFOs' potential to discharge. The Clean Air Act offers an avenue for limiting CAFO air pollution, but this law, too, is rarely enforced against CAFOs, despite various attempts from citizen groups to expand air pollution coverage of this industry. The Resource Conservation and Recovery Act, however, presently offers a promising angle for addressing CAFO groundwater pollution, due in large part to a citizen group lawsuit that was ongoing at the time of writing. Finally, federal reporting laws for the release of hazardous pollutants offer another approach for at least collecting information about CAFO pollution and warning the public about harmful emissions. But EPA has hindered application of those laws to CAFOs. Federal environmental regulation of CAFOs thus leaves much to be desired.

6.2.1.1 Clean Water Act

The Clean Water Act (CWA) provides primary federal authority for protecting water quality, with the stated intention to "restore and maintain the chemical, physical, and biological integrity of the Nation's waters."[37] The CWA prohibits the discharge of a pollutant from any point source into national waters unless such discharge is authorized by a permit under the National Pollutant Discharge Elimination System (NPDES).[38] NPDES permits authorize some water pollution but significantly restrict its type and quantity, and they require permit holders to take steps to prevent pollution.[39] These permits are therefore critical to the successful implementation of the CWA.[40]

Under this law, Congress specifically identified CAFOs as "point sources,"[41] potentially subjecting them to permitting requirements. However, EPA estimates that just over 40 % of CAFOs have CWA permits.[42] Additionally, the promising CWA "point source" designation for CAFOs excludes a huge portion of factory farms: of the 450,000 industrial U.S. animal factories, only 15 % are considered large enough to qualify as CAFOs.[43] The "small" industrial farms, which EPA defines as housing up to, for example, 3000 pigs or 25,000 laying hens, thus presumptively evade CWA oversight.[44]

Further, EPA's permitting jurisdiction over CAFOs has been severely circumscribed. More than a decade ago, recognizing that CAFOs routinely discharge pollutants into national waters, EPA promulgated rules requiring that CAFOs apply for permits if they were likely to discharge, and establishing that CAFOs that discharge without permits were liable both for the discharge and for failing to apply for a permit.[45] EPA promulgated those rules to improve CAFO

---

[37] 33 U.S.C. § 1251(a) (2012).

[38] *Id.* §§ 1311(a), 1342. NPDES permits are issued either by EPA, the federal agency tasked with administering the CWA, or by states that participate in a federally approved permitting system. *Id.* § 1342.

[39] *Id.* § 1251(a)(1). For example, NPDES permits establish "effluent restrictions" that limit the "quantities, rates, and concentrations of chemical, physical, biological, and other constituents which are discharged from point sources" into national waters. 40 C.F.R. § 401.11(i).

[40] *Waterkeeper v. U.S. Envtl. Prot. Agency*, 399 F.3d 486, 492 (2d Cir. 2005) (internal citation marks omitted).

[41] 33 U.S.C. § 1362(14).

[42] National Pollutant Discharge Elimination System (NPDES) Concentrated Animal Feeding Operation (CAFO) Reporting Rule, 76 Fed. Reg. 65,431, 65,447 (Oct. 21, 2011).

[43] U.S. Envtl. Prot. Agency (2015).

[44] U.S. Envtl. Prot. Agency, Regulatory Definitions of Large CAFOs, Medium CAFO, and Small CAFOs.

[45] Revised National Pollutant Discharge Elimination System Permit Regulation and Effluent Limitations Guidelines for Concentrated Animal Feeding Operations in Response to the Waterkeeper Decision, 73 Fed. Reg. 70, 418 (Nov. 20, 2008); National Pollutant Discharge Elimination System Permit Regulation and Effluent Limitations Guidelines and Standards for Concentrated Animal Feeding Operations (CAFOs), 68 Fed. Reg. 7176, 7266 (Feb. 12, 2003).

designs and operations, through permitting requirements, in ways that would prevent CAFOs from polluting the water. However, two cases rejected EPA's preventative approach, largely eviscerating the CWA's applicability to CAFOs by precluding EPA from requiring a permit until after a CAFO has already polluted and its pollution has been proven.[46]

The regulations implementing the CWA—which once held great promise as the United States' only environmental law with provisions explicitly applicable to livestock production—are thus currently broken.

6.2.1.2 Clean Air Act

The Clean Air Act (CAA) seeks to improve air quality and promote public health and welfare.[47] Under the CAA, any AFO that exceeds established air emission thresholds for certain pollutants can be regulated,[48] and there are several CAA programs with the potential to address CAFO pollution: national air quality standards, construction and operating permits, industry-specific technology-based regulations, and limitations on hazardous air pollutants. However, CAFOs have historically escaped regulation under this statute.

The federal government could address CAFO air pollution under national air quality standards. EPA establishes a national ambient air quality standard for each of the ubiquitous "criteria" pollutants it identifies and imposes limitations on emissions of these pollutants.[49] Under this rubric, "criteria" pollutants are those that "may reasonably be anticipated to endanger public health or welfare."[50] However, EPA has not designated as criteria pollutants either of two key CAFO pollutants—ammonia and hydrogen sulfide, which both pose numerous health risks. In 2011, citizen groups petitioned EPA to designate ammonia as a criteria pollutant, primarily as an effort to address CAFO emissions.[51] As of early 2015, EPA still had not responded to the ammonia CAA petition, so the citizen groups brought a lawsuit under the Administrative Procedure Act (APA) asserting that the agency had unlawfully delayed its response.[52] However, the court dismissed that lawsuit, deciding that the groups must sue under the CAA instead of the APA, and the issue was ongoing at the time of writing. If EPA grants this petition, states would have to evaluate and ultimately address ammonia emissions. In 2013, citizen

---

[46] *Nat'l Pork Producers Council v. U.S. Envtl. Prot. Agency*, 635 F.3d 738, 744–45 (5th Cir. 2011); *Waterkeeper*, 399 F.3d at 504.

[47] 42 U.S.C. § 7401(b)(1) (2012).

[48] U.S. Gov't Accountability Office (2008), pp. 2–3.

[49] 42 U.S.C. § 7409 (National Primary and Secondary Ambient Air Quality Standards).

[50] *Id.* § 7408(a)(1)(A).

[51] Envtl. Integrity Project, Petition for the Regulation of Ammonia as a Criteria Pollutant Under Clean Air Act Sections 108 and 109 (Apr. 2011).

[52] *Envtl. Integrity Project v. U.S. Envtl. Prot. Agency*, No. 15-cv-139, Compl. (D.D.C. Jan 28, 2015).

groups also brought suit against EPA for failing to list ammonia and hydrogen sulfide as a criteria pollutant,[53] but that suit was dismissed in 2014.[54]

The federal government could also address CAFO air pollution through construction and operating permits. Under the CAA, EPA can require permits for the construction or modification of facilities,[55] as well as for ongoing operation.[56] Through permits, the agency can require reductions in emissions of various pollutants, including those commonly emitted by CAFOs: ammonia, hydrogen sulfide, and greenhouse gases such as methane.[57] Citizen groups have brought suits against several CAFO dairies operating without CAA permits.[58] Such groups also brought a lawsuit against EPA for failing to require permits for CAFOs' operations, but that case was dismissed.[59] Overall, EPA and states have been reluctant to impose or enforce CAA permitting requirements for CAFOs.

A third method for the federal government to reduce CAFO air pollution is through technology-based regulations. Under the CAA, EPA can establish such regulations on an industry-by-industry basis.[60] The agency could use this authority to create standards for pollutant emissions that are specific to the CAFO industry, but it has not. In 2009, citizen groups petitioned EPA to designate CAFOs as a new industrial category under this authority, and to regulate CAFOs' emissions of numerous pollutants, including ammonia, hydrogen sulfide, and greenhouse gases.[61] The agency still had not responded to that petition by early 2015, so the citizen groups brought suit alleging unlawful delay,[62] and that suit was ongoing at the time of writing.

Finally, the federal government could restrict CAFOs' emissions of certain pollutants by designating those pollutants as hazardous and establishing standards for emissions. Under the CAA, EPA designates pollutants with the most serious health effects—generally those anticipated to be carcinogenic, mutagenic, neurotoxic, chronically toxic, or a threat to reproductive function—as hazardous air pollutants (HAPs).[63] However, the agency has not yet regulated CAFOs under its

---

[53] *Zook v. U.S. Envtl. Prot. Agency*, No. 1:13-cv-01315-RJL, Compl. (D.D.C. Aug 29, 2013).

[54] *Zook v. U.S. Envtl. Prot. Agency*, No. 1:13-cv-01315-RJL, Order (D.D.C. June 30, 2014); *Zook v. McCarthy*, No. 1:13-cv-01315-RJL, Notice of Appeal (D.D.C. July 29, 2014).

[55] 42 U.S.C. §§ 7475(a)(1), 7479(1), 7602(j) (New Source Review Program).

[56] *Id.* § 7661c (CAA permits for major stationary sources).

[57] *Id.* §§ 7401–7515, 7661–7661f.

[58] *Ass'n of Irritated Residents v. Fred Schakel Dairy*, No. 1:05-CV-00707 (E.D. Cal. 2008); *Ass'n of Irritated Residents v. C & R Vanderham Dairy*, No. 1:05-CV-01593 (E.D. Cal. 2008).

[59] *Zook v. U.S. Envtl. Prot. Agency*, No. 1:13-cv-01315-RJL, Order (D.D.C. June 30, 2014); *Zook v. McCarthy*, No. 1:13-cv-01315-RJL, Notice of Appeal (D.D.C. July 29, 2014).

[60] *See generally* 42 U.S.C. § 7411(b), (d), (f) (New Source Performance Standard Program).

[61] Humane Soc'y U.S., Petition to List Concentrated Animal Feeding Operations Under Clean Air Act Section 111(b)(1)(A) of the Clean Air Act, and to Promulgate Standards of Performance Under Clean Air Act Sections 111(b)(1)(B) and 111(d) (Sept 21, 2009).

[62] *Humane Soc'y U.S. v. U.S. Envtl. Prot. Agency*, No. 15-cv-0141, Compl. (D.D.C. Jan 28, 2015).

[63] 42 U.S.C. § 7412 (Hazardous Air Pollutants).

program for HAPs, even though the toxicity of ammonia, hydrogen sulfide, and other CAFO emissions is comparable that of other regulated HAPs. In 2009, citizen groups petitioned EPA to designate hydrogen sulfide as a HAP,[64] but at the time of writing, the agency had not responded.

Further underscoring EPA's general reluctance to regulate CAFOs under the CAA, the agency entered into a sweeping consent agreement with thousands of CAFO operators in 2005 under which it agreed to work with CAFOs to monitor and study their air emissions but, in trade, issued a moratorium on suits for those CAFOs' violations of permitting requirements and emissions restrictions.[65] Citizen groups challenged that consent agreement, but a court dismissed their lawsuit in 2007,[66] and the agreement remains in place. Consequently, EPA still is not enforcing CAA requirements for the approximately 14,000 industrial animal factories that entered into the agreement.

6.2.1.3 Resource Conservation and Recovery Act

The Resource Conservation and Recovery Act (RCRA) governs the treatment, storage, and disposal of solid and hazardous waste "to minimize the present and future threat to human health and the environment."[67] RCRA prohibits management or disposal of solid waste that contaminates groundwater or otherwise jeopardizes the environment or public health. Under this law, "discarded material" from agricultural operations qualifies as "solid waste."[68] RCRA prohibits management or disposal of solid waste in a manner that constitutes "open dumping,"[69] which it defines to include placing solid waste on land such that it contaminates groundwater by exceeding EPA-determined contaminant levels.[70] The law also prohibits handling, management, and disposal of solid waste that causes or contributes to the creation of an imminent and substantial endangerment to human health or the environment.[71]

Unlike the CWA, RCRA does not explicitly govern CAFO pollution. However, as discussed below, citizen groups have successfully argued that a CAFO dairy's

---

[64]Letter from Neil J. Carman, Sierra Club, to EPA Administrator Lisa Jackson, Hydrogen Sulfide Needs Hazardous Air Pollutant Listing Under CAA Title III (Mar. 30, 2009).

[65]Animal Feeding Operations Consent Agreement and Final Order, 70 Fed. Reg. 4958, 4958 (Jan. 31, 2005).

[66]*Ass'n of Irritated Residents v. U.S. Envtl. Prot. Agency*, 494 F.3d 1027, 1028–37 (D.C. Cir. 2007).

[67]42 U.S.C. § 6902(b) (2012).

[68]*Id.* § 6903(27).

[69]*Id.* § 6945(a).

[70]*See Community Ass'n for Restoration of the Env't, Inc. (CARE) v. Cow Palace, LLC*, No. CV-13-3016-TOR, Order re: Cross Mots. Summary Judgment, 81–82 (E.D. Wash. Jan 14, 2015) (discussing and outlining these RCRA regulations).

[71]42 U.S.C. § 6972(a)(1)(B).

management and disposal of farm animal waste constitutes open dumping of solid waste under RCRA, and that nitrates from this facility pollute groundwater in excess of federal contaminant levels, causing or contributing to public and environmental endangerment.[72]

6.2.1.4 Reporting the Release of Hazardous Substances

The Comprehensive Environmental Response, Compensation, and Liability Act (CERCLA) requires facilities to immediately notify federal authorities upon knowledge of the release of a hazardous substance that equals or exceeds EPA's reportable quantity within a 24-hour period.[73] Under CERCLA, reportable releases include emissions into ambient air, surface water, and groundwater,[74] and several CAFO pollutants, including ammonia, are characterized as hazardous substances.[75]

Similarly, the Emergency Planning and Community Right-to-Know Act (EPCRA) requires facilities to notify of local or state emergency planning committees or commissions after release of a reportable quantity of any hazardous substance subject to notification requirements under CERCLA.[76] Under EPCRA, the facility's emergency notice must include the chemical name of the released substance, an estimate of the amount released, and the time and duration of the release.[77]

In 2008, however, EPA issued a rule exempting CAFOs from certain CERCLA and EPCRA reporting requirements for air emissions,[78] bowing to the same industry pressures that resulted in the agency's 2005 consent agreement not to enforce CAA provisions against thousands of CAFOs during an undetermined (and still ongoing) period of monitoring and studying CAFO air emissions.[79] Under this rule, the agency removed all CERCLA reporting requirements for the air release of hazardous pollutants from animal manure.[80] EPA's rule also removed EPCRA air emissions reporting requirements for releases from animal manure for all but the largest CAFOs.[81] As small consolation, EPA's CERCLA/EPCRA exemption does not apply to reporting for the release of hazardous pollutants into water.[82]

---

[72] *CARE v. Cow Palace, LLC*, No. CV-13-3016-TOR, Compl. 2–3 (E.D. Wash. Feb. 14, 2013).

[73] 42 U.S.C. § 9603(a) (2012); 40 C.F.R. § 302.6(a).

[74] 42 U.S.C. § 9601(8), (22).

[75] 40 C.F.R. § 302.4.

[76] 42 U.S.C. §§ 11001(c), 11004(b)(1) (2012); 40 C.F.R. § 355.40.

[77] 42 U.S.C. § 1104(b)(2).

[78] CERCLA/EPCRA Administrative Reporting Exemption for Air Releases of Hazardous Substances From Animal Waste at Farms, 73 Fed. Reg. 76,948, 76,950 (Dec. 18, 2008).

[79] Animal Feeding Operations Consent Agreement and Final Order, 70 Fed. Reg. 4958, 4958 (Jan. 31, 2005).

[80] 73 Fed. Reg. at 76,950.

[81] *Id.*; 40 C.F.R. § 355.31(g).

[82] 73 Fed. Reg. at 76,953.

Citizen groups challenged the exemption in 2009,[83] and the court ordered EPA to reconsider its rule.[84] However, the court left the exemption rule in place, and 6 years later the agency still had not revised it, so in April 2015 citizen groups petitioned an appellate court to either review the merits of the CERCLA/EPCRA exemption rule or order EPA to revise the rule within 9 months,[85] and a case was ongoing in the appellate court at the time of writing. Nevertheless, the CERCLA/EPCRA exemptions for CAFO air emissions presently remain in effect.

6.2.2 *Animal Welfare Laws*

Federal law provides almost no protection for farm animals. The Animal Welfare Act is the central piece of federal legislation aimed at animal protection.[86] However, this law explicitly exempts farm animals.[87] In fact, with the exception of certain requirements for the comparatively small number of farm animals reared under federal organic standards,[88] no federal law governs the treatment of animals on farms or in CAFOs.[89]

Paradoxically, though, some farm animals do receive limited protection at the time of death: the federal Humane Methods of Slaughter Act (HMSA) mandates that livestock be rendered insensible to pain before slaughter.[90] But HMSA exempts poultry,[91] which constitute more than 95 % of animals slaughtered for food.[92] Farm animals also receive minimal protection during transport under the federal Twenty Eight Hour Law, which prohibits carriers from transporting animals for more than 28 h without unloading them for food, water, and rest but has significant exceptions.[93] Beyond those two laws, all other attempts to pass federal protections for farm animals have been unsuccessful.[94]

---

[83] *Waterkeeper Alliance v. U.S. Envtl. Prot. Agency*, No. 09-1017, Petition for Review (D.D.C. Jan. 15, 2009).

[84] *Waterkeeper Alliance v. U.S. Envtl. Prot. Agency*, No. 09-1017, Order (D.D.C. Oct. 19, 2010).

[85] *Waterkeeper Alliance v. U.S. Envtl. Prot. Agency*, No. 09-1017, Motion to Recall the Mandate or, in the Alternative, Petition for Writ of Mandamus (D.C. Cir. Apr. 15, 2015).

[86] 7 U.S.C. §§ 2132–2159 (2012).

[87] *Id.* § 2132(g).

[88] Organic Foods Production Act, 7 U.S.C. §§ 6501–6522 (2012); 7 C.F.R. § 205.238(a)(4) (requiring the "provision of conditions which allow for exercise, freedom of movement, and reduction of stress appropriate to the species" for animals raised under the federal organic program).

[89] Wolfson (1999), p. 14.

[90] 7 U.S.C. § 1902 (2012).

[91] *Id.*

[92] Humane Soc'y U.S. (2014).

[93] 49 U.S.C. § 80502(a) (2012).

[94] Pew Charitable Trusts (2008), p. 38.

Like the federal government, states provide only extremely limited protections for farm animals. Many states exempt farm animals from anticruelty statutes, often by expressly tolerating all "customary," "common," or "established" farming practices.[95] In essence, these exemptions allow the CAFO industry to self-determine how it treats farm animals, and it does so on the basis of economic interest rather than animal welfare. In fact, the very existence of farm animal exemptions in state anticruelty laws indicates recognition that much treatment of farm animals would otherwise qualify as unlawful animal cruelty.[96] Further, states' criminal anticruelty laws require government enforcement, and even where statutes do not exempt farm animals, prosecutors face significant hurdles, and cases involving farm animals are rarely pursued.[97]

Animal advocates have sought to extend state laws by publicizing farm animal abuse and promoting policy reform. In California, for example, after advocates released an undercover video demonstrating that employees at a slaughter facility in that state had horribly mistreated "downer" animals (those too sick or injured to stand or walk),[98] the state amended its penal code to ban the slaughter of nonambulatory animals and require slaughterhouses to euthanize them.[99] However, the U.S. Supreme Court overturned California's protection for disabled farm animals on the basis that it was preempted by a federal law governing mean inspection.[100]

Animal advocates have also sought novel legal routes around the failure of state anticruelty laws to protect farm animals. In Texas, for example, citizen groups brought a lawsuit alleging that the state had failed to enforce provisions of its health code that affected the treatment of chickens, resulting in tolerance for CAFOs' squalid conditions and thus creating a public health risk.[101] Although the court acknowledged that egg facilities' unsanitary conditions and poor hen health can jeopardize public health, it dismissed the suit on the ground that the state had discretion in enforcing its health code.[102]

Animal advocates have, however, had far greater success in regulating CAFOs' farm animal abuse through passage of state anti-confinement laws.

---

[95]Wolfson and Sullivan (2004), p. 212.

[96]Wolfson (1999), p. 10.

[97]Wolfson and Sullivan (2004), pp. 209–12.

[98]Humane Soc'y U.S. (2011).

[99]*Nat'l Meat Ass'n v. Harris*, 132 S.Ct. 965, 969–70 (2012).

[100]*Id.* at 968.

[101]*Ctr. for Food Safety v. Lakey*, No. 03-13-00094-cv, Memo. Opinion 1–2 (Tex. Ct. App., 3d Dist., Feb. 19, 2014).

[102]*Id.* at 11–12.

6.2.3 State Anti-Confinement Laws

In CAFOs, millions of pigs, calves, and laying hens spend their lives confined in spaces that do not even allow them to turn around or extend their limbs. For these animals, gestation crates (which severely restrict female breeding pigs for the majority of their pregnancies), veal crates, and hen battery cages prevent natural behaviors, resulting in injuries and substantial physical and psychological distress. To combat these abuses, animal advocates have increasingly promoted state-level anti-confinement legislation. And they have been successful. In 2002, Florida voters passed the first ballot measure banning the use of gestation crates.[103] Since then, seven additional states have banned both gestation crates and veal crates,[104] and two of those seven states have also banned battery cages.[105] California's anti-confinement law, which passed in 2008, was a landmark victory, ultimately resulting in benefits for millions of egg-laying hens across the country, as discussed in detail below.

6.2.4 "Ag Gag" and "Veggie Libel" Laws

"Ag gag" and "veggie libel" laws are state-level efforts to keep the public from documenting and discussing CAFOs' harms to animals, the environment, and human health. These laws aptly demonstrate that the CAFO industry has something to hide, and they arguably tread on constitutionally protected speech.

6.2.4.1 "Ag Gag" Laws

"Ag gag" laws criminalize activities that expose the inhumane, unsafe, and illegal conditions at industrial animal factories. These laws prohibit one or more of the following: recording a facility without consent[106]; falsifying a resume to gain employment, as in undercover operations[107]; or failing to immediately report animal cruelty to the police, which necessarily ends any ongoing undercover

---

[103]Fla. Const. art. 10, § 21.

[104]Ariz. Rev. Stat. §§ 13-2910.07–13-2910.08 (Arizona); Cal. Health & Safety Code §§ 25990, 25995–97 (California); Colo. Rev. Stat. §§ 35-50.5-101–103 (Colorado); Me. Rev. Stat. Ann. tit. 7, § 4020 (Maine); Mich. Rev. Stat. § 287.746 (Michigan); Or. Rev. Stat. § 600.150 (Oregon); R.I. Rev. Stat. chap. 4-1.1 (Rhode Island).

[105]Cal. Health & Safety Code §§ 25990, 25995–97 (California); Mich. Rev. Stat. § 287.746 (Michigan).

[106]*E.g.*, Kan. Stat. Ann. § 47-1827; N.D. Cent. Code § 12.1-21.1-02; Utah Code Ann. § 76-6-112.

[107]*E.g.*, Iowa Code § 717A.3A; Utah Code Ann. § 76-6-112.

investigation.[108] Those who violate ag gag laws risk criminal prosecution, fines, and jail time, and liability can extend even to organizations that support undercover investigations.[109] Three states passed ag gag laws in the 1990s.[110] In recent years, largely in response to undercover CAFO investigations demonstrating horrific conditions and animal abuse, ag gag laws have gained momentum, passing in four other states,[111] and more than two dozen additional ag gag bills were introduced in the last 5 years but ultimately failed.[112]

As written and by design, ag gag laws effectively curtail documentation of CAFO abuses and conditions. However, scholars have roundly criticized these laws as unconstitutional, and public interest organizations challenged Utah's ag gag law in 2013 and Idaho's ag gag law in 2014, asserting, among other things, that these laws unconstitutionally abridge free speech.[113] A court overturned Idaho's law in 2015, and the case is currently on appeal. The ag gag lawsuit in Utah was ongoing at the time of writing.

6.2.4.2 "Veggie Libel" Laws

"Veggie libel" laws, which are also known as "food disparagement" laws, make it easier for the food industry to sue members of the public who speak out about CAFOs. These laws have passed in 13 states.[114] Under them, a company can file a lawsuit for economic damages caused by public criticism of a perishable food—for example, if consumers reject the food after learning about production practices.[115] Notably, veggie libel laws lower the burden of proof for the food industry, with some even establishing strict liability (liability absent a finding of fault)[116] or

---

108 *E.g.*, Missouri Rev. Code § 578.013.1 (requiring anyone who witness animal cruelty to report it to law enforcement within 24 h).

109 *See, e.g.*, Utah Code Ann. § 76-6-112 (providing for a fine of up to $2500 and up to a year in jail).

110 Kan. Stat. Ann. § 47-1827 (Kansas); N.D. Cent. Code § 12.1-21.1-02 (North Dakota); Mont. Code Ann. § 81-30-103 (Montana).

111 Idaho Rev. Stat. § 18-7042 (Idaho); Iowa Code § 717A.3A (Iowa); Missouri Rev. Code § 578.013.1 (Missouri); Utah Code Ann. § 76-6-112 (Utah).

112 *See* Am. Soc'y for the Prevention of Cruelty to Animals (2015) (outlining the introduction of state ag gag laws).

113 *Animal Legal Def. Fund v. Otter*, No. 1:14-cv-104, Compl. (D. Idaho Mar. 17, 2014); *Animal Legal Def. Fund v. Herbert*, No. 2:13-CV-00679-RJS, Compl. (D. Utah July 22, 2013).

114 Ala. Code § 6-5-620 et seq. (Alabama); Ariz Rev. Stat. § 3-113 (Arizona); Colo. Rev. Stat. § 35-31-101 (Colorado); Fla. Stat. § 865.065 (Florida); Ga. Code § 2-16-1 et seq. (Georgia); Idaho Code § 6-2003(4) (Idaho); La. Stat. § 3:4501 et seq. (Louisiana); Miss. Code Ann. § 69-1-251 (Mississippi); N.D. Cent. Code § 32-44-01 et seq. (North Dakota); Ohio Code § 2307.81 (Ohio); Okla. Stat. § 5-100 et seq. (Oklahoma); S.D. Codified Laws § 20-10A-1 et seq. (South Dakota); Tex. Civ. Prac. & Rem. Code § 96.001 et seq. (Texas).

115 *Id.*

116 Ala. Code § 6-5-623.

requiring a defendant to prove that a statement about food was based on reliable scientific information.[117]

Like ag gag laws, veggie libel laws have been heavily criticized as unconstitutional limitations on protected free speech.[118] However, as intended, the threat of prosecution under these laws has effectively silences some public discussion of food production practices.[119] A handful of cases have been brought under veggie libel laws,[120] but none has yet reached constitutional questions.

6.2.5 Antimonopoly Laws

Many CAFO abuses stem from the structure of this industry. CAFOs—particularly those raising chickens and pigs—are often vertically integrated, meaning that farmers (the "growers") have contracts with corporate meat packing companies (the "integrators") to raise animals until slaughter.[121] Under these contracts, the growers do not own the animals, and the integrators control all phases of animal rearing, including details such as equipment requirements and the timing and content of feed, although the growers are responsible for managing and disposing of animal waste.[122] This structure results in a handful of large corporations controlling the majority of the U.S. market for chicken, eggs, and pork.[123]

Antimonopoly laws could reduce consolidation in the CAFO industry, but they generally are not enforced. In his 2008 campaign, now-President Barack Obama acknowledged the problems of consolidation within the livestock industry, criticizing the industry's anticompetitive behavior and vowing to strengthen antimonopoly laws and ensure that farm programs benefit small farms, rather than vertically integrated corporate structures.[124] Those promising statements led to a 2012 report by antitrust officials in the federal Justice Department in which the department assured the public that it had "redoubled its efforts to prevent anticompetitive agricultural mergers and conduct."[125] The Justice Department challenged several

[117]*See* Ala. Code § 6-5-621(1) (a statement is "deemed to be false if it is not based on reasonable and reliable scientific inquiry, facts, or data."); Ga. Code Ann. § 2-16-2(1) (similar); La. Rev. Stat. Ann. § 4502(1) (similar).

[118]*See, e.g.*, Jones (2000–2001), p. 839; Wasserman (2000), p. 334; Semple (1996), p. 411.

[119]*See generally* Nomai (1999).

[120]*See, e.g.*, *Tex. Beef Grp. v. Winfrey*, 201 F.3d 680 (5th Cir. 2000); *Action for Clean Env't v. Georgia*, 457 S.E.2d 273 (Ga. Ct. App. 1999).

[121]Pew Charitable Trusts (2008), p. 5.

[122]*Id.*, pp. 5–6, 42.

[123]*Id.*, p. 6.

[124]Farmers for Obama, Ensuring Economic Opportunity for Family Farmers.

[125]U.S. Dep't of Justice (2012), p. 16.

proposed company acquisitions by agribusinesses,[126] but no significant change to CAFO industry consolidation has resulted.

6.3 Clean Water Act CAFO Regulation

CAFOs are a significant source of water pollution, and the CWA expressly brings CAFOs within EPA's regulatory ambit as "point sources."[127] Beginning more than a decade ago, EPA has attempted to increase CWA oversight of CAFOs through rules that required more NPDES permits for CAFOs and restricted CAFOs' land application of manure.[128] However, the CAFO industry challenged those rules, and, in a saga of events, two cases severely undermined EPA's preventative approach, substantially eroding CWA applicability to CAFOs.

6.3.1 2003 CWA CAFO Rule

In 2003, EPA tried to improve CWA oversight of CAFOs by promulgating a rule that expanded the number of CAFOs needing NPDES permits and added requirements for CAFOs' land application of manure.[129] According to EPA, improvements to CAFO CWA oversight were urgently needed because "[i]mproper management of manure from CAFOs is among the many contributors to remaining water quality problems. Improperly managed manure has caused serious acute and chronic water quality problems throughout the United States."[130] The agency explained that its rule would ensure that the 15,500 CAFOs it targeted would more safely manage the 300 million tons of manure they produce each year.[131]

Under EPA's 2003 rule, CAFOs either had to seek permits or demonstrate that they had "no potential to discharge."[132] The agency established this "duty to apply" for a permit on a presumption that CAFOs are likely to discharge pollutants into national waters.[133] EPA's rule sought to greatly increase CWA coverage of CAFO

---

[126]*See id.*, pp. 17–19 (describing Justice Department challenges to livestock industry acquisitions on antitrust grounds).

[127]33 U.S.C. § 1362(14).

[128]National Pollutant Discharge Elimination System Permit Regulation and Effluent Limitations Guidelines and Standards for Concentrated Animal Feeding Operations (CAFOs), 68 Fed. Reg. 7176, 7266 (Feb. 12, 2003).

[129]*Id.* at 7176.

[130]*Id.*

[131]*Id.*

[132]*Id.* at 7182.

[133]*Id.* at 7201.

pollution and thus pollution prevention strategies—through permits—and provide the agency with much-needed information about CAFO discharges.

6.3.2 *Waterkeeper v. EPA*

In *Waterkeeper Alliance, Inc. v. U.S. Environmental Protection Agency*, various groups challenged EPA's 2003 rule.[134] Among other arguments, industry groups asserted that the agency had exceeded its CWA jurisdiction by requiring CAFOs to either obtain permits or demonstrate that they had "no potential" to discharge.[135] Under the industry groups' reasoning, the CWA explicitly applies only to the actual "discharge of any pollutant," so there is no CWA violation, and thus no permit obligation, until a discharge has in fact already occurred.[136]

The court recognized that EPA's new permit requirement promoted the CWA's goal of preventing water pollution, concluding that EPA had demonstrated that this requirement was likely necessary to effectively regulate CAFOs' water pollution, given that CAFOs "are important contributors to water and pollution and . . . have, historically at least, improperly tried to circumvent" the CWA permitting process.[137] Nevertheless, the court agreed with the CAFO industry groups: "the [CWA] gives the EPA jurisdiction to regulate and control only *actual* discharges—not potential discharges, and certainly not point sources themselves."[138] The court thus vacated the CAFO permit requirement in EPA's 2003 rule.

6.3.3 2008 CWA CAFO Rule in Response

In 2008, in response to *Waterkeeper*, EPA again tried to strengthen CWA CAFO oversight, issuing a revised rule that eliminated the automatic duty to apply for a permit but required CAFOs to obtain permits if they either discharge or propose to discharge.[139] The 2008 rule clarified that CAFOs "propose to discharge" if they are "designed, constructed, operated, or maintained such that a discharge would occur."[140] Under the rule, EPA required CAFOs to self-determine whether they

---

[134] *Waterkeeper Alliance, Inc. v. U.S. Envtl. Prot. Agency*, 399 F.3d 486 (2d Cir. 2005).

[135] *Id.* at 504.

[136] *Id.* at 504–05 (discussing CWA requirements applicable to "the discharge of any pollutant").

[137] *Id.* at 505–06, 506 n. 22.

[138] *Id.* at 505 (emphasis in original).

[139] Revised National Pollutant Discharge Elimination System Permit Regulation and Effluent Limitations Guidelines for Concentrated Animal Feeding Operations in Response to the Waterkeeper Decision, 73 Fed. Reg. 70,418, 70,423 (Nov. 20, 2008).

[140] *Id.*

proposed to discharge based on, *inter alia*, uncorrected past discharges, local weather patterns, aspects of CAFO construction such as waste storage quality and capacity, and whether operational and management procedures were designed to prevent discharges.[141] Where CAFOs discharged without a permit, the 2008 rule established that the CAFOs were potentially liable both for failing to obtain a permit and for the discharge.[142]

6.3.4 National Pork Producers Council v. EPA

In *Nat'l Pork Producers Council (NPCC) v. U.S. Environmental Protection Agency*, both citizen groups and CAFO industry groups challenged EPA's 2008 CWA rule.[143] As in *Waterkeeper*, CAFO industry groups argued, among other things, that EPA had exceeded its CWA authority, this time by requiring CAFOs to obtain permits if they "propose" to discharge.[144]

The court agreed with industry that EPA's "propose to discharge" requirement once again extended beyond the agency's CWA authority over "the discharge" of a pollutant by requiring CAFOs to get permits due only to a likelihood of discharge, rather than an actual, historical discharge event.[145] According to the court, "there must be an actual discharge into navigable waters to trigger the CWA's requirements and the EPA's authority. Accordingly, the EPA's authority is limited to the regulation of CAFOs that discharge. Any attempt to do otherwise exceeds the EPA's statutory authority."[146] Consistent with that conclusion, the court vacated the requirement in the 2008 rule that CAFOs that "propose" to discharge must obtain permits, and also held that CAFOs are not liable under the CWA for failing to apply for permits.[147]

In response to that second defeat, in 2012, EPA promulgated a CAFO CWA rule that eliminated any pre-discharge CAFO permit requirement.[148] A second critical outcome from *NPPC v. EPA* was a settlement agreement between EPA and the environmental groups.[149] Pursuant to that agreement, the agency committed to

[141] *Id.* at 70,423–24.

[142] *Id.* at 70,424.

[143] *Nat'l Pork Producers Council v. U.S. Envtl. Prot. Agency*, 635 F.3d 738 (5th Cir. 2011).

[144] *Id.* at 749.

[145] *Id.* at 750.

[146] *Id.* at 751.

[147] *Id.* at 751–52.

[148] National Pollutant Discharge Elimination System Permit Regulation for Concentrated Animal Feeding Operations: Removal of Vacated Elements in Response to 2011 Court Decision, 77 Fed. Reg. 44494 (July 30, 2012).

[149] National Pollutant Discharge Elimination System (NPDES) Concentrated Animal Feeding Operation (CAFO) Reporting Rule, 76 Fed. Reg. 65,431, 65,435 (Oct. 21, 2011) (explaining the settlement agreement).

propose a CWA reporting rule that required CAFOs to provide EPA with basic information about their locations, number of animals, and manure disposal practices, regardless of whether the CAFOs had permits.[150] EPA agreed to propose the rule in 2011, and to take final action on the rule the following year.[151]

6.3.5 EPA Reporting Rule Proposal and Withdrawal

As agreed, EPA proposed a CAFO reporting rule in 2011, pursuant to CWA section 308, which expressly authorizes information collection from "point sources," including CAFOs, in order to prevent water pollution.[152] The rule offered two options through which EPA would require either all CAFOs or a subset of CAFOs—those in "focus watersheds where CAFO discharges may be causing water quality concerns"—to submit "necessary information" to the agency that included the CAFOs' contact information, locations, CWA permitting status, number and type of animals, and number of acres available for land application of manure.[153]

As EPA explained, the information it proposed to collect would enable the agency to catalog and locate CAFOs and also inform decisions on how best to prevent ongoing water pollution through developing and enforcing CWA requirements.[154] EPA asserted that the rule was necessary to promote transparency and provide "a comprehensive body of data that would serve as a basis for sound decisionmaking about EPA's CAFO program."[155]

Informing EPA's proposed rule was a 2008 U.S. government accountability report concluding that despite EPA's long-term regulation of CAFOs, the agency had neither the requisite information to assess CAFO water pollution nor the data to ensure CAFO compliance with the CWA.[156] In fact, the report found, no federal agency collects accurate and consistent data on CAFOs,[157] and it recommended that EPA develop a national inventory of CAFOs.[158] In its proposed reporting rule, EPA confirmed that it would establish such an inventory.[159]

---

[150] *Id.*

[151] *Id.*

[152] *Id.* at 65,431. EPA clarified that its reporting rule did not contravene *Waterkeeper* and *NPPC v. EPA* because it imposed only information disclosure, not permitting requirements. *Id.*

[153] *Id.* at 65,435.

[154] *Id.* at 65,436.

[155] *Id.* at 65,435.

[156] U.S. Gov't Accountability Office (2008), p. 48.

[157] *Id.* at 4.

[158] *Id.* at 48.

[159] 76 Fed. Reg. at 65,435.

In 2012, however, rather than finalizing the proposed CAFO reporting rule, EPA succumbed to industry pressure and withdrew it.[160] The agency asserted that instead of requiring CAFOs to report the necessary information, it would seek that information voluntarily and from existing government sources.[161] Citizen groups challenged EPA's withdrawal of the reporting rule as unlawful, arguing that the agency had unreasonably disregarded its own acknowledgment of deficiencies in existing government information about CAFOs.[162] However, the court upheld EPA's withdrawal.

6.3.6 Discouraging CWA Outcomes

By categorizing CAFOs as point sources, the CWA seemed a likely candidate for comprehensive oversight and prevention of CAFO water pollution. And EPA tried to strengthen this law's preventative approach through rules reasonably requiring CAFOs to preemptively obtain permits on the basis that they commonly discharge pollutants into national waters. However, *Waterkeeper* and *NPPC v. EPA* severely undercut EPA's CWA authority, effectively prohibiting the agency from requiring CAFOs to obtain permits until they have already discharged pollutants into national waters.

Further, despite EPA's initially fairly bold stand against CAFO water pollution in its 2003 and 2008 rules, the agency eventually caved to industry influence, withdrawing the proposed reporting rule under which it could have gathered information about CAFO discharges to more effectively regulate this industry. As it stands, EPA lacks even basic knowledge of the amounts and types of CAFO water pollution, let alone effective CWA regulatory control.

6.4 CAFO New Legal Angles

Although the current U.S. regulatory system for CAFOs is failing to adequately prevent or address pollution, public health threats, and farm animal abuse, citizen groups have recently used novel legal and public policy strategies to rein in CAFO harms. At the federal level, citizen groups brought a RCRA challenge that is poised to prevent CAFO dairy pollution of groundwater. At the state level,

[160]National Pollutant Discharge Elimination System (NPDES) Concentrated Animal Feeding Operation (CAFO) Reporting Rule, 77 Fed. Reg. 42,679 (July 20, 2012) (withdrawing 2011 proposed CAFO reporting rule).

[161]*Id.* at 42,681.

[162]*Envtl. Integrity Project v. U.S. Envtl. Prot. Agency*, No.: 1:13-cv-1306, Compl. 2 (D.D.C. Aug. 28, 2013).

California's anti-confinement laws have improved the lives of millions of egg-laying hens and extensively introduced and informed public discussion of CAFO industry practices.

6.4.1 RCRA CAFO Dairy Challenge

In 2013, citizen groups brought a lawsuit asserting that a Washington CAFO dairy, Cow Palace, had violated RCRA through improper management and disposal of animal waste that endangers public health and the environment.[163] Fundamentally, the citizen groups argued that Cow Palace's management and land application of manure, which the dairy touted as useful fertilization, was far in excess of any rate beneficial to crops, and thus hazardous waste disposal.[164] In early 2015, a court agreed, holding that Cow Palace's stored and land-applied manure constituted solid waste under RCRA.[165]

Cow Palace has more than 11,000 cows, annually generating over 100 million gallons of manure.[166] The dairy manages and disposes of the manure by holding it onsite to transform it into compost, storing it in dirt waste impoundments (euphemistically called "lagoons"), and spraying it on nearby fields.[167] Cow Palace disingenuously characterized its yearly 100 million gallons of manure as a "valuable product" that it gifted to third parties, sold as compost, and used to fertilize fields around the dairy.[168] However, the citizen groups demonstrated that Cow Palace's lagoons and compost piles leached contaminants into the soil, and that the dairy sprayed manure on fields at levels far higher than crops and the soil could usefully absorb, in fact applying tens of millions of gallons of manure to fields that were not in need of fertilization.[169] The citizen groups also showed that Cow Palace had contributed to nitrate pollution in surrounding areas in excess of safety levels set by EPA.[170]

In 2014, the citizen groups moved for summary judgment on several key RCRA issues, including (1) that animal waste that leaks into groundwater and is over-applied to fields is solid waste; (2) that such improper waste management and disposal violates RCRA's ban on open dumping; and (3) that these conditions may

[163]*Community Ass'n for Restoration of the Env't, Inc. (CARE) v. Cow Palace, LLC*, No. CV-13-3016-TOR, Compl. 2 (E.D. Wash. Feb. 14, 2013).

[164]*Id.* at 15, 86.

[165]*Id.* at 109.

[166]*Id.* at 4–5.

[167]*Id.* at 15, 22, 30.

[168]*Id.* at 5.

[169]*Id.* at 18.

[170]*Id.* at 31.

cause or contribute to public endangerment.[171] Under RCRA, "solid waste" includes, among other things, any "discarded material" from agricultural operations.[172] This law exempts from the "solid waste" definition manure that is returned to the soil as fertilizer.[173] However, animal waste applied to fields in excess of its use as fertilizer does not qualify for this exemption.[174] RCRA prohibits management or disposal of solid waste in a manner that constitutes "open dumping."[175] Under RCRA regulations, open dumping includes placing solid waste on land such that it contaminates groundwater by exceeding EPA-determined contaminant levels for, *inter alia*, nitrates.[176] RCRA also prohibits facilities from causing or contributing to the creation of an imminent and substantial endangerment to human health or the environment through handling, management, or disposal of solid waste.[177]

The court granted the citizen groups' motion for summary judgment on all three grounds.[178] Regarding Cow Palace's assertion that it was fertilizing crops, the court found that the dairy had applied manure at levels exceeding agronomic nutrient uptake rates, discarding the manure and thus transforming it into solid waste under RCRA.[179] The court hinged its determination of whether land-applied manure is solid waste on the issue of whether the manure is productively used as fertilizer, rather than an unwanted material in need of disposal. The court also held that the manure Cow Palace stored in lagoons constituted solid waste under RCRA because the lagoons' leakage, which was due to "the poorly designed temporary storage features of the lagoons," converted the manure into waste by abandoning it in underlying soil.[180] Similarly, the court ruled that manure in Cow Palace's unlined composting area "is both knowingly abandoned and accumulating in dangerous quantities and thus a solid waste."[181] Finally, concerning the endangerment of public health and the environment, the court concluded that there was "no triable issue as to whether the Dairy's operations are contributing to the high nitrate levels

[171] *Id.* at 78–79.

[172] 42 U.S.C. § 6903(27).

[173] 40 C.F.R. § 257.1(c)(1).

[174] *CARE v. Cow Palace, LLC*, No. CV-13-3016-TOR, Order re: Cross Mots. Summary Judgment, 85–86 (E.D. Wash. Jan. 14, 2015) (explaining this exemption for useful fertilizer).

[175] 42 U.S.C. § 6945(a).

[176] *See CARE v. Cow Palace, LLC*, No. CV-13-3016-TOR, Order re: Cross Mots. Summary Judgment 81–82 (discussing and outlining these RCRA regulations).

[177] 42 U.S.C. § 6972(a)(1)(B); *see CARE v. Cow Palace, LLC*, No. CV-13-3016-TOR, Order re: Cross Mots. Summary Judgment at 80 (describing what plaintiffs must establish to show liability under this provision).

[178] *CARE v. Cow Palace, LLC*, No. CV-13-3016-TOR, Order re: Cross Mots. Summary Judgment at 109.

[179] *Id.* at 88.

[180] *Id.* at 93.

[181] *Id.* at 95.

in the groundwater,"[182] demonstrating the potential for imminent and substantial endangerment to the public and the environment.[183]

Given the court's RCRA rulings, the parties settled the case in 2015, with Cow Palace agreeing to line lagoons and limit nutrient application, among other improvements. Going forward, the court's strong rulings on essential RCRA issues suggest that citizen groups may also be successful in holding other CAFO dairies accountable for their groundwater pollution under RCRA.

6.4.1.1 CERCLA/EPCRA Challenge

In addition to asserting RCRA violations, the citizen groups also argued that Cow Palace had violated two federal reporting statutes—CERCLA and EPCRA—by failing to notify agencies of the release of certain hazardous pollutants, including ammonia.[184] CERCLA requires facilities to immediately notify federal authorities upon knowledge of the release of a hazardous substance that equals or exceeds EPA's reportable quantity within a 24-hour period.[185] EPA has established that ammonia is a hazardous substance with a reportable quantity of 100 pounds per day.[186] Similarly, EPCRA requires notification of local and state agencies after release of a reportable quantity of a hazardous substance subject to notification requirements under CERCLA.[187]

As noted above, a 2008 EPA rule exempts CAFO air emissions from CERCLA reporting requirements and restricts EPCRA air reporting requirements only to the largest CAFOs.[188] However, the rule did not exempt CAFOs from CERCLA and EPCRA reporting for the release of hazardous pollutants into water, including nitrogen pollution attributable in part to ammonia.[189] Further, Cow Palace is too large to be eligible for the EPCRA air emissions reporting exemption.[190]

The citizen groups alleged that Cow Palace releases more than 100 pounds of ammonia per day but fails to provide notice of this reportable release, contravening

---

[182] *Id.* at 97, 101–02.

[183] *Id.* at 104–05.

[184] *CARE v. Cow Palace, LLC*, No. CV-13-3016-TOR, Compl. 2–3 (E.D. Wash. Feb. 14, 2013); *see* 42 U.S.C. § 9603(a) (establishing CERCLA reporting requirement); 42 U.S.C. § 11004 (establishing EPCRA reporting requirement).

[185] 42 U.S.C. § 9603(a); 40 C.F.R. § 302.6(a).

[186] 40 C.F.R. § 302.4.

[187] 42 U.S.C. §§ 11001(c), 11004(b)(1); 40 C.F.R. § 355.40; 40 C.F.R. pt. 355, app. A.

[188] CERCLA/EPCRA Administrative Reporting Exemption for Air Releases of Hazardous Substances From Animal Waste at Farms, 73 Fed. Reg. 76,948, 76,950 (Dec. 18, 2008).

[189] *Id.* at 76,953.

[190] *See id.* at 76,952 (limiting the EPCRA exemption to CAFOs containing fewer than 700 mature dairy cows); *CARE v. Cow Palace, LLC*, No. CV-13-3016-TOR, Order re: Cross Mots. Summary Judgment at 4 (stating that Cow Palace has more than 7372 milking cows).

both CERCLA and EPCRA.[191] According to the citizen groups, the dairy in fact releases nearly 1600 pounds of ammonia per day, at minimum, and possibly more than 4700 pounds per day.[192] As noted, that case settled in 2015.

6.4.1.2 Potential Impacts

This groundbreaking RCRA CAFO case could have far-reaching consequences, as CAFOs have previously treated stored and land-applied manure as exempt from regulation as a solid waste under RCRA. Going forward, this case could force CAFOs that do not have CWA permits to either seek such permits or ensure that their waste management, storage, and disposal meet RCRA requirements, including preventing the leakage of contaminants into groundwater. Given that the court's initial rulings were robustly favorable to the citizen groups, RCRA is now potentially an important new legal angle for addressing CAFO pollution.

6.4.2 California Anti-Confinement Law: Proposition 2

In 2008, California voters approved Proposition 2, an anti-confinement ballot initiative.[193] Proposition 2, which went into effect in 2015, prohibits California CAFOs from confining farm animals in a manner that does not allow them to turn around freely, lie down, stand up, and fully extend their limbs.[194] The farm animals generally subjected to the intensive confinement practices targeted by this law include veal calves, gestating pigs, and laying hens. However, since few industrial veal and pig operations exist in California, Proposition 2 primarily affects the California egg CAFOs that confine more than 19 million egg-laying hens in battery cages too small for the hens to spread their wings. Proposition 2 effectively bans the use of battery cages.

In 2010, recognizing that Proposition 2 imposes stricter standards on California egg producers than egg CAFOs elsewhere in the country, the state legislature passed AB 1437, which extended Proposition 2's space requirements to the hens producing all eggs sold within California.[195]

The Proposition 2 campaign was a well-orchestrated and far-sighted collaboration between diverse nonprofits to combat the massive lobbying and financial sway

[191] *CARE v. Cow Palace, LLC*, No. CV-13-3016-TOR, Compl. at 25–26.

[192] *Id.* at 26–27.

[193] Cal. Health & Safety Code §§ 25990–25994.

[194] *Id.* § 25990 ("[A] person shall not tether or confine any covered animal, on a farm, for all or the majority of any day, in a manner that prevents such animal from: (a) [l]ying down, standing up, and fully extending his or her limbs; and (b) [t]urning around freely.").

[195] *Id.* §§ 25990, 25995–97.

of the CAFO egg industry. Even before introducing Proposition 2, advocates publicized a 2007 undercover investigation of a California slaughter facility where employees abused farm animals too sick or injured to stand on their own.[196] That investigation and resulting public uproar caused a massive meat recall. With public awareness of CAFO animal abuses at a zenith, advocates introduced Proposition 2, inviting public discussion of the conditions for industrially raised farm animals.

During the Proposition 2 campaign, the industrial egg industry assured the public that laying hens were well treated and their practices environmentally sound, and that family farmers were the ones opposing this ballot initiative. To combat those misrepresentations, Proposition 2 advocates aired the industry's secrets, disseminating undercover footage from a 2008 investigation in an egg CAFO that showed horrendous conditions for the hens,[197] bringing several environmental lawsuits against California industrial egg producers,[198] and firing off legal complaints demonstrating that the egg industry had laundered more than $4.5 million in out-of-state agribusiness money through an unregistered ballot committee,[199] as well as a successful federal lawsuit demonstrating that a federal egg-promotion agency had illegally set aside $3 million to support advertising in opposition to Proposition 2.[200]

With CAFOs' positive framing of their industry in shambles, they resorted to a financial appeal: Proposition 2, they claimed, would cause egg prices in California to soar. In response, Proposition 2 advocates filed federal antitrust petitions unearthing evidence that the egg industry had artificially inflated egg prices, eliminated competition, and defrauded consumers. A subsequent federal investigation of egg industry price-fixing was extensively covered in national news, from the front page of the Wall Street Journal to Business Week magazine.[201]

California voters got the picture, approving Proposition 2 by a large majority. However, the CAFO industry did not concede its fight against even these small

[196] Humane Soc'y U.S. (2012).

[197] Blume (2008).

[198] *Aliva v. Olivera Egg Ranch*, No. 08-1220 (E.D. Cal., filed Oct. 20, 2008); *In re The Humane Soc'y U.S.*, Cal. Regl. Water Quality Control Board (Oct. 7, 2008).

[199] Complaint Against the United Egg Producers, Inc., California for SAFE Food, a Coalition of Family Farmers, Veterinarians, and Consumers, No on Proposition 2, and the U.S. Poultry & Egg Association, Cal. Fair Political Pract. Comm'n (Sept. 3, 2008); Complaint Against the United Egg Producers, Inc., California for SAFE Food, a Coalition of Family Farmers, Veterinarians, and Consumers, No on Proposition 2, and the U.S. Poultry & Egg Association, Cal. Fair Pract. Comm'n (Sept. 11, 2008); Complaint Against the United Egg Producers, Inc., California for SAFE Food, a Coalition of Family Farmers, Veterinarians, and Consumers, No on Proposition 2, and the U.S. Poultry & Egg Association, Cal. Fair Pract. Comm'n (Oct 1, 2008).

[200] *Humane Soc'y U.S. v. Schafer*, No. 08-3843, Compl. (N.D. Cal. Aug. 12, 2008).

[201] *See, e.g.*, Wilke (2008).

improvements for farm animal welfare, and instead brought legal challenges to both Proposition 2 and AB 1437.

6.4.2.1 Industry Challenges

In 2012, an industrial egg producer challenged Proposition 2 as unconstitutionally vague, asserting that because the law does not specify minimum cage sizes for egg-laying hens, egg producers cannot determine which types of housing are lawful.[202] The court, however, easily dismissed that suit: "All Proposition 2 requires is that each chicken be able to extend its limbs fully and turn around freely. This can be readily discerned using objective criteria. Because hens have a wing span and a turning radius that can be observed and measured, a person of reasonable intelligence can determine the dimensions of an appropriate confinement that will comply with Proposition 2."[203]

Next, in 2014, the state of Missouri filed suit, arguing that AB 1437 unconstitutionally regulates commerce outside California.[204] The states of Nebraska, Oklahoma, Alabama, Kentucky, and Iowa—home to some of the largest industrial egg producers in the country—joined the case.[205] Proposition 2 advocates moved to dismiss the states' case for lack of standing, arguing that the states had failed to allege any interest apart from that of private egg CAFOs.[206]

In October 2014, the court dismissed the states' case on standing grounds.[207] According to the court, the states had failed to show that they represented the interests of their citizens, rather than primarily the economic interests of a small group of industrial egg producers.[208] The states appealed,[209] and that appeal is currently pending. In the meantime, Proposition 2 and AB 1437 have gone into effect.

6.4.2.2 The Promise of Proposition 2

Proposition 2 was a landmark victory in state-level efforts to improve CAFO regulation. The ballot initiative itself opened a public conversation about the misery CAFOs impose on laying hens and the public health effects of food produced

---

[202] *Cramer v. Harris*, No. 2:12-cv-03130-JFW- JEM, Order, at *1 (9th Cir. Feb. 4, 2015).

[203] *Id.* at *3.

[204] *State of Missouri v. Harris*, No. 2:14-cv-00341-KJM-KJN, Order, at *2 (E.D. Cal. Oct. 2, 2014).

[205] *Id.*

[206] *Id.* at *10.

[207] *Id.*

[208] *Id.* at *15.

[209] *State of Missouri v. Harris*, No. 2:14-cv-00341-KJM-KJN, Notice of Appeal (E.D. Cal., Oct. 24, 2014).

through intensive confinement. The regional and national media and pre-ballot litigation around Proposition 2 also informed millions of Americans about the environmental harms of CAFOs, including air and water pollution. Compounding the success of Proposition 2, which liberated California's 19 million laying hens from battery cages, this law set in motion AB 1437, which leveled the playing field by requiring other producers that sell eggs in California to meet the space requirements in Proposition 2, extending to many more millions of hens throughout the country the freedom to stretch their wings. Proposition 2 paved the way for further CAFO regulation by illuminating for the public the conditions in this industry.

6.5 Conclusion

The CAFO industry has significant influence over nearly every aspect of its own regulation, from academic research to legal and policy development and the enforcement of legislation.[210] Consequently, state and federal laws have largely been stripped of efficacy in overseeing this industry, causing unacceptable harms to public health, the environment, and the farm animals themselves. As a recent national report warned, "Our diminishing land capacity for producing food animals, combined with dwindling freshwater supplies, escalating energy costs, nutrient overloading of soil, and increased antibiotic resistance, will result in a crisis unless new laws and regulations go into effect in a timely fashion."[211]

To adequately address these harms, the federal government needs to prevent air and water pollution by requiring CAFOs to get permits under the CAA and CWA; creating standards under the CAA for hazardous air pollutants; enforcing provisions of the CAA, CWA, and RCRA where CAFOs are polluting without permits; reinstating and enforcing reporting for CAFOs' releases of hazardous air pollutants under CERCLA and EPCRA; placing limits on CAFO industry consolidation through antitrust methods; and establishing a federal farm animal welfare law that is enforceable by citizens. States, for their part, can assist with enforcement of these federal laws and set higher standards for air and water pollution; eliminate exemptions for criminal farm animal cruelty; and establish their own standards for farm animal welfare with citizen enforcement. In the meantime, citizen groups will continue to monitor and confront the worst CAFO abuses, as best they can, through creative and diligent use of legal and policy avenues.

---

[210]Pew Charitable Trusts (2008), p. viii.

[211]*Id.*, p. 77.

References

Am. Soc'y for the Prevention of Cruelty to Animals (2015) Ag-gag bills at the state level. https://www.aspca.org/fight-cruelty/advocacy-center/ag-gag-whistleblower-suppression-legislation/ag-gag-bills-state-level. Accessed 12 Apr 2015

Berry W (1977) The unsettling of America: culture and agriculture. Avalon Books, New York

Blume H (Oct 14, 2008) Footage of mistreated hens released in support of Proposition 2. http://articles.latimes.com/2008/oct/14/local/me-chickens14. Accessed 12 Apr 2015

Copeland C (2010) Air quality issues and animal agriculture: a primer. Research Serv., RL 32948

Doorn MRJ et al. (2002) Review of emissions factors and methodologies to estimate ammonia emissions from animal waste handling. U.S. Envtl. Prot. Agency

Farmers for Obama, Ensuring Economic Opportunity for Family Farmers. http://farmersforobama.org/?p=50. Accessed 12 Apr 2015

Gurian-Sherman D (2008) CAFOs uncovered: the untold costs of confined animal feeding operations. Union of Concerned Scientists

Humane Soc'y U.S. (2010) An HSUS Report: The welfare of animals in the pig industry

Humane Soc'y U.S. (Nov. 9, 2011) High court considering case on sick, injured livestock law. http://www.humanesociety.org/news/blog/2011/11/supreme_court_downers_11092011.html. Accessed 12 Apr 2015

Humane Soc'y U.S. (Oct. 9, 2012) Another Hallmark/Westland investigation milestone. http://blog.humanesociety.org/wayne/2012/10/hallmark-investigation-video.html. Accessed 12 Apr 2015

Humane Soc'y U.S. (Sept 15, 2014) Farm animal statistics: slaughter totals. http://www.humanesociety.org/news/resources/research/stats_slaughter_totals.html. Accessed 12 Apr 2015

Jones EG (2000–2001) Forbidden fruit: talking about pesticides and food safety in the era of agricultural product disparagement laws. Brook Law Rev 66:823

Nat'l Res. Def. Council, Facts about pollution from livestock farms. http://www.nrdc.org/water/pollution/ffarms.asp. Accessed 12 Apr 2015

Nomai AJ (1999), Food disparagement laws: a threat to us all. Free Heretic Pubs

Pew Charitable Trusts & Johns Hopkins Bloomberg Sch. Pub. Health (2008) Putting meat on the table: industrial farm animal production in America: A Report of the Pew Commission on Industrial Farm Animal Production

Rollin B (2010) Farm factories. In: Imhoff D (ed) The CAFO reader: the tragedy of industrial animal factories. University of California Press, Berkeley

Semple M (1996) Veggie libel meets free speech: a constitutional analysis of agricultural disparagement law. VA Environ Law J 15:403

U.S. Dep't of Justice (2012) Competition and agriculture: voices from the workshops on agriculture and antitrust enforcement in our 21st century economy and thoughts on the way forward

U.S. Envtl. Prot. Agency (2002) State compendium: programs and regulatory activities related to animal feeding operations

U.S. Envtl. Prot. Agency (2012) Inventory of U.S. greenhouse gas emissions and sinks: 1990–2010

U.S. Envtl. Prot. Agency (2014) Animal feeding operations. http://www.epa.gov/agriculture/anafoidx.html. Accessed 12 Apr 2015

U.S. Envtl. Prot. Agency (2015) What is a CAFO?. http://www.epa.gov/region07/water/cafo/. Accessed 12 Apr 2015

U.S. Envtl. Prot. Agency, Animal waste: what's the problem? http://www.epa.gov/region9/animalwaste/problem.html. Accessed 12 Apr 2015

U.S. Envtl. Prot. Agency, Regulatory definitions of large CAFOs, Medium CAFO, and small CAFOs. http://www.epa.gov/npdes/pubs/sector_table.pdf. Accessed 12 Apr 2015

U.S. Gov't Accountability Office (2008) Concentrated animal feeding operations: EPA needs more information and a clearly defined strategy to protect air and water quality from pollutants of concern

Wasserman HM (2000) Two degrees of speech protection: free speech through the prism of agricultural disparagement laws. Wm Mary Bill Rights J 8:323
Wilke JR (Sept 23, 2008) Federal prosecutors probe food-price collusion. Wall St. J. http://online.wsj.com/article/SB122213370781365931.html?mod=googlenews_wsj
Wolfson DJ (1999) Beyond the law: agribusiness and the systemic abuse of animals raised for food or food production. Farm Sanctuary, Watkins Glen
Wolfson DJ, Sullivan M (2004) Foxes in the hen house—animals, agribusiness, and the law: a modern American Fable. In: Sunstein CR, Nussbaum MC (eds) Animal rights: current debates and new directions

Chapter 7
The Political Ecology of the Dairy Industry

Clare Gupta

Abstract This chapter provides an overview of the structure of the dairy industry, including the laws and policies that influence and regulate this critical sector of the global agricultural economy. This chapter begins with an introduction to the bi-lateral and multi-lateral trade policies that guide the international market for milk and other dairy-related products, as well as food safety and environmental regulations. The chapter then addresses the nature of the U.S. dairy industry—its structure, key support policies, and changes to the industry over the past several decades. The impacts of the U.S. dairy industry on the environment, animal welfare, and human health are also highlighted. The chapter concludes with a case study of the dairy industry in Hawaii, which reflects many of the larger trends occurring nation-wide.

7.1 International Dairy Policy

7.1.1 Bi-Lateral and Multi-Lateral Trade Policies

Out of all the cow's milk produced globally, only 8 % is traded on international markets, and consists primarily of butter, cheese, and dry milk powders, with limited trade in fluid milk products. Currently, Organization of Economic Cooperation and Development (OECD) member countries account for over 80 % of world dairy exports,[1] and many governments lend a high level of support to domestic dairy production.[2] Among all OECD countries, there has been a significant increase in the number of cows per farm and a rise in production intensity due to improved technology (and this has resulted in associated environmental effects: see below).[3]

[1]See OECD (2004), p. 31.
[2]Ibid., p. 93.
[3]Ibid., pp. 58, 63, 66, 70.

C. Gupta (✉)
University of California, Davis, New Haven, CT, USA
e-mail: claregupta@gmail.com

G. Steier, K.K. Patel (eds.), *International Farm Animal, Wildlife and Food Safety Law*, DOI 10.1007/978-3-319-18002-1_7

The major global exporters of dairy products include the European Union (EU), New Zealand, and Australia.[4] Since 2002, the U.S. has been a net importer of dairy products, primarily from the EU and New Zealand.[5] The principal document governing the multilateral trade liberalization of dairy products is the Agreement on Agriculture of the Uruguay Round (UR).[6]

The dairy industry is one of the most regulated sectors in developed countries, for several reasons.[7] First, milk is one of the most basic foods, and therefore countries want to ensure an adequate supply. Due to its importance, countries prefer to be self-sufficient and would rather avoid importing large quantities of milk. Second, milk is bulky and highly perishable. Historically, the balance of market power rested with milk handlers: dairy farmers had little choice other than to accept the handler's price or dump the milk. This was a significant problem for dairy farmers before governments put in place policies and institutions to protect them. Improvements in transportation have also enabled milk to stay fresh for longer.

A third reason for the heavy regulation of the dairy industry in developed countries is due to the seasonal imbalance between milk demand and supply. In North America, milk production is highest in spring and summer and lowest in fall, whereas milk consumption is generally highest in the fall and lowest in the summer. Nevertheless, milk producers face consumer pressure to keep the supply constant year-round. Fourth, minor changes in the supply or demand of milk can cause severe changes in milk prices due to the high inelasticity of supply and demand.[8] Fifth, milk markets are oligopsonistic (many sellers relative to very few buyers of raw milk). Indeed, milk markets have been historically highly localized, and dairy farmers have little control over pricing and outlets.[9] Lastly, there are different types (grades) of milk that are used for various purposes, depending on certain sanitary health standards—for instance, somatic cell count, bacterial count, and conditions of farm facilities. Because of these considerations, government regulations exist to grade milk accordingly. The number of farmers and milk handlers is decreasing, however, as the same time as dairy cooperatives are growing and consolidating. As a result, some critics argue that dairy farmers have more market power today relative to when the government intervention programs were legislated, and argue for trade liberalization.[10]

While milk is one of the most highly supported agricultural commodities, there are significant variations between countries in the level of support provided to milk producers. Market price support has traditionally been the most dominant support category in all OECD countries except New Zealand, followed by payments based

[4]See Hadjigeorgalis (2005), p. 4.

[5]Ibid., p. 6.

[6]Ibid., p. 3.

[7]See Suzuki and Kaiser (2005), p. 1901.

[8]Ibid., p. 1902.

[9]Ibid., pp. 1902–1903.

[10]See Suzuki and Kaiser (2005), p. 1903.

on input use, which apply in all OECD countries.[11] Market price support policies are designed to protect producers from low prices, thus insulating them from market changes; they have been widely effective in reaching this objective.[12] Trade measures (for example, tariffs, import quotas, and export subsidies) have also been used historically in many OECD countries to protect dairy producers from traded products and to enable domestic pricing arrangements.[13] Overall however, there has been a downward trend in support prices in OECD member countries since the early 1990s.[14]

Agricultural support policies have been affected by the World Trade Organization (WTO) Uruguay Round Agreement on Agriculture (URAA) commitments of 1994 to reduce the level of support provided through trade measures such as quotas, tariffs, and export subsidies and other production-distorting support.[15] The purposes of this agreement are to avoid dairy surpluses and shortages while maintaining prices at an equitable level, as well as to improve cooperation in the dairy products sector to attain objectives pertaining to the expansion and liberalization of world trade.[16] For example, under the UR Agreement, U.S. tariff rate quotas replaced dairy import quotas. Unlike an import quota, which limits the total amount of product that may enter an importing country in a given year, a tariff rate quota establishes a two-tier tariff for imports. While imports below a set limit may enter the country duty free or at a reduced tariff rate, imports above this limit enter at a higher, generally prohibitive rate.[17]

The WTO agreements from the Uruguay Rounds were terminated in 1997, and during the 2004 Doha Round of WTO negotiation member countries reached an agreement on a framework for reducing agricultural supports. This suggests that reducing dairy supports is inevitable. The agreements call for the most trade-distorting supports to be substantially reduced, with product-specific capping of spending.[18] This agreement directly impacts the U.S. Dairy Export Incentive Program (see below).[19] There has also been an increase in the number and strength of policies to address environmental issues in agriculture.[20]

There are also several bilateral trade agreements that apply to dairy products. The most important of these include NAFTA, the Closer Economic Relations (CER) Agreement between Australia and New Zealand, and the U.S. and Australia Free Trade Agreement. NAFTA entered into force in 1994, and led to

[11]See OECD (2004), p. 96.

[12]Ibid., p. 102.

[13]Ibid., p. 98.

[14]Ibid., pp. 91–92.

[15]Ibid., p. 91.

[16]See World Trade Organization (1994), p. 1.

[17]See Hadjigeorgalis (2005), p. 4.

[18]See Suzuki and Kaiser (2005), pp. 1903–1904.

[19]Ibid., p. 1906.

[20]See OECD (2004), pp. 19–21.

the phasing out of all tariffs for trade with Mexico (the Canadian portion of NAFTA excluded dairy products). As part of the CER Agreement, all dairy trade was liberalized between Australia and New Zealand. Under the U.S.-Australia Free Trade Agreement, which entered into force on January 1, 2005, Australia guarantees duty-free tariff treatment of all U.S. dairy products. Australia also gains additional access to the U.S. market under this agreement through several duty-free tariff-rate quotas on dairy products not previously imported from Australia.[21]

Ultimately, while the dairy industry is one of the most regulated sectors in developed countries, liberalization is proceeding at a steady pace through advances in tariff reduction and quota elimination in the NAFTA and U.S.-Australia Free Trade Agreements. Globalization of the milk market is having a major role in shaping the U.S. dairy industry through innovations such as refrigeration, cheap transport, increased communication, and reduced trade barriers and tariffs.[22]

7.1.2 Safety Regulation of Dairy by International Organizations

International organizations also attempt to regulate the international food product trade. The most notable of these organizations include the World Health Organization (WHO), the Food and Agriculture Organization (FAO), the World Trade Organization (WTO), the Codex Alimentairus Commission (CAC), and the International Epidemic Animal Disease Office. One of the goals of these organizations is to globally standardize food safety applications. Some important programs and administration bodies include the WTO's Sanitary and Phytosanitary Measures (SPS), as well as the Agreement on Technical Barriers to Trade (TBT), which established hygiene rules and standards.

Every country that signs the SPS agreement may determine its own measures for food safety, but the agreement requires each country to meet certain hygiene standards to be at a minimum acceptable risk level. For dairy trade, the Codex Committees determine international standards. The committees include the Codex Committee on Milk and Milk Products (CCMMP), the Codex Committee on Food Additives (CCFA), the Codex Committee on Food Hygiene (CCFH), the Codex Committee on Pesticide Remains (CCPR), and the Codex Committee on Residues of Veterinary Drugs in Foods (CCRVDF). The most important of these is the CCMMP, which aims to determine the principle safety codes related to dairy products and prepare the international standards, codes, and other guidelines for dairy production. The main food safety criterion for dairy products is having a low bacteria count: low or zero levels of pathogens that could negatively affect human

[21]See Hadjigeorgalis (2005), p. 3.

[22]See von Keyserlingk et al. (2013), p. 5417.

health; low or zero amounts of veterinary drug remnants; and minimum contamination from chemical pollutants and microbial toxins.

The private sector of the world dairy industry is widely applying advanced hazard management and control processes due to market demands and legal regulations. Additionally, as a result of international agreements, both developed and developing countries now extensively employ Hazard Analysis at Critical Control Points (HACCP). Regardless of such efforts, there are practical problems with the dairy industry and food safety regulations. In countries that have a weak dairy industry, it is difficult to implement these laws in practice. This is further complicated because the formation and application of HACCP plans can be difficult and time-consuming, even in developed countries.[23] Implementing HACCP protocols is particularly challenging in the dairy industry because milk and milk products are highly perishable and require monitoring on the farm and in production facilities.[24]

7.1.3 Environmental Regulation of Dairy by International Organizations

Nutrient pollution and greenhouse gas emissions due to dairy production are common and have risks for human and environmental health. Current environmental policies relevant to milk production thus tend to focus on water pollution and ammonia, and, more recently, on greenhouse gas (GHG) emissions and biodiversity. While there are relatively few environmental policies specific to the dairy industry, there are broader agricultural policies that are nevertheless applicable to dairy producers; all tend to be regulatory in nature. Some policies are the result of international environmental agreements (a trend likely to continue), while others are government-specific. Certain regulations are in place to limit point-source pollution, while others are beginning to address non-point pollution by controlling factors such as manure quantity and disposal. Overall, the number of environmental regulations is increasing in OECD countries. Due to the high costs of new regulations on dairy farmers, many countries have introduced payments to farmers in the form of grants or interest/tax concessions. These financial supports tend to be available for a limited time after the introduction of the regulation.[25]

In some countries, particularly in Europe, policy instruments have been used to encourage organic dairy farming. Many OECD countries in Europe provide financial support in the form of annual per-hectare payments for the conversion to, and maintenance of, organic milk production. In North America, the government provides dairy producers with some assistance to offset the costs of organic

[23]See Demirbas et al. (2006), p. 238.
[24]Ibid., p. 240.
[25]See OECD (2004), p. 20.

certification. Globally there has been a significant increase in organic dairy farmers in the past decade, although organic production still remains a small share of total milk production in most countries.[26]

7.2 The United States Dairy Industry

7.2.1 The Structure of the Dairy Industry in the United States

In the United States, the move towards industrial dairying emerged in the early 1900s, during which there was increasing public concern for dairy sanitation, along with demands for milk to be produced by scientists ("bacteriologists").[27] Americans' views of agricultural production shifted towards a marginal cost perspective that reduced farming to three factors: land, labor, and capital. Under this framework, the farmer would use expert advice to manage the land and labor with the goal of increasing productive efficiency.[28] As such, the industrialization of dairy production focused on consolidating dairy agriculture into the management of the three factors, or inputs (land, labor, and capital) in order to create the most efficient and clean form of production.[29]

The industrial vision of dairying emphasized worker efficiency, the technical education of the farmer, the ability to produce high-quality milk year round, a favorability of large herds, and a high level of production per cow.[30] Newly industrialized farms used fewer acres per cow and used each acre more intensively through the cultivation of high-protein feeds.[31] Dairy industrialization, however, came with added risks and costs. As market milk farms made higher investments and carried larger debt loads, they began to bear greater risk. Furthermore, market milk farms took on increased risks of disease as they pushed a biological milk production system (the cow) to greater production.[32] In sum, the perfect dairy farm, based on theoretical work by Cornell economists, was one that had a larger herd of cows, good valley cropland, high milk yields, substantial amounts of machinery (including a milking parlor and a bulk tank), and a manger farmer with trained employees.[33]

Despite this defined ideal, the "perfect dairy farm" based on industrialization was largely unattainable, with most dairy farms remaining small and centered

---

[26]Ibid., p. 165.

[27]See DuPuis (2002), p. 125.

[28]Ibid., p. 128.

[29]Ibid., p. 130.

[30]Ibid., pp. 131, 141–142.

[31]Ibid., pp. 137–138.

[32]Ibid., p. 139.

[33]Ibid., p. 141.

around labor provided by the family.[34] As a result, the industrialization of dairy farming in the Northeast and Midwest was very slow.[35] Today, the fully industrial and efficient dairy farm has only come close to realization in California, largely due to the state's unique environmental conditions.[36] Unlike farms in the Northeast and the Midwest, Western dairy farms closely fit the neoclassical model of industrial efficiency, with dry-lot farms in the West buying most of their feed, employing the latest milking technologies, and providing labor by employees rather than a family. These farms are able to outcompete other dairy regions and now account for a significant proportion of U.S. milk production.[37]

While all 50 states produce milk, the ten states of California, Wisconsin, New York, Pennsylvania, Idaho, Minnesota, New Mexico, Michigan, Texas, and Washington together produce 71 % of U.S. milk. Over 70 % of dairy farms in the U.S. were family-owned or family corporations in 2002, although there has been a recent increase in the number of large-scale dairy operations. The U.S. dairy industry is also characterized by its heavy reliance on foreign-born workers. Dairy farms are labor-intensive and the majority of jobs are filled with immigrant laborers, many of whom are undocumented.[38] Researchers estimate that foreign labor currently represents 41 % of the dairy workforce.,[39,40] Most U.S. dairy farmers belong to producer-owned cooperatives, which aggregate members' milk and move it to processors and manufacturers. The U.S. dairy industry receives a significant amount of help from the federal government through programs such as federal milk marketing orders (where processors must pay a set minimum to farmers for their milk), a price support program (the Dairy Product Price Support Program), direct payments to producers (the Milk Income Loss Contract program), and the Dairy Export Incentive Program.[41] The two primary objectives of U.S. dairy policy are (1) to provide price supports to establish a minimum farm income, and (2) to incorporate counter-cyclical price stabilization systems so as to ensure an orderly supply and marketing of farm milk.[42]

[34]Ibid., pp. 141–142.

[35]Ibid., p. 142.

[36]Ibid., p. 142.

[37]Ibid., p. 160.

[38]See von Keyserlingk et al. (2013), p. 5406.

[39]See Susanto et al. (2010), p. 1776.

[40]Higher amounts of hired foreign labor relative to the total hired labor increase the probability of exit intentions from dairy farming. Essentially, an expected labor shortage in the future (due to tightened immigration policy) increases the probability of exiting dairy farming. The effects of herd size, however, seem to supersede the effects of the ratio of foreign labor to total labor in influencing the probability of exit from dairy farming (see Susanto et al. 2010, pp. 1778, 1780).

[41]See Hadjigeorgalis (2005), p. 2.

[42]See Bozic and Gould (2009), p. 238.

7.2.2 *Federal Support Policies for the Dairy Sector*

The U.S. dairy industry has undergone more government intervention and regulation than almost any other domestic industry. Both federal and state governments subsidize milk production and regulate dairy prices. Dairy programs have been a major expenditure contributing to U.S. budget deficits for the past several decades.[43] These programs stimulate additional milk output, raise the U.S. price of milk, and shift income from taxpayers and consumers to support producers in the dairy industry.[44] Past studies by the USDA's Economic Research Service (ERS), however, suggest that the impact of U.S. dairy programs on producer returns over the past 20 years has increased the farm price of milk by only 1 %, and has had a limited impact on the financial viability of dairy farms.[45] As a result, dairy programs have been the subject of heavy criticism over the years.

Three important government dairy programs are import quotas on foreign dairy products (i.e. border measures), federal milk marketing orders (FMMOs), and the dairy price support program (as well as associated government purchases of manufactured dairy products). As the federal government's role in milk marketing has evolved, the emphasis of these programs has shifted. Specifically, the indirect assistance to producers through the price support program has been replaced by more direct involvement in the form of limited duration innovative programs.[46] Today, government programs assist with milk marketing through export enhancement, low-income feeding programs, dairy research, advertising, and promotion.[47]

What follows is a brief review of the three main government dairy programs: border measures, federal milk marketing orders, and the dairy price support program.

7.2.2.1 Border Measures

Border measures create import barriers for most foreign dairy products, and create export subsidies for a few manufactured dairy products. The U.S. relies on import controls to prevent the U.S. market from being flooded with inexpensive products from other countries.[48] Dairy products that are imported to the U.S. are subject to tariff-rate quotas (TRQs). These TRQs impose a relatively low tariff on imports up to a determined quota, and set a relatively high tariff on any quantity above this quota.[49] Such efforts have limited U.S. imports of dairy products subjected to less

[43]See Stukenberg et al. (2006), p. 1198.

[44]See Sumner and Balagtas (2002), p. 7.

[45]See Blayney et al. (2006), p. 1.

[46]See Stukenberg et al. (2006), p. 1205.

[47]See Price (2004); Stukenberg et al. (2006), p. 1205.

[48]See Manchester and Blayney (2001), p. 6.

[49]See Sumner and Balagtas (2002), p. 2.

than 6 % of U.S. consumption. The TRQs have also contributed to the higher price of milk and dairy products in the U.S. when compared to dairy products traded on the world market.[50] At the same time, the U.S. government provides small amounts of direct financial subsidies to U.S. exporters of dairy products to encourage the disposal of dairy products acquired in the dairy support program.[51]

7.2.2.2 Federal Milk Marketing Orders

FMMOs were authorized by Congress under the Agricultural Agreements Act of 1937 and are administered by the United States Department of Agriculture (USDA). This legislation states that the purposes of FMMOs are (1) to establish orderly marketing, rather than rely on chaotic marketing conditions; (2) to establish pricing mechanisms that are fair to farmers, distributors, and consumers; and (3) to provide consumers with an adequate supply of high-quality milk.[52] FMMOs arose because improvements in transportation and technology throughout the twentieth century (for example, refrigerated transport) led to a market-boundary struggle between dairy farmers, as milk products were no longer confined to a specific spatial area.[53] The government thus established milk market order legislation to recreate orderly markets because milksheds could no longer be based on physical transportation considerations alone. FMMOs gave the federal government the power to organize and regulate markets,, and they essentially led to the establishment of dairy territories.[54] Ultimately, milk market order legislation was influenced by ideas of economic efficiency, boundary struggles influencing the reach of local milksheds, and improvements in transportation and production technology.[55]

FMMOs jointly established farm, wholesale, and retail prices for milk and manufactured dairy products. The federal milk order program provides dairy producers with a means of equally sharing revenues generated by a classified pricing system. The classified pricing system requires dairy handlers (processors) to pay a higher price for milk used for fluid consumption (as opposed to milk used in manufactured dairy products).[56] The USDA supports the milk price to dairy producers by purchasing storable dairy products, which processors can sell to the USDA at an established rate.[57] FMMOs divide the country into geographic regions, and the manufacturers or processors in each region are required to pay farmers at least the minimum price for the four milk classes. The intent was for the FMMOs to

[50]Ibid., p. 1.

[51]Ibid., p. 2.

[52]See Bartlett (1972), p. 18.

[53]See DuPuis and Block (2006), p. 5.

[54]Ibid., p. 6.

[55]See DuPuis and Block (2006), pp. 4–5.

[56]See Shields (2009), p. 6.

[57]See Stukenberg et al. (2006), p. 1195.

provide market stability, but in reality they allow the federal government, acting for milk producers, to price discriminate.[58]

FMMOs have three key effects: price discrimination, revenue pooling, and regionalization. Price discrimination happens because the minimum processor prices require fluid milk plants to pay a higher price for farm milk than other types of dairy processors. Revenue pooling happens because the regulated farm milk price is an average of minimum prices in various uses. This eliminates the incentive for farmers to compete for the high-value fluid market.[59] In other words, plants that are part of the milk market order pool are the only ones required to pay order prices, and these market-order plants pay into the "pool," which blends the various prices of milk according to the various uses to which that milk has been put, leading to a "blend price" received by farmers.[60] Finally, regional differences in minimum prices and prices received by farmers are maintained through marketing orders that use restrictions on cross-region milk shipments.[61] Because each federal marketing order pertains to a geographically distinct region of the U.S., this discourages the transport of milk across regions as each order may rely on different means of price discrimination and minimum price setting for end products.[62] Today, approximately two-thirds of the nation's fluid milk is regulated under FMMOs. Dairy farmers generally support FMMOs, as they set minimum prices and help balance the marketing power traditionally held by processors.[63]

The Federal Agriculture Improvement and Reform Act of 1996 mandated reforms to the FMMO program. Specifically, the Act changed the way the minimum prices paid to farmers were determined. The Act also consolidated the number of FMMOs.[64]

7.2.2.3 Dairy Price Support Program

Dairy price supports have been part of U.S. dairy policy since 1949 and historically have operated as a market intervention program, where the government offers to purchase nonperishable dairy products from manufacturers at a specified intervention price level.[65] The dairy price support program was legislated by the Agricultural Act of 1949, which specified that farm milk prices must be supported at between 75 and 90 % of parity. The Act authorizes the Secretary of Agriculture to determine the specific price support level within this range. Parity was based on the

[58] See Chouinard et al. (2010), p. 62.

[59] See Cakir and Balagtas (2012), p. 648.

[60] See DuPuis and Block (2006), p. 4.

[61] See Cakir and Balagtas (2012), p. 648.

[62] See Sumner and Balagtas (2002), pp. 3, 5.

[63] See Shields (2009), p. 6.

[64] See Chouinard et al. (2010), p. 59.

[65] See Chang and Mishra (2011), p. 2945.

index of prices paid by farmers for commodities, services, interest, taxes, and wages relative to the base period 1910–1914. Additionally, the Act called for farm milk prices to be supported indirectly by the Commodity Credit Corporation (CCC) through governmental purchases of butter, cheese, and nonfat dry milk from the processors of these products. This process was meant to ensure that farm prices of manufactured milk remained above the legislated support price. Despite this intention, the Act ultimately led to large dairy surpluses held by the government.

To address the problem of dairy surpluses, the Agriculture and Food Act of 1981, the 1983 Dairy and Tobacco Adjustment Act, and the Food Security Act of 1985 all lowered the support price. The Acts also allowed for additional price reductions if government surpluses of manufactured dairy products continued to remain high. Since then, the support price has remained steady.[66] U.S. Farm Bills, however, have been moving towards a less regulated agricultural sector, which would eventually mean reformation of federal milk marketing orders and the gradual elimination of dairy price supports. Nonetheless, such efforts have been postponed due to low commodity prices in recent years; most recent policy efforts have aimed to provide ad-hoc emergency assistance to farmers to address this issue.[67] For example, the Milk Income Loss Contract (MILC) program, implemented in 2000, provides payments to dairy farm operators to partially reimburse their forgone income when price of Class I milk falls below a predefined level.[68]

7.2.2.4 Recent Changes to Dairy Policy Resulting from the Agricultural Act of 2014

The recent 2014 Agricultural Act, also known as the Farm Bill, outlines dairy provisions to update the safety nets that have been in place since the mid-twentieth century. Specifically, the policies pertaining to dairy in the 2014 Agricultural Act include the prescription of two new programs to benefit dairy producers and low-income populations, as well as the repeal or reauthorization of several other dairy-related provisions.[69]

Several dairy programs were terminated in the 2014 Act. First, the Dairy Product Price Support (DPPSP) program was discontinued (although the more permanent Dairy Price Support Program from the 1949 Agricultural Act was retained, but suspended for the duration of the new Farm Bill). Second, the Act called for the termination of the Milk Income Loss Contract (MILC) program once the new Margin Protection Program (MPP) begins.[70] This latter program pays dairy

[66]See Chouinard et al. (2010), p. 61.

[67]See Sumner and Balagtas (2002), p. 7.

[68]See Bozic and Gould (2009), p. 5.

[69]See U.S. Congress (2014).

[70]See Stephenson and Novakovic (2014), p. 1.

operators a margin protection payment when actual dairy production margins fall below the threshold levels for the margin protection payment.[71] The MPP is linked with another new program, the Dairy Production Donation Program, which calls for the government to purchase dairy products to provide nutritional assistance to low-income individuals.[72] Third, the Act terminated the Dairy Export Incentive Program (DEIP).[73] Finally, the Act repealed the Federal Milk Marketing Order Review Commission, which was charged with conducting a comprehensive review and evaluation of the federal and non-federal milk marketing order systems currently in effect.[74]

At the same time as the programs above were eliminated, a number of existing dairy programs were continued. The Act kept the Dairy Forward Pricing Program, which allows non-cooperative buyers of milk who are regulated under Federal Milk Marketing Orders (FMMOs) to offer farmers forward pricing on Class II, III, or IV milk, instead of the minimum FMMO blend price for pooled milk. The Act also kept the Dairy Indemnity Program, which provides payments to dairy producers if a public regulatory agency directs them to remove their raw milk from the commercial market because of contamination issues. Finally, the Act retained some provisions to increase the development of export markets under the National Dairy Promotion and Research Program.[75] Combined, the programs ensure that farmers are able to sell their products and maintain their income.[76]

7.2.3 State-Level Dairy Policy and Regulations

Certain state regulations apply to the dairy market and industry. While state regulations were important during the New Deal period, they have declined in recent years, although many states retain the authority to control milk markets (for example, California and Pennsylvania).[77] States have attempted to regulate producer prices through a variety of ways: direct measures of production costs or changes in these costs; connections to prices in nearby Federal Milk Marketing Orders; economic formulas; and hearing processors. On the consumer level, several states regulate either the wholesale or the retail prices of fluid milk products—or both—although they differ in the regulation of resale prices. Today, retailers generally exert strong control over pricing, and therefore resale price control has

[71] Ibid., pp. 2–3.

[72] Ibid., p. 3.

[73] Ibid., p. 1.

[74] See U.S. Congress (2014).

[75] See Stephenson and Novakovic (2014), p. 1.

[76] See Sumner and Balagtas (2002), p. 7.

[77] See Manchester and Blayney (2001), p. 11.

been reduced in importance and strength. Finally, many states have the authority to regulate trade practices.[78]

7.2.4 The Role of Cooperatives in the Dairy Sector

In addition to state and federal government regulations and policies that influence dairy pricing, there are also nongovernment (market) pricing instruments at play. The most important of these is cooperatives, which have been in existence for decades. They tend to be regional organizations as opposed to local, especially due to a suite of mergers, consolidations, and acquisitions in recent years.[79] Additionally, the purpose of cooperatives has changed over the past century: today, they represent member interests in the rulemaking processes of federal and state regulated markets, sell raw milk to buyers, and process or manufacture raw milk in cooperative plants.[80]

In many cases, cooperatives have assumed the operation of a complete milk procurement and distribution system that reduces costs to individual handlers. They can achieve significant economies of scale by coordinating supply with demand using full-supply arrangements. This supply–demand coordination will also reduce the uncertainties for handlers, fluid milk processors, and dairy product manufacturers. However, if members cannot produce enough milk to meet commitments, the cooperative may have to buy milk from other sources, which is an added cost to members. Cooperatives are also important because they influence price-making in regulated milk markets.[81]

Overall, dairy cooperatives help dairy farmers counter market power of dairy product processors and manufacturers in a variety of ways. These include: improving the milk transportation system; balancing seasonal fluctuations; providing dependable supplies of milk to milk handlers; processing milk into fluid and manufactured products; providing market information; and devoting significant resources to lobbying governments for policies to help dairy farmers.[82] As they have become more consolidated and efficient, cooperatives have lowered their operating costs. The lower costs in turn have helped improve the cooperatives' ability to bargain for and obtain payments above the minimum prices established in the Federal Milk Marketing Orders to help defray some of the costs of servicing those markets.[83] The power of cooperatives is reinforced by government regulations.[84] The 1922 Capper-Volstead Act, for example, partially exempts U.S. farm

[78]Ibid., p. 12.
[79]See Manchester and Blayney (2001), p. 13.
[80]Ibid., pp. 13–14.
[81]Ibid., p. 14.
[82]See Suzuki and Kaiser (2005), p. 1903.
[83]See Manchester and Blayney (2001), p. 14.
[84]See Suzuki and Kaiser (2005), p. 1903.

cooperatives from anti-trust laws. This exemption allows farms to coordinate milk marketing and input purchases.[85]

7.2.5 Structural Changes to the U.S. Dairy Sector: 1970s to Present

The dairy sector accounts for 12 % of the gross value of U.S. agricultural production, and since the 1970s it has undergone significant globalization and structural change. Some farm-level changes include increased farm size; technological evolution; shifting production locations away from those used traditionally; and a decline in productive milk cow numbers.[86] In particular, there has been a decrease in the number of U.S. farms with milk cows, and in the number of dairies.[87] Researchers have found that the combination of older farmers, higher off-farm income, lower returns over variable cost, and a diversification of farm income are associated with the decision to leave dairy farming.[88] Melhim et al., for instance, found that while all farm groupings with fewer than 500 milk cows exhibited negative growth rates in recent years, the number of farms with 500–999 milk cows grew by 36 %. The number of farms with 1000 or more milk cows more than doubled in this same period. Dairy farms and producers are also becoming more geographically concentrated, specifically in western states.[89] This concentration in the dairy industry is expected to lead to further environmental degradation and adverse impacts to rural communities, as discussed below.[90]

Changes to the dairy manufacturing industry include new value-added dairy products; new uses for by-products of dairy production; improved productivity through new technologies; expansion of products to displace those typically imported; and an increase in average processing plant size.[91] Overall, dairy farms are becoming larger, more specialized, and more productive. At the same time, processors and retailers have become more concentrated, raising the possibility for non-competitive behavior in these industries.[92] Dairy policies have become more market-oriented and rely increasingly on international dairy market exports. This is

[85]See Cakir and Balagtas (2012), p. 647.

[86]See Stukenberg et al. (2006), pp. 1202–1203.

[87]See Susanto et al. (2010), p. 1774.

[88]See Bragg and Dalton (2004), p. 3097; Susanto et al. (2010), p. 1775.

[89]See Melhim et al. (2009), p. 2.

[90]Ibid., p. 4.

[91]See Bozic and Gould (2009), p. 1.

[92]See Cakir and Balagtas (2012), pp. 647–649.

partly due to declining domestic milk prices, but it is also a contributing factor to price decline.[93] Foreign direct investment in the U.S. dairy industry has also increased significantly in recent years.[94]

Particularly in the past decade, the dairy industry has faced economic stress due to continued growth in milk production, weakening demand, and high feed costs. During 2007 and 2008, strong demand raised the price of dairy products and the farm price of milk. In 2008, however, feed prices rose rapidly due to greater corn demand for ethanol, strong global demand for grain, heightened investment in commodity markets with uncertain prospects for U.S. corn and soybean yields, and flooding in the Midwest. Together these factors contributed to the rapid decline of farm milk prices while feeding costs remained high, putting dairy farmers in financial danger. At the same time, dairy productivity has become increasingly efficient,[95] but this trend has not been met by a corresponding decrease in the number of dairy cows in farmers' herds. Instead, the dairy cow numbers in farmers' herds has been increasing alongside the rise in efficiency of milk production. These two conditions have led to the expansion of milk supplies at the same time as domestic demand has weakened. Similarly, U.S. dairy exports—especially of cheese—have been growing, and yet overall demand has decreased due to the trifecta of the global economic recession, higher dairy production abroad, and a stronger dollar.[96]

The structural changes associated with the concentration of the dairy industry have also resulted in changes to land-use, including reduced cropland, reduced crops for feed (as they have been replaced by corn for ethanol production), and reduced water usage.[97] Furthermore, there has been a net export of nutrients from major crop-producing areas to areas with a high concentration of animal agriculture. This is the result of the specialization and concentration of livestock and crop production in different geographical locations within the U.S.[98] There has also been the trend of dairy farms migrating westward, resulting in a higher proportion of milk production in the western half of the U.S., which in turn is closely intertwined with increased farm size and cow numbers.[99] This is problematic because the majority of dairy production is now in areas with fewer water resources.[100] In this way, the U.S. dairy industry's push for increased efficiency

[93]See Bozic and Gould (2009), p. 1.

[94]See Stukenberg et al. (2006), p. 1204.

[95]Advancements in genetics, nutrition, and herd management have significantly contributed to the fourfold increase in milk yields between 1944 and 2007, and well as the associated reduction in the number of farms and cows (von Keyserlingk et al. 2013, p. 5406).

[96]See Shields (2009), p. 2.

[97]See von Keyserlingk et al. (2013), pp. 5406, 5408.

[98]Ibid., pp. 5408–5409.

[99]Ibid., p. 5409.

[100]Ibid., pp. 5409, 5412.

and production through technology, productive assets, and economies of scale has come with a heavy environmental price tag.

7.2.6 Environmental Impacts of the Dairy Industry

Environmental concerns related to dairy production in the United States include impacts to water quality and supply, impacts to soil and land use, climate change, energy, nutrient management, and air emissions. Agricultural production is one of the greatest consumers of water (for example, through irrigation and feed production), and this has profound effects on food production—effects that are likely to increase in the face of a growing global population. Another related concern is the pollution of drinking water by nutrients, and particularly water contamination by nitrates.[101] Competition for the land currently used for dairy productions will continue to come from biofuel producers, who in turn are funded by government corn subsidies.[102] The dairy industry is also relies extensively on nonrenewable resources, such as petroleum, for cropping and feeding. It thus has a large carbon footprint.[103] These environmental problems could be addressed through greater pasture-based production, which may be a suitable alternative to increased land use for feed production.[104]

Concentrated animal agriculture is a significant source of nitrogen and phosphorous contamination of surface water. Animal manure is typically applied to land to supply nutrients for crop growth, however excess application can result in nutrient losses and the contamination of groundwater, surface water, and the air. The federal government now requires states to develop watershed implementation plans (WIP) to reduce nutrient losses from farms over a planned and monitored time course.[105] Dairy production also results in air emissions—including greenhouse gases (GHG), volatile organic compounds, and air pollutants—some of which must meet National Ambient Air Quality Standards and are regulated by the U.S. Clean Air Act and the Environmental Protection Agency (EPA). Major sources of emissions include feeding systems, animal housing, manure collection, treatment and storage structures, and land application.[106]

Finally, increased animal numbers and changing animal production systems have led to more stringent federal regulations for concentrated animal feeding operations (CAFOs), including a permit process and a Total Maximum Daily Load program. The EPA manages the CAFO permit program in seven states,

[101] Ibid., p. 5411.
[102] Ibid., p. 5408.
[103] Ibid., p. 5413.
[104] Ibid., p. 5412.
[105] See von Keyserlingk et al. (2013), p. 5415.
[106] Ibid., p. 5415.

while the remaining states have their own regulatory programs under EPA oversight, which operate to regulate effluent discharge.[107] While these programs do address some of the environmental costs of dairy industrialization, they do not directly address dairy cow welfare, as discussed below.

7.2.7 Animal Welfare Impacts of the Dairy Industry

Animal welfare is a growing area of concern for the U.S. dairy industry, especially as production practices intensify and become increasingly concentrated. Animal welfare concerns are fundamentally grounded in the belief that humans have a moral responsibility to maintain an acceptable standard of life care and welfare for all animals, including those used for food. More specifically, animal welfare concerns revolve around issues of pain and suffering, and capacity to exhibit normal behaviors. The U.S. has minimal federal welfare regulation for food production animals.[108] Specific legal requirements for assuring animal welfare are the responsibility of each state, and there are no national regulations governing farm animal management.[109]

With a lack of government regulation, the responsibility for promoting animal welfare in the dairy industry has largely fallen on corporations, and this has been on a voluntary basis and in response to consumer demand. For example, in 2014 Nestlé announced a Commitment on Farm Animal Welfare that aims to improve farm animal welfare across their global supply chain.[110] This announcement followed a company report called "Nestlé in Society" that detailed their Responsible Sourcing

[107]Ibid., p. 5409.

[108]Ibid., pp. 5419–5420.

[109]See Bergman et al. (2014), p. 4275.

[110]Nestlé also makes several commitments in this announcement. First, they write that they will ensure that all animal-derived materials derived used in the manufacturing of Nestlé, products fully comply with applicable local laws and regulations on farm animal welfare. Second, they support the development and implementation of science-based international standards and guidelines by World Organization for Animal Health (OIE), and will contribute to the development of an International Organization for Specification (ISO) technical specification to support the implementation of these OIE guidelines. More specifically, they plan to engage with supply chain partners to establish traceability of the animal-derived materials they source. They will also undertake a monitoring program to understand the current status of farm animal and welfare practices, and the materiality of the use of such practices in their supply chains. Third, they will support and implement actions to promote animal health and welfare, and eliminate practices which contravene the "Five Freedoms" for animals, which they outline in their report. Fourth, they will engage with suppliers, farmers, industry associations, governments, international organizations, NGOs, scientists, and other relevant stakeholders to improve their understanding of farm animal issues; adapt commitments and practices to achieve the goal of improving farm animal welfare in supply chains; develop awareness of farm animal health and welfare in the food supply chain; and implement collective actions to address gaps. Finally, Nestlé commits to regularly and publicly reporting its progress in meeting this Commitment (Nestlé 2014).

Guideline Assessment protocol, which includes farm animal welfare standards. Similarly, in 2012, Dean Foods, one of the nation's leading dairy processors and distributors, highlighted the formation of an industry-leading Animal Welfare Council in its Corporate Sustainability Report. While these developments are signs of potential progress towards a more humane treatment of farm animals, there are limitations to a corporate-led rather than government-initiated approach. Such commitments may be difficult to enforce and standardize, and may lend themselves to a form of corporate greenwashing.

7.2.8 *Human Health Impacts of the Dairy Industry*

7.2.8.1 Current Regulatory Framework

In the dairy industry and beyond, consumers are increasingly demanding information about how their food is produced. Yet there is a mismatch between consumer demand for more regulation and labeling and the current regulatory approach.[111] Currently, the FDA interprets food safety and health from the perspective of acute effects of contamination and long-term effects of nutritional deficiencies. It does not consider long-term effects of exposure to foods containing genetically engineered technologies, antibiotics, hormones, and other chemicals used in production, processing, or packaging as relevant to food safety and health considerations. Ultimately, the FDA and USDA view foods produced using new methods as not differing materially (i.e. not differing in their composition) from traditional products and therefore automatically consider them safe and without the need for regulation.[112]

The U.S. currently lacks any regulations pertaining to the use of milk from cloned animals and their progeny. Such milk may be sold to consumers without labeling, preapprovals, or post-market monitoring under current U.S. law.[113] The FDA also does not require labeling of genetically modified products or products containing ingredients from genetically modified organisms.[114] This differs from the regulatory approach taken by the EU, which bases its food regulation policies on the precautionary principle.[115] The responses by the FDA and the USDA to significant food production controversies have often been delay, inaction, and avoidance of regulatory responsibility.[116] The FDA continues to assert that it lacks the authority to regulate food processing and production.[117] As a result,

[111]See Dragich (2013), p. 405.

[112]Ibid., p. 391.

[113]See Strauss (2011), p. 355.

[114]Ibid., pp. 373–374.

[115]See Dragich (2013), p. 391.

[116]Ibid., p. 423.

[117]Ibid., p. 407.

producers and consumers have taken to voluntary industry action and state and local regulation, which have produced mixed results.[118]

7.2.8.2 rBST

In the dairy industry, one prominent subject of consumer concern about the human health impacts of current dairy production methods is the use of rBST, a synthetically produced hormone that combines with naturally occurring bovine somatotropin (bST) to increase milk production in cows by up to 10 %.[119]

When it was first developed, the FDA required Monsanto to determine whether rBST was biologically absorbed into the human body. Monsanto performed one study and found that it is not absorbed; the company thus concluded that milk from cows treated with rBST is safe for human consumption. Monsanto did not, however, consider whether the drug might have negative health risks to the animals, which could lead to human safety risks as a result. In particular, cows treated with rBST could be more likely to develop an udder infection, which is treated with antibiotics that can remain in milk.[120] Nevertheless, in 1993, the FDA approved Monsanto's controversial application for their version of rBST, finding that there was no significant compositional difference between milk from cows treated with the drug and those that were not.[121]

The FDA concluded that the hormone was safe for cows and that milk produced from such cows was safe for human consumption.[122] Nevertheless, many consumers remain concerned that the use of rBST harms dairy cows; leads to the increased use of antibiotics that wind up in the food supply; cause a number of health problems in humans, including cancer; and impairs milk quality. As such, several consumer groups have pushed for labeling to reflect that milk has been produced with the use of rBST.[123] Several court suits regarding state regulations that both prohibited (Ohio) and required (Vermont) the labeling of milk derived from rBST have resulted in court orders concluding that requiring a producer to label—or to not label—a product goes against first-amendment rights.[124] Therefore, producer labeling can only occur if it is voluntary and not misleading.[125]

---

[118]Ibid., pp. 423–424.

[119]Ibid., p. 398.

[120]See Beyranevand (2012), p. 10.

[121]Ibid., pp. 1–2.

[122]See Dragich (2013), p. 398.

[123]Ibid., p. 399.

[124]Ibid., p. 415.

[125]Ibid., pp. 415–416.

7.2.8.3 Modified Atmosphere Packing

The second production concern related to dairy products is the food industry's use of Modified Atmosphere Packaging (MAP) to help extend the shelf life of packaged foods, including dairy products (specifically cheese).[126] In MAP packaging, the air in the package is replaced with either a single gas or a mix of gases, including carbon monoxide. This approach extends shelf life by slowing respiration; maintaining the appearance, texture, and quality of the food; slowing the growth of some microorganisms; and preserving flavor. However, MAP does not inhibit the growth of many bacteria responsible for food-borne illnesses. This is an issue because MAP packaging makes food products look fresher than they actually are, which may cause consumers to buy food products they would have avoided had they not been treated to look more well-preserved. This packaging technique can thus lead to consumption of older food products that contain harmful bacteria.[127] Nevertheless, the FDA and USDA argue that the "indirect" additives used in food packaging migrate into food in negligible quantities and have little effect on the food itself, and consequently do not require regulation.[128]

7.2.8.4 Raw Milk

The United States Public Health Service (USPHS) prescribed the Pasteurized Milk Ordinance (PMO) in 1924 as a model regulation to help states and municipalities create an effective program to prevent milk-borne diseases. It contains provisions governing the production, processing, and sale of Grade "A" milk and milk products, and calls for only safely pasteurized or processed milk and milk products to be sold to the final consumer or distributors (such as restaurants and grocery stores). The PMO is used as a basic standard in the Voluntary Cooperative State-USPHS/FDA Program for the Certification of Interstate Milk Shippers, in which all U.S. states and territories participate.[129]

Three percent of the U.S. population drinks unpasteurized milk, but this number is growing and contributes to the increased public debate between public health authorities and consumers, particularly because of the significant risk of foodborne illness associated with the consumption of raw milk and raw milk products.[130] While the FDA regulations prohibit the interstate sale of unpasteurized milk for human consumption,[131] 30 states allow raw milk sales within their borders.[132]

[126]See Dragich (2013), pp. 403–404.

[127]Ibid., p. 404.

[128]Ibid., p. 419.

[129]See Kennedy (2004), p. 1.

[130]See David (2012), p. 598.

[131]See Langer et al. (2012), p. 386.

[132]See David (2012), p. 598.

These states have legalized the sale or distribution of raw milk and raw milk products through statutes, administrative rules or regulations, and policies.[133]

7.2.9 Organic Dairy Production

Organic milk production is one of the fastest-growing segments of organic agriculture in the U.S., especially among small dairy operations with the goal of improving farm profitability. Certified organic milk production systems rely on ecologically-based practices that prohibit antibiotics and hormones in the cow herd, as well as the use of synthetic chemicals in dairy feed production. Organic producers also attempt to accommodate the animals' natural nutritional and behavioral requirements (for instance, access to pasture). These requirements can add to production costs and create obstacles to adoption due to additional managerial challenges, time commitment, and space required.

It is difficult and costly to transition from conventional to organic production for several reasons.[134] The USDA's National Organic Program (NOP) requires that all pasture and cropland providing feed for organic dairies be managed organically for a minimum of 36 months. Dairy cattle must be fed 100 % organic feed and receive organic health care for 12 months before their milk can be certified, thus adding to the cost of the transition.[135] Producers must keep adequate records and maintain a detailed, verifiable audit trail so that each animal can be traced back to the farm where it lived. The records must include information on the amount and sources of all medications administered, as well as details on all feeds and feed supplements bought and fed.[136] Organic milk production costs are consequently higher than those for conventional, especially during the transition period. Despite the initially higher overhead, however, returns above operating and capital costs on these small organic operations compare favorably with those of small conventional operations. Most notably, organic milk production is competitive in terms of the ability of a farm to use its pasture for a significant portion of dairy feed. This may incentivize pasture-based dairies to transition to organic production. It may also encourage the development of startup organic dairies in situations where pasture can be suitably managed as organic dairy feed.[137]

Organic and non-rBGH milk production represent one of the few opportunities for dairy producers to market differentiated identity-preserved products.[138] Yet ironically, the organic milk industry is more concentrated than its conventional

---

[133] See Kennedy (2004), p. 1.

[134] See McBride and Greene (2009), p. 793.

[135] Ibid., p. 794.

[136] See U.S. Congress (1990), pp. 25–10.

[137] See McBride and Greene (2009), p. 811.

[138] See DuPuis (2002), p. 225.

counterpart, with only three firms (Horizon Organic, Alta Dena, and Organic Valley) serving 95 % of the organic milk market.[139] Horizon is a vertically integrated diary company that provides organic milk mainly by transporting it from its own two centralized, large herds. In contrast, most conventional dairy companies, including the largest national firms (for example Suiza and Dean Foods), do not tend to own farms.[140]

7.3 The Dairy Industry in Hawaii: A Case Study

The evolution of the dairy industry in the state of Hawaii is an excellent example of the way in which, over the past several decades, the U.S. dairy industry has become consolidated in the name of efficiency—regardless of the potential costs to the quality of milk produced, and in turn the environmental and human health impacts of milk consumption. Like elsewhere in the United States, Hawaii consumers are now beginning to push back against this economic transformation, and call for greater protection of local, smaller-scale dairy farms. The following section highlights the history and political economy of milk production in Hawaii in order to demonstrate both the challenges and potential opportunities for reforming current dairy policy in Hawaii, and in the United States more broadly.

7.3.1 History of Dairy in Hawaii

Cattle were first introduced in Hawaii in 1793, and the first commercial dairy began operating in 1869. During the Second World War, dairy farms played an important role in the health of the military, as milk was given freely to injured military personnel. By 1955, there were 86 dairies in Hawaii and by 1965 cow populations had peaked with 15,100 head. The rapid growth of dairy operations, coupled with a limited number of processors, led to discriminate purchasing practices and prices that caused turmoil within the industry.[141] Protests by dairy producers in the state capital of Honolulu led to the creation of the Milk Act. One of the goals of the Act was to ensure order in the market place—it was similar to the Federal Milk Marketing Orders throughout the rest of the country, however these did not apply to Hawaii.[142]

Like other national dairy policies, the Milk Act was aimed at ensuring a sufficient supply of fluid milk by providing a reasonable return on investment to

[139]Ibid., pp. 223, 225.

[140]Ibid., p. 223.

[141]See Lee (2007), p. 2.

[142]Ibid., p. 3.

producers. This legislative support was justified according to the logic that fluid milk must be given special consideration as an essential nutritional component in the average American diet.[143] The Milk Act established quotas, which represented the fluid milk demand within a milk shed. It also set the baseline for milk production within a farm. Milk quotas are based on production levels for Grade A, Class I milk—in other words, the quotas are based on fluid milk only. Any surplus milk receives a lower price for Class II utilization (milk that becomes, for instance, a dairy product such as yogurt or cottage cheese). Quotas are administered by the Hawaii Department of Agriculture via the Milk Control Branch, though they may be traded among farmers without the interference of any state entity. Milk quotas provide assurance that milk produced within the quota will receive the Class I price, since the quota ostensibly represents the fluid milk demand.[144]

Despite the price support of the quota system, by 1974 dairy operations in Hawaii began to decrease, while the number of cows per farm continued to increase. Hawaii remained self-sufficient in milk production until 1982, when the milk supply was tainted with Heptachlor, a chemical found in the pineapple-based roughage given to dairy cows.[145] During the period of time in which Hawaii-produced milk was recalled, consumers shifted to powdered milk and reconstituted fluid milk imported from the mainland. This unfortunate event opened the gateway for processors to import milk from the mainland on a regular basis, even when the Heptachlor scare was over. Starting in 1984, when Safeway first imported processed milk into the marketplace, processors began to look outside of Hawaii for milk. While Hawaiian processors initially remained loyal to local milk, they would import milk as "filler" during the summer months when local production was depressed. Nevertheless, when bulk shipment became possible in the mid-1990s, processors began purchasing milk from the mainland on an even greater basis.[146]

In the past several decades since the Heptachlor incident, environmental issues, feed costs, improvements in transportation, lower milk prices to farmers, and an aging ownership in the industry, along with changing dynamics of the market place, have all led to the decline of Hawaii's dairy industry.[147] One particular point of contention for dairy producers in Hawaii is the milk pricing system. From the inception of the Milk Act in 1967 and up until 1998, milk was priced according to cost of production. In 1997, feed prices were at an all time high and producers petitioned for a milk price increase. While a cost of production study performed by Hawaii's Department of Agriculture determined that this price increase was warranted, milk processors insisted that producers agree to a more competitive pricing system. They wanted a system that better reflected the marketplace for fluid

[143]Ibid., p. 4.

[144]Ibid., p. 5.

[145]See Lee (2007), p. 2.

[146]Ibid., p. 15.

[147]Ibid., p. 2.

milk; one that could make Hawaiian milk producers competitive with mainland imports.[148] As a result, the Oahu milk shed moved from prices based on cost of production to prices based on a formula that consisted of the California Class I price, plus shipping, plus a premium. The Milk Act was amended to reflect this change in 1998.[149]

While the pricing formula aimed to set a premium for locally produced milk (approximately 20 cents per gallon at the farm gate), in practice this premium became negligible. This was because rising oil prices beginning in 2005 increased the shipping cost, which effectively erased the premium for local milk. Furthermore, there was a loophole in the existing Milk Act that allowed a processor to declare Class II utilization. Because Class II milk was only paid about 32 % of the Class I price, processors took advantage of this loophole and purchased a higher proportion of Class II than Class I milk from local producers.[150] This move drove a number of local dairy operations out of business since they ended up receiving a much lower blend price for Class I and Class II milk, rather than the higher price intended for exclusively Class I milk.

Meanwhile, milk imports to Hawaii via bulk containers increased to fill the void created by local producers as they left the industry. With the new pricing formula, Hawaii had essentially removed a method of ensuring a reliable supply of fluid milk and replaced it with a system that was much more variable and dependent on external forces.[151] Given that the goal of state and federal dairy programs is to provide a stable supply of milk for the fluid market, processors might be expected to source Class I/fluid milk—the demand for which is highly *inelastic*—from local dairies first. Only then might they source milk for Class II products—the demand for which is much more *elastic*—from milk producers further afield. Yet processors have chosen to do exactly the opposite: instead, they buy local milk for Class II utilization and import fluid Class I milk in order to avoid paying a premium for local fluid milk. While the Hawaiian processors' approach makes sense from a purely economic bottom-line perspective, it raises questions on the quality of imported products and the sustainability of local milk markets. Would Hawaii want to be totally dependent on imports?[152] From the perspective of Hawaiian milk producers, the answer is certainly no.

[148]Ibid., p. 9.

[149]Ibid., p. 9.

[150]Ibid., p. 9.

[151]Ibid., p. 10.

[152]See Lee (2007), pp. 14–15.

7.3.2 *Trade-Offs Between Efficiency and Quality*

Quality is an important issue in the debate over Hawaii's current milk sourcing policies. Imported milk travels great distances to reach the islands and thus there is more of an opportunity for health- and nutrition-related problems to arise. Milk quality can be measured in two broad categories: (a) bacteria-related issues; and (b) nutrient changes in milk following handling. Bacteria-related issues result in spoilage and potential illness. Handling, such as through the extended heating of milk, can denature proteins or amino acids and as a result change the nutrient values of the milk. The pasteurization process itself does alter some proteins and enzymes. Hawaiian milk is unique in that it is pasteurized *twice*—first before being shipped to Hawaii and then a second time prior to bottling for retail in Hawaii.[153]

One recent study shows that milk imported to Hawaii often exceeds federal regulatory limits for bacterial counts as early as 5 days prior to expiration.[154] While the source or site of this contamination is still undetermined, the author of the study notes a lengthy duration of time between milking the cow on the mainland and the milk's arrival at a retail location in Hawaii. He estimates that at its expiration date, the age of milk from a California cow is at minimum 24.7 days, and at maximum 30 days. It is unlikely that any other state in the nation has 25–30 day old milk sold to consumers unless it is ultrapasteurized. Furthermore, no law is actually broken with this excessive duration between milking and retail because no states regulate shelf-life.[155] Hawaii's current milk sourcing policies evidently raise concerns about the quality and safety of the milk it imports.

7.3.3 *The Value of Local Milk*

Since there is no law that forces local processors to use locally produced milk, the remaining local dairies in Hawaii (of which there are two) are vulnerable to the whims of market forces and processor decisions. Increasingly, however, consumers in Hawaii are exhibiting preferences for locally-produced food.[156] This new trend stems directly from concerns about the environmental, economic, and human/animal health impacts of the current industrialized and globalized food system. In a recent study of consumer food preferences on Oahu, for example, consumers stated that if they understood all the characteristics of local milk—including freshness and level of pasteurization—they would be willing to pay up to $1.25 more per quart of local milk.[157]

[153]It is worth noting that California specifically prohibits the repasteurization of fluid milk for fluid purpose. Lee (2007), p. 16.

[154]Ibid., p. 17.

[155]Ibid., p. 18.

[156]See OmniTrak Group Inc. (2011), p. 1.

[157]Ibid., p. 2.

Already, there are some Hawaiian branding programs that have helped to identify local milk and allowed it to receive a premium at the retail level, in part because of consumer willingness to pay slightly higher prices. The Mountain Apple brand, for instance, is sold at regional chain grocery stores and has had great success in marketing local products, and local milk in particular. The growing sales of the Mountain Apple brand, coupled with the introduction of Whole Food's branding of local milk, attest to consumer preference to buy local when it is available. At the same time, Hawaii's two remaining local dairies face significant hurdles in maintaining their businesses, especially as feed costs continue to increase and make it difficult to compete with mainland milk imports. The dairies' viability rests on their capacity to gain a premium for their milk from consumers who perceive benefits of purchasing and consuming local milk—such as, support for the local agricultural economy, freshness, and greater nutritional value.

7.4 Conclusion

The dairy industry is one of the most regulated sectors in developed countries, largely due to milk's unique characteristics. The United States in no exception: the U.S. dairy industry has undergone more government intervention and regulation than almost any other domestic industry. Federal and state governments subsidize milk production and manage dairy prices. While dairy industries have historically been heavily protected by nation-states, economic liberalization in recent years has proceeded at a steady pace through policies such as tariff reduction and quota elimination. In the U.S. in particular, recent 2014 Agricultural Act (Farm Bill) dairy provisions revamp the safety nets that have been in place since the mid-twentieth century.

Another important trend in the U.S. dairy industry is the great intensification of U.S. dairy production over the past few decades. While this has led to greater "efficiency" in the eyes of some, it has also led to greater external costs, as noted by critics of the industrial agriculture model. Namely, current milk production techniques pose potential threats to human and environmental health, such as increased exposure to antibiotics, as well as impacts on water quality and air emissions.

Compounding the environmental and health issues, there is a mismatch between consumer demand for more regulation and labeling and the current regulatory approach. Because the FDA interprets food safety and health only in terms of acute effects of contamination and nutritional deficiencies, it does not consider relevant to food safety the long-term effects of exposure to food containing genetically engineered technologies, antibiotics, hormones, and other chemicals used in food production, processing, or packaging. As a result, producers and consumers have taken to voluntary industry action in addition to pushing for more state and local regulation—the ultimate results of which remain to be seen.

References

Bartlett RW (1972) Are federal milk orders operating in the public interest? Ill Agric Econ 12(1): 18–22
Bergman MA, Richert RM, Cicconi-Hogan KM et al (2014) Comparison of selected animal observations and management practices used to assess welfare of calves and adult dairy cows on organic and conventional dairy farms. J Dairy Sci 97(7):4269–4280
Beyranevand LJ (2012) Milking it: reconsidering the FDA's refusal to require labeling of dairy products produced from rBST treated cows in light of International Dairy Foods Association v Boggs. Fordham Environ Law Rev 23(1):102–138
Blayney D, Gehlhar M, Bolling CH et al (2006) U.S. dairy at a global crossroads. U.S. Department of Agriculture, Economic Research Service, Economic Research Report Number 28. http://www.ers.usda.gov/media/868595/err28_002.pdf. Accessed 16 Dec 2014
Bragg LA, Dalton TJ (2004) Factors affecting the decision to exit dairy farming: a two-stage regression analysis. J Dairy Sci 87(9):3092–3098
Bozic, M, Gould BW (2009) The dynamics of the U.S. milk supply: implications for changes in U.S. dairy policy. University of Wisconsin-Madison Department of Agricultural & Applied Economics, Staff Paper No. 540. http://www.aae.wisc.edu/pubs/sps/pdf/stpap540.pdf. Accessed 16 Dec 2014
Cakir M, Balagtas JV (2012) Estimating market power of U.S. dairy cooperatives in the fluid milk market. Am J Agric Econ 94(3):647–658
Chang HH, Mishra AK (2011) Does the Milk Income Loss Contract Program improve the technical efficiency of U.S. dairy farms? J Dairy Sci 94(6):2945–2951
Chouinard HH, Davis DE, LaFrance JT et al (2010) Milk marketing order winners and losers. Appl Econ Perspect Policy 32(1):56–76
David SD (2012) Raw milk in court: implications for public health policy and practice. Public Health Rep 127(6):598–601
Demirbas N, Karahan O, Kenanoglu Z et al (2006) The evaluation of the developments in food safety systems formation in the world for dairy industry from the standpoint of Turkey. Agric Econ-Czech 52(5):236–243
Dragich M (2013) Do you know what's on your plate? The importance of regulating the processes of food production. J Environ Law Litig 28(3):385–445
DuPuis EM (2002) Nature's perfect food: how milk became America's drink. New York University Press, New York
DuPuis EM, Block D (2006) Sustainability and scale: U.S. milk-market orders as relocalization policy. Environ Plann A 40(8):1987–2005
Hadjigeorgalis E (2005) The U.S. dairy industry and international trade in dairy products. New Mexico State, College of Agriculture and Home Economics, Agricultural Experiment Station, Cooperative Extension Service Technical Report 42. http://ageconsearch.umn.edu/bitstream/23950/1/tr050042.pdf. Accessed 16 Dec 2014
Kennedy P (2004) An overview of U.S. state milk laws. http://www.realmilk.com/state-updates/raw-milk-statutes-and-codes-page-1/. Accessed 16 Dec 2014
Langer AJ, Ayers T, Grass J et al (2012) Nonpasteurized dairy products, disease outbreaks, and state laws – United States, 1993-2006. Emerg Infect Dis 18(3):385–391
Lee CN (2007) Issues related to Hawaii's dairy industry. Department of Human Nutrition, Food, and Animal Science, College of Tropical Agriculture and Human Resources, University of Hawaii, Manoa
Manchester AC, Blayney DP (2001) Milk pricing in the United States. U.S. Department of Agriculture, Market and Trade Economics Division, Economic Research Service. Agriculture Information Bulletin No. 761. http://www.ers.usda.gov/publications/aib-agricultural-information-bulletin/aib761.aspx. Accessed 16 Dec 2014
McBride WD, Greene C (2009) Costs of organic milk production on U.S. dairy farms. Appl Econ Perspect Policy 31(4):793–813

Melhim A, O'Donoghue EJ, Shumway CR (2009) Do the largest firms grow and diversify the fastest? The case of U.S. dairies. Appl Econ Perspect Pol 31(2):284–302

Nestle (2014) Nestle commitment on farm animal welfare. http://www.nestle.com/asset-library/documents/creating%20shared%20value/rural_development/nestle-commitment-farm-animal-welfare.pdf. Accessed 16 Dec 2014

OmniTrak Group Inc. (2011) Local food market demand study of O'ahu shoppers executive summary. Ulupono Initiative, Honolulu

Organisation for economic cooperation and development (2004) Agriculture, trade, and the environment: The dairy sector. http://www.oecd-ilibrary.org/docserver/download/5104051e.pdf?expires=1418751308&id=id&accname=ocid177224&checksum=6959B94F8EBAB3E964561F79ABAD8EC4. Accessed 16 Dec 2014

Price MJ (2004) Effects of U.S. dairy policies on markets for milk and dairy products. U.S. Department of Agriculture, Economic Research Service. Technical Bulletin Number 1910. http://www.ers.usda.gov/media/880661/tb1910.pdf0x-_ls-.pdf. Accessed 16 Dec 2014

Shields DA (2009) Dairy market and policy issues. Congressional Research Service 7-5700, R40205. http://congressional.proquest.com/congressional/result/pqpresultpage.gispdfhitspanel.pdflink/http%3A$2f$2fprod.cosmos.dc4.bowker-dmz.com$2fapp-bin$2fgis-congresearch$2f4$2f5$2fd$2fd$2fcrs-2009-rsi-0126_from_1_to_13.pdf/entitlementkeys=1234|app-gis|congresearch|crs-2009-rsi-0126. Accessed 16 Dec 2014

Stephenson MW, Novakovic AM (2014) The dairy subtitle of the Agricultural Act of 2014. Choices 29(1):1–4

Strauss D (2011) An analysis of the FDA food safety modernization act: protection for consumers and boon for business. Food Drug Law J 66(3):353–376

Stukenberg D, Blayney D, Miller J (2006) Major advances in milk marketing: government and industry consolidation. J Dairy Sci 89(4):1195–1206

Sumner DA, Balagtas JV (2002) United States' agricultural systems: an overview of U.S. dairy policy. In: Roginski H, Fuquay J, Fox P (eds) Encyclopedia of dairy sciences. Elsevier Science Ltd, Oxford

Susanto D, Rosson CP, Anderson DP et al (2010) Immigration policy, foreign agricultural labor, and exit intentions in the United States dairy industry. J Dairy Sci 93(4):1774–1781

Suzuki N, Kaiser HM (2005) Impacts of the Doha Round Framework Agreements on dairy policy. J Dairy Sci 88(5):1901–1908

U.S. Congress (1990) Organic Foods Production Act of 1990, Title XXI of the Food, Agriculture, Conservation, and Trade Act of 1990, Public Law 101-624. U.S. Congress, Washington, D.C.

U.S. Congress (2014) Act of February 7, 2014, Public Law 113-79, 128 STAT. 649, to provide for the reform and continuation of agricultural and other programs of the Department of Agriculture through fiscal year 2018, and for other purposes. U.S. Congress, Washington, D.C.

von Keyserlingk MAG, Martin NP, Kebreab E et al (2013) Invited review: sustainability of the US dairy industry. J Dairy Sci 96(9):5405–5425

World Trade Organization (1994) International dairy agreement. http://www.wto.org/english/docs_e/legal_e/ida-94_01_e.htm. Accessed 16 Dec 2014

Chapter 8
Live Export of Farm Animals

Lewis Bollard

Abstract Every year livestock producers send millions of cows, pigs, sheep, and other farm animals on international journeys to slaughter in foreign lands. This "live export trade" is valuable and widespread—worth roughly $21 billion a year to the estimated 109 countries that engage in it. But, as the trade has expanded over the last century, it has also become a source of increasing public controversy due to the abuses that animals often suffer in the trade. This chapter documents the history of the trade, its current state, and the nascent legal framework that regulates it. I argue that the legal framework has failed to address the animal welfare problems associated with the trade. Several more promising international legal solutions exist to ameliorate problems associated with the trade—particularly working through international institutions and adopting a live export treaty. But only one reform will end the cruelty associated with the trade: abolishing the live export trade itself.

8.1 Introduction

In 2011, the Australian television news program, ABC's Four Corners, aired an undercover investigation by advocacy group Animals Australia at Indonesian slaughterhouses exposing the treatment of Australian cattle.[1] The presenter began the show by noting that he was personally repulsed by the "gross, horrible abuse."[2] The footage showed Australian cattle, who had been shipped for live Halal slaughter in Indonesia, screaming in pain as they were whipped, kicked, dragged with ropes, and slowly cut apart while alive.[3] Some cattle took 33 neck cuts to die, while others were still kicking in pain minutes after being cut.[4] All of these abuses

[1]ABC Four Corners (2011).

[2]ABC Four Corners (2011).

[3]ABC Four Corners (2011).

[4]ABC Four Corners (2011).

L. Bollard (✉)
Open Philanthropy Project, Washington, DC, USA
e-mail: lewis@openphilanthropy.org

G. Steier, K.K. Patel (eds.), *International Farm Animal, Wildlife and Food Safety Law*, DOI 10.1007/978-3-319-18002-1_8

occurred in a slaughterhouse approved and inspected by Meat and Livestock Australia, the Australian industry body.[5]

The screening of the hour-long report created a public outcry over the live export trade. Dr. Temple Grandin, the most prominent animal welfare consultant to the U.S. meat industry, called the abuses "absolutely terrible," stating "this is clearly absolutely not acceptable for a developed country to be sending those cattle in there."[6] A week later, the Australian Government suspended all live exports to Indonesia after backbench members of Parliament threatened to revolt over the issue.[7] In the ensuing several years, the live export trade became a subject central to Australian politics as Animals Australia released over 20 undercover investigations documenting the abuse of Australian cattle and sheep across Southeast Asia and the Middle East.[8]

But the subsequent years also showed the resiliency of the live export trade. Within a month of the indefinite suspension of the trade to Indonesia, the Australian Government had authorized live exports to resume.[9] When a new Australian prime minister came to power in 2013, he even apologized to the Indonesian government for the short suspension of the trade "in panic over a TV program."[10] And in late 2014, the Government announced that it was close to finalizing a new agreement with China that would allow for the live export of another million cattle annually, worth an estimated $1 billion.[11] Animal welfare groups promptly condemned the move.[12]

This chapter explores the political and legal controversy surrounding the international live export trade—the shipment of millions of cattle, sheep, and other farm animals every year across national borders. Although this trade has origins in the nineteenth century, it has only become an object of heated debate in the last few decades as the animal welfare movement has exposed abuses associated with the trade. And it has only become the subject of international law and regulatory systems even more recently. This chapter explores the origins of the trade, its current state, and the current legal frameworks that govern it. Much of the chapter focuses on the case study of Australia's live export trade to Asia and the Middle East, since that trade has been the focus of the most controversy—and the most attempts at regulation.

The chapter concludes with proposed legal solutions for the trade. I argue that the trade in its current state is cruel and the source of significant animal suffering. Reforms to the trade—and in particular tougher international oversight—could

[5]ABC Four Corners (2011).

[6]Ferguson (2011).

[7]Coorey and Allard (2011).

[8]Animals Australia (2015).

[9]Ferguson and Masters (2012).

[10]News.com.Au (2013).

[11]Medhora (2014).

[12]Medhora (2014).

significantly decrease the suffering associated with the trade. But the experience of reforms to date suggests that no amount of reform could stop the suffering of animals in live export. Only the abolition of the live export trade could achieve that.

8.2 The History of the Live Export Debate

The modern live export trade's origins in the exports of live cattle by nineteenth century American and Australian cattlemen to European and Asian consumers. From its inception, the trade has been controversial. As early as 1879, anti-cruelty advocates drew attention to overcrowding and inhumane deaths in the transport of cattle from America to Europe. Since the 1980s, the issue has become a central focus of anti-cruelty advocates. These advocates have had success in reforming aspects of the trade, but have seldom achieved their ultimate goal—stopping the trade entirely.

8.2.1 The History of the Live Export Trade

The trade in live animals for meat likely dates back to the first trade between early pastoralists. Prior to the invention of refrigeration, the trade in live animals was the primary trade in meat. But most live export was likely local, since when animals are forced to walk trade routes they require more feed and lose weight rapidly. Long-distance transnational live export trade routes only became possible with the advent of large sailing ships regularly covering trade routes in the nineteenth century.

In the early nineteenth century, Britain was rapidly urbanizing, and domestic cattle producers, perennially struck by food and mouth disease, could no longer satisfy the demand for beef from Britain's new city-dwellers.[13] America's new meat packing titans, who had built huge slaughterhouses in Chicago after the Civil War to slaughter range cattle imported from as far afield as Texas and the Dakotas, had experimented with exporting dressed beef to Britain.[14] But the beef often spoiled, so the meat packers began experimenting with sending live cattle.[15] The trade only became profitable, though, in 1877, when Britain halted all imports of European cattle after they were struck with an outbreak of pleuro-pneumonia.[16] American meat-packers pounced on the opportunity and in just the two years between 1876

[13]Zimmerman (1962).
[14]Zimmerman (1962).
[15]Zimmerman (1962).
[16]Zimmerman (1962).

and 1878, the number of live cattle exported from American to Britain ballooned from around 250 animals a year to around 25,000.[17] The nineteenth century American-Britain live export trade resembled many aspects of the modern live export trade. The railroads brought cattle from Missouri, Kansas, Nebraska and other range-states to the two major export ports: New York and Boston, where they were shipped to the major import ports of London and Liverpool.[18] By August, 1878, the Liverpool meat markets were selling more American beef than British beef.[19]

In Australia, the live export trade began in 1885, with a shipment of cattle from the Northern Territory to Hong Kong.[20] At the time both places were British colonies, and the lack of refrigeration meant that the live export trade was the only viable way to ship meat. But in 1889 the trade ceased due to an outbreak of disease, and only re-emerged in the late 1940s, becoming a significant industry in the late 1970s.[21] In the meantime, Taiwan had become the primary exporter of live pigs to Hong Kong.[22] Taiwan extended that trade in live pigs to Japan and Malaysia in the 1980s and 1990s—until the outbreak of food-and-mouth disease in 1997 stopped the trade.[23]

In the late twentieth century, a reduction in trade barriers and resulting increase in all forms of trade, caused a significant rise in live exports, especially within the European Union (EU). In Britain alone, the number of cattle exported live each year rose from 162,000 in 1986 to two million in 1993, the year that the common European market came into force, eroding the last remaining barriers to trade between Britain and Europe.[24] Within the EU, the number of pigs transported across national borders every year increased from 16 million to 28 million between 2005 and 2009 alone, mostly in the export of pigs from farms in Denmark and the Netherlands to slaughter in Germany.[25]

8.2.2 *The History of Efforts to Reform or Abolish the Trade*

From the start, the live export trade came under scrutiny from animal welfare advocates. In 1879, the recently formed American Society for the Prevention of Cruelty to Animals (ASPCA) conducted an investigation into the "life of exported

[17]Zimmerman (1962).

[18]Zimmerman (1962).

[19]Zimmerman (1962).

[20]Livestock Export Review (2003).

[21]Livestock Export Review (2003).

[22]Brooke et al. (2008), p. 305.

[23]Brooke et al. (2008), p. 305.

[24]Howkins and Merricks (2000).

[25]European Commission (2011).

cattle."[26] The investigator found that cattle in the transatlantic trade were typically overcrowded into narrow stalls, where the only space to lie down was in their own manure.[27] He also found that veterinarians seldom accompanied the ships, and euthanasia methods were rudimentary. In one notable exchange, the ASPCA investigator asked a sailor how he had killed an injured steer on board:

> Investigator: "How did you kill the steer?"
> Sailor: "Oh I did it with this" (showing a small knife)
> Investigator: "You could hardly have killed him immediately with that knife."
> Sailor: "No, but I gave him such a cut that he couldn't get over it."[28]

The average export ship lost 1–10 % of its cattle in transit, but there were stories of "ghost ships," like one that left Boston with 400 cattle, was battered by a storm, and arrived in Liverpool with just one live cow on board.[29] At one point, Lloyd's of London, which insured live export ships, reportedly wrote to the Privy Council, urging it to compel improvements to the "most unhealthy and disgraceful manner" in which cattle were shipped.[30] Ultimately conditions slowly improved after British courts made the shipping firms liable for lost cattle (previously exporters had been), and as shippers realized that avoiding winter sailings and installing better ventilation reduced their losses.[31] In the ensuing decades, animal welfare advocates largely diverted their focus to other issues, although the Humanitarian League waged a short-lived campaign around 1897 against the live export of cattle across the Irish Sea.[32]

But the live export issue gained prominence again in 1950s Britain as the public learned of abuses in the live cattle trade to Europe. In 1956, a new group called the Protection of Livestock for Slaughter Association started a campaign to regulate the trade.[33] This led the British Parliament the following year to establish the Balfour Committee on the Export of Live Cattle to investigate the trade.[34] The Committee's report, published in 1957, concluded that the live export trade was undesirable for animal welfare, but that the alternative of the carcass trade was not economically viable.[35] The Committee, though, recommended an extensive set of animal welfare reforms to the trade—from restrictions on the types of ships to be used and the conditions on board to the weather conditions under which boats could sail.[36] The

[26]Zimmerman (1962).
[27]Zimmerman (1962).
[28]Zimmerman (1962).
[29]Zimmerman (1962).
[30]Zimmerman (1962).
[31]Zimmerman (1962).
[32]Zimmerman (1962).
[33]Howkins and Merricks (2000).
[34]Howkins and Merricks (2000).
[35]*See* comments of Baroness Stocks, House of Lords debate, Jan. 21, 1971, in Hansard, http://hansard.millbanksystems.com/lords/1971/jan/21/export-of-live-animals.
[36]See comments of Baroness Stocks, House of Lords debate, Jan. 21, 1971, in Hansard, http://hansard.millbanksystems.com/lords/1971/jan/21/export-of-live-animals.

British Government largely adopted these recommendations in 1957. The Government pledged to only allow the export of British cattle to European countries that agreed to a set of conditions that came to be known as the "Balfour assurances"—that the cattle not be re-exported, that their journey after embarkation be no longer than 60 miles, and that the cattle be slaughtered humanely.[37]

In the following decades, the newly active animal welfare groups exposed numerous violations of the Balfour assurances, igniting new calls for abolition of the live export trade. The Royal Society for the Prevention of Cruelty to Animals (RSPCA), in particular, conducted a number of undercover investigations, which documented trucks transporting animals excessive distances and foreign slaughterhouses killing British animals in inhumane conditions. The RSPCA drew on these investigations to mount a national "Stop the Live Export of Food Animals" campaign.[38] In an impassioned debate in the British House of Lords in 1971, several lords and baronesses argued that the live export trade is inherently cruel. Lord Somers, reporting on the RSPCA's most recent investigation, argued that the Balfour assurances were inadequate to stop animal cruelty.[39] Baroness Stocks even compared the potential abolition of the live export trade to the House of Lords' abolition of the slave trade a century and a half earlier (coincidentally both were centered on the same port in Liverpool), noting "[i]n history there have been occasions when profitable interests have been subjected to considerations of humanity."[40] But the government initially resisted banning the live export trade, noting the trade's economic importance.[41]

In the following two years, though, Britain did temporarily ban the trade under increasing public pressure. In 1972, the newly formed group Compassion in World Farming (CIWF) presented to Parliament a petition with half a million signatures calling for an end to the live export trade.[42] The following year, the British House of Commons voted to ban the live export trade based on evidence of the cruelty of the trade and foreign slaughterhouses.[43] But the Government set up a committee to investigate the trade, which concluded that instead of banning the trade, the Government should regulate it with common European transport and slaughter standards.[44] Following the presentation of the committee's report, in 1975, the

---

[37] See comments of Mr. Amory, Ministry of Agriculture, Fisheries, and Food, House of Commons debate, July 1, 1957, in Hansard, http://hansard.millbanksystems.com/commons/1957/jul/01/export-of-live-cattle.

[38] Howkins and Merricks (2000).

[39] See comments of Lord Somers, House of Lords debate, Jan. 21, 1971, in Hansard, http://hansard.millbanksystems.com/lords/1971/jan/21/export-of-live-animals.

[40] See comments of Baroness Stocks, House of Lords debate, Jan. 21, 1971, in Hansard, http://hansard.millbanksystems.com/lords/1971/jan/21/export-of-live-animals.

[41] See comments of Lord Denham, House of Lords debate, Jan. 21, 1971, in Hansard, http://hansard.millbanksystems.com/lords/1971/jan/21/export-of-live-animals.

[42] Howkins and Merricks (2000).

[43] Barclay (2000).

[44] Barclay (2000).

House of Commons voted narrowly—232 votes to 191—to resume the trade.[45] Over the coming decade there were repeated efforts to ban what Members of Parliament called the "horrible export trade," but all fell short, even as the live export trade grew in scale.[46]

In 1993, the RSPCA and CIWF launched a publicity campaign to end live export entirely, after the creation of the EU caused a relaxation of the few existing live export regulations.[47] Confronted by the power of the farm lobby in the House of Commons, CIWF turned to new targets: the EU, companies involved in the trade, and celebrities like actress Joanna Lumley who could influence public opinion. In April 1994, CIWF presented the European Commission with a petition with a million signatures calling for an end to the live export trade.[48] That August, British Airways pledged to stop carrying live animals after CIWF and Tory MP Sir Teddy Taylor highlighted a shipment of 50 sheep sent in the hold of a jumbo jet.[49] Soon after, the shipping line P&O, which carried 60 % of the trade, pledged to stop carrying live animals after receiving hundreds of thousands of consumer complaints.[50] The major ferry company Stena soon followed suit.[51]

But exporters simply moved their business to different shipping companies and ports, resulting in an escalation of live export protests in Britain. In January, 1995, over 1300 police were deployed at Shoreham port to keep protesters from stopping the boarding of sheep onto a vessel bound for Europe.[52] In February, 1995, a driver transporting British veal calves to planes at Baginton airport drove over a live export trade protester, killing her.[53] Within months, the firm that flew the veal calves to Europe had gone bankrupt, ending the trade through the airport.[54] When exporters tried to send sheep out of the port at Brightlingsea later that year, a poll showed that an incredible 40 % of the town's residents participated in the protests against the trade.[55]

---

[45]Barclay (2000).

[46]*See*, *e.g.*, Dalyell (1977).

[47]Howkins and Merricks (2000).

[48]Howkins and Merricks (2000).

[49]Howkins and Merricks (2000).

[50]Howkins and Merricks (2000).

[51]Howkins and Merricks (2000).

[52]Howkins and Merricks (2000).

[53]Honigsbaum (2005).

[54]Honigsbaum (2005).

[55]Howkins and Merricks (2000).

8.3 The Modern Live Export Trade

Today a majority of the world's nations likely participate in the live export trade, although reliable numbers are hard to come by. Meat and Livestock Australia, the body that promotes the Australian live export trade, claims that 109 nations export live farm animals—a number that is impossible to verify.[56] The largest trade routes are in live pigs and poultry between European nations, but these routes lack the worst characteristics of the long-distance live export trade: days in transit followed by slaughter under weaker standards than the exporting countries. The biggest long-distance live export trades are in cattle, pigs, and sheep, especially from Australia. This section summarizes the trends in the global live export trade, and then covers Australia's trade in more depth.

8.3.1 The Major Global Live Export Trade Routes

The trade is worth roughly $21 billion dollars a year, according to the United Nations.[57] (Although the UN collates national statistics on the value of the trade, it does not collate statistics on the number of animals involved.) The five largest exporters of live animals in 2013, by value, were the Netherlands, France, Canada, Germany, and Denmark.[58] The five largest importers of live animals in 2013, by value, were the United States, Germany, Italy, the Netherlands, and Saudi Arabia.[59] The presence of European nations in these lists is deceptive, since the European open market and the close proximity of these nations, has fostered a large European cross-border flow in animals. Although this short-distance trade involves large numbers of animals, it lacks many of the problems associated with the long-distance live export trade, both because of the short travel times and because of the EU's stringent animal welfare regulations.

In the long-distance live export trade, the biggest players are quite different. Australia, Brazil, and the United States are likely the biggest long-distance exporters of live animals.[60] The Australian trade alone, which primarily sends cattle and sheep to Indonesia and the Middle East, was worth $1.46 billion in 2014, up from $1.01 billion in 2013.[61] China was the biggest exporter in Asia, with a trade in live animals worth $580 million in 2013, Ethiopia was the biggest exporter in Africa with a trade worth $340 million, and Jordan was the biggest in

[56]"Australia's live export trade," ABC, May 29, 2013, http://www.abc.net.au/news/2013-05-30/about-live-export-trade/4719636.

[57]UN ComStat Database (2015).

[58]UN ComStat Database (2015).

[59]UN ComStat Database (2015).

[60]UN ComStat Database (2015).

[61]UN ComStat Database (2015).

the Middle East, with a trade worth $230 million.[62] The biggest long-distance importers were likely Saudi Arabia ($970 million), China and Hong Kong ($590 million), Russia ($410 million), Turkey ($350 million), Indonesia ($340 million), Lebanon ($290 million), Qatar ($220 million), and Jordan ($200 million).[63]

In Europe, over three million live animals are exported each year to non EU countries, according to CIWF.[64] The group documented the shipment of bulls from Hungary and Slovakia to Turkey, finding that the bulls were kept inside overheated trucks without access to clean water for over 60 hours.[65] In 2009, more than one billion chickens and turkeys were also transported across national borders in the EU, along with 37 million pigs, cattle, and sheep.[66]

There is little good data on the live export trades within Asia. China is likely responsible for the world's largest long-distance domestic trade in live animals. Beef cattle raised in Zhangbei near Inner Mongolia are shipped over 1500 miles to slaughter in Guangdong.[67] Sheep and cattle raised in China's inland provinces are shipped almost as far to ports, before being shipped over 7000 miles to the Middle East.[68] Taiwan, by contrast, has not been involved in the international live export trade since the outbreak of foot-and-mouth disease in 1997.[69] But there is still a significant domestic trade in live animals because of the popularity of live animal markets.[70]

A recent study estimated that producers send six million animals a year to Saudi Arabia for religious festivities alone.[71] In particular, large numbers of live animals are slaughtered during the Haj and Ramadan.[72] Of these six million animals, about 42 % come from the Horn of Africa (Djibouti, Ethiopia, Somalia and Eritrea) and Sudan, with another 43 % and 16 % from Australia and Eastern Europe, respectively.[73] The Horn of Africa live export trade is not well documented, but the limited evidence that does exist suggests appalling abuses. A recent paper by Bahraini and Sudanese researchers into health issues in the Horn of Africa trade noted in passing that camels are typically sent inside boats that are too short for them to stand up in.[74] The exporters have solved this problem by tying down the

[62]UN ComStat Database (2015).
[63]UN ComStat Database (2015).
[64]Compassion in World Farming (2015).
[65]Compassion in World Farming (2015).
[66]European Commission (2011).
[67]Brooke et al. (2008), p. 302.
[68]Brooke et al. (2008), p. 302.
[69]Brooke et al. (2008), p. 304.
[70]Brooke et al. (2008), p. 304.
[71]Abbas et al. (2014).
[72]Abbas et al. (2014).
[73]Abbas et al. (2014).
[74]Abbas et al. (2014).

camels, resulting in many arriving at the quarantine facility with bruises, fractures, myositis and pneumonia.

The most valuable global live export trade, by far, at a value of $8.6 billion in 2013, is in cattle.[75] The biggest volumes in that trade are within Europe and North America, although Venezuela, Turkey, and Russia are also major importers of live cattle.[76] The second most valuable trade, worth $5.3 billion in 2013, is in live pigs.[77] The majority of this volume is the sale of pigs from Denmark and the Netherlands into Germany and Poland, although China and Hong Kong are increasingly large importers through what is likely a long-distance trade.[78] The trade in live poultry is the third most valuable globally, at $3.1 billion in 2013.[79] The biggest trade in poultry takes place over short distances between Germany, the Netherlands, and Belgium, and between Malaysia and Singapore, with a far smaller long-distance trade.[80] By comparison, the trade in live sheep and other farm animals is relatively small, at just over $1 billion each in 2013, but generates more attention because a larger portion of the trade is long-distance.[81] In particular, the largest live sheep trade is between Australia and the Middle East—a controversy to which we now turn.

8.3.2 *The Australian Live Export Trade*

No trade in live animals has generated more public controversy than the Australian trade to Indonesia and the Middle East. Partly this is a nature of its size: Australia conducts the world's largest long-distance cattle and sheep trade. The most recent Australian government figures show that the country's livestock industry exported 1.9 million live sheep and 778,000 cattle in 2013, with 14,000 sheep and 800 cattle dying on the ships.[82] Both numbers are falling though: in 2008, Australian producers exported 4.2 million live sheep, of whom 40,000 died at sea.[83] The trade in sheep is almost exclusively to the Middle East, with 98 % of the sheep in the early 2000s sent to Saudi Arabia, Kuwait, Jordan, the United Arab Emirates, Bahrain, Oman, Qatar, the Palestinian Territories, Israel and Lebanon.[84] The vast majority of Australian sheep exported today are still sent to the Middle East for religious

[75]UN ComStat Database (2015).
[76]UN ComStat Database (2015).
[77]UN ComStat Database (2015).
[78]UN ComStat Database (2015).
[79]UN ComStat Database (2015).
[80]UN ComStat Database (2015).
[81]UN ComStat Database (2015).
[82]Australia Department of Agriculture (2013a).
[83]Australia Department of Agriculture (2013a).
[84]Livestock Export Review (2003).

slaughter.[85] The trade in cattle is primarily to Southeast Asia, especially Indonesia, where they are sent to feedlots to fatten up before slaughter.[86]

In 1985, the Australian Senate Select Committee on Animal Welfare concluded that "if a decision were to be made on the future of the trade solely on animal welfare grounds, there is enough evidence to stop the trade."[87] The Committee explained that "[t]he trade is, in many respects, inimical to good animal welfare, and it is not in the interests of the animal to be transported to the Middle East for slaughter."[88] But the Committee also acknowledged that for economic reasons the trade would not end in the short term.[89] As such, it recommended that the government implement a series of animal welfare improvements to the live trade while encouraging a transition to a trade solely in refrigerated meat in the long term.[90]

Thirty years later, though, the Australian live export trade shows no signs of converting to a refrigerated meat trade. The live export trade persists in part because of the lack of export-approved slaughterhouses in the Northern Territory and Western Australia, where most live export producers are concentrated, according to a recent Australian Parliamentary briefing report.[91] It also persists because Middle Eastern consumers prefer to buy live cattle and sheep than refrigerated meat.[92] This is due to high fodder, water, and meat subsidies provided by a number of Middle Eastern governments, the preference of local souks to sell live animals, and the demand for the animals to be slaughtered according to Halal tradition.[93] And in some South East Asian nations—particularly Indonesia, which is now the largest recipient of live Australian cattle—the cold storage supply chain is not reliable enough to keep packaged meat refrigerated.[94]

But the controversy around the Australian live export trade has not abated. In 2003, 5,691 sheep (out of a shipment of 57,937) died at sea when Saudi Arabia rejected a shipment of sheep aboard the *Cormo Express* infected with "scabby mouth."[95] After two months at sea, the sheep were finally unloaded in Eriteria, where they were slaughtered in makeshift slaughterhouses.[96] In the ensuing controversy, two live export firms went out of business and People for the Ethical Treatment of Animals launched a global boycott of Australian wool, persuading

---

[85] ACIL Tasman (2009).
[86] Livestock Export Review (2003).
[87] Australian Senate Select Committee on Animal Welfare (1985).
[88] Australian Senate Select Committee on Animal Welfare (1985).
[89] Australian Senate Select Committee on Animal Welfare (1985).
[90] Australian Senate Select Committee on Animal Welfare (1985).
[91] Coombs and Gobbett (2013).
[92] Coombs and Gobbett (2013).
[93] Coombs and Gobbett (2013).
[94] Coombs and Gobbett (2013).
[95] Coombs and Gobbett (2013).
[96] The Age (2004).

Abercrombie and Fitch to ditch Australian merino wool.[97] The Australian government commissioned another review of the trade after the disaster, which concluded that a more stringent regulation of the trade was needed.[98] This resulted in Australia's Standards for the Export of Livestock (see below).

Since then, Australia has been repeatedly rocked by investigations of the live export trade by Animals Australia, an advocacy group. Animals Australia has conducted 33 investigations into Australian live export trade markets over the last decade, from Israel and Gaza to Indonesia and Malaysia.[99] Each of the investigations has documented widespread abuses of professed Australian industry standards, even in slaughter plants specifically certified and designed by Australian export groups.[100] For example, the 2011 investigation in Indonesia, mentioned in the introduction to this chapter, filmed abuses occurred in 11 different Australian certified slaughterhouses.[101] Several of these slaughterhouses featured equipment branded "Meat and Livestock Australia" and had recently received Australian inspectors to train the staff in humane slaughter techniques.[102] Several of these investigations have led the Australian Government to briefly suspend live export trade routes—to Saudi Arabia throughout the 1990s, to Egypt in 2006, and to Indonesia in 2011.[103] But, as of the time of this article's publication, the Australian live export trade was growing and showed no signs of abating.

8.4 Current Legal Framework for Live Export

The live export trade remains largely unregulated in most of the world. But there are exceptions. The World Organization for Animal Health (OIE) has adopted standards that govern certain aspects of the trade. The EU has regulated the trade, while the US has largely failed to. And Australia has tried to regulate the trade through an extensive industry self-regulation scheme, that appears to have largely failed in practice.

8.4.1 OIE Live Export Standards

The OIE has developed the first globally applicable standards for the live export trade. The OIE (the acronym refers to the organization's historic French name) is the intergovernmental organization tasked with improving animal health

[97]Coombs and Gobbett (2013).

[98]Livestock Export Review (2003).

[99]Animals Australia (2015).

[100]Animals Australia (2015).

[101]ABC Four Corners (2011).

[102]ABC Four Corners (2011).

[103]Hastreiter (2013), pp. 184–185.

worldwide. In 2004, the OIE integrated animal welfare guiding principles into its Terrestrial Animal Health Code, its set of guidelines for the transportation and handling of animals, which until then had largely concerned animal health and zoonotic diseases. Since then, OIE member nations have adopted 12 global animal welfare standards, covering more than 100 pages of specific guidelines, which have been integrated into the Terrestrial Animal Health Code.[104] The first three sets of animal welfare recommendations concern the transport of animals by sea, land, and air—showing the importance that the OIE attaches to transport conditions as an influence on animal welfare.[105]

The OIE standards for live export speak largely in vague generalities, avoiding specific requirements. For example, the standards for transport by sea state that: the journey length "should be kept to a minimum," though no maximum length is set; animals' desires "should be taken into account," though what conditions satisfy those desires is not explained; "[c]alculations for the space allowance for each animal should be carried out in reference to a relevant national or international document," though no relevant documents are referenced.[106] Similarly, the standards for transport by land call for water and feed to be "available as appropriate," but do not state when it is appropriate, and advise drivers to "avoid[] group sizes which are too large," but do not state how many animals is too many for one truck.[107]

Some of the OIE standards also appear to be divorced from the reality of the live export trade. For example, the land transport standards recommend that once in motion "[s]ick or injured animals should be segregated," but do not explain how a driver could segregate a sick or injured animal in a crowded transport truck, even if the driver knew an animal had become sick or injured in transit.[108] Similarly, the standards recommend removing urine and faeces from the floor of the truck, but do not explain how a driver could do this in a crowded truck or where he would remove the waste to.[109] And they suggest positioning animals to "enable each animal to be observed regularly during the journey to ensure their safety and good welfare," but do not explain how this is possible in modern slaughter trucks, where one truck driver may transport as many as 2000 pigs at a time.[110]

But the standards nonetheless recommend a number of sensible ideas in each area. Ship loading facilities should have smooth floors, curved passages leading to the ship, and lack distracting noises, air currents, and shiny objects—much as modern American slaughterhouses are meant to.[111] Vessels should contain

[104]OIE (2014), Chapter 7: Animal Welfare.

[105]OIE (2014), Chapter 7: Animal Welfare.

[106]OIE (2014), Chapter 7.2: Transport of Animals by Sea.

[107]OIE (2014), Chapter 7.2: Transport of Animals by Land.

[108]OIE (2014), Chapter 7.2: Transport of Animals by Land.

[109]OIE (2014), Chapter 7.2: Transport of Animals by Land.

[110]*See, e.g.*, Taylor (2014).

[111]OIE (2014), Chapter 7.2: Transport of Animals by Sea.

non-slip flooring, a backup power supply to provide ventilation to the animals, and a design that prevents faeces or urine from animals on upper levels soiling animals on the lower levels.[112] Animals should not be goaded with electric prods or sticks, should have enough room to lie down comfortably at all times, and should be humanely euthanized if severely ill.[113] Truck drivers should limit the duration of their rest stops when carrying animals and spray water on pigs in hot weather.[114]

The OIE standards are also the only live export animal welfare standards to enjoy international acceptance. The OIE standards are set by consensus by all 180 OIE member states, which include all of the largest live exporting nations and many developing nations that are yet to adopt any domestic animal welfare laws.[115] In particular, in some of the biggest nations in the Middle Eastern live export trade—like Sudan, Ethiopia, and Saudi Arabia—the OIE standards are the only protections that animals have. And although compliance with the OIE standards is currently limited, the standards provide an aspirational benchmark for these countries. The OIE helps nations to reach this benchmark through workshops and conferences to build the capacity of national regulators to adopt and enforce the OIE standards.[116] And the OIE has developed animal welfare strategic plans to implement the standards in each of its five regions, while appointing an implementation liaison in each of its 180 member states.[117]

Moreover, the OIE standards are a presumptively valid basis for regulating trade under the World Trade Organization Agreement on the Application of Sanitary and Phytosanitary Measures, which formally recognizes the OIE standards as the international reference for animal health.[118] This is critical given the live export trade inherently involves international trade. Any other animal welfare standards that nations tried to impose on the international trade in live animals would be vulnerable to challenge and invalidation before the WTO dispute settlement body.

8.4.2 European and U.S. Live Export Regulations

The EU has proactively regulated the welfare of animals in the live export trade since 1977 (initially through the EU's predecessor, the European Communities).[119]

[112]OIE (2014), Chapter 7.2: Transport of Animals by Sea.

[113]OIE (2014), Chapter 7.2: Transport of Animals by Sea.

[114]OIE (2014), Chapter 7.2: Transport of Animals by Land.

[115]OIE (2013).

[116]*See*, *e.g.*, OIE (2012).

[117]IFC (2014).

[118]*See* Agreement on Sanitary and Phytosanitary Measures: Preamble, Article 3, and Annex A, paragraph 3(b).

[119]European Council (1977).

The EU's current live export regulations were adopted in 2004.[120] The regulation defines any journey longer than 8 h—which is standard in all transnational live export routes—as a "long journey" subject to stricter regulation.[121] It provides for detailed record-keeping and animal welfare checks at all border crossings by official veterinarians. The technical annex to the regulations forbids the transport of young animals for more than 100 km, "downer" animals who are too sick or injured to walk, and animals with open wounds or prolapses.[122] Both mammals and birds must be watered at least every 12 h and fed at least every 24 h in transport.[123]

In the U.S., the nation's oldest animal welfare law governs the welfare of farm animals in transport. The Twenty-Eight Hour Law of 1873 provides that any transporter "may not confine animals in a vehicle or vessel for more than 28 consecutive hours without unloading the animals for feeding, water, and rest," unless the vehicle or vessel provides animals with food, water, space, and an opportunity for rest.[124] For the first half-century following the law's passage, it was rigorously enforced on rail carriers, who established about 900 feed, water, and rest stations for transported animals across the nation.[125]

But since the 1960s, the U.S. Department of Agriculture (USDA) has almost completely failed to enforce the humane transport law, especially in the cross-border trade in live farm animals. For decades, the USDA contended that the Twenty-Eight Hour Law only applied to rail, and not truck, transports of animals, despite the law's clear application to all "vehicle[s] or vessel[s]."[126] And even after the USDA dropped this position, there is little evidence that it has brought any cases over violations of the law in the last decades (by contrast, it brought almost 400 cases under the law in 1967 alone).[127] The USDA did apparently investigate one complaint of 135 pigs who died in transit between the US and Mexico in 2006, but failed to respond to a complaint alleging that more than 80 pigs died in transit between Canada and Hawaii the following year.[128] The USDA has also only adopted the most minimal regulations governing the transport of live animals, requiring merely that animals have a clean space, free of obvious hazards, and with enough air to breathe.[129]

---

[120] European Council (2004).

[121] European Council (2004).

[122] European Council (2004).

[123] European Council (2004).

[124] 49 U.S.C. §§ 80502(a)(1) and 80502 (c).

[125] Animal Welfare Institute (2010).

[126] Animal Welfare Institute (2010).

[127] Animal Welfare Institute (2010).

[128] Animal Welfare Institute (2010).

[129] *See* 9 C.F.R. § 3.138.

8.4.3 Australia's Exporter Self-Regulatory Scheme

Since 2004, the Australian government and industry groups have collaborated on Australian Standards for the Export of Livestock (ASEL).[130] To export live farm animals, an exporter must secure a government license, which in turn requires compliance with the ASEL.[131] The government regulates ASEL compliance through monitoring livestock exporters and government-accredited veterinarians, and premises used in the live export trade.[132] Every six months, the government reports on mortality figures in the live export trade to the Australian Parliament.[133] The government claims that mortality rates in the trade have fallen since the ASEL system has been in place.[134]

The ASEL outline a comprehensive set of animal welfare standards for live export. Standard five of the ASEL requires immediate reporting of any incident that has the potential to cause serious harm to the health and welfare of animals, end-of-voyage health and welfare reporting, and daily reporting on the health and welfare of livestock on voyages of greater than ten days' duration, and defines acceptable upper levels of mortality during an export voyage.[135]

But the ASEL appear to be widely disobeyed and woefully under-enforced. For example, in 2012, Bahrain refused to allow an Australian ship to unload a shipload of sheep on health grounds—a direct violation of the ASEL requirement.[136] The Australian Government then authorized the ship to move on to Pakistan, where there were no ASEL-approved slaughterhouses.[137] The Australian Government hastily approved a Pakistani slaughterhouse under ASEL standards but did not inform the Pakistani Government that the shipment had been previously rejected by Bahrain.[138] When the local Pakistani health authorities found out, they ordered that most of the 20,000 sheep be killed.[139] Whistleblower footage showed local workers attempting to bludgeon the sheep to death before throwing them into huge pits—with many sheep still alive and moaning in pain.[140] Yet the Australian Government not only failed to stop this abuse, but approved a shipment of cattle to Pakistan the following week.[141]

---

[130]Schipp and Sheridan (2013).

[131]Schipp and Sheridan (2013).

[132]Schipp and Sheridan (2013).

[133]Schipp and Sheridan (2013).

[134]Schipp and Sheridan (2013).

[135]Schipp and Sheridan (2013).

[136]Ferguson and Masters (2012).

[137]Ferguson and Masters (2012).

[138]Ferguson and Masters (2012).

[139]Ferguson and Masters (2012).

[140]Ferguson and Masters (2012).

[141]Ferguson and Masters (2012).

And in yet another recent incident, in November 2013, the authorities allowed producers to board 112 sheep into a non-ventilated forward cargo hold of a plane flying from Perth to Kuala Lumpur, Malaysia, resulting in 44 of the sheep overheating or suffocating to death.[142] The Department of Agriculture subsequently presented a report, which put no blame on the airline or exporters, and required only that the airline in future stack slightly fewer sheep into the hold and only use planes with functioning air conditioning systems—conditions that the Department dropped a month later.[143]

Since 2011, the Australian government has also required exporters to take part in the industry-operated Exporter Supply Chain Assurance System, in response to widely publicized abuses in the live export trade.[144] The ESCAS standards focus on achieving certain animal welfare outcomes, although they allow for flexibility in the approach to those outcomes, depending on the importing country.[145] The ESCAS standards start with the framework of the minimal animal welfare standards of the OIE Terrestrial Code.[146] They flesh out these standards with a more detailed checklist of how animals should be treated at all six stages in the supply chain, from leaving Australian shores to slaughter in a foreign nation.[147] This checklist is designed to be audited to determine if individual exporters are complying with the ESCAS.[148] The ESCAS standards apply to all Australian export markets except Egypt.[149]

The ESCAS standards' most glaring omission is that they do not require the stunning of animals prior to slaughter. The Australian government justifies this omission on the basis that the OIE Terrestrial Code does not require animals to be stunned before slaughter, and that Australia's humane slaughter laws do not require stunning for religious slaughter.[150] But even the Australian Veterinary Association, which actively supports the Australian live export trade, states that all animals live exported should be stunned before slaughter.[151] And this omission undercuts the argument of live export promoters that their participation in the trade gives them the leverage to raise animal welfare standards in the receiving nations.[152] The omission reflects the lack of power that Australian authorities have to demand humane treatment of animals once they arrive on foreign soil—a key critique leveled by animal advocates against the live export trade in general.

[142] Australia Department of Agriculture (2013b).

[143] Australia Department of Agriculture (2013b).

[144] Schipp and Sheridan (2013).

[145] Schipp and Sheridan (2013).

[146] Schipp and Sheridan (2013).

[147] Schipp and Sheridan (2013).

[148] Schipp and Sheridan (2013).

[149] Schipp and Sheridan (2013).

[150] Schipp and Sheridan (2013).

[151] Australian Veterinary Association (2015).

[152] See, e.g., Australian Veterinary Association (2015).

The ESCAS standards appear to also be widely ignored because no meaningful penalties are attached to their violation. For example, one of the ESCAS standards provides that "[a]nimals must be protected from exposure to adverse weather conditions or alternative arrangements must be made to alleviate heat/cold stress."[153] But in August 2013, 4,179 merino sheep died of heat exhaustion on a boat transporting 75,508 sheep from Adelaide and Freemantle to Qatar and the United Arab Emirates.[154] The government report found that almost all of the sheep died on day 21 of the voyage, when the boat docked in Qatar, where temperatures peaked at 38 °C (100 °F).[155] The report also found that the deaths were mostly of "wether" merino sheep which have heavier coats than others, and were almost exclusively in the poorly ventilated lower decks of the boat.[156] Yet the report did not fault the exporter. [157]And the Department of Agriculture allowed the exporter to send another consignment of 77,095 sheep to the Middle East just two months later, subject to the sole condition that the exporter provide the sheep with 10 % more space on board.[158]

8.5 Possible Legal Solutions to Live Export

Most stakeholders in the live export trade acknowledge the need for reforms. Governments, animal welfare groups, and even export associations have all accepted the need to improve animal welfare standards in the trade. The disagreement is over what reforms are needed. This section explores five possible solutions to animal welfare abuses in the live export trade: (1) strengthen existing OIE standards governing the trade; (2) implement memoranda of understanding between exporting and importing countries; (3) limit the trade to countries with animal welfare provisions already in place; (4) sign a treaty governing the live export trade; or (5) abolish the live export trade. Although all five solutions could significantly improve animal welfare in the trade, only abolishing the trade will ensure an end to the unnecessary suffering of exported animals.

---

[153]Australia Department of Agriculture (2015b).
[154]Australia Department of Agriculture (2013c).
[155]Australia Department of Agriculture (2013c).
[156]Australia Department of Agriculture (2013c).
[157]Australia Department of Agriculture (2013c).
[158]Australia Department of Agriculture (2013c).

8.5.1 Strengthen the OIE Standards

The most promising multilateral solution is for the OIE to strengthen both the text and enforcement of its existing standards governing the live export trade. As noted above, the OIE has already developed extensive standards concerning the live export trade. These standards have a number of advantages as the basis for regulating the live export trade. They are comprehensive: they cover the treatment of animals in all major forms of transport—by sea, air, and land—from the point of leaving the farm right through slaughter. They enjoy widespread acceptance: the OIE's 180 member nations have already signed off on them. And they are presumptively valid under the WTO, making them an appropriate basis to regulate this inherently international trade.

But there are two primary limitations to the OIE standards as the solution to live export abuses. First, the standards contain no enforcement mechanism. As a result, there is little evidence that the OIE standards have independently raised animal welfare standards anywhere—except, perhaps, for where they have been tied to binding commitments, as in Australia's ESCAS program. Second, the standards do not restrict what are likely the two cruelest aspects of the live export trade: the long-distance transport of animals, and the slaughter of animals without prior stunning. Although the standards insist that journey length "should be kept to a minimum," they contain no guidance on what that means.[159] And the insistence of the Australian government that it is complying with the OIE standards while condoning over 8,000 mile journeys to the Middle East suggests that this guidance is meaningless.[160] Moreover, the omission of any stunning requirement for ritual slaughter—while a common omission even in developed nation's animal welfare laws—can result in horrendous suffering in countries where the vast majority of slaughter is done without stunning and with little concern for the length of time it takes the animal to die.[161]

Still, while the live export trade persists, strengthened OIE standards present one of the more appealing options for its regulation. These standards at least acknowledge the importance of animal welfare to the trade, and make a number of sensible recommendations around flooring, low stress environments, and adequate ventilation in transit. They also set the basis for a system of continuous improvement. The OIE's commitment to cooperating with governments to improve enforcement of the standards is particularly encouraging. When applied in short distance trades where ritual slaughter is not at issue—for example, in the cattle trade between EU nations and Belarus and Ukraine—the OIE standards may adequately ensure animal welfare. And with relatively small changes—for instance, mandating stocking densities, limiting journey durations, and requiring use of modern ritual slaughter techniques—the OIE standards might ensure decent levels of animal welfare in

[159] OIE (2014), Chapter 7: Animal Welfare.

[160] Australia Department of Agriculture (2015a).

[161] *See, e.g.*, ABC Four Corners (2011).

longer-distance trades. The challenge for the OIE is to ensure that these standards are actually enforced, and strengthened continuously in the years ahead.

8.5.2 Reach Memoranda of Understanding on Animal Welfare

One bilateral solution to live export trade abuses is for exporting nations to reach memoranda of understanding (MOUs) on animal welfare with importing nations. This has been Australia's solution of choice. Australia has coupled regulation of live animal exporters through the ACEL and ESCAS programs with MOUs with ten countries in Africa and the Middle East.[162] These MOUs aim to raise animal welfare standards in the importing countries. They require, for example, that all animals be unloaded on arrival regardless of their health status, to prevent a recurrence of the *Cormo Express* disaster where diseased Australian sheep wallowed at sea for over two months because no nation would accept them.[163] The primary advantage of these MOUs is that they bind importing countries to specific animal welfare provisions and achieve a degree of animal welfare enforcement on foreign soil that the exporting country could not achieve alone.

But in practices these MOUs appear to have been largely disobeyed because of the lack of repercussions attached to them. For example, in 2012 Bahrain refused to unload a shipment of animals because of concerns about their health status—a direct violation of the MOU between Australia and Bahrain.[164] Yet Australia did not cut off the live export trade with Bahrain following this incident, presumably because of pressure from Australian exporters to not cut off any markets.[165] And Australia has had MOUs with Middle Eastern importing nations since 2004, yet egregious abuses have continued in the Middle Eastern trade since then.[166] It seems unlikely that the MOUs will have any real force until Australia indicates that it is willing to stop the trade, or otherwise sanction importing nations, when the MOUs are violated.

In 2014, Australian Minister of Agriculture Barnaby Joyce, a loud proponent of the live export trade, conceded that "MOUs have mixed success as they are statements of intent between governments and are not legally binding."[167] Although Mr. Joyce made this argument in the self-serving context of justifying why his government was opening up live export shipments to countries lacking MOUs, he identified their primary limitation. Given they lack any legally binding

[162] Australia Department of Agriculture (2015a).
[163] Australia Department of Agriculture (2015a).
[164] Ferguson and Masters (2012).
[165] Ferguson and Masters (2012).
[166] Animals Australia (2015).
[167] Joyce (2014).

quality, they are effective only if accompanied by a credible threat that the trade will stop if the MOU is violated. Since no exporting country has so far shown that commitment, MOUs are limited in their effect.

8.5.3 *Only Export to Nations with Animal Welfare Protections*

A stronger bilateral version of an MOU would be a policy by an exporting country that it will only allow the export of live animals to countries that already have strong animal welfare laws in place. Australia, for instance, could insist that its 31 live export trading partners[168] each adopt meaningful animal welfare laws within a set period of time, for example 5 years. In theory, this approach could not only ensure adequate welfare standards in the live export trade, but also boost animal welfare laws across the developing world.

The evidence, though, from the one nation where this has occurred is not promising. In 2010, Indonesia enacted its Farm and Animal Welfare Law, in part to ensure the continuance of the live export trade with Australia.[169] But just 18 months later, ABC's Four Corners found flagrant violations of the law at 11 slaughterhouses across Indonesia.[170] According to the Indonesian Meat Importers Association, the problem was that the government hadn't yet introduced regulations enforcing the law, meaning that there were no sanctions for its breach.[171] But even if there had been regulations in place, it is unclear whether Indonesia had the resources to properly enforce the law. And the bigger problem may be the insistence of countries like Indonesia on conducting ritual slaughter without stunning. After the Four Corners investigations, Animals Australia insisted that the only slaughterhouses in Indonesia capable of meeting minimal animal welfare standards were the four slaughterhouses that require stunning.[172] Those slaughterhouses, though, represent just a fraction of the approximately 100 Indonesian slaughterhouses that kill imported cattle.[173]

[168]Joyce (2014).
[169]Alford (2011).
[170]Alford (2011).
[171]Alford (2011).
[172]Alford (2011).
[173]Alford (2011).

8.5.4 Sign a Live Export Animal Welfare Treaty

A more promising approach to ensuring importing nations follow animal welfare standards would be to sign a treaty governing live export. The treaty could impose clear standards for the treatment of animals in the live export trade—for example, that mortality rates cannot exceed set levels in transit and that all sick animals must be immediately humanely euthanized. In return for market access—and perhaps other trade incentives—importing nations would pledge to adopt binding animal welfare standards. The failure to adopt and enforce these standards would then become a treaty violation, subject to sanctions or penalties. Ideally, such a treaty would provide a private enforcement mechanism, so that animal welfare groups and export associations could pursue violations where a government was unwilling to act.

A live export treaty could be bilateral or multilateral. A bilateral treaty—for example, between Australia and Indonesia—would likely have the best chance of initial passage. But it is not hard to imagine such a treaty becoming multilateral, if other nations like Malaysia, Singapore, and Thailand wanted to join under similar conditions. In the longer run, it is possible to imagine a multilateral treaty governing the live export trade across an entire region like the Middle East—or even globally. While the benefits of such a treaty are still largely speculative, the success of the Council of Europe and the European Union in raising animal welfare standards suggests that a treaty is the most promising option short of abolishing the live export trade.

8.5.5 Abolish the Live Export Trade

The simplest solution to the cruelty of the live export trade would be to abolish it. For decades, animal advocates have advocated for this solution, arguing that the live export trade is inherently cruel and incapable of meaningful reform. In Australia, animal protection groups have advocated for the abolition of the live export trade since a Senate Select Committee found in 1985 that this would be the best outcome for animal welfare.[174] The Australian RSPCA commissioned a report in 2009 that claimed that the live sheep trade could be abolished in five years at a cost of $200 million.[175]

The strongest argument against abolishing the live export trade is that exporting nations would lose their leverage to improve animal welfare standards in importing countries. The Australian Department of Agriculture argues that "[o]ur ongoing involvement in the livestock export trade provides an opportunity to influence

[174] Australian Senate Select Committee on Animal Welfare (1985).

[175] ACIL Tasman (2009).

animal welfare conditions in importing countries."[176] The Department notes that Australia has signed MOUs with ten countries in the Middle East and Africa, raising animal welfare standards in those countries.[177] Although self-serving, this is not a frivolous argument. Many of the countries where Australia exports live cattle and sheep have only rudimentary animal welfare laws. If Australia and other exporting nations with high animal welfare standards can leverage their involvement in the live export trade to force animal welfare improvements across these countries, that could raise global animal welfare standards.

But there are three problems with this argument. First, it is unclear how much leverage exporting nations really have over animal welfare standards in the importing country. Australia has tried harder than any other exporting nation to raise animal welfare standards in importing countries. Yet it has still failed to secure progress on the most important animal welfare issues—the absence of stunning before slaughter, the transport of animals through hot climates without adequate ventilation, and the repeated abuse of animals in foreign slaughterhouses. Indeed, Australia has been left pleading with importing countries to follow the OIE standards—standards they were already legal obliged to follow.

Second, animal welfare improvements seldom extend beyond the live export trade that is the subject of protests. For example, in Indonesia, the Australian meat industry has worked with the slaughterhouses involved in the trade to raise their animal welfare standards (seemingly unsuccessfully). But it has done nothing with slaughterhouses not involved in the trade. The leverage argument only makes sense if Australia can secure improvements across Indonesian slaughterhouses. If Australia is only securing improvements in the treatment of Australian cattle, there is no leverage—and the Australian cattle would still have been better off slaughtered in a regulated Australian slaughterhouse.

Third, developed nations have other levers beyond the live export trade that they could use to improve animal welfare in other nations. For example, the EU has negotiated significant animal welfare improvements in developing nations by requesting the inclusion of animal welfare provisions in all new free trade agreements.[178] Tying animal welfare improvements to market access to the EU appears to have already improved farm animal treatment in Chile, Uruguay, and South Korea.[179] Australia could use free trade agreements or other diplomatic avenues to improve animal welfare in Asia and the Middle East, as it claims to be doing through the live export trade. It is telling—and undercuts Australia's leverage argument—that it has never sought to include farm animal welfare in its free trade negotiations.

---

[176] Australia Department of Agriculture (2015a).

[177] Australia Department of Agriculture (2015a).

[178] Cabanne (2013).

[179] Cabanne (2013).

8.6 Conclusion

The live export trade is a major animal welfare issue, affecting the wellbeing of hundreds of millions of animals every year. The lack of regulation of the trade causes animals to suffer—especially through long voyages in hot climates and inhumane deaths in ill-regulated slaughterhouses. Most exporters have largely ignored regulating the trade. Australia has tried, and failed. There are more promising solutions—particularly strengthening the OIE standards and adopting a live export treaty. But ultimately only one reform will guarantee an end to the unnecessary suffering of animals in the live export trade: the abolition of the trade itself.

References

Abbas B, Yousif M, Nur H (2014) Animal health constraints to livestock exports from the Horn of Africa. Rev Sci Tech Off Int Epiz 33(3)

ABC Four Corners (2011) A bloody business. Aired May 30, 2011. http://www.abc.net.au/4corners/special_eds/20110530/cattle/. Accessed 15 March 2015

ACIL Tasman (2009) The value of live sheep exports from Western Australia. ACIL Tasman, March, 2009. http://www.rspca.org.au/sites/default/files/website/Campaigns/Live-export/Live-exports-vs-the-meat-trade/ACIL%20Tasman%202009%20-%20The%20value%20of%20live%20sheep%20exports%20from%20Western%20Australia.pdf. Accessed 14 March 2015

Alford P (2011) Indonesian slaughter laws can't stop abattoir torture. The Australian, June 2, 2011. http://www.theaustralian.com.au/national-affairs/indonesian-slaughter-laws-cant-stop-abattoir-torture/story-fn59niix-1226067530577. Accessed 22 March 2015

Animal Welfare Institute (2010) Legal protections for farm animals during transportation. https://awionline.org/sites/default/files/uploads/legacy-uploads/documents/FA-LegalProtectionsDuringTransport-081910-1282577406-document-23621.pdf. Accessed 10 Apr 2015

Animals Australia (2015) Live export investigations. http://www.banliveexport.com/investigations/. Accessed 29 March 2015

Australia Department of Agriculture (2013a) Reports to Parliament on livestock mortalities. http://www.agriculture.gov.au/export/live-animals/livestock/regulatory-framework/compliance-investigations/investigations-mortalities. Accessed 14 March 2015

Australia Department of Agriculture (2013b) Mortality investigation report 49: sheep exported by air to Malaysia in November 2013. http://www.agriculture.gov.au/export/live-animals/livestock/regulatory-framework/compliance-investigations/investigations-mortalities/sheep-malaysia-report-49. Accessed 14 March 2015

Australia Department of Agriculture (2013c) Mortality investigation report 46: sheep exported to Qatar and the United Arab Emirates - September 2013. http://www.agriculture.gov.au/export/live-animals/livestock/regulatory-framework/compliance-investigations/investigations-mortalities/report-46. Accessed 14 March 2015

Australia Department of Agriculture (2015a) Live animal export trade. http://www.agriculture.gov.au/animal/welfare/export-trade. Accessed 22 March 2015

Australia Department of Agriculture (2015b) ESCAS animal welfare standards. http://www.agriculture.gov.au/export/live-animals/advisory-notices/2015/2015-05. Accessed 5 Apr 2015

Australian Senate Select Committee on Animal Welfare (1985) Export of Live Sheep from Australia, Australian Government Printing Service, Canberra, 1985. http://www.aph.gov.au/Parliamentary_Business/Committees/Senate/significant%20reports/animalwelfarectte/

exportlivesheep/~/media/wopapub/senate/committee/history/animalwelfare_ctte/export_live_sheep/00contents.ashx. Accessed 14 March 2015

Australian Veterinary Association (2015) Position statement on live animal export. http://www.ava.com.au/policy/151-live-animal-export. Accessed 14 March 2015

Barclay C (2000) Research Paper 00/11: The export of farm animals bill. House of Commons, London

Brooke PDB et al (2008) "Asia" in Cussen V & Garces L. Long distance transport and welfare of farm animals. CABI 2008

Cabanne C (2013) The EU-Chile free trade agreement—a boost for animal welfare. Eurogroup for Animals, Brussels

Compassion in World Farming (2015) Live animal exports from the EU. http://www.ciwf.org.uk/our-campaigns/investigations/transport-investigations/investigation-live-animal-exports-from-the-european-union/. Accessed 10 Apr 2015

Coombs M, Gobbett H (2013) Live animal exports. In: Parliamentary Library briefing book. http://www.aph.gov.au/About_Parliament/Parliamentary_Departments/Parliamentary_Library/pubs/BriefingBook44p/AnimalExports. Accessed 15 March 2015

Coorey P, Allard T (2011) Live cattle ban to stay. Sydney Morning Herald. June 8, 2011. http://www.smh.com.au/environment/animals/live-cattle-ban-to-stay-20110607-1fr8b.html. Accessed 29 March 2015

Dalyell T (1977) Westminster scene: Cruelty to animals, in New Scientist, May 12, 1977, p. 348

European Council (1977) Council Directive 77/489/EEC of 18 July 1977 on the rules on the protection of animals during international transport; O J L 200

European Council (2004) Council Regulation (EC) No 1/2005 of 22 December 2004 on the protection of animals during transport and related operations and amending Directives 64/432/EEC and 93/119/EC and Regulation (EC) No 1255/97 OJ L 3

European Commission (2011) Report from the Commission to the European Parliament and the Council on the impact of Council Regulation (EC) No 1/2005 on the protection of animals during transport. Brussels. http://ec.europa.eu/food/animal/welfare/transport/docs/10112011_report_en.pdf

Ferguson S (2011) Transcript of an interview with Dr Temple Grandin. ABC Four Corners. May 30, 2011. http://www.abc.net.au/4corners/content/2011/s3230885.htm. Accessed 29 March 2015

Ferguson S, Masters D (2012) Another bloody business. ABC Four Corners. Nov. 7, 2012. http://www.abc.net.au/4corners/stories/2012/11/02/3623727.htm. Accessed 29 March 2015

Hastreiter M (2013) Animal welfare standards and Australia's live exports industry to Indonesia: creating an opportunity out of a crisis. Wash Univ Global Stud Law Rev 12(1):181

Honigsbaum M (2005) Woman who died in veal protest becomes martyr of wider cause. The Guardian. Feb. 4, 2005. http://www.theguardian.com/uk/2005/feb/05/animalwelfare.world

House of Commons debate, July 1, 1957, in Hansard. http://hansard.millbanksystems.com/commons/1957/jul/01/export-of-live-cattle

House of Lords debate, Jan. 21, 1971, in Hansard. http://hansard.millbanksystems.com/lords/1971/jan/21/export-of-live-animals

Howkins A, Merricks L (2000) Dewy-Eyed Veal Calves. Live animal exports and middle-class opinion, 1980–1995. Agric Hist Rev 48(1):85–103

Joyce B (2014) Live trade set to recommence with Bahrain. Media Release. Feb. 28, 2014. http://www.agricultureminister.gov.au/Pages/Media%20Releases/live-trade-set-to-recommence-with-bahrain.aspx. Accessed 22 March 2015

Livestock Export Review (2003) A Report to the Minister for Agriculture, Fisheries and Forestry. http://www.australiananimalwelfare.com.au/app/webroot/files/upload/files/keniry_review_jan_04.pdf. Accessed 15 March 2015

Medhora S (2014) Cattle live export deal with China condemned by animal welfare groups. The Guardian. November 6, 2015. http://www.theguardian.com/business/2014/nov/07/cattle-live-export-deal-with-china-condemned-by-animal-welfare-groups. Accessed 15 March 2015

News.com.Au (2013) Abbott backs live exports to the hilt. Oct. 31, 2013. http://www.news.com.au/national/breaking-news/aust-sheep-brutally-slaughtered-in-jordan/story-e6frfku9-1226750086643. Accessed 29 March 2015

OIE (2014) Terrestrial Animal Health Code. http://www.oie.int/index.php?id=169&L=0&htmfile=titre_1.7.htm. Accessed 14 March 2015

Schipp M, Sheridan A (2013) Applying the OIE Terrestrial Animal Health Code to the welfare of animals exported from Australia. Rev Sci Tech Off Int Epiz 32(3)

Taylor K (2014) 700 pigs dead after crash on interstate, Wish TV, Nov. 20, 2014. http://wishtv.com/2014/11/20/truck-full-of-pigs-overturns-on-interstate/. Accessed 14 March 2015

The Age (2004) Cormo Express disaster haunts industry. The Age, Oct. 29, 2014. http://www.theage.com.au/articles/2004/10/29/1098992281427.html. Accessed 15 March 2015

UN Comstat Database (2015) United Nations Commodity Trade Statistics Database. http://comtrade.un.org/db/ce/ceDefault.aspx. Accessed 14 March 2015

Zimmerman WD (1962) Live export cattle trade between United States and Great Britain. Agric Hist 36(1)

Chapter 9
Harmonized Approaches in Intensive Livestock Production Systems in Europe

Kea Ovie

Abstract Animal protection in general and as well as the protection of farm animals in particular is neither a value nor a target of the European Union (EU) and its Common Agricultural Policy (CAP). However, farm animal protection within the EU has become increasingly important within the last few years. Since 2009, contract law includes a horizontal clause for the protection of the welfare of animals as sentient beings in Article 13 of the Treaty on the Functioning of the European Union (TFEU). For the EU and its Member States the horizontal clause is a commandment of consideration and optimization with respect to the determination and implementation of the agricultural policy. In order to draw attention to the protection of farm animals and primarily to secure the targets of the CAP, the EU harmonizes farm animal protection based on the ancillary competence of the CAP. Especially in the areas of animal keeping, transport and slaughtering, the national law systems shall be harmonized to avoid distortions due to economical competition. While adopting such law acts of harmonization, a balance between interests of agriculture on the one hand and as well as animal protection on the other hand has to be ensured. Adopting animal protection laws at the EU level is limited, though. These limitations cover among others the caveat for culture. Moreover, the EU legislator cannot be obliged to adopting harmonization laws, because of a wide margin of discretion. Due to the balance of interests among the Member States and the protection of farm animals, most law acts include a certain minimum harmonization from which the Member States can deviate by stricter animal protection measures. Finally, in some areas of animal keeping European harmonization processes are still missing completely. This chapter gives an overview of the status quo regarding the harmonization in farm animal protection based on the current secondary law situation in the EU. Several case studies are used to illustrate different EU law acts and their level of harmonization. General problems with respect to harmonization are explained and, finally, an outlook on farm animal protection in the EU is given.

K. Ovie (✉)
Georg-August-University Göttingen, Institute for Agricultural Law, Göttingen, Germany
e-mail: kea.ovie@jura.uni-goettingen.de

G. Steier, K.K. Patel (eds.), *International Farm Animal, Wildlife and Food Safety Law*, DOI 10.1007/978-3-319-18002-1_9

9.1 Introduction

Animal protection in a broader sense includes all legal guidelines related to (farm) animals to protect their welfare.[1] The context of this article concentrates mainly on keeping, transport and slaughtering of farm animals. It thereby focuses on regulations that aim to achieve the prevention of keeping- or handling-associated pain, suffering or damages to farm animals.

Animal protection has increasingly gained public interest.[2] On the one hand, consumers become increasingly aware that their purchasing behavior might influence market structures and production conditions of animal products.[3] On the other hand, more and more consumers expect politics to support campaigns for higher animal protection standards in the production of animal products.[4] However, animal protection in the EU cannot be regulated separately by each Member State at the national level, entirely detached from the EU Law, since the EU Law and national laws are increasingly linked to each other.[5] Therefore, the EU Law has an essential influence on national legislation, administration and jurisdiction. To guarantee that all participants in the supply chain of animal products can produce under the same conditions and thus not to jeopardize neither free trade of goods within the EU, the national legal systems ought to be harmonized by the use of European legal acts.[6]

To illustrate the harmonization of farm animal protection in the EU, Sect. 9.2 of this article summarize the statutory framework of EU law and particularly the CAP. The standardization of animal protection in the European primary law is presented as well as the institute of harmonization as the "third pillar"[7] of the CAP. Section 9.3 illustrates the legal basis for animal protection and the existing frequency of law acts of harmonization and their level of harmonization. Some critical points regarding the harmonization of farm animal protection are discussed in Sect. 9.4. Finally, the last section (9.5) gives an outlook on future developments in farm animal protection in the EU.

[1] Kluge, von Loeper. In: Bergmann (2015) p. 921.

[2] Weinberger, Knorr. In: Dombert and Witt (2011), § 22, n. 96.

[3] Cf. Härtel. In: Ruffert (2013), vol. 5, § 7, n. 81.

[4] Apel (2010), p. 215.

[5] Härtel (2012), chapter 25, n. 13.

[6] Cf. Härtel. In: Ruffert (2013), vol. 5, § 7, n. 76.

[7] Holzer (2014), p. 153.

9.2 Background

9.2.1 European Union and the Common Agricultural Policy

9.2.1.1 Framework of the European Union

The EU as a "predominantly supranationally organized association of states"[8]—currently consisting of 28 Member States (stand July 2016)—has an own legal entity (Article 47 Treaty on European Union - TEU) and international personality.[9] The EU can be subject of rights and duties.[10] In this context one can speak of the "European Integration",[11] i.e. the process "[...] of the association between the European states in an organizationally established framework".[12] For improving integration the Member States transfered parts of their sovereign powers to the EU.[13]

The EU and its self-concept are based on its values.[14] The primary objective of the EU is to enforce these values.[15] The EU and the Member States are obliged to respect and promote them.[16] A violation of them can be sanctioned.[17] After World War II the values were characterized by "[...] the "magical triangel of values", comprising freedom, economic efficiency and integration, which is captured in the slogan "peace through economic integration"[...]".[18] Over time the EU developed more and more to a community of social values.[19] The respect of human dignity, freedom, democracy, equality, the principle of rule of law, the respect of human rights including the rights of persons belonging to minorities are central values of the EU (Article 2 TEU). But, the protection of animals in general and the protection of farm animals in particular do not belong to these written values.

---

[8]Geiger. In: Geiger et al. (2010) Art. 1 TEU, n. 5; also Calliess. In: Calliess and Ruffert (2011), Art. 1 TEU, n. 27.

[9]Geiger. In: Geiger et al. (2010) Art. 47 TEU, n. 4.

[10]Geiger. In: Geiger et al. (2010) Art. 47 TEU, n. 1; Ruffert. In: Calliess and Ruffert (2011), Art. 47 TEU, n. 5 ff.

[11]Calliess. In: Calliess and Ruffert (2011), Art. 1 TEU, n. 5, 9 ff.

[12]Bieber et al. (2015), § 1 n. 1.

[13]Piepenschneider. In: Bergmann (2015), p. 538.

[14]Calliess. In: Calliess and Ruffert (2011), Art. 2 TEU, n. 2; Geiger. In: Geiger et al. (2010) Art. 2 TEU, n. 1, 9, 32.

[15]Calliess. In: Calliess and Ruffert (2011), Art. 2 TEU, n. 32; Geiger. In: Geiger et al. (2010) Art. 2 TEU, n. 6.

[16]Geiger. In:Geiger et al. (2010) Art. 2 TEU, n. 6 f.

[17]Calliess. In: Calliess and Ruffert (2011), Art. TEU, n. 32; Geiger. In: Geiger et al. (2010) 2 TEU, n. 7.

[18]Calliess. In: Calliess and Ruffert (2011), Art. 2 TEU, n. 2.

[19]Calliess. In: Calliess and Ruffert (2011), Art. 2 TEU, n. 3.

The EU pursues different objectives, which are laid down in the so-called target-trinity[20] of Article 3 TEU. The original main objective of integration was to secure the European peace by communitarization.[21] Additional aims of the EU are the promotion of the values of the Union and the peoples' well-being. Today's tasks of the EU focus on integration and the establishment of an internal market as part of the peoples' well-being.[22] The internal market is an area without internal borders in which the free circulation of goods, persons, services and capital resources ("four freedoms") is ensured (Article 26(2) TFEU).[23] The topographical boundaries of the internal market are the borders of the Member States with third countries. For the purpose of free trade between all Member States, there are no tariffs, "non-tariff-related trade barriers are reduced and the foreign trade with third countries is subject to a common foreign trade tariff (customs union)".[24] Hence, the European integration is primarily an economic integration.[25] The protection of farm animals itself is not a target of the Union.

The EU does not have a "competence-competence". That means, it is only allowed to act within the limits of the legal competences conferred upon it by the Member States in the Treaties (Principle of Conferral, Article 5(2) TEU). The Member States remain "Masters of the Treaties".[26]

To conclude, superficially the EU is an economic union. Its main objectives are economic integration and the establishment of an internal market. Animal protection is neither a value nor a target of the EU.

9.2.1.2 Common Agricultural Policy

The internal market also comprises agriculture, fisheries and trade in agricultural products (Article 38(1) subpara. 2 sentence 2 TFEU). The term "agricultural products" captures products of the soil, of stockfarming, and of fisheries as well as directly related products of first-stage processing (Article 38(1) subpara. 2 sentence 2, (3) TFEU in conjunction with Annex I TFEU). Furthermore, living animals (Chapter 1), meat (Chapter 2), fish, crustaceans and molluscs (Chapter 3), milk and

[20]Ruffert. In: Calliess and Ruffert (2011), Art. 3 TEU, n. 13.

[21]Ruffert. In: Calliess and Ruffert (2011), Art. 3 TEU, n. 15; cf. ibid. In: Ruffert (2013), vol. 5, § 1, n. 2.

[22]Ruffert. In: Calliess and Ruffert (2011), Art. 3 TEU, n. 22.

[23]Synonymous "Common Market", see Kahl. In: Calliess and Ruffert (2011), Art. 26 TFEU, n. 8 ff.

[24]Piepenschneider. In: Bergmann (2015), p. 170.

[25]Ruffert. In: Ruffert (2013), vol. 5, § 1, n. 1.

[26]Calliess. In: Calliess and Ruffert (2011), Art. 5 TEU, n. 6.

dairy products, birds' eggs natural honey (Chapter 4) and other animal products belong to agricultural products and therefore are considered goods.[27] Thus, speaking of farm animals consequently means that they are products and goods in terms of the internal market.

A special policy of the EU for the realization of the internal market in this economic sector is the CAP (Article 38(1) subpara 2 sentence 1 TFEU), which was shaped in its initial phase (from 1962 to 1980) by food shortage in the aftermath of World War II.[28] In this phase, the CAP had been nearly exclusively oriented towards foodstuff production, which should guarantee the self-sufficiency of the EU with agricultural goods and a sufficient income for the farmers.[29] Furthermore, the CAP has the following targets (Article 39(1) TFEU), which bind the Union legislator and have to be used for the interpretation of agricultural policy legal requirements made by the EU[30]: Increasing agricultural productivity (a), ensuring a fair living standard for the agricultural community (b), stabilizing markets (c), assuring the availability of supplies (d) and ensuring reasonable prices for consumers (e). Even though the social and economic conditions in the EU have changed significantly after World War II, these written objectives remained unchanged.[31] However, the protection of farm animals does not belong to the strongly economically[32] shaped and final[33] written objectives of the CAP.

The CAP has a specific relevance[34] within the EU, which becomes apparent for the following considerations: Just highlighting its status as a primary law shows that the CAP is recognizable as a "special case of the free movement of goods".[35] In comparison to other economic sectors, an own particular policy is practiced beyond the use of internal market instruments.[36] General requirements of the internal market can only be applied if not stated otherwise in the Articles 38 to 44 (Article 38(2) TFEU). The chapter of the Treaty on competition rules may only be applied if the European Parliament and the Council have determined this (Article 42 TFEU).

[27]See the name of the column 2 "Description of products" Annex I List referred to in Art. 38 TFEU, EU (2007).

[28]Holzer (2014), p. 128; Düsing. In: Dombert and Witt (2011), § 26, n. 118; Härtel. In: Ruffert (2013), vol. 5, § 7, n. 2; Thiele. In: Calliess and Ruffert (2011), Art. 39 TFEU, n. 2; development of the CAP, see Martínez (2014), § 6, n. 24 ff.

[29]Härtel. In: Ruffert (2013), vol. 5, § 7, n. 2.

[30]Khan. In: Geiger et al. (2010) Art. 39 TFEU, n. 3; cf. Busse. In: Lenz and Borchardt (2010), Art. 39 TFEU, n. 1.

[31]Busse. In: Lenz and Borchardt (2010), Art. 39 TFEU, n. 1; ibid. In: Lenz and Borchardt (2010), Art. 38 TFEU, n. 3; Khan, in: Geiger et al. (2010) Art. 39 TFEU, n. 2; Martínez (2014), § 6, n. 72.

[32]Martínez (2014), § 6, n. 73.

[33]Bittner. In: Schwarze (2012), Art. 39 TFEU, n. 4.

[34]Düsing. In: Dombert and Witt (2011), § 26, n. 119.

[35]Martínez (2014), § 6, n. 27.

[36]Thiele. In: Calliess and Ruffert (2011), Art. 38 TFEU, n. 3.

The CAP regulates both cross-border issues and the agricultural economy as a whole within the Member States.[37] Additionally, the CAP is characterized by planned economy patterns: According to the principle of Union preference agricultural goods from the EU take precedence over imports from third countries[38] in order to protect the agricultural internal market against low-price imports and fluctuations of prices on the world market.[39] Further, regarding the design of the CAP and the realization of its objectives, the particular nature of agricultural activities and the close linkage between agriculture and the economy as a whole have to be taken into account (Article 39(2) TFEU). Among other things, it is necessary to effect the appropriate adjustments by degrees. (Article 39(2) lit. b TFEU).

To conclude, the protection of farm animals is neither a value or a target of the EU, nor a value or target of the CAP. In fact, the EU in general and the CAP in particular pursue primarily economical targets. Farm animals are defined as agricultural products within the CAP and for that reason they are treated by law like other agricultural goods. Lifestock farming and the production of animal products are assigned to the law of agricultural production[40] and have to be seen in the context of the realization of the targets of the CAP and the internal market.

9.2.2 Harmonization of Law

The harmonization of law as means of realizing the internal market[41] is the "third pillar" of the CAP, besides the instrument of common organization of the market ("first pillar") and rural development ("second pillar").[42] Harmonization of law "means convergence of the national legal systems across the Member States"[43] "with the aim to ensure a co-development and to avoid disturbances in the internal market".[44] In the agricultural sector the harmonization of law is considerably important for the realization of the free internal market of agricultural products.[45] Furthermore, "legal acts [in the sector of agricultural policy] shall achieve and

[37]Düsing. In: Dombert and Witt (2011), § 26, n. 119.

[38]Düsing. In: Dombert and Witt (2011), § 26, n. 119; Martínez (2014), § 6, n. 89.

[39]Martínez (2014), § 6, n. 89; Oppermann et al. (2014), § 24, n. 9.

[40]Busse. In: Schulze et al. (2015) § 25, n. 43 f.; see table of contents Norer (2012).

[41]Ruffert. In: Ruffert (2013), vol. 5, § 1, n. 47.

[42]Martínez (2014), § 6, n. 91; Norer, Bloch. In: Dauses (2014), G n. 166.

[43]Remien. In: Schulze et al. (2015), § 14, n. 2.

[44]Grupp. In: Bergmann (2015), p. 508; cf. Fischer, in: Lenz and Borchardt (2010), preliminary note Art. 114 TFEU, n. 1 f.

[45]Norer, Bloch. In: Dauses (2014), G n. 166.

protect a high standard of agricultural basic and final products on behalf of a productive agriculture committed to technical progress".[46] A legal harmonization by measures of an approximation of law can only be performed by the EU, if it has the legal competence.[47] Although there can be fluent transitions,[48] approximation of law shall not be confused with unification of law, which strives for a uniform jurisdiction.

Approximation of law can be carried out by all acts of secondary law based on primary law, such as directives, regulations, decisions, recommendations and opinions (Article 288 TFEU). The individual legal acts can be differentiated regarding their addressees and their legal effects. Thereby, directives and regulations have a special importance in the field of harmonization of the protection of farm animals. The addressees of directives are the Member States.[49] Directives are only binding with respect to their objectives.[50] The choices of form as well as methods of realization are left to each of the Member States (Article 288(3) TFEU). Nevertheless, the Member States have the duty to implement the instructions of these directives into national legislation entirely and within the given time limit.[51] Regulations, in contrast to directives, are directly applicable and constitute direct rights and obligations to the Member States and their citizens.[52] They do not need a national legislation act for their effectiveness.

An interpretation of the secondary law has always to occur in the light of the primary law.[53] Due to this hierarchical relation the secondary law may not contradict the primary law.[54] In case of conflicts, the primary and the secondary law take precedence in its application over national legislation and constitutional law of the Member States.[55] Additionally, the prohibitive effect of the EU-Law forbids an independent legislation by the Member States in areas, in which the EU has introduced binding law (Article 2(2) sentence 2 TFEU).[56] However, in so-called

---

[46]Norer, Bloch. In: Dauses (2014), G n. 166; also Härtel. In: Ruffert (2013), vol. 5, § 7, n. 76.

[47]Ruffert. In: Ruffert (2013), vol. 5, § 1, n. 10; the terms of approximation of law and harmonization are often used interchangeably, see Kahl. In: Calliess and Ruffert (2011), Art. 114 TFEU, n. 13.

[48]Remien. In: Schulze et al. (2015) § 14, n. 2; auch Fischer. In: Lenz and Borchardt (2010), Vorb. Art. 114 TFEU, n. 1; Kahl. In: Calliess and Ruffert (2011), Art. 114 TFEU, n. 1.

[49]Biervert. In: Calliess and Ruffert (2011), Art. 288 TFEU, n. 24.

[50]Crit. Calliess (2012) p. 820.

[51]Biervert. In: Calliess and Ruffert (2011), Art. 288 TFEU, n. 23. In Germany, this implementation is usually carried out by national laws and regulations; to the direct effect of directives see ibid. In: Calliess and Ruffert (2011), Art. 288 TFEU, n. 27.

[52]Biervert, in: Calliess and Ruffert (2011), Art. 288 TFEU, n. 20.

[53]Borchardt, in: Schulze et al. (2015), § 15, n. 51; cf. Schwarze. In: Schwarze (2012), Art. 19 TEU, n. 31.

[54]König. In: Schulze et al. (2015), § 2, n. 2.

[55]ECJ (1964), n. 1269 ff.; Ehlers. In: Schulze et al. (2015), § 11, n. 39.

[56]In detail Calliess. In: Calliess and Ruffert (2011), Art. 2 TFEU, n. 2; Lachmayer and Bauer (2008), p. 746.

clauses for enacting higher levels of protection the Union legislator can admit a remaining scope of action to the Member States,[57] which is determined by the degree of harmonization of the secondary law. The range of harmonization varies between minimum harmonization to full harmonization. Minimum harmonization implies that the Member States are allowed to enact more stringent laws as far as they comply with the Union Law.[58] Such minimal clauses are located at the end of the directives. The minimum harmonization is a specific case of partial harmonization, in which the EU just partially regulates a certain subject area.[59] In the case of optional harmonization the Member States are free to maintain their own legal system with respect to national issues, although the national legal system has to provide at least the same protection as the corresponding directive.[60] Full harmonization means that legislation acts do not principally exhibit a clause for mandating higher levels of protection due to their exhaustive character.[61]

9.2.3 Cross-Sectional Task Protection of Animals

Even though the protection of animals is neither a declared value nor an objective of the EU and the CAP, the primary law includes some animal protective provisions. Especially the horizontal clause and cross-sectional task of Article 13 TFEU as well as the possibility of quantity restriction due to animal protection in Article 36 TFEU are of particular importance for legislations in the fields of production, transport and slaughtering farm animals. Article 36 TFEU is a justification-reason for particular national restrictions of the fundamental freedoms. However, since the primary scope of this contribution is the role of the EU, in the following only the horizontal clause Article 13 TFEU will be discussed. It states:

> In formulating and implementing the Union's agriculture, fisheries, transport, internal market, research and technological development and space policies, the Union and the Member States shall, since animals are sentient beings, pay full regard to the welfare requirements of animals, while respecting the legislative or administrative provisions and customs of the Member States relating in particular to religious rites, cultural traditions and regional heritage.

[57] Calliess. In: Calliess and Ruffert (2011), Art. 2 TFEU, n. 2.

[58] Lachmayer and Bauer (2008), p. 746; Remien. In: Schulze et al. (2015) § 14, n. 42.

[59] Lachmayer and Bauer (2008), p. 746; Remien. In: Schulze et al. (2015) § 14, n. 42.

[60] Remien. In: Schulze et al. (2015) § 14, n. 45; Herrnfeld. In: Schwarze (2012), Art. 114 TFEU, n. 60.

[61] Herrnfeld. In: Schwarze (2012) Art. 114 TFEU, n. 57; cf. Remien. In: Schulze et al. (2015) § 14, n. 42.

9.2.3.1 Historical Development of the Protection of Animal Welfare in the EU

As described above, while establishing the EU (or the previous organizations) economic interests and peace-keeping issues were of foremost importance. Consequently, there was no animal protective legislation in the founding treaties. Not until 1992 a declaration on the protection of animals was incorporated into the Final Act of the Treaty of Maastricht (Declaration N° 24[62]). This declaration only had an "appellative"[63] character for the EU and its Member States due to its lack of binding legal force and therefore could only be used for interpretation of contract law.[64] In 1997, the Treaty of Maastricht was followed by the Treaty of Amsterdam to which a Protocol (Prot. N° 33) on protection and welfare of animals was annexed.[65] This protocol was supposed to improve the protection of animals and to increase the level of animal protection in the EU. This amendment to the protocol resembled an increasing legal importance of the protection for animals.[66] It is especially supported by the fact that protocols have the same legal status as Contract Law (Article 51 TEU) and therefore protocols produce legal effects.[67] Henceforth, infringement proceedings (Article 258 TFEU), action for annulment (Article 263 TFEU) as well as preliminary ruling procedures (Article 267 TFEU) before the European Court of Justice (ECJ) were possible.[68] Since then, animal protection must be taken into consideration when balancing legally protected interests for planned legislative acts, also in the field of agriculture. Since then, when legislative acts in the field of agriculture are planned animal protection must be taken into consideration and must be balanced with other legally protected interests.[69]

In 2009, the protocol was followed by the Treaty of Lissabon[70] with the Article 13 TFEU. With the implementation of this Article on animal welfare into the treaty itself, animal protection gained increasing political importance.[71] Comparing Prot. N° 33 with the Article 13 TFEU some changes of wording are apparent: The horizontal clause Article 13 TFEU was extended by the political fields of fisheries, technological development, and space policies. Furthermore, it now includes the reference of animals as "sentient beings", which was taken over from the justification declaration of the protocol.

---

[62]EU (1992), p. 103.

[63]Schmidt. In: Schwarze (2012), Art. 13 TFEU, n. 1; also Kluge, von Loeper. In: Bergmann (2015) p. 992.

[64]Becker. In: Schwarze (2012), Art. 51 EUV, n. 9 f.; Geiger. In:Geiger et al. (2010) Art. 51 TEU, n. 5.

[65]EU (1997), p. 110.

[66]Cf. Härtel (2012) chapter 25, n. 12; critical Caspar (2001), p. 7 f.

[67]Becker. In: Schwarze (2012), Art. 51 EUV, n. 2; Härtel (2012) chapter 25, n. 12.

[68]Becker. In: Schwarze (2012), Art. 51 EUV, n. 2; Caspar (2001), p. 75.

[69]Nettesheim. In: Grabitz et al. (2014), Art. 13 TFEU, n. 3.

[70]EU (2007).

[71]Calliess. In: Calliess and Ruffert (2011), Art. 13 TFEU, n. 1.

9.2.3.2 Protection of Welfare

The term *welfare* is not explicitly defined in the TFEU.[72] It is assumed that animals have to be protected from avoidable pain and harmful effects, from non-behaviourally appropriate keeping as well as from species-inadequate nutrition or care.[73] Furthermore welfare means more than just the absence of disease and injuries. Additionally, animals "[should be protected from] mental suffering as it is particularly associated with the suppression of physiological and ethological needs".[74] The special characteristics of animals, their sensory perceptions and their feelings are to be accepted and to be respected.[75] Basic needs of animals should fully be ensured.[76] In the field of farm animal welfare, general principles apply—such as sufficient freedom of movement, prevention of heat, sufficient food, and water.[77]

9.2.3.3 Ethical Protection of Animals

Animal protection according to Article 13 TFEU is based on an ethical motivated understanding of animal protection,[78] which regards animals as fellow creatures and sentient beings.[79] This is also apparent in the wording of animals as "sentient beings". Animals therefore have an (self-) interest of avoidance of pain, suffering, and damage, which is worthwhile protecting.[80] Every individual animal shall be

[72]Breier. In: Lenz and Borchardt (2010), Art. 13 TFEU, n. 4; Schmidt. In: Schwarze (2012), Art. 13 TFEU, n. 7.

[73]Cf. as already to the Prot. N° 33 Hirt et al. (2007), introduction, n. 35; cf. Breier. In: Lenz and Borchardt (2010), Art. 13 TFEU, n. 4; Calliess. In: Calliess and Ruffert (2011), Art. 13 TFEU, n. 3 and. Martínez (2014), § 6, n. 21, speak of "ensuring a welfare state".

[74]Hirt et al. (2007), introduction, n. 35.

[75]Frenz (2011), p.(106).

[76]von Loeper. In: Kluge (2002), intro. n. 109.

[77]Breier. In: Lenz and Borchardt (2010), Art. 13 TFEU, n. 4.

[78]Schmidt. In: Schwarze (2012), Art. 13 TFEU, n. 7; as already to the Prot. N° 33: Hirt et al. (2007), introduction, n. 35; Lorz and Metzger (2008), intro., n. 80; von Loeper and Kluge (2002), introduction, n. 109.

[79]Cf. as already to the Prot. N° 33: Hirt et al. (2007), introduction, n. 21; Lorz and Metzger (2008), intro., n. 26; von Loeper. In: Kluge (2002), introduction, n. 48 f.; referring to the term "fellow creature" see Blanke (1959) p. 198.

[80]Nettesheim. In: Grabitz et al. (2014), Art. 13 TFEU, n. 12; Schmidt. In: Schwarze (2012), Art. 13 TFEU, n. 7.

protected for its own sake.[81] Nevertheless, an unrestricted protection of animal welfare[82] or an inviolable inherent value of animals[83] cannot be awarded from the wording.

9.2.3.4 Content

Article 13 TFEU includes an obligation for the EU institutions and the Member States to consider animal protection in the sense of a binding legal bid.[84] While formulating and implementing the listed policies in Article 13 TFEU, welfare requirements of animals have to be fully taken into account.[85] Furthermore, animal protection is to be taken into consideration as a balancing concern in discretionary decision. This leads to the fact that the Article 13 TFEU resembles also a binding guideline for the realization of animal protection.[86] "As a parallel to other horizontal principles [e.g. environmental protection in Article 11 TFEU], issues on animal welfare do not gain absolute priority, but they have to be brought in a balanced relation to conflicting interests [mostly of human nature] in a matter of practical concordance"[87] As already shown, farm animals are classified as goods according to the EU law. Their production is intended to provide income for the farmers as well as to guarantee sufficient food supply for the consumers. Therefore, the ethical animal protection of Article 13 TFEU has always to be seen in the context of anthropocentric and economic interests of farmers and consumers as well. Additionally, Article 13 TFEU functions as an interpretation aid for undefined legal terms and in the context of discretionary decisions.[88]

---

[81]Schmidt. In: Schwarze (2012), Art. 13 TFEU, n. 7; cf. as already to the Prot. N° 33: von Loeper. In: Kluge (2002), introduction, n. 52; crit. in case of farm animals ibid. In: Kluge (2002), introduction, n. 109.

[82]Epiney. In: Vedder and Heintschel von Heinegg (2012) Art. 13 TFEU, n. 3, Fn. 6.

[83]Nettesheim. In: Grabitz et al. (2014), Art. 13 TFEU, n. 12; cf. Schmidt. In: Schwarze (2012), Art. 13 TFEU, n. 7.

[84]Calliess (2012), p. 819, synonymous in German language: Berücksichtigungsgebot und Rücksichtnahmepflicht in: Müller-Graff. In: Vedder and Heintschel von Heinegg (2007) Art. III-121; Berücksichtigungsgebot: Schmidt. In: Schwarze (2012), Art. 13 TFEU, n. 7; Rücksichtnahmegebot: ibid. In: Schwarze (2012), Art. 13 TFEU, n. 2; Nettesheim. In: Grabitz et al. (2014), Art. 13 TFEU, n. 9; Maisack (2012), 5 (5); cf. as already to the Prot. N° 33: Hirt et al. (2007), introduction, n. 37.

[85]Schmidt. In: Schwarze (2012), Art. 13 TFEU, n. 7; cf. ECJ (2001), n. 85.

[86]Calliess. In: Calliess and Ruffert (2011), Art. 13 TFEU, n. 7.

[87]Calliess. In: Calliess and Ruffert (2011), Art. 13 TFEU, n. 7.

[88]Calliess 2012, p820; ibid.In: Calliess and Ruffert (2011), Art. 13 TFEU, n. 10.

9.2.3.5 Consequences

Addressees of the legal bid are the EU and its Member States.[89] As already shown, while formulating and implementing the listed policies in Article 13 TFEU, welfare requirements of animals have to be fully taken into account. Formulating means a definition of other policies "by specification and definition of measures [. . .], be it through formal measures, such as abstract general acting in the field of legislation, or through non-binding legal acts. Therefore, the formulating of other policy areas is meant, if the requirements of animals as sentient beings can be read into the respective treaty provisions as part of these other policies.[90] Implementing other policies means the "operational phase",[91] i.e. adopting secondary law in those policy areas listed in Article 13 TFEU.[92] It is the obligation of the Member States to pay full regard to the welfare requirements of animals, by legislative implementation, by administration, and by judice.[93] While implementing EU law, the Member States have to consider the taken evaluations by the EU during determination of its policies.[94] In purely national issues, the Member States do not have to adhere to this consideration bid.[95]

9.2.3.6 Limitations

By the wording of Article 13 TFEU, the commandment of consideration has limitations.[96] The horizontal animal protection clause includes a caveat for culture in the second half sentence ("[. . .] while respecting the legislative or administrative provisions and customs of the Member States relating in particular to religious rites, cultural traditions and regional heritage."). This caveat can relativize farm animal protection standards in the EU. Thus, due to cultural reasons, the Member states are allowed to maintain acts allowing certain activities, which might be problematic with respect to animal welfare.[97]

---

[89]Calliess. In: Calliess and Ruffert (2011), Art. 13 TFEU, n. 5; Schmidt. In: Schwarze (2012), Art. 13 TFEU, n. 2.

[90]Breier. In: Lenz and Borchardt (2010), Art. 13 TFEU, n. 7.

[91]Breier. In: Lenz and Borchardt (2010), Art. 13 TFEU, n. 7.

[92]Ibid., in: Lenz and Borchardt (2010), Art. 11 TFEU, n. 9; cf. Schmidt. In: Schwarze (2012), Art. 13 TFEU, n. 5.

[93]Cf. Käller. In: Schwarze (2012), Art. 11 TFEU, n. 10.

[94]In detail Schmidt. In: Schwarze (2012), Art. 13 TFEU, n. 5.

[95]Schmidt. In: Schwarze (2012), Art. 13 TFEU, n. 2.

[96]At this point it will not be discussed whether the policies described in Art. 13 TFEU are listed exhaustively, since agriculture (and thus the production of animal products) belongs to the listed policies.

[97]Calliess, in Calliess and Ruffert (2011), Art. 13 TFEU, n. 9; Schmidt. In: Schwarze (2012), Art. 13 TFEU, n. 8.

9.2.3.7 Judicial Control

The European Council of Justice (ECJ) verifies whether the commandment of consideration of animal protection was complied by the competent authority.[98] Indeed during the realization of the commandment of consideration, the Union institutions have a wide scope for discretion.[99] The annulment of harmonizing legal norms by the ECJ due to violations of Article 13 TFEU is subject of high requirements.[100] During examining of whether Article 13 TFEU was injured, the ECJ is limited to control "[. . .] if the measure [. . .] is vitiated by [. . .] an obvious error or an abuse of discretion, whether the authority has clearly exceeded the bounds of its discretion",[101] and whether the measure has been obviously unsuitable.[102] The requirement of consideration was sufficiently recognized "[. . .] if the animal welfare aspects are recognizable and the assessment between the different objectives is comprehensible".[103]

9.2.3.8 Further Obligations

Article 13 TFEU expresses the binding mandate to act to prioritize animal protection as high as possible.[104] In this sense it is spoken of the commandment of optimization[105] for further policies. Due of this broad discretion, further concrete mandates to act—as included in Article 191, 192 TFEU as part of the environmental policy—cannot be deduced.[106] Rather the commandment of optimization is already realized if the EU guarantees a minimum of animal protection.[107] Thus, an

[98]Calliess, in Calliess and Ruffert (2011), Art. 13 TFEU, n. 7.

[99]Schmidt. In: Schwarze (2012), Art. 13 TFEU, n. 7.

[100]Cf. Caspar (2001), S. 32 für das Prot. N° 33.

[101]Schmidt. In: Schwarze (2012), Art. 13 TFEU, n. 7.

[102]ECJ (2001), n. 80.

[103]Schmidt. In: Schwarze (2012), Art. 13 TFEU, n. 7.

[104]Calliess, in Calliess and Ruffert (2011), Art. 13 TFEU, n. 7.

[105]Cf. so already for Prot. N° 33 the German terms: Lorz and Metzger (2008), intro., n. 80 ("Optimierung des Tierschutzes"); cf. Caspar (2001), p. 16 ("rechtsverbindliche Gestaltungsaufgabe zur Optimierung des Tierschutzes"; ibid., p. 32 und 77 ("Optimierungsgebot"). cf. for Art. 20a of the German constitutional law as an "Optimierungsgebot": Schulze-Fielitz. In Dreier (2006), vol. 2, Art. 20a, n. 26.

[106]cf. Nettesheim. In: Grabitz et al. (2014), Art. 13 TFEU, n. 4; cf. so already for Prot. N° 33 Caspar (2001), p. 77.

[107]Cf. Caspar (2001), p. 78 for Prot. N° 33.

obligation of performing optimizing acts does not apply as well as the adoption of protective duties[108] or a "[. . .] comprehensive mandate to protect the animals".[109]

9.3 Status Quo of the Harmonization of Farm Animal Protection

9.3.1 The Legal Situation of Farm Animal Protection

As already explained, the EU needs a legal competence for enacting harmonization laws. In the area of (farm) animal protection the EU has no obvious legal competence.[110]

9.3.1.1 Legal Competence Outside the CAP

It needs to be noted that Article 13 TFEU is not a legal competence as it refers to other competences.[111] Article 2 TEU and Article 3 TEU also represent no legal competences due to dogmatic reasons. As far as animal welfare is not recognized as a target according to Article 3 TEU, it cannot fall in the "flexibility clause" of Article 352 TFEU.[112] As written above, also Article 36 TFEU refers to animal protection. This prescription regulates the competences of the Member States to impose restrictions on the freedom of goods protecting the health and life of animals. But, Article 36 TFEU is not a legal competence for enacting secondary law by EU institutions.

9.3.1.2 Legal Competence Inside the CAP

The legal competence in Article 43(2) TFEU is the correct legal competence in cases regarding the regulation of agricultural products and the achievement of the

[108]Schmidt. In: Schwarze (2012), Art. 13 TFEU, n. 7; critically to enforceable legal obligations Müller-Graff. In: Vedder and Heintschel von Heinegg (2007) Art. III-121; Nettesheim. In: Grabitz et al. (2014), Art. 13 TFEU, n. 13.

[109]Cf. Caspar (2001), p. 77 for Prot. N° 33.

[110]Schmidt. In: Schwarze (2012), Art. 13 TFEU, n. 3.

[111]Härtel. In: Ruffert (2013), vol. 5, § 7, n. 84; Schmidt. In: Schwarze (2012), Art. 13 TFEU, n. 3.

[112]Härtel. In: Ruffert (2013), vol. 5, § 7, n. 84.

CAP's targets.[113] However, recourse to Article 43(2) TFEU is not possible if the law acts only relate to agricultural products but do not pursue the targets of the agricultural policy [114] However, animal protection, on its own, is not a target of the CAP. Thus, the second requirement of the legal competence cannot be readily affirmed regarding to provisions on handling farm animals. Nevertheless, directives and regulations for harmonizing animal protection are based on this legal competence because they serve—and probably primarily—agricultural targets.[115] Besides improving animal protection, these measures are supposed to substantially contribute to equal conditions of competition between producers in the individual Member States.[116] This fact can also be deduced from the recitals of the directives and regulations. A reduction of unequal conditions of competition will particularly be strived if different national animal protection laws impede free trade with animal products. Without European minimum standards the Member States could conduct animal protection dumping[117] in order to gain significant competitive advantages. "[For that reason], individual animal protection plays just an indirect role in terms of an ancillary competence within the framework of the CAP or under economic aspects, respectively."[118]

9.3.1.3 Principle of Subsidiarity

The derivation of a legal competence from Article 43(2) TFEU is in conformity with the principle of subsidiarity according to Article 5(3) TEU. This article implies: "Under the principle of subsidiarity, in areas which do not fall within its exclusive competence, the Union shall act only if and in so far as the objectives of the proposed action cannot be sufficiently achieved by the Member States, either at central level or at regional and local level, but can rather, by reason of the scale or effects of the proposed action, be better achieved at Union level."[119] Therefore, the principle of subsidiarity includes a criterion of necessity and a criterion of efficiency. A predictive decision has to be made for both criteria. In this connection, EU institutions have a wide margin of assessment and a wide scope of design.[120]

Farm animal protection is embedded in agricultural policy and various economic linkages within the internal market. Due to the freedom of movements within the EU transferring animal production from one Member State to another is easily

[113]von Rintelen. In: Grabitz et al. (2014) Art. 43 TFEU, n. 1.

[114]von Rintelen. In: Grabitz et al. (2014), Art. 43 TFEU, n. 2.

[115]Caspar (2001), p. 18; Härtel. In: Ruffert (2013), vol. 5, § 7, n. 85; Lachmayer and Bauer (2008) p. 863; Calliess. In: Calliess and Ruffert (2011), Art. 13 TFEU, n. 12.

[116]ECJ (1988a); ECJ (1988b).

[117]Nentwich (1994), p. 88.

[118]Schulze-Fielitz. In: Dreier (2006) vol. 2, Art. 20a, n. 18.

[119]Lienbacher. In Schwarze (2012), Art. 5 TEU, n. 15.

[120]Lienbacher. In Schwarze (2012), Art. 5 TEU, n. 23, 26 f.

possible. Normally, higher animal protection standards are accompanied by increasing production costs.[121] Therefore, the Member States raising animal protection standards are exposed to the risk of impairing the own economy location. Likewise, it is not forseeable that all 28 Member States are in condition to enact legal acts for the protection of farm animals coordinated in line with each other without using the EU framework. One example of this is that housing pregnant sows in crates is not prohibited in all Member States.[122] This is based on different perceptions of animal protection across the Member States[123] and no identical states of science. Thus, a uniform animal protection level is only possible at EU level.[124]

9.3.1.4 Relation to General Legal Competences of the Internal Market

Compared to the general harmonization legal competence of the internal market (Article 114 TFEU), Article 43(2) TFEU is the more specific legal competence.[125] Hence, recourse to Article 114 TFEU is not possible.[126] Due to the ancillary competence recourse to Article 352 TFEU is prevented.[127]

9.3.2 Selected Examples for Existing Harmonized Rules

In its resolution of 20 February 1987 on animal welfare, the European Parliament invited the Commission to make proposals about Community rules covering general aspects of the rearing of livestock.[128] Following this request, harmonizing directives and regulations have been enacted in the areas of keeping, transporting, and slaughtering. Additionally to the specification of the primary law target definition in Article 13 TFEU,[129] the following secondary law serves the principle of the single market:

Council Directive 98/58/EC of 20 July 1998 concerning the protection of animals kept for farming purposes (CD 98/58/EC)[130]

121 Caspar (1999) p. 209; animal protection as a selling point, see Apel (2010), p. 216.

122 Wollenteit and Lemke (2013) p. 178, fn. 4.

123 Apel (2010), p. 217.

124 Cf. Caspar (2001), p. 18 to the Prot. N° 33.

125 Khan. In: Geiger et al. (2010), Art. 43 TFEU, n. 8.

126 Cf. Bittner. In:Schwarze (2012), Art. 43 TFEU, n. 11.

127 Geiss. In: Schwarze (2012), Art. 352, n. 20; cf. Busse. In: Schulze et al. (2015) § 25, n. 32: Art. 352 TFEU as a legal competence for non- Annex I products.

128 Recital CD 98/58/EC.

129 Nettesheim. In: Grabitz et al. (2014) Art. 13 TFEU, n. 16.

130 EC (1998).

Council Directive 2007/43/EC of 28 June 2007 laying down minimum rules for the protection of chickens kept for meat production (CD 2007/43/EC)[131]

Council Directive 1999/74/EC of 19 July 1999 laying down minimum standards for the protection of laying hens (CD 1999/74/EC)[132]

Council Directive 2008/120/EC of 18 December 2008 laying down minimum standards for the protection of pigs (CD 2008/120/EC)[133]

Council Directive 2008/119/EC of 18 December 2008 laying down minimum standards for the protection of calves (CD 2008/119/EC)[134]

Council Regulation (EC) No 1/2005 of 22 December 2004 on the protection of animals during transport and related operations and amending Directives 64/432/EEC and 93/119/EC and Regulation (EC) No 1255/97 (CR (EC) N° 1/2005)[135]

Council Regulation (EC) No 1099/2009 of 24 September 2009 on the protection of animals at the time of killing (CR (EC) N° 1099/2009)[136]

9.3.2.1 General Minimum Standards

The CD 98/58/EC defines minimum standards for the protection of animals kept for farming purposes (Article 1(1)). In general, CD 98/58/EC applies to all farm animals (Article 1(2)). The owners or keepers take all appropriate measures to ensure the welfare of animals under their care and to ensure that those animals are not subject to any unnecessary pain, suffering, or injury (Article 3). Furthermore the conditions under which animals (excluding fishes, reptiles, or amphibians) are bred or kept, have to meet the requirements set out in the appendix. Thereby, the species, its degree of development, its adaption and domestication, and its physiological and ethological needs in accordance with established experiences and scientific knowledge need to be taken into account (Article 4). It must be emphasized that no animal must be kept for farming purposes unless it can reasonably be expected (on the basis of its genotype or phenotype) that it can be kept without detrimental effects on its health or welfare (N° 21 in the Annex).

[131]EC (2007).

[132]EC (1999).

[133]EC (2008a).

[134]EC (2008b).

[135]EC (2005).

[136]EC (2009).

In addition to the relevant provisions of the general CD 98/58/EC, special requirements apply for several farm animals according to the aforementioned directives. In the following, some issues that are discussed in public[137] are outlined and assessed in a legal evaluation:

9.3.2.2 Stocking Density for Broilers

A fattening period of broilers takes about 4 weeks. During this period the animal litter will not be changed and the excrements of the animals accumulate. Thus, a higher stocking density leads to a higher degree of pollution of the animal litter.[138] Thereby, burns could be formed at the foot pads of the animals. These are associated with pain for them. Thus, the animals might have problems walking and through lesions bacterias can easily enter the body.[139] Studies show that at the time of slaughter the proportions of broilers with severe foot pad lesions ranged between 7 and 25 %.[140] Generally it cannot be said that the stocking density is the only decisive factor for animal health.[141] However, several studies found out that the frequency and the intensity of foot pad lesions increase with higher stocking densities.[142] Thus, a higher stocking density can have negative effects on animal health.[143] This view is also confirmed by the Panel on Animal Health and Welfare (AHAW) of the European Food Safety Authority (EFSA).[144] Particularly, exceeding a stocking density of 25 kg/m^2 is seen in a very critical light.[145]

According to the CD 2007/43/EC, the maximum stocking density must not exceed 33 kg/m^2 at any time (Article 3(1), (2) and Annex I). By derogation from these paragraphs, broilers can be kept at a higher stocking density if further criteria are met (Article 3(3), Annex II). These criteria are for example notification and record keeping obligations; additionally, the stalls must be equipped with a ventilation system and heating and cooling systems to comply with special requirements on concentration of ammoniak and carbon dioxid, room temperature, and air humidity if necessary.

In case of complying with the following additional criteria, the maximum stocking density can be increased from 39 kg/m^2 to a maximum of 42 kg/m^2 (Article 3(4), (5), Annex V). These criteria are: a) the monitoring of the holding carried out by the competent authority within the last 2 years did not reveal any

---

[137]A general overview of problems in the livestock sector can be found in Caspar (1999), p. 209 ff.

[138]Spindler and Hartung (2010), p. 11.

[139]Keppler et al. (2009), p. 33; Spindler and Hartung (2010), p. 11.

[140]Kamphues (2014), p. 2, who refers to a study carried out in Germany.

[141]De Jong et al. (2012) p. 39.

[142]Spindler and Hartung (2010), p. 17 f.

[143]Spindler and Hartung (2010), p. 6.

[144]EFSA (2000); so probably still de Jong, Berg, Butterworth, Estevéz (2012), p. 39, 74.

[145]EFSA (2000) p. 64, 66, 107, N° 25.

deficiencies with respect to the requirements of this directive, b) the monitoring by the owner or keeper of the holding is carried out using the guides to good management practice referred to in Article 8 and c) in at least seven consecutive, subsequently checked flocks from a house the cumulative daily mortality rate was below 1 % + 0.06 % multiplied by the slaughter age of the flock in days. By the way of derogation from the last criteria, the competent authority may decide to increase the stocking density when the owner or keeper has provided sufficient explanation for the exceptional nature of a higher daily cumulative mortality rate or has shown that the causes lie beyond his sphere of control.

Hence, the stocking density allowed by the EU law is markedly higher compared to the stocking density required by the EFSA. By approving a stocking density of 42 kg/m^2 more animal protection problems are to be expected. However, lower stocking densities can be associated with higher costs for the producers.

9.3.2.3 Beak Trimming of Laying Hens

In a lot of laying hens farms so-called feather-pecking is a severe problem as the laying hens peck out feathers from each other. In the worst case, laying hens continue pecking at hens already attacked until fatal cannibalism. Feather-pecking and cannibalism are unwanted behaviors that have multifactorial causes.[146] As a consequence, the beaks are trimmed in order to avoid these behaviors and also to limit the damages to laying hens. In doing so the tips of the beaks are removed with tongs, with the so-called "hot knife", or partly by the usage of the infrared treatment. The infrared beam does not detach the tips of the beaks but interferes in the tissue structure of the beaks. Finally, the treated area falls off by rubbing during feed intake after approximately 10–14 days. Since feather-pecking and cannibalism exist in laying hens stock regardless of the production method, there are a lot of different opinions about the need of beak trimming.[147] But it is indisputable that the beak trimming is damage to every single animal, irrespective of whether the animals experience pain or suffering by that measure[148] and that animals are adapted to the husbandry system.

The appendix of the CD 1999/74/EG implies in N° 8 regarding operations on laying hens that without prejudice to the provisions of N° 19 of the Annex to Directive 98/58/EC, all mutilations shall be prohibited. In order to prevent feather pecking and cannibalism, however, the Member States may authorize beak trimming, provided it is carried out by qualified staff and on chickens that are less than 10 days old and intended for laying. Therefore, causing such irreversibly damages to animals are legal in the EU law. Further, the EU law does no prescribe certain kinds of procedures for beak trimming.

[146]Telle (2011), p. 12 ff.

[147]Telle(2011), p. 14 f.

[148]Windhorst (2013), p. 1 f.

9.3.2.4 Keeping Sows in Crates

Pigs are social animals and live together in groups. They separate their environment in several areas, such as feaces and lying areas. Additionally, the possibility to nest-building is of primary importance.[149] In practice, however, sows are kept in so-called crates over a long time period. Objectives of this breeding system are increasing the probability of a successful insemination as well as protecting piglets against being crushed by the sow. Crates are fixed steel boxes that prevent the sow to turn around. As a result, they can neither separate the environment according to their natural behavior nor build a nest for their piglets. Moreover, their urge to move and to explore is significantly restricted. This kind of keeping is a limitation of their behaviours and means stress for the animals. This might lead to stereotypes such as idle chewing or physical problems.[150] There is much discussion about the argument that fewer piglets are crushed compared to other keeping forms.[151] In several Member States, like Sweden, Denmark, and Austria for example, keeping in crates depends on more specific criteria and is just allowed over a clearly limited time period. These countries focus on other systems for farrowing, which are still being studied.[152] In principle since 1 January 2013, keeping sows in groups is a general obligation in the EU (Article 3(1a)) CD 2008/120/EC). Sows and gilts have to be kept in groups during a time period starting four weeks after the service to one week before the expected time of farrowing (Article 3(4) subparagraph 4 sentence 1). Furthermore, in the time between—that means for the time of farrowing and lactation—keeping sows in crates is allowed. Assuming that a sow farrows twice a year and she spends 11 weeks per farrowing in the crates results in approximately 5 months per year in which the sow has not the possibilty to move, to build a nest, and is incapable to separate the area for feeding and defecating purposes. In that time sows and gilts have permanent access to manipulable material (Article 3(5), Annex I Chapter 1 N° 4) and in the week before the expected farrowing time they must be given suitable nesting material in sufficient quantity (Annex I Chapter B N° 3). This requirement only applies unless it is not technically feasible for the slurry system used in the establishment. Thus, the existing slurry system decides whether the sows receive nesting material and finally also influences the sows' health and welfare.

---

[149] Wechsler (1997) p.175.

[150] Wechsler (1997), p. 176 f.; EFSA (2007a), p. 3; EFSA (2007b) p. 29, 39 f.

[151] So in the year 1997 Wechsler (1997), p. 177.

[152] As one of ongoing research projects in Germany: Schrey "Studies on animal health, behavior and the performance of sows and piglets in a group housing system with free farrowing", University of Veterinary Medicine Hannover.

9.3.2.5 Castration of Pigs

When male fattening pigs become sexually mature they increasingly produce specific hormones that can have a negative effect on the flavor of the meat. However, only 2–10 % of the boar meat have these characteristics and not all costumers can smell this special "boar meat".[153] Nevertheless, 80 % of the pigs are castrated when they are young piglets to avoid the so-called boar odour.[154] Additionally, castrations shall lead to fewer ranking fights and thus to easier handling of the animals. Even though the rates of castrated animals differ across the Member States (e.g. 11 % in Portugal, 33 % in Spain[155]), and some Member States even renounce that practice (Great Britain, Ireland) or apply different alternatives,[156] castration is generally not forbidden in the EU. Indeed, CD 2008/120/EC, Annex I N° 8 implies that all procedures intended as an intervention carried out for other than therapeutic or diagnostic purposes or for the identification of the pigs and resulting in damage to or the loss of a sensitive part of the body or the alteration of bone structure are prohibited. An exception of this is the castration of male pigs. Tearing of issues is the only illegal castration procedure. To the seventh day of life, the castration shall only be performed by a veterinarian or a trained person experienced in performing the applied techniques with appropriate means and under hygienic conditions. Although, it is commonly accepted that castrations without anaesthetics cause pain and suffering for young piglets, the EU law does not prescribe the usage of these kinds of substances in that case.[157] Solely castrations after the seventh day of life have to be conducted by a veterinarian and using anaesthetica and additional prolonged analgesia under hygienic conditions. However, this means higher costs for the lifestock owners and thus most piglets are castrated in the first days of life. On the other hand, generally a castration in the first seven days of life has positive effects. In these days the animals receive their earmarks for marking and a syringe with iron for a good physical development. Conducting these measures simultaneously means less stress for the animals because they do not receive multiple treatments.

9.3.2.6 Dehorning of Calves

Horns of bovine animals serve to clarify the hierarchy among themselves. Often the size of the horns or threatening with them is sufficient to determine the ranking order. Physical confrontations happen only rarely. Due to animal housing techniques and safety purposes, however, female and male bovine animals are dehorned

---

[153] LfL (2015).

[154] MSD (2015).

[155] MSD (2015).

[156] MSD (2015).

[157] Wechsler (1997), p. 182.

when they are calves. One such method is using a cautery to cauterize the horn system. Cauterization means that bleedings are stopped and disinfected. Normally, these wounds close unproblematically and heal up quickly. Another method is the etching pen. The etching pen burns and destroys the horn system with a corrosive medium (sodium hydroxide for example). This method is not as safe as cauterizing because the corrosive medium can trickle down and thus violate the skin or eyes of the animals. It is proven that dehorning without anaesthesia leads to pain and suffering for the calves. Additionally, dehorning causes harm to the animals.[158] In practice there is much discussion on different alternatives that would make dehorning completely unnecessary.[159] Keeping horned bovines in modern play-pens, as one alternative, could pose an acceptable risk for both humans and animals. Also the breed of bovines without a horn is subject of further research. However, this kind of breeding is critisized because it means that the animals are adapted to a certain lifestock system instead of adapting the lifestock system to the animals' requirements.

In the EU law no special requirements for dehorning exists, i.e. for example there is no duty to anaesthetize. Thus, the Member States' requirements must be applied (Article 3 CD 98/58/EC, Annex I N° 19). In Germany there is a general prohibition of amputations regarding to vertebrates (§ 6 (1) sentence 1 German Tierschutzgesetz—TierSchG[160]). Exceptions are possible in the case of dehorning calves under the age of six weeks if it is essential for the intended animal use, its own protection, or the protection of other animals (§ 6(1) sentence 2 N° 3). Also surgeries on animals without anaesthesia are generally illegal (§ 5(1) sentence 1). Here again, dehorning of calves under the age of six weeks represents an exception (§ 5(3) sentence 2). Accordingly, dehorning of calves without anaesthesia, which may cause the animal pain, is legal in Germany. Only, some federal states in Germany plan to phase out dehorning without anaesthesia.[161]

9.3.2.7 Long Journey Times

The CR (EC) N° 1/2005 codifies some requirements to protect animals during transport. For example, no person shall transport animals or cause animals to be transported in a way likely to cause injury or undue suffering to them (Article 3(1)). Additionally, all necessary arrangements have been made in advance to minimize the length of the journey and meet animals' needs during the journey (Recital N° 5, Article 3(2) lit. a). Despite these legal requirements, long journey times of more than eight hours (Article 2 lit. m) are legal if further conditions are met (Article 11, particularly Annex I Chapter VI). The EU legislator does not prohibit long

[158] Sambraus (1997), p. 122.
[159] As one of many: TVT (2012).
[160] BT (2006).
[161] ML (2011), p. 7.

journey times in general. Indeed it is proved that animals are exposed to high stress during these transports: Animals are taken from their normal housing into a new environment and experience strange sounds and unfamiliar lighting conditions. At least bovine animals are separated from their known conspecifics and have to deal with unknown conspecifics. They also have no place to retreat during transport. Particularly loading and unloading means high stress for the animals. The longer the journey time, the more the welfare is affected and the more animals suffer damages.[162] Based on these animal protection problems the question of ethical justifiability of long journey times has been discussed for years.[163] Particularly long journey times occur, if the producer exploits price differences by the transport in a Member State[164] or if animals have to be transported after a road transport by sea and therefore they have to be transported to a remote port.[165] Although the majority of all consignments of live animals took less than 8 h during the time period from 2005 to 2009, at least one third lasted longer than 8 h.[166] Further, an increasing trend towards longer transport times can be found for that time period.[167] In addition, in practice maximum transport times prescribed are often violated.[168] The EU legislator has failed to adopt a ban on long journey times during the creation of the CR (EC) N° 1/2005. Hence, the free movement of goods was given priority over the welfare of animals in Article 13 TFEU.

9.3.2.8 Slaughtering Without Stunning

The regulation (EC) N° 1099/2009 contains general requirements for killing farm animals and related activities. The key premise is to spare animals to be slaughtered any avoidable pain, stress, or suffering (Article 3(1)). Therefore, they must be stunned before the actual killing. For this reason, the regulation prescribes anesthetic procedures in order to ensure that the loss of consciousness and sensibility are maintained until death of the animal (Article 4, Annex I). Likewise killing methods are required that have to be made after the anesthetic procedure, which do not result in instantaneous death (so called "simple stunning"). These procedures are for example bleeding, pithing, electrocution, or prolonged exposure to anoxia (Article 4(1)). In the case of animals slaughtered by religious rites, this requirement shall not apply provided that the slaughter takes place in a slaughterhouse (Article 4(4)).

[162]Recital N° 18 CR (EC) N° 1/2005.
[163]Fikuart (1997), p. 496 (496); EFSA (2004).
[164]COM (2011), p. 4.
[165]For example through Austria see Fikuart (1997), p. 496.
[166]COM (2011), p. 4, table 2.
[167]COM (2011), p. 4, table 2.
[168]COM (2011), p. 11.

Under Jewish law or Islamic law, meat is only "kosher"or "halal"("allowed" or "permitted"), if the animal has not been stunned before killing.[169] According to the beliefs of these religions it is not allowed to consume meat of animals which have been harmed before the religious slaughtering. In order to ensure this, the neck of the (unstunned) animal has to be cut by knife. Usually only one sharp cut is needed to cut off large blood vessels as well as trachea and esophagus. Afterwards the animal bleeds to death. The objective is a residue-free bleeding of the animals, since the consumption of blood is forbidden in Islam and Judaism. According to the Scientific Report of the EFSA animals (cattle here) lose their consciousness only gradually and thus not immediately. The loss of consciousness occurs between the time of cutting the carotid artery and the moment where the cerebral circulation becomes insufficient to maintaining normal brain functions. This state of consciousness, and thus possibly also pain and suffering, may persist until death.[170]

Based on EU law no obligations can be derived for stunning while slaughtering in case of religious rites (e.g. electric shocks). However, the Member States may enact stricter national animal protection laws (Article 26(2) letter c). Several Member States (Sweden, Norway, Iceland, Liechtenstein, the Netherlands, and Denmark) have made use of this opportunity and banned slaughtering animals without stunning. In contrast, Lithuania and Poland again allowed this procedure in 2014. These differences in legal requirements point out that many conflicts exist between interest in religious freedom and animal welfare, and also that there is no uniform understanding of animal welfare and protection within the EU.

9.3.3 Harmonization Degrees in Farm Animal Protection

9.3.3.1 Minimum Harmonization

The EU directives for keeping lay down a minimum harmonization. Several recitals of the directives as well as individual articles refer to this degree of harmonization. Thus, the Member States may, in compliance with the general rules of the Treaty, maintain or apply stricter provisions within their territories for the protection of animals kept for farming purposes than those laid down in these directives. They shall inform the Commission about any of such measures and shall communicate to the Commission the texts of the main provisions of national law which they adopt.[171]

---

[169] In some circles of Islam it is recognized that stunning by electric shock is reversible. Therefore, a stunning for a short time by electrocution is allowed.

[170] EFSA (2013), p. 11.

[171] Recitals and Art. 10 (2), (3) CD 98/58/EC; CD 1999/74/EC recital N° 10, Art. 13(2), (3); CD 2008/119/EC Art. 11 and CD 2008/120/EC Art. 12; CD 2007/43/EC Art. 1(2) subpara 2.

9.3.3.2 Full Harmonization

In the light of difficulties that have arisen during the implementation of the Directive 91/628/EC at national level the directive has been replaced by the regulation CR (EC) N° 1/2005 (Recital N° 6, 10). Replacing the former directive with a directly applicable regulation has intensified the level of harmonization. The CR (EC) N° 1/2005 is an example for almost complete full harmonization. Stricter national measures aimed at improving the animals' welfare may only persist or be enacted if the respective animals are entirely transported within the territory of a Member State or by ship departing from the territory of a Member State (Article 1 (3)). Particularly for the transport of cows and pigs the Member States are allowed to define time limitations of eight hours (Annex I Chapter. V. Clause 1.9). Consequently in this segment of animal transport an optional harmonization exists. However, the Member States are not allowed to restrict the time for cross-border transports to 8 h.

Also in the area of slaughtering the previous CD 93/119/EG was replaced by the regulation CR (EC) N° 1099/2009. As in the area of transport, the reasons for this were considerable differences in the implementation of the former directive by the Member States. There were significant doubts with respect to animal protection and competition between companies (Recital N° 3). In order to avoid undermining pre-existing animal protection standards in certain areas in the Member States, certain flexibility is left to the Member States. As a consequence the Member States are allowed to maintain or even extend existing national rules in certain areas (Article 26(1)). Thereby it has to be ensured that the Member States do not use the respective regulations in such a way that they affect the functioning of the internal market (Recital N° 57). Areas in which Member States can enact requirements with higher animal protection standards are very limited and refer only to the killing and related operations of animals outside of a slaughterhouse (Article 26 (2a)), the slaughtering and related operations of special farmed game including reindeer (Article 26(2b)), and the slaughtering and related operations of animals in accordance with Article 4(4) (Article 26(c)). When, on the basis of new scientific evidence, a Member State has the opinion that it is necessary to take measures at enacting more extensive protection of animals at the time of killing in relation to the methods of stunning, it shall notify the Commission of the envisaged measures. After a mandatory procedure the Commission can approve or refuse the measures and can propose changes for the requirements of methods of stunning (Article 26 (3)). Altogether, the CR (EC) N° 1099/2009 as well as CR (EC) N° 1/2005 is designed as a full harmonization with the possibility of a partially optional harmonization.

If the Member States have no room for manoeuvre in the directives or the regulations, they are not allowed to interpret the European guidelines of the harmonizing secondary law too unilaterally in favor of animal protection from Article 13 TFEU.[172] In these cases, they are bound to the decisions of consideration of the enacted secondary legislation. If they have the opinion that animal protection from Article 13 TFEU is not sufficiently respected within the consideration of interests of lifestock owners, carriers, and slaughterhouses, only the judicial procedures as mentioned in Sect. 9.2.3 remain to them.

9.4 Harmonization in Practice

In case of using animals as goods, general problems and also specific tensions may arise between ethical animal protection according to Article 13 TFEU and interests of lifestock keepers, carriers, and slaughterhouses. The following sections provide an overview of some of these problematic fields.

9.4.1 General Pros and Cons of Harmonization

The Harmonization of national legal systems can have pros and cons for animal protection itself as well as producers of animal products.

9.4.1.1 Increasing Animal Protection Standards

Adopting harmonization law acts might have positive effects on animal protection standards in the EU. This is the case when previous animal protection standards in the Member States were lower than those implemented by the new secondary law. Since all participants within the EU have to meet the same standards, harmonization law acts increase the animal protection standards in this case. This includes positive effects on lifestock keepers, transporters, and slaughterhouses, too, because the same competitive conditions apply to actors within the supply chain. Thus, individual actors are not able to obtain economic advantages due to lower standards.[173]

9.4.1.2 Reducing Animal Protection Standards

On the other hand this mechanism can have negative effects on animal protection standards. In case of enacting directives or regulations that require lower animal protection standards, compared to existing standards set by the national legislator,

[172]Schmidt. In: Schwarze (2012), Art. 13 TFEU, n. 5; cf. ECJ (2011).
[173]BR (2011).

the Member State is allowed to accept the Union law with a lower protection standard. This similarly happened in the case of stocking densities of broilers in Germany.[174] There have been no national legal requirements for stocking density in Germany. However, with the recognition of the "National Benchmarks for a Voluntary Agreement for Keeping Broilers and Turkeys"[175] the sector has expressed not to exceed a stocking density of 35 kg/m^2. Although the aforementioned expert opinion of the EFSA pleads for a stocking density of 25 kg/m^2, after adopting RC 2007/42/EC a stocking density of up to 42 kg/m^2 is allowed at Union level. RC 2007/42/EC is just a minimum harmonization and the Member States are allowed to enact stricter measures. Indeed Germany has advocated that the average of three consecutive fattening runs may not exceed a stocking density of 35 kg/m^2 (§ 19(4) Tierschutznutztierhaltungsverordnung—TierSchNutztV[176]). However, this legal requirement only applies if the average final fattening weight is lower than 1600 g, which is only the case for the so-called "short-fattening" (i.e. the fattening phase lasts 29–32 days). In the cases of a fattening phase of 36–38 days (final fattening weight: 2000–2200 g) and a fattening phase of 41–43 days (with a final fattening weight of approximately 2.5 kg) it is allowed to increase the stocking density up to 39 kg/m^2 (§ 19(3), (4) TierSchNutztV). These two kinds of fattening phases are preferred by farmers based in the German federal state Lower Saxony[177] where a lot of broilers are kept (with a total of 6.3 million in 2011[178]). In summary, it can be said that after adopting the directive and implementing it into national law a higher stocking density is allowed than before. Thus, adopting the directive has not necessarily improved animal protection in that point, because the enactment was a sign for Germany that broilers can be kept at higher stocking densities.

9.4.1.3 Replacing Directives by Regulations

A tendency in the case of the enactment of harmonizing secondary law acts is discernible to the effect that directives are replaced by regulations.[179] Thus, the question arises whether enacting directives with a minimum harmonization results in a better animal protection than enacting regulations, which naturally provide less room for manoeuvre for the Member States. Regulation as a stricter instrument is meant to solve the problems arising in conjunction with the implementation of directives. By introducing regulations the EU provides a more rigid framework for the Member States in order to realize a common level of rights in all Member

[174] Apel (2010), p.217; in detail Drossé (2010).

[175] BML (1999).

[176] BMEL (2006).

[177] LWK (2014).

[178] Wing (2015).

[179] Especially in the areas of transport and slaughter see 9.3.3; Härtel (2012) chapter 25, n. 13.

States.[180] However, this rigid framework has only a positive effect on animal protection if the animal protection standards are sufficiently high. Therefore it is generally unclear what type of harmonization is preferable. Thus, due to competitive reasons, all Member States need to consider whether they enact stricter standards than the EU standards as this can lead to higher costs and competitive disadvantages.[181]

9.4.2 *Orienting the Secondary Law Towards Minimum Harmonization*

Harmonizing law acts often define only a minimum standard. This has many reasons. The predominantly economic targets of the CAP can be an obstacle to the animal needs regarding welfare or species-appropriate keeping.[182] However, these aspects have to be brought in line while taking into account all various interests.[183] Due to their legal competence, the existing harmonizing law acts primarily pursue economic targets according to the CAP.[184] As explained above, the commandment of optimization is already fulfilled if the EU guarantees a minimum of animal protection.[185] Additionally, the majority of the 28 Member States must agree to the law acts on which they have to vote. Hence, it can be argued that a balance between the interests can only be achieved when rules in the area of animal protection have to be accepted and implemented by all Member States. This will only be the case in recognition of minimum standards.[186] Within the framework of the clauses for mandating higher levels of protection the Member States

[180]Holzer (2014), p. 476.

[181]Differenziated by Weikard (1992), p. 109 ff.

[182]Cf. Caspar (2001), p. 22.

[183]CD 2007/43/EC recital N° 10; In contrast to CD 2007/43/EC, the CD 1999/74/EC states that a balance between the interest "must" be fulfilled instead of "should" be fulfilled in creating a balance (recital N° 9); CD 2008/120/EC recital N° 12; the CD 2008/119/EC does not mention such compensation in recitals.

[184]Lachmayer and Bauer (2008), p. 863.

[185]Caspar (2001), p. 78.

[186]Recital CD 98/58/EC; CD 2007/43/EC recital N° 6; CD 1999/74/EC recital N° 6; CD 2008/120/EC recitals N° 6, 7; CD 2008/119/EC recitals N° 5, 6; Caspar (2001), p. 23.

have to decide for themselves whether they enact further and stricter animal protection guidelines. Positive examples include heightened requirements for space[187] and better designs of keeping systems with fully slatted floors[188] for fattening pigs in Germany. A negative example is that unlike other Member States keeping in crates is not generally forbidden in Germany.[189]

9.4.3 *Deficient Obligation for Optimization*

9.4.3.1 Nonexistent Specification

However, existing directives or regulations dealing with handling farm animals do not include all kinds of farm animals. For example, there is no secondary law for dairy cows, beef cattles, turkeys, water fowl, fur-bearing animals, rabbits, and fishes. For these species the general CD 98/58/EC applies, which comprises only a minimum standard of protection. Likewise, there are no guidelines for slaughtering fishes and crustaceans.[190] As aforementioned, Article 13 TFEU does no allow deriving an obligation for the EU for enacting optimizing law acts. It follows that the EU cannot be obliged to enact harmonization instruments for these species. In addition, the EU does not violate against obligations to protect or obligations to act, in so far as it does not enact detailed secondary law. However, any new animal protection law acts by the EU have to be based on latest scientific findings, opinions, and practical experiences.

9.4.3.2 Restrictions of Animal Protection for Cultural Reasons

The inactivity of the EU in areas of cultural reasons (cultural caveats) can be seen as another gap of harmonization. Particularly the production of foie gras is severely criticized. The animals experience verifiably pain and suffering when being force-fed by the insertion of a metal rod in the throats.[191] Thereby the liver grows from approximately 300 g to 1000–2000 g and becomes fatty. Especially in France these products are culinary delights and a specialty of the French cuisine. Foie gras is a part of the French culture. The production is not prohibited by EU law. Prohibitions only partly exist at national level. On the one hand, there are voices in the literature

[187] Art. 3(1a) CD 2008/120/EC compared with § 29 (2) TierSchNutztV.

[188] Art. 3(2b) CD 2008/120/EC compared with §§ 29(2) sentence 2; 22(3) N° 8 TierSchNutztV.

[189] Art. 3(4) subpara 1 sentence 1 CD 2008/120/EC and § 24(4) TierSchNutztV.

[190] Apel (2010), p. 216.

[191] EFSA (1998).

that there is no harmonization banin cultural areas.[192] But, on the other hand, it remains difficult to derive obligations to protect for enacting animal protection laws in the cultural areas: If there is no obligation to protect (referring to Article 13 TFEU) regarding to areas in which a prohibition on harmonisation does not exist, this applies even more for areas of cultural caveats. In the absence of the obligation to protect, the EU cannot be forced to enact secondary law acts in the area of foie gras. Even within the EFSA there is no agreement about appropriate measures to be taken.[193] Further, it is not possible to enact national trade barriers like bans on imports of foie gras because Article 36 TFEU only protects animals which are located on the own national territory.[194]

9.5 Outlook

The various directives and regulations regarding to animal protection reveal that the EU was not inactive in this area. A so-called animal welfare strategy for the protection of animals exists for the period 2012–2015.[195] In November 2015 the European Parliament voted for an "Animal Welfare Strategy 2016-2020 (joint motion for a resolution, 11/25/2015). The resolution calls on the European Commission to present a new animal welfare stretegy for that period after 2015. Stand July 2016 no Commission decision is known. However, at this stage, it is not obvious that the EU plans further optimization measures for farm animal protection in the near future: On the one hand the Commission would like to promote a production of high-quality food products and a variety of high-level food products in consideration of the requirements in terms of animal health and animal protection.[196] On the other hand, currently there are many other international issues on the agenda of the EU. Due to the fact that no obligation to enact more specific rules can be derived from the primary law, further harmonizing and detailed law acts cannot be expected. Additionally, there are no specific plans to improve animal protection standards in existing law acts. A summary of the existing law acts by a framework law for animal protection will not take place in the short-run. In the last contract amendments neither the farm animal protection has become a specific EU policy nor received its own written legal competence. Furthermore, farm animal protection will only play a role as an ancillary competence in connection with the realization of economic interests. Thus, the future development of farm animal protection at Union level cannot be predicted. Animal protection has increasingly gained

[192]Breier. In: Lenz and Borchardt (2010), Art. 13, n. 11; so probably also Calliess. In: Calliess and Ruffert (2011), Art. 13 TFEU, n. 9; another view: Kotzur. In Geiger et al. (2010) Art. 13 TFEU, n. 3.; Nettesheim. In: Grabitz et al. (2014), Art. 13 TFEU, n. 14.

[193]EFSA (1998), p. 69.

[194]Maisack (2012), p. 8.

[195]COM (2012).

[196]COM (2010); Holzer (2011), p. 99 f.

publicity in the consumer's mind. The EU decides within its wide discretion if farm animal protection will be realized in further harmonizing and detailed law acts.

References

Apel W (2010) Ziel- und richtungslos, Die Europäische Union und der Tierschutz in der Landwirtschaft. In: AgrarBündnis e.V (ed) Kritischer Agrarbericht 2010, Schwerpunkt "Boden", Konstanz, 215-221

Bergmann J (ed) (2015) Handlexikon der Europäischen Union, 5th edn. Nomos, Baden-Baden

Bieber R, Epiney A, Haag M (2015) Die Europäische Union: Europarecht und Politik, 11th edn. Nomos, Baden-Baden

Blanke F (1959) Unsere Verantwortlichkeit gegenüber der Schöpfung. In: Vogelsanger P, Der Auftrag der Kirche in der modernen Welt: Festgabe zum 70. Geburtstag von Emil Brunner, Zwingli Verlag Zurich and Stuttgart, pp 193–198

BMEL (2006) Tierschutz-Nutztierhaltungsverordnung in der Fassung der Bekanntmachung vom 22. August 2006 (BGBl. I S. 2043), die zuletzt durch Art. 1 der Verordnung vom 5. Februar 2014 (BGBl. I S. 94) geändert worden ist. Bundesministerium für Ernährung und Landwirtschaft. Available via http://www.gesetze-im-internet.de/tierschnutztv/. Accessed 16 Mar 2015

BML (1999). Bundeseinheitliche Eckwerte für eine freiwillige Vereinbarung zur Haltung von Jungmasthühnern (Broiler, Masthähnchen) und Mastputen. Bundesministerium für Landwirtschaft. Az. 321-3545/2; 23. September 1999, 4157/3659. Available via PROVIEH http://www.provieh.de/downloads_provieh/eckwerte.pdf. Accessed 26 Mar 2015

BR (2011) Agrarbericht 1990. Agrar- und ernährungspolitischer Bericht der Bundesregierung. 8.2.1990. BT Drs. 11/6387, Chapter 6.3, No. 286 "Tierschutz". Available via http://dipbt.bundestag.de/doc/btd/11/063/1106387.pdf. Accessed 27 Mar 2015

BT (2006) Tierschutzgesetz in der Fassung der Bekanntmachung vom 18. Mai 2006 (BGBl. I S. 1206, 1313), das zuletzt durch Artikel 3 des Gesetzes vom 28. Juli 2014 (BGBl. I S. 1308) geändert worden ist. Deutscher Bundestag. Available via http://www.gesetze-im-internet.de/tierschg/BJNR012770972.html. Accessed 27 Mar 2015

Callies C (2012) Tierschutz zwischen Europa- und Verfassungsrecht – Überlegungen am Beispiel der Tierversuchsrichtlinie. NuR 2012:819–829

Calliess C, Ruffert H-J (eds) (2011) EUV/AEUV, Das Verfassungsrecht der Europäischen Union mit Europäischer Grundrechtecharta Kommentar, 4th edn. Beck, Munich

Caspar J (1999) Tierschutz im Recht der mordernen Industriegesellschaft, Eine rechtliche Neukonstruktion auf philosophischer und historischer Grundlage. In: Hoffmann-Riem W, Koch H-J, Ramsauer U, Forum Umweltrecht, Schriftenreihe der Forschungsstelle Umweltrecht der Universität Hamburg Bd. 31, Zugl. Hamburg, Univ., Habil-Schr., 1st edn. Nomos, Baden-Baden, 1998

Caspar J (2001) Zur Stellung des Tieres im Gemeinschaftsrecht. In: Caspar J, Harrer F (eds) Das Recht der Tiere und der Landwirtschaft, 1st edn. Nomos, Baden-Baden

COM (2010) European Commission. Communication from the Commission to the European Parliament, the Council, the European Economic and Social Committee and the Committee of the Regions. The CAP towards 2020: Meeting the food, natural resources and territorial challenges of the future. 18 November 2010 COM (2010) 672 final. Available via http://ec.europa.eu/agriculture/cap-post-2013/communication/com2010-672_en.pdf. Accessed 27 Mar 2015

COM (2011) European Commission. Report from the Commission to the European Parliament and the Council on the impact of Council Regulation (EC) No 1/2005 on the protection of animals during transport. 10 November 2011. COM (2011) 700 final. Available via http://ec.europa.eu/food/animal/welfare/transport/docs/10112011_report_en.pdf. Accessed 27 Mar 2015

COM (2012) European Commission. Communication from the Commission to the European Parliament, the Council and the European Economic and Social Committee on the European Union Strategy for the Protection and Welfare of Animals 2012–2015. 15 February 2012 COM (2012) 6 final/2. Available via http://ec.europa.eu/food/animal/welfare/actionplan/docs/aw_strategy_19012012_en.pdf. Accessed 27 Mar 2015

Dauses MA (ed) (2014) Handbuch des EU-Wirtschaftsrechts, vol. 1, 36. Erg.-Lfg., Stand Oktober, Beck, Munich

de Jong I, Berg C, Butterworth A, Estevéz I (2012) 'Scientific report updating the EFSA opinions on the welfare of broilers and broiler breeders. Supporting Publications 2012:EN-295. [116 pp.]. Available via http://www.efsa.europa.eu/en/supporting/doc/295e.pdf. Accessed 19 Mar 2015

Dombert M, Witt K (eds) (2011) Münchener Anwaltshandbuch Agrarrecht. Beck, Munich

Dreier H (ed) (2006) Grundgesetz Kommentar, vol II, Article 20-82, 2nd edn. Mohr Siebeck, Tübingen

Drossé I (2010) Wenn das Huhn auf den Hund kommt, Tierschutzprobleme bei der intensiven Hühnermast – und das Versagen der Politik. In: AgrarBündnis e.V (ed) Kritischer Agrarbericht 2010, Schwerpunkt "Boden", Konstanz, pp 233–238

EC (1998) Council Directive 98/58/EC of 20 July 1998 concerning the protection of animals kept for farming purposes (CD 98/58/EC). (OJ L 221, 08/08/1998 P. 0023–0027). Available via http://eur-lex.europa.eu/legal-content/EN/TXT/HTML/?uri=CELEX:31998L0058&qid=1427735600126&from=DE. Accessed 30 Mar 2015

EC (1999) Council Directive 1999/74/EC of 19 July 1999 laying down minimum standards for the protection of laying hens (CD 1999/74/EC). (OJ L 203, 03/08/1999 P. 0053–0057). Available via http://eur-lex.europa.eu/legal-content/EN/TXT/HTML/?uri=CELEX:31999L0074&qid=1427735732083&from=DE. Accessed 30 Mar 2015

EC (2005) Council Regulation (EC) No 1/2005 of 22 December 2004 on the protection of animals during transport and related operations and amending Directives 64/432/EEC and 93/119/EC and Regulation (EC) No 1255/97 (CR (EC) No 1/2005) (OJ L 3/1). Available via http://eur-lex.europa.eu/legal-content/EN/TXT/HTML/?uri=CELEX:32005R0001&qid=1427735845601&from=DE. Accessed 30 Mar 2015

EC (2007) Council Directive 2007/43/EC of 28 June 2007 laying down minimum rules for the protection of chickens kept for meat production (CD 2007/43/EC) (OJ L 182/19). Available via http://eur-lex.europa.eu/legal-content/EN/TXT/HTML/?uri=CELEX:32007L0043&qid=1427735948652&from=DE. Accessed 30 Mar 2015

EC (2008a) Council Directive 2008/120/EC of 18 December 2008 laying down minimum standards for the protection of pigs (CD 2008/120/EC) (OJ L 47/5). Available via http://eur-lex.europa.eu/legal-content/EN/TXT/HTML/?uri=CELEX:32008L0120&qid=1427736005531&from=DE. Accessed 30 Mar 2015

EC (2008b) European Council. Council Directive 2008/119/EC of 18 December 2008 laying down minimum standards for the protection of calves. (OJ L 10/7). Available via http://eur-lex.europa.eu/legal-content/EN/TXT/HTML/?uri=CELEX:32008L0119&qid=1427734787194&from=DE. Accessed 30 Mar 2015

EC (2009) Council Regulation (EC) No 1099/2009 of 24 September 2009 on the protection of animals at the time of killing (CR (EC) No 1099/2009) (OJ 303/1). Available via http://eur-lex.europa.eu/legal-content/EN/TXT/HTML/?uri=CELEX:32009R1099&qid=1427736082241&from=DE. Accessed 30 Mar 2015

ECJ (1964) Judgment of the Court of 15 July 1964 - Flaminio Costa v E.N.E.L. - Reference for a preliminary ruling: Giudice conciliatore di Milano - Italy - Case 6/64 - Available via http://eur-lex.europa.eu/legal-content/EN/TXT/HTML/?uri=CELEX:61964CJ0006&from=EN. Accessed 30 Mar 2015

ECJ (1988a) Judgment of the Court of 23 February 1988 - United Kingdom of Great Britain and Northern Ireland v Council of the European Communities - Minimum standards for the protection of laying hens kept in batter cages. - Case 131/86. European Court reports 1988 Page 00905. Available via http://eur-lex.europa.eu/legal-content/EN/TXT/HTML/?uri=CELEX:61986CJ0131&from=DE. Accessed 30 Mar 2015

ECJ (1988b) Judgment of the Court of 23 February 1988 - United Kingdom of Great Britain and Northern Ireland v Council of the European Communities - Case 68/86. European Court reports 1988 Page 00855. Available via http://eur-lex.europa.eu/legal-content/EN/TXT/HTML/?uri=CELEX:61986CJ0068&from=DE. Accessed 30 Mar 2015

ECJ (2001) Judgment of the Court of 12 July 2001 - Jippes - Minister van Landbouw, Natuurbeheer en Visserij - Case 189/01. Available via http://curia.europa.eu/juris/document/document.jsf?text=&docid=46530&pageIndex=0&doclang=EN&mode=req&dir=&occ=first&part=1. Accessed 30 Mar 2015

ECJ (2011) Judgment of the Court of 21 Decembre 2011 - Rs. C-, Danske Svineproducenter / Justitsministeriet. Case 316/10. Available via http://curia.europa.eu/juris/celex.jsf?celex=62010CJ0316&lang1=de&type=TXT&ancre=. Accessed 30 Mar 2015

EFSA (1998) European Food Safety Authority. Report of the Scientific Committee on Animal Health and Animal Welfare 'Welfare Aspects of the Production of Foie Gras in Ducks and Geese'. Available via http://ec.europa.eu/food/animal/welfare/international/out17_en.pdf. Accessed 19 Mar 2015

EFSA (2000) European Food Safety Authority. Report of the Scientific Committee on Animal Health and Animal Welfare 'The Welfare of Chickens Kept for Meat Production (Broilers)´ SANCO.B.3/AH/R15/2000. Available via http://ec.europa.eu/food/fs/sc/scah/out39_en.pdf. Accessed 19 Mar 2015

EFSA (2004) European Food Safety Authority. Scientific Report of the Scientific Panel on Animal Health and Welfare on a request from the Commission related to the welfare of animals during transport. Available via http://www.efsa.europa.eu/de/scdocs/doc/44ax1.pdf. Accessed 27 Mar 2015

EFSA (2007a) European Food Safety Authority. Scientific Opinion of the Panel on Animal Health and Welfare 'Animal health and welfare aspects of different housing and husbandry systems for adult breeding boars, pregnant, farrowing sows and unweaned piglets'. The EFSA Journal 572:1-13. Available via http://www.efsa.europa.eu/de/scdocs/doc/572.pdf. Accessed 19 Mar 2015

EFSA (2007b) European Food Safety Authority, 'Scientific Report on animal health and welfare aspects of different housing and husbandry systems for adult breeding boars, pregnant, farrowing sows and unweaned piglets'. Annex to The EFSA Journal 572:1-13. Available via http://www.efsa.europa.eu/de/scdocs/doc/572.pdf. Accessed 19 March 2015

EFSA (2013) European Food Safety Authority. 'Scientific Opinion on monitoring procedures at slaughterhouses for bovines. EFSA J 11(12):3460, 65 pp. Available via http://www.efsa.europa.eu/de/efsajournal/doc/3460.pdf. Accessed 19 Mar 2015

EU (1992) European Union. Treaty on European Union signed at Maastricht. Declaration No. 24 on the protection of animals, (OJ. C 191/01 29.07.1992, p. 103). Available via http://eur-lex.europa.eu/legal-content/EN/TXT/PDF/?uri=OJ:C:1992:191:FULL&from=EN. Accessed 27 Mar 2015

EU (1997) European Union. Treaty of Amsterdam amending the Treaty on European Union, the Treaties establishing the European Communities and certain related acts. Protocol No. 33 on protection and welfare of animals. (OJ. No. C 340, 2. 10. 1997, p. 110). Available via http://www.europarl.europa.eu/topics/treaty/pdf/amst-en.pdf. Accessed 27 Mar 2015

EU (2007). European Union. Consolidated versions of the Treaty on European Union and the Treaty on the Functioning of the European Union - Consolidated version of the Treaty on the Functioning of the European Union - Protocols - Annexes - Declarations annexed to the Final Act of the Intergovernmental Conference which adopted the Treaty of Lisbon, signed on 13 December 2007 - Tables of equivalences. (OJ. C 326, 26/10/2012 P. 0001–0390). Available via http://eur-lex.europa.eu/legal-content/EN/TXT/HTML/?uri=CELEX:12012E/TXT&from=EN. Accessed 27 Mar 2015

Fikuart K (1997) Tiertransporte. In: Sambraus HH, Steiger A (eds) Das Buch vom Tierschutz. Ferdinand Enke, Stuttgart, pp 496–509

Frenz W (2011) Umwelt- und Tierschutzklausel im AEUV. NuR 2011:103–107

Geiger R, Khan D-E, Kotzur M (2010) EUV/AEUV Vertrag über die Europäische Union und Vertrag über die Arbeitsweise der Europäischen Union Kommentar, 5th edn. Beck, Munich

Grabitz E, Hilf M, Nettesheim M (ed) (2014) Das Recht der Europäischen Union, Ergänzungslieferung, Stand, vol 54. Beck, Munich

Härtel I (ed) (2012) Handbuch des Fachanwalts Agrarrecht. Luchterhand, Cologne

Hirt A, Maisack C, Moritz J (2007) Tierschutzgesetz: Kommentar, 2nd edn. Vahlen, Munich

Holzer G (2011) Agrarrecht, 2nd edn. Neuer wissenschaftlicher Verlag, Vienna

Holzer G (2014) Agrarrecht, 3rd edn. Neuer wissenschaftlicher Verlag, Vienna

Kamphues J (2014) Zur Bedeutung von Fütterung und Haltung für die Fußballengesundheit beim Mastgeflügel. Available via wing. http://www.wing-vechta.de/themen/fussballengesundheit/zur_bedeutung_von_f_tterung_und_haltung_f_r_die_fu_ballengesundheit_beim_mastgefl_gel.html. Accessed 19 Mar 2015

Keppler C, Vogt-Kaute W, Knierim U (2009) Tiergesundheit von langsam wachsenden Masthühnern in Öko-Betrieben - Eine Feldprüfung. In: Rahmann G, Schumacher U (eds) Praxis trifft Forschung – Neues aus der ökologischen Tierhaltung, pp 31–46. Available via http://literatur.ti.bund.de/digbib_extern/dk042683.pdf. Accessed 19 Mar 2015

Kluge H-G (ed) (2002) Tierschutzgesetz Kommentar, 1st edn. Kohlhammer, Stuttgart

Lachmayer K, Bauer L (2008) Praxiswörterbuch Europarecht. Springer, Vienna

Lenz CO, Borchardt K-D (eds) (2010) EU-Verträge Kommentar nach dem Vertrag von Lissabon, 5th edn. Bundesanzeiger Cologne

LFL (2015) Bayrische Landesanstalt für Landwirtschaft. Available via http://www.lfl.bayern.de/schwerpunkte/tierwohl/068541/index.php. Accessed 7 Mar 2015

Lorz A, Metzger E (eds) (2008) Tierschutzgesetz mit Allgemeiner Verwaltungsvorschrift, Rechtsverordnungen und Europäischen Übereinkommen sowie Erläuterungen des Art. 20a GG, 6th edn. Beck, Munich

LWK (2014) Landwirtschaftskammer Niedersachsen. Gute Ergebnisse in der Hähnchenmast. Meldung vom 18.12.2014. Landwirtschaftskammer Niedersachsen, Available via http://www.lwk-niedersachsen.de/index.cfm/portal/1/nav/229/article/26509.html. Webcode 10127795. Accessed 9 Mar 2015

Maisack C (2012) Was tut sich in Sachen "Nutztierschutz"auf EU-Ebene?. In: Nutztierschutztagung Raumberg-Gumpenstein. Lehr- und Forschungszentrum für Landwirtschaft Raumberg-Gumpenstein. pp 5–12. Available via http://www.google.de/url?sa=t&rct=j&q=&esrc=s&source=web&cd=1&ved=0CCIQFjAA&url=http%3A%2F%2Fwww.raumberg-gumpenstein.at%2Fcm4%2Fde%2Fforschung%2Fpublikationen%2Fdownloadsveranstaltungen%2Ffinish%2F556-nutztierschutztagung-2012%2F4983-was-tut-sich-in-sachen-nutztierschutz-auf-eu-ebene.html&ei=yQYLVbKHIMXYU7uNgZgP&usg=AFQjCNGZG_KZ0XaPs8Lsy8qmCZ2MuLDm0w&bvm=bv.88528373,d.d24&cad=rja. Accessed 19 Mar 2015

Martínez J (2014) Die Gemeinsame Agrar- und Fischereipolitik der Union. In: Niedobitek M (ed) Europarecht – Politiken der Union. DE GRUYTER, Berlin/Boston, pp 709–774

ML (2011) Niedersächsisches Ministerium für Ernährung, Landwirtschaft und Verbraucherschutz. Tierschutzplan Niedersachen. Available via http://www.ml.niedersachsen.de/download/72939/_Tierschutzplan_Niedersachsen_.pdf. Accessed 27 Mar 2015

MSD (2015) MSD Tiergesundheit in Deutschland. Intervet Deutschland GmbH. Available via. http://www.msd-tiergesundheit.de/News/Fokusthemen/Saugferkelkastration/Status_Quo_EU.aspx. Accessed 27 Mar 2015

Nentwich M (1994) Die Bedeutung des EG-Rechts für den Tierschutz. In: Harrer F, Graf G (eds) Tierschutz und Recht: Tierschutz im Straf- und Zivilrecht – verfassungsrechtliche Grundlagen

– Tierschutz in der EG – rechtliche Grenzen des tierschützerischen Aktionismus – rechtshistorische und philosophische Wurzeln der geltenden Regelungen. Orac, Vienna
Norer R (ed) (2012) Handbuch des Agrarrechts, 2nd edn. Springer Österreich
Oppermann T, Classen C D, Nettesheim M (2014) Europarecht Ein Studienbuch, 6th edn. Beck, Munich
Ruffert M, (ed) (2013) Europäisches sektorales Wirtschaftsrecht. In: Hatje A, Müller-Graff P-C (eds) Enzyklopädie Europarecht, vol 5, 1st edn. Nomos, Baden-Baden
Sambraus HH (1997) Rind. In: Sambraus HH, Steiger A (eds) Das Buch vom Tierschutz. Ferdinand Enke, Stuttgart, pp 107–126
Schulze R, Zuleeg M, Kadelbach S (eds) (2015) Europarecht Handbuch für die deutsche Rechtspraxis, 3rd edn. Nomos, Baden-Baden
Schwarze J (ed) (2012) EU-Kommentar, 3rd edn. Nomos, Baden-Baden
Spindler B, Hartung J (2010) Abschlussbericht - Untersuchungen zur Besatzdichte bei Masthühnern entsprechend der RL 2007/43/EG. Stiftung Tierärztliche Hochschule Hannover. Available via http://www.tierschutz-landwirtschaft.de/Gutachten_Hartung_Spindler_2010.pdf. Accessed 19 Mar 2015
Telle M (2011) Verhaltensbeobachtungen bei der Kleingruppenhaltung von Legehennen (LSL). Dissertation, Ludwig-Maximilian-Universität München. Available via http://edoc.ub.uni-muenchen.de/13429/1/Telle_Monika.pdf. Accessed 19 Mar 2015
TVT (2012) Zur Enthornung von Rindern Merkblatt Nr. 86. Tierärztliche Vereinigung für Tierschutz e.V. Available via http://www.google.de/url?sa=t&rct=j&q=&esrc=s&source=web&cd=1&ved=0CCIQFjAA&url=http%3A%2F%2Fwww.tierschutz-tvt.de%2Findex.php%3Fid%3Dmerkblaetter%26eID%3Dtx_rtgfiles_download%26tx_rtgfiles_pi1%255Buid%255D%3D8&ei=TzEVVZahIon3OrvogZAI&usg=AFQjCNEDuS_pZShZ3TER3JSBaI-Xai8JHw&bvm=bv.89381419,d.ZWU&cad=rja. Accessed 27 Mar 2015
Vedder C, Heintschel von Heinegg W (2007) (eds) Europäischer Verfassungsvertrag Handkommentar, 1st edn. Nomos Baden-Baden
Vedder C, Heintschel von Heinegg W (eds) (2012) Europäisches Unionsrecht EUV AEUV Grundrechte-Charta Handkommentar mit den vollständigen Texten der Protokolle und Erklärungen des EAGV, 1st edn. Nomos, Baden-Baden
Wechsler B (1997) Schwein. In: Sambraus HH, Steiger A (eds) Das Buch vom Tierschutz. Ferdinand Enke, Stuttgart, pp 173–185
Weikard H-Peter (1992) Der Beitrag der Ökonomik zur Begründung von Normen des Tier-und Artenschutzes - Eine Untersuchung zu praktischen und methodologischen Problemen der Wirtschaftsethik. In: Volkwirtschaftliche Schriften, issue 419. Duncker & Humblot, Berlin
Windhorst H-W (2013) Kann in der Legehennen- und Mastputenhaltung schon bald auf die Schnabelbehandlung verzichtet werden?. Available via wing. http://www.wing-vechta.de/themen/schnabelbehandlung/schnabelbehandlung_vor_dem_aus_1.html. Accessed 19 Mar 2015
Wing (2015) Daten und Fakten zur Geflügelwirtschaft – Jungmasthühnerhaltung. Wissenschafts- und Informationszentrum Nachhaltige Geflügelwirtschaft Universität Vechta. Available via http://www.wing-vechta.de/pdf_files/dokumente/materialien_broiler.pdf. Accessed 19 Mar 2015
Wollenteit U, Lemke I (2013) Die Vereinbarkeit der Haltung von abferkelnden Sauen in Kastenständen mit dem Tierschutzrecht und die Zulässigkeit eines Verbots dieser Haltungsform. NuR 2013:177–183

Chapter 10
Meat Production and Antibiotics Use

Meghan Davis and Lainie Rutkow

Abstract Debate over how regulation can address the growing public health crisis of antimicrobial resistance has addressed both the regulatory framework for intervention and the political choice to intervene, balancing control of the public health risk from agricultural use of antimicrobials and economic benefit to agribusiness from such use. This chapter reviews current U.S. laws and regulations pertaining to non-therapeutic use of antimicrobials in livestock and to surveillance of antimicrobial-resistant pathogens of food animal origin. Regulatory efforts in the United States and Europe are compared, with an emphasis on the scientific evidence for public health success or failure of these policy interventions. The chapter also provides the scientific context that informs regulatory efforts in the U.S. and global efforts to address the problem of antimicrobial resistance. Recommendations for combined regulatory, surveillance, and research strategies are offered, with a focus on science-based regulatory approaches and mechanisms for evaluation of the public health benefits of regulation.

10.1 Introduction

People must eat. As the population has grown, global food production has increased to meet this demand, driven first by the Green Revolution of the 1930s and refined during the intervening decades by continued industrialization and intensification of agricultural practices. What is most remarkable about the increase in world food production is that it has met a growing demand for diets rich in calories, including from animal products.[1] One of the foundations of this achievement is the growing dominance of concentrated animal feeding operations (CAFOs), typified by raising a large number of animals, crowded together in barns or small areas of land.[2]

A version of this chapter was originally published in Volume 25 of the Tulane Environmental Law Journal 2011–2012.

[1]Kastner et al. (2012), pp. 6868–6872.

[2]Silbergeld et al. (2008a).

M. Davis (✉) • L. Rutkow
Johns Hopkins University, Baltimore, MD, USA
e-mail: mdavis65@jhu.edu

G. Steier, K.K. Patel (eds.), *International Farm Animal, Wildlife and Food Safety Law*, DOI 10.1007/978-3-319-18002-1_10

Simultaneous with the development of this approach was the discovery of antibiotics, including the potential uses of these drugs to enhance the growth or increase the output[3] of food-producing animals.[4]

Antimicrobial drugs, including antibiotics,[5] are important to human and veterinary medicine for the treatment of infectious diseases.[6] Indeed, global trends have demonstrated increasing consumption of antimicrobials in humans and animals.[7] However, bacteria may develop resistance to one or more classes of antibiotics, allowing them to survive and reproduce even in the presence of these drugs.[8] When antibiotic-resistant pathogens cause infection, the human and economic costs are high.[9] For example, in the United States, human health care costs associated with treating diseases resistant to antibiotics are estimated at over $4 billion annually[10] and may reach $7 billion.[11] Patients infected with resistant bacteria generally have higher mortality, higher morbidity, longer hospital stays, and higher rates of sequelae than those with susceptible infections.[12]

---

[3]Rusoff (1951), pp. 652–655; Stokstad and Jukes (1950), pp. 523–528.

[4]Food-producing animals, also known as livestock or food animals, include all animals raised for meat, milk, or eggs for human consumption. Pigs, poultry ("layer" chickens which produce eggs, "broiler" chickens raised for meat, and turkeys), dairy cows, beef cattle, and farmed fish (e.g., catfish) are examples of the most common food-producing animals raised in the United States. See U.S. Dep't Agric., Census of Agriculture (2007), available at http://www.agcensus.usda.gov/Publications/2007/Full_Report/usv1.pdf.

[5]In this chapter, the terms "antimicrobial" and "antibiotic" may occasionally appear to be used interchangeably, as antibiotics are, by some definitions, considered to be antimicrobials. Not all antimicrobials are antibiotics, however. Some regulations may apply to all antimicrobials broadly (used to treat infections with viruses, bacteria, parasites, and fungal organisms), and others to drugs used to treat bacterial infections specifically. Technically, the term antibiotic refers only to chemicals naturally produced by microorganisms that kill or impair other microorganisms; otherwise, synthetic "antibiotics" are considered antimicrobials. For a lay definition of these terms, see Ctrs. for Disease Control & Prevention, Antibiotic/Antimicrobial Resistance, www.cdc.gov/drugresistance/index.html (last visited Sept. 13, 2011). See also Luca Guardabassi & Patrice Courvalin, Modes of Antimicrobial Action and Mechanisms of Bacterial Resistance, in Antimicrobial Resistance in Bacteria of Animal Origin 1, 1 (Frank M. Aarestrup ed., 2006) (concerning use and misuse of the terms antimicrobial and antibiotic).

[6]Peter Lees et al., Drug Selection and Optimization of Dosage Schedules to Minimize Antimicrobial Resistance, in Antimicrobial Resistance in Bacteria of Animal Origin, (Frank M. Aarestrup ed., 2006), at 49.

[7]Van Boeckel et al. (2014), pp. 742–750.

[8]Id. at 49.

[9]Am. Soc'y Microbiology, Report of the ASM Task Force on Antibiotic Resistance (1995), available at http://www.asm.org/images/docfilename/0000005962/antibiot[1].pdf; Oguz Resat Sipahi (2008), pp. 523–526.

[10]Am. Soc'y Microbiology, (1995), at 3.

[11]Coast and Smith (2003), pp. 241–242.

[12]Sipahi (2008), p. 526.

Bacteria can acquire genes for resistance from other bacteria, and this process of genetic exchange can occur in microorganisms carried by humans and animals, or present in the environment.[13] Because of this complex ecology, use of antibiotics in one setting, such as agriculture, can drive emergence of resistant bacteria capable of causing disease in humans.[14] Even if only a modest fraction of antimicrobial-resistant infections in humans are caused directly by use of antimicrobials in food-producing animals, the population burden and economic cost from such use potentially is high.

In food-producing animals, antimicrobials either may be administered to treat disease or used at low levels in feed to promote animal growth, which the industry presents in terms of improved feed efficiency and control of pathogens.[15] However, this latter use of antimicrobial drugs (for growth promotion)[16] typically involves feeding them to animals at levels that result in doses that are not high enough to kill or inhibit all target bacteria[17] (*i.e.*, at drug concentrations below those required to treat clinical infection).[18] This drives emergence of resistant organisms in those animals and in the environment.[19] Use of antimicrobial drugs in agriculture exceeds that in human clinical settings nearly eightfold.[20] In 2012, the U.S. Food and Drug Administration (FDA) reported that 13.2 million kilograms (over 28 million pounds) of antimicrobials were sold or distributed domestically for use in food-

[13]Antimicrobial Resistance in Bacteria of Animal Origin, (Frank M. Aarestrup ed., 2006), at 26 (adapted from Alan H. Linton, Antibiotic Resistance: The Present Situation Reviewed, 100 Veterinary Rec. 354 (1977) and modified by R. Irwin from a model sometimes referred to as the "confusogram").

[14]Silbergeld et al. (2008a), p. 151; Gilchrist et al. (2007), pp. 313–314; Angulo et al. (2004a), pp. 485, 487–490; McEwen and Fedorka-Cray (2002), p. S99.

[15]Prescott (2006), p. 22.

[16]The practice of feeding antimicrobials at levels below that which treat clinical infection, alternately termed "non-therapeutic" or "sub-therapeutic" use, originated in the late 1940s and early 1950s. During that era, this use was shown to hasten animal weight gain and, at times, reduce mortality in herds or flocks. In the United States, "subtherapeutic levels" sometimes are defined as concentrations of antimicrobials that are less than 200 g per ton of feed. The degree to which this use remains an economic incentive for an individual farmer or industrial producer depends on many factors, including the underlying health and environmental living conditions of the animals. See id. at 19–23.

[17]Antimicrobial drugs differ in their ability to kill (bacteriocidal drugs) or inhibit (bacteriostatic drugs) different kinds of bacteria. For example, fluoroquinolone drugs (e.g., ciprofloxacin and enrofloxacin) are broad-spectrum and are active against gram-negative bacteria (e.g., E. coli) and gram-positive cocci (e.g., Staphylococcus aureus), but have only weak activity against anaerobic bacteria (e.g., Clostridium).

[18]Prescott (2006), pp. 22–23.

[19]Silbergeld et al. (2008a), pp. 151–169; Gilchrist et al. (2007), pp. 313–314; Angulo et al. (2004b), p. 78; Levy et al. (1976), pp. 40–42.

[20]Margaret Mellon et al., Hogging It!: Estimates of Antimicrobial Abuse in Livestock xiii (2001).

producing animals.[21] Agricultural uses typically represent 75–80 % of the antimicrobial drug sales in the United States, with over 90 % of these antimicrobials administered in animal feed or water.[22] In addition, in the early 2010s, 97–98 % of antimicrobial drugs sold or distributed for use in food-producing animals were approved for sale over-the-counter (OTC).[23]

The regulation of antimicrobial use in agriculture has received attention at the national and global levels in recent years. In 1997, the World Health Organization held the first of many conferences on antimicrobial resistance,[24] and designated certain antimicrobials "critically important"[25] to human health during a later conference in Canbarra.[26] In 1998, the European Union (EU) passed a commission ruling banning the use of a number of antimicrobials in animal feed.[27] A study of the impact of a ban on certain antimicrobial drugs in Denmark showed little economic impact to that country's broiler chicken industry, although the swine industry experienced a one percent increase in overall costs of production.[28] Offsetting this minor cost was a tremendous decrease in the percentage of bacteria from swine and broiler chickens that were resistant to the banned antimicrobials.[29] This suggests that regulation may offer an effective public health strategy to combat antimicrobial resistance of agricultural origin.

---

[21]This is the fourth report on such uses produced in response to requirements of the Center for Veterinary Medicine of the FDA under § 512 of the Animal Drug User Fee Amendments of 2008 (ADUFA) 21 U.S.C. § 360b(l) (2009). This estimate includes all uses in food-producing animals for all purposes (growth promotion, prophylaxis, or therapy), and regardless of route of administration (via injection, oral administration, or in medicated feed). See Ctr for Veterinary Med., U.S. Food and Drug Administration, 2012. Summary Report on Antimicrobials Sold or Distributed for Use in Food-Producing Animals (2014), available at http://www.fda.gov/downloads/ForIndustry/UserFees/AnimalDrugUserFeeActADUFA/UCM416983.pdf.

[22]Letter from Karen Meister, Supervisory Congressional Affairs Specialist, Food and Drug Administration, to Rep. Louise A. Slaughter, U.S. House of Representatives (Apr. 19, 2011), available at http://www.louise.house.gov/images/stories/FDA_Response_to_Rep._Slaughter.pdf.

[23]OTC drugs are sold or dispensed without requirement for human or veterinary prescription; Ctr for Veterinary Med., U.S. Food and Drug Administration, 2012, at 5.

[24]World Health Org., The Medical Impact of Antimicrobial Use in Food Animals (1997), available at http://whqlibdoc.who.int/hq/1997/WHO_EMC_ZOO_97.4.pdf.

[25]Two criteria were used by WHO to determine the importance of antibiotics that may be used in food-producing animal production for human health. The first criterion was the importance of the drug in human health, i.e., whether or not the drug was the only or one of few available to treat a given disease. The second criterion was the use of a given antibiotic to treat specifically zoonotic disease, i.e., a disease that can be transmitted from an animal to a human. These were given higher weight. See World Health Org., Critically Important Antibacterial Agents for Human Medicine for Risk Management Strategies of Non-Human Use 4–5 (2005), available at www.who.int/foodborne_disease/resistance/amr_feb2005.pdf.

[26]World Health Org., Critically Important Antibacterial Agents, (2005).

[27]Aarestrup et al. (2001), p. 2054.

[28]Emborg and Wegener (2005), pp. 168–169.

[29]Id. at 163–67.

Scientists,[30] professional organizations,[31] public health advocates[32] and the U.S. Government Accountability Office[33] have argued that the U.S. government's current oversight of antimicrobial use in agriculture—and indeed, efforts by many countries on a global scale—are insufficient to address the problem of rising antimicrobial resistance. Within the last 40 years, the FDA has developed primarily non-binding guidance about the use of non-therapeutic antimicrobials in livestock in the United States.[34] Congressional efforts to give legal effect to the principles of appropriate antimicrobial use described in this guidance have failed, and FDA's guidance continues to lack enforceability.[35] Agribusiness has opposed legislation requiring the reduction or elimination of non-therapeutic use of antimicrobials in livestock, as illustrated in an example later in the chapter.[36] This slow progress of the

[30]Silbergeld et al. (2008a); Gilchrist et al. (2007); McEwen and Fedorka-Cray (2002).

[31]Am. Soc'y Microbiology, (1995), pp. 7–8; John G. Bartlett et al., Statement of the Infectious Diseases Society of America before the Food and Drug Administration Part 15 Hearing Panel on Antimicrobial Resistance (2008), available at http://www.idsociety.org/uploadedFiles/IDSA/Policy_and_Advocacy/Current_Topics_and_Issues/Advancing_Product_Research_and_Development/Antimicrobials/Statements/ee434daf62ba4fedac689288741635704.pdf#search=%22Statement of the Infectious Diseases Society of America before the Food Drug Administration Part 15 Hearing Panel on Antimicrobial Resistance%22.

[32]Pew Comm'n Indus. Food Animal Prod., Putting Meat on the Table: Industrial Farm Animal Production in America 15–16 (2008), available at http://www.pewtrusts.org/uploadedFiles/wwwpewtrustsorg/Reports/Industrial_Agriculture/PCIFAP_FINAL.pdf; Mellon et al., 17.

[33]U.S. Gen. Accounting Office, Antibiotic Resistance: Federal Agencies Need to Better Focus Efforts to Address Risk to Humans from Antibiotic Use in Animals (2004), available at www.gao.gov/cgi-bin/getrpt?GAO-04-490.

[34]U.S. Food & Drug Admin., Guidance for Industry #152 – Evaluating the Safety of Antimicrobial New Animal Drugs with Regard to Their Microbiological Effects on Bacteria of Human Health Concern (2003), available at www.fda.gov/cvm/guidance/fguide152.pdf; Guidance for Industry #209 – The Judicious Use of Medically-Important Antimicrobial Drugs in Food-Producing Animals (2012), available at http://www.fda.gov/downloads/AnimalVeterinary/GuidanceComplianceEnforcement/GuidanceforIndustry/UCM216936.pdf; U.S. Food & Drug Admin., Guidance for Industry #213 - New Animal Drugs and New Animal Drug Combination Products Administered in or on Medicated Feed or Drinking Water of Food- Producing Animals: Recommendations for Drug Sponsors for Voluntarily Aligning Product Use Conditions with GFI #209 (2013), available at http://www.fda.gov/downloads/AnimalVeterinary/GuidanceComplianceEnforcement/GuidanceforIndustry/UCM299624.pdf.

[35]Donald Kennedy, Cows on Drugs, N.Y. Times, Apr. 18, 2010, at WK11; Preservation of Antibiotics for Medical Treatment Act of 2009: Hearing before the H. Comm. on Rules, 111th Cong. (2009) (statement of Joshua Sharfstein, Principal Deputy Comm'r, U.S. Food and Drug Admin.).

[36]Kennedy, (2010); Animal Health Inst., Political Bans on Antibiotics are Counterproductive: European Test Case: Increased Animal Disease, Mixed Human Health Benefit (2006), available at www.ahi.org/content.asp?contentid=715; Am. Veterinary Med. Ass'n, 111th Congress Legislative Agenda: H.R. 1549/S. 619 Preservation of Antibiotics for Medical Treatment Act -- Active Pursuit of Defeat (2010), available at http://www.avma.org/advocacy/avma_advocate/apr10/aa_apr10_all.asp; Eric Gonder, Letter to the Editor, Poultry Veterinarians' Perspectives on Antimicrobial Resistance, 237 J. Am. Veterinary Med. Ass'n 258 (2010); Becky Tilly, Letter to the Editor, Poultry Veterinarians' Perspectives on Antimicrobial Resistance, 237 J. Am. Veterinary Med. Ass'n 258 (2010).

U.S. towards new regulatory approaches and enforcement of existing regulations governing uses of antimicrobial drugs in food-producing animals is illustrative of global challenges faced in policy efforts to combat the rise of antimicrobial-resistant pathogens broadly.[37] This chapter presents a comparison of the U.S. approach to a more progressive approach used in the European Union and places regulation of antimicrobial drug use in food-producing animals in a scientific context.

10.2 Overview of Antimicrobial Resistance

While microorganisms may produce antibacterial chemicals naturally,[38] the first documented use of antimicrobial agents by humans was in Egypt in the sixteenth century B.C.[39] Mass production of the first antibiotic, penicillin, began in 1941 to treat wounded soldiers during World War II.[40] The use of antimicrobials quickly became common in both humans and animals to reduce morbidity and suffering by speeding recovery from infection and to cure patients whose natural immune response alone could not eliminate an infection.[41] As use of antimicrobials became more common, so too did selection for organisms resistant to them.

10.2.1 Selection for Resistance

Because of the abundant natural sources of antibiotic substances within ecosystems, resistance to antibiotics predates human use of antimicrobial chemicals by many millennia.[42] Antimicrobial resistance in any given microbe may develop through a process of genetic exchange or mutation, where acquisition of a resistance gene or changes to the bacteria's genetic code provide a mechanism for a given bacterium to survive in the presence of a given antimicrobial or group of antimicrobial drugs.[43] The basic mechanisms of antimicrobial resistance are, in most cases, well-understood.[44] Antimicrobials typically attack one of four bacterial targets: peptidoglycans important to the structure of bacterial cell walls; ribosomes that synthesize important bacterial proteins; enzymes involved in bacterial genome replication; or bacterial cytoplasmic membranes.[45] Resistance genes encode

[37]Laxminarayan et al. (2013), pp. 1057–1098.

[38]Baltz (2008), p. 557.

[39]Forrest (1982), pp. 198–200 (describing uses of copper, mercury, honey, and resins).

[40]Keyes et al. (2003), pp. 45–46.

[41]Id. at 45–46.

[42]Vanessa M. D'Costa et al., Antibiotic Resistance is Ancient, Nature, (accessed ahead-of-print, August 31, 2011).

[43]Keyes et al. (2003), pp. 45–46.

[44]Guardabassi and Courvalin (2006) pp. 1–18.

[45]Id. pp. 8–12.

proteins that allow bacteria to evade attack, typically by providing target-specific evasion from the antimicrobial, by inactivating the drug, or by removing the drug from the bacterium.[46] Therefore, in the presence of an antimicrobial chemical, a susceptible bacterium will die and a resistant bacterium will survive to reproduce. As a result, resistant strains will quickly dominate the population of bacteria present in a human, an animal, or the environment.[47] This process is known as *selection*.[48]

Bacteria may acquire genes for antimicrobial resistance from other bacteria through a process called *horizontal gene transfer*.[49] Such transfers can occur between bacteria of different species.[50] An example is the acquisition of the *vanA* gene, which confers resistance to the critically-important antibiotic vancomycin, by methicillin-resistant *Staphylococcus aureus* (the "superbug" MRSA) from vancomycin-resistant *Enterococcus* (VRE, another "superbug").[51] Of clinical concern, multiple resistance genes may travel together, conferring multidrug resistance with a single genetic transfer event.[52]

Just as humans may live together in communities, so too do microbes, including both "good" commensal bacteria that do not cause disease and "bad" pathogens.[53] Such communities are termed *microbiomes*, and the environments in which these microbes live are *microbial ecosystems*.[54] The concept of the ecosystem, in which all living beings and non-living constituents of an area influence each other,[55] is important to understanding how antimicrobial drugs influence bacterial communities.[56] An example of a microbiome is the collection of microorganisms that comprise the human intestinal flora, and this population of bacteria and other microbes plays an important role in digestion and other gastrointestinal functions.[57]

---

[46]Mechanisms of resistance vary among bacteria according to the specific antibiotic or class of antimicrobials under consideration. For example, the mecA gene in Staphylococcus aureus, making this pathogen methicillin-resistant (MRSA), alters a target protein normally used by the class of penicillin drugs (including methicillin) to inhibit cell wall synthesis. This altered protein, PBP2a, does not bind well to penicillin drugs, and thus MRSA evades penicillin attack. See id.

[47]Wright (2007), pp. 175, 183–184.

[48]Keyes et al. (2003), p. 51.

[49]This typically occurs on a mobile genetic element (e.g., plasmid), which is a piece of genetic material capable of being transferred between bacteria, usually via a process called bacterial conjugation. See Andremont (2000), p. S178.

[50]Ito et al. (2003), pp. 41–49.

[51]Id. at 49.

[52]Wright (2007), pp. 175–186.

[53]Dethlefsen et al. (2007), pp. 811–812.

[54]Davis et al. (2011), pp. 244–245.

[55]Loreau (2010), pp. 49–55.

[56]Davis et al. (2011), pp. 244–245.

[57]Turnbaugh et al. (2007), p. 804.

Microbial ecosystems are dynamic; they change in response to new components.[58] A small number of resistant bacteria in a microbiome may occur through natural processes, such as mutation.[59] However, when antimicrobials are added to a microbial ecosystem (*e.g.*, by administering drugs to sick humans or by feeding antimicrobials to broiler chickens in an industrial poultry production environment), these drugs increase selective pressure in the feed itself, in the animal's intestine, and in the manure or litter. This, in turn, may drive increases in the populations of resistant bacteria.[60] As resistant bacteria multiply, the number of genes for resistance also multiplies.[61] The sum of all the diverse genes for resistance in a community of microbes is called the *resistome*, or reservoir of resistance.[62] When a new bacterium, such as a pathogen, enters a microbial community under the influence of antimicrobials, it may more easily acquire the "information," or resistance gene, that will allow it to survive.[63] Even a "good" bacterium may develop resistance and transfer this information to a pathogen, making consideration for resistance in both commensal and pathogenic bacteria (*i.e.*, consideration of the entire resistome) important to any discussion of antimicrobial regulation. Further, resistant bacteria may protect susceptible members of their microbial community (including potential pathogens) from antimicrobial effects, although the mechanisms of such "altruistic" behavior are not yet well characterized.[64] This underscores the importance of considering entire microbial communities, not just specific pathogens, in designing strategies to retain clinical efficacy of antimicrobial agents.

10.2.2 *Judicious Use*

Physicians and researchers typically have associated the recent increase in infections caused by drug-resistant pathogens with poor medical practices and overuse of antimicrobials in the environment of a hospital or clinic.[65] Hospital environments may promote selection for and transmission of resistant bacteria.[66] To help

[58] Davis et al. (2011), pp. 244–245.

[59] Wright (2007), p. 176.

[60] Id.

[61] Multiplication of resistance genes may occur through expansion of resistant populations of bacteria (one resistant bacterium becomes two, etc.), and also through horizontal gene transfer, in which the plasmid that contains the gene itself is copied and shared with a formerly susceptible bacterium.

[62] Wright (2007) 49, p. 176.

[63] Davis et al., An Ecological Perspective, (2011), at 256; Skippington and Ragan (2011), p. 707.

[64] Lee et al. (2010), p. 82.

[65] Silbergeld et al. (2008b), pp. 1391–1392.

[66] Kola et al. (2010), p. 46; Fishman (2006), pp. S53–S61.

reduce this phenomenon, good medical practice dictates that a patient who is infected with a resistant organism should be identified through medical follow-up, and another antimicrobial drug should be prescribed to effectively eliminate the resistant organism.[67] The following hypothetical example illustrates this practice: Sam enters an outpatient clinic because she has developed an abscess on her hand following a sports injury. Her physician cultures the wound and starts Sam on amoxicillin, a type of antimicrobial related to penicillin. Two days later, the laboratory reports that the wound is infected with methicillin-resistant *Staphylococcus aureus* (MRSA), a microbe resistant to the entire beta-lactam class of antimicrobials that includes penicillin. In light of this information, Sam's physician follows up with her and prescribes clindamycin, an antimicrobial more likely to treat the infection based on the resistance profile (*i.e.*, culture and sensitivity report) provided by the laboratory. This is an example of *antimicrobial stewardship* or *judicious use*.[68]

Veterinary use of antimicrobials to treat clinical infection in individual animals, such as pets,[69] also falls under judicious use guidelines similar to those employed by physicians who treat humans.[70] For treatment of an individual animal, a veterinarian may follow a similar model as presented above, seeking laboratory culture and sensitivity testing of suspected infections.[71] For food-producing animals, a veterinarian instead may seek laboratory confirmation of a suspected disease by testing a representative sample of animals in the flock, school, or herd.[72] Antimicrobial use in livestock may be under veterinary supervision to treat a diagnosed infection, and drugs for disease treatment often are administered by injection.[73] The majority of antimicrobial use in food-producing animals in the United States, however, historically has not been for disease treatment but instead for growth promotion or other purposes.[74] Without needing a veterinary prescription, food animal producers have been able to purchase antimicrobial supplements to add to the feed of the animals they raise for either growth promotion purposes or for prevention or control of disease in animals exposed to or at risk of exposure to particular pathogens (also termed *prophylaxis*).[75]

---

[67]Dellit et al. (2007), p. 159; Ctrs. Disease Control & Prevention, Get Smart: Know When Antibiotics Work, (2011), available at www.cdc.gov/getsmart/index.html.

[68]Dellit et al. (2007), pp. 159–160.

[69]Pets, or companion animals, include dogs, cats, horses, rabbits, and other animals that might be kept in or near the household.

[70]Am. Veterinary Med. Ass'n, AVMA Guidelines for Judicious Therapeutic Use of Antimicrobials (2010), available at http://www.avma.org/issues/jtua/jtua_poultry.asp.

[71]Id.

[72]Radostits et al. (1985).

[73]Parenteral use (injection) is common for disease treatment except some uses in poultry production and aquaculture due to difficulty of injection or the muscle damage an injection could cause in these smaller species. See id. at 85.

[74]Ctr for Veterinary Med., Summary Report, (2012).

[75]Love et al. (2011a), p. 279.

10.2.3 Antimicrobials as Pollutants

Antimicrobial use at non-therapeutic levels in food-producing animals (livestock), primarily for growth promotion and other production purposes,[76] is of increasing concern.[77] Because food-producing animals excrete 75 % of the antimicrobials they consume unchanged or as active metabolites of the drug,[78] antimicrobials not only apply selective pressure on the intestinal microbial community of the food-producing animal, but also on the microbial community of the animal's environment, such as the barn, pasture, and fields where manure is applied.[79] Spillage of medicated feed may contaminate local soils and waters.[80] The presence of antimicrobial drugs from these sources can influence the local microbial ecology, allowing resistant organisms to survive and to become more common in bacterial communities in and around CAFOs.[81] Further, the CAFO environment,[82] marked by crowding of animals in small, often indoor spaces, intensifies the spread of bacteria among animals and increases pathogen contamination of their barns or pens.[83] This led scientist Dr. Jose Luis Martinez to coin the term *antibiotic pollution*, which may refer to either the antimicrobial chemicals themselves (which, like other chemical pollutants, may degrade over time) or the resistance genes they foster (which may, in fact, multiply through horizontal gene transfer and reproduction of resistant bacteria).[84] Residents of rural communities may be exposed to antimicrobial pollution through air and water contaminated by manure waste,[85] and consumers nationwide (and globally) can be exposed through the

---

[76]In addition to medication of animals, antimicrobials also may be used in agricultural environments, in environmental sanitation, and crop treatment; these latter uses are regulated by the Environmental Protection Agency. See U.S. Envtl. Prot. Agency, Pesticide Registration Manual: Chapter 18 - Other Federal or State Agency Requirements (2010), available at www.epa.gov/pesticides/bluebook/chapter18.html#antimicrobial.

[77]World Health Org., The Medical Impact, (1997), pp. 1–6; Silbergeld et al. (2008a); McEwen and Fedorka-Cray (2002).

[78]Elmund et al. (1971), pp. 129–131.

[79]Davis et al. (2011), pp. 246–248.

[80]Love et al. (2011a), p. 279.

[81]Halling-Sørensen et al. (1988), pp. 357–359; Sengeløv et al. (2003), pp. 587, 590–592; Diarra et al. (2007), p. 6566.

[82]CAFOs, otherwise known as industrial food animal production facilities, are typified by high-throughput methods designed to achieve a uniform product (meat, milk, or eggs) in a standardized period of time to accommodate mechanized harvest methods. High animal density, waste (manure) concentration, and use of antimicrobials, often in medicated feed, are hallmarks of these systems. See Davis et al. (2011), pp. 244–245; Love et al. (2011a), p. 279; Silbergeld et al. (2008a), p. 123.

[83]Silbergeld et al. (2008a), p. 123.

[84]Martinez (2009), p. 2893.

[85]Davis et al. (2011), p. 247; Graham and Nachman (2010), pp. 646–654; Chapin et al. (2004), p. 137.

retail meat,[86] seafood[87] or other products they contact, such as fertilizer derived from contaminated animal products.[88]

Both national surveillance and independent research data support the existence of these pathways of exposure to resistant pathogens and genes for resistance.[89] Antimicrobial resistance patterns in bacteria cultured from humans have been shown to follow resistance trends in food and food-producing animals for bacteria that can be transmitted between animals and humans, termed *zoonoses*.[90] In the U. S., studies have reported that resistance genes and resistant *Salmonella* bacteria from food-producing animals matched those found in humans.[91] Similar associations for ceftiofur resistance[92] were identified in a national surveillance program, the National Antimicrobial Resistance Monitoring System (NARMS),[93] which is a joint effort of the FDA, the Centers for Disease Control and Prevention (CDC), and United States Department of Agriculture (USDA).[94] Food is an important route for

[86]U.S. Food and Drug Administration, NARMS 2008 Executive Report 1–3 (2009), available at http://www.fda.gov/downloads/AnimalVeterinary/SafetyHealth/AntimicrobialResistance/NationalAntimicrobialResistanceMonitoringSystem/UCM253024.pdf; McEwen and Fedorka-Cray (2002), pp. S99–S101.

[87]Love et al. (2011b), p. 7232.

[88]David C. Love et al., Poultry Feather Meal from the United States and China Contains Residues of Multiple Pharmaceuticals and Personal Care Products (PPCPs). (2012; on file with author).

[89]Aarestrup et al. (2008), pp. 733–738; McEwen et al. (2010), p. 561.

[90]U.S. Food and Drug Admin., NARMS 2008, (2009); McEwen et al. (2010), pp. 561–562. Effects from use of fluoroquinolones, virginiamycin, and other drugs will be discussed, infra.

[91]M'ikanatha et al. (2010), p. 929; Alexander et al. (2008), p. 191.

[92]Of note, the finding of an association between use of cephalosporins, including ceftiofur, in food-producing animals and cephalosporin resistance in human isolates was the basis for an attempt by the FDA to restrict extra-label use of these antimicrobials in food-producing animals. New Animal Drugs; Cephalosporin Drugs; Extralabel Animal Drug Use; Order of Prohibition, 73 Fed. Reg. 38,110–38,113 (July 3, 2008) (to be codified at 21 C.F.R. pt. 530). The initial order was revoked before it took effect. See U.S. Food & Drug Admin., FDA Revokes Order Prohibiting Extra-label Use of Cephalosporin (2008), available at www.fda.gov/AnimalVeterinary/NewsEvents/CVMUpdates/ucm054431.htm. A new order to prohibit certain extra-label uses of certain cephalosporins was published in early 2012. New Animal Drugs; Cephalosporin Drugs; Extralabel Animal Drug Use; Order of Prohibition, 77 Fed. Reg. 4,735–745 (January 6, 2012) (to be codified at 21 C.F.R. pt. 530). Extra-label use by veterinarians is use in a species or at a dosage or via a route not specifically included in the approval (label) of that animal drug. The Animal Medicinal Drug Use Clarification Act (AMDUCA) of 1994, as implemented by FDA regulation (21 C.F.R. § 530), authorizes the veterinarian to prescribe an animal drug for extra-label use under certain conditions. This extra-label use is, in part, a response to the many species veterinarians need to treat which may not have specifically been tested during the drug approval process.

[93]U.S. Food & Drug Admin., National Antimicrobial Resistance Monitoring System - Enteric Bacteria: 2004, Human Isolates Final Report (2004), available at http://www.cdc.gov/narms/annual/2004/NARMSAnnualReport2004.pdf.

[94]FDA's Role in Antimicrobial Resistance: Hearing Before the Subcomm. on Livestock, Dairy, and Poultry of the H. Comm. On Agriculture, 110th Cong. (2008) (statement of Bernadette Dunham, Director, Ctr. Veterinary Med.).

transmission of zoonotic pathogens from food-producing animals to humans because of its broad impact on potentially all citizens.[95]

10.3 Regulation of Antimicrobial Drugs for Food-Producing Animals in the United States and European Union

Antimicrobial resistance has threatened human health globally for over half a century.[96] The history of policies to address antimicrobial resistance of agricultural origin began in England.[97] In 1960, the Netherthorpe Committee was established[98] to consider whether feeding antimicrobials to food-producing animals was potentially hazardous to human or animal health.[99] Although the Netherthorpe Committee did not find evidence of risk from such practices, later scientific evidence regarding the development of multiple drug resistance from animal feeding of antimicrobials re-opened the issue.[100] A new committee, dubbed the Swann Committee, was formed in 1968, leading to the first European report on the topic.[101] Commissioned by the English Parliament and delivered to the House of Lords in 1969, the Swann Report warned against using the same classes of antimicrobials for growth promotion in animals that were used in human therapy.[102] Although this report recommended the formal establishment of a committee to oversee regulation on the subject, this did not materialize in Britain until 1998.[103] Subsequent efforts in Britain have included the development of a chapter[104] of the Alliance for the

[95]Even vegetarians and vegans may be impacted by zoonotic bacteria through the food they eat, because vegetables may be contaminated by water or dust containing bacteria of food animal origin. Examples include E. coli 0157:H7 outbreaks traced to animal manure spread in apple orchards and irrigation water for spinach crops. See Gerba and Smith (2005), p. 42.

[96]Sherris and Florey (1951), p. 309.

[97]Kiser, p. 1058.

[98]England's Netherthorpe Committee was established in response to a 1955 meeting of the Agricultural Research Institute of the National Academy of Sciences (NAS) held on October 17–18 in Washington, D.C. in which, although resistance in animal microbes to in-feed antimicrobials was found, a conclusion of no hazard to human health was made. Id.

[99]Id.

[100]Id. at 1058–1059.

[101]Id. at 1059–1060.

[102]Soulsby (2007), p. i77; House of Lords, Use Of Antibiotics In Animal Husbandry And Veterinary Medicine (Swann Report) (1969), http://hansard.millbanksystems.com/commons/1969/nov/20/use-of-antibiotics-in-animal-husbandry.

[103]Id. at i77.

[104]Id. at i78 (concerning UK involvement).

Prudent Use of Antibiotics[105] and participation in the international Reservoirs of Antibiotic Resistance (ROAR) network.[106] The ROAR network of scientists, which includes federally-funded U.S. researchers, has focused attention on the environmental spread of resistant bacteria and the ecology of pathogenic and non-pathogenic (commensal) organisms in regard to the transfer of resistance genes.[107]

In 1996, the World Organisation [*sic*] for Animal Health (OIE)[108] established an international body, the International Cooperation on Harmonization of Technical Requirements for Registration of Veterinary Medicinal Products (VICH), to synchronize the registration standards for veterinary products and surveillance standards for post-marketing evaluation of approved veterinary drugs internationally.[109] Both the United States and the European Union have adopted many VICH standards.[110] Further, the United States has acknowledged the importance of global efforts to combat antimicrobial resistance.[111]

Despite this attempt at harmonization, European Union members and other countries have progressed ahead of the United States in regulatory and surveillance efforts for non-therapeutic uses of antimicrobials in livestock.[112] Although federal agencies first proposed restriction of antimicrobial use in food animals in 1977,[113] the first enforceable action to limit such use did not take place for almost three decades.[114] Global and U.S. federal efforts to combat antimicrobial resistance can be divided into three broad categories: programs to support research and surveillance of antibiotic resistance to better describe the problem,[115] bans, restrictions, or approval limitations for antimicrobial use in food-producing animals,[116] and

[105]The Alliance for the Prudent Use of Antibiotics (APUA) is an international advocacy organization based at Tufts University in the United States and sponsors the ROAR network of scientists. Alliance for the Prudent Use of Antibiotics, available at http://www.tufts.edu/med/apua/about_us/what_we_do.shtml (last visited Nov. 21, 2011).

[106]Soulsby (2007), p. i78.

[107]Reservoirs of Antibiotic Resistance, available at http://www.roarproject.org/ROAR/html/index.htm (last visited Nov. 21, 2011) (describing research activities and U.S. funding mechanisms); Soulsby (2007), p. i78 (concerning UK involvement).

[108]At the time, the World Organisation [sic] for Animal Health was called the Office International des Epizooties (OIE). The OIE is a global reference body, headquartered in Paris with 178 member countries, dedicated to international cooperation to combat animal diseases. The United States is a member of this 80-year old world organization.

[109]Id.

[110]Id.

[111]Interagency Task Force on Antimicrobial Resistance, A Public Health Action Plan, 2011 Revision, 130, at 15.

[112]Centner, at 6–7.

[113]Falkow and Kennedy (2001), p. 397.

[114]21 C.F.R. §§ 520, 522, 556 (2001); 21 C.F.R. §§ 520, 556 (2005) (concerning the withdrawal of FDA approval for uses in poultry of veterinary fluoroquinolones).

[115]42 U.S.C. § 247d-5 (2011).

[116]21 C.F.R. §§ 520, 556.

guidance statements for industry to inform self-regulation and best management practices.[117]

10.3.1 Surveillance Programs

Research and surveillance efforts through European Union systems and U.S. programs provide information that may inform additional, post-approval regulation of antimicrobials, but do not provide a legal mechanism to restrict use of the drugs.[118] Nonetheless, these programs are critical, not only to create a knowledge base on which to build or modify regulatory approaches, but also to evaluate the success of existing regulatory or policy strategies.

10.3.1.1 Surveillance Programs in the European Union

Many European countries have developed national surveillance systems for testing foodborne and other bacterial agents, and the efforts of these agencies are being harmonized.[119] Although many aspects of these programs are similar to the United States NARMS program and related surveillance networks,[120] a few scientifically-appealing characteristics distinguish European systems. In Denmark, development of the DANMAP surveillance program integrated bacterial and antimicrobial surveillance data with detailed surveys of antimicrobial use and geocoded[121] information on farm locations and human and animal cases of disease.[122] Collecting addresses, GPS points, or other geocoded information allows integration of surveillance systems for human and animal pathogens through a spatial matrix, allowing better linkage of outbreaks that occur in temporal and spatial proximity. When funding for expensive molecular testing of isolates is limited, selection of candidate isolates to test may be guided by this kind of epidemiologic evidence. Sweden's Strategic Program for the Rational Use of Antimicrobial Agents and Surveillance of Resistance (STRAMA) program combined surveillance across

[117]U.S. Food & Drug Admin., Guidance for Industry #152, (2003).

[118]U.S. Food and Drug Administration, NARMS 2008, (2009), at 2.

[119]Monnet (2000), p. 91.

[120]See supra.

[121]Geocoding is a technique for converting an address into a point on a map on the basis of latitude and longitude. Researchers can use this information to conduct spatial data analysis comparing sources of antimicrobial contamination with patterns of resistance in human, animal, and environmental bacteria. See Beth Feingold et al., Spatial Analysis of Livestock Associated MRSA, (conference abstract) Ass'n Am. Geographers (conf. abstract, Apr. 13, 2011), available at http://meridian.aag.org/callforpapers/program/AbstractDetail.cfm?AbstractID=39548.

[122]Danish Integrated Antimicrobial Resistance Monitoring and Research Program.

human and veterinary clinical testing (including companion animals) with education both on resistance trends and also judicious use practices.[123]

Efforts at coordination across countries within the European Union may provide a useful model for international efforts for resistance surveillance involving the United States. EU countries and the U.S. participate in the international SENTRY surveillance program, but this surveillance network focuses exclusively on human clinical isolates.[124] The European Food Safety Authority (EFSA), established in 2002 as part of Europe's food safety program,[125] and the European Centre for Disease Prevention and Control (ECDC), founded in 2005 to coordinate European health agencies,[126] manage a European-wide program for surveillance of zoonoses and foodborne bacteria.[127] Multi-national studies on antimicrobial resistance trends have been conducted since 1999 by the European Antimicrobial Susceptibility Surveillance in Animals (EASSA) program through the European Animal Health Study Centre (CEESA).[128] These studies have demonstrated general trends of decreasing resistance in bacteria isolated from animals following the ban, but also found a few paradoxic plateaus or increases in resistance.[129] Numerous scientists and stakeholders have noted the importance of pre- and post-regulation monitoring.[130] This will be addressed in greater detail later in the chapter.

10.3.1.2 United States National Antimicrobial Resistance Monitoring Program

The National Antimicrobial Resistance Monitoring Program (NARMS), part of the Emerging Infections Program, was launched in 1996 as the primary U.S. surveillance program for antimicrobial resistance in foodborne pathogens.[131] This is a multi-agency effort involving, within the USDA, the Food Safety and Inspection Service (FSIS), Agricultural Research Service (ARS) and Animal and Plant Health Inspection Service (APHIS); and, within the Department of Health

[123] Andreasen et al., at 41–42.

[124] JMI Laboratories: Surveillance, available at http://www.jmilabs.com/surveillance/ (last visited Nov. 21, 2011).

[125] Euro. Food Safety Auth., About EFSA, www.efsa.europa.eu/en/aboutefsa.htm (last visited Sept. 17, 2011).

[126] Euro. Ctr. Disease Prevention & Control, Mission, www.ecdc.europa.eu/en/aboutus/Mission/Pages/Mission.aspx (last visited Sept. 17, 2011).

[127] Council Directive 2003/99/EC, available at http://eur-lex.europa.eu/LexUriServ/LexUriServ.do?uri=OJ:L:2003:325:0031:0040:EN:PDF.

[128] de Jong et al. (2009), p. 733.

[129] Id. (noting that few resistance patterns following the bans returned to zero, and also that some resistance patterns (e.g., to streptogramins and fluoroquinolones) remain higher than expected); see infra.

[130] Halpern, at 16; Hawkey (2008), p. i1; John and Fishman (1997), p. 471.

[131] FDA's Role in Antimicrobial Resistance: Hearing, (2008) (statement of Bernadette Dunham).

and Human Services (HHS), the FDA, including the Center for Veterinary Medicine (CVM), and the CDC.[132] Specifically, NARMS microbiologists test four groups of foodborne bacteria—*Salmonella, Campylobacter, Enterococcus*, and *E. coli*—[133]for resistance to certain antimicrobials, they bank strains for future testing, and they perform molecular strain typing of certain isolates.[134] This work currently is implemented in a growing number of states that comprise the FoodNet surveillance program for diseases of foodborne origin.[135] Currently, NARMS is an umbrella program for three distinct entities: PulseNet (CDC), the "human arm" of the program which is a database of isolates from human foodborne infections; VetNet (USDA), the "animal arm" of the program which parallels PulseNet for isolates of animal origin; and the "retail arm," which is an active surveillance program for meats from federally-inspected slaughterhouses and is a collaboration between CVM, CDC, and FoodNet, although most of the laboratory work is performed by branches of the USDA.[136]

As the primary surveillance network for antimicrobial resistance of animal origin, NARMS is limited in its focus on antimicrobial resistance in foodborne bacteria. While food is an important pathway for transmission of zoonotic diseases between animals and humans, other pathways, such as occupational health risks and rural community exposure to industrial agricultural environments, are not captured by this surveillance system.[137] Although U.S. surveillance programs targeting antimicrobial resistance of animal origin do not provide the same level of scientific data as programs like DANMAP, these activities have had similar per-capita costs: $8 per capita for DANMAP and $6 per capita for U.S. programs.[138] In 2012, the Infectious Disease Society of America suggested that "[t]he U.S. is far behind other

[132]USDA/HHS Response to the House and Senate Reports: Agriculture, Rural Development, Food and Drug Administration, and Related Agencies Appropriations Bill 1 (2000), available at http://www.fda.gov/downloads/AnimalVeterinary/SafetyHealth/AntimicrobialResistance/UCM134733.pdf.

[133]Salmonella spp. and Campylobacter jejuni are human enteric pathogens, while Enterococcus and E. coli may be present commensally or may cause disease opportunistically.

[134]The antimicrobials are: Azithromycin, Ciprofloxacin, Clindamycin, Erythromycin, Florfenicol, Gentamicin, Nalidixic Acid, Telithromycin, and Tetracycline.

[135]FoodNet was launched with five states, and additional states were added slowly through a state application/selection process. The current FoodNet states are Connecticut, Georgia, Maryland, Minnesota, New Mexico, Oregon, Pennsylvania, Tennessee, California (selected counties), Colorado (selected counties), and New York (selected counties). See Samantha Yang, FoodNet and Enter-net: Emerging Surveillance Programs for Foodborne Diseases, 4 Emerging Infectious Diseases 457 (1998); Ctrs. for Disease Control & Prevention, FoodNet – Foodborne Diseases Active Surveillance Network, http://www.cdc.gov/foodnet/ (last visited Sept. 15, 2011).

[136]U.S. Food & Drug Admin., NARMS Program (2010), http://www.fda.gov/AnimalVeterinary/SafetyHealth/AntimicrobialResistance/NationalAntimicrobialResistanceMonitoringSystem/ucm059089.htm.

[137]Id.

[138]Sorensen et al. (2014), p. 2.

countries in collecting and benefiting from data on antibiotic consumption and resistance."[139]

Despite its limitations, NARMS exemplifies recent governmental success to improve surveillance for antimicrobial-resistant pathogens. It was strengthened in the past decade, not only through the 1997 FSI, but also through the work of a collaborative interagency task force, detailed later in this chapter, which added VetNet, and expanded the testing program for retail meat products.[140] In addition, NARMS data were used to support the only current ban on use of an antimicrobial drug in a food-producing animal species that was based on a risk assessment evaluating the effects on antimicrobial resistance in people, discussed in the next section.

10.3.2 *Bans on Antimicrobial Uses*

Globally, restrictions on uses of antimicrobials in food-producing animals fall into three categories: (1) bans or approval limitations, (2) voluntary guidances, and (3) no or minimal restriction of use. European Union countries are among the only nations worldwide to rely strongly on bans, the first category of restrictions.

10.3.2.1 European Bans on Antimicrobial Use

Avoparcin, an antimicrobial drug related to the critically-important human drug vancomycin, was introduced for use as a growth-promoting antimicrobial (GPA) in food animal production during the 1970s in Europe.[141] Such use rapidly led to the emergence of a large community reservoir of vancomycin-resistant enterococcus (VRE) in both animal and human populations.[142] Further, VRE strains were found in food-producing animals only in countries in which avoparcin was used in animal feed,[143] and not in Sweden or the United States, where avoparcin was not used.[144]

---

[139]IDSA (2012).

[140]Interagency Task Force on Antimicrobial Resistance, Progress Report: Implementation of a Public Health Action Plan to Combat Antimicrobial Resistance, Progress Through 2007 3 (2008), available at http://www.cdc.gov/drugresistance/actionplan/2007_report/ann_rept.pdf.

[141]Witte (2000), p. S19.

[142]Bonten et al. (2001), p. 314; Silbergeld et al. (2008b); Aarestrup et al. (2000), pp. 63–68; Letter from Frank M. Aarestrup to Rep. Nancy Pelosi, 244.

[143]The countries are: Belgium, Denmark, Finland, France, Germany, Great Britain, the Netherlands, and Norway. Wegener et al. (1999), pp. 329–331.

[144]Sweden banned all growth promoters in 1986, and avoparcin was not approved as a growth promoter in the United States due to concerns about carcinogenicity. See Wegener et al. (1999), pp. 330–331.

Virginiamycin was introduced to Europe during the same period as avoparcin, but unlike avoparcin, it was used in the United States as well.[145] Not only did resistance to this antimicrobial emerge in animals fed the growth promoter,[146] but resistance was found in human clinical isolates *prior* to the release of Synercid®,[147] a human drug in the same class as virginiamycin.[148] This initial finding, combined with later molecular evidence,[149] strongly suggested that use of virginiamycin in food-producing animals contributed to human disease. This conclusion was further supported by how rarely human physicians prescribed streptogramin antimicrobial drugs before and after Synercid's release.[150]

In Denmark, avoparcin use in livestock for growth promotion was banned in 1994, and virginiamycin use was banned in 1997.[151] Danish food safety authorities considered the ban a public health success.[152] Overall use of antimicrobials[153] in livestock decreased by over 50 %, although therapeutic use did increase slightly.[154] Prevalence of antimicrobial resistance in animal isolates dropped quickly.[155] At the same time, the ban had little negative economic or animal welfare effect on the Danish pig industry.[156] Despite a brief, one-percent increase in the mortality of weaner pigs[157] following bans of non-therapeutic use of antimicrobials in this

[145]Witte (2000), pp. S19–S20.

[146]Id.; Aarestrup et al., Associations Between the Use of Antimicrobial Agents, (2000), pp. 68–69.

[147]Synercid®, also known as quinupristin/dalfopristin, was the first streptogramin drug widely released for human use, but its final approval in 1999 came decades after use of virginiamycin began in food-producing animals. See B. Pavan, Synercid Aventis, 1 Current Opinion Investigational Drugs 173 (2000). Q/D remains a drug of last resort for certain highly-resistant infections, in part due to side effects. See Welte and Pletz (2010), pp. 391–393.

[148]Werner et al. (1998), p. 401.

[149]Werner et al. (2002), p. 81.

[150]The multitude of potential sources of antimicrobial use in both veterinary and human clinical environments for other drugs makes assessment of cause more difficult. See infra. In the case of virginiamycin, human uses of related streptogramins did not significantly contribute to antimicrobial pollution for that class of drugs, making this an unusual case and one scientifically useful to consider.

[151]Letter from Frank M. Aarestrup to Rep. Nancy Pelosi; Centner.

[152]Letter from Frank M. Aarestrup to Rep. Nancy Pelosi,; Preservation of Antibiotics for Medical Treatment Act of 2009: Hearing, (2009) (statement of Frank M. Aarestrup & Henrik Wegener), available at http://www.livablefutureblog.com/wp-content/uploads/2009/08/testimony-of-dr-frank-moller-aarestrup-1.pdf.

[153]This is determined according to milligrams of antibiotic used per kilogram of meat produced.

[154]Letter from Frank M. Aarestrup to Rep. Nancy Pelosi.

[155]Id.

[156]Id. at 1.

[157]In industrial animal production, animals often are sectioned into age groups, sometimes called production stages, because these animals will need to be fed differently according to weight and age. "Weaner" pigs are piglets that recently have been moved away from their mothers and a milk diet and onto other foods. Conventionally, this is done at 3–5 weeks of age. This process is stressful and weaner pigs, like many young food animals, are more susceptible than other age groups to diseases to which they might be exposed.

production group, the overall rate of swine production in Denmark has continued to increase.[158] Management changes on Danish farms also may have contributed to the improvements in pig weaner mortality,[159] similar to results found in Sweden following its GPA ban.[160] The Danish chicken industry experienced improvements in production.[161] In broiler chickens, feed-conversion efficiency[162] increased following the ban, and the percent mortality decreased.[163]

Based in part on the bans in Denmark and Sweden,[164] the European Union first imposed an EU-wide GPA ban[165] in 1997, withdrawing approval for the antimicrobial drug avoparcin.[166] In 1998, it withdrew GPA approval for four additional antimicrobials,[167] including virginiamycin.[168] In the same year, the United Kingdom's Parliament updated the 1969 Swann Report to recommend further limits on non-therapeutic use of antimicrobials in food-producing animals and to establish the "Swann Committee."[169] In 2001, the World Health Organization

---

158Letter from Frank M. Aarestrup to Rep. Nancy Pelosi, at 1; Emborg and Wegener (2005); Aarestrup et al., Effect of Abolishment, (2001).

159Letter from Frank M. Aarestrup to Rep. Nancy Pelosi, at 1.

160Andreasen et al. at 42.

161Letter from Frank M. Aarestrup to Rep. Nancy Pelosi, at 1.

162Feed conversion is a measure of how much weight an animal gains as a function of the amount of feed it consumes. With efficient feed conversion, most of the feed consumed is used for weight gain. With poor feed conversion, feed (energy) may be used for other purposes (e.g., activity). An analogy is the difference between a human who has a sedentary lifestyle and gains weight rapidly and a human who is very active and, despite having a similar caloric intake, does not gain weight rapidly.

163Letter from Frank M. Aarestrup to Rep. Nancy Pelosi, at 1.

164Andreasen et al., at 41–42; Castanon (2007), pp. 2466, 2469–2470 (concerning the legal grounds for permitting antimicrobials in animal feeds in the European Union, particularly the harmonization of restrictions in certain member countries established before accession into European Union membership).

165These regulatory efforts have not gone unchallenged. Both Alpharma and Pfizer, major pharmaceutical companies that make and market drugs for non-therapeutic use in livestock in the United States and Europe, attempted to overturn the European bans on the basis of (1) alleged errors of risk assessment relating to the scientific evidence, and (2) alleged misapplication of powers, in this case: the application of the precautionary principle, which allows for regulation to proceed when evidence exists for harm but data are incomplete. European Courts dismissed the cases brought by Alpharma and Pfizer on the grounds that the European Commission, in mandating the original and amended legislation concerning restrictions on feed additives, had proper authorization to do so pursuant to its directive for the protection of animal or human health or the environment. See Case T-70/99, Alpharma, Inc. v. Council Euro. Union, 2002 E.C.R. II-03495; Case T-13/99, Pfizer Animal Health SA v. Council Euro. Union, 2002 E.C.R. II-3305; Council Directive 70/524, 1970 O.J. (L270) (EC).

166Council Directive 97/6, 1197 O.J. (L272) (EEC).

167The antimicrobials were: spiramycin, tylosin, bacitracin zinc, and virginiamycin. Soulsby (2007), p. i78.

168Council Regulation 2821/98, 1998 O.J. (L351) (EEC).

169Centner, at 2; Soulsby (2007), p. i77; Goforth & Goforth, at 49; see supra.

recommended international bans or global management strategies on use of certain classes of antimicrobials for growth promotion where it concludes that use in food-producing animals selects for resistance to antimicrobials of importance to human medicine.[170] In 2005, the EU banned the four remaining antimicrobials[171] used in growth promotion.[172] Shortly after the EU-wide bans, decreases in streptogramin (quinupristin-dalfopristin) and glycopeptide (vancomycin) resistance in bacteria isolated from both humans and animals were found across Europe.[173]

Voluntary bans also have been attempted in the EU, specifically in Denmark, for use of critically-important 3rd and 4th generation cephalosporins.[174] Unlike many of the EU banned drugs, which were used typically in feed for growth promotion indications, these cephalosporins were given by injection to prevent disease in newborn piglets. In response to finding extended-spectrum cephalosporinase (ECS)-producing isolates among slaughter pigs, the Danish Agriculture and Food Council promulgated a voluntary ban on use of these drugs in pig production, which led to a major reduction in ESC-producing *E. coli* bacteria.[175]

10.3.2.2 U.S. Ban on Fluoroquinolone Antimicrobials in Poultry

In the United States, veterinary antimicrobials, like those intended for use in humans, are regulated by the FDA through delegated authority from the Federal Food, Drug and Cosmetic Act.[176] Most limitations on use of veterinary antimicrobials occur through the drug approval process. Before drug companies can market a new animal drug (including antimicrobials), FDA must review scientific documentation on the safety and efficacy of the drug's proposed use and approve its label, which contains information about doses, species, and indications for use.[177] Many antimicrobials for non-therapeutic use in medicated animal feed received FDA approval by the early 1970s.[178] However, federal documents outline concerns with promotion of antimicrobial resistance from approved veterinary drugs in

---

[170]Collignon et al. (2009), p. 132; World Health Org., WHO Global Strategy for Containment of Antimicrobial Resistance (2001), http://www.who.int/csr/resources/publications/drugresist/WHO_CDS_CSR_DRS_2001_2_EN/en/.

[171]The antimicrobials were: monensin, avilamycin, salinomycin, and flavomycin. Soulsby (2007), p. i78.

[172]Id.

[173]Werner et al., Molecular Analysis, (2002) at 90; van den Bogaard et al., at 146–48.

[174]Agerso and Aarestrup (2013), p. 569.

[175]Id, at 572.

[176]21 U.S.C. §§ 301–399.

[177]Id.

[178]Kiser et al. (1971), pp. 55–56; Kiser (1976), pp. 1058–1059; Prescott (2006), pp. 24–25 (describing first FDA task force (1972) on use of antibiotics in animal feeds, which cited public health concerns with promotion of resistance).

food-producing animals as early as the 1970s.[179] Selection for resistance by approved antimicrobials, particularly with increases in drug use, indicates the importance of public health monitoring for antimicrobial resistance after a drug has been approved.

Fluoroquinolones, as a class of antimicrobials, were introduced to human clinical use in the mid-1980s. In the mid-1990s, these drugs, including enrofloxacin,[180] were approved for use in food-producing animals.[181] Enrofloxacin, as a chemical, is metabolized in animals to the human drug ciprofloxacin.[182] In poultry, enrofloxacin may be administered to a whole flock as a water additive, which may lead to variation in the dose each chicken receives.[183] Such in-feed or in-water administration is known to select for resistance in bacteria that colonize treated chickens.[184]

The NARMS surveillance network recorded no ciprofloxacin resistance among *Campylobacter jejuni*[185] isolates from poultry products in 1989 and 1990, before the approval for use in food-producing animals.[186] After the authorization, scientists found rising trends of resistance to ciprofloxacin in *Campylobacter* strains using data from NARMS.[187] Among humans, eating chicken products was found to be a risk factor for a human having a ciprofloxacin-resistant *Campylobacter*.[188] In addition, particular *Campylobacter* strains causing disease in humans were matched to strains found in retail chicken products.[189] Based on this evidence, the FDA proposed restrictions on fluoroquinolones in 2000 by publishing its intent to withdraw approval of the New Animal Drug Application (NADA) for use of enrofloxacin in poultry.[190]

Both approval and withdrawal of approval can occur through FDA action.[191] Withdrawal of drug approval carries a different regulatory burden than the approval mechanism. Specifically, drug manufacturers must prove efficacy and safety for

---

[179]Tollefson et al. (1997), pp. 709–710 (citing concerns with food animal use of antimicrobials, particularly in animal feed).

[180]Baytril®, or enrofloxacin, is a relative of the human drug ciprofloxacin used to treat humans exposed to the bioterrorism agent anthrax. Ciprofloxacin also is used to treat humans with other clinically-important infections.

[181]Zhao et al. (2010), p. 7949; Animal Drugs, Feeds, and Related Products; Enrofloxacin for Poultry; Withdrawal of Approval of New Animal Drug Application, 70 Fed. Reg. at 44, 048.

[182]Nielsen and Gyrd-Hansen (1997), p. 246.

[183]Love et al. (2011a), pp. 279–283.

[184]Randall et al. (2006), p. 4030; van Boven et al. (2003), p. 719.

[185]C. jejuni is a food-borne enteric pathogen that may be found in poultry at high rates (90–100 % of birds) without causing signs of disease in the birds. See McCrea et al. (2006a), p. 2908.

[186]Zhao et al. (2010), p. 7949; Gupta et al. (2004), p. 1102.

[187]Zhao et al. (2010), p. 7949; Gupta et al. (2004), p. 1107 (figure is of particular note, demonstrating trends of rising resistance after approval of fluoroquinolones for use in poultry).

[188]Kassenborg et al. (2004), p. S279.

[189]Smith et al. (1999), p. 1525.

[190]Enrofloxacin for Poultry; Opportunity for a Hearing, 65 Fed. Reg. 64,954 (Oct. 31, 2000).

[191]Tollefson et al., Regulation of Antibiotic Use in Animals, (1997) at 418–23; see supra.

drug approval, but the FDA, not the drug manufacturer, has the initial burden of raising questions about the safety of drugs already on the market.[192] Once this initial burden of production has been met, the burden of persuasion shifts to the drug manufacturer to prove that the drug indeed remains safe.[193] This "safety clause" allows for review of drugs when new evidence, beyond that provided with the initial application, becomes available.[194] This is in contrast to regulatory efforts in other industries, such as chemical production (similar in that most antimicrobials are chemical compounds), in which the burden of proof at all stages is on the producer to demonstrate safety.[195]

Guidance #78, finalized in 1999, states that FDA believes it is necessary to consider the potential human health impact of the microbial effects associated with all uses of all classes of antimicrobial new animal drugs intended for use in food-producing animals when approving such drugs.[196] This expansion of "safety" to include both direct toxic effects from a chemical, as well as indirect effects on human health from antimicrobial resistance, was an important regulatory step that allowed the FDA to justify restriction of fluoroquinolones.[197]

NARMS data informed a risk assessment performed by the FDA in 2000, in which the agency quantified the increased risk to human health from fluoroquinolone use in poultry production.[198] After prolonged administrative litigation with Bayer, the company that produces enrofloxacin,[199] the FDA ultimately succeeded in banning fluoroquinolone use in poultry in 2005.[200] At the time of writing, the fluoroquinolone ban was the only national risk assessment-based restriction[201]

---

[192]21 U.S.C. § 360b(e)(1)(B) (2010); Tollefson et al., Regulation of Antibiotic Use in Animals, (1997) at 418–23.

[193]Briceno, at 5–6 (Of note, this occurs in the context of a regulatory hearing before a hearing officer under Part 16 of the regulations, and can be appealed to the Commissioner).

[194]21 U.S.C. § 360b(e)(1)(B); Tollefson et al., Regulation of Antibiotic Use in Animals, (1997) at 423.

[195]Toxic Substances Control Act, 15 U.S.C. § 2603 (2010).

[196]Ctr. Veterinary Med., U.S. Food & Drug Admin., Human Health Impact of Fluoroquinolone Resistant Campylobacter Attributed to the Consumption of Chicken 2 (2000), available at http://www.fda.gov/downloads/AnimalVeterinary/SafetyHealth/RecallsWithdrawals/UCM152308.pdf.

[197]Id. at 2.

[198]Id.

[199]Ramanan Laxminarayan & Anup Malani, Extending the Cure: Policy Responses to the Growing Threat of Antibiotic Resistance 106 (2007), available at http://www.rwjf.org/files/research/etcfullreport.pdf.

[200]Zhao et al. (2010) at 7949; Animal Drugs, Feeds, and Related Products; Enrofloxacin for Poultry; Withdrawal of Approval of New Animal Drug Application, 70 Fed. Reg. at 44,048.

[201]The uncertainties inherent to any risk assessment, which are particularly profound for microbial risk assessment, were attacked by Bayer, the company that produces Baytril®, during its effort to stop the FDA's withdrawal of approval. See Briceno, at 5–6. These risk assessment techniques have also been hotly debated in the scientific community. See Feingold et al. (2010), p. 1170; Toze et al. (2010), p. 1038.

designed to address public health concerns with antimicrobial resistance of agricultural origin in the United States.[202]

10.3.3 *U.S. Guidance on Antimicrobial Use in Food-Producing Animals*

In the past two decades, regulations have been promulgated and bills have been introduced to provide additional oversight of veterinary and human antimicrobial use in the U.S.[203] The NARMS surveillance program, begun in 1996, was strengthened in 1997 through the President's Food Safety Initiative (FSI).[204] The FSI introduced risk assessment[205] as a tool to address the potential for animal drugs to promote antimicrobial resistance. This was later formalized in the "Framework Document" that became FDA Guidance #152.[206] The 1997 FSI led to the formation of the President's Council on Food Safety in 1998, which then appointed the Interagency Task Force on Antimicrobial Resistance (hereafter, "Task Force") in 1999.[207] In 2012, the FDA finalized Guidance #209 as a formal statement of its opinion that the use of antimicrobials to promote growth in food-producing animals runs counter to public health goals[208] and, in 2013, provided Guidance #213 to provide guidance to drug sponsors to align their efforts with Guidance #209. Additionally, the Task Force provides periodic updates of its Action Plan, which presents a framework for future activity by federal agencies to address the larger problem of antimicrobial resistance, including activities related to uses of antimicrobials in food-producing animals.

---

[202]Preservation of Antibiotics for Medical Treatment Act: Hearing before the H. Comm. on Rules, 111th Cong. 27 (2009) (statement of Margaret Mellon) (noting FDA's failure to use its authority to restrict antibiotic use except in the case of fluoroquinolones in poultry).

[203]See infra Appendix I: Regulatory Timeline.

[204]President's National Food Safety Initiative, 62 Fed. Reg. 13,589 (Mar. 21, 1997) (which improved coordination among agencies by clarifying their roles in prevention and emergence of resistant pathogens).

[205]Risk assessment is a process used by government agencies and other groups, including industry, to characterize and quantify hazards associated with certain activities. Originally designed for assessment of toxicants, risk assessment more recently has been applied to hazards of microbial origin, including concerns with antimicrobial resistance.

[206]U.S. Food & Drug Admin., A Proposed Framework for Evaluating and Assuring the Human Safety of the Microbial Effects of Antimicrobial New Animal Drugs Intended for Use in Food-producing Animals (1998), available at www.fda.gov/AdvisoryCommittees/CommitteesMeetingMaterials/VeterinaryMedicineAdvisoryCommittee/ucm126607.htm; U.S. Food & Drug Admin., Guidance for Industry #152, (2003).

[207]U.S. Food & Drug Admin., A Proposed Framework, (1998).

[208]Ctr. Veterinary Med., Draft Guidance #209.

10.3.3.1 A Public Health Action Plan to Combat Antimicrobial Resistance

In 1999, the U.S. government convened the Interagency Task Force on Antimicrobial Resistance (henceforth "Task Force") in response to a February 25 congressional hearing.[209] The goal was to unify strategies among the disparate federal agencies to reduce the burden of antimicrobial resistance and relieve the impacts of antimicrobial resistance on human health.[210] Three agencies—the CDC, the FDA, and the National Institutes of Health (NIH)—were assigned to jointly chair the Task Force.[211] Additional members of the Task Force included the USDA and the Environmental Protection Agency (EPA), among others.[212]

The Task Force published "A Public Health Action Plan to Combat Antimicrobial Resistance" (henceforth "Action Plan") in 2001.[213] This document, its 2011 update,[214] and its 2015 update[215] (the latter in response to Executive Order 13676[216]) detailed the domestic and international goals of U.S. federal agencies with regard to antimicrobial resistance and use of antimicrobials in humans and animals. Although these documents provide a framework for additional regulatory

[209]Ctrs. for Disease Control & Prevention, Interagency Task Force on Antimicrobial Resistance (2011), www.cdc.gov/drugresistance/actionplan/taskforce.html; James M. Hughes, Statement on Antimicrobial Resistance: Solutions to a Growing Public Health Threat (1999), available at www.hhs.gov/asl/testify/t990225c.html. The Task Force began work before formal Congressional action to organize and fund it was passed in 2000 through H.R. 2498. See Resources for the Future, Policy Responses to the Growing Threat of Antibiotic Resistance. Extending the Cure: Policy Brief 9 (May 2010), available at www.extendingthecure.org.

[210]Interagency Task Force on Antimicrobial Resistance, A Public Health Action Plan to Combat Antimicrobial Resistance (2001), available at http://www.cdc.gov/drugresistance/actionplan/aractionplan-archived.pdf.

[211]Id. at 2.

[212]Initial members included the Agency for Healthcare Research and Quality, the Health Care Financing Administration, the Health Resources and Services Administration, the Department of Agriculture, the Department of Defense, the Department of Veterans Affairs, and the Environmental Protection Agency. Later, the Centers for Medicare and Medicaid Services and the Department of Health and Human Services Office of the Assistant Secretary for Preparedness and Response were added.

[213]Interagency Task Force on Antimicrobial Resistance (2001).

[214]Interagency Task Force on Antimicrobial Resistance, A Public Health Action Plan to Combat Antimicrobial Resistance: 2011 Revision (2011), available at http://www.cdc.gov/drugresistance/pdf/public-health-action-plan-combat-antimicrobial-resistance.pdf.

[215]Interagency Task Force on Antimicrobial Resistance, National Action Plan for Combating Antimicrobial Resistant Bacteria (2015), available at https://www.whitehouse.gov/sites/default/files/docs/national_action_plan_for_combating_antibotic-resistant_bacteria.pdf.

[216]Issued by President Barack Obama on September 18, 2014; The renamed "National Action Plan" also supports the World Health Assembly resolution 67.25, endorsed in May 2014, concerning antimicrobial resistance.

action on the part of agencies, they neither provide a legal mandate for enforcement nor any penalty for noncompliance.[217]

To date, this collaborative effort has led to the initiation or enhancement of a number of projects involved in controlling antimicrobial resistance, including the expansion of the NARMS surveillance network as previously described,[218] mandates for new research on use of antimicrobials in food-producing animals as part of the USDA's National Animal Health Monitoring System (NAHMS),[219] and the evaluation of fluoroquinolone resistance from poultry and poultry products[220] that led to the subsequent FDA ban on fluoroquinolone use in poultry.[221] Surveillance systems beyond NARMS have been bolstered, including the National Healthcare Safety Network (NHSN), although this latter program focuses exclusively on infections in healthcare settings and is not currently linked to animal monitoring systems.[222] A USDA program, the Collaboration in Animal Health, Food Safety, and Epidemiology,[223] was developed to track *Salmonella*, *Campylobacter*, *E. coli*, and *Enterococci* on sentinel swine farms, and to conduct pilot programs in New York state and the midwest for dairy herd risk assessment.[224]

Since its inception, non-governmental stakeholders have criticized the Task Force for its lack of progress towards implementation of some of the goals outlined in the Action Plan.[225] In *Hogging It*, the Union of Concerned Scientists (UCS) called for a faster implementation of Priority Action 5 of the Action Plan, regarding improved monitoring systems.[226] In addition, UCS has advocated for mandates about companies' reports of the quantities and types of antimicrobials employed for therapeutic and non-therapeutic uses as feed additives in greater detail than previously provided.[227] While the Animal Drug User Fee Amendments of 2008,[228] which required the FDA to provide annual summary reports on sale and distribution

[217]Id.

[218]See supra Part 10.3.1.2.

[219]Interagency Task Force on Antimicrobial Resistance (2011), at 17 (some reports are pending publication).

[220]Id. at 65.

[221]Animal Drugs, Feeds, and Related Products; Enrofloxacin for Poultry; Withdrawal of Approval of New Animal Drug Application, 70 Fed. Reg. 44,048 (Aug. 1, 2005) (to be codified at 21 C.F.R. pts. 520 & 556); see infra.

[222]Interagency Task Force on Antimicrobial Resistance (2011), at 6. According to a presentation by the Task Force at a public meeting for comment (November 15, 2011 in Washington, D.C.), at the time of writing, the Task Force plans to expand NHSN further, including collection of data on geographic distribution of infections in healthcare settings.

[223]Interagency Task Force on Antimicrobial Resistance (2008), at 18.

[224]Id. at 19.

[225]Jones (1996), p. 153.

[226]Mellon et al. (2009), p. 65.

[227]Id. at 65–66.

[228]21 U.S.C. § 360b(l) (2010).

of antimicrobials for use in food-producing animals, partially succeeded in addressing these concerns, the lack of refinement of the information provided engendered further criticism.[229] Opposition to implementation of regulatory and research efforts related to the Action Plan also has come from the pharmaceutical industry, agribusiness, and allied professionals who may benefit economically from non-therapeutic use of antimicrobials.[230] Heated debate centered around a top priority action item in the Action Plan: to "refine and implement the proposed FDA framework for approving new antimicrobial drugs for use in food-animal production and, when appropriate, [to re-evaluate] currently approved antimicrobial drugs."[231]

The most recent update incorporates critiques from both sides of the issue and calls for a one health approach to integrated surveillance systems spanning human and animal monitoring programs, specifically pointing to integration of the National Animal Health Laboratory Network (NAHLN) and the Veterinary Laboratory Investigation and Response Network (Vet-LIRN), as well as improved coordination with global organizations, WHO and OIE. The national goal will be to "create a regional public health laboratory network to strengthen national capacity to detect resistant bacterial strains and a specimen repository to facilitate development and evaluation of diagnostic tests and treatments."[232]

A key goal in the most recent update is to address the "[m]isuse and over-use of antibiotics in healthcare and food production [that] continue to hasten the development of bacterial drug resistance, leading to loss of efficacy of existing antibiotics."[233] Notably, this mandate includes elimination of antimicrobial uses for growth promotion and supervision of antimicrobial uses for treatment, prevention and control under veterinary oversight. The guidance documents that govern how the FDA will respond to this mandate are detailed in the next sections. In particular, Guidance #209 and Guidance #213 provide a roadmap for how FDA expects industry to implement the voluntary guidance aimed at reducing or eliminating growth promotion uses and bringing other uses of antimicrobials under veterinary oversight.

---

[229]Dave Love, Drug Amounts for Food Animals Now Reported by FDA: Thanks, It's About Time!, Johns Hopkins Center for a Livable Future Blog (Dec. 13, 2010), www.livablefutureblog.com/2010/12/drug-amounts-for-food-animals-now-reported-by-fda-thanks-it%E2%80%99s-about-time (regarding the need to report amounts by specific drug and also by use in food-producing animals).

[230]Letter from Am. Ass'n Avian Pathologists et al., to Michael B. Enzi, Ranking Member, Senate Comm. on Health, Education, Labor & Pensions (Nov. 18, 2009), available at http://www.meatami.com/ht/a/GetDocumentAction/i/55364 (urging defeat of the bill); Kennedy, (2010); Food Marketing Inst., Low-Level Use of Antibiotics in Livestock and Poultry, available at http://www.fmi.org/docs/media/bg/antibiotics.pdf (last visited Sept. 16, 2011); Timothy S. Cummings, Stakeholder Position Paper: Poultry, 73 Preventive Veterinary Medicine 209 (2006).

[231]Interagency Task Force on Antimicrobial Resistance (2001), at 29.

[232]See supra ITFAR 2015, at 9.

[233]See supra ITFAR 2015, at 5.

10.3.3.2 FDA Guidance #152

First proposed in 1998 as the "Framework Document,"[234] the FDA published "Evaluating the Safety of Antimicrobial New Animal Drugs with Regard to their Microbiological Effects on Bacteria of Human Health Concern" (FDA Guidance #152) in 2003.[235] This document provided FDA's recommendation that consideration of indirect effects on human health through antimicrobial resistance pathways be included when evaluating the safety of new animal drugs.[236] FDA Guidance #152 offers instruction to drug sponsors on conducting qualitative risk assessments for new drugs under consideration for approval to assess their abilities to pose risks to human health through the development of antimicrobial resistance.[237] FDA then uses the submitted risk assessments to inform safety assessment for the drugs in question.

According to the testimony of the then Director of the FDA's Center for Veterinary Medicine, Bernadette Dunham,[238] before the House Committee on Agriculture in 2008 and the testimony of then Principal Deputy Commissioner for the FDA, Joshua Sharfstein,[239] to the House Committee on Rules in 2009, the FDA, at the time, had slowly begun voluntary application of these criteria to currently approved antimicrobial drugs. Both Dunham and Sharfstein cited the 2001 *Public Health Action Plan to Combat Antimicrobial Resistance* as a key document guiding this and similar FDA regulatory efforts. However, like all voluntary guidances, Guidance #152 was not designed to be legally binding, and the FDA permitted industry to use alternate methods (other than risk assessment) to assess the microbial food safety of some proposed drugs.[240]

10.3.3.3 FDA Guidance #209

In June 2010, the FDA issued Draft Guidance #209,[241] which it finalized in 2012.[242] In this document, the FDA stated that it "believes that the weight of scientific

[234]U.S. Food & Drug Admin., A Proposed Framework, (1998).

[235]U.S. Food & Drug Admin., Guidance for Industry #152, (2003).

[236]Id.

[237]U.S. Food & Drug Admin., Guidance for Industry #152, (2003); Tollefson (2004), p. 415.

[238]FDA's Role in Antimicrobial Resistance: Hearing, (2008) (statement of Bernadette Dunham).

[239]Preservation of Antibiotics for Medical Treatment Act of 2009: Hearing, (2009) (statement of Joshua Sharfstein).

[240]Alternative methods are not detailed expressly in the document; instead, industry is urged to discuss possible alternatives with FDA officials. U.S. Food & Drug Admin., Guidance for Industry #152, (2003), at 1–2.

[241]Comments were solicited through the end of August 2010, and the FDA has stated that it intends to issue a final document. At the time of writing, the timeline for the final document is unknown. See Ctr. Veterinary Med., Draft Guidance #209.

[242]U.S. Food & Drug Admin., The Judicious Use of Medically-Important Antimicrobial Drugs in Food-Producing Animals (2012), available at http://www.fda.gov/downloads/AnimalVeterinary/GuidanceComplianceEnforcement/GuidanceforIndustry/UCM216936.pdf.

evidence supports the recommendations outlined in this guidance document."[243] Specifically, the agency advanced two guiding principles for antimicrobial use in animals: (1) "medically-important antibiotics," meaning those with demonstrated human clinical uses,[244] should be restricted to disease treatment uses in animals in response to specific pathogens and not be used for production purposes such as growth promotion; and (2) antimicrobials should be used under the supervision of a veterinarian, whether through direct oversight or after consultation.[245] This guidance, like Guidance #152, did not provide a legal means of enforcement of these principles, and the FDA again explicitly allowed for consideration of alternative approaches to accomplish its stated goals.[246]

In this document, the FDA outlined differences between animal drugs approved before Guidance#152, which did not have to meet microbiological safety standards, and those approved after the guidance was issued.[247] New Animal Drug Applications (NADAs) submitted since 2003 must incorporate risk assessment for drug safety by analyzing potential harm through selection for antimicrobial resistance, or must use alternative methods to evaluate microbiological safety. This change in the drug approval process, although not legally binding (like Guidance #152), shifted the burden of demonstrating human microbiological safety of new antimicrobials to the drug manufacturer. In contrast, to remove a drug approved before 2003 from the market or to amend its approval, the FDA must raise concerns and provide evidence for risk from antimicrobial resistance to humans for these drugs.[248] Many of these drugs have been on the market for decades,[249] longer than the surveillance systems have been in existence. This limits the ability of the FDA to provide data on trends in resistance before and after drug approval, which would hinder any FDA effort to justify a drug's withdrawal.

While the new drug approval process is more rigorous in considering antibiotic resistance explicitly, the dichotomy between recommending higher standards for new drug approval and applying lesser standards for existing drugs may serve as a disincentive to drug manufacturers to develop and market new antimicrobials. The Infectious Diseases Society of America (IDSA), representing clinicians and

---

[243]Id. at 17.

[244]Id. at 3.

[245]Id.

[246]Id.

[247]Id. at 13–15.

[248]U.S. Food & Drug Admin., Proposal to Withdraw Approval of the New Animal Drug Application for Enrofloxacin for Poultry 5 (Docket no. 00 N-1571, Mar 16, 2004); Ctr. Veterinary Med., Draft Guidance #209, at 11 ("However, initiating action to withdraw an approved new animal drug application (NADA), in whole or in part, based on the results of a post-approval safety review would require the agency to make the showing required under section 512(e)(1) of the [Food Drug and Cosmetic] Act.").

[249]Love et al. (2011a), p. 280.

researchers on the front line of antimicrobial resistance, has campaigned for years to address the dwindling pipeline of new antimicrobial drugs needed to combat human and animal disease from highly drug resistant pathogens.[250]

Of note, the American Veterinary Medical Association (AVMA) provided comment on the guidance in its draft form, stating that it "is concerned that mandating veterinary oversight of veterinary antimicrobials may not guarantee improved veterinary involvement or a valid veterinarian-client-patient relationship," in part due to the availability of medication over-the-counter and in part due to the established shortage[251] of food animal veterinarians.[252] The AVMA also speculated that "antimicrobials used for production purposes may have unknown mechanisms of action which may actually be therapeutic," going further to suggest that medically-important antimicrobials used for production purposes (*i.e.*, for growth promotion) be relabeled for therapeutic use instead, avoiding FDA's stated recommendations to limit growth promotion use.[253] Both in the finalized document for Guidance #209, and in the issuance of Guidance #213, discussed next, the FDA has incorporated concerns raised by stakeholders.

10.3.3.4 FDA Guidance #213

Guidance #213 (henceforth "GFI#213"), released in late 2013, is designed to provide drug sponsors with recommendations for compliance with Guidance #209 within a 3-year time frame.[254] It specifically targets medically-important antimicrobial drugs administered in the feed or water as the FDA deems such uses to "pose higher risk to public health."[255] Of note, as the FDA acknowledges in this document, any extralabel uses of antimicrobial drugs approved for use in feed or water already are illegal under Sections 512(a)(2) and (a)(4)(A) of the Food, Drug and Cosmetic Act.[256]

GFI #213 explicitly references the three types of marketing status it uses for approvals of new animal drugs or drug combinations: (1) over-the-counter (OTC), (2) veterinary prescription (Rx), or (3) veterinary feed directive (VFD), the latter discussed further in the next section. In this document, the FDA expands on the

[250]Boucher et al. (2009), p. 1.

[251]U.S. Dep't Agric., Veterinary Medicine Loan Repayment Program (2011), www.csrees.usda.gov/nea/animals/in_focus/an_health_if_vmlrp.html.

[252]Am. Veterinary Med. Ass'n, AVMA Responds to Federal Register Request for Comments (2010), available at http://www.avma.org/advocacy/federal/regulatory/public_health/judicious_use_antimicrobial_drugs.asp.

[253]Id.

[254]i.e. by the end of 2016, or longer, if the FDA deems this necessary.

[255]Supra FDA, Guidance#213 at 8.

[256]Id.

tension between the allowance of OTC labeling versus the requirement of veterinary oversight, stating that it believes that the "judicious use of medically important antimicrobial new animal drugs in the feed or water of food-producing animals needs the scientific and clinical training of a licensed veterinarian."[257] In this, the FDA allows for legal use of OTC labeling but suggests instead that drug sponsors align with judicious use recommendations for Rx or VFD, both for new drug approvals and also for existing approvals.

In issuing this document, the FDA provides a framework to voluntarily phase out production uses with regard to existing OTC drug approvals using the following process: (1) voluntary withdrawal of production claims, or (2) change in marketing status to Rx or VFD for products without production claims. For drug sponsors, the submission would only require revised labeling. Drug sponsors wishing to add therapeutic indications to existing products would have additional requirements, including information on risk of antimicrobial resistance, as outlined in GF1#213 and consistent with GFI#152 for new drug approvals.

10.3.3.5 Veterinary Feed Directive Drugs

Medicated feeds containing antimicrobial drugs have been used since the 1940s and 1950s,[258] and this use became widespread in the United States and globally during the 1960s and 1970s.[259] In the United States, the FDA regulates medicated animal feeds, which deliver non-prescription antimicrobials, differently than it regulates pharmaceutical grade antimicrobials typically used for therapeutic indications.[260] New drugs for use in animal feed, including antimicrobials, are divided into two categories on the basis of withdrawal period (*i.e.*, the length of time required between cessation of drug delivery and harvesting of milk or meat from the animal).[261] Category I drugs require no withdrawal period.[262] Category II drugs

---

[257] Supra FDA, Guidance#213 at 5.

[258] Use of growth promoting antibiotics in medicated animal feed was shown to be associated with increased rates of animal weight gain. Love et al. (2011a), p. 280. However, the use of antibiotics in feed was coupled with industrialization of the animal production process, in which high-throughput techniques were combined with single-species raising in small spaces (barns or feedlots) and commodity feed supplementation, see Pew Comm'n Indus., (2008). As a result, disentangling the exact mechanism of action of the antibiotics used for growth promotion has been difficult; scientists and others speculate that bacterial metabolic effects, host microbial ecology effects, and effects from treatment of sub-clinical disease may play roles independently or in combination. See Kiser, at 1063. Further, in some settings, use of growth promoting antibiotics has been shown to have little or no positive effect on animal growth and no economic benefit. See Graham et al. (2007), p. 79.

[259] Love et al. (2011a), p. 280.

[260] 21 C.F.R. § 558 (2010).

[261] Id.

[262] Id.

require a withdrawal period. These categories are each subdivided into Type A, Type B, and Type C medicated feeds, on the basis of manufacturing guidelines.[263] The length of withdrawal typically depends on the amount of a drug that could remain in milk or meat at the time of harvesting, otherwise known as the drug's "residue."[264] Antimicrobial drug residues are considered potentially harmful to human health either through human drug sensitivity (*i.e.*, allergic reaction), through promotion of antimicrobial resistance, or through disruption of normal microflora in the intestinal microbiome[265] of humans who consume the residues inadvertently in food products.[266]

Recent FDA efforts have amended this regulatory structure to provide veterinary oversight of this historically non-prescription process. In 1996, the FDA added a new class of medications for addition to animal feeds, known as "veterinary feed directive" (VFD) drugs.[267] VFD drugs are antimicrobials or other drugs for which the FDA considers the risks too high for over-the-counter marketing.[268] VFD drugs require a written statement by a licensed veterinarian, akin to a prescription written in the context of a valid veterinary-patient-client relationship,[269] which orders the use of the VFD drug in animal feed.[270] Although the AVMA has advocated for a number of changes to the logistic structure of the process, veterinary professional

[263]Type A medicated articles are used for manufacture of another Type A medicated article or for production of Type B or Type C medicated feed. Type B medicated feeds are used for the manufacture of other medicated feeds and contain nutrients (e.g., minerals or vitamins). Type C medicated feeds are complete feeds (i.e., contain all nutrients needed) or are "top-dressed" feeds (often literally placed on top of other feed). These are offered as free-choice supplements, meaning that animals choose how much of the medicated feed--and therefore the drug--to consume. These contain nutrients (e.g., vitamins, minerals) and other nutritional ingredients, and are produced by diluting Type A medicated articles or Type B medicated feeds. Certain licenses are required for manufacturers, or feed mills, of Type B or Type C medicated feeds. See id.

[264]Mitchell et al. (1998), pp. 742–743.

[265]Tollefson et al. (2006), p. 421.

[266]van Houweling and Gainer (1978), p. 1413.

[267]21 C.F.R. § 558.

[268]21 C.F.R. § 512(b).

[269]Unlike actual prescriptions, VFD orders circumvent state pharmacy laws while providing for a higher degree of professional control than the typical, over-the-counter labels approved for the majority of medicated animal feeds. See 21 U.S.C. § 354. At the time of writing, this category only had been used for one new antimicrobial, Schering-Plough's Aquaflor®, or florfenicol (a drug related to chloramphenicol), approved in 2005 (NADA 141–246; a Type A medicated feed article used to make Type C medicated feed for catfish). See Appendix II: Critically-Important Antimicrobials. Chloramphenicol is rarely employed for human clinical use due to toxicity concerns. See Editorial, Fatal Aplastic Anemias from Chloramphenicol, 247 New Eng. J. Med. 183 (1952).

[270]Veterinary Feed Directive: Final Rule, 65 Fed. Reg. 76,924 (Dec. 8, 2000) (to be codified at 21 C.F.R. pts. 510, 514, 558).

groups, including food animal practitioners, generally have supported the VFD requirements.[271]

10.3.4 Conclusion on Regulatory Approaches

Collectively, experiences with banning or restricting antimicrobials used for growth promotion in food-producing animals suggest that, although the United States pork and poultry industries may experience minor economic impacts from similar bans (such as those proposed under the Preservation of Antimicrobials for Medical Treatment Act[272]), management strategies may help overcome some of these costs in animal mortality and feed conversion. In addition, when considering the cost to society from disease and death caused by antimicrobial-resistant bacteria, the marginal financial efficiencies from growth promotion uses of antimicrobials pale in comparison.[273] Further, the European experience shows that a ban can be successful from a public health perspective in reducing the percent of bacteria isolated from animals and foods that are resistant to antimicrobials.[274] Even bans as broad as those implemented in Europe, however, may improve but will not fully eliminate the problem of antimicrobial resistance, particularly considering the global nature of antimicrobial use in multiple industries. In particular, strong, global surveillance programs are central to creating a scientific base of evidence from which to build policy or regulation, and also to measurement of the effectiveness of such approaches in providing benefits to society. The next section will detail how regulatory and legislative bodies must consider that any use of antimicrobials can select for resistance, and will provide science-based recommendations on building a regulatory framework and supporting public health efforts to better address this global problem.

---

[271] Am. Veterinary Med. Ass'n, AVMA Submitted Comments Regarding the Veterinary Feed Directive (Aug. 26, 2010), available at http://www.avma.org/advocacy/federal/regulatory/practice_issues/drugs/Veterinary_Feed_Directive.asp; Greg Cima, Antimicrobial Oversight Could Increase Through VFDs, JAVMA News (November 15, 2011), available at http://www.avma.org/onlnews/javma/nov11/111115p_pf.asp (regarding the American Association for Bovine Practitioners's support of VFD oversight of over-the-counter antimicrobial drugs for food animal use).

[272] H.R. 1150, re-introduced by Rep. Louise Slaughter (D-NY, a microbiologist) on March 14, 2013. This act originally was introduced by Representative Brown (D-OH) to the 106th Congress in 1999, and most recently had been introduced to the House of Representatives by Rep. Slaughter and to the Senate by Sen. Diane Feinstein (C-DA) in 2011. PAMTA would amend Sections 201 and 512 of the Federal Food, Drug and Cosmetic Act to rescind approval for certain critically-important antimicrobials for production uses in food-producing animals.

[273] Teillant and Laxminarayan (2015).

[274] U.S. Gen. Accounting Office, (2004).

10.4 Impact of Regulation on the Problem of Antimicrobial Resistance

Because antimicrobial-resistant infections pose an urgent and global public health threat, the question that remains is not whether action should be taken on a regulatory front, but how best to accomplish the goal of restricting the spread and impact of antimicrobial resistance. Addressing judicious use in human clinical settings is important, and furthering development of novel antimicrobial drugs and alternative therapies is critical.[275] In addition, as the case of virginiamycin demonstrated,[276] new antimicrobials intended for human clinical use should neither be first nor concurrently licensed for growth promotion uses. Further, given the economic disincentives to research and development for new antimicrobials, regulatory effort is needed urgently to protect the current arsenal of drugs.[277] Addressing use of antimicrobials in agriculture presents an opportunity for scientific evidence-based intervention through regulation and policy.[278]

10.4.1 Critically-Important Antimicrobials

Both European and proposed U.S. regulatory strategies to address antimicrobial resistance focus on "critically-important antimicrobials," also known as "medically-important antibiotics," or those antimicrobials used in human clinical settings to treat known pathogens (see Table 10.1).[279] Some have called this a "one bug, one drug" model.[280] However, this approach has several critical limitations.

First, antimicrobial resistance is not limited to pathogens, and resistance in commensal (non-pathogenic) bacteria can spread to pathogens in bacterial communities.[281] Because both pathogens and non-pathogens may acquire and exchange genes that confer resistance, surveillance systems like NARMS, limited to a few bacteria, primarily foodborne pathogens, may miss important pools of resistant commensal bacteria and non-tested pathogens (*e.g.*, *Staphylococcus aureus*).[282]

---

[275]See supra.

[276]See supra.

[277]Boucher et al. (2009), p. 1.

[278]Aarestrup et al., Resistance in Bacteria, (2008) (reviewing options for strategies to control antimicrobial resistance and their anticipated effectiveness from a scientific perspective).

[279]Collignon et al. (2009).

[280]Silbergeld et al. (2008a), p. 156.

[281]Keyes et al. (2003), pp. 45–51; Skippington and Ragan (2011), pp. 3–5.

[282]In an October 12, 2010 letter to Rep. Louise Slaughter, the Food and Drug Administration noted that NARMS personnel are exploring the possibility of adding S. aureus to the list of tested organisms. Letter from Jeanne Ireland, Assistant Comm'r Legislation, U.S. Food & Drug Admin., to Rep, Louise Slaughter, U.S. House of Representatives (Oct. 12, 2010), available at http://www.keepantibioticsworking.com/new/KAWfiles/64_2_107766.pdf.

Table 10.1 Table of selected antibiotics by class according to human and veterinary use[a]

Antimicrobial class	WHO classification[b]	Human antimicrobials	Veterinary antimicrobials		
		Drug example(s)	Use	Drug example(s)	Use
Beta-Lactams					
Penicillins	Critically important	Amoxicillin Ampicillin	Tx	Amoxicillin Ampicillin	Tx (FA, C) P (FA, C) GPA (FA)
Cephalosporins	Critically important (3rd & 4th generation)	Ceftriaxone	Tx	Ceftiofur	Tx (FA, C) P (C) Extra-label FA use restricted
	Highly important (1st & 2nd generation)	Cefazolin (Ancef) Cephalexin (Keflex)	Tx	Cephalexin	Tx (FA, C) Extra-label FA use restricted
Glycopeptides	Critically important	Vancomycin	Tx	Avoparcin[c]	GPA (FA)
Fluoroquinolones	Critically important	Ciprofloxacin	Tx P (anthrax)	Enrofloxacin	Tx (FA, C) Extra-label FA use restricted
Streptogramins	Highly important	Synercid (quinupristin-dalfopristin)	Tx	Virginiamycin[d]	P (FA) GPA (FA)
Oxazolidinones	Critically important	Linezolid	Tx	Linezolid	C use limited
Tetracyclines	Highly important	Oxtetracycline Doxycycline	Tx	Oxytetracycline Chlortetracycline	Tx (FA, C) P (FA, C) GPA (FA)
Macrolides	Critically important	Azithromycin Erythromycin Tylosin	Tx	Erythromycin Tylosin	Tx (FA, C) P (FA) GPA (FA)
Sulfonamides	Highly important	Trimethoprim-sulfamethoxazole	Tx	Trimethoprim-sulfamethoxazole	Tx (FA, C) P (C)
Lincosamides	Highly important	Clindamycin Lincomycin	Tx	Clindamycin Lincomycin	Tx (FA, C) P (FA) GPA (FA)

Aminoglycosides	Critically important	Amikacin Gentamicin Kanamycin	Tx	Amikacin Gentamicin Streptomycin Apramycin	Tx (C) P (FA) GPA (FA)
Phenicols	Highly important	Chloramphenicol	Limited uses (toxicity)	Florfenicol Chloramphenicol[e]	Tx (FA, C) P (FA) GPA (FA)

Use Codes

Tx—therapeutic uses

P—prophylaxis (treat individuals known or believed to be exposed to an infectious agent, or to prevent emergence of infection in food-producing animals, *e.g.*, dairy dry cow treatment to prevent mastitis)

GPA—growth promotion

Species Codes

FA—livestock (food-producing animals)

C—companion animals (dogs, cats, horses, *etc.*)

[a]Table adapted from Collignon et al., (2009) at 139–40; Guardabassi & Courvalin, (2006) at 6–7; Angelo A. Valois et al., Geographic Differences in Market Availability, Regulation, and Use of Veterinary Antimicrobial Products, in Guide to Antimicrobial Use in Animals (Luca Guardabassi et al., eds., 2008), at 70–71

[b]Food & Agric. Org., United Nations, Joint FAO/WHO/OIE Expert Meeting on Critically Important Antimicrobials. Report of the FAO/WHO/OIE Expert Meeting (2011), available at http://www.who.int/foodsafety/publications/antimicrobials-third/en/

[c]Not typically used in the United States

[d]Banned in EU

[e]Banned in U.S. (and EU) for use in food-producing animals (never approved due to human health hazard)

Hence, global and U.S. regulatory and policy efforts should expand surveillance systems to include both commensal and pathogenic bacteria, and to include non-food pathways, such as occupational health monitoring,[283] for potential transmission of resistant zoonoses to humans.[284] Occupational transmission of MRSA to veterinarians,[285] farmers,[286] and slaughter workers[287] has been demonstrated. A particular MRSA strain, ST398,[288] was found in food animals, especially pigs, and may be transmitted to humans.[289] This strain commonly carried a plasmid encoding for multiple resistance genes to different classes of antimicrobials, including tetracycline.[290]

In addition, reservoirs for resistant bacteria may occur in many species.[291] This includes humans, food-producing animals, companion animals (*e.g.*, dogs, cats, and horses), and occasionally exotic or wild animals.[292] Companion animals, to date, have not been part of routine monitoring programs for antimicrobial resistant bacteria,[293] despite research evidence that demonstrates trends of sometimes high rates of resistant bacteria in these populations.[294] Many human families consider companion animals as part of their households,[295] and antimicrobial-resistant infections may spread between humans and their animal companions.[296] National recommendations for harmonization between human medical and veterinary practice for community surveillance, treatment for antimicrobial-resistant infections within and among households, and judicious use of antimicrobial drugs are warranted.[297] At minimum, integrating healthcare surveillance networks,

---

[283]Ricardo Castillo et al., Antimicrobial Resistant Bacteria: An Unrecognized Work-Related Risk in Food Animal Production, Safety & Health at Work (invited paper, in submission, 2011; on file with author).

[284]Scott Weese (2006).

[285]Loeffler et al. (2010), p. 282.

[286]Smith et al. (2009), p. e4258.

[287]Mulders et al. (2010), p. 743; van Cleef et al. (2010), p. 756.

[288]This strain designation, ST398, is based on genetic methods using a process known as multilocus sequence typing. Other typing mechanisms, such as pulsed-field gel electrophoresis (PFGE) may generate a different "name." ST398 originally was known as a PFGE non-typable (NT-) MRSA.

[289]Harper et al. (2010), p. 101.

[290]Kadlec and Schwarz (2010), p. 3589.

[291]Halpern, at 4–5.

[292]Weese (2006), p. 445.

[293]Id.

[294]Chomel and Sun (2011), pp. 167–170; Loeffler and Lloyd (2010), p. 595; van Duijkeren et al. (2010), p. 96.

[295]Chomel and Sun (2011), p. 167.

[296]Bramble et al. (2011), p. 617.

[297]The agenda of the Antimicrobial Resistance Summit (2011) was integrating surveillance and regulation with infection prevention activities in multiple settings, and with education and research efforts. See Gottlieb and Nimmo (2011), p. 281.

such as NHSN,[298] with other national databases like NARMS would allow better tracking of the movement of resistance determinants and resistant pathogens between the community and the hospital.[299] Ideally, establishment of a veterinary clinical surveillance system, integrated with human healthcare networks, would help quantify the role of companion animal antimicrobial therapies in selecting for household-level resistance. This information could guide recommendations for judicious use practices in both veterinary and human medicine. In addition, expansion of monitoring systems to include rural community hospitals, which typically do not participate in antimicrobial stewardship programs or surveillance networks,[300] would allow better tracking of potential community exposure to antibiotic pollution that may occur through environmental pathways in rural areas.

Finally, and most important, use of antimicrobials not considered "medically-important" may co-select for bacteria resistant to drugs used in human clinical settings.[301] In other words, the use of one allowed antimicrobial in livestock may drive resistance to an antimicrobial restricted to human use.[302] This is a key limitation of regulatory approaches that focus exclusively on "critical antimicrobial animal drugs."[303] For example, both penicillin and cephalosporin antimicrobials are known to select for beta-lactam-resistant bacteria (*e.g.*, the "superbug" MRSA).[304] Recent action by the FDA, however, limited certain extra-label uses of cephalosporins in food-producing animals.[305] For example, a banned extra-label[306] agricultural use of cephalosporins noted by the FDA to be of great concern is the routine injection into chicken eggs prior to hatch.[307] To be effective at limiting selective pressure for beta-lactam resistance, both penicillins and cephalosporins need to be restricted simultaneously.

All antimicrobials, including those not considered critically important by the World Health Organization, should be evaluated for the potential to induce resistance to a broad spectrum of antimicrobial drugs in a range of bacteria.

---

[298]See supra.

[299]Silbergeld et al. (2008b), pp. 1392–1393 (concerning movement of pathogens and resistant bacteria between the hospital and the community).

[300]Johannsson et al. (2011), pp. 367–372 (regarding the need to include small community hospitals in computerized networks and provide other incentives for participation in antimicrobial stewardship programs).

[301]Gottlieb and Nimmo (2011), p. 282.

[302]This may occur because the genes for resistance may co-locate to the same mobile genetic element. See Silbergeld et al. (2008b), p. 1394.

[303]H.R. 965.

[304]S. 1211.

[305]Supra note 94.

[306]Davis et al. (2009), p. 528 (concerning prohibited extra-label uses; others may be allowed).

[307]McReynolds et al. (2000), pp. 1524–1525.

Some mechanisms of resistance may be broad.[308] Even more concerning, some metals (*e.g.*, zinc),[309] and non-antimicrobial pharmaceuticals (*e.g.*, aspirin, a salicylate),[310] also may play important roles in selecting for resistant organisms or promoting resistance mechanisms. While the extent of the ability of non-antimicrobials to select for antimicrobial-resistant bacteria is not yet well-characterized, improved reporting of all drugs (not just certain antimicrobial drugs) used in food-producing animals will allow better monitoring of this potential phenomenon.

10.4.2 Anticipated Impact of Regulation on Resistance

Surveillance and regulation do not occur in a vacuum; the intent of these programs is to produce a beneficial effect for society. Understanding how changes in regulation of antibiotics will impact the epidemic of antimicrobial resistance requires a scientific understanding of the microbial ecology of resistance. While the experience of regulatory authorities in Europe offers a model for a generally successful public health intervention, other antimicrobial restrictions, such as the fluoroquinolone ban in poultry in the United States, have achieved less success in the short term from a public health perspective.

Resistance to ciprofloxacin has persisted despite the ban on fluoroquinolones.[311] Data from the NARMS surveillance program demonstrated a lack of immediate improvement in ciprofloxacin resistance in chickens, chicken breasts, and human isolates of the important foodborne pathogen *Campylobacter jejuni*[312] following the 2005 ban on fluorquinolone use.[313] A simple analysis of these data reveals a three percent increase, on average, of ciprofloxacin resistance in *C. jejuni* isolates from these sources after the ban (2006–2009) compared to before the ban

[308]Certain drug efflux pumps will provide resistance to multiple families of antibiotics. In addition, other characteristics, such as the thickness of a cell wall, may help exclude antibiotics from a bacterium, conferring partial resistance. The latter is one mechanism of action for partial vancomycin resistance in some MRSA isolates. See Howden et al. (2010), pp. 99, 107–109.

[309]Cavaco et al. (2011), p. 344.

[310]Shen et al. (2011) pp. 7128–7133.

[311]Nannapaneni et al. (2009), p. 1348; see supra.

[312]Campylobacter jejuni may colonize chickens at high rates without causing disease, making contamination of food products more likely. See McCrea et al. (2006b), pp. 136–143. Campylobacter is the leading cause of foodborne illness in the United States, responsible for an estimated 2 million human infections annually. See Samuel et al. (2004), p. S165.

[313]U.S. Food & Drug Admin., National Antimicrobial Resistance Monitoring System: 2009 Executive Report 82 (2011), available at http://www.fda.gov/AnimalVeterinary/SafetyHealth/AntimicrobialResistance/NationalAntimicrobialResistanceMonitoringSystem/ucm268951.htm; Zhao et al., (2010) at 7951 (noting trend in ciprofloxacin resistance in figure 1).

(2002–2005).[314] United States researchers also have noted the lack of reduction in fluoroquinolone resistance in poultry isolates following the ban.[315] NARMS retail data from 2009 support a statistically-significant 6 % increase in fluoroquinolone resistant *C. jejuni* from retail meats between 2002 and 2009.[316] In Europe, even after the bans on growth promoters, high rates of fluoroquinolone resistance were found in *Campylobacter* and other bacterial species in both humans and poultry, but in Australia, where fluoroquinolones never were approved for food-producing animal use, cases of domestically-acquired human ciprofloxacin-resistant *Campylobacter* have been rare.[317]

Multiple potential mechanisms may explain this persistence. First, in the United States, fluoroquinolones were restricted only in poultry, and use was allowed to continue in other species, such as cattle.[318] Second, international shipment of food products and global human travel may spread resistant strains and resistance determinants beyond the boundaries of regulation. A pandemic ciprofloxacin-resistant clone of *Salmonella enterica* Serotype Kentucky was found in both humans and chickens, and use of fluoroquinolones in poultry production in Nigeria and Morocco was implicated in the rapid international spread of the pathogen.[319] Third, contrary to historical scientific belief that resistance genes are burdensome to bacteria,[320] certain genes may not be jettisoned quickly once selective pressure is reduced.[321] Finally, as noted above,[322] cross-resistance within bacteria to multiple drugs may allow non-target antimicrobials to provide selective pressure. Of note,

---

[314]This simple analysis was performed by the author (MFD). Methods: Briefly, data on the proportion of resistant isolates, by type and year, were adapted from the NARMS 2009 (360) report to Stata 11 (College Station, TX). A linear regression model was run on the proportion of ciprofloxacin resistance compared to a dichotomous variable (after vs. before the ban) for time trend, and clustering within type of isolate (human, chicken breast, and chickens). Results: After the ban, on average, the proportion of ciprofloxacin resistance increased 0.029 (~3 %), and this estimate was statistically significant ($p = 0.008$). No statistical differences were seen by type of isolate, controlling for year ($p = 0.36$). Overall averages for percentage of ciprofloxacin resistance found since the ban (for humans, chicken breasts, and chickens combined) were: 21.3 % (2009), 23.0 % (2008), 21.5 % (2007), and 14.9 % (2006). See U.S. Food & Drug Admin., National Antimicrobial Resistance Monitoring System: 2009 Executive Report, 362, at 82.

[315]Price et al. (2007), p. 1035; Nannapaneni et al., 360, at 1348–53; Silbergeld et al. (2008a), pp. 156–157.

[316]U.S. Food & Drug Admin., NARMS 2009 Retail Meat Report 9 (2009), available at http://www.fda.gov/AnimalVeterinary/SafetyHealth/AntimicrobialResistance/NationalAntimicrobialResistanceMonitoringSystem/ucm257561.htm.

[317]Lohren et al. (2008), pp. 132–133.

[318]U.S. Food & Drug Admin., FDA Approves Fluoroquinolone Product for Use in Cattle (1998), available at http://www.fda.gov/AnimalVeterinary/NewsEvents/FDAVeterinarianNewsletter/ucm089486.htm.

[319]Le Hello et al. (2011), p. 679.

[320]Laxminarayan and Malani (2007), p. 50.

[321]Silbergeld et al. (2008a), pp. 156–157.

[322]See supra.

tetracycline drugs, used widely in food animal production, are known to select for fluoroquinolone resistance.[323] Data from NARMS in 2009 show 50 % tetracycline resistance in chicken isolates of *Campylobacter*, 46 % resistance in chicken product isolates, and 43 % resistance in human isolates.[324] The degree to which other pharmaceutical products, such as aspirin, promote fluoroquinolone resistance is unknown.[325] Whether resistance (*e.g.*, to fluoroquinolones) that is easy to induce is more likely to persist also is unknown.

An additional concern with the fluoroquinolone ban was, paradoxically, the strength of the scientific evidence used for its support. The risk assessment conducted by the FDA demonstrated a strong connection between use of a particular antimicrobial in poultry and emergence of resistance patterns in the same family of antimicrobial in humans.[326] Industry[327] and members of Congress[328] have suggested that, for regulation to occur, regulatory authorities must prove that use of antimicrobials at non-therapeutic levels caused resistance in a particular bacterium, and that this specific bacterium was transmitted to humans.

Causation is difficult to prove in science, particularly in as dynamic a setting as antimicrobial resistance. Multiple sources can contribute to the problem, but proof that any one pathway was the cause for a particular case of disease in a particular individual is challenging.[329] Any and all uses of antimicrobials may contribute to selective pressure, including therapeutic uses in both human and veterinary hospital environments.[330] In addition, soil organisms and other microbes[331] may produce

---

[323]Cohen et al. (1989), p. 1318.

[324]U.S. Food & Drug Admin., National Antimicrobial Resistance Monitoring System: 2009 Executive Report, 362, at 82 (Data on cross-resistance, however, are not available in published reports, which provide only prevalences of resistance in particular pathogens by source, i.e., food animals, retail meat, or humans).

[325]Shen et al. (2011).

[326]Ctr. Veterinary Med., Human Health Impact, (2000) at 2.

[327]Kiser, at 1062.

[328]Letter from Rep. Tom Latham, U.S. House of Representatives, to Lester M. Crawford, Acting Comm'r, U.S. Food & Drug Admin. (Sept. 1, 2004), available at http://www.fda.gov/ohrms/dockets/dockets/00n1571/00n-1571-m000006-vol403.pdf (in which Representative Latham suggests that FDA should have "scientific certainty" to ban fluoroquinolone use).

[329]This is similar to the burden of ascribing a "cause" for cancer in a particular individual suffering from its effects, particularly when the cancer is potentially linked to many sources (e.g., diet, smoking habits, chemical exposures, and genetics). However, chemicals and commercial products (e.g., cigarettes) have been regulated despite this difficulty. Further, for chemicals, in vitro (cell culture) and in vivo (laboratory animal) assays demonstrating carcinogenicity in the laboratory prove sufficient for risk assessment purposes. On the contrary, similar laboratory and field assays demonstrating the ability of antibiotics to select for resistance and promote transfers of genetic material in bacteria conferring resistance are attacked by opponents of regulation as insufficient evidence of harm. Cummings, (2006) at 209–12.

[330]Muto et al. (2003), p. 362.

[331]After all, Fleming discovered penicillin by isolating it from the mold Penicillium.

antibiotics at very low concentrations,[332] although public health impacts from these natural sources may be limited. Both humans and animals may carry bacteria, including zoonotic pathogens, that harbor genes for antimicrobial resistance.[333] Isolating agriculture as the specific cause of any given human case of MRSA or *Salmonella* requires expensive molecular testing at all stages of transmission, which typically is not performed in either surveillance or clinical settings.[334] For the fluoroquinolone ban, molecular evidence was provided that linked strains of fluoroquinolone-resistant bacteria in food products to the same strains in human cases of disease.[335] This "high bar" set by the fluoroquinolone ban offers a barrier to regulation of antimicrobials whose effects are harder to demonstrate. In its recent order of prohibition for certain extra-label uses of cephalosporins, the FDA addressed this perception of a need to conduct a risk assessment and prove that an adverse event has occurred in humans in order to take regulatory action, noting instead that "it is not limited to making risk determinations based solely on documented scientific information, but may use other suitable information as appropriate."[336]

The complex ecology of bacterial resistance also impacts interpretation of the public health success or failure of regulation. In some cases, broad use of an antimicrobial, such as in medicated animal feed or water, may open a veritable "Pandora's box" of resistance.[337] Subsequent attempts to reduce usage, particularly when the reduction in use is limited to one or several countries, or limited only in a single species of food-producing animal, may be less successful than anticipated. In these cases, broader restrictions may be needed, and restrictions on multiple drugs, not just the target antimicrobial, should be explored. In the case of fluoroquinolones, some evidence links tetracycline to selection for fluoroquinolone resistance,[338] suggesting the potential need to restrict more than one class of antimicrobial to achieve the public health target effect.

---

[332]Wright (2007), pp. 183–184.

[333]Lloyd (2007), p. S148; Cuny et al. (2010), p. 109.

[334]Mullner et al. (2009), p. 1311. In this study, surveillance and laboratory data were combined, and isolates tested using molecular techniques, to determine that most cases of human campylobacteriosis were attributable to poultry. Government intervention in poultry production practices led to a decline in human cases. New Zealand's relative isolation--as an island country--likely enhanced the determination of cause. See id.

[335]Smith et al. (1999), pp. 1525–1532.

[336]Supra note 94, New Animal Drugs; Cephalosporin Drugs; Extralabel Animal Drug Use; Order of Prohibition at 743 (61 FR 57732 and 57738, November 7, 1996). (At the time of writing, this order of prohibition was still in public comment and was scheduled to take effect in April, 2012).

[337]Goforth & Goforth, at 12.

[338]Cohen et al. (1989), pp. 1318–1325.

10.4.3 Environmental Pollution

Many bacteria and their genes for resistance can survive in the environment.[339] Industries involved in antimicrobial manufacture, trade, and usage—from pharmaceutical companies to agribusiness to medical enterprises—are connected through environmental pathways. Effluent into surface waters from an antimicrobial manufacturing plant was found to drive selection for antimicrobial resistance in bacteria found downstream.[340] Use of antimicrobials in food-producing animals on farms has been tied to contamination of local and regional soils and waters.[341] Human and animal use may result in discharge of drugs into sewage,[342] leading to contamination of surface water.[343] Manures and animal by-products that contain antimicrobial residues may enter other industries through sale or trade.[344] As a result, both animals and humans may be exposed to unintended doses of antimicrobials through drinking water or other sources.[345] This evidence makes antimicrobial pollution in the environment important to consider as a future regulatory target.

The Task Force was a multi-agency effort, and included participation by the Environmental Protection Agency (EPA), the agency most likely to spearhead future regulation of antimicrobials in the environment. Indeed, recommendations in the Action Plan included plans to consider environmental impacts.[346] Strategies to address antimicrobial chemical pollution discharged into the environment, however, need to account for the diverse reservoir of resistance genes found in native soil microorganisms.[347] Because genes for resistance can be found broadly in the environment, attempts to reduce environmental antimicrobial pollution may need to be equally broad, targeting both point and non-point sources of antimicrobial discharge simultaneously. Consideration of antimicrobial pollution may require novel risk assessment techniques. In contrast to most regulated chemicals, which do not multiply in the environment, even low concentrations of antimicrobials may drive selective pressure for antimicrobial resistance, expanding the local reservoir of resistance genes.[348] This is in contrast to traditional EPA assessment, which often assumes a threshold below which adverse effects are assumed to be negligent.[349]

[339]Davis et al. (2011), pp. 247–248.

[340]Li et al. (2010), p. 3444.

[341]Love et al. (2011a), p. 279; Davis et al. (2011), pp. 246–248.

[342]Graham et al. (2011), p. 418.

[343]Ji et al. (2010), p. 641.

[344]Graham and Nachman (2010), p. 653; Love et al. (2012).

[345]Ji et al. (2010).

[346]Interagency Task Force on Antimicrobial Resistance (2001), at 30 (discussing role of EPA in antibiotic and antibiotic pesticide registrations).

[347]Wright (2007).

[348]Love et al. (2011a), p. 280 (particularly figure 1).

[349]Nat'l Res. Council, Toward a Unified Approach to Dose-Response Assessment, in Science and Decisions: Advancing Risk Assessment 128 (2009).

Current scientific evidence is insufficient to quantify the role of environmental antimicrobial pollution in driving the epidemic of antimicrobial resistance, and it is equally insufficient to allow accurate prediction of the scope of regulation that might be needed to achieve a public health benefit. As a result, a first step toward consideration of how this reservoir might be regulated should involve expansion of surveillance programs and research funding, followed by testing potential regulatory efforts in carefully-chosen ecosystems through pilot intervention programs at the local or state level. In the meantime, educational efforts could target reduction of antimicrobial contamination that occurs at known sources, such as on CAFOs. For example, researchers recently have shown that poultry farms that transitioned from conventional to organic (no antimicrobial use) practices had significantly lower prevalence of resistance in *Enterococci* bacteria found in litter, feed, and water compared to conventional farms that used antimicrobials.[350] Strategies could include incentives to support organic practices and regulatory support of improved veterinary oversight of antimicrobial use in food-producing animals, particularly use in medicated feed and water.

10.4.4 Veterinary Oversight

Historically, in the United States and elsewhere, veterinary involvement in antimicrobial use on farms has been low, and this lack of oversight may contribute to inappropriate use of antimicrobials by producers.[351] Research through the National Animal Health Monitoring System (NAHMS) for dairy operations showed that producers consulted a veterinarian only 46 % of the time before choosing an antimicrobial, and they based their antimicrobial choice on culture and sensitivity results only 20 % of the time.[352]

The current U.S. Action Plan calls for increased veterinary oversight of antimicrobial uses in food-producing animals. As previously noted,[353] the AVMA has voiced concerns with the burden such oversight would place on the inadequate food-producing animal veterinary workforce.[354] This demonstrates the need to harmonize regulations and legislation addressing antimicrobial usage with support for the scientific expertise and occupational resources needed to accomplish the

[350]Amy R. Sapkota et al. Lower Prevalence of Antibiotic-Resistant Enterococci on U.S. Conventional Poultry Farms That Transitioned to Organic Practices, Envtl. Health Persps. (accessed ahead-of-print, 2011).

[351]Goforth & Goforth, at 70.

[352]Animal & Plant Health Inspection Serv., U.S. Dep't Agric., Dairy 2007 Part III: Reference of Dairy Cattle Health and Management Practices in the United States (2008), available at http://www.aphis.usda.gov/animal_health/nahms/dairy/downloads/dairy07/Dairy07_dr_PartIII_rev.pdf.

[353]See supra.

[354]Am. Veterinary Med. Ass'n., AVMA Responds, (2010).

goals of any federal directive.[355] Veterinary training systems, such as the National Veterinary Accreditation Program (NVAP) through USDA,[356] could be one venue through which national recommendations are harmonized. In return, food animal veterinarians could serve as consultants to large producers (CAFOs) to assist with programs to reduce antimicrobial usage and also to help these producers accurately report such usage to state and federal authorities.

Even within the Action Plan, the FDA makes concessions regarding disease diagnosis by the veterinarian providing oversight. Specifically, the FDA states:

> Numerous risk factors have been documented to increase susceptibility to bacterial disease, including environmental factors (such as temperature extremes and inadequate ventilation), host factors (such as age, nutrition, genetics, immune status), and other factors (such as stress of animal transport). From FDA's standpoint, the administration of a drug to animals when a veterinarian determines that there is a risk of a specific disease, based on the presence of such risk factors, could be considered judicious prevention use.[357]

While these provisions undoubtedly will prevent some animal morbidity and mortality and are consistent with veterinary clinical practices related to herd management, they do not require a diagnosis of current bacterial disease to be present prior to prescription, nor do they require that individual animals (rather than an entire cohort) undergo targeted treatment. Adoption of best practices among producers to reduce stress and improve environmental engineering controls to foster animal health is another route to reduce the need for antimicrobial interventions, one that could be further supported by policy efforts.

10.4.5 Surveillance for Antimicrobial Usage

Improving transparency of antimicrobial usage, particularly in the livestock and pharmaceutical industries, is critical for future regulatory and surveillance efforts on a global scale. Public provision of antimicrobial drug distribution, consumption, indication for usage (*i.e.* disease conditions by species), time of use, and location of

[355]At the time of writing, only one specific federal incentive existed to support entry of veterinarians into food animal practice, public practice, and research. The Veterinary Medicine Loan Repayment Act (VMLRP) is a small program to help provide partial repayment of educational loans, but only in specific, designated shortage areas that require nomination by state health officials. See U.S. Dep't Agric., Veterinary Medicine Loan Repayment Program, (2011) . The average veterinary student loan burden is $130,000, and the average starting salary is $65,000 and may be lower in rural areas. See Scott R. Nolen, JAVMA News: Student Loan Subsidy's End Raises Concerns, Sept. 15, 2011), available at http://www.avma.org/onlnews/javma/sep11/110915u.asp.

[356]U.S. Dep't Agric., Animal Health: National Veterinary Accreditation Program (2011), http://www.aphis.usda.gov/animal_health/vet_accreditation/.

[357]Supra Guidance #213, at 7.

use could enhance global, regional, and local surveillance efforts.[358] Even poultry industry veterinarians acknowledge the limitations this lack of data imposes on clinical, research, surveillance, and policy efforts.[359] Further, although USDA provides some public information on farm locations and farming practices in the United States, its database is incomplete.[360] This lack of information hinders regulatory, research and surveillance efforts by limiting the evidence base for public health conclusions. Open access to information could be used, not just as evidence to support antimicrobial restriction, but also as evidence to support a decision not to restrict certain individual antimicrobials or drug classes. European surveillance systems may offer models for expansion of data reporting in the United States and elsewhere in the world.[361]

10.5 Example: NRDC Lawsuit

The first uses of antimicrobials in animal feed to enhance or promote growth were approved by the FDA in 1951. Subsequent concern over the potential for these uses to drive antimicrobial resistance led the FDA to appoint a Task Force in 1970, which proposed withdrawal of the approvals of certain antimicrobial drugs for production purposes. This led to issuance of notices of opportunity for hearing for two antimicrobial drugs: penicillins and tetracyclines.[362] In 1977, these notices concluded, particularly for penicillin and tetracycline, that production uses in animal feed were not safe for the public.[363] Despite this assertion, the FDA did not move to act to withdraw or limit any approvals for these antimicrobials for indications for use in animal feed.

In 1999 and again in 2005, the plaintiffs—Center for Science in the Public Interest, the Food Animal Concerns Trust, Public Citizen, and the Union of Concerned Scientists—petitioned the FDA to withdraw approvals for use of penicillins and tetracyclines in animal feed. The FDA initially ignored, and later denied

---

[358]Meghan Davis & Tyler Smith. More Data, Better Data: How FDA Could Improve the Animal Drug User Fee Act, Center for a Livable Future Blog (Nov. 15, 2011), available at: http://www.livablefutureblog.com/2011/11/adufa-more-data-better-data (providing details of comments by the author (MFD) given during a public meeting at FDA in Rockville, Maryland on Nov. 7, 2011 regarding reauthorization of ADUFA).

[359]Cummings (2006), pp. 209–212.

[360]U.S. Dep't Agric., Census of Agriculture, (2007).

[361]See supra.

[362]Penicillin-Containing Premixes: Opportunity for Hearing, 42 Fed. Reg. 43772 (Aug. 30, 1977) ("Penicillin NOOH"); Tetracycline NOOH, 42 Fed. Reg. 56264 (Oct. 21, 1977).

[363]Penicillin Containing Premixes; Opportunity for Hearing, 42 Fed. Reg. 43,772, 43,772 (Aug. 30, 1977).

these petitions.[364] Therefore, in May 2011, the National Resources Defense Council (NRDC) filed a lawsuit against the FDA alleging that FDA's failure to act to withdraw or limit approvals for penicillin and tetracycline constituted an unlawful violation of the Administrative Procedure Act and the Food, Drug and Cosmetic Act.[365] In March and June 2012, respectively, judges from the U.S. District Court for the Southern District of New York sided with NRDC on this lawsuit and a supplemental complaint alleging that the FDA acted capriciously in denying the original petitions.[366] This decision required the FDA to proceed with hearings regarding the withdrawal of approval for use of penicillins and tetracycline antimicrobial drugs in animal feed.

The FDA appealed these decisions to the U.S. Second Circuit Court of Appeals, which ruled in July 2014 to reverse the prior decisions in favor of the FDA.[367] The majority opinion concluded that a hearing was not required by the Administrative Procedure Act and, therefore, the FDA was within its statutory authority to undertake—or not—withdrawal of drug approval.[368] Chief Judge Katzmann dissented, siding with the lower court's conclusion that "21 U.S.C. §360b(e)(1) requires the FDA to continue the proposed withdrawal proceedings."[369] As a result of this decision, the FDA continued with voluntary withdrawals of approvals of antimicrobials added to animal feeds for growth-promotion uses, but was not required to specifically initiate hearings for the withdrawal of approval of any use of penicillin or tetracycline antimicrobials in animal feeds.

10.6 Conclusion

In 2011, the World Health Organization dedicated its World Health Day to the global issue of antimicrobial resistance. Perhaps serendipitously, 2011 also marked World Veterinary Year.[370] Veterinarians are at the forefront of current regulatory efforts to address the problem of antimicrobial resistance—aligned with multiple stakeholders, among them agencies attempting to promulgate regulations (*i.e.* FDA's CVM), and agribusiness interests (*e.g.* AVMA and others) attempting to limit regulatory restrictions on use of antimicrobials in food-

[364] Docket No. FDA-1999-P-1286 and Docket No. FDA-2005-P-0007.

[365] NRDC et al. v. FDA et al. 11 Civ 3562 (RMB).

[366] NRDC et al. v. FDA et al. 11 Civ 3562 (THK), http://www.hpm.com/pdf/blog/NRDC%20-%20SDNY%20CP%20Decision.pdf.

[367] United States Court of Appeals for the Second Circuit, August Term, 2012. Docket Nos. 12-2106-cv(L), 12-3607-cv(CON), decided July 24, 2014. http://docs.nrdc.org/health/files/hea_14072401a.pdf.

[368] http://docs.nrdc.org/health/files/hea_14072401a.pdf.

[369] http://docs.nrdc.org/health/files/hea_14072401b.pdf, at 3.

[370] World Veterinary Year 2011 (2011), www.vet2011.org/.

producing animals. Many stakeholders call for a science-based approach to regulation.[371]

Understanding the science, specifically the ecology of antimicrobial resistance, underscores the need to better regulate non-therapeutic use of antimicrobials in food-producing animals.[372] Since movement of resistance genes can occur across national boundaries,[373] international strategies, and perhaps global regulatory authorities,[374] are needed to address the emergence and transmission of antimicrobial resistance. Within the United States, integration and harmonization of federal agency efforts, expansion of regulation of non-therapeutic antimicrobial use in food-producing animals, increased funding for research and surveillance of antimicrobial resistance, and mandates for public reporting of information critical to these programs will further domestic efforts to combat antimicrobial resistance. Failure of the current system to address growth promotion and similar non-therapeutic uses of antimicrobials in agriculture undermines federal and global efforts to control antimicrobial resistant infections in people, leading to a high economic cost and human burden of disease.[375] Although FDA currently has authority to regulate antimicrobial use in food animals, proposed legislation and existing regulatory efforts only partially address these public health concerns.[376]

Existing EU regulations and surveillance programs offer possible options for U.S. and global efforts to limit the non-therapeutic use of antimicrobials in livestock. Ultimately, efforts that consider the global ecosystem of resistance, including pathogenic and non-pathogenic bacteria and gene transfer among populations of bacteria,[377] are critical to U.S. and global strategies to curb the rise of antimicrobial resistance. On November 3, 2009, the White House released a joint US-EU declaration, which established:

> a transatlantic task force on urgent antimicrobial resistance issues focused on appropriate therapeutic use of antimicrobial drugs in the medical and veterinary communities, prevention of both healthcare- and community-associated drug-resistant infections, and strategies for improving the pipeline of new antimicrobial drugs, which could be better addressed by intensified cooperation between us.[378]

---

[371]Ctr. Veterinary Med., Draft Guidance #209, 37, at 4; Am. Veterinary Med. Ass'n, AVMA Responds, (2010).

[372]Possible exceptions could include antibiotics that have been tested for resistance and cross-resistance by multiple, independent researchers and proven not to be a threat to public health.

[373]O'Brien (1997), p. S2.

[374]A key conclusion of the Australian Society for Infectious Diseases/Australian Society for Antimicrobials' Antimicrobial Resistance Summit (Feb. 7-8, 2011) was the need for "a national interdisciplinary body . . . to manage the looming antimicrobial resistance crisis." See Gottlieb & Nimmo, (2011) at 281.

[375]Coast and Smith (2003), p. 242.

[376]U.S. Food & Drug Admin., Guidance for Industry #152, (2003); Ctr. Veterinary Med., Draft Guidance #209, 37; H.R. 965; S. 1211.

[377]Wright (2007), pp. 175–186.

[378]Press Release, Office Press Sec'y, White House, U.S.-EU Joint Declaration and Annexes (Nov. 3, 2009), available at http://www.whitehouse.gov/the-press-office/us-eu-joint-declaration-and-annexes.

Bacteria do not respect national boundaries.[379] Scientific evidence should inform both the national regulatory strategies and the domestic and international surveillance systems that are important, not just to monitor the problem, but also to evaluate the impacts of regulation. The regulatory process itself should be guided by evidence of success, but such evidence should not be required *a priori* for new regulatory effort, nor should incremental regulations be delayed.[380] Instead, policy-makers should focus on crafting regulation based on scientific evidence and providing for mechanisms of iterative evaluation of the public health impact of regulation.

Intervene we must. The human, societal, and economic costs of drug-resistant infections are high. Given the complexity of the issue, a single regulation—a single target—is unlikely to be broadly successful. Imposing restrictions on use of antimicrobials in food-producing animals for growth promotion is one of many targets, and one that is scientifically easier to justify than it is politically feasible. Long-term efforts grounded in scientific evidence are needed to harmonize use and restriction of use of antimicrobials internationally, and across multiple industries, particularly food animal production.

The World Health Organization (WHO) has classified antimicrobial drugs according to their relative importance to human health for purposes of stewardship and risk assessment to combat the growing global problem of antimicrobial resistance. Critically important antimicrobials are those antimicrobials recommend by WHO to receive top priority for risk management, surveillance, and other strategies for preservation of human use. They adhere to two criteria: (1) that the antimicrobial agent is the only or one of few available drugs to treat serious human disease, and (2) that the antimicrobial agent is used to treat infectious agents transmitted between humans and animals. Highly important antimicrobials are ranked below critically important and are so designated based on meeting one or the other of the above criteria. Important antimicrobials are used in human medicine but adhere to neither criteria; these are ranked as the lowest relative priority for stewardship initiatives. This table provides an overview of antimicrobials according to drug class, WHO classification, human drug examples and uses, and veterinary drug examples, species and uses. This overview illustrates that many classes of antimicrobial drugs deemed critically important by WHO for human use have extensive veterinary uses in both companion and food animal species, with indications ranging from growth promotion to prophylaxis or prevention to treatment.

[379]Monecke et al. (2011), p. e17936 (demonstrating international movement of clones of MRSA).

[380]Incremental regulations should not be held to the same standards of evaluation as more comprehensive, multi-agency regulatory efforts, since partial or limited restrictions may be equally limited in their ability to achieve the desired public health effect.

References

Aarestrup FM et al (2000) Associations between the use of antimicrobial agents for growth promotion and the occurrence of resistance among enterococcus faecium from broilers and pigs in Denmark, Finland, and Norway. Microb Drug Resist 6:63–68

Aarestrup FM et al (2008) Resistance in bacteria of the food chain: epidemiology and control strategies. Exp Rev Anti-Infect Ther 6:733–738

Agerso Y, Aarestrup FM (2013) Voluntary ban on cephalosporin use in Danish pig production has effectively reduced extended-spectrum cephalosporinase-producing Escherichia coli in slaughter pigs. J Antimicrob Chemother 68(3):569

Alexander KA et al (2008) Antimicrobial resistant salmonella in dairy cattle in the United States. Vet Res Commun 33:191

Andremont A (2000) Effect of antibiotics on bacterial resistance ecology: role of the digestive tract. Med Mal Infect 30:S178

Angulo FJ et al (2004) Antimicrobial resistance in zoonotic enteric pathogens. Sci Technol Rev 23:485, 487–490

Angulo FJ et al (2004b) Antimicrobial use in agriculture: controlling the transfer of antimicrobial resistance to humans. Semin Pediat Infect Dis 15:78

Baltz RH (2008) Renaissance in antibacterial discovery from actinomycetes. Curr Opin Pharmacol 8:557

Bonten MJ et al (2001) Vancomycin-resistant enterococci: why are they here, and where do they come from? Lancet Infect Dis 1:314

Boucher HW et al (2009) Bad bugs, no drugs: No Eskape! an update from the Infectious Diseases Society of America. Clin Infect Dis 48:1

Bramble M et al (2011) Potential role of pet animals in household transmission of methicillin-resistant Staphylococcus aureus: a narrative review. Vector Borne Zoonotic Dis 11:617

Castanon JIR (2007) History of the use of antibiotic as growth promoters in European poultry feeds. Poult Sci 86: 2466, 2469–2470

Cavaco LM et al (2011) Zinc resistance of Staphylococcus aureus of animal origin is strongly associated with methicillin resistance. Vet Microbiol 150:344

Chapin A et al (2004) Airborne multidrug-resistant bacteria isolated from a concentrated swine feeding operation. Environ Health Perspect 113:137

Chomel BB, Sun B (2011) Zoonoses in the Bedroom. Emerg Infect Dis 17:167–170

Coast J, Smith RD (2003) Antimicrobial resistance: cost and containment. Exp Rev Anti-Infect Ther 1:241–242

Cohen SP et al (1989) Cross-resistance to fluoroquinolones in multiple-antibiotic-resistant (Mar) Escherichia coli selected by tetracycline or chloramphenicol: decreased drug accumulation associated with membrane changes in addition to OmpF reduction. Antimicrob Agents Chemother 33:1318

Collignon P et al (2009) World Health Organization ranking of antimicrobials according to their importance in human medicine: a critical step for developing risk management strategies for the use of antimicrobials in food production animals. Clin Infect Dis 49:132

Cuny C et al (2010) Emergence of methicillin-resistant Staphylococcus aureus (MRSA) in different animal species. Int J Med Microbiol 300:109

Davis JL et al (2009) Update on drugs prohibited from extralabel use in food animals. J Am Vet Med Ass'n 235:528

Davis MF et al (2011) An ecological perspective on U.S. industrial poultry production: the role of anthropogenic ecosystems on the emergence of drug-resistant bacteria from agricultural environments. Curr Opin Microbiol 14:244–245

de Jong A et al (2009) A Pan-European survey of antimicrobial susceptibility towards human-use antimicrobial drugs among zoonotic and commensal enteric bacteria isolated from healthy food-producing animals. J Antimicrobial Chemother 63:733

Dethlefsen L et al (2007) An ecological and evolutionary perspective on human-microbe mutualism and disease. Nature 449:811–812
Diarra MS et al (2007) Impact of feed supplementation with antimicrobial agents on growth performance of broiler chickens, clostridim perfringens and enterococcus counts, and antibiotic resistance phenotypes and distribution of antimicrobial resistance determinants in Escherichia coli isolates. Appl Environ Microbiol 73:6566
Elmund GK et al (1971) Role of excreted chlortetracycline in modifying the decomposition process in feedlot waste. Bull Environ Contam Toxicol 6:129–131
Emborg H-D, Wegener HC (2005) The effect of banning antibiotics for growth promotion in poultry and swine production in Denmark. In: Miranowski JA, Scanes CG (eds) Perspectives in world food and agriculture, vol 2. Blackwell Publishing, Oxford, UK, pp 168–169
Falkow S, Kennedy D (2001) Antibiotics, animals, and people--again! Science 291:397
Feingold B et al (2010) A Niche for infectious disease in environmental health: rethinking the toxicological paradigm. Environ Health Perspect 118:1170
Fishman N (2006) Antimicrobial stewardship. Am J Med 119:S53–S61
Forrest RD (1982) Early history of wound treatment. J R Soc Med 75:198–200
Gerba CP, Smith JE (2005) Sources of pathogenic microorganisms and their fate during land application of wastes. J Environ Q 34:42
Gottlieb T, Nimmo GR (2011) Antibiotic resistance is an emerging threat to public health: an urgent call to action at the antimicrobial resistance summit 2011. Med J Aust 194:281
Graham JP, Nachman KE (2010) Managing waste from confined animal feeding operations in the United States: the need for sanitary reform. J Water Health 8:646–654
Graham JP et al (2007) Growth promoting antibiotics in food animal production: an economic analysis. Pub Health Rep 122:79
Graham DW et al (2011) Antibiotic resistance gene abundances associated with waste discharges to the Almendares River near Havana Cuba. Environ Sci Technol 45:418
Gupta A et al (2004) Antimicrobial resistance among campylobacter strains, United States, 1997–2001. Emerg Infect Dis 10:1102
Halling-Sørensen B et al (1988) Occurrence, fate and effects of pharmaceutical substances in the environment--a review. Chemosphere 36:357–359
Harper AL et al (2010) An overview of livestock-associated MRSA in agriculture. J Agromedicine 15:101
Hawkey PM (2008) The growing burden of antimicrobial resistance. J Antimicrobial Chemother 62:i1
Howden BP et al (2010) Reduced vancomycin susceptibility in Staphylococcus aureus, including vancomycin-intermediate and heterogeneous vancomycin-intermediate strains: resistance mechanisms, laboratory detection, and clinical implications. Clin Microbiol Rev 23:99, 107–109
IDSA (2012) Statement of the infectious diseases society of America presented at the Interagency Task Force on Antimicrobial Resistance (ITFAR) Meeting. Washington DC
Ito T et al (2003) Insights on antibiotic resistance of Staphylococcus aureus from its whole genome: genomic Island SCC. Drug Resist Updat 6:41–49
Ji K et al (2010) Influence of water and food consumption on inadvertent antibiotics intake among general population. Environ Res 110:641
Johannsson B et al (2011) Improving antimicrobial stewardship: the evolution of programmatic strategies and barriers. Infect Cont Hosp Epidemiol 32:367–372
John JF, Fishman NO (1997) Programmatic role of the infectious diseases physician in controlling antimicrobial costs in the hospital. Clin Infect Dis 24:471
Jones RN (1996) The emergent needs for basic research, education, and surveillance of antimicrobial resistance: problems facing the report from the american society for microbiology task force on antibiotic resistance. Diag Microbiol Infect Dis 25:153

Kadlec K, Schwarz S (2010) Novel ABC transporter gene, vga(C), located on a multiresistance plasmid from a porcine methicillin-resistant Staphylococcus aureus ST398 strain. Antimicrob Agents Chemother 54:3589

Kassenborg HD et al (2004) Fluoroquinolone-resistant campylobacter infections: eating poultry outside of the home and foreign travel are risk factors. Clin Infect Dis 38:S279

Kastner T, Rivas M, Koch W (2012) Global changes in diets and the consequences for land requirements for food. PNAS 109(18):6868–6872. available at doi:10.1073/pnas.1117054109/-/DCSupplemental

Keyes K et al (2003) Antibiotics: mode of action, mechanisms of resistance, and transfer. In: Torrence ME, Isaacson RE (eds) Microbial food safety in animal agriculture: current topics. Iowa State Press (Blackwell), Ames, Iowa, pp 45–46

Kiser JS (1976) A perspective on the use of antibiotics in animal feeds. J Anim Sci 42:1058–1059

Kiser JS et al (1971) Antibiotics as feedstuff additives: the risk-benefit equation for man. CRC Crit Rev Toxicol 1:55–56

Kola A et al (2010) Is there an association between nosocomial infection rates and bacterial cross transmissions? Crit Care Med 38:46

Laxminarayan R, Duse A, Wattal C, Zaidi AKM, Wertheim HFL, Sumpradit N et al (2013) Antibiotic resistance--the need for global solutions. Lancet Infect Dis 13(12):1057–1098. doi:10.1016/S1473-3099(13)70318-9

Le Hello S et al (2011) International spread of an epidemic population of Salmonella enterica serotype Kentucky ST198 resistant to ciprofloxacin. J Infect Dis 204:679

Lee HH et al (2010) Bacterial charity work leads to population-wide resistance. Nature 467:82

Levy SB, FitzGerald GB, Macone AB (1976) Spread of antibiotic-resistant plasmids from chicken to chicken and from chicken to man. Nature 260(5546):40–42

Li D et al (2010) Antibiotic resistance characteristics of environmental bacteria from an oxytetracycline production wastewater treatment plant and the receiving river. Appl Environ Microbiol 76:3444

Lloyd DH (2007) Reservoirs of antimicrobial resistance in pet animals. Clin Infect Dis 45:S148

Loeffler A, Lloyd DH (2010) Companion animals: a reservoir for methicillin-resistant Staphylococcus aureus in the community? Epidemiol Infect 138:595

Loeffler A et al (2010) Meticillin-resistant Staphylococcus aureus carriage in UK veterinary staff and owners of infected pets: new risk groups. J Hosp Infect 74:282

Lohren U et al (2008) Guidelines for antimicrobial use in poultry. In: Guardabassi L et al (eds) Guide to antimicrobial use in animals. Blackwell Publishing, Oxford, UK, pp 132–133

Loreau M (2010) Linking biodiversity and ecosystems: towards a unifying ecological theory. Phil Trans R S Lond Ser B Biol Sci 365:49–55

Love DC et al (2011a) Dose imprecision and resistance: free-choice medicated feeds in industrial food animal production in the United States. Environ Health Perspect 119:279

Love DC et al (2011b) Veterinary drug residues in seafood inspected by the European Union, United States, Canada, and Japan from 2000 to 2009. Environ Sci Technol 45:7232

M'ikanatha NM et al (2010) Multidrug-resistant salmonella isolates from retail chicken meat compared with human clinical isolates. Foodborne Pathog Dis 7:929

McCrea BA et al (2006a) A longitudinal study of salmonella and campylobacter jejuni isolates from day of hatch through processing by automated ribotyping. J Food Prot 69:2908

McCrea BA et al (2006b) Prevalence of Campylobacter and salmonella species on farm, after transport, and at processing in specialty market poultry. Poult Sci 85:136–143

McEwen SA, Fedorka-Cray PJ (2002) Antimicrobial use and resistance in animals. Clin Infect Dis 34:S99

McEwen SA et al (2010) Antibiotics and poultry - a comment. Can Vet J 51:561

McReynolds JL et al (2000) The effect of in ovo or day-of-hatch subcutaneous antibiotic administration on competitive exclusion culture (PREEMPT^{TM}) establishment in neonatal chickens. Poult Sci 79:1524–1525

Mitchell JM et al (1998) Antimicrobial drug residues in milk and meat: causes, concerns, prevalence, regulations, tests, and test performance. J Food Prot 61:742–743
Monecke S et al (2011) A field guide to pandemic, epidemic and sporadic clones of methicillin-resistant Staphylococcus aureus. PLoS One 6, e17936
Monnet DL (2000) Toward multinational antimicrobial resistance surveillance systems in Europe. Int J Antimicrob Agents 15:91
Mulders MN et al (2010) Prevalence of livestock-associated MRSA in broiler flocks and risk factors for slaughterhouse personnel in the Netherlands. Epidemiol Infect 138:743
Mullner P et al (2009) Assigning the source of human Campylobacteriosis in New Zealand: a comparative genetic and epidemiological approach. Infect Genet Evol 9:1311
Muto CA et al (2003) SHEA guideline for preventing nosocomial transmission of multidrug-resistant strains of Staphylococcus aureus and enterococcus. Infect Cont Hosp Epidemiol 24:362
Nannapaneni R et al (2009) Ciprofloxacin-resistant campylobacter persists in raw retail chicken after the fluoroquinolone ban. Food Addit Contam 26:1348
Nielsen P, Gyrd-Hansen N (1997) Bioavailability of enrofloxacin after oral administration to fed and fasted pigs. Pharmacol Toxicol 80:246
O'Brien TF (1997) The global epidemic nature of antimicrobial resistance and the need to monitor and manage it locally. Clin Infect Dis 24:S2
Prescott J (2006) History of antimicrobial usage in agriculture: an overview. In: Aarestrup FM (ed) Antimicrobial resistance in bacteria of animal origin. ASM Press, Washington, DC, p 22
Price LB et al (2007) The persistence of fluoroquinolone-resistant campylobacter in poultry production. Environ Health Perspect 115:1035
Radostits OM et al (1985) General principles. In: Radostits OM et al (eds) Herd health: food animal production medicine, 2nd edn. W.B. Saunders, Philadelphia, p 16
Randall LP et al (2006) Modification of enrofloxacin treatment regimens for poultry experimentally infected with salmonella enterica serovar typhimurium DT104 to minimize selection of resistance. Antimicrob Agents Chemother 50:4030
Rusoff L (1951) Antibiotic feed supplement (aureomycin) for dairy calves. J Dairy Sci 34:652–655
Samuel MC et al (2004) Epidemiology of sporadic campylobacter infection in the United States and declining trend in incidence, FoodNet 1996–1999. Clin Infect Dis 38:S165
Scott Weese J (2006) Prudent use of antimicrobials. In: Giguère S et al (eds) Antimicrobial therapy in veterinary medicine, 4th edn. Iowa State Press (Blackwell), Ames Iowa, p 445
Sengeløv G et al (2003) Bacterial Antibiotic Resistance Levels in Danish Farmland as a Result of Treatment with Pig Manure Slurry. Environ Int 28:587, 590–592
Shen Z et al (2011) Salicylate functions as an efflux pump inducer and promotes the emergence of fluoroquinolone-resistant mutants in Campylobacter jejuni. Appl Environ Microbiol 77 (20):7128–7133
Sherris JC, Florey ME (1951) Relation of penicillin to sensitivity in staphylococci to clinical manifestations of infection. Lancet 1:309
Silbergeld EK et al (2008a) Industrial food animal production, antimicrobial resistance, and human health. Annu Rev Public Health 29:151
Silbergeld EK et al (2008b) One reservoir: redefining the community origins of antimicrobial-resistant infections. Med Clin N Am 92:1391–1392
Sipahi OR (2008) Economics of antibiotic resistance. Expert Rev Anti-Infect Ther 6:523–526
Skippington E, Ragan MA (2011) Lateral genetic transfer and the construction of genetic exchange communities. FEMS Microbiol Rev 35:707
Smith KE et al (1999) Quinolone-resistant campylobacter jejuni infections in Minnesota, 1992–1998. New Eng J Med 340:1525
Smith TC et al (2009) Methicillin-Resistant Staphylococcus aureus (MRSA) Strain ST398 is Present in Midwestern U.S. Swine and Swine Workers. PloS One 4:e4258

Sorensen AC, Lawrence RS, Davis MF (2014) Interplay between policy and science regarding low-dose antimicrobial use in livestock. Front Microbiol 2. epub (doi:10.3389/fmicb.2014.00086/full)

Soulsby L (2007) Antimicrobials and animal health: a fascinating nexus. J Antimicrobial Chemother 60:i77

Stokstad ELR, Jukes TH (1950) Further observations on the "animal protein factor.". Exp Biol Med 73(3):523–528

Teillant A, Laxminarayan R (2015) Economics of antibiotic use in US swine and poultry production. Choices 30(1), epub

Tollefson L (2004) Developing new regulatory approaches to antimicrobial safety. J Vet Med Ser B 51:415

Tollefson L et al (1997) Therapeutic antibiotics in animal feeds and antibiotic resistance. Rev Sci Tech (Int Office Epizootics) 16:709–710

Tollefson L et al (2006) Regulation of antibiotic use in animals. In: Giguère S et al (eds) Antimicrobial therapy in veterinary medicine, 4th edn. Iowa State Press (Blackwell), Ames, Iowa, p 421

Toze S et al (2010) Use of static quantitative microbial risk assessment to determine pathogen risks in an unconfined carbonate aquifer used for managed aquifer recharge. Water Res 44:1038

Turnbaugh PJ et al (2007) The human microbiome project. Nature 449:804

Van Boeckel TP, Gandra S, Ashok A, Caudron Q, Grenfell BT, Levin SA, Laxminarayan R (2014) Global antibiotic consumption 2000 to 2010: an analysis of national pharmaceutical sales data. Lancet Infect Dis 14(8):742–750. doi:10.1016/S1473-3099(14)70780-7

van Boven M et al (2003) Rapid selection of quinolone resistance in campylobacter jejuni but not in Escherichia coli in individually housed broilers. J Antimicrob Chemother 52:719

van Cleef B et al (2010) High prevalence of Nasal MRSA carriage in slaughterhouse workers in contact with live pigs in the Netherlands. Epidemiol Infect 138:756

van Duijkeren E et al (2010) Methicillin-resistant Staphylococcus aureus in horses and horse personnel: an investigation of several outbreaks. Vet Microbiol 141:96

van Houweling CD, Gainer JH (1978) Public health concerns relative to the use of subtherapeutic levels of antibiotics in animal feeds. J Anim Sci 46:1413

Wegener HC et al (1999) Use of antimicrobial growth promoters in food animals and enterococcus faecium resistance to therapeutic antimicrobial drugs in Europe. Emerg Infect Dis 5:329–331

Welte T, Pletz MW (2010) Antimicrobial treatment of nosocomial meticillin-resistant Staphylococcus aureus (MRSA) pneumonia: current and future options. Int J Antimicrob Agents 36:391–393

Werner G et al (1998) Association between quinupristin/dalfopristin resistance in glycopeptide-resistant enterococcus faecium and the use of additives in animal feed. Euro J Clin Microbiol Infect Dis 17:401

Werner G et al (2002) Molecular analysis of streptogramin resistance in enterococci. Int J Med Microbiol 292:81

Witte W (2000) Selective pressure by antibiotic use in livestock. Int J Antimicrob Agents 16:S19

Wright GD (2007) The antibiotic resistome: the nexus of chemical and genetic diversity. Nat Rev Microbiol 5:175, 183–184

Zhao S et al (2010) Antimicrobial resistance of campylobacter isolates from retail meat in the United States between 2002 and 2007. Appl Environ Microbiol 76:7949

Gilchrist MJ et al (2007) The potential role of concentrated animal feeding operations in infectious disease epidemics and antibiotic resistance. Environ Health Perspect 115:313–314

Aarestrup FM et al (2001) Effect of abolishment of the use of antimicrobial agents for growth promotion on occurrence of antimicrobial resistance in fecal enterococci from food animals in Denmark. Antimicrob Agents Chemother 45:2054

Dellit TH et al (2007) Infectious Diseases Society of America and the society for healthcare epidemiology of America guidelines for developing an institutional program to enhance antimicrobial stewardship. Clin Infect Dis 44:159

Chapter 11
Food Production and Animal Welfare Legislation in Australia: Failing Both Animals and the Environment

Alex Bruce and Thomas Faunce

Abstract In this chapter, we explore how animals in Australia are raised and processed domestically for food and exported internationally. We trace the rise of corporate domination of farm animal production in Australia as a response to increasing domestic and international demand for meat products and describe the systematic exclusion of State and Territory *Animal Welfare Acts* to farm animals intended to be processed for food. In doing so, we illustrate the complexity of Australia's regulatory framework governing Australian farm animals by taking the poultry industry (chicken meat and eggs) as our case study. We then explore Australia's highly controversial live animal export industry; its highly visible failures and more recent attempts by the Australian government to introduce traceability and accountability into the live export supply chain. We then discuss the major environmental externalities associated with intensive farm animal operations; identifying the energy inefficiencies associated with raising and processing animals for human consumption. We note the contribution of intensive animal farming to atmospheric CO_2 emissions as well as the water degradation caused by waste matter runoff. In response to these environmental externalities, we propose the use of artificial photosynthetic technology as a means of transforming the farm animal industry from one of net energy and resource taker to one approaching energy and waste neutrality. We conclude by examining four major problems with the troubled relationship between farm animals and the Australian domestic food and live export industries. The chapter is purposefully written for the well informed reader, interested in farm animal welfare in the different countries of the world, but who is not necessarily informed about the industry and regulatory framework in Australia. Accordingly, the chapter is written in a way that is not overly-technical, but nevertheless leads the reader through the sometimes complex and contradictory nature of farm animal regulation in Australia.

A. Bruce (✉) • T. Faunce
Australian National University, College of Law, Canberra, ACT, Australia
e-mail: alex.bruce@anu.edu.au; thomas.faunce@anu.edu.au

G. Steier, K.K. Patel (eds.), *International Farm Animal, Wildlife and Food Safety Law*, DOI 10.1007/978-3-319-18002-1_11

11.1 Introduction

11.1.1 Animal Welfare Protection in Australia

In this chapter, we explore how animals in Australia are raised and processed domestically for food and exported internationally. We trace the rise of corporate domination of farm animal production in Australia as a response to increasing domestic and international demand for meat products and describe the systematic exclusion of State and Territory *Animal Welfare Acts* to farm animals intended to be processed for food.

In doing so, we illustrate the complexity of Australia's regulatory framework governing Australian farm animals by taking the poultry industry (chicken meat and eggs) as our case study. We then explore Australia's highly controversial live animal export industry; its highly visible failures and more recent attempts by the Australian government to introduce traceability and accountability into the live export supply chain.

We then discuss the major environmental externalities associated with intensive farm animal operations; identifying the energy inefficiencies associated with raising and processing animals for human consumption. We note the contribution of intensive animal farming to atmospheric CO_2 emissions as well as the water degradation caused by waste matter runoff.

In response to these environmental externalities, we propose the use of artificial photosynthetic technology as a means of transforming the farm animal industry from one of net energy and resource taker to one approaching energy and waste neutrality.

We conclude by examining four major problems with the troubled relationship between farm animals and the Australian domestic food and live export industries.

The chapter is purposefully written for the well informed reader, interested in farm animal welfare in the different countries of the world, but who is not necessarily informed about the industry and regulatory framework in Australia. Accordingly, the chapter is written in a way that is not overly-technical, but nevertheless leads the reader through the sometimes complex and contradictory nature of farm animal regulation in Australia.

11.1.2 Commonwealth (Federal) and State Laws

Our discussion in this chapter proceeds in the following manner. Following this introduction, Sect. 11.2 discusses the socio-historical transition from small, family-operated farming concerns to large, corporate-owned agricultural enterprises. It explores the reasons for this shift and identifies several consequences, particularly the physical and emotional separation of farmers and the animals they tended. It traces the gradual domination of the Australian agricultural industry by corporations whose concerns are for profit maximisation and efficiency rather than farm animal welfare.

In Sect. 11.3, we explain how and why the regulation of farm animals in Australia involves a complex relationship between Commonwealth (Federal) and State laws. Lack of express Commonwealth Constitutional power has resulted in Australian States and Territories assuming principal responsibility for the welfare of animals. However, we also illustrate the process by which farm animals are systematically excluded from State and Territory *Animal Welfare* laws.

Our case study for these processes and difficulties is introduced in Sect. 11.4 in which we explore the Australian poultry industry. The complex "layering" of *Statutes*, *Codes of Practice*, *Standards and Guidelines*, *Regulations* and *Subordinate Instruments*, reveals that the welfare of both chickens and laying hens is subordinated to economically efficient animal husbandry practices and the profit motive.

In Sect. 11.5 we explain how these same dynamics operate in Australia's live export industry. Although it is one of the world's largest exporters of live animals, Australia's export industry has been mired in controversy. Highly visible media campaigns in recent years have exposed systematic animal welfare abuses both domestically and in Australia's export markets. We provide an overview of the Australian government's response to these welfare abuses; the *Exporter Supply Chain Assurance Scheme*.

The transition from small family operated farms to large intensive animal farming enterprises has also resulted in significant environmental externalities. In Sect. 11.6 we discuss these externalities; the energy inefficiencies associated with maintaining large farm animal industries and the implications for atmospheric and water table degradation. As a response, we argue that artificial photosynthetic technology has a significant rôle to play in mitigating these environmental externalities.

We conclude in Sect. 11.7 by evaluating Australia's farm animal industry and identify four significant criticisms that collectively ensure that the industry continues to subordinate the welfare of animals to the efficient and profitable maintenance of Australia's primary industries.

11.2 Transition from Animal Husbandry to Corporate Agricultural Enterprises

11.2.1 The Evolution of Intensive Animal Farming

"In terms of impact on the planet, animal agriculture is second only to nuclear war"[1]

[1]Alex Hershaft, *Environmental Action Magazine* 1990.

Before the industrial revolution, most agricultural enterprises were family-operated concerns. Small tenet farmers or private land owners cultivated relatively small plots of land, tended small herds of animals and then sold fresh produce at weekly town markets. The mechanisation of industry initiated by the industrial revolution was focussed on producing automobiles, weapons and ships. The British "Agricultural Revolution" was not principally initiated by mechanisation so much as new methods of land use such as crop rotation, land enclosure and animal husbandry techniques such as selective breeding.[2]

In these circumstances, there was a closer physical and emotional relationship between people and the animals they kept. It was in the interests of farmers to ensure optimal welfare for their animals since the economic security of their families often depended on the sale of meat, milk and eggs at regional markets. Likewise, smaller family-farms were unlikely to use environmentally damaging production methods. Since farmers and their families often worked directly with their animals and were in physical contact with their crops, they tended to avoid the use of chemicals and pesticides. Animal wastes were used to fertilize land deleted of the nutrients need to grow crops used to feed *both* the animals and the famer's family.[3] Family farmers were therefore "motivated to raise their crops and animals in the most environmentally sounds and healthy way".[4]

However, in the wake of the economic recovery following the Second World War, rural agriculture transformed from family owned farms to the dominance of corporate-owned highly vertically integrated concentrated animal feeding operations ("CAFOs").[5] Prior to the Second World War, most consumers in Western societies did not consumer large amounts of meat products.

However, post-war prosperity, generated by massive employment for returning service men and the growth of the consumer economy resulted in a significant increase in consumer demand for red meat and chicken meat products.[6] Australian consumers' increasing demand for meat products in the era mirrored similar demand in most Western countries.[7]

In the decades following the Second World War, consumers' disposable incomes rose and used their increased purchasing power to diversify their diets to include more meat-based products. In order to meet this demand, food animal industries, informed by neo-classical economics transformed the way in which the success or failure of agricultural enterprises was measured.

---

[2]Allen (1992), p. 209. For a more animal-centric analysis, see Thomas (2005), p. 71.

[3]Cheever (2000), p. 43.

[4]Wender (2011), p. 141.

[5]Follmer (2009), pp. 45, 51–52.

[6]McAllister et al. (2006), p. 41.

[7]Tim O'Brien, *Compassion in World Farming Trust, Factory Farming: The Global Threat*, 1998, Compassion in World Farming.

Now, success is measured in terms of efficiency and profits per unit as corporations embraced techniques of mass-production of food animal products.[8] Wealth-maximisation, that holy grail of neo-classical economics, was the key to successful animal farming practices.[9] Pursuing economically efficient farming/processing practices necessitated characterising animals as units of production; capital to be exploited as efficiently as possible in the process of being turned into meat products.[10] Farmers then began to explore more efficient methods of manipulating animal "products" in order to increase profits. These methods included artificial feed-cycles to ensure year-round meat production and the increasing use of antibiotics to stimulate growth and ward off infection.[11]

One of the many significant consequences of the transformation of the agricultural industry from small family-oriented enterprises to corporate controlled agri-businesses was the physical and emotional alienation of people from animals and the environment.[12] When corporations manage large CAFOs and when processing animals becomes industrialised, the connection that people have enjoyed with animals for millennia is broken.

What it means to kill an animal for food, or to live on a farm sprayed with chemical pesticides becomes theoretical when animal husbandry and agricultural production becomes industrialised. In these circumstances, the term "animal husbandry" is certainly more ironic than referential.[13]

Like other Western industrialised nations, the production of animals and agriculture in Australia is heavily corporatised and mechanised. The reality is that most of the animals in Australia that are slaughtered for their meat or farmed for their eggs do not see the sun or feel the earth, they do not socialise with other animals, they do not able to express their natural instincts but are confined in mass-factories before being slaughtered or their eggs harvested. This process of factory farming is described as[14]:

> a system of raising animals using intensive production line methods that maximise the amount of meat produced while minimising costs. Industrial animal agriculture is characterised by high stocking densities and/or close confinement, forced growth rates, high mechanisation and low labour requirements.

[8]Ibrahim (2007), p. 86.

[9]Evans (2006), p. 167.

[10]Gunderson (2011), p. 1, 3–4.

[11]Mason and Singer (1990).

[12]Casuto (2012), pp. 73–75.

[13]Cassuto (2007), p. 59.

[14]*From Nest to Nugget: An Expose of Australia's Chicken Factories*, Voiceless, November 2008 at 9.

11.2.2 Corporate Domination of Farm Animal Production in Australia

Because of its relatively small population, Australia does not possess the economies of scale necessary for all industries to be competitive.[15] Large industries in Australia are often characterised by natural monopolies, by duopolies or oligopolies. The food animal industry is no exception as most of the animal meat produced in Australia for both domestic consumption and export is processed by a few dominant corporations.

The Australian chicken meat industry is a virtual duopoly. According to the Australian Chicken Meat Federation ('ACMF'): 'the two largest (companies) Baiada Poultry and Inghams Enterprises, supply more than 80 % of Australia's chicken meat'.[16] The beef industry is dominated by four producers. Swift Australia, Cargill Australia, Teys Brothers and Nippon Meats supply almost 50 % of meat products in Australia.[17]

And in 2011, the Australian Competition and Consumer Commission ('the ACCC') cleared a proposed acquisition of Teys Brothers by Cargill Beef Australia; an acquisition that permitted the merger of Australia's second and fourth largest beef processors leading to a further concentration of corporate production of animal food products.[18]

11.2.3 The Scale of Australian Domestic Farm Animal Production and Consumption

Like other Western nations, the industrialisation of the food animal industry in Australia is a response to demand. With a population of nearly 24 million people[19] most of whom are located in cities, Australia simply would not be able to produce sufficient food if the agricultural industry were not industrialised.

Population growth therefore inevitably requires larger or more efficient means of production, or sometimes both. When population growth is also underpinned by higher living standards, consumer demand for previously unavailable products

[15]Alex Bruce, *Australian Competition Law*, 2013, LexisNexis Butterworths, Sydney Australia at 7 ff.

[16]Australian Chicken Meat Federation Inc, *The Australian Chicken Meat Industry: An Industry in Profile*, 2012 at 13. <http://www.chicken.org.au/industryprofile/> Accessed on 29 March 2012.

[17]*Top 25 Red Meat Processors*, 'Feedback', Meat & Livestock Industry Journal Supplement, October 2005.

[18]*ACCC will not Oppose Teys Bros and Cargill Beef Australia Proposed Merger*, ACCC Media Release dated 6 July 2011.

[19]http://www.abs.gov.au/ausstats/abs@.nsf/Web+Pages/Population+Clock?opendocument.

increases. This is evidenced in Australia by the significant increase in consumer demand for red meat and chicken meat products since the Second World War.[20]

What this means is that Australians consume a significant amount of meat and meat products each year. This demand is reflected in the gradual increase in Australian meat production for both domestic and export markets over the last 10 years. The Australian Bureau of Statistics' livestock products report for the September 2011 quarter indicated that total red meat production in Australia increased by one per cent to 751,000 tonnes compared with the previous quarter while chicken meat production for the quarter amounted to 251,000 tonnes.[21]

To give these figures some perspectives, and in relation to the 2010 figures, Meat and Livestock Australia[22] reports:

> Over the 12 months to September 2010, fresh meat purchases increased 3 % to about 133 million serves/week. Contributing to the trend was a rise in beef (by 4 %), lamb (up 2 %) and chicken purchases (up 6 %) to 52 million serves/week, 22 million serves/week and 38 million serves/week, respectively.[23]

The number of animals that are raised each year in Australia solely for the purpose of slaughter for human consumption is truly staggering. Yet most Australians are unaware of the way in which this process occurs. Australian animal advocacy group "Voiceless" states that:

> More than 5 million pigs, 13 million hens and 420 million meat or 'broiler' chickens are raised for food production in Australia every year. Most of these animals spend their lives crammed together in giant factory farms.[24]

By 2050, the United Nations Population Division predicts that the world's population will reach somewhere between 8 and 11 billion people.[25] Much of this population growth is expected in developing countries where a growing middle class, with more disposable income generating increasing demand for meat products as part of their diet.[26] This is particularly so in China and India where demand for meat products is quickly growing.[27]

---

[20]McAllister et al. (2006), p. 41.

[21]Australian Bureau of Statistics, *Livestock Products, Australia*, December 2011 (cat no 7215.0), p 4, on <www.abs.gov.au> at *Statistics* (cited 27 March 2012).

[22]Meat and Livestock Australia (MLA) is a corporation whose members are Australian cattle producers. MLA is the corporate entity that acts as the cattle farmer's advocate in the development of Commonwealth primary industry policies. It also provides marketing and research on behalf of its member cattle farmers.

[23]Meat and Livestock Australia, *Australian Fresh Meat Consumption Increases*, 3 December 2010, on <www.mla.com.au> at *Prices & Markets*, then *Market News* at *Dec 2010* (cited 21 August 2011).

[24]Voiceless, *Lifting the Veil of Secrecy: The Animal Behind your Food*, May 2007.

[25]United Nations Department of Economic and Social Affairs; *World Population Prospects - The 2010 Revision* <http://www.un.org/popin/> (accessed 29 March 2012).

[26]Thornton (2010), pp. 2854–2855.

[27]Hocquette and Chatellier (2011), p. 20.

In order to meet this expected demand for food generally and meat products particularly, the United Nations Food and Agricultural Organisation estimates that agricultural output will need to increase by 70 % but must do so in circumstances of a world-wide decline in agricultural land because of climate change, dwindling fossil fuel supplies and the general movement of people off the land and into cities, urban and sub-urban areas..[28]

Most of the suggestions for meeting these challenges involve increasing the efficiency of CFAO's through more efficient breeding and production techniques rather than advocating plant-based diets or even artificially grown meat products.[29] In these circumstances, the challenge for most Western countries will be to increase the efficiency of existing CFAO's in order to produce sufficient meat products for domestic consumption and the emerging foreign demand for meat products.

For example, the National Farmers Federation ("the NFF"), the peak industry representative body for farmers in Australia has specifically noted the strategic advantages available to Australian meat and grain producers in satisfying future demand from developing countries. In its *NFF Farm Facts: 2012* Report, the NFF observes[30]:

> The prospects for agriculture are huge, with the need to feed, clothe and house a booming world population. Expanding Asian societies need food and fibre like never before and, due to their growing affluence, are demanding produce of the highest quality. The challenge for Australian agriculture and our famers will be in meeting this booming need for food and fibre through increasing production. Agriculture has an enormous uptake of new technology.

Western industrialised countries generally, and Australia particularly are therefore proposing to meet the expected increase in world demand for meat products by increasing the efficiency and productivity of agricultural practices generally and CAFOs specifically.[31]

In the process, attention is being drawn to the suffering that food animals will inevitably experience as a result of the growth in corporate exploitation of food animals.[32] This is an issue that requires investigating whether and to what extent Australia's *Animal Welfare* legislation protects farm animals.

---

[28]Hume et al. (2011), p. 1 at 2.

[29]Galyean et al. (2011), pp. 29–32.

[30]National Farmers Federation, *NFF Farm Facts: 2012* at 3. < http://www.nff.org.au/farm-facts.html> accessed on 29 March 2012.

[31]Vinnari and Tapio (2009), p. 269.

[32]Winders and Nibert (2004), p. 76.

11.3 The Role and Function of Australian Animal Welfare Legislation

11.3.1 Constitutional Limitations on a Comprehensive Animal Welfare Regime

Animal law in Australia is a complex and multi-faceted area.[33]

There is a bewildering amount of regulation across the nine different Australia jurisdictions concerning farm animals generally and the production of beef, pork and chicken meat specifically. Despite its lack of Constitutional power the Commonwealth through the *Primary Industries Standing Committee* ("the PISC") has created *Model Codes of Practice* ("MCOPs") issued under the auspices of the *Primary Industries Ministerial Council* that relate to the management of the chicken, pork and beef industries.

Each State and Territory has incorporated the content of these MCOPs to a greater or lesser extent in jurisdiction-specific regulation in the form of their own *Codes* and legislation. However, the difficulty is that there is no consistency in the extent to which these MCOPs are incorporated into the legal framework of the States and Territories. There is no consistency in the legal effect of these MCOPs even if they are adopted. And there is no consistency in the coverage of these MCOPs.

Before exploring the regulation of chicken, pork and beef meat production, it is necessary to understand the legal and regulatory complexity of the Australian animal law framework. The principal reason for this complexity is the lack of a national regulatory regime that applies consistently throughout Australia. In turn, the lack of a national regulatory regime is the result of the absence in the Commonwealth Constitution of an express power permitting the Commonwealth government to nationally regulate animals and economic activity involving animals.

11.3.2 Lack of Constitutional Power

Before Federation in 1901, various colonies in Australia had already enacted different forms of animal welfare legislation. Beginning with Van Dieman's Land (Tasmania) and then New South Wales these forms of animal welfare legislation were implemented throughout the late 1800s.[34]

These early forms of colonial legislation, principally in the *Police Acts*, were broadly based on English animal welfare legislation that was intended to prohibit cruelty to animals. Throughout the 1850s and 1860s, colonial animal welfare

[33]Plowman et al. (2008), pp. 25–29.

[34]Jamieson (1991), p. 238.

legislation was amended so that it not only became more focussed but also carved out exemptions for a number of practices including the extermination of rabbits, foxes and wild dogs as well as hunting, trapping or shooting any wild animal.[35]

Upon Federation, Australia's Constitution came into effect. However, during the Constitutional Conventions that preceded the drafting of the Constitution, there was no direct discussion of a power with respect to animals but indirect discussion within the context of freedom of trade between States.[36]

Consequently, the Constitution does not directly address the issue of animal welfare with the result that the federal government does not possess Constitutional power to legislate for the provision of animal welfare.

Despite this lack of express power, the Constitution does provide the Federal Government with several indirect powers to regulate animals. The quarantine power in 51(ix); fisheries power in s 51(x), the Trade & Commerce power in s 51 (i) and the external affairs power in s 51(xx) of the Constitution provide the Federal Government with the capacity to indirectly regulate animals.[37]

Accordingly, the Federal Government indirectly regulates animals in international trade, treaties that involve animals (such as the *Convention on International Trade in Endangered Species*), the export of animals, biosecurity, customs and imports and management of feral animals or other invasive species.

While there is no express power in the Constitution permitting the direct regulation of animals, it should be noted that Kirby J in *ABC v Lenah Game Meats Pty Ltd*[38] held that free discussion of governmental and political issues of animal welfare is protected by the implied freedom of political communication in the Constitution.

The law regulating animals is therefore found mostly in State and Territory legislation. Although the common law classifies animals as property[39] in reality, the regulation of animals and animal welfare involves a complex network of Commonwealth, State and Territory legislation, Codes of Practice, Regulations and Subordinate legislation. At the local government level, regulations exist concerning the registration of domestic pets, animal control and other issues.

This lack of express Constitutional power carries the following consequences for any systematic exploration of how animals generally and farm animals particularly, are regulated and treated in Australia:

First, principal regulatory authority for animals and animal welfare rests with the States and Territories. However, the eight different States and Territories throughout Australia have separate and often inconsistent regimes regulating animals and animal welfare.

---

[35]Jamieson (1991), p. 238.

[36]*Official Record of the Debates of the Australia Federal Constitution*, Sydney, 22 September 1897 at 1059–1064.

[37]Cao (2010), p. 100.

[38]*ABC v Lenah Game Meats Pty Ltd* (2001) 208 CLR 199 at.

[39]*Saltoon v Lake* [1978] 1 NSWLR 52.

Second, in the States, there is an added layer of government in the form of local councils. Pursuant to State *Local Government Acts*, local councils have been given responsibility to manage animals in their jurisdiction.

Third, despite a lack of express Constitutional power, the Commonwealth government has attempted to provide both legal and policy leadership on issues of animal management and animal welfare. However the various Strategies, Model Codes of Practice and Animal Welfare Standards that it has created do not have the status of law.

Fourth, as a result, these Commonwealth initiatives have been implemented to a greater or lesser extent by the States and Territories. In some States, these initiatives have been incorporated into animal-specific legislation. In others they have not. In some States, compliance with a Commonwealth Code is mandatory, in others it is not. The result is an inconsistent and often patch-work regime of animal management and regulation.

Fifth, even where all States and Territories have similar legislation concerning a particular issue of animal regulation (such as animal welfare or pest regulation), there are often many differences amongst the legislation making it difficult to identify a consistent regulatory strategy.

Sixth, this means that in almost all cases, in order to understand the way that, for example, animals as pests, are regulated it will be necessary to excavate many different layers of Acts, Regulations, Codes, Standards and Subordinate legislation in each individual State and Territory.

Finally, the many different Acts, Regulations, Codes, Strategies and Subordinate Legislation have necessitated an equally difficult and complex bureaucratic structure across the Commonwealth, States and Territories. Many different and related government departments are responsible for the design, implementation and review of animal regulations. Identifying them all can be extremely difficult.

11.3.3 At the Commonwealth Level

The Commonwealth *Department of Agriculture, Fisheries and Forestry* ("DAFF") is the principal regulatory agency at the Commonwealth level with portfolio responsibility for animals.[40] DAFF is responsible for developing a general overarching strategy toward the regulation of animals in Australia. This strategy is called the *Australian Animal Welfare Strategy* ("the AAWS") and forms the basis for State and Territory approaches to animal welfare in Australia.

The AAWS was initially endorsed by the *Primary Industries Ministerial Council* in May 2004. A revised edition titled *Australian Animal Welfare Strategy and National Implementation Plan 2010–2014* was issued in August 2011.[41]

[40]http://www.daff.gov.au/.

[41]http://www.daff.gov.au/__data/assets/pdf_file/0004/1986223/cic-102054-aaws.pdf.

Advising the Minister for DAFF is the *National Consultative Committee on Animal Welfare* ("NCCAW").[42] The NCCAW's task is to develop general guidelines for animal welfare. Once developed, the Minister for DAFF 'formally reports' them to state and territory ministers responsible for animal welfare for 'their consideration and appropriate action'.

In 2005, the Federal Government gave 4 years of funding to enable the AAWS to be implemented. To do that the *Australian Animal Welfare Strategy Advisory Committee* ("the AAWSAC") was established. Its task is to oversee the gradual implementation of the AAWS. The process of implementing the AAWS is guided by the *National Implementation Plan of the Australian Animal Welfare Strategy* ("National Implementation Plan") which was endorsed by SCOPI, in April 2006.

The overall coordination of the National Implementation Plan is under the regulatory oversight of the *Primary Industries Standing* Committee (PISC), which in turn, reports back to the SCOPI.

11.3.4 Farm Animals are Exempt from Animal Cruelty Prohibitions

At common law animals are classified as property: *Saltoon v Lake*[43] and at least in theory may be treated as chattels by their owners. Despite their status as property, animals are provided with a prima facie measure of protection against cruelty by State and Territory legislation.[44] For the purposes of discussion this article will refer to the provisions of the New South Wales *Prevention of Cruelty to Animals Act 1979* (NSW) ("the POCTA Act").

The difficulty is that while these *Animal Welfare Acts* do prohibit acts of cruelty, they also exempt conduct that is permitted under a MCOP such as the Commonwealth *Poultry Code.*

For example, s. 13 of the *Animal Care and Protection Act 2001* (Qld) provides that the *Regulation* may make a Code of Practice about animal welfare. Part 2 of Schedule 1 of the *Animal Care and Protection Regulation 2002* (Qld) designates most existing Commonwealth MCOPs as "Voluntary Codes of Practice" including the MCOP relating to *Domestic Poultry*.

However, this MCOP is *voluntary* and its standards are not compulsory for farmers producing chickens to be processed for meat. Not only are Queensland chicken farmers exempt from compliance with the standards in the *Domestic*

---

[42] http://www.daff.gov.au/animal-plant-health/welfare/nccaw.

[43] *Saltoon v Lake* [1978] 1 NSWLR 52.

[44] *Animal Care and Protection Act 2001* (Qld); *Prevention of Cruelty to Animals Act 1979* (NSW); *Animal Welfare Act 1992* (ACT); *Prevention of Cruelty to Animals Act 1986* (Vic), *Animal Welfare Act* (Tas); *Animal Welfare Act 1993* (SA), *Animal Welfare Act 2002* (WA); *Animal Welfare Act* (NT).

Poultry MCOP, s. 40 of the *Animal Care and Protection Act 2001* (Qld) creates an "offence exemption" for an offence if the offence was constituted by doing an act permitted by a code of conduct.

Why are conditions like these not considered to breach relevant *Animal Welfare Acts*? Because as we noted above (for example) s. 40 of the *Animal Care and Protection Act 2001* (Qld) creates an "offence exemption" for an offence if the offence was constituted by doing an act permitted by a code of conduct. And the current Commonwealth MCOP for *Domestic Poultry* permits this treatment of chickens in Australia.

Likewise, the slaughter of animals at abattoirs necessarily involves conduct calculated to destroy their lives[45] However, animals intended for slaughter are generally exempted from scrutiny under State and Territory animal welfare statutes. These exemptions also extend to practices involving the religious slaughter of animals.[46] How are farm animal exempted from the more general *Animal Welfare Acts*?

The animal welfare legislation in each State and Territory prohibit acts of cruelty toward animals[47] Sections 5 and 6 of the POCTA Act prohibit acts of cruelty and aggravated acts of cruelty toward animals. Section 5 provides;

> 5 Cruelty to Animals
>
> (1) A person shall not commit an act of cruelty upon an animal.
>
> (2) A person in charge of an animal shall not authorise the commission of an act of cruelty upon the animal.
>
> (3) A person in charge of an animal shall not fail at any time:
>
> (a) to exercise reasonable care, control or supervision of an animal to prevent the commission of an act of cruelty upon the animal,
>
> (b) where pain is being inflicted upon the animal, to take such reasonable steps as are necessary to alleviate the pain, or
>
> (c) where it is necessary for the animal to be provided with veterinary treatment, whether or not over a period of time, to provide it with that treatment.

Section 6 provides:

> 6 Aggravated Cruelty to Animals
>
> (1) A person shall not commit an act of aggravated cruelty upon an animal. Maximum penalty: 1000 penalty units in the case of a corporation and 200 penalty units or imprisonment for 2 years, or both, in the case of an individual.

Additional provisions in both the POCTA Act and the *Crimes Act 1990* (NSW) prohibit other forms of conduct toward animals that would cause pain and

---

[45] Welty (2007), pp. 176–182.

[46] *Australian Standard for the Hygienic Production & Transportation of Meat and Meat Products for Human Consumption* (AS 4696:2007); *National Animal Welfare Standards for Livestock Processing Establishments Preparing Meat for Human Consumption 2009-2010* (2nd Ed), *Model Code of Practice for the Welfare of Animals: Livestock at Slaughtering Establishments 2002*.

[47] Cao et al. (2010), p. 115.

distress.[48] These include protection from being transported in a way that causes unreasonable, unnecessary or unjustifiable pain and protection from being mutilated in a certain way.[49]

The legislation also exempts techniques by which animals are slaughtered for religious purposes.

Although the POCTA Act prohibits acts of cruelty, defences are available for conduct directed toward the slaughtering of animals for food generally and for religious rituals specifically. Section 24(1)(b)(ii) provides that

> 24 Certain defences
>
> (1) In any proceedings for an offence against this Part or the regulations in respect of an animal, the person accused of the offence is not guilty of the offence if the person satisfies the court that the act or omission in respect of which the proceedings are being taken was done, authorised to be done or omitted to be done by that person:
>
> (b) in the course of, and for the purpose of:
>
> (ii) destroying the animal, or preparing the animal for destruction, for the purpose of producing food for human consumption, in a manner that inflicted no unnecessary pain upon the animal,

In a similar manner, a specific defence under the POCTA Act is created for the slaughter of animals according to the religious rituals of both the Jewish and Islamic traditions. Section 24(1)(c)(i) of the POCTA provides:

> **24 Certain defences**
>
> (1) In any proceedings for an offence against this Part or the regulations in respect of an animal, the person accused of the offence is not guilty of the offence if the person satisfies the court that the act or omission in respect of which the proceedings are being taken was done, authorised to be done or omitted to be done by that person:
>
> (c) in the course of, and for the purpose of, destroying the animal, or preparing the animal for destruction:
>
> (i) in accordance with the precepts of the Jewish religion or of any other religion prescribed for the purposes of this subparagraph.

In this way, the destruction of animals generally for the purposes of domestic food consumption undertaken at commercial abattoirs is not characterised as an act of cruelty.

11.4 Feedlots and Fattened Hens: Farm Animal Welfare Concerns

11.4.1 Present and Future Demand Chicken Meat and Eggs

Australians love to eat chicken meat and eggs. The statistics above indicated that in 2010–2011 almost 500 million chickens were slaughtered for their meat while

[48] *Prevention of Cruelty to Animals Act 1979* (NSW); sections 4, 5, 7, 8, 9, 10, 12 and 16; *Crimes Act 1990* (NSW) s 530.

[49] Cao et al. (2010), pp. 192–194.

hundreds of millions of eggs are consumed by Australian's each year; largely produced from CFAOs. There is a plethora of Commonwealth, State and Territory regulatory instruments relevant to the poultry industry.

A general overview of these instruments illustrates the complexity.

11.4.2 Poultry Regulation Generally

It is difficult to identify with any degree of certainty all of the *Acts, Regulations, Codes of Practice, Standards, Industry Guidelines* and *Recommendations* in each State and Territory that regulate the chicken meat and egg industries. Most States have multiple "layers" of regulation concerning the same practice (such as egg laying fowls) making it both time-consuming and difficult to identify all of the overlapping Commonwealth and State regulatory requirements.

The starting point is the Commonwealth *Model Code of Practice for the Welfare of Animals—Domestic Poultry* (4th Ed) ("MCOP - Poultry") issued by the SCOPI in 2002. The MCOP-Poultry is "intended to help people involved in the care and management of poultry to adopt standards of husbandry that are acceptable."[50]

However, supplementing the MCOP - Poultry are the Commonwealth *National Animal Welfare Standards for the Chicken Meat Industry* (2008), the *National Animal Welfare Standards—Manual for Chicken Meat Farming* (2008), *Model Code of Practice for the Welfare of Animals—Land Transport of Poultry*—2nd Ed (2006) the *National Biosecurity Manual for Contract Meat Chicken Farming* (2008).

All States and Territories except South Australia have attempted to incorporate the MCOP—Poultry into their jurisdiction in one form or another.

However, some States have chosen not to adopt the MCOP—Poultry but have instead created their own Poultry Codes. For example the Victorian *Department of Primary Industries* issued a *Code of Accepted Farming Practice for the Welfare of Poultry* in December 2003 that is based upon the MCOP—Poultry. The Victorian government also amended its *Planning and Environment Act 1987* (Vic) to include the *Victorian Code for Broiler Farms* (2009) as part of the *Victorian Planning Provisions* of that Act.

Supplementing these Codes are the *Prevention of Cruelty to Animals (Domestic Fowl) Regulations 2006* (Vic).

In Western Australia, broiler chicken farms were until recently, regulated by the *Chicken Meat Industry Act* 1977 (WA) and associated *Chicken Meat Industry Act (Participation in Growth Expansion) Regulations 1978* (WA). In addition, the Western Australian Department of Local Government and Regional Development has also published its own Code. The *Code of Practice for Poultry in Western*

---

[50]*Model Code of Practice for the Welfare of Animals – Domestic Poultry* 2002, 4th ed, Primary Industries Standing Committee, Canberra "Preface".

Australia issued in March 2003 is also based on the MCOP—Poultry. However, in 2010 the above Acts and Regulations were repealed. From 2011, the Australian Competition and Consumer Commission authorised WA broiler chicken farmers to collectively bargain with their contracted processors and the above Acts were repealed.[51]

Complicating matters is the fact that even where some States have explicitly adopted the MCOP—Poultry, they have enacted poultry—specific legislation or subordinate regulations that operate along-side the MCOP—Poultry.

For example, NSW has enacted the *Poultry Meat Industry Act 1986* (NSW), the *Poultry Meat Industry Regulation 2008* (NSW) and a *Code of Practice for the Conduct of Negotiations between Processors and Contract Growers* issued by the NSW Poultry Meat Industry Committee. In NSW, these instruments are in turn supplemented by Part 2A of the *Prevention of Cruelty to Animals (General) Regulation 2006* (NSW) titled "Confinement of Fowl for Egg Production".

Queensland has adopted the MCOP—Poultry and has also enacted very specific regulations concerning "domestic fowl" in Chapter 3 of the *Animal Care and Protection Regulation 2002* (Qld). Likewise, the ACT has adopted the MCOP—Poultry but has also created offences relating to "Commercial Egg Production" in Part 6 of the *Animal Welfare Regulation 2001* (ACT).

Yet, other States have neither specific *Codes* nor legislation but "Guidelines" of uncertain legal status. For example, Primary Industries and Resources SA promotes a document called *Guidelines for the Establishment and Operation of Poultry Farms in South Australia* dated March 1998 and prepared by the SA Government in conjunction with private enterprise. However, Part 3A of the *Animal Welfare Regulations 2000* then creates regulations relating to "domestic fowls".

All of these different *Guides, Guidelines, Codes of Practice* and *Standards* embody the MCOP—Poultry to a greater or lesser degree. The extent to which the MCOP—Poultry has been adopted by the different Australian States and Territories and the extent to which it is then given legal effect is indicated by the table at the conclusion of this chapter.

11.4.3 Broiler Chicken Meat Production

The processing of chickens for their meat and eggs involves very different processing techniques and processes. At chick hatcheries throughout Australia day-old chicks are firstly separated by gender. Strong female chicks may be used for either egg or meat production.

Weak female chicks or male chicks are not suitable for either egg production and usually not exploitable for their meat. Instead they are killed. Section 14.1 of the

[51] http://www.accc.gov.au/content/index.phtml/itemId/974882.

current *Model Code of Practice for the Welfare of Animals—Domestic Poultry* (2002) provides:

> Culled or surplus hatchlings awaiting disposal must be treated as humanely as those intended for retention or sale. They must be destroyed humanely by a recommended method such as carbon dioxide gassing or quick maceration...

Each day around Australia, hundreds of thousands of live male chicks and weak female chicks are separated out by sorters and fed by conveyer belt into a machine that grinds them into slurry in little under a second.[52] The separated out strong female chicks are then used for the production of meat or eggs.

Chickens intended to be raised for their meat are referred to as "broiler hens". They are bred to grow in size and weight very quickly. Most of these chickens are stored under artificial light and temperature conditions in huge sheds providing an artificial environment and containing up to 60,000 chickens.[53] Through intensive feeding practices, these chickens are "harvested" sometimes within 30–35 days of arrival.

The commercial structure of the chicken meat industry in Australia involves a large processor/supplier such as Baiada Poultry or Inghams Chickens entering into supply contracts with corporations that grow the chickens for them. The structure of the industry was generally described by the Australian Competition Tribunal in *Re VFF Chicken Meat Growers' Boycott Authorisation* (2009) ACompT 2 (21 April 2006) in the following terms:

> In Victoria, as elsewhere in Australia, chicken meat processors deliver day old chicks to growers and collect the grown chickens after about five to eight weeks. The processors provide feed and veterinary requirements but otherwise the growers care for the chickens and manage their growth. Processors and growers enter into contracts under which growers' services are supplied.[54]

Managing this corporate relationship often involves a complex supply chain. For example, in *Baiada Poultry Pty Ltd v The Queen*[55] the High Court considered an appeal involving an industrial accident at a CAFO operated by a company operated by the Houbens. The Houbens were contract growers for Baiada Poultry Pty Ltd.

After the Houbens's chickens reached 32 days of age, they were ready for harvesting. Baiada Poultry Pty Ltd then engaged another contractor company called DMP Poultech Pty Ltd to catch and pack the chickens into crates. However, Baiada Pty Ltd then engaged yet another contractor company, Azzopardi Haulage Pty Ltd to then transport the chickens to its slaughter facility at Laverton in Victoria.

After taking delivery of the day old chicks, the growers ensure the chickens reach optional harvesting weight through a combination of selective breeding and

---

[52] Sharman (2008), pp. 46–48.

[53] Australian Chicken Meat Federation inc: *Growing Meat Chickens* < http://www.chicken.org.au/page.php?id=6> (Accessed 30 March 2012).

[54] *Re VFF Chicken Meat Growers' Boycott Authorisation* (2009) ACompT 2 (21 April 2006) at 1.

[55] *Baiada Poultry Pty Ltd v The Queen* [2012] HCA 14 (30 March 2012).

forced feeding regimes. There are corporations that specialise in breeding animals with genetic characteristics enabling them to convert feed into weight within a very short timeframe.

For example, US corporation Cobb-Vantrass Inc has created for export to world-wide chicken meat markets, a broiler chicken marketed under the brand "Cobb 700". The Cobb 700 is marketed in the following terms[56]:

> The Cobb 700 has been selected to achieve meat yield more efficiently than any other breed. In a high yield market the primary function of a broiler is to produce meat, and most importantly, breast meat. A new measurement to evaluate the efficiency of producing breast meat yield is **Breast Meat Feed Conversion** – the amount of feed required to produce a pound or kilogram of breast meat. The Cobb 700 consistently provides the lowest Breast Meat Feed Conversion offering more saleable meat per bird at a lower cost of production, a new standard in high yield.

The emphasis is on efficiency of feed-weight conversion. The principal goal of owners of chicken-meat factories is economic efficiency. The owner wants the most weight from its chickens for the expenditure made on feed for the chickens.

In order to increase efficiency, the chickens housed in giant sheds are exposed to artificial lighting cycles. The use of fluorescent or other artificial lights are intended to distort natural sleeping and feeding patterns in order to maximise weight gain and to control aggression.[57] The permanent confinement in high-density sheds combined with constant exposure to artificial feeding and lighting regimes causes suffering to the chickens for the month or so that they have to live before being harvested.

In terms of overcrowding, or stocking densities, Stephanie Buijs et al.[58] explained the reason broiler chickens tended to stand or lie close to the walls of indoor sheds or pens. They noted that this behaviour increased as more and more chickens were crowded into the sheds. Their conclusion was that this behaviour was a fear-based response as an adaptation to violence associated with over-crowding.

This violence often takes the form of cannibalism and other self-destructive practices. Glatz et al note[59] "cannibalism, egg eating, feather picking and vent picking are common traits where birds are housed together under high light intensity. beak trimming is an animal husbandry practice commonly carried out in the poultry industry involving the removal of part of the top and bottom beak of a bird to blunt the beaks enough so that pecking cannot do any damage."

Beak trimming and toe removal, intended to mitigate these destructive behaviours are therefore permitted by the MCOP-Poultry and described below.[60]

---

[56] *Cobb 700 – The New Standard in High Yield*; Cobb-Vantrass Inc, 2007. <http://www.cobb-vantress.com/Products/ProductProfile/Cobb700_Sales_Brochure_2007.pdf> (Accessed 1 April 2012).

[57] Olanrewaju et al. (2006), p. 301.

[58] Buijs et al. (2010), p. 102.

[59] Glatz et al. (2009), p. 3.

[60] *Model Code of Practice for the Welfare of Animals – Domestic Poultry* 2002, 4th ed, Primary Industries Standing Committee, Canberra, Chapter 13, paragraphs 13.2 and 13.4 at p. 19.

In terms of artificial lighting regimes, Deep and Schwean-Lardner et al.[61] demonstrated how low light regimes exerted a negative effect on broiler chickens in the form of increased ulcerative foot-pad lesions and underdeveloped eye size.

The strategic manipulation of light, feed and genetic structure to ensure accelerated growth rates is the principal cause of deformities, diseases and death in broiler chickens. In 2000, the European Commission Scientific Committee on Animal Health and Animal Welfare issued its Report: *The Welfare of Chickens Kept for Meat Production (Broilers)*.[62] The Report concluded that "it is obvious that rapid growth which is the result of genetic selection and intensive feeding and management systems is the main cause of various skeletal disorders and metabolic diseases that have become important causes of mortality."[63]

Accelerated growth rates are achieved in chicken muscle and fatty tissues and not in bone density or internal structural support. Consequently, by the time many broiler chickens reach optimal harvesting weight their leg and bone structures are under-developed or fragile.

Often many chickens are lame suffering ruptured tendons and other metabolic disorders.[64] The constant indoor confinement often leads to respiratory diseases and death.[65]

The management practices that cause these deformities, diseases and death are permitted by the MCOP—Poultry. For example, clause 12.5 advises that in the event of feather picking (a stress indicator) or cannibalism, adjustments should be made, including the elimination of shafts of natural sunlight.[66]

And while antibiotics are not used to accelerate growth, the Australian poultry industry uses a large quantity of antibiotics to combat bacterial infection caused by high-density confinement and prophylactically to prevent the spread of infection. When used this way, antibiotics are included with chicken feed to ensure that birds do not get sick.[67]

When it is time for harvesting, broiler chickens are captured by clean-up crews that are legally permitted to carry up to 5 chickens in each hand; usually dangling by their legs.[68] The chickens are then packed into cages of sometimes up to 28 chickens per square metre.[69] After they arrive at a processing factory, the chickens are

---

[61]Deep et al. (2010), p. 2326.

[62]European Commission Scientific Committee on Animal Health and Animal Welfare *The Welfare of Chickens Kept for Meat Production (Broilers)*, Adopted 21 March 2000.

[63]Ibid at 30.

[64]*From Nest to Nugget*, November 2008, Voiceless at 12.

[65]Sirri et al. (2007), p. 734; Bilgili et al. (2009), p. 583.

[66]*Model Code of Practice for the Welfare of Animals – Domestic Poultry* 2002, 4th ed., Primary Industries Standing Committee, Canberra at 17.

[67]Turnidge (2001), p. 26.

[68]*Model Code of Practice for the Welfare of Animals – Land Transport of Poultry*, 2006 Primary Industries Standing Committee, Canberra, clause 4.5.2 at p. 9.

[69]Ibid at clause 4.2.3 at p 7.

quickly killed. The process is described by Animals Australia in the following terms[70]:

> When the trucks arrive at the slaughterhouse, chickens are pulled from the crates and shackled upside down by their feet into metal stirrups on an overhead conveyor. The conveyor carries them into the killing room where their heads pass through an electrified water bath intended to stun them. As they pass along further, an automatic knife cuts their throat, and then they proceed into a scalding tank to loosen their feathers before plucking.
>
> Unfortunately things do not always run smoothly. Some birds lift their heads and miss the electrified water bath and they are therefore still fully conscious when they reach the automatic knife. Some birds may also miss the knife and are then lowered into the 50 degree scalding tank while still alive. Back-up people are supposed to cut the throats of the chickens that miss the automatic knife, but due to the emphasis on speed in the processing plants this may not always occur. There are no animal welfare inspectors onsite to ensure that the slaughter process is humane.

The difficulties involved in this process and associated suffering experienced by the animals were explored by Hindle et al.[71] who noted suffering experienced by broilers, hens, and ducks as a result of variations in the electrical current used and in the resistance caused by multiple animals being stunned simultaneously.

11.4.4 Battery Hens: Egg Production

In 2010 Australia's flock size of egg producing chickens was almost 21 million birds. During the same period, some 345 *million* dozen eggs were produced with a gross production value at market of $1.5 billion.[72] The chicken egg industry maintains extremely efficient methods of production. Chickens that are used to produce eggs for human consumption are also housed, fed and treated in ways that maximise the economic efficiency of each chicken.[73]

Similar to broiler chicken processing, corporations use intensive factory farming methods such as artificial lighting and feed cycles to maximise egg production. Several hens are stored in tiny cages that are stacked on top of each other in tiers. These cages are sloped to facilitate feeding but in the process, result in foot and claw problems.[74]

Being confined in this way causes hens to develop adverse behaviour such as pecking and cannibalism. In order to prevent or at least limit the damage caused by

---

[70] Animals Australia: *Broiler Chickens Fact Sheet* (undated). <http://www.animalsaustralia.org/factsheets/broiler_chickens.php#toc5>.

[71] Hindle et al. (2010), p. 402.

[72] Australian Egg Corporation Limited, 2010 Annual Report at 3.

[73] Ibid at 12.

[74] Sharman (2008), pp. 49–50.

this behaviour, the MCOP-Poultry permits chicken farmers to de-beak chicks.[75] Because the beak is a chicken's principal sensory organ, many egg-producing chickens live in a permanent state of disfigurement and disorientation.

A further motivation for de-beaking chickens is to avoid injury caused by overcrowding. Appendix 1 to the MCOP—Poultry permits multiple chickens to be housed in cages of 550 square centimetres per chicken.

To provide some perspective, an A4 sheet of paper has an area of 625 square centimetres. A standard laying hen is at least 40-cm high when she stands erect and is approximately 45-cm long and 18-cm wide, without her wings extended. Her body space takes therefore takes up an area of about 810 square centimetres.

The chickens in these cages are entirely female. After hatching, chicks are separated according to gender and health. Pursuant to the MCOP-Poultry, unhealthy female chicks and male chicks can be fed into a high-speed grinding machine while still alive.[76] Millions of chicks are slaughtered in this way before they produce either eggs or meat. They are simply regarded as unwanted by-products.

Several attempts have been made to prohibit battery-cage farming of eggs. In 1997, the ACT government enacted the *Animal Welfare (Amendment) Act 1997* (ACT) banning battery cages. However, the Act was never enforced because of allegations by other States that it beached *National Competition Policy Principles.*

Another attempt by the ACT Greens to introduce the *Eggs (Cage Systems) Legislation Amendment Bill 2009* (ACT) was also defeated. Similar legislation in Tasmania was defeated in 2010. In 2011, NSW Greens MP Dr John Kay introduced the *Truth in Labelling (Free-range Eggs) Bill 2011* into the NSW Parliament. The Bill passed the Legislative Council but was defeated in the Legislative Assembly in 2013.

The current debate in Australia over what level of stocking density constitute "free-range" remains extremely important. In September 2011, the Australian Egg Corporation proposed new *Draft Egg Standards* that would permit the label "free range" to be applied to eggs produced by farms that permit a stocking density of up to 20,000 chickens per hectare.[77] This stands in stark contrast to the MCOP-Poultry that permits a stocking density of 1500 birds per hectare for non-cage meat chicken farms.[78] Although it was not passed, the *Truth in Labelling (Free-range Eggs) Bill 2011* (NSW) would have confined the use of "free range" to chickens produced on farms using a stocking density consistent with the MCOP-Poultry.

The relationship of this debate to the protection offered by State *Animal Welfare* Acts is difficult and does not benefit the hens in many respects.

---

[75] *Model Code of Practice for the Welfare of Animals – Domestic Poultry* 2002, 4th ed, Primary Industries Standing Committee, Canberra cl 12.5 at 17.

[76] *Model Code of Practice for the Welfare of Animals – Domestic Poultry* 2002, 4th ed, Primary Industries Standing Committee, Canberra, cl 14 at 21.

[77] <http://www.aecl.org/system/attachments/454/original/Draft%20Egg%20Standards%20of%20Australia.pdf?1316400975> (Accessed 1 April 2012).

[78] *Model Code of Practice for the Welfare of Animals – Domestic Poultry* 2002, 4th ed, Primary Industries Standing Committee, Canberra, Appendix 2, A2.1.4 at p. 28.

11.4.5 *Relationship with State and Territory* Animal Welfare Acts

The difficulty in identifying the numerous and often over-lapping Commonwealth, State and Territory MCOPs, Acts and Regulations relevant to the poultry industry is complicated by the uncertain relationship that these MCOPs, Acts and Regulations have with State and Territory *Animal Welfare Acts* and the *Regulations* that are made pursuant to those Acts.

Understanding this relationship is important because these *Animal Welfare Acts* are the principal source of protection for animals from acts of cruelty. The difficulty is that while these *Animal Welfare Acts* do prohibit acts of cruelty, they also exempt conduct that is permitted under MCOPs such as the MCOP-Poultry. Sharman notes[79]:

> As each jurisdiction's animal welfare law purports to apply to all animals, prima facie, chickens appear to be protected from cruelty. Despite this, any close examination of State and Territory animal welfare legislation reveals that chickens, like many other animals used for food production purposes, fall largely outside the reach of the law when it comes to the most meaningful of protections.

For example, s. 13 of the *Animal Care and Protection Act 2001* (Qld) provides that the *Regulation* may make a Code of Practice about animal welfare. Part 2 of Schedule 1 of the *Animal Care and Protection Regulation 2002* (Qld) designates most existing MCOPs as "Voluntary Codes of Practice" including the MCOP—Poultry.

However, the MCOP-Poultry is *voluntary* and its standards are not compulsory for farmers producing chickens to be processed for meat. Not only are Queensland chicken farmers exempt from compliance with the standards in the MCOP-Poultry but s. 40 of the *Animal Care and Protection Act 2001* (Qld) creates an "offence exemption" for an offence if the offence was constituted by doing an act permitted by a code of conduct.

How does this work in practice?

Assume that a person in Queensland confines their pet chicken in a dark garage, inside a cramped cage without access to sunlight, dirt or the capacity to scratch, turn around or interact with other chickens. That person would surely have breached their duty of care toward the chicken imposed by s. 17(1) of the *Animal Care and Protection Act 2001*.

Section 17(3) states that a person breaches the duty of care provision if he or she fails to take reasonable steps to provide for the animal's needs for accommodation for the animal or to permit the animal to display normal patterns of behaviour – s 17 (3)(a)(ii) and (iii).

[79] Sharman (2008), pp. 46–48.

However, if that same person also owned a battery hen farm, raising chickens for egg production in the manner and in the conditions discussed earlier, the same conduct would be legal.

Why are conditions like these not considered to breach relevant *Animal Welfare Acts*? Because (for example) s. 40 of the *Animal Care and Protection Act 2001* (Qld) creates an "offence exemption" for an offence if the offence was constituted by doing an act permitted by a code of conduct. And the current MCOP—Poultry permits this treatment of chickens in Australia.

In addition, each State and Territory *Animal Welfare Act* creates specific exemptions for conduct associated with the slaughter of animals for food. For example, although the *Prevention of Cruelty to Animals Act 1979* (NSW) prohibit acts of cruelty, defences are available for conduct directed toward the slaughtering of animals for food generally and for religious rituals specifically. Section 24(1)(b)(ii) of the Act provides that

> **24 Certain defences**
>
> (1) In any proceedings for an offence against this Part or the regulations in respect of an animal, the person accused of the offence is not guilty of the offence if the person satisfies the court that the act or omission in respect of which the proceedings are being taken was done, authorised to be done or omitted to be done by that person:
>
> (b) in the course of, and for the purpose of:
>
> (ii) destroying the animal, or preparing the animal for destruction, for the purpose of producing food for human consumption, in a manner that inflicted no unnecessary pain upon the animal,

In this way, industrial scale egg and chicken meat farming in all States and Territories of Australia are systematically exempt from scrutiny under relevant *Animal Welfare Acts*. It might be thought that the existence of MCOPs are drafted to ensure that the production of chicken meat and eggs (as well as other meat-animals) protects the welfare of the animals.

However, it is important to remember that these MCOPs are created by the *Primary Industries Standing Committee*, part of the *Primary Industries Ministerial Committee*. These committees are in fact constituted by representatives of the corporations that profit from the commercial exploitation of animals. And as is clear from an examination of the MCOP—Poultry, animal welfare is in fact subordinated to economic efficiency and profit.

Similar conclusions can be drawn about the application (or lack of application) of State and Territory *Animal Welfare* Acts; "the animal welfare statutes of each jurisdiction permit a series of encroachments on bodily liberty and bodily integrity in the interests of maximising production".[80]

[80] Sharman (2008), pp. 46–48.

11.5 Australia's Live Export Industry

11.5.1 The Economic Significance of Australia's Live Animal Export Industry

Australia is one of the world's largest, if not *the* largest exporter of live animals, principally, cattle, sheep and goats. In 2013, the combined value of cattle, sheep and goat exports was almost AUD $700 million (€ 460 million).[81] Both the quantity and therefore the economic value of these live animal exports is about to significantly increase following the signing of a Free Trade Agreement between Australian and China during 2014. Amongst other initiatives, the Australia-China Free Trade Agreement ("FTA") anticipates a phased removal of current import tariffs on beef, beef offal and sheep. According to *Meat and Livestock Australia Limited* ("MLA"), the peak research and lobby group for Australia's beef, sheep and goat producers, the removal of these import tariffs has the potential to boost the value of beef production alone to AUD $3.3 billion by 2030.[82]

These statistics underscore the importance of the live animal export to the Australian economy and following the FTA initiatives, the industry is likely to be increasingly important to the future of trade in the Asia-Pacific region. Not surprisingly, given the sheer quantities of animals that are exported, the live animal export industry has attracted continual controversy and criticism. Indeed a 1985 Senate Inquiry into the export of sheep to the Middle East concluded[83]:

> If a decision were to be made on the future of the trade purely on animal welfare grounds, there is enough evidence to stop the trade. The trade is, in many respects, inimical to good animal welfare, and it is not in the interests of the animal to be transported to the Middle East for slaughter.

Clearly successive Australian governments have decided that animal welfare is to be subordinated to the economic imperative; live animal exports have not only increased since 1985, but are about to dramatically surge following the implementation of the Australia-China FTA and the eventual removal of animal import tariffs.

---

[81]Australian Bureau of Agricultural and Resource Economics and Sciences (ABARES), *Live Export Trade Assessment*, July 2014 at (vii).

[82]Colin Bettles, *China Deal a "Cracking" Win: NFF*, The Land, 17 November 2014. http://www.farmonline.com.au/news/agriculture/agribusiness/general-news/china-deal-a-cracking-win-nff/2717567.aspx (Accessed 12 December 2014).

[83]Senate Select Committee on Animal Welfare, Parliament of Australia, *Export of Live Sheep from Australia* (1985) at 185.

11.5.2 Underlying Legislative Framework of the Live Export Industry

The regulation of live animal exports from Australia is complex, involving overlapping *Statutes, Codes, Schemes and Regulations*. The industry is principally regulated by the *Australian Meat and Live-stock Industry Act 1997* (Cth) and the *Export Control Act 1982* (Cth). Pursuant to these statutes, the Secretary for Agriculture may issue a licence to companies wishing to export animals. However, following live animal export scandals in Indonesia and Egypt (discussed below), the Australian Government created the *Exporter Supply Chain Assurance Scheme* with the intention of improving the welfare of animals being exported.

The ESCAS regime is discussed below.

11.5.3 Export Animal Welfare Crises

The concerns expressed by the Senate Committee were dramatically proven in 2011 when the Australian Investigative Television Program "Four Corners" aired footage of Indonesian abattoir workers inflicting gross abuses on Australian export cattle. The public outrage following the program convulsed the Australian Government into suspending the entire live export trade to Indonesia for 6 months while it addressed the issues.[84]

This was not the first time the Australian government had suspended live exports after cruel animal handing practices were exposed. Exports of live sheep and cattle to Saudi Arabia were suspended between 1991 and 2000 and in 2006 export of live sheep to Egypt was suspended after shocking footage of animal abuse was reported.

Part of the difficulty lies in the inability of the Australian government to mandate and enforce animal welfare standards in export destination countries. Australian domestic *Animal Welfare* legislation does not protect sheep as they are being processed in Cairo street markets. Instead, the Australian government relies upon *Memoranda of Understanding* ("MOU"). Pursuant to these MOUs the export destination country agrees to comply with certain minimum welfare standards in the handling of live animals.[85] These MOUs are aspirational documents and do not have the status of domestic law in the signatory countries. Because they are unenforceable, breaches of MOUs are regularly observed; as one Egyptian veterinarian noted[86]:

---

[84]A large, multi-million dollar class action suit has been initiated against the Australian government by beef export companies seeking compensation for losses suffered during the period of suspension. http://www.farmonline.com.au/news/agriculture/agribusiness/general-news/live-ex-class-action-moves-forward/2716030.aspx?storypage=0 (Accessed 12 December 2014).

[85]Hastreiter (2013), p. 181.

[86]Coghlan (2014), p. 49.

> Egyptians don't care – and our government doesn't care about animal welfare. We only care about meat inspection. Before the animal is killed, we don't care. So no-one orders the workers to stop these bad actions and there is no punishment. So it continues.

Not surprisingly, continued exposure of animal welfare abuses in Egypt again led to the suspension of the live export trade in 2013. The Egyptian attitude of indifference toward the suffering of exported animals was apparently reflected in Australia. In November 2012, the then president of the Meat Division of the Western Australian Farmers Federation expressed approval of Egyptian abattoirs and criticised the suspension of live exports as "another blow to beef producers reeling from falling prices caused by a lack of demand for live exports."[87]

Live exports to Egypt were resumed in May 2014 after the Australian Government introduced the *Exporter Supply Chain Assurance System* ("ESCAS").

11.5.4 The "Exporter Supply Chain Assurance System" (ESCAS)

In an attempt to reform the live export industry following the Indonesian and Egyptian abuses, the Australian government introduced the *Exporter Supply Chain Assurance System* ("ESCAS").

By making the *Australian Meat and Live-stock Industry (Conditions on live-stock export licences) Order 2012* ("the Order"), the Australian government required exporters of live animals to comply with the *Export Control (Animals) Oder 2004* ("the Animals Order"). The Animals Order was the actual legislative instrument that contains the elements of the ESCAS regime.

From 2012, a company wishing to export live animals from Australia must obtain an ESCAS approval from the Secretary for Agriculture. The exporter's ESCAS must include details of how animals are to transition through all parts of the supply chain to the point of slaughter. The ESCAS must include details of how the handling of livestock will comply with the animal welfare standards of the World Organisation for Animal Health as well as how the animals may be traced through the supply chain to the point of slaughter.[88]

The effectiveness of ESCAS in protecting the welfare of animals exported for slaughter and its relationship to wider international trade law is a matter of some debate.[89] However, whether animals are exported for slaughter or simply processed domestically, there are significant environmental externalities associated with intensive animal farming.

---

[87] Brad Thompson and Andrew Tillett, *Call to Make Live Export Ban Permanent*: "The West Australian" 7 May, 2013.

[88] Black (2013), p. 80.

[89] Ibid. See also; Chaudhri (2014), p. 279.

11.6 Farm Animal Industries and Environmental Degradation

11.6.1 From Family Farms to Corporate Agribusiness

Earlier, we noted that the transition from small, sustainable family-operated farms to large scale, corporate-owned industrial animal farms has resulted in the alienation of people from natural agricultural cycles.[90] Consumer demand for cheap and plentiful meat, eggs and other animal products is met through artificial manipulation of animal growth and feeding cycles and even genetic content. Meeting and then sustaining this consumer demand is both intensively energy consuming and polluting. Concentrated Animal Feedlot Operations require enormous quantities of feed that, in turn, require enormous quantities of fertilizer and water.

Likewise, the outputs from CAFO enterprises is more than just meat and other animal products. There is a very significant and damaging environmental cost. Most beef animals are fed grain that is grown specifically for that purpose. The amount of water and fertilizer need to produce the necessary quantities of grain are immense. Producing 1 kilo of meat requires almost 20,000 l of water whereas producing 1 kilo of wheat requires only 2000 l.[91]

11.6.2 Animal Agriculture and the Environmental Domino-Effect

This intensity of resource use produces an environmental domino-effect. In order to grow the crops needed to feed the animals intended to be meat products, huge amounts of land needs to be cleared. Approximately 40 % of the world's land area is devoted to such food production, resulting in deforestation, loss of bio-diversity and water and soil degradation.[92] Consumer preferences for meat also results in higher CO_2 emissions. The annual beef consumption by an average family of four in the United States requires 1100 l of fossil fuel to be burned. This process then releases an amount of CO_2 that is equivalent to the amount produced by a car in six months.[93] Likewise, the creation of pastureland by clearing forests release trapped CO_2 and methane as trees are burnt. Land clearing also results in fewer trees to capture atmospheric CO_2 for conversion into oxygen.

Intensive animal farming also results in massive increases in methane. Cows have a large body mass and a ruminant digestive system and produce is the third largest source of atmospheric methane.[94]

---

[90]Casuto (2012), pp. 75.

[91]Tao (2002–2003), p. 15.

[92]Elferink and Nonhebel (2007), p. 1778.

[93]Tao (2002–2003), p. 333.

[94]Cheever (2000), p. 44.

11.6.3 Large-Scale Intensive Animal "Agriculture"

Large-scale, corporate animal facilities produce enormous quantities of waste. In the United States, CAFOs produce over 100 times more excrement than the human population, with a mid-size piggery producing waste equivalent to a city with a population of 12,000.[95] When this amount of waste leeches into the water table, it results in devastating environmental destruction. For example, in 1995 animal feedlots in the US State of North Carolina discharged 242 *million* litres of animal excrement into rivers, lakes and ponds resulting in the destruction of approximately 10 million fish.[96] Since less than 1 % of the world's water is suitable for human consumption, the environmental and physical externalities associated with these intensive animal farming practices is dire.

11.6.4 Artificial Photosynthesis as Alternative Agribusiness Technology

More solar energy strikes the Earth's surface in 1 hour of each day than the energy used by all human activities in 1 year. At present the average daily power consumption required to allow a citizen to flourish with a reasonable standard of living is about 125 kWh/day. Much of this power is devoted to transport (~40 kWh/day), heating (~40 kWh/day) and domestic electrical appliances (~18 kWh/day), with the remainder lost in electricity conversion and distribution (McKay 2009). Global energy consumption is approximately 450 EJ/year, much less than the solar energy potentially usable at ~1.0 kW per square metre of the earth—3.9×10^6 EJ/year even if we take into the earth's tilt, diurnal and atmospheric influences on solar intensity (Pittock 2009).

Photosynthesis as a natural process is equally important with DNA in the progress of humanity. Photosynthesis provides the fundamental origin of our oxygen, food and the majority of our fuels; it has been operating on earth for over 2 billion years. Photosynthesis can be considered as a process of planetary respiration: it creates a global annual CO_2 flux in from the atmosphere and an annual O_2 flux out to atmosphere. In its present nanotechnologically-unenhanced form, photosynthesis globally already traps around 4000 EJ/year solar energy in the form of biomass (Kumar, Jones and Hann 2009). The global biomass energy potential for human use from photosynthesis as it currently operates globally is approximately equal to human energy requirements (450 EJ/year) (Hoogwijk, Faaij, van den Broek, Berndes, Gielen, Turkenburg 2003).

[95]Tao (2002–2003), p. 334.

[96]Lee Nardo (2000), p. 83.

In its most basic form, AP technology is therefore about replicating the biochemical process employed by plants to convert sunlight into energy during the day and to capture atmospheric carbon dioxide in the evenings for the same process. The science is complex and a more detailed explanation of the process can be found elsewhere.[97]

However, for the purposes of this article, the process can be described fairly succinctly. Scientists hope to develop cost-effective and efficient technology that will capture photons from within the solar spectrum and then use those photons to split water molecules into its component elements of hydrogen and oxygen.

The hydrogen gas may then be distributed for use as fuel or combined with carbon dioxide to create carbon-based fuels. In the "dark-reaction" cycles, atmospheric carbon dioxide may be captured and then combined with other elements to make carbon sugars.[98] Ultimately, a potential future may be imagined where all cities, freeways and building might be fitted with AP technology that uses hydrogen gas to make fresh water and absorbing carbon dioxide to make fertilizers and basic foods.[99]

Artificial photosynthesis can facilitate other energy options H_2–based fuels, the most promising perhaps by combining it with atmospheric nitrogen to make ammonia. Ammonia is already is shipped, piped, and stored in large volumes in every industrial country around the world as an agricultural fertilizer. As a fuel, ammonia has been proven to work efficiently in a range of engine types, including internal combustion engines, combustion turbines, and direct ammonia fuel cells. Due to its high energy density and an extensive, existing ammonia delivery infrastructure, ammonia is ready for the market today as an alternative to gasoline.

If such artificial photosynthetic technology is incorporated into every building, road and vehicle on the earth's surface than the positive outcome will be that humanity's structure will be producing abundant safe, low carbon fuels and fertilizers. In such a world it will be much more feasible for communities and families to support many of their basic food needs off–grid through organic farming rather than relying on distant sourced food provide by large corporate marketing chains.

The emergence of artificial photosynthesis technology ("AP technology") as a potential alternative mechanism of energy production has therefore attracted considerable interest in recent years.[100] The potential of AP technology to alleviate anthropogenic destruction of the environment is vast, particularly if the technology can be engineered into the rapidly increasing urban environment. Urban structures capable of utilizing photon energy in the form of sunlight to catalyse water into hydrogen fuel for human use is as revolutionary as it is beneficial for the future preservation of humanity and the environment.[101]

---

[97]Faunce (2011), p. 276 ff.

[98]Ibid.

[99]Bruce and Faunce (2015).

[100]Faunce (2012).

[101]Faunce et al. (2013), p. 695.

The demonstrated beneficial potential for AP technology, outlined in the other articles in this collection and elsewhere more than justifies establishing a Global Artificial Photosynthesis Project dedicated to facilitating the introduction of AP technology.[102]

11.6.5 Transitioning from Corporate Control to New Ethic of Farm Animal Welfare Based on "Sustainocene" Principles

In this way the *process* of making the hard decisions necessary in confronting anthropogenic environmental degradation, population growth and resource scarcity must necessarily involve bracketing short-term selfish desires in favour of exercising deeper virtues. In this sense, AP technology promises to be not just a technical solution to environmental problems, but as a vehicle for the exercise of the virtues necessary for the wider "ecocentric transformation of human consciousness sustained by contemplative traditions (that) are our collective destiny", a transformation that necessarily embraces animal welfare.[103]

11.7 Conclusion: Evaluating Australia's Animal Law Regime

11.7.1 Australian Animal Law Regulatory Regime

The complexity of Australia's animal law regulatory regime has been repeatedly criticised for its failure to both protect the welfare of the vast majority of animals in Australia and to achieve meaningful advances in animal welfare.[104]

There are several reasons for these criticisms but they are commonly distilled by the literature[105] into four major faults:

1. The complexity of the regulatory regime;
2. The existence of an inherent conflict of interest in the bodies responsible for drafting animal welfare Codes and Standards;
3. The inconsistencies and often contradictory language and structure of animal welfare laws;

[102]Faunce et al. (2015).

[103]Faunce (2015), p. xxxii.

[104]White (2007), p. 347; Ellis (2010), p. 4.

[105]Ibid.

4. The lack of a coherent and adequately resourced strategy to enforce animal welfare laws.

These alleged deficiencies are interrelated. *Because* Australia's regulatory regime is so complex, *inconsistencies and contradictions* often arise. This complexity also means that most animal welfare policies are not created by Parliament, but by governmental departments through committees that are composed of industry representatives.

Thus a *conflict of interest* is often at work. The lack of a national or overarching animal welfare law or regulator means that enforcement of existing animal welfare laws is left to a charitable organisation; the RSPCA; a private charitable organisation that is vastly underfunded.[106]

11.7.2 The Complexity of the Regulatory Regime

Earlier this chapter explored the way in which animal laws across Australia have been created and administered through a complex network of Departments, Councils, Committees and Working Groups. It noted that although the Commonwealth government does not have direct Constitutional power to create laws relating to animals, it has assumed a leadership role in creating animal welfare and related policies. It attempts this through the creation of the *Australian Animal Welfare Strategy* and *Model Codes of Practice* that are progressively being translated into *Welfare Standards*.

None of these *Strategies* or *Model Codes of Practice* or *Animal Welfare Standards* has the status of law; they are not legislative instruments like statutes. It is not possible for any authority to institute legal proceedings alleging that a person has, for example, breached a provision of the *Australian Animal Welfare Strategy* or any of the *Model Codes of Practice or Standards*.

It is the responsibility and discretion of the States and Territories to give legal effect to these Commonwealth *Strategies* and *Model Codes of Practice* through State or Territory laws that apply them. In theory this co-operative approach overcomes the lack of Constitutional power in the Commonwealth government to directly legislate. However in practice, the universal application of these Commonwealth *Strategies* and *Model Codes of Practice* has been inconsistent and in some case, duplicated in State legislation and Codes.

Not all of the Commonwealth *Model Codes of Practice* and *Welfare Standards* have been adopted and applied across the States and Territories. And even where these documents have been adopted, they are inconsistently applied. In some States compliance is mandatory and in others, it is voluntary.

[106] Cao (2010), p. 140.

Even where a State or Territory has specifically incorporated a Commonwealth *Code of Practice* or *Animal Welfare Standard*, the extent of the incorporation and its legal effect are often very confused.

11.7.3 Conflicts of Interest

Commonwealth *Model Codes of Practice* and *Animal Welfare Standards* are created by a variety of non-statutory entities and issued by different Committees. Most are issued under the auspices of the former *Primary Industries Ministerial Council* (now the *Standing Council of Primary Industries*—SCoPI) committee composed of the Primary Industries Ministers from each State and Territory. The objective of SCoPI/SCOPI is "to develop and promote sustainable, innovative and profitable agriculture, fisheries/aquaculture, and food and forestry industries"[107]

In other words, the interests of the SCoPI lie in the profitable development of primary industries and not in the welfare of animals. Where there is a conflict of interest, the welfare of animals is subordinated to efficient industry practices and market forces.[108]

Conflicts of interests are evident from the very beginning of the creation of *Model Codes of Practice* and *Animal Welfare Standards*. The body responsible for actually creating the Codes and Standards was the *Animal Welfare Working Group* ("the AWWG") that was a sub-committee of the *Animal Health Committee*.[109]

In 2011, the AAWG became the *Animal Welfare Committee* ("the AWC") as a sub-committee of the *Animal Welfare and Product Integrity Task Force* within the Commonwealth Department of Agriculture, Fisheries and Forestry.[110] Since 2014, these bodies have been disbanded, along with Australia's *Animal Welfare Strategy*.

Attempting to identify the relevant Committee, Working Group or Department responsible for developing animal welfare related documents can therefore be confusing. The composition of the AAWG/AWC includes members of government departments whose focus is not animal welfare but industry productivity.

The reality is that Commonwealth *Model Codes of Practice* and *Animal Welfare Standards* are created by Committees composed of representatives of both Government departments whose principal focus is not animal welfare and primary industry representatives whose principal focus is the economic and profitable development of primary industries.[111]

Elizabeth Ellis provides the example of NSW representatives being drawn from the Animal Welfare Branch of the Department of Primary Industries whose stated

[107]http://www.mincos.gov.au/about_SCoPI.
[108]Ellis (2010), p. 13.
[109]http://www.dfat.gov.au/facts/animal_welfare.html.
[110]http://www.daff.gov.au/animal-plant-health/animal/committees/ahc/awwg.
[111]Dale (2009), pp. 185–186.

goal is to act "in partnership with industry and other public sector industries in New South Wales."[112]

11.7.4 Confused and Inconsistent Language and Structure

Even where a *Model Code of Practice* or *Animal Welfare Standard* has been incorporated into State or Territory legislation, in many cases there are inconsistencies in the relationship of the adopted Code or Standard with other State or Territory *Animal Welfare Acts*.

For example, each State and Territory has enacted *Animal Welfare Acts* that prohibit acts of cruelty toward animals. However, these same Acts also create specific exemptions or defences for conduct toward animals that is permitted under a *Model Code of Practice* or *Animal Welfare Standard*.

The effect is to place most of the animals in Australia beyond the reach of *Animal Welfare Acts* even though those Acts are specifically intended to provide for animal welfare. For example, the recitals of the *Animal Care and Protection Act 2001* (Qld) ("the Queensland Act") states:

> ....An Act to promote the responsible care and use of animals and to protect animals from cruelty, and for other purposes.

Section 15(1) of the Queensland Act provides that a regulation may require a person to comply with the whole or part of a *Model Code of Practice*.

However the *Animal Care and Protection Regulation 2002* (Qld) ("the Regulation") provides that almost all of the *Codes of Practice* and *Animal Welfare Standards* are completely voluntary; they are not enforceable in Queensland. In addition, s 40 of the Queensland Act then creates a specific offence exemption for conduct that was permitted by a Code of Practice (that is not enforceable in any case).

11.7.5 Lack of Effective Centralised Enforcement

There is no single enforcement authority responsible for animals and animal welfare. In reality, enforcement of animal welfare and animal related issues is fragmented across different government departments and private associations such as the RSPCA. The RSPCA is in fact a collection of 8 different private charitable associations incorporated under State and Territory *Associations Incorporation Legislation*. The RSPCA's in each State and Territory have limited enforcement powers and funding.[113]

---

[112]Ellis (2010), p. 13.

[113]Deborah. Cao, *Animal Law in Australia and New Zealand*, 2010 Lawbook Co, Sydney, NSW at 140.

The difficulty in detecting and prosecuting contraventions of Australia's animal laws is exacerbated by State and Territory initiatives such as the *Prevention of Cruelty to Animals (Prosecution) Act 2007* intended to limit the standing for those persons entitled to prosecute contraventions of animal welfare laws.

Complexity, confusing and inconsistent language, lack of comprehensive enforcement and an inherent conflict of interest in Australia's regulatory regime combine to ensure that food animals remain some of the most vulnerable sentient beings in contemporary society.

And food animals are likely to remain so given an expanding world population, an enlarging middle class in Asia and higher incomes driving an increase in the demand for animal food products. Agricultural producers are predicting significant economic gains to be made through the export of animals, food products and other agricultural resources. In fact the National Farmers Federation quotes former Australian Prime Minster Julia Gillard as anticipating "the potential for a new golden era of Australian agriculture given the rise of Asia."[114]

Economic efficiency is the key to realising this potential golden era. Large vertically integrated corporations managing CAFOs throughout Australia will attempt to realise even higher productivity gains from the animals they process. Under the profit-oriented SCOPI, existing MCOPs and future Animal Standards will continue to permit the handling and exploitation of animals in ways that involve cruel practices in pursuit of efficiency.

Animals in these CFAOs are commodities, units of production intended to maximise the profit of agricultural corporations as they exploit emerging world markets, satisfying increasing consumer demand for animal food products.

Constitutionally unable or perhaps politically unwilling to mandate Australia-wide application of MCOPs and Animal Standards, the Commonwealth government continues to preside over a complex and inefficient regime that attempts contradictory policy objectives.

Until 2014, the Australian government expressed its policy of encouraging animal welfare through the *Australian Animal Welfare Strategy*. In the 2014 Budget, the government eliminated the *Strategy* and diverted its funds to the live export industry.

What little policy oriented toward farm animal welfare protection remains lies in the hands of State and Territory governments working through the *Standing Committee on Primary Industry* to produce MCOPs and Animal Standards that permit farming practices that would be illegal under State and Territory *Animal Welfare* legislation. Of course, these exemptions assist SCoPI in fulfilling its stated policy objective of ensuring efficient and profitable primary industries.

Farm animals in Australia are thus legally characterised as property, ethically characterised as utilitarian inputs to corporate profit and, to ensure "efficient"

---

[114]National Farmers Federation, *NFF Farm Facts: 2012* at 3. < http://www.nff.org.au/farm-facts.html> accessed on 29 March 2012.

treatment, placed largely beyond the reach of most State and Territory *Animal Welfare Acts*.

References

Allen R (1992) Tracking the industrial revolution in England. Econ Hist Rev 52(2):209

Bilgili S, Hess J, Blake J, Macklin K, Saenmahayak B, Sibley J (2009) Influence of bedding materials on footpad dermatitis in broiler chickens. J Appl Poultry Res 18:583

Black C (2013) Live export and the WTO: considering the exporter supply chain assurance system. Macquarie Law J 11:77–80

Bruce A, Faunce T (2015) Nanotechnology-based artificial photosynthesis: food security and animal rights in the sustainocene. Chapter 12 In: Faunce T (ed) Nanotechnology toward the Sustainocene. Pan Stanford Publishing, Singapore, p 259

Buijs S, Keeling L, Vangestel C, Baert J, Vangeyte J, Tuyttens F (2010) Resting or hiding? why broiler chickens stay near walls and how density affects this. Appl Anim Behav Sci 124:97–102

Cao D (2010) Animal Law in Australia and New Zealand. Lawbook Company Limited, Australia, p 100

Cao D, Sharman K, White S (2010) Animal Law in Australia and New Zealand. Lawbook Co, Thomson Reuters, Sydney, p 115

Cassuto D (2007) Bred meat: the cultural foundation of the factory farm. Law Contemp Probl 70:59

Casuto D (2012) Old McDonald's had a farm: the metaphysics of factory farming. J Anim Ethics 2 (1):73–75

Chaudhri R (2014) Animal welfare and the WTO: the legality and implications of live export restrictions under international trade law. Fed Law Rev 42:279

Cheever H (2000) Concentrated animal feeding operations: the bigger picture. Alberta Law Environ Outlook 5:43

Coghlan S (2014) Australia and live animal export: wronging nonhuman animals. J Anim Ethics 4 (2):45–49

Dale A (2009) Animal welfare codes and regulations – the devil in disguise? In: Sankoff P, White S (eds) Animal law in Australasia. The Federation Press, Sydney, pp 174, 185–186

Deep A, Schwean-Larder K, Crowe T, Fancher B, Classen H (2010) Effect of light intensity on broiler production. Process Charact Welf Poultry Sci 89:2326

Elferink EV, Nonhebel S (2007) Variation in law requirements for meat production. J Cleaner Prod 15:1778

Ellis E (2010) Making sausages & law: the failure of Australian animal welfare laws to protect both animals and fundamental tenets of Australia's legal system Aust Anim Prot Law J 4:4

Evans G (2006) To what extent does wealth maximisation benefit farmed animals? Anim Law 13:167

Faunce T (2011) Global artificial photosynthesis project: a scientific and legal introduction. J Law Med 19:275–276 ff

Faunce F (2012) Nanotechnology for a sustainable world. Global artificial photosynthesis as the moral culmination of nanotechnology. Edward Elgar, Cheltenham

Faunce T (2015) Preface in nanotechnology toward the sustainocene. Pan Stanford Publishing, Singapore, p xxxii

Faunce TA, Lubitz W, Rutherford AW et al (2013) Energy and environment policy case for a global project on artificial photosynthesis. Energy Environ Sci 6(3):695

Faunce TA, Bruce A, Donohoo AM (2015) Toward the sustainocene with global artificial photosynthesis, Chapter 13. In Faunce T (ed) Nanotechnology toward the sustainocene. Pan Stanford Publishing

Follmer J (2009) Whatever happened to old Macdonald's farm....concentrated animal feeding operation, factory farming and the safety of the nation's food supply. J Food Policy (2009) 5, 45:51–52

Galyean M, Ponce C, Schultz J (2011) The future of beef production in North America. Anim Front 1(2):29–32

Glatz P, Critchley K, Hill M, Lunman C (2009) The Domestic Chicken, ANZCCART Fact Sheet A11:3. <http://www.adelaide.edu.au/ANZCCART/publications/A11_DomesticChickenFactSheet.pdf> (Accessed 1 April 2012)

Gunderson R (2011) From cattle to capital: exchange value, animal commodification, and barbarism. Crit Sociol 1:3–4

Hastreiter M (2013) Animal welfare standards and Australia's live exports industry to Indonesia: creating an opportunity out of a crisis. Wash Univ Glob Stud Law Rev 12:181

Hindle V, Lambooij E, Reimert H, Workel L, Gerritzen M (2010) Animal welfare concerns during the use of water bath for stunning broiles, hens and ducks. Poultry Sci 89:401–402

Hocquette J-F, Chatellier V (2011) Prospects for the European beef sector over the next 30 years. Anim Front 20

Hume DA, Whitelaw CBA, Archibald AL (2011) The future of animal production: improving productivity and sustainability. J Agric Sci 1:2

Ibrahim D (2007) A return to descartes: property, profit and the corporate ownership of animals. Law Contemp Probl 70:86

Jamieson P (1991) Duty and the beast: the movement in reform of animal welfare law. Univ Queensl Law J 16(2):238

Lee Nardo M (2000) Feedlots - Rural America's sewer. Anim Law 83(2000):6

Mason J, Singer P (1990) Animal factories. Harmony Books

McAllister R et al (2006) Australian pastoralists in time and space: the evolution of a complex adaptive system. Ecol Soc 11(2):41

Olanrewaju H et al (2006) A review of lighting programs for broiler production. Int J Poultry Sci 5 (4):301

Plowman K, Pearson A, Topfer J (2008) Animals and the Law in Australia: a livestock industry perspective. Aust Law Reform Comm 91:25–29

Sharman K (2008) Putting the chicken before the egg: layer egg housing laws in Australia. Anim Prot Law J 1:46–48

Sirri F, Minelli G, Folegatti E, Lolli S, Meluzzi A (2007) Foot dermatitis and productive traits in broiler chickens kept with different stocking densities, litter types and light regimen. Ital J Anim Sci 6(suppl 1):734

Tao B (2002–2003) A stitch in time: addressing the environmental, health and animal welfare effects of China's expanding meat industry. Georgetown Int Environ Law Rev 15:321

Thomas R (2005) Zooarcheology, improvement and the British agricultural revolution. Int J Hist Archeol 9(2):71

Thornton P (2010) Livestock production: recent trends, future prospects. Philos Trans R Soc 365:2854–2855

Vinnari M, Tapio P (2009) Future images of meat consumption in 2030. Futures 41:269

Welty J (2007) Humane slaughter laws. Law Contemp Probl 70:176–182

Wender M (2011) Goodbye family farms and hello agribusiness. Villanova Environ Law J 22:141

White S (2007) Regulation of animal welfare in Australia and the emergent commonwealth: entrenching the traditional approach of the state and territories or laying the ground for reform? Fed Law Rev 35:347

Winders B, Nibert D (2004) Consuming the surplus: expanding 'meat' consumption and animal oppression. J Soc Policy 24(9):76

Chapter 12
Textbox: The Farm Bill

Patty Lovera

Abstract Whether consumers get their food at a supermarket, a farmers market, a restaurant or a food bank, the Farm Bill had some impact on what they were eating. With the Farm Bill, a major piece of legislation that is revised and renewed about every 5 years, Congress sets the policies and programs that shape what food is available to the public, how it is produced and where it is sold. The Farm Bill covers government support for farmers, agricultural research and marketing, trade policies, energy issues, rural land use and conservation programs, and the Supplemental Nutrition Assistance Program, the primary government assistance program to help low-income families purchase food.

Whether consumers get their food at a supermarket, a farmers market, a restaurant or a food bank, the Farm Bill had some impact on what they were eating. With the Farm Bill, a major piece of legislation that is revised and renewed about every 5 years, Congress sets the policies and programs that shape what food is available to the public, how it is produced and where it is sold. The Farm Bill covers government support for farmers, agricultural research and marketing, trade policies, energy issues, rural land use and conservation programs, and the Supplemental Nutrition Assistance Program, the primary government assistance program to help low-income families purchase food.

12.1 A Broken Food System

The meatpackers, food processors and retailers who buy crops and livestock have gotten bigger and more powerful in the last several decades. Because there are so few competitors, they do not bid up the price of crops and livestock; instead, they tend to push down the prices farmers receive. Farmers facing long-term declining prices have been forced to specialize in one or two crops or a single stage of life for

P. Lovera (✉)
Food & Water Watch, Washington, DC, USA
e-mail: Plovera@fwwatch.org

G. Steier, K.K. Patel (eds.), *International Farm Animal, Wildlife and Food Safety Law*, DOI 10.1007/978-3-319-18002-1_12

a single type of livestock and scale up to recoup their losses with more sales. This trend is often explained to farmers as "get big or get out." Over the same period, agribusinesses and market-oriented "reformers" chipped away at the farm policies that ensured that farmers were paid more for their crops than it cost to grow them. The big corporate buyers wanted to pay as little as possible for farm products like corn, cattle and milk and changed farm policy to do away with programs designed to help farmers avoid the cycle of overproduction that drives down crop prices.

Federal farm programs were developed to provide a safety net for farmers to blunt the effects of wild price swings that are unique to agriculture. While the demand for food remains fairly steady—people do not become hungrier when food is cheap or less hungry when it is expensive—the supply of food is vulnerable to droughts, floods, pests or unusually good seasons with high yields. All of these factors can create volatility in the price farmers are paid for their crops.

While low crop prices hurt farmers, they are a boon to the agribusinesses that buy these commodities. The agribusiness processors and grocery stores that buy farm products have taken advantage of the savings from cheap input prices to consolidate into larger operations. Since the 1990s, every segment of the agriculture and food industry—from seeds to grain companies to meatpackers to food processors to grocery stores—has become considerably more concentrated as a wave of mega-mergers increased the size and dominance of the largest players.

This consolidation hurts farmers when they buy supplies and when they sell their farm products. At the beginning of the food chain, there are very few companies supplying farmers with inputs like seed and fertilizer, and the lack of competition drives up costs for farmers. There are also few companies buying crops and livestock, so farmers and ranchers are essentially forced to sell at whatever prices these agribusiness giants offer.

Consumers also feel the pinch of consolidation at the grocery store. The number of brands and food varieties at the supermarket creates the illusion of abundant choice, but most food is manufactured by only a handful of firms that sell their products under many brand names. Supermarket chains themselves are very concentrated, with half of sales going to four companies. On the local level, the top four chains can control more than 70 % of the marketplace.[1] Walmart alone controls more than 50 % of the grocery market in 29 markets across the country.[2]

Because a few agribusiness and grocery companies wield most of the buying power in the food system, they can pay farmers a low price at one end of the food chain and charge consumers a high price for their groceries at the other. Since the mid-1980s, the cost of a typical basket of groceries, adjusted for inflation, has risen

[1]"*Supermarket News*'s top 75 retailers for 2009." *Supermarket News*. June 2009; Martinez, Steve W. USDA Economic Research Service. "The U.S. Food Marketing System: Recent Developments 1997–2006." Economic Research Report Number 42. May 2007 at note 11 at 18.

[2]United Food & Commercial Workers. "Ending Walmart's rural stranglehold." 2010 at 6.

relatively steadily.[3] In contrast, the farmer's share of consumer dollars spent on this market basket of groceries fell from 35 % in 1984 to 23 % in 2008.[4]

12.2 How Did We Get Here?

U.S. farm policy was not always set up to favor large agribusiness.

12.2.1 The New Deal

The federal role in agriculture dates to before the Civil War but was expanded after the agricultural collapse during the Great Depression.[5] Farm prices were sky high after World War I scorched European farmland, and American farmers planted far and wide to take advantage of the high prices. However, as European production recovered, overproduction led U.S. prices to plummet in the 1920s.[6] When a severe drought hit in the 1930s, much of the farmland dried up and blew away, creating a dustbowl in the Great Plains. Farm bankruptcies exploded, and many farmers lost their land when the topsoil disappeared in dust storms. Farmers attempted to organize voluntary and cooperative reductions in supply to try to balance out prices, but were unable to do so without government support.[7]

President Franklin D. Roosevelt's New Deal aimed to balance out the wild market fluctuations and provide a safety net during years of low farm prices. In effect, the government was ensuring that agribusinesses buying the farmers' commodity crops paid farmers a decent price for their crops that at least covered the cost of producing them. And without the rampant corporate concentration that exists today, the number of competitive buyers for crops effectively bid prices upward. Farmers earned their income from selling their crops for a fair return when farm policies ensured that volatility did not undermine the viability of farm households. These programs worked pretty well for farmers and consumers for decades.

[3]USDA National Agricultural Statistics Service, Agricultural Statistics Annual (1994 to 2010) at Table 9-34.

[4]*Ibid. at* Table 9-34.

[5]*See* Effland, Anne B. W. USDA Economic Research Service. "U.S. Farm Policy: The First 200 Years." Agricultural Outlook. March 2000 at 22, 24.

[6]*Ibid.* at 24.

[7]USDA ERS. "History of Agricultural Price-Support Programs, 1933–1984." (AIB 485). December 1984 at 1–2.

12.2.2 Export Promises

In the 1970s, policymakers began a shift in farm policy that continues to reverberate to this day. First, agribusiness-friendly politicians contended that global demand for U.S. exports could replace the policies of the New Deal.[8] The Cold War thaw during the Nixon administration presented the prospect for new exports to the Soviet Union and worldwide.[9] Farmers were encouraged to plant "fencerow-to-fencerow" to feed the promised ever-increasing demand for their farm products around the world.[10] The export proponents argued that excess farm production could be exported, which would prevent over-supply in the U.S. market and prevent crop prices from falling.

12.2.3 Deregulation

The agricultural boom of the 1970s—with high crop prices, newfound export markets and farm expansion fueled by low interest rates—was sharply reversed in the early-1980s farm crisis. Crop prices fell and farmers paid more for seeds, fertilizer and other inputs than they received for their crops.[11] Net farm income fell by half between 1981 and 1983, and farmland values slid by almost a third between 1982 and 1985.[12] There was a higher rate of farm bankruptcies in 1987 than during the Great Depression, and more than 9500 farms filed for bankruptcy between 1987 and 1989.[13] However, instead of recognizing that the export boom of the seventies was a bubble that had burst, policymakers kept their faith in trade and turned their attention to unraveling farm safety nets.

During the 1980s, market-driven agricultural policy initiatives began to replace the programs of the New Deal. Policies that prevented farm prices from collapsing were increasingly viewed as limiting U.S. export opportunities.[14] Policymakers insisted that pushing crop prices lower would help exporters undercut foreign competition and sell more, essentially "dumping" U.S. crops on foreign markets for a price that was lower than the local cost of production.[15] Overseas farmers

---

[8] *Ibid.* at 27.

[9] *Ibid.* at 29; *see* President Richard Nixon, Annual Message to the Congress on the State of the Union, January 30, 1974.

[10] Wyant, Sara. "Memories of Agriculture Secretary Earl Butz." *Agri-Pulse*. February 10, 2008.

[11] U.S. General Accounting Office. "Farm Finance: Financial Condition of American Agriculture as of December 31, 1986." GAO/RCED-88-26BR. October 1987 at 49.

[12] *Ibid.* at 13, 33.

[13] USDA ERS. "Farmer Bankruptcies and Farm Exits in the United States, 1899-2002." (AIB-788). March 2004 at 13.

[14] Effland. 2000 at 24.

[15] Ray et al. (2003), p. 16.

could not compete with the flood of U.S. imports that were cheaper than locally produced goods.[16] And U.S. farmers tried to make up for their low prices by producing more, which drove down prices even further.

12.2.4 *Freedom to Farm = Freedom to Fail*

Despite the initial deregulation of the 1980s, some programs that could stabilize prices for commodity crops still existed. However, all this changed with the 1996 "Freedom to Farm" bill, which ended the structural safety nets that had protected farmers during lean years for decades. At the behest of the giant agribusiness corporations that purchase commodity crops, the 1996 Farm Bill completely eliminated the requirement to keep some land idle, which encouraged farmers to plant as much as they could.

The failure of the 1996 Farm Bill led to record-level government farm payments. Although the legislation was designed to completely phase out farm program payments, dramatically falling farm prices led to direct government payments to farmers. Critics from the left and right pointed to the direct payments as a poster-child for wasteful agricultural policy. Nearly two decades later, the 2014 Farm Bill formally ended direct payments and reallocated federal funds to subsidize premiums for crop insurance instead.

12.3 Who Benefits from Bad Policy?

Farmers lose when crop prices collapse, but buyers of those crops win. With lower cost inputs of corn and soybeans, agribusiness can produce processed foods and high-fructose corn syrup much more cheaply. And instead of raising livestock on pasture, animals can be crammed into factory farms and fed artificially cheap corn- and soybean-based animal feed. A Tufts University study found that factory farms saved $34.8 billion between 1997 and 2005 because they were able to buy feed at below-production cost.[17] The buyers were silent when crop prices fell for decades. When commodity prices rose in 2007 and 2008, meatpackers and poultry processors saw significant drops in profit as the cost of their major input—feed—started to rise.[18]

---

[16] *Ibid.* at 11, 13.

[17] Starmer, E., and T.A. Wise. Global Development and Environment Institute of Tufts University. "Feeding at the Trough: Industrial Livestock Firms Saved $35 Billion from Low Feed Prices." GDAE Policy Brief No. 07-03. December 2007.

[18] Simon, Ellon. "Pork, chicken prices may rise in next wave of food inflation." *The Associated Press.* May 5, 2008; Purdue University. [Press release]. "Pork industry facing twin horrors, says Purdue expert." March 7, 2008.

12.4 Are Subsidies Really the Problem?

It has become quite common for every problem in the food system to be blamed on misguided farm subsidy programs. But no matter how often it is repeated, it's not that simple. The 2014 Farm Bill's end to direct government payments will not fix the problems in our food supply because the payments were the result, not the cause, of the low prices farmers received for their crops.

Farm program payments are not the main reason that U.S. farmers grow lots of corn and soybeans. Farmers plant crops that are in demand by the largest buyers: grain-trading companies, meatpackers and factory farms, and food manufacturers.

Because of decades of corporate-controlled farm policy and consolidation of agribusiness crop buyers, commodity crops are the only option for farmers in many parts of the country. A rural wheat farmer with a few thousand acres of wheat cannot suddenly switch to growing tomatoes to sell directly to consumers at the farmers market. The infrastructure needed to sustain this type of transition away from intensive commodity crop production no longer exists.

12.5 Better Agriculture Policy

Instead of encouraging overproduction and maintaining farm programs that really benefit the big agribusiness companies, it is time to restore supply management policies and price safety nets that make agribusiness, not taxpayers, pay farmers fairly for the food they grow. This means bringing back strategic grain reserves, requiring that farmers leave a portion of land fallow and maintaining minimum price floors for crops to ensure that, at the very least, farmers are paid for the cost of producing their crops.

12.5.1 Reining in Corporate Control

The consolidated market power of meat and poultry companies has reduced the earnings of livestock producers, forced them to become significantly larger and encouraged them to adopt the more intensive practices used on factory farms. The supposed efficiency gains from larger operations ignore the considerable cost to communities and the environment from this type of industrialized agriculture. The intensive methods come with a host of environmental and public health costs such as air pollution, contamination of water with manure and increases in antibiotic resistant bacteria in the environment.

The 2008 Farm Bill included the first-ever livestock title that made some progress in addressing the lack of competition in the livestock sector, including a provision that directed USDA to develop new rules to ensure that livestock producers are treated fairly by meatpackers and poultry companies.[19] In 2010, USDA issued proposed rules to prohibit unfair and abusive contract terms for poultry and hog growers, including banning retaliation against growers who speak out against unfair treatment, allowing growers to opt out of binding arbitration clauses in contract disputes and ensuring that growers who make significant investments in their farms receive long enough contracts to repay the loans.[20] In addition, the rules as originally proposed ensured that favored livestock producers were not rewarded with sweetheart deals from the meatpackers while others received lower prices for the same number and quality of livestock.

By late 2011, the rulemaking process had ground to a standstill under pressure from the meatpacking and poultry industry and many needed reforms were indefinitely delayed. Fully implementing the original proposed rules as well as additional reforms are still needed, like addressing captive supply arrangements including prohibiting meatpackers from owning livestock and thus manipulating market prices, and ensuring all contracts are based on pre-agreed, set prices and firm dates of delivery. This would prohibit meatpackers from using a pricing system that could provide unfair advantage to some producers and disadvantage others.

12.5.2 Regional Food Systems

The Farm Bill triggers hundreds of millions of dollars of USDA spending on rural development, ranging from grants to local governments and community organizations to government-backed loans to businesses.[21] Unfortunately, many past bills have focused funding only on larger projects like broadband Internet access or businesses that don't help rebuild food systems, like hotels or convenience stores selling processed food.[22] What has been sorely lacking is investment in agricultural-related industries and infrastructure that would support the vegetable, grain, dairy and livestock farmers who need distribution, packing and processing facilities before they can bring their products to market.

The 2014 Farm Bill expanded funding for the Farmers Market and Local Food Promotion Program to $30 million per year, nearly tripling the funding provided by the previous bill.[23]

---

[19]PL 110-246 Food, Conservation, and Energy Act of 2008. June 18, 2008 at §11002.

[20]75 Fed. Reg 35338. June 22, 2010.

[21]PL 110-246 Food, Conservation, and Energy Act of 2008. June 18, 2008 at Title VI Rural Development.

[22]Pates, Mikkel. "Vilsack appears with Peterson in Minnesota." *AgWeek*. February 23, 2010.

[23]USDA. "2014 Farm Bill Highlights." March 2014 at 2.

12.5.3 Promoting Environmental Stewardship

While a major portion of funding generated by the Farm Bill has historically gone to conservation programs that either encourage farmers to take vulnerable land out of production, the 2014 Farm Bill cut funding for conservation programs.[24]

Several Farm Bills have redirected conservation to subsidize short-term, technology-heavy "fixes" to the pollution problems of industrial livestock operations such as manure digesters for large scale factory farms. Instead, funds should facilitate the transition to and maintenance of farm management strategies that improve biodiversity, minimize air and water pollution and conserve soil, water and other essential resources. Conservation programs should support the transition to organic farming and help farmers identify crops and techniques appropriate to their region's water resources and climate.

12.6 Conclusion

U.S. farm policy, for decades driven by agribusiness's desire for cheap raw materials, should instead focus on ensuring that farmers and farmworkers who grow our food can earn a decent living, that farmers can sell their goods in genuinely competitive regional markets and that consumers are able to access sustainably grown, regionally produced food.

Reference

Ray D et al (2003) Rethinking US agricultural policy: changing course to secure farmer livelihoods worldwide. Agricultural Policy Analysis Center, University of Tennessee, p 16

[24]Clausen, Roger. "Conservation." USDA Economic Research Service (ERS). April 11, 2014.

Part III
Marine Animals and Fishing

Chapter 13
Overfishing and Bycatch

Anastasia Telesetsky

Abstract Humans have been consuming seafood since the genesis of *Homo sapiens*. Today, marine fisheries are the most important source of wild food in the world, providing the primary source of protein for millions of people particularly in developing countries. Yet, marine fisheries are vastly overexploited due to a variety of factors including overcapitalization in the industry, increasing levels of technology, illegal fishing, and reckless harvesting. The collapse of fisheries reflects a double jeopardy for many individuals and communities. In addition to the immediate losses of food resources, there are also associated costs in the form of lost livelihoods for both this generation and future generations. These losses may be particularly acute for developing countries since half of the world's fish trade is sourced from developing countries. This chapter describes two related phenomena associated with marine fisheries law—overfishing and bycatch—and outlines the existing legal regimes to address these phenomena.

13.1 Introduction

Humans have been consuming seafood since the genesis of *Homo sapiens*.[1] Today, marine fisheries are the most important source of wild food in the world, providing the primary source of protein for millions of people particularly in developing countries.[2] Yet, marine fisheries are vastly overexploited due to a variety of factors including overcapitalization in the industry, increasing levels of technology, illegal

[1]Marean et al. (2007) (Documenting shell middens dating back 164,000 years containing brown mussels, giant periwinkles, and whelks.)

[2]World Health Organization, Global and Regional Food Consumption Patterns and Trends, Available: http://www.who.int/nutrition/topics/3_foodconsumption/en/index5.html (A billion people depend on fish as a primary source of protein).

A. Telesetsky (✉)
University of Idaho College of Law, Natural Resources and Environmental Law Program, Moscow, ID, USA
e-mail: atelesetsky@uidaho.edu

G. Steier, K.K. Patel (eds.), *International Farm Animal, Wildlife and Food Safety Law*, DOI 10.1007/978-3-319-18002-1_13

fishing, and reckless harvesting. The collapse of fisheries reflects a double jeopardy for many individuals and communities. In addition to the immediate losses of food resources, there are also associated costs in the form of lost livelihoods for both this generation and future generations.[3] These losses may be particularly acute for developing countries since half of the world's fish trade is sourced from developing countries.[4] This chapter describes two related phenomena associated with marine fisheries law—overfishing and bycatch—and outlines the existing legal regimes to address these phenomena.

13.2 Overfishing

Historically, fisheries have been the classic open-access resources. The sustainability of the resource was not questioned until the advent of fishing technology in the form of trawls and steam powered ships and their toll on inland fishery stocks and later on coastal stocks. With the passing of each decade, humans became increasingly efficient in their predatory fishing practices culminating in what are dubbed today "supertrawlers" measuring over an American football field in length with a 300 m long net that could almost contain the Queen Mary 2 cruise liner (311 m), and a 275 metric ton cargo capacity.[5]

One of the most perplexing questions for fisheries biologist is "how many fish are in the sea?" The answer depends on complex and competing models. Regardless of exact number of fish in the sea, the overall trends of modern exploitation can be sharply contrasted with the historical promise of boundless fishing to satisfy the collective freedom to fish. Once internationally important harvest species included the Indian sardine (1940s), Japanese sardine (1940s and 1950s), South African pilchard (1965–1966), Greenland cod (1968), Georges Bank haddock (1968), Namibian pilchard (1970–1971), Peruvian anchovy (1972–1973), Gulf of Guinea sardine (1973–1974), and Canadian Atlantic cod (1990s) have all been severely overfished.[6] According to the Food and Agriculture Organization (FAO) estimate, today 87 % of marine fisheries are fully-exploited (operating at or close to an optimal yield level, with no ability to expand the fishery) or overexploited (operating beyond optimal yield level and including recovering and depleted stocks).[7]

---

[3]The classic story of a community catastrophe based on the collapse of a fishery is the Grand Banks, Canada cod fishery. Kurlansky M (1997) Cod: A Biography of the Fish that Changed the World, Walker Books.

[4]Food and Agriculture Organization, Committee on Fisheries, Sub-Committee on Fish Trade, COFI:FT/XI/2008/3, June 2008: para. 10 (Seafood export value for developing countries is $25 billion per year).

[5]*See* Ross (2012).

[6]Garcia (2003), p. 71. The Ecosystem Approach to Fisheries, FAO Fisheries Technical Paper 43.

[7]Food and Agriculture Organization (2012), p. 11 (57 % of species are fully exploited and 29.9 % are overexploited).

Using similar data, consisting largely of fishery data from Europe and North America, fishery scientists conclude that as of 2009, one-third of "all stocks can be classified as overfished".[8] Other scientists using broader data sets than those used by FAO estimate that as much as two-thirds of all global fisheries are overfished.[9]

Overfishing is a complex phenomenon engaging a variety of actors ranging from artisanal fishing fleets operating in territorial waters to industrial fleets operating on the high seas. The underlying cause of overfishing is poor management whether it is at the local level for sedentary stocks or at a global level for migratory stocks. Compounding the management issues are two distinct but significant problems that institutions have legally addressed with only limited success: illegal, unreported, and unregulated fishing (IUU fishing) and overcapacity of fishing effort. While these problems also occur within artisanal fleets, this chapter will look at both of these phenomenona with an emphasis on industrial fleets.[10]

13.3 IUU Industrial Fishing

IUU industrial fishing is the result of a number of independent but related factors including overcapacity of boats (discussed in the next sub-section), the provision of government subsidies to distant water fishing fleets, inadequate fishery management, and corruption in enforcement efforts. With the existing pressure on marine resources from legal fishing interests who seek management decisions that favor larger but potentially ecologically riskier catch quotas over ecologically conservative smaller quotas, IUU fishing activities may become the tipping point for the viability of some marine fisheries. The presence of IUU fishing may determine in a well-managed fishery whether a fishery can continue operation for future seasons or will need to be precautionarily closed. In an unmanaged fishery, IUU fishing may result in the eventual irreversible collapse of a stock.

Adopted by the international community as a term of art, the negotiated term "IUU fishing" is an attempt to provide a bright-line between sustainable and unsustainable fishing. In general, "Illegal" fishing refers to fishing activities that violate national laws or international measures, which are supposed to establish a maximum sustainable yield for commercial fisheries. "Unreported" fishing refers to the failure of vessels to accurately report catches. This undesirable fishing practice is particularly problematic because future species quotas are decided based on catch registers. The final category of "unregulated" fishing was created to respond to the

[8]Hilborn and Hilborn (2012), p. 123. (Noting that historical data has not been collected in Asia and Africa but that similar trends are expected in these regions as fishing pressure increases to meet food security demands).

[9]Costello et al. (2012).

[10]Artisanal fisheries refer to subsistence fishing using smaller vessels, generally serving local markets. Industrial fisheries refer to commercial fishing using a combination of small and large vessels to participate in a global fisheries market.

reality that some activities that are otherwise legal, may still be damaging in terms of conservation outcomes.[11] In practice, "unregulated" fishing has been applied to activities of vessels in the area under the jurisdiction of a regional fisheries management organization ("RFMO") where the flag state of a given vessel is not a member of the RFMO.

Another way of thinking about the operation of IUU industrial fishing vessels is to distinguish between rule-obedient fleets, "grey fleets", and "black fleets". Rule-obedient fleets are properly registered and adhere to the rules and regulations associated with their fishing permits. Rule-obedient fleets can still contribute to overfishing but the responsibility for this overfishing must be assigned to the fishing agency setting total allowable catches ("TAC" referring to a harvest cap on a given stock) and allocating permits among individuals or companies. A "grey fleet" includes vessels that are flagged to a State where at least one of the beneficial owners resides who have permits for a certain quota but make the choice to harvest resources beyond the quota. A "black fleet" includes vessels that are either stateless or have been registered to a flag of convenience in order to avoid monitoring or inspection regimes. Ultimately, the nature of a given fishing fleet or vessel will determine what type of control a coastal state may be able to exercise to prevent IUU fishing or seek compensation for lost stocks.

In addition to ecological costs illustrated by the collapse of stocks such as the Atlantic Bluefin tuna, IUU fishing also has economic and social costs. From an economic perspective, IUU fishing undervalues fishery resources and harms legitimate businesses. For example, the American crab industry may have lost $560 million over the last decade in market opportunities to illegal Russian crabbers whose black market is 3–4 times larger than their legal market.[12] Global estimates of losses from illegal and unreported fishing range from $10 to 23.5 billion annually.[13] From a social cost perspective, IUU fishing has attracted the interest of organized crime syndicates. Because IUU fishing for certain species such as tuna or swordfish is highly profitable and difficult to detect due to the commingling of legal and illegal products, IUU fishing may become just another profit generating strategy for groups who were formerly trading drugs, arms, or people.[14] This global interest in pursuing IUU fishing has resulted in additional social harms in the form of serious labor abuses of crewmembers that have been coerced into dangerous working conditions.[15]

[11]There is no single definition of IUU fishing. There may be a consensus on definition: the International Plan of Action to Prevent, Deter and Eliminate Illegal, Unreported and Unregulated Fishing, FAO Committee on Fisheries, March 2001 (endorsed by FAO Council June 2001).

[12]Joyce (2014).

[13]Agnew et al. (2009).

[14]UN Office of Drugs and Crime, Transnational Organized Crime in the Fishing Industry (2011), http://www.unodc.org/documents/human-trafficking/Issue_Paper_-_TOC_in_the_Fishing_Industry.pdf.

[15]International Labor Organization, Caught at Sea: Forced Labour and Trafficking in Fisheries (2013), http://www.ilo.org/wcmsp5/groups/public/---ed_norm/---declaration/documents/publication/wcms_214472.pdf.

13.3.1 Overcapacity of Vessels in Industrial Fleets

After a heartbreak, the clichéd response is that there are plenty of fish in the sea. The more accurate statement would be there are plenty of boats on the ocean. Driven by a number of factors including heavily subsidized boat construction, the Food and Agriculture Organization estimates that there are approximately 4.36 million fishing vessels of which 3.23 million vessels operate within marine waters.[16] There are so many boats in the water that the EU industrial tuna fleet reached its quota in one week in 2010.[17] The economic efficiency of harvesting quota in one week comes at the price of adequate recruitment since many of the tuna were gathered right before the spawning season began. In the U.S., a 2008 report by the National Marine Fisheries Service found that 12 of the 25 U.S. commercial fishing operations it reviewed had 50 % more boats than needed to bring in its total fish catch for the year.[18] Overcapacity is not just about the number of vessels but also the size of the vessels deployed and the technical efficiency of the vessels.

Subsidy programs have fueled the growth of the industry with programs such as the Capital Construction Fund and the Fisheries Obligation Guarantee Program leading to more vessels being constructed than can sustainably fish in a given fishery. When there are too many large vessels deployed, there are typically less fish to catch per vessel and less profits for all of the actors in the fishery. In a region with too many boats, the only politically attractive options to manage overcapacity have been either buyback programs or allocations of individual fishing quotas.

13.3.2 Bycatch

The general term, bycatch refers to fishing discards, which tend to be non-target fish (both commercial and non-commercial) and may include birds, sea mammals, and turtles that are accidentally trapped in nets. Bycatch refers to generally undesired species or unintended species that fishermen dispose of overseas either because they

[16]See the State of the World Fisheries and Aquaculture 2012. (Food and Agriculture Organization of the United Nations: Rome, Italy, 2012): 10–11 (73 % of all fishing vessels [marine and inland] are registered in Asia, 11 % in Africa, 8 % Latin America and the Caribbean, 3 % North America, and 3 % in Europe. 60 % of these vessels are motorized. Of these motorized vessels, 83 % were shorter than 12 m. On 2 % of the motorized vessels are 24 m or longer).

[17]World Wildlife Fund, EU Industrial Tuna Fishing Boats Reaching Quota in a Week is a Sign of Massive Overcapacity, June 9, 2010, http://wwf.panda.org/?193767/EU-industrial-tuna-fishing-boats-reaching-quota-in-a-week-is-sign-of-massive-overcapacity-WWF.

[18]National Marine Fisheries Service, Excess Harvesting Capacity in U.S. Fisheries. (2008) Available: www.nmfs.noaa.gov/msa2007/docs/042808_312_b_6_report.pdf.

do not want to be accused of damaging already endangered species (regulatory discards) or because keeping the discards would impact economic bottom lines by taking valuable cargo space (economic discards). In terms of non-target fish, the phenomenon of bycatch becomes a question of vast mountains of wasted protein. UNEP estimates that 30 million tons of fish are discarded annually accounting for 25 % of the marine catch.[19] Some fisheries are more prone to bycatch than other fisheries. The shrimp fisheries have notoriously high levels of bycatch with discard rates of 5 tons of fish for every ton of shrimp harvested.[20]

Certain types of fishing gear are particularly problematic. For example, when gillnets extending up from 5 to 11 km are deployed, these nets may capture large numbers of sharks.[21] Whales and other cetaceans are commonly captured as bycatch. On the whole the extent of bycatch is not fully understood since many States do not have comprehensive statistics. One estimate identifies bycatch rates of 60,000 cetaceans per year for fisheries in Iran, India, Sri Lanka, Pakistan, Oman and Yemen.[22]

13.4 Legal Frameworks to Address Global Overfishing and Bycatch

There is no dearth of international law focused on addressing the issues of overfishing and bycatch. This section reviews some of the major instruments including both binding legal instruments such as treaties and voluntary instruments such as codes and plans.

13.4.1 Binding Legal Instruments

13.4.1.1 Law of the Sea

States negotiated the United Nations Convention on the Law of the Sea ("LOS") in order to create "a legal order for the seas and oceans which will facilitate international communication, and will promote the peaceful uses of the seas and oceans, the equitable and efficient utilization of their resources, the conservation of the living resources, and the study, protection and preservation of the marine

[19]Stockhausen et al. (2012), pp. 90–95.

[20]Hilborn and Hilborn (2012), p. 116.

[21]Indian Ocean Tuna Commission–2012–WPEB08–13 Status Report on Bycatch of Tuna Gillnet Operations in Pakistan.

[22]Anderson (2014), p. 4. Cetaceans and Tuna Fisheries in the Western and Central Indian Oceans (2014).

environment."[23] Even though there are no direct references to IUU fishing, overcapacity, or bycatch in the treaty, these topics are implicitly addressed in the portions of the treaty creating obligations for conservation and management. Two sections of the lengthy treaty are of particular relevance: Part V and Part VII. Part V covers State obligations in relation to the Exclusive Economic Zone (EEZ), the marine area under the jurisdiction of the coastal State, which is measured 200 nautical miles (230.1 miles) from base-points along the coast of either a mainland or an island. When the LOS treaty was negotiated, the creation of the EEZ marked a radical departure from existing law by enlarging the jurisdiction of coastal States and requiring flag States to comply with coastal States fishing laws. The EEZ as a concept was a resource control boon for coastal States since most of the valuable commercial fish are found within EEZs and coastal States are empowered to evaluate their fishing capacity and decide whether or not to allow other States to fish within its EEZ. In terms of optimal utilization of the living marine resources, the EEZ changed a formerly open-access resource into a resource that theoretically would be better managed because of oversight from the coastal State. Part VII covers the high seas and refers to areas beyond national jurisdiction.[24] The flag state is the primary authority over high seas fishing activities.

Issues of overcapacity and reducing bycatch are not directly mentioned in the LOS. Within the EEZ, the coastal state is expected to promulgate measures that should in theory address the overcapacity issue by focusing on best management of stocks to produce a maximum sustainable yield. Reducing bycatch is not directly referenced in the Law of the Sea unless the bycatch is either a "species associated with or dependent upon a harvested species."[25] In the case of associate or dependent species, coastal States are expected to maintain or restore these species in order to protect targeted harvest species. In practice, the Food and Agriculture Organization has called for a more ecologically reliable ecosystem approach to fisheries.[26]

While the LOS has empowered coastal States to better manage EEZ resources for at least target species, it may have also inadvertently incentivized coastal States to build fleets. Article 62(2) provides that the coastal State, "shall determine its capacity to harvest the living resource of the exclusive economic zone." There are no further provisions describing how a State should determine its capacity. When a

[23]United Nations Convention on the Law of the Sea, December 10, 1982, 1833 U.N.T.S. 3, 397; 21 I.L.M. 1261 (1982).

[24]The acronym ABNJ is frequently used for "areas beyond national jurisdiction" by intergovernmental organizations.

[25]UNCLOS, Article 61.

[26]Food and Agriculture Organization, FAO Technical Guidelines on the Ecosystem Approach to Fisheries Vol. 4(Suppl. 2) (2003): 6. (Ecosystem Approach to Fisheries is defined as an approach that takes into "account the knowledge and uncertainties about biotic, abiotic and human components of ecosystems and their interactions" and applies "an integrated approach to fisheries within ecologically meaningful boundaries.").

State does not "have the capacity to harvest the entire allowable catch", then it is expected "through agreements or other arrangements. . .[to] give other States access to the surplus of the allowable catch."[27]

While a State may design its policies to primarily maximize resource usage which would include extending fishing opportunities to other States, it may also decide to prioritize State interests and limit access to EEZ waters for purposes of fishing to coastal State boats. In order to prioritize State interests while also complying with its obligation to promote "optimum utilization of the living resources", a coastal State must make a decision either to set a lower allowable catch that matches its actual fleet capacity or to increase its fishing capacity to meet the maximum sustainable production levels. A government decision to restrict allowable catch to match existing fleet capacity might raise issues of good faith compliance and inefficient resource protectionism. The latter decision of increasing a coastal State fleet's capacity seems less problematic under the treaty as a means of asserting broader control over the EEZ waters. To understand the potential impact of Article 62(2) on fleet overcapacity, one hypothesis that might be empirically tested is whether coastal State fishing fleets increased in overall numbers after a State declared an EEZ to maximize domestic harvest of an EEZ living marine resource by a coastal state or whether fishing vessel cargo capacity increased across the domestic fleet. After New Zealand declared its Exclusive Economic Zone in 1978, there appears to some correlation with capacity growth of the domestic fleet. In the period from 1978 to 1996, the number of New Zealand's domestic vessels over 24 m grew while vessels under 12 m decreased leading to an overall increase of 7000 t of domestic capacity.[28]

LOS Article 62(4) requires foreign nationals to comply with the conservation measures and fishing laws of the coastal state when fishing. In this treaty section, the term "national" may refer to individual persons, companies, and vessels flagged to a specific State. In maritime law, any vessel that travels internationally must be registered with a State and fly that State's flag. While Article 62 does not address the potential for illegal fishing activity by coastal state nationals in coastal state waters or overcapacity by a coastal state fleet, it does provide a legal means for a coastal State to ensure that vessels from neighboring states or distant water fishing vessels comply with conservation and management measures for EEZ stocks. In practice, this means not only the coastal State should exercise conservation oversight in its EEZ waters, but flag states should also exercise some degree of due diligence over its registered vessels. This means that if a State such as China contracts to deploy a distant water fishing fleet to fish in the exclusive economic zone of another State such as Guinea, then China should undertake some measures to ensure that its flagged vessels are not just following Chinese laws but also local conservation laws.

[27]UNCLOS, Article 62(2).

[28]Connor (2001), pp. 151–186. Available at http://www.fao.org/docrep/005/y2498e/y2498e12.gif.

In Part VII of the LOS involving the high seas, there is no specific mention of how States should manage resources to avoid overfishing or bycatch. Instead, the same general standards about attaining maximum sustainable yield and protecting species associated or dependent on harvested species are applied to high seas fishing activities.[29] States are further expected to "co-operate with each other in the conservation and management of living resources in the areas of the high seas" and "as appropriate, co-operate to establish sub-regional or regional fisheries organization to this end."[30] The threat of overfishing is potentially even more acute on the high seas because the law of the sea recognizes the right of States to pursue "freedom of fishing." While this freedom is subject to states cooperating to set allowable catch limits and establishing conservation measures, there are few means of enforcing cooperation under the treaty.[31]

Ongoing cooperation in management of open-access fisheries is critical. Environmental economists observe that overexploitation becomes more acute with the increase in parties harvesting a stock that crosses boundaries. When two countries fish a single stock that straddles a border, it is 9 % more likely to be overharvested and 19 % more likely to be depleted than if the stock only occurred within the waters of one country. For a stock that is shared by five countries, it is 36 % more likely to be overharvested and 82 % more likely to be depleted than if the stock only occurred within the waters of one country.[32]

13.4.1.2 UN Fish Stocks Agreement

The UN Fish Stocks Agreement was negotiated to address problems specific to highly migratory species and straddling stocks including free ridership.[33] A highly migratory species such as tuna is a species that may cross from one States' EEZ in the Western Atlantic through the high seas to another States' EEZ in the Eastern Atlantic. A straddling stock is a species such as cod whose population lives both within an EEZ and the high seas. Thus, the challenge for management of both migratory and straddling stocks is that these species may be governed very differently depending on the given location of the species. For example, within an EEZ, there may be a strict conservation regime; in contrast, there may be no regime on

---

[29]UNCLOS, Article 119(1).

[30]UNCLOS, Article 118.

[31]UNCLOS, Article 87(e).

[32]Also McWhinnie (2009).

[33]See UNCLOS, Article 64 ("The coastal State and other States whose nationals fish in the region for the highly migratory species listed in Annex 1 shall cooperate directly or through appropriate international organizations with a view to ensuring conservation and promoting the objective of optimum utilization of such species throughout the region, both within and beyond the exclusive economic zone.") and Article 116 (States have the right for their nationals to engage in fishing on the high seas subject to the rights, duties, and interests of coastal States as defined by Article 64); Bjorndal and Munro (2003).

the high seas since an appropriate regional fisheries management organization has not been formed.[34]

The Fish Stocks Agreement is an implementing agreement for Article 64 and Article 116 of the Law of the Sea. In terms of addressing concerns regarding overfishing and reducing bycatch, the Fish Stocks Agreement requires compatibility between conservation and management measures, which are created either within the EEZ or on the high seas.[35] As a general conservation and management principle, the treaty calls for coastal States and States fishing on the high seas to "minimize. . .waste, discards. . .catch of non-target species, both fish and non-fish species. . .and impacts on associated or dependent species, in particular endangered species, through measures including, to the extent practicable, the development and use of selective, environmentally safe and cost-effective fishing gear and techniques." The Fish Stocks Agreement has 82 members. Unfortunately, a number of States who have historically been accused of not using selective gear—resulting in the capture of associated species such as sea turtles, are still not members of the Fish Stocks Agreement.[36] Most managed fisheries have some sort of bycatch provisions. These provisions usually require parties fishing to use specific technology such as turtle excluder devices or may impose bycatch limits on vessels when there are high levels of bycatch in the fishery.

If the concept of the EEZ was the new idea for the LOS, the most powerful new idea within the Fish Stocks agreement was the introduction of the "precautionary approach", which if properly implemented, should address at least the overcapacity component of the overfishing problem. A typical precautionary approach requires that States not undermine sustainability objectives in permitting fishing within their waters or by their nationals. Under the treaty, coastal States and States fishing on the high seas are expected to apply the precautionary approach that includes setting reference points to correspond "to the state of the resource and of the fishery, and which can be used as a guide for fisheries management."[37] Where there is inadequate information about the status of the resource, States can set provisional reference points based on similar stocks and then readjust the point after additional monitoring. It is particularly important for States to include management measures that "can be implemented when precautionary reference points are approached."[38]

---

[34]See e.g. Bjørndal and Munro (2003).

[35]Agreement for the Implementation of the Provisions of the United Nations Convention of the Law of the Sea of December 10, 1982 Relating to the Conservation and Management of Straddling Fish Stocks and Highly Migratory Fish Stocks (UN Fish Stock Agreement) 34 I.L.M. 1542 (1995), Article 7(2).

[36]Before the WTO, the U.S., in the Shrimp-Turtle dispute alleged since India, Malaysia, Pakistan, and Thailand were failing to use turtle excluder devices in their fisheries, the U.S. could exclude shrimp products from these States. The WTO Dispute Panel disagreed with the U.S. position. India, Malaysia, Pakistan, and Thailand have declined to become parties to the Fish Stocks Agreement.

[37]UN Fish Stock Agreement, Annex II(1).

[38]Id.

While individual States may designate reference points, they are more likely to be implemented collectively through a regional fisheries management organization.[39]

13.4.1.3 Agreement to Promote Compliance with International Conservation and Management Measures by Fishing Vessels on the High Seas

The text of the Law of the Sea has not provided sufficient and specific guidance to forestall the continued tragedy of open-access resources in places such as the high seas. States adopting high seas conservation and management measures that might constrain the fishing activity of its nationals observed an exodus of vessel registrations favoring the so-called flags of convenience which are registrations in States that are unlikely to pass strict laws on conservation and management or to systematically enforce any such laws. Semi-notorious flags of convenience particularly associated with IUU fishing include States such as Panama and Mongolia (a land-locked State).

Drafted in 1993, the core obligation of the Agreement is the clarification of flag State responsibility over registered vessels to ensure that any vessel registered to the State does not undermine international conservation and management standards.[40] Specifically, the Agreement clearly states any vessel that will be fishing on the high seas must be in possession of an authorization from the appropriate Flag State authorities.[41]

To provide for some degree of coordination across States, the Agreement required a "record of fishing vessels" to be compiled by the FAO. At a minimum the record must include the name of the fishing vessel, flagging history of the vessel, name and address of owners, and physical information about the vessel.[42] In theory, the record would be particularly useful for regional monitoring and enforcement efforts. In reality, however the record has been less than complete for those States who are members. For example, FAO has access to fishing license information for only 13.5 % of the total 6300 ships in the database.[43]

This chronic absence of information is particularly problematic since States are obliged (under the Law of the Sea) to cooperate in managing high seas marine living resources. When some States fail to provide any information about the conditions attached to their fishing authorizations, cooperation is very difficult. This problem will be addressed in the final chapter describing additional law and policy interventions to address overfishing and bycatch.

---

[39]See e.g. Indian Ocean Tuna Commission, Resolution 13/10 On interim target and limit reference points and a decision framework (2013).

[40]Article III(1).

[41]Article III(2).

[42]Article VI(1).

[43]See FAO Fishery Records Collections, High Seas Vessels Authorization Record, http://www.fao.org/fishery/collection/hsvar/4/en.

13.4.1.4 Convention for the Prohibition of Fishing with Long Driftnets in the South Pacific (Wellington Convention)

By the late 1980s, the issue of bycatch had reached crisis proportions. Policymakers recognized that high seas driftnets extending over ten nautical miles were capturing large numbers of marine mammals, birds, turtles, and non-target fish species. In 1987, the United States in the North Pacific had been negotiating with drift netting States such as Japan, Korea, and Taiwan under its Driftnet Impact Monitoring, Assessment and Control Act for more responsible fishing practices.[44] South Pacific States sought to control drift netting practices and negotiated a treaty banning driftnets and preventing drift netting vessels from accessing ports within signatory States.[45] The objectives of this Convention were further reinforced with United Nations General Assembly resolutions approved by some of the States that were formerly permitting high seas drift-netting in their fleet. Drift netting is noticeably reduced because large fish importers such as the U.S. have been engaged in enforcing the prohibition and by leveraging its enforcement capacity through Memorandums of Understanding on enforcement concerns with countries such as China.[46]

13.4.1.5 FAO Agreement on Port State Measures

The International Plan of Action on IUU Fishing (discussed below) called upon States to consider developing port state measures that would prohibit landings and transshipment of catches unless a vessel could carry the burden of proof that its catch was properly harvested. In response to frustration over the continued negligence of flag States—especially flag of convenience States that failed to exercise adequate control and jurisdiction over their fleets, key port States, in 2005, negotiated a Model Port State Measure Scheme. This Scheme eventually became the basis for the FAO Agreement on Port State Measures. Under this binding agreement Parties have agreed to institute a series of basic port measures requiring vessels to request permission for port entry, requiring the use of designated ports, restricting landing/transhipment of fish, imposing port inspections, and other specific measures if a vessel is deemed to be engaged in IUU fishing.[47] Two of the largest importers of fish, the United States and the European Union are members of the Agreement.

---

[44] 16 U.S.C. 1822.

[45] Convention for the Prohibition of Fishing with Long Driftnets in the South Pacific, 29 I.L.M. 1449 (1990).

[46] See National Marine Fisheries Service, 2012 Driftnet Report: 7 and 17. http://www.nmfs.noaa.gov/ia/iuu/driftnet_reports/2012_driftnet_report.pdf (In the U.S., reports of drift-netting in the North Pacific were reduced from 98 sightings in 2006 (mostly Japan and Canada) to 1 sighting in 2012 (a United States vessel.)

[47] See Agreement on Port State Measures to Prevent, Deter and Eliminate Illegal, Unreported and Unregulated Fishing, Food and Agriculture Organization (2009).

13.4.1.6 Additional Treaties

States have negotiated a number of other treaties to provide some institutional capacity for tackling overfishing and bycatch by focusing collective fishing management efforts on a particular geographical region or on a particular stock.[48] Typically, the regional fisheries management organizations (RFMOs) attempt to address overfishing by establishing conservation and management measures for all members and by allocating stock among members based on maintaining a maximum sustainable yield. Rules of reducing bycatch are typically embedded in conservation measures. Historically, these organizations have been weak. Still, the FAO has been exploring the possibility of rights-based fishing for migratory stocks, which as suggested later in the chapter would provide an incentive for better fishery monitoring and enforcement by participants.[49]

On the issue of bycatch RFMOs have attempted to systematically address some of the recurring involving interactions between fishing vessels and protected species. At least three of the tuna RFMOs have binding measures on longline seabird bycatch.[50] Three of the tuna RFMOs have specific measures on turtle bycatch by purse seiners.[51] None of the tuna RFMOs have measures to reduce the catch of sharks, though there are restrictions on finning. Other measures are also in place for juvenile tuna and for dolphins.

13.4.2 Soft Law

13.4.2.1 FAO Code of Conduct for Responsible Fishing

In 1995 FAO parties drafted a voluntary code of conduct to be adopted in the context of pre-existing obligations under the Law of the Sea and applied globally by "members and non-members of FAO, fishing entities, sub-regional, regional and global organizations, whether governmental or non-governmental, and all persons concerned with the conservation of fishery resources and management and development of fisheries, such as fishers, those engaged in processing and marketing of

[48]See e.g. Convention for the Conservation of Atlantic Tunas, May 14, 1966, 20 U.S.T. 2887 (creating ICCAT body that sets quotas regularly); Convention on the Conservation of Antarctic Marine Living Resources, May 20, 1980, 33 U.S.T. 3476. (Creating a Commission to set regular conservation measures).

[49]Squires et al. (2013).

[50]International Commission for the Conservation of Atlantic Tuna, Resolution 07-07 (2007); Indian Ocean Tuna Commission Resolution 10-06 (2010); Western and Central Pacific Fisheries Commission Resolution 2007-04 (2007).

[51]Indian Ocean Tuna Commission Resolution 09-06 (2009); Inter-American Tropical Tuna Commission, Resolution 07-03 (2007); Western and Central Pacific Fisheries Commission Conservation and Management Measure 2008-03 (2008).

fish and fishery products, and other users of the aquatic environment in relation to fisheries."[52] The Code is particularly significant in light of international food law. Where other documents recognize the significance of the fishing industry, the code provides—as one of its objective, the promotion of fisheries for "food security and food quality, giving priority to the nutritional needs of local communities."[53]

Just as the code calls for "responsible fishing", it also offers specific principles for implementation. In keeping with the commitments under the Fish Stocks Agreements—also negotiated in 1995, the Code recommends the application of the precautionary approach by States and RFMOs along with the use of selective fishing gear that reduces waste, bycatch, and negative impacts on the environment.[54] Explicitly, the Code calls for States to "prevent overfishing and excess fishing capacity" by implementing measures that "ensure that fishing effort is commensurate with the productive capacity of the fishery resources and their sustainable utilization."[55] This is a departure from previous multilateral documents where there is no explicit recognition that "overfishing" is a chronic problem for open access resources.

On the topic of reducing bycatch, mentioned several times throughout the Code, not only States but also industry groups are expected to develop technologies that reduce discard, limit use of technology that increases discard rates, and promote "gear and practices that increase survival rates of escaping fish."[56] States can provide additional incentives by requiring the use of certain gear, methods, or practices.[57]

While perhaps discussion of the topics of overfishing and bycatch are not surprising for a code of responsible fishing, the code extends its reach to "responsible international trade." A practice of responsible fishing should further sustainable development and not just for the affluent parties in the world. The code urges States and intergovernmental organizations such as multilateral banks to ensure that their policies do not "adversely impact the nutritional rights and needs of people for whom fish is critical to their health and well being and for whom other comparable sources of food are not readily available or affordable."[58] This language suggests that fish must be protected as more than just commodity products. The Code makes it clear that it matters where fish originate and who are the beneficiaries of fisheries products at the place of origin.

[52]Code of Conduct for Responsible Fisheries, 28th Session of the FAO Conference on 31 October 1995: Article 1.2.

[53]Id. at Article 2.

[54]Id. at Articles 6.5 and 6.6.

[55]Id. at Article 6.3.

[56]Id. at Article 8.4.5.

[57]Id. at Article 8.5.1.

[58]Id. at Article 11.2.15.

13.4.2.2 FAO-IPOA-IUU Fishing

Building on the Code of Conduct for Responsible Fisheries, the FAO Committee for Fisheries decided that compliance with the voluntary objectives could be improved through the development of "international plans of action" (IPOA). Like the Code, the IPOAs are voluntary and are intended in part to create a common language concerning international fisheries management. Under the IPOA-IUU Fishing, States must "prevent, deter, and eliminate" IUU fishing practice.[59] FAO State Parties distinguished between illegal fishing, unreported fishing, and unregulated fishing.[60] Curiously, the parties did not define "fishing". So it is unclear whether IUU fishing for the purposes of the document is limited to a plain meaning usage of the term or encompasses a broader set of activities (e.g. at-sea processing).[61] Building on existing international instruments, the IPOA-IUU Fishing "address in an effective manner all aspects of IUU fishing."[62] States are expected to take a number of steps to end IUU fishing including exercising responsibility over nationals who are registered under their flag, discouraging nationals from flagging with flags of convenience, creating appropriate sanctions "of sufficient severity", ending subsidies to IUU actors and conducting monitoring, control, and surveillance schemes.[63] Specific advice is provided for flag States to avoid flagging vessels with a history of non-compliance, and to consider domestic processes that integrate the registration of a vessel with issuing authorizations to fish.[64] In order to improve application of the plan to each nation's unique situation, each State was expected to develop a National Plan of Action within 3 years of adopting the IPOA-IUU Fishing.[65] By 2014, only 11 states had posted their National Plan of Action publicly on the Food and Agriculture Organization page; only one of these States is considered to be a Flag of Convenience.[66]

---

[59] International Plan of Action to Prevent, Deter, and Eliminate Illegal, Unreported and Unregulated Fishing (2001) Paragraph 8.

[60] Id. at Paragraph 3.

[61] Flag states are expected to provide oversight for their fishing, transport and support vessels. Id. at Paragraph 48.

[62] Id. at Paragraph 16.

[63] Id. at Paragraphs 16–24.

[64] Id. at Paragraphs 36 and 40.

[65] Id. at Paragraphs 25–26.

[66] See FAO Fisheries and Aquaculture—IUU National Plan; http://www.fao.org/fishery/ipoa-iuu/npoa/en (Providing plans for Belize, Ghana, South Korea, Argentina, Fiji, Australia, Canada, United States, Chile, Japan, New Zealand; Belize is considered to be a Flag of Convenience state and under recent pressure from the European Union has become active in combatting IUU fishing by its fleet).

13.4.2.3 FAO–IPOA Capacity

States drafted the IPOA on Capacity in response to the Code of Conduct for Responsible Fishing. Similarly, the IPOA on Capacity is a voluntary plan encouraging States to develop National Plans of Action focusing on assessment and monitoring of fishing capacity to tackle overfishing and economic waste. The IPOA on Capacity was expected to achieve “efficient, equitable and transparent management of fishing capacity” by 2005 at national, regional, and global scales after an initial assessment, adoption of preliminary management measures, and full implementation of capacity management.[67] However, in practice calling for a global record of fishing vessels has proven to be difficult to execute in practice.

By 2002, States were expected to draft national plans, make these plans public, and be prepared to “reduce fishing capacity in order to balance fishing capacity with available resources on a sustainable basis.”[68] This might involve creating new livelihood for fishing communities so as to reduce pressure on fishing stocks.[69] Simultaneously, States were expected to reduce subsidies and other economic incentives contributing to overcapacity. States are expected to cooperate in reducing fishing capacity particularly for overfished high seas stocks such as tuna.[70]

As part of the IPOA on Capacity, problems associated with reflagging and IUU fishing undermining the global fish stock management are acknowledged. Regarding IUU fishing, the IPOA calls for States to cooperate multilaterally to ensure action by flag States that “do not exercise effectively their jurisdiction and control over their vessels which may operate in a manner that contravenes or undermines the relevant rules of international law and international conservation and management measures.”[71] Where there may not be time to cooperate because a stock is in danger of imminent collapse, States are expected to “individually, bilaterally, and multilaterally, as appropriate, to reduce substantially the fleet capacity” to pursue stocks especially “trans-boundary, straddling, highly migratory, and high seas stocks which are significantly overfished.”[72] But, the IPOA on Capacity has not generated strong national responses. As of 2014, only three States have submitted national plans of action to be published by the Food and Agricultural Organization.[73]

[67] International Plan of Action for the Management of Fishing Capacity (1999) Paragraphs 7 and 9.

[68] Id. at 21.

[69] Id. at paragraph 22.

[70] Id. at paragraph 31.

[71] Id. at paragraph 33.

[72] Id. at paragraphs 39–40.

[73] See IPOA-Capacity National Plans, http://www.fao.org/fishery/ipoa-capacity/npoa/en (Plans have been submitted by Indonesia, Namibia, and the United States).

13.4.2.4 FAO–IPOA for Seabirds and FAO-IPOA for Sharks

In 1999 based on the Code of Responsible Conduct, States negotiated a voluntary IPOA to protect seabird populations from longline fisheries and to regulate shark fisheries. Regarding seabirds, the FAO observed that albatross, gulls, petrels, and fulmars are common bycatch in a number of global fisheries.[74] States who fish using longlines are expected to assess their fisheries to determine whether there is a seabird bycatch problem. In that case, they should implement a National Plan of Action by 2001, consider the unique qualities of each of the long-line fisheries under their jurisdiction and use technical measures to mitigate the problem.[75] As of 2014, only nine countries have publicly submitted National Plans of Action to the Food and Agriculture Organization even though far more States operate longline fisheries.[76]

Sharks provide key food resources for low-income food deficit regions. But they have been widely overharvested beyond replacement rates in many fisheries.[77] States directly overharvesting sharks or catching sharks as bycatch, should adopt a national plan of action by 2001.[78] Waste may be minimized so that sharks captured for their fins are also retained rather than discarded.[79] Only 10 states have national plans; six of which are considered to be among the leading shark fishing States responsible for 80 % of the global shark harvests.[80] Indonesia, the largest shark fishing state in terms of reported tonnage has not provided to FAO a public national plan of action.[81]

[74]International Plan of Action for Reducing Incidental Catch of Seabirds in Longline Fisheries (1999), paragraph 2.

[75]Id. at paragraphs 12, 16, 17, and Technical Note on Some Optional Technical and Operational Measures for Reducing the Incidental Catch of Seabirds in Longline Fisheries (providing specific mitigation techniques such as *inter alia* weighting longline gear, increasing line sinking rate, setting lines under water, or discharging offal in a manner that does attract seabirds).

[76]Also FAO Fisheries and Aquaculture—Seabirds National Plan, http://www.fao.org/fishery/ipoa-seabirds/npoa/en (Providing plans from Argentina, Japan, Canada, Uruguay, South Africa, Brazil, New Zealand, Australia, and the United States).

[77]International Plan of Action for the Conservation and Management of Sharks (1999) paras. 2 and 15.

[78]Id. at paragraphs 18 and 20.

[79]Id. at paragraph 22.

[80]FAO Fisheries & Aquaculture—Sharks National Plan; http://www.fao.org/fishery/ipoa-sharks/npoa/en (Including plans from Japan, Argentina, Uruguay, Canada, Seychelles, Malaysia, Ecuador, Australia, Taiwan, United Kingdom and United States); FAO Fisheries & Aquaculture—National and Regional Plans of Actions, http://www.fao.org/fishery/topic/18123/en (Canada, United States, United Kingdom, Malaysia, Taiwan, and Japan are leading shark fishing States).

[81]FAO Fisheries & Aquaculture—National and Regional Plans of Actions, http://www.fao.org/fishery/topic/18123/en.

13.5 National Efforts

In addition to implementing obligations under international agreements, a number of States have also adopted law and policy measures to combat bycatch, IUU fishing or overcapacity particular to their State. For example, in Norway fishing rules require any bycatch to be used as food.[82] In the United States, the President in June 2014 in response to reports of rampant mislabeling of fish products introduced the U.S. Comprehensive Framework to Combat Illegal, Unreported and Unregulated Fishing and Seafood Fraud.[83] As part of this framework, a cross-agency task force—involving participation from 14 agencies, provided recommendations to ensure IUU fishing does not contribute to additional economic losses or undermine U.S. efforts to reduce global hunger.[84] To combat IUU fishing, the Task Force proposed additional attention on ending subsidies contributing to overfishing, implementation of the global database on fishing vessels called for in several international agreements, and use of free trade agreements to pressure trading partners to address IUU fishing. The U.S. also proposed reliance on Customs Mutual Assistance Agreements to improve information gathering, improvement of at-sea enforcement efforts, and creation of a seafood traceability program particularly for fish likely to be part of IUU fishing effort.[85]

In Indonesia, the government in response to reports of 5400 fishing vessels operating illegally in Indonesian waters, decided to sink foreign fishing vessels determined by a court to be engaged in illegal fishing activities, specifically to sanction violators and create a deterrent to future IUU fishing activity.[86] The European Union is responsible for common fisheries policies for its individual member States. A community wide system is created by the EU to tackle IUU EU states prohibit trading in IUU fisheries products, enforced through a port inspection process.[87] In addition to identifying individual IUU vessels, the EU may also identify non-cooperating third countries that are not permitted to import fisheries products into the EU.[88]

Summary A broad framework of law has been created to tackle overfishing with a general international support. Until recently, little has been done on the domestic front to minimize the threats that overfishing or bycatch pose to the renewability of

[82]Waldman (2014), p. 360.

[83]Presidential Memorandum Establishing a Comprehensive Framework to Combat Illegal, Unreported, and Unregulated Fishing and Seafood Fraud (June 17, 2014) http://www.whitehouse.gov/the-press-office/2014/06/17/presidential-memorandum-comprehensive-framework-combat-illegal-unreporte.

[84]Department of Commerce, National Oceanic and Atmospheric Administration, Recommendations of the Presidential Task Force on Combating Illegal, Unreported and Unregulated Fishing and Seafood Fraud, Notice and request for Comments, FR Doc. 2014-29628 (December 18, 2014).

[85]Id.

[86]Parlina (2014).

[87]European Union Council Regulation 1005/2008 (September 29, 2008): Chapter II.

[88]Id. at Article 38.

fishery resources. While trade embargos from large importers may impact some IUU fishing activity, more comprehensive domestic efforts are necessary to address the open-access nature of fishing, criminal network of IUU fishing, and the lack of comprehensive information about fish harvests.

13.5.1 Challenges with Existing Domestic Law and Policy

While international law exists urging States to combat overfishing and bycatch, there remains an urgent need for domestic law to respond to these problems in order to ensure long-term food security. Even though most commercial fishing is in domestic waters e.g. the Exclusive Economic Zone, domestic fishing codes often remain woefully out of date in terms of their management approaches with insufficient catch limitations set to conserve stocks. International law obligations have not been effectively translated into domestic law and many States have failed to develop or review national plans of actions that systematically address fishing capacity, bycatch, or IUU fishing.

13.5.2 Lack of Domestic Definitions of Resource Problems

In general, most individual States do not have specific plans to end overfishing and reverse losses beyond the general concept that States should not permit overfishing. Among the larger fishing nations, only the United States has formally defined overfishing in its laws and used that definition to regulate its registered fleets and activities in its jurisdictional waters. The U.S. approach is useful and might serve as a model for other States seeking a legal strategy to tackle overfishing and bycatch. In the amended Magnuson-Stevens act and its implementing regulations, the United States defines "overfishing" as "a rate or level of fishing mortality that jeopardizes the capacity of a fishery to produce the maximum sustainable yield on a continuing basis."[89] Fisheries management in the U.S. requires fisheries management plans to "prevent overfishing while achieving, on a continuing basis, the optimum yield from each fishery for the United States fishing industry."[90] Since 2010, U.S. fisheries management plans must "specify objective and measurable criteria for identifying when the fishery to which the plan applies is overfished (with an analysis of how the criteria were determined and the relationship of the criteria to the reproductive potential of stocks of fish in that fishery) and, in the case of a fishery which the Council or the Secretary has determined is

[89] Magnuson Stevenson Fishery Conservation and Management Act 16 U.S.C. 1802(34).
[90] Id. 16 U.S.C. 1851(a)(1).

approaching an overfished condition or is overfished, contain conservation and management measures to prevent overfishing or end overfishing and rebuild the fishery."[91]

In order to monitor progress in ending overfishing within the U.S. waters and vessels, the Secretary of Commerce has reporting duties. In the case of fisheries within U.S. waters, the Secretary is expected to report annually to both Congress and Regional Fishery Management Councils on whether fisheries are overfished or "are approaching a condition of being overfished" based on "trends in fishing effort, fishery resource size, and other appropriate factors."[92] Councils are expected to respond to these reports by creating new fisheries management plans or amending existing plans to "(A) to end overfishing immediately in the fishery and to rebuild affected stocks of fish; or (B) to prevent overfishing from occurring in the fishery whenever such fishery is identified as approaching an overfished condition."[93] If Councils fail to prepare or amend management plans, the Secretary of Commerce will respond with measures to prevent overfishing.[94]

Additionally, the Secretary is expected to report on fisheries managed under an international agreement using criteria provided in the agreement.[95] In a case where a species is being overfished or is likely to be overfished "due to excessive international fishing pressure" the U.S. does not have any obligations under an existing agreement. Thus, the Secretary of Commerce and the appropriate Councils will be expected to submit recommendations for both domestic regulations and for international actions that can be taken to end overfishing.[96]

Still, the Secretary of Commerce is not required to consult with a Council in instances deemed to be overfishing. The Secretary may create interim measures to reduce overfishing for any fishery independent of any existing fishery management plan.[97] Also the Secretary has the ability when it "is necessary to prevent or end overfishing" to buyback fishing vessels and fishing permits.[98]

The legally defined prohibition on overfishing, while not necessarily popular with commercial or recreational fishing interests, has made a difference in resource management. In 2011, the U.S. former chief scientist at the National Oceanic and Atmospheric Administration announced that overfishing in federally managed fisheries had come to an end.[99] Catch limits continue to be enforced in the U.S. waters and additional regulations have been introduced to reduce both

[91]Id. 16 U.S.C. 1853(a)(10).

[92]Id. 16 U.S.C 1854 (e)(1).

[93]Id. 16 U.S.C. 1854(e)(3).

[94]Id. 16 U.S.C. 1854(e)(5).

[95]Id. 16 U.S.C. 1854(e)(1).

[96]Id. 16 U.S.C. 1854 (i)(2).

[97]Id. 16 U.S.C. 1855(c).

[98]Id. 16 U.S.C. 1861a(b)(1) and (2).

[99]Lindsay (2011).

overfishing for migratory species and bycatch of threatened species e.g. the Bluefin tuna.[100]

13.5.3 *Difficulties with Prosecuting Violations*

Even in the presence of political will to end certain types of overfishing, it can be very difficult to prosecute for overfishing. For example, the Border Guard Service of Russia pursued the Cambodian flagged Iskander through the Sea of Okhotsk where it had been harvesting live king crab without a permit in spite of the crab populations in the Far East being in jeopardy. Before the vessel was seized, the crew dumped the catch overboard.[101] Because there was no evidence to base a poaching claim, the crew was not prosecuted but released to pursue similar illegal activity most likely in some other part of the ocean.

In the years to come, there may be opportunities to use new cost-effective technology such as conservation drones or satellite data in order to enhance the ability to prosecute IUU fishing actors. In order for such technology to actually serve a deterrent effect, fishery management laws must have adequate sanctions. Existing legal sanctions for major fishing States generally only involve civil penalties or loss of the cargo rather than criminal charges.[102] IUU industrial fishing should be prosecuted as an organized crime.

13.5.4 *Inadequate Enforcement of Laws*

Even if domestic laws properly define the resource problem, set adequate catch limits, and assign appropriate sanctions, there often remains the recurring problem of enforcement due to lack of vessels and enforcement personnel. While the ministry or Department of Fisheries are expected to enforce the fisheries laws, these governmental agencies may have no independent enforcement authority. Fisheries agencies must rely upon their national military or coast guard to enforce fishery laws. Underenforcement occurs when there are ineffective communications between competing institutions or when the enforcement agency perceives fisheries enforcement to be a low priority. Some States have attempted to address this gap by

---

[100]National Oceanic and Atmospheric Agency Fisheries, Final Amendment 7 to the 2006 Atlantic High Migratory Species Fishery Management Plan, August 2014 (Providing for reallocation of Atlantic bluefin tuna quotas; regulating the pelagic longline fishery with individual quotas and gear usage, requiring pelagic longline vessels and purse seines to report via Vessel Monitoring System, and requiring automated catch reporting system for general and harpoon fishing entities).

[101]Joyce (2014).

[102]Telesetsky (2014), pp. 132–142.

creating fishery specific regional enforcement capacity. For example, the South Pacific Niue Treaty provides for potential regional coordination on fisheries surveillance and law enforcement over a broad stretch of the Pacific Ocean. For example, the State parties to the Niue Treaty have agreed to assist each other with extradtion of individuals who violate the fisheries laws.[103] These types of coordination treaties might be effective for other regions where overfishing and lack of enforcement threatens fishery resources.

13.6 Future Legal Developments

Humans have been overexploiting marine resources for centuries. In 1376, a complaint was made to King Edward of England regarding overfishing using a trawl where the author noted that "[T]he fishermen take such quantity of small fish that they do not know what to do with them; and that they feed and fat their pigs with them, to the great damage of the commons of the realm and the destruction of the fisheries".[104] While the degree of damage to the inland fisheries was notable, the oceans had not yet experienced the full extent of man as a predator. Today, with increasing demands for protein to feed a growing global population, it is essential that overfishing and bycatch be systematically addressed to protect marine living resources as a future reliable food source. Tackling these age-old problems will require both political commitment and a comprehensive strategy and improvements to existing legal frameworks described earlier. The remainder of this chapter explores several law and policy ideas that should improve the odds of long term survival of living marine resources and livelihoods dependent on these resources.

13.6.1 Marine Reserves

There are no simple policy interventions for reversing the trends of overfishing since excess fishing does not only damage commercial stocks but may transform otherwise productive ecosystems by changing natural patterns of competition between species. The removal of older and larger fish from the ecosystem—the more successful spawners, may reach a tipping point for the ultimate regeneration

[103]Niue Treaty on Cooperation in Fisheries Surveillance and Law Enforcement in the South Pacific Region (http://www.ffa.int/system/files/Niue%20Treaty_0.pdf) (Negotiated by Australia, Cook Islands, Federated States of Micronesia, Fiji, Kiribati, Marshall Islands, Nauru, New Zealand, Niue, Palau, Papua New Guinea, Solomon Islands, Tokelau, Tonga, Tuvalu, Vanuatu, and Western Samoa).

[104]Roberts (2007), p. 132 (Quoting from G.L. Alward (1932) The Sea Fisheries of Great Britain and Ireland).

of the fishery. In an overfished system, overall abundance of fish stocks may be as low as 10 % of the pre-fishing numbers, which depending on the species may not be sufficient to sustain the stock.[105] A decline in fish will also impact related species such as birds and marine mammals.

A robust framework of marine reserves is one possible insurance for healthier ecosystems. In some fisheries such as the Caribbean, marine managers have already proven the value of marine reserves for the local economy as harvest rates have increased adjacent to marine reserves due to spillover from the marine reserve. Arguably with the current degraded state of marine fisheries, we should globally exercise a paradigm shift where reserves operate not as the exception to fishery management but as the rule. Today, marine reserves are treated largely as an exception with most of ocean area open to intensive fishing activities. As a result, there is little chance for fisheries to ecologically rebound. Industrial fishing has become largely about how well a captain can use fish finding equipment and not their knowledge of the ocean.

If marine reserves were to become the rule, there will need to be a great deal more rule of law on the oceans to ensure conservation and management measures for the reserves are respected. The challenge will be that some parties (e.g. small-scale community fishermen) may consider themselves winners because marine reserves would salvage their disappearing livelihoods. Meanwhile, other parties such as multinational fishing companies will view themselves as losers since marine reserves will limit both their mobility and flexibility. To avoid the potential that marine reserves will be perceived as zero sum games, political leaders may need to offer legal fishing industries temporary financial relief if certain fisheries were to be closed and fleets historically operating in those fisheries had to forego annual profits. While it may be possible for some fishing vessels to become marine reserve patrols in lieu of their fishing activities, this opportunity would likely be available to only a narrow subset of the existing fleet.

Creation of a vast network of marine reserves with no-fishing or extremely limited fishing is really a best-case scenario for food security interests because it will create a bank of resources to be drawn on in the decades to come. The more reserves there are the better, since climate trends suggest that future fisheries may not be geographically identical to current fisheries. However, achieving a dispersed marine network that is relatively comprehensive in terms of covering spawning areas, may come at great direct costs to fishing fleets. Even if States were willing to forego present interests in return for a larger dividend in the future—with a possible greater payoff in terms of a food security strategy, States might find it extremely challenging to prevent the fishing industry from ramping up contemporary production in areas likely to be designated as future reserve sites.

[105]Hilborn and Hilborn (2012), pp. 110–111. (Emphasizing that ecosystem will experience change for sustainable fishing practices, but only overfishing "completely transforms the ecosystem.").

13.6.2 Property Interests

To address particularly the overcapacity challenges, some States have introduced individual fishing quotas (catch shares). Some advantages of these programs are reduction of fishing capacity, decreased competition, increased safety, and reduction in the amount of lost gear. While number of States already have these programs—e.g. the United States, New Zealand, Canada, and Iceland, there are many more coastal States and Regional Fisheries Management Organizations designing programs to promote a secure tenure right in a fishing quota. Theoretically, secure tenure rights might provide some degree of self-enforcement because participants may be willing to fund enforcement activities to protect their property interests.[106] Yet, the challenge with designing these programs is ensuring they are transparent and fair in how they allocate quotas. If fishermen believe a catch share system will not give them a reasonable opportunity to secure their livelihood, then they will not willingly participate in the program and may even sabotage efforts to create rationale management. One possible means of assigning rights in a shared fishery would be through a central auction mechanism.[107]

One interesting test case for property rights is being designed in the Eastern Pacific Ocean with the support of the Interamerican Tropical Tuna Commission (IATTC), World Wildlife Fund, and the International Seafood Sustainability Forum to create a property regime for migratory tuna fisheries. Recognizing that 85 countries are involved in harvesting tuna valued at around $10 billion, the IATTC, WWF, and ISSF are exploring the possibility of addressing excess harvesting capacity by creating enforceable and tradable fishing entitlements for the tuna purse seine fisheries.[108]

13.6.3 Financing New Technology

Efforts to systematically change fishing practices in a given fishery, e.g. purchasing new equipment to eliminate some unintended bycatch may be financially burdensome. Programs are needed to assist fishermen maximize economic yield.[109]

[106]Cunningham (2013). (Describing a program where fishers invest approximately $1 million to protect rights to harvest lobsters).

[107]Telesetsky (2013).

[108]World Bank (August 6, 2014) Ocean Partnerships for Sustainable Fisheries and Biodiversity Conservation- Models for Innovation and Reform, Report No. PAD 962, Annex 6: Regional Sub-Project for a Global Think Tank and Tuna Fisheries in the Eastern Pacific Ocean.

[109]Achieving Maximum Economic Yield (MEY), which is calculated based on cost of fishing versus value of catch, requires less fishing effort than fishing to the level of Maximum Sustainable Yield (MSY).

Governments or non-governmental organizations dedicated to environmental sustainability might establish lending program to assist with financing regulatory transitions. In the United States, the Environmental Defense Fund (EDF) established the California Fisheries Fund providing loans to fishermen affected by regulatory changes so they can acquire permits, equipment, and working capital to further sustainable fishery goals. Under the revolving fund, the EDF was willing to accept fishing quota as collateral for the loans as long as the fishermen agreed to comply with certain management conditions.[110] Notably, these are not grants but business loans. A similar business loan is offered by Verde Ventures invested in businesses such as a Mexican fishing cooperative with property rights in spiny lobster quota, but has also voluntarily agreed to establish no-take zones in the area where it has its tenure rights.[111]

Fishery wide loan programs for technology installation may dissipate some resistance to using technology. The United States has been introducing some technology into certain fisheries with wider application. For example, in the groundfish industry in the Gulf of Alaska, Bering Sea, and Aleutian Islands, the US requires catcher/processor vessels and motherships to have at-sea scales that can be used to estimate groundfish catch in support of the large-scale catch share program in this region.[112] Drawing on the model of the California Fisheries Fund, costs for installing the new at-sea scales technology could be financed with low-interest or no-interest loans that can be collateralized based on a specific catch share right. Theoretically, the at-sea scale technology including the relay of daily test results for flow scales and video to monitor scale results can eventually become more widely disseminated to include other fisheries requiring accurate catch accounting.

13.7 Information and Education

The last policy idea should perhaps be the easiest to implement because it does not require re-envisioning access to the commons, rethinking parts of the existing property law regime or developing appropriate financing contracts. Yet, efforts to the use the law to shine some light on the increasingly murky global industry of fishing have faced resistance by States and players. As recent studies have exposed commercial seafood fraud with rampant mislabeling,[113] it is obvious that more information is needed about the actual source of our food so wholesalers, retailers, and consumers can make better decisions.

[110] Holmes et al. (2014), p. 70.

[111] Id. at 71.

[112] Department of Commerce, NOAA (November 18, 2014): Fisheries of the Exclusive Economic Zone off Alaska; Monitoring and Enforcement; At-Sea Scales Requirements, Federal Register Volume 79, Number 222: 8610-68619.

[113] Oceana (2013) (87 % of the samples labeled as snapper and 59 % of those samples labeled as tuna were mislabeled).

Future policy is likely to focus on catch documentation. Here, the starting point for information is assigning unique vessel numbers to all vessels participating in the global industrial fishing fleet and not just 100 GT or larger vessels.[114] IMO numbers would form a sounder basis for the creation of a Global Record of Fishing Vessels, which for decades, the FAO members have tried to implement. States have not submitted adequate public information to allow other States to identify vessels' beneficial owners or determine whether a given vessel is authorized to fish because of concerns of protecting their nationals' fishing strategies. Open access to the information in fishing authorizations must be approached as a public good because these authorizations are the foundation of rational fishing management to ensure that too many vessels operating within a given range do not overexploit fishery stock.

Finally, there is room for the power of education. While some fishermen have an education in formal fishery management, many do not. Particularly for industrial fleets, States prior to issuing registrations to vessels might mandate some lessons in basic fishery management so that fishing fleets will better understand the implications of some of their gear choices or fishing strategies. Of particular importance for fishing-fleet-owners is a working understanding that maximum economic yield for a fishery requires less harvesting than maximum sustainable yield.

This is a critical juncture for marine fisheries. In less than half a century, global fleets have nearly depleted a number of fisheries that humans have depended upon for millennia. Because it may be dangerous to depend entirely on a technical fix to deliver food security e.g. scaled-up aquaculture, fisheries laws, particularly at the domestic level, must become more ambitious in targeting recovery and be more comprehensively implemented across the industry. If archaeologists of the future were to excavate our trash middens, what might they find? Bones of marine fish and shells similar to those found in vicinity of our 164,000 year old fish-consuming ancestors? Or will they find our famished bodies with pockets full of cash but with nothing to eat? These are the law and policy choices for this generation.

References

Agnew D et al (2009) Estimating the worldwide extent of illegal fishing. PLoS One 4

Agreement for the Implementation of the Provisions of the United Nations Convention of the Law of the Sea of December 10 1982 Relating to the Conservation and Management of Straddling Fish Stocks and Highly Migratory Fish Stocks (UN Fish Stock Agreement) (1995) 34 I.L.M. 1542

[114] The Fish Site (December 6, 2013) Exemption on Fishing Vessels Having IMO Identification Numbers Lifted, http://www.thefishsite.com/fishnews/21963/exemption-on-fishing-vessels-having-imo-identification-numbers-lifted (Indicates the two RFMOs: International Commission for the Conservation of Atlantic Tunas (ICCAT) and the Commission for the Conservation of Antarctic Marine Living Resources (CCAMLR) require for its members to have unique vessel identification numbers).

Anderson R (2014) Cetaceans and Tuna Fisheries in the Western and Central Indian Oceans
Bjørndal T, Munro GR (2003) The management of high seas fisheries. In Folmer H, Tietenberg T (eds) The International Yearbook of Environmental and Resource Economics 2003/2004, pp 1–35
Connor R (2001) Changes in fleet capacity and ownership of harvesting rights in New Zealand fisheries. In: Shotton R (ed) Case studies on the effects of transferable fishing rights on fleet capacity and concentration. FAO Fisheries Technical Paper, No. 412, pp 151–186
Costello C et al (2012) Status and solutions for the world's unassessed fisheries. Science 338:517–520
Cunningham E (2013) Catch shares in action: Mexican Baja California FEDECOOP benthic species territorial use rights for fishing system. Environmental Defense Fund
Food and Agriculture Organization (2012) The State of the World Fisheries and Aquaculture 2012
Garcia S (2003) The ecosystem approach to fisheries, FAO Fisheries Technical Paper 43
Hilborn R, Hilborn U (2012) Overfishing. Oxford University Press
Holmes L et al (2014) Towards investment in sustainable fisheries: a framework for financing the transition. Environmental Defense Fund and The Prince of Wale's International Sustainability Unit
Joyce S (2014) Searching for the Russian Crab Mafia. Bloomberg Businessweek, June 19
Kurlansky M (1997) Cod: a biography of the fish that changed the world. Walker Publishing Company, New York
Lindsay J (2011) Steve Murawski: overfishing to stop in U.S. for first time in a century. 10 January 2011. http://www.huffingtonpost.com/2011/01/10/steve-murawski-overfishing_n_806679.html
Marean CW et al (2007) Early human use of marine resources and pigment in South Africa during the Middle Pleistocene. Nature 449:905–908
McWhinnie S (2009) The tragedy of the commons in international fisheries: an empirical examination. J Environ Econ Manag 57:321–333
Oceana (2013) National Seafood Fraud Testing Results http://oceana.org/sites/default/files/National_Seafood_Fraud_Testing_Results_Highlights_FINAL.pdf
Parlina I (2014) RI to sink foreign vessels fishing illegally. The Jakarta Post, December 4
Roberts C (2007) The unnatural history of the sea. Island Press, Washington
Ross M (2012) Australian broadcasting corporation, super trawler: destructive or sustainable? 12 September 2012. http://www.abc.net.au/news/2012-08-15/super-trawler-debate/4200114
Squires D, Allen R, Restrepo V (2013) Rights based management in International Tuna Fisheries. Food and Agriculture Organization http://www.fao.org/docrep/018/i2742e/i2742e.pdf
Stockhausen B et al (2012) Discard mitigation – what we can learn from waste minimization practices in other natural resources? Marine Policy 36:90–95
Telesetsky A (2013) Going once, going twice--sold to the highest bidder: restoring equity on the high seas through centralized high seas fish auctions. Law of the Sea Institute Publication. Available at http://www.law.berkeley.edu/files/Telesetsky-final.pdf
Telesetsky A (2014) Laundering fish in the global undercurrents: illegal, unreported, and unregulated fishing and transnational organized crime. Ecol Law Q 41:101–159
Waldman J (2014) How Norway and Russia made a cod fishery live and thrive. Yale Environ, p 360

Chapter 14
Perspectives and Predicaments of GMO Salmon

Nicole Negowetti

Abstract AquAdvantage Salmon, produced by AquaBounty Technologies, Inc., is the first genetically engineered (GE) fish to be considered for commercial production and human consumption in the United States. Its application is currently under review by the Food and Drug Administration (FDA). Although private companies around the world are working on developing at least 35 species of GE fish and shellfish—including catfish, carp, oysters, and trout—no country has yet approved any of them for commercial production or human consumption.

14.1 Introduction

AquAdvantage Salmon, produced by AquaBounty Technologies, Inc., is the first genetically engineered (GE)[1] fish to be considered for commercial production and human consumption in the United States. Its application is currently under review by the Food and Drug Administration (FDA). Although private companies around the world are working on developing at least 35 species of GE fish and shellfish—including catfish, carp, oysters, and trout—no country has yet approved any of them for commercial production or human consumption.[2]

AquAdvantage Salmon is engineered to produce an insulin-like growth factor hormone (IGF-1) year-round, which gives it the ability to grow to market size in less time than conventional farmed salmon.[3] The engineered genetic construct combines a growth hormone protein from the unrelated Pacific Chinook salmon with regulatory sequences from an antifreeze protein gene found in ocean pout.

[1] Although transgenic and genetically modified are other terms used interchangeably, this chapter refers only to genetically engineered (GE) animals.

[2] Center for Food Safety (2015a).

[3] AquaBounty Technologies (2010), p. 13 [noting that AquAdvantage salmon have an "enhanced growth rate compared to non-transgenic Atlantic salmon"].

N. Negowetti, M.A.; J.D. (✉)
Valparaiso University School of Law, Valparaiso, IN, USA
e-mail: nicolen@gfi.org

G. Steier, K.K. Patel (eds.), *International Farm Animal, Wildlife and Food Safety Law*, DOI 10.1007/978-3-319-18002-1_14

AquaBounty then inserts this construct into the genome of Atlantic salmon.[4] Unlike conventional salmon, which produce growth hormone only during the spring and summer, the ocean pout promoter acts like a switch, keeping the growth hormone protein from turning off and allowing for continued growth of the fish.[5] The purpose of this GE fish is to significantly decrease the time from birth to market and "improv [e] the economics of land-based production."[6] These modifications could mean that the salmon reaches the market sooner, with less feed and overhead costs for farmers.[7]

Demand for fish products is expected to increase in response to the rising world population and improvements in the standard of living over the next 25 years. As a result, fish supplies are under ever-increasing pressure. According to the United Nations Food and Agriculture Organization (FAO), approximately 80 % of global fish stocks are overexploited, depleted or endangered.[8] Global fish production and aquaculture remains one of the fastest-growing food producing sectors, providing almost half of all fish for human food.[9] This share is projected to rise to 62 % by 2030, as catches from wild capture fisheries level off and demand from an emerging global middle class substantially increases.[10] Fish remains among the most traded food commodities worldwide. In 2012, approximately 200 countries reported exports of fish and fishery products. For some developing countries, fish accounted for more than half of the total value of traded commodities.[11] Due to the rising demand for meat and the loss of agricultural land, there is pressure to use biotechnology to improve productivity in animal agriculture.[12] Worldwide, the demand for fish continues to increase at a higher rate than wild fish populations can support on their own.[13] The FAO estimates that by 2030, annual commercial production will need to increase by an additional 28.8 million metric tons in order to maintain per capita fish consumption at current levels.[14] Beyond traditional aquaculture, GE fish provide a possible means of meeting seafood demand more efficiently.[15]

[4]AquaBounty Technologies (2010), p. 12.

[5]See Food and Drug Administration Center for Veterinary Medicine (2010), p. 110.

[6]AquaBounty Technologies (2010), p. 12.

[7]Bratspies (2008), p. 5.

[8]FAO (2014) and United Nations (2010).

[9]FAO (2014), p. iii.

[10]Ibid., pp. iii–iv.

[11]Ibid., p. 7.

[12]National Research Council (2002), p. 16.

[13]Logar and Pollock (2005), p. 17.

[14]www.fao.org/newsroom/en/news/2008/1000930/index.html, accessed 11/10/2010.

[15]Logar and Pollock (2005), p. 18.

14.1.1 *Overview of GE Animal Regulation in the United States*

Although several federal statutes apply indirectly to GE animals, there are currently no U.S. laws specifically governing their production and sale. Neither the U.S. Department of Agriculture (USDA) nor the U.S. Food and Drug Administration (FDA) have developed standards for GE animals that might be imported into the United States.[16] A patchwork of federal laws, dividing authority among different federal agencies, address the environmental issues raised by conventional aquaculture. For example, the U.S. Environmental Protection Agency (EPA) manages water pollution issues posed by aquaculture facilities, while the Army Corps of Engineers considers environmental impacts when it issues permits for aquaculture facilities in navigable waters.[17] No federal agency, however, has clear legal authority to regulate aquaculture facilities to avoid potential harm to wild fish communities, unless the wild fish are already threatened or endangered under the Endangered Species Act.[18] Furthermore, fisheries and aquaculture are primarily state-regulated.[19] Biotechnology has added further complicated an already problematic regulatory regime. Under a federal policy developed in the 1980s, products created through the use of biotechnology are treated the same as conventionally produced products. The development of biotechnology requires are different approach, however, particularly as it raises questions about which federal law applies.[20]

The FDA regulates GE animals under the New Animal Drug Application (NADA) provisions of the Federal Food, Drug, and Cosmetic Act (FDCA).[21] GE animals are regulated as animal drugs because the FDA considers the transfer of genetic information as a means of delivering a drug (hormone, protein, etc.) to the tissue of the animal. In the case of the AquAdvantage Salmon, the promoter is viewed as a drug that delivers growth hormone to the tissues of the fish. To receive FDA approval to sell a GE fish for human food, producers must complete a New Animal Drug Application.

On December 20, 2012, the FDA released its draft environmental assessment of the proposed conditions of use for GE fish, declaring a preliminary finding of no significant impact (FONSI) for the AquAdvantage Salmon application.[22] The FDA concluded that food from AquAdvantage Salmon is "as safe as food from conventional Atlantic salmon"[23] and that the "development, production, and grow-out of

[16]Food and Drug Administration Center for Veterinary Medicine (2010), p. 128.

[17]Pew Initiative on Food & Biotechnology (2004), preface.

[18]Ibid.

[19]Ibid.

[20]Ibid.

[21]21 U.S.C.A. §§ 360(a)(1), 360(a)(3).

[22]Food and Drug Administration Center for Veterinary Medicine (2012).

[23]Food and Drug Administration (2012), p. 3.

AquAdvantage Salmon . . . will not result in significant effects on the quality of the human environment in the United States."[24] On March 13, 2014, FDA Commissioner Dr. Margaret Hamburg stated that the FDA was still evaluating Aquabounty's application.[25] If approved, the salmon will become the first genetically engineered animal approved for human consumption in the United States.[26]

14.1.2 Predicaments of GE Salmon

The pending application of AquaBounty's GE AquAdvantage salmon, which has been subjected to one of the longest regulatory assessments in FDA history, has raised significant concerns regarding the FDA's regulatory oversight. The FDA's regulation has been heavily criticized primarily because of the confusion created by the outdated regulatory framework for biotechnology. As Professor Rebecca Bratspies of the City University of New York so aptly states, "transgenic salmon are a clear case of science outpacing policy."[27] The regulation of animal drugs is criticized as being antiquated; formulated before GE animals were thought possible. Furthermore, as Professor Bratspies explains, "scientific development of transgenic fish has taken place without a corresponding development of policies for managing the unique risks posed by these fish."[28] Scientists also widely debate the potential environmental risks and possible adverse effects on human health, such as increased allergenicity. Opponents of GE fish question whether the FDA possesses the expertise and authority necessary to adequately assess the risks posed by GE fish. They have also criticized the FDA's review process under the NADA as lacking transparency.[29] In general, opponents are concerned that there are no federal regulations under the NADA to prevent or minimize the environmental and human health risks posed by GE animals.[30] This concern is indeed warranted: once the United States approves these fish, they will enter the ocean ecosystem with very little chance of ever being removed.[31]

Although there is much opposition to the approval of the AquAdvantage salmon, some observers suggest that rejecting AquaBounty's application could signal that

[24]Ibid., p. 4.

[25]Upton and Cowan (2014), p. 13 citing U.S. Congress, Senate Committee on Health, Education, Labor, and Pensions, *Protecting the Public Health: Examining FDA's Initiatives and Priorities*, 113th Cong., 2nd sess., March 13, 2014.

[26]Voosen (2010).

[27]Bratspies (2008), p. 6.

[28]Ibid., p. 10.

[29]McEvilly (2013), pp. 415–416.

[30]Bratspies (2008), p. 10.

[31]Bratspies (2008), p. 11.

science-based regulatory oversight is subject to political intervention.[32] Furthermore, "regulatory uncertainty makes [the] commercialization [of GE fish] in the United States prohibitively expensive and is moving development of this technology offshore to countries with more predictable policy environments."[33] According to AquaBounty Technologies chief executive Ron Stotish, efforts by critics to create uncertainty in the fish, the company, and the regulatory process are a form of "harassment" that has "basically arrested an entire generation of technology" which is vital for the "U.S., future projects and new opportunities."[34] There is also a concern, as stated by the FDA, that denying the AquAdvantage application may drive AquaBounty and other biotech companies to produce GE salmon in foreign countries like China—and thus possibly in less-secure facilities than AquaBounty has proposed—which could increase the likelihood of environmental damage.[35]

14.1.3 Chapter Roadmap

This chapter provides an overview of the FDA's regulation of GE animals. It first examines the FDA's review of AquaBounty's pending application and discusses the key criticisms of the FDA's approach to oversight. Section 14.2 explores the international implications of GE salmon approval. Section 14.3 provides an overview of the regulation of GE salmon, while Sect. 14.4 analyzes the FDA's regulatory approach. Section 14.5 examines issues surrounding the labeling of GE fish, and Sect. 14.6 briefly explains the impact of the FDA's review on future GE fish. Section 14.7 explores some of the U.S. state and Congressional measures that have attempted to impose additional requirements for, and limits on, the cultivation of GE fish. Section 14.8 provides recommendations for reforming the regulatory process. Finally, Sect. 14.9 offers concluding thoughts.

14.2 International Implications of GE Salmon Approval by the United States

The international scope of AquaBounty's plan for the production of GE salmon has environmental, economic, and safety implications for countries across the world. According to its proposal for FDA approval, AquaBounty's intention is to produce the eggs on Prince Edward Island, Canada. It will then transport them to inland

---

[32]Van Eenennaam et al. (2013).

[33]Ibid.

[34]AquaBounty Technologies (2014c).

[35]Schwab (2013), p. 13; See Food and Drug Administration Center for Veterinary Medicine (2012), pp. 3, 4, 23–24.

facilities in Panama, and raise the AquAdvantage salmon to market size. Following that, AquaBounty will harvest and ship the fish back to U.S. markets.[36] Therefore, in addition to its pending application for FDA approval, AquaBounty is also subject to Canadian and Panamanian[37] regulations. In the November 2013 decision that marked the first time any government had approved the commercial scale production of a GE food animal, the government of Canada granted AquaBounty Technologies Inc. permission to export up to 100,000 GE fish eggs a year from a hatchery in Prince Edward Island to a site in the Panamanian rainforest. In addition to Canada's expected role in producing the AquAdvantage eggs, AquaBounty has also identified China, Chile and Argentina as target countries for GE salmon production,[38] which would further expand distribution of the fish.

14.2.1 Cartagena Biosafety Protocol to the Convention on Biological Diversity

Salmon do not recognize national borders, and thus GE salmon may implicate the Cartagena Biosafety Protocol to the Convention on Biological Diversity. The Protocol is an international agreement with the objective of ensuring the safe transfer, handling, and use of GMOs that "may have adverse effects on the conservation and sustainable use of biological diversity, taking also into account risks to human health, and specifically focusing on transboundary movements."[39] The Protocol is the primary international agreement has set rules for the transboundary movement of living modified organisms (LMOs). Furthermore, the rules are legally binding under international law.[40] It defines an LMO as any "living organism that possesses a novel combination of genetic material obtained through the use of modern biotechnology."[41] The Protocol is a supplement to the UN Convention on Biological Diversity signed at the UN Rio Conference in 1992.[42] Contained within are measures to protect genetic resources from the potential risks that the release of genetically modified organisms might pose.[43]

The Protocol divides LMOs in three categories: LMOs for voluntary introduction into the environment, such as seeds for planting, live fish for release, or

---

[36]AquaBounty Technologies (2010), p. 15.

[37]Based on a 2012 inspection Panama recently fined AquaBounty for failing to secure necessary permits, particularly around its use of water and pollution of the local environment. See Biron (2014).

[38]Schwab (2013), p. 12.

[39]Cartagena Protocol on Biosafety to the Convention on Biological Diversity (2000), p. 3.

[40]Ibid., p. 4.

[41]Ibid., p. 4.

[42]Ibid., p. 1.

[43]Ibid., pp. 1–19.

micro-organisms for bioremediation; LMOs destined for contained use,[44] which Article 3(b) of the Protocol defines as instances where LMOs "are controlled by specific measures that effectively limit their contact with, and their impact on, the external environment;"[45] and, lastly, LMOs intended for direct use as food or feed, or for processing.[46] This international agreement, ratified or acceded to by 168 countries and the European Union, guarantees states the power to control the movement of LMOs, like GE salmon, within their territory if such organisms are introduced into the environment.[47] The Protocol, therefore, gives states the authority to ban or regulate aquaculture of GE fish.[48] Despite its regulatory power, however, several countries have refused to ratify, including those with high levels of agricultural exports such as the United States, Argentina, Australia and Canada.[49] Furthermore, the Protocol does not guarantee states the power to control the import of such organisms for use as food.[50]

14.2.2 *Different Approaches to GMO Regulation*

Worldwide, countries have taken different approaches to managing genetically modified organisms, and thus it is possible that a transgenic fish could be approved in the United States but banned elsewhere. In the EU, for example, "GMOs can only be placed on the market after having undergone a stringent science-based risk assessment on a case-by-case basis."[51] Six European states–Austria, France, Germany, Hungary, Luxembourg, and Greece–have outright banned genetically modified organisms.[52]

Among countries that have not banned GMOs, state practices often differ significantly from international agreements. The Codex Alimentarius Commission Guidelines (Codex Guidelines), created by the World Health Organization (WHO) and the Food and Agriculture Organization of the United Nations (FAO), were designed to establish and promote the coordination of international food standards.[53] Although the FDA has indicated that it would adhere to these guidelines,[54] its

[44]Cartagena Protocol on Biosafety to the Convention on Biological Diversity (2000), p. 14.

[45]Ibid., p. 4.

[46]Ibid., p. 14.

[47]Ibid., p. 1.

[48]Bratspies (2008), p. 5.

[49]GMO Compass (2015).

[50]Bratspies (2008), p. 5.

[51]EUROPA (2006) (explaining the European Union's approach).

[52]Ibid.

[53]Codex Alimentarius, (2015), http://www.fao.org/fao-who-codexalimentarius/about-codex/en/

[54]Food and Drug Administration (2009) ("[t]he information needed to establish food safety for food from GE animals under an NADA is consistent with that described in the Codex Guidelines.")

procedure for evaluating GE animals runs contrary to them. Under the Codex Guidelines, the safety assessment method recommended for whole foods requires a strict multidisciplinary approach.[55] When genetic modification changes the characteristic of a food, additional data or information may be necessary, along with "…the use of appropriate conventional toxicology or other studies on the new substance."[56,57] A majority of countries require toxicity tests for whole foods that have undergone genetic modification.[58] Indeed, some nongovernmental organizations have warned that the FDA's regulation of AquAdvantage salmon under the NAD approach, which adopts a substantial equivalence test,[59] contradicts the Codex Guidelines' more rigorous safety assessment approach.[60]

14.2.3 Different Approaches to GMO Labeling

Unlike the United States, the EU and other countries have imposed stringent labeling and traceability rules on GMOs—requirements that might be impossible for GE fish producers to meet.[61] In 2004, for example, the EU enacted regulations to mandate labeling for all food products that make direct use of genetically modified organisms at any point in their production.[62] Labeling on GE foods is mandatory under certain circumstances in 64 countries,[63] including Australia, New Zealand,[64]

[55]World Health Organization & Food and Agriculture Organization of the United Nations (2009), p. 60.

[56]Ibid., p. 70.

[57]World Health Organization & Food and Agriculture Organization of the United Nations (2009), p. 67.

[58]Food and Water Watch, Consumers Union, and The Center for Food Safety (2012), p. 7.

[59]"The food and feed safety step of the hierarchical review process addresses the issue of whether food or feed from the GE animal poses any risk to humans or animals consuming edible products from GE animals compared with the appropriate non-transgenic comparators." Food and Drug Administration (2010d), p. 5.

[60]Food and Water Watch, Consumers Union, and The Center for Food Safety (2012), p. 8; See World Health Organization & Food and Agriculture Organization of the United Nations (2009), p. 57.

[61]European Union Deliberate Release Directive 2001/18/EC and Regulation 1829/2003.

[62]Commn. Reg. 641/2004/EC on genetically modified food and feed, OJ L 102/1 (2004); Reg. 1829/2003/EC on genetically modified food and feed, OJ L 268/6-7 (2003); Reg. 1830/2003/EC concerning the traceability and labeling of genetically modified organisms, OJ L 268/24–25 (2003).

[63]Center for Food Safety (2015a, b).

[64]Austrl. N.Z. Food Stands. Code—Stand. 1.5.2 (2011), http://www.comlaw.gov.au/Details/F2011C00118.

Taiwan,[65] South Korea,[66] Russia,[67] and Japan.[68] The lack of labeling on AquaBounty salmon could put at risk US salmon exports to the EU, as well as to other countries where GE animals have not been authorized.[69]

14.3 Regulation of GE Salmon

14.3.1 Defining Genetic Engineering

Genetic modification broadly refers to changes in an organism's genetic makeup that do not occur in nature. As far back as 8000 years ago, traditional farmers and scientists modified the genetics of animals by selecting those individuals with desirable traits for further breeding. Since the discovery of the genetic code in the 1960s, however, modern biotechnology and genetic engineering have made it possible to take a gene (or genes) for a specific trait from one organism and transfer it to another organism of a different species.[70]

The FDA defines genetically engineered (GE) animals as those animals modified by recombinant DNA (rDNA) techniques.[71] These techniques expand the range of traits that may be transferred to another organism and increase the speed and efficiency through which desirable traits, such as herbicide-resistance, are incorporated into organisms. Research programs in China, Cuba, India, Korea, the Philippines, and Thailand[72] have subjected approximately 50 species of fish to genetic modification and have developed more than 400 fish/trait.[73] Fish, along with other marine animals, have been the focus of many genetic engineering programs in attempt to reduce the production costs of human food as well as to produce pharmaceuticals.[74] Additionally, fish are good candidates for genetic engineering because they produce large quantities of eggs that can be fertilized and developed externally.[75]

---

[65]Ibid., p. 1.

[66]Ibid., pp. 1–2.

[67]Ibid., p. 9.

[68]Carter and Gruere (2003), p. 2 (see discussion on labeling guidelines for GM foods in Japan, the United States, Australia, New Zealand, European Union, Canada, Argentina, South Korea, and Indonesia).

[69]Varela (2013), p. 526.

[70]National Research Council (2002), p. 4.

[71]Food and Drug Administration Center for Veterinary Medicine (2009). Note that the FDA's definition is similar to the Cartagena Protocol's definition of an living modified organism is any "living organism that possesses a novel combination of genetic material obtained through the use of modern biotechnology." Cartagena Protocol on Biosafety to the Convention on Biological Diversity (2000), p. 4.

[72]Upton and Cowan (2014), p. 5.

[73]Cowx et al. (2010), p. 3.

[74]Upton and Cowan (2014), p. 5.

[75]Ibid.

14.3.2 Overview of the United States Regulatory Process for Biotechnology

In 1986, the White House Office of Science and Technology Policy (OSTP) promulgated the Coordinated Framework for Regulation of Biotechnology (Coordinated Framework).[76] The Coordinated Framework, while not legally binding, distributed regulatory responsibilities to three agencies based on pre-existing statutory mandates: the U.S. Department of Agriculture (USDA), the Environmental Protection Agency (EPA),[77] and the FDA.[78] The Coordinated Framework maintained that existing statutes were sufficient enough to provide agencies with jurisdiction and authority to ensure adequate regulation of biotechnology, although the agencies could take legislative action as the field advanced.[79] The Coordinated Framework's working group concluded, "For the most part [existing laws] as currently implemented would address regulatory needs adequately."[80] As the Pew Initiative has noted, however, this has encouraged agencies to reinterpret old statutes in order to fit new biotechnology products into decades-old legal frameworks.[81] Furthermore, the policy stated that a commercial product should be regulated based on the product's composition and intended use, regardless of its manner of production. This policy therefore has the effect of regulating genetically engineered food in the same manner as other conventionally produced foods.[82] The result is that no single statute and or single federal agency specifically governs the regulation of biotechnology products, and the division of responsibility among the agencies is unclear. Indeed, some newer applications of biotechnology did not exist when the current regulatory framework was articulated, leading the National Research Council to report that the Coordinated Framework "might not be adequate

---

[76]51 Fed. Reg. 23302.

[77]The EPA reviews the potential effects on human health and the environment of pesticidal substances produced by plants, under the Federal Insecticide, Fungicide, and Rodenticide Act (FIFRA; 7 USC § 136 et seq.) and the pesticide residue provisions of the federal Food, Drug, and Cosmetic Act (FDCA; 21 USC § 346a). The role of the EPA, although relevant to the regulation of GE plants, is beyond the scope of this chapter. However, the FDA's 2009 Guidance recognized that the EPA may also assert jurisdiction over certain GE animals, such as insects, and that FDA was discussing with it "the best approach for oversight." Food and Drug Administration Center for Veterinary Medicine (2009), p. 5, fn. 1.

[78]See Coordinated Framework for Regulation of Biotechnology, 51 Fed. Reg. at 23,304 (stating that the agencies involved have "extensive experience with products that involve living organisms" and that new developments will be reviewed by the FDA, USDA, and EPA in the same manner for safety).

[79]Pew Initiative on Food & Biotechnology (2001), p. 6 ("[T]he application of existing statutes to biotechnology led to significant questions about overlapping authorities among the agencies, as well as uncertainties about whether the agencies would follow consistent approaches in using these authorities.").

[80]Coordinated Framework for Regulation of Biotechnology, 51 Fed. Reg. at 23,303.

[81]*See* Pew Initiative on Food & Biotechnology (2004), pp. 10–11.

[82]Ibid.

to address unique problems and characteristics associated with animal Biotechnologies."[83]

14.3.3 FDA Regulatory Framework

The FDA regulates food; animal feed additives; and human and animal drugs, including those from biotechnology; to ensure that they pose no human health risks under the Federal Food, Drug, and Cosmetic Act (FFDCA)[84] as well as under the Public Health Service Act.[85] The FFDCA requires all food and feed manufacturers to ensure that their food products are safe and properly labeled, which includes those developed through genetic engineering,

In 2009, the FDA issued a final guidance to the food industry explaining that gene-based modifications of animals for production or therapeutic claims fall within the purview of the agency's Center for Veterinary Medicine (CVM), and it would thus regulate genetically engineered food animals as New Animal Drugs (NAD).[86] Under the FFDCA, drugs are defined in Section 201(g) as "articles intended for the use in the diagnosis, cure, mitigation, treatment, or prevention of disease in man or other animals; and articles (other than food) intended to affect the structure or any function in the body of man or other animals."[87] The FDA determined that an rDNA construct is an animal drug because its insertion into the animal's genome alters the animal's structure or function, which therefore grants the FDA authority over the resulting GE animal, as well as all offspring that descend from the original transgenic animal as a result of breeding.[88] Under the statute, NADs are generally deemed unsafe for human consumption unless the FDA has approved a New Animal Drug Application (NADA) for a particular use—such as in the case of certain Investigational New Animal Drugs (INADs).[89]

The FDA's 2009 guidelines explain how the NADA requirements in 21 C.F.R. 514.1 apply to GE animals. It outlines seven categories of information required during the pre-market approval process: product identification, molecular characterization of the construct, molecular characterization of the GE animal lineage, phenotypic characterization of GE animal, genotypic and phenotypic durability assessment, food/feed safety and environmental safety assessments, and effectiveness/claim validation.[90] The FDA's Center for Veterinary Medicine

[83]National Research Council (2002).

[84]21 U.S.C. §301 et seq.

[85]42 U.S.C. §201 et seq.

[86]21 U.S.C. §321; Food and Drug Administration Center for Veterinary Medicine (2009), p. 5.

[87]Food and Drug Administration Center for Veterinary Medicine (2009), p. 4.

[88]Ibid., pp. 4 and 6.

[89]21 U.S.C.A. §§ 360(a)(1), 360(a)(3).

[90]Food and Drug Administration Center for Veterinary Medicine (2009), pp. 20–25.

(CVM) will then conduct a food safety assessment that includes examining both the potential of direct toxicity from a GE animal, such as allergenicity, as well as any indirect toxicity.[91] Food and feed will be considered safe if the composition of edible materials from the GE animal can be shown to be "substantially equivalent" to that from a non-GE animal.[92] Therefore, if animals of the same or comparable type are commonly and safely consumed, there is a presumption that food from the GE animal is safe and the product will not have to be labeled.[93]

The National Environmental Policy Act (NEPA)[94] requires federal agencies to consider the environmental consequences of an action before proceeding, and to involve the public in its decision-making process. To demonstrate compliance with NEPA, federal agencies must prepare an environmental impact statement (EIS) for federal actions anticipated to have "significant" impacts on the environment. The EIS is a detailed evaluation of the proposed action and provides opportunity for the public, other federal agencies, and outside parties to provide input into the process. The FDA decision regarding an NADA is subject, however, to the NEPA's evaluation of the application. The FDA then determines whether the decision would have a "significant" impact on the environment, and thus, whether or not an EIS is necessary.[95]

The FDA examines the documentation submitted by an applicant for an NAD in order to evaluate the potential for environmental impacts. This process also includes determining the potential of the inadvertent release or escape of the GE animal and/or its products into the environment, and whether certain measures may mitigate any potential significant impacts that would adversely affect the human environment.[96] Following this step, the FDA must create a preliminary environmental assessment (EA).[97] "The EA is a public document that provides sufficient information to allow FDA to either prepare an environmental impact statement (EIS) or issue a finding of no significant impact (FONSI)."[98] The agency issues a finding of no significant impact (FONSI) when the NAD would have no significant environmental impacts. If the EA determines that the environmental consequences of an NAD are anticipated to be significant, an EIS is prepared.[99]

[91]Ibid., p. 24.

[92]Ibid.

[93]Ibid.

[94]42 U.S.C. 4321.

[95]Section 514.1(b)(14).

[96]Food and Drug Administration Center for Veterinary Medicine (2009), p. 12.

[97]Actions that, based on an agency's past experience with similar actions, have no significant impacts are categorically excluded from the requirement to prepare an EA or EIS. 21 C.F.R. §514.1(b)(14). Because approval of GE salmon is unprecedented, the FDA was required to prepare an EA.

[98]2009 Guidance, p. 18. The EA becomes public when the animal is approved.

[99]In cases where significant impacts are anticipated, the federal agency may decide to prepare an EIS without first preparing an EA. CFR at 4.

Under the NAD regulatory protocols and the Trade Secrets Act, the FDA must keep confidential all information acquired through the New Animal Drug approval process in order to protect the applicant's trade secrets.[100] The FDA cannot even disclose whether an application exists until after the approval is published in the Federal Register.[101] The NAD application process has been the subject of much criticism, however, because the above restrictions can limit the opportunity for public input before the final decision. This issue will be further discussed in Sect. 14.4.3.

14.3.4 United States Department of Agriculture Regulatory Framework

Although several USDA agencies could be involved in the regulation of GE animals, the USDA has not established a policy or made a decision regarding whether it intends to exercise its statutory authority to regulate GE animals.[102] The USDA's Animal and Plant Health Inspection Service (APHIS) has the broad authority, under the Animal Health Protection Act (AHPA),[103] to regulate animals and their movement so as to control the spread of diseases and pests to farm-raised animals. Based on that authority, APHIS focuses more on the potential effects of GE animal traits on the health of the US livestock population, while the FDA is more concerned with the direct effects of genetic engineering on individual animals.[104] APHIS also administers the Viruses, Serums, Toxins, Antitoxins, and Analogous Products Act,[105] which ensures the safety and effectiveness of animal vaccines and other biological products, including those that have been genetically engineered. In addition, it oversees the Animal Welfare Act,[106] portions of which govern the humane treatment of several kinds of warm-blooded animals used in research (but does not generally include cold-blooded animals such as fish).

In 2007, APHIS established an Animals Policy Branch within its Biotechnology Regulatory Services to determine APHIS's role in the regulation of genetically engineered animals.[107] In the September 19, 2008 Federal Register, APHIS requested information from the public and scientists on how GE animals might

[100]*See* 21 U.S.C. § 331(j) (2006) (prohibiting information acquired under the authority of several sections related to the Animal Drug approval process from being revealed).

[101]21 C.F.R. § 514.11 (2012).

[102]Upton and Cowan (2014), p. 5.

[103]7 U.S.C. §8301 et seq.

[104]Food and Drug Administration (2008), p. 54408.

[105]21 U.S.C. §151–159.

[106]7 U.S.C. §2131 et seq.

[107]United States Department of Agriculture, Office of Inspector General (2011), pp. 10–11.

affect US animal health.[108] Scientists indicated the need for APHIS and the FDA to collaborate in order to ensure adequate regulations and safeguards regarding GE animals and insects.[109] The public expressed concerns that, because of the overlap in APHIS' and the FDA's regulatory authority over GE animals, animals that the FDA does not review may not be referred to APHIS.[110] In its 2009 Guidance, the FDA stated its intention to develop a memorandum of understanding with APHIS to determine each agency's role in the comprehensive oversight of GE animals.[111] To date, APHIS has not decided how it should proceed with regulating GE animals and insects, but it has acknowledged the need to publicly clarify the APHIS regulatory framework for GE animals and insects.[112]

14.3.5 Overview of AquaBounty's Proposed Operations

In its application to the FDA, AquaBounty indicated that it plans to produce salmon eggs at a specific facility on Prince Edward Island (PEI), Canada. Eggs would then be shipped to Panama and reared to market size in land-based facilities based in the Panamanian highlands, which would reduce the risk of GE salmon escaping and interacting with wild salmon populations. The salmon would be processed in Panama before being shipped to the United States for retail sale, and no live fish would be imported into the United States. AquaAdvantage salmon would be produced and imported into the United States under specified conditions proposed by AquaBounty.[113] AquaBounty claims that they would only produce sterile female GE Atlantic salmon through a process that renders the animal incapable of reproducing.[114] Measures to prevent the fish and eggs from escaping into the wild include various passive and active forms of containment.[115] The PEI facility would also have high security to stop unauthorized or unintentional access. The river that supplies the PEI facility runs into several other tributaries before discharging into the Pacific Ocean. To supply the PEI facility's grow-out tanks, water is diverted from the river and into a basin.[116] Screens, cover nets, and jump fences are used to prevent the salmon from escaping.[117] The grow-out facility in Panama, on the other hand, is located in a remote highland area at an elevation of 5000 ft.[118] The Panama

[108] Food and Drug Administration (2008), p. 54408.
[109] United States Department of Agriculture, Office of Inspector General (2011), p. 11.
[110] Ibid.
[111] Food and Drug Administration Center for Veterinary Medicine (2009), p. 6, fn 1.
[112] United States Department of Agriculture, Office of Inspector General (2011), pp. 12 and 14.
[113] AquaBounty Technologies (2010), p. 10.
[114] Ibid.
[115] Food and Drug Administration Center for Veterinary Medicine (2010), p. 117.
[116] AquaBounty Technologies (2010), p. 46.
[117] Ibid., p. 63.
[118] Food and Drug Administration Center for Veterinary Medicine (2010), p. 123.

site is geographically isolated from wild salmon species, and environmental conditions in the river's estuary and the Pacific Ocean are unfavorable for salmon survival anyway.[119]

14.3.6 *AquaBounty's AquAdvantage Salmon Application Process*

AquaBounty first approached the FDA in 1993 concerning the commercial use of GE salmon, and in 1995 they formally applied for approval. In 2009, AquaBounty provided the FDA with the last required study of AquAdvantage Atlantic salmon for their New Animal Drug Application (NADA). The FDA then issued a final Guidance in 2009 to industry explaining that it would regulate GE food animals using the New Animal Drug Application (NADA), which treats the inserted gene construct as an animal drug.[120] The FDA also indicated in the Guidance that it "is interested in increasing the transparency of its deliberations and actions," and therefore, "we intend to hold public advisory committee meetings prior to approving any GE animal."[121] However, it also stated that it "may revisit that policy in the future as we gain more experience with reviews of GE animals."[122] To provide it with expert advice on science, technology, and policy, the FDA relies on advisory committees and panels. In order to obtain specific advice on the safety and effectiveness of the AquAdvantage Salmon's NADA, the FDA's Veterinary Medicine Advisory Committee (VMAC) met on September 19–20, 2010.[123] The VMAC is composed of members with technical expertise in areas such as veterinary medicine, animal science, microbiology, biostatistics, and food sciences.[124] In the interest of transparency, the committee gave the public the opportunity to provide written submissions and oral testimony.[125]

The VMAC Chairman's report provided the following four conclusions regarding its evaluation of the AquAdvantage Salmon NADA. First, the committee found

---

[119]Ibid.

[120]See Food and Drug Administration Center for Veterinary Medicine (2009), p. 5.

[121]Ibid., p. 12.

[122]Ibid.

[123]Food and Drug Administration (2010d). See Food and Drug Administration (2010b), asking the VMAC to address four questions: (1) Do the data and information demonstrate that the rDNA construct is safe to AquAdvantage salmon? (2) Do the data and information demonstrate that there is a reasonable certainty of no harm from consumption of foods derived from AquAdvantage salmon? (3) Do the data indicate that AquAdvantage Salmon grow faster than their conventional counterparts? (4) Are any potential environmental impacts from AquAdvantage Salmon production adequately mitigated by AquaBounty Technologies' proposed conditions of use?

[124]Opponents of the AquaBounty application have argued that more experts in fisheries and ecology should have been included on the committee. Upton and Cowan (2014), p. 14.

[125]Food and Drug Administration (2010) Background Document.

no evidence that the rDNA construct is unsafe to AquAdvantage salmon. Second, it concluded that the data was appropriate and sufficient to establish similarities and equivalence between AquAdvantage Salmon and Atlantic salmon, which thus demonstrate that there is a reasonable certainty of no risk to humans in consuming foods derived from AquAdvantage salmon. Third, the committee found evidence that the AquAdvantage Salmon grow faster than their conventional counterparts. Lastly, it recognized that although the risk of escape from the Prince Edward Island and Panamanian facilities could never be zero, both locations have extensive barriers in place that make such an event unlikely. However, because part of the containment strategy is dependent on the management of standard operating procedures, the committee felt that both sites would need rigorously adhere to the safety policies in order to sustain the barriers.[126]

On December 20, 2012, the FDA announced the opportunity for the public to comment on the draft environmental assessment of impacts associated with the NADA submitted by AquaBounty[127] and the FDA's preliminary finding of no significant impact (FONSI).[128] In addition, the agency determined that:

> It is reasonable to believe that approval of the AquAdvantage Salmon NADA (New Animal Drug Application) will not have any significant impacts on the quality of the human environment of the United States (including populations of endangered Atlantic salmon) when produced and grown under the conditions of use for the proposed action. FDA preliminarily concludes that the development, production, and growout of AquAdvantage Salmon under the conditions proposed . . . will not result in significant effects on the quality of the human environment in the United States.[129]

To assess potential environmental impacts, the FDA consulted with the Fish and Wildlife Service and the National Marine Fisheries Service (NOAA Fisheries).[130] The FDA has made a "no effect" determination under the Endangered Species Act (ESA)[131] and has concluded that approval of the AquAdvantage Salmon NADA will not jeopardize the continued existence of Atlantic salmon—which is listed as threatened or endangered—nor will it result in the destruction or adverse modification of their critical habitat. A 60-day public comment period initially ran through February 25, 2013, but was extended through to April 26, 2014.[132] On November 23, 2013, Environment Canada granted AquaBounty permission to export up to

[126]Senior (2010).

[127]Food and Drug Administration Center for Veterinary Medicine (2012).

[128]Food and Drug Administration (2012).

[129]Ibid.

[130]In 2007, legislation was passed to require the FDA to consult with the National Marine Fisheries Service and to produce a report on any environmental risks associated with genetically engineered seafood products, including the impact on wild fish stocks.21 U.S.C. §2106. According to FDA, the two agencies have consulted on this matter, but this report has not been developed and no target date for its completion has been specified.

[131]16 U.S.C.1531 et seq.

[132]78 Federal Register 10620–10621 (February 14, 2013).

100,000 eggs a year from a hatchery in Prince Edward Island to Panama.[133] On March 13, 2014, the FDA Commissioner, Dr. Margaret Hamburg, stated that the AquaBounty NAD application is still under consideration and that the FDA will be moving forward with it in a deliberate science-driven way.[134] Currently, AquaBounty's research facility is operating and raising GE salmon in Panama. Since 1989, AquaBounty's breeding program has produced 10 generations of AquAdvantage Salmon.[135]

14.4 Analysis of the FDA's Regulatory Approach to GE Salmon

14.4.1 Potential Human Health and Safety Issues

As a National Research Council study has concluded, because genetic engineering can introduce new protein into a food product, there is a low to moderate food safety risk in consuming GE seafood.[136] Several consumer groups have expressed food safety concerns and have questioned whether GE salmon could pose public health hazards, such as introducing previously unknown allergens into the food supply or introducing a known allergen into a "new" food.[137,138] Although the VMAC concluded that test results established similarities and equivalence between AquAdvantage salmon and non-GE Atlantic salmon, the Chairman's Report added that the data submitted by AquaBounty does not conclusively show that AquAdvantage Salmon would be more or less allergenic than Atlantic salmon.[139] The FDA has maintained that people who are allergic to Atlantic salmon will likely be allergic to AquAdvantage Salmon because it is likewise a finfish, but not because it has been genetically engineered.[140] In its preliminary finding of no significant impact released on December 20, 2012, the FDA reiterated that food from AquAdvantage salmon is as safe as food from non-GE salmon and that there are

[133]Goldenberg (2013a, b). Two Canadian environmental groups—Ecology Action Centre (NS) and Living Oceans Society (BC)—filed a judicial review application with the Federal Court on Dec. 23, 2013 to decide if the federal government violated its own law when it permitted the manufacture of the AquAdvantage salmon. The legal challenge asserts that the approval is unlawful because it failed to assess whether the GE salmon could become invasive, potentially putting ecosystems and species such as wild salmon at risk. Ecology Action Centre (2014).

[134]Upton and Cowan (2014), p. 13 citing U.S. Congress, Senate Committee on Health, Education, Labor, and Pensions, Protecting the Public Health: Examining FDA's Initiatives and Priorities, 113th Cong., 2nd sess., March 13, 2014.

[135]AquaBounty Technologies (2014a).

[136]National Research Council (2004).

[137]Food and Water Watch, Consumers Union, and The Center for Food Safety (2012), p. 3.

[138]Ibid.

[139]Senior (2010).

[140]Food and Drug Administration (2013).

no significant food safety hazards or risks associated with consuming AquAdvantage salmon.[141]

On February 7, 2012, three non-governmental organizations petitioned the FDA's Office of Food Additive Safety (OFAS) to review the AquaBounty application under the FFDCA food additive provisions.[142] The petitioners argued that the gene expression product (GEP) of the genetic construct that creates the AquAdvantage salmon is a food additive under FFDCA.[143] AquAdvantage salmon exhibit an elevated level of Insulin Growth Factor-1(IGF-1), which some studies have linked to breast, colon, prostate, and lung cancers.[144] It may also have increased allergen potential.[145] The petition asserts that IGF-1 is a novel food additive and constitutes a "material fact" about the GE salmon, compared to its non-GE counterpart which does not contain the additive. In the petition, the non-governmental organizations requested that the FDA make a statement that neither the AquAdvantage salmon nor the gene expression product (GEP) used to create it is "generally regarded as safe". In particular, the petition asked for extensive pre-market testing, arguing that "[t]he Agency's general classification of rDNA constructs as new animal drugs does not displace or override the Agency's regulations and guidelines, and nothing precludes the Agency from also regulating GE salmon and its components as food additives."[146]

Independent scientists have criticized the studies submitted by AquaBounty for lacking sufficient rigor and appropriate methodology. For example, attendees at the VMAC meeting raised concerns about the small sample size for the allergenicity tests used in AquaBounty Technologies' research.[147] They also critiqued the lack of independent, peer-reviewed studies to support some data.[148] Committee member Dr. Jodi Ann Lapidus, for instance, characterized the food safety testing process as "a bit ad hoc." She argued, "some of the studies that were done could have been thought through . . . .more carefully, [and could have] employed more rigorous study design method."[149] Given the skepticism surrounding the studies that have already taken place, more research needs to be done to determine what long term risks, if any, come with consuming AquAdvantage salmon.

---

141 Ibid. (2012).

142 Food and Water Watch, Consumers Union, and The Center for Food Safety (2012).

143 21 U.S.C. §201(s), 21 U.S.C. §321.

144 Food and Water Watch, Consumers Union, and The Center for Food Safety (2012), p. 16.

145 Ibid., p. 6.

146 Ibid., p. 4.

147 Food and Drug Administration (2010d), pp. 290–291 (noting that risk of potential allergenicity was 20 % higher in AquAdvantage Salmon, but that this figure was not statistically significant because the sample size was only six fish).

148 Food and Drug Administration (2010d), p. 293 (noting that the analysis of IGF levels looked at only two studies—a peer-reviewed publication from 1992 and an AquaBounty study from 2004—and the analysis of allergen potency focused on a 2006 study furnished by AquaBounty).

149 Food and Drug Administration (2010d), p. 366.

14.4.2 Environmental Concerns

The potential environmental harm that might be caused by GE fish is of great concern to several scientists and environmental groups. A National Research Council report, for instance, states that transgenic fish pose the "greatest science-based concerns associated with animal biotechnology, in large part due to the uncertainty inherent in identifying environmental problems early on and the difficulty of remediation once a problem has been identified."[150] However, as discussed in Sect. 14.2.4 above, the FDA concluded in its draft environmental assessment that it "found no evidence that approval of an NADA for AquaAdvantage Salmon would result in significant impacts on the environment in the United States."[151] Furthermore, according to the FDA, it has verified that AquAdvantage Salmon would be produced and grown in secure facilities. The FDA considers the likelihood of the escape and survival of GE salmon to be extremely remote. Indeed, the FDA describes the environment around the egg-producing facility in PEI and the grow-out facility in Panama as inhospitable to fish. In the event that any fish do escape and survive, reproduction in the wild would be unlikely because the AquAdvantage salmon will be all female triploid fish, nearly all of which will be sterile.[152]

Despite the FDA's assertion that AquAdvantage salmon production is environmentally sound, opponents challenge these conclusions.[153] Some sources, for example, critique the security of AquaBounty's facilities. While the FDA has withheld from the public any visual documentation, an independent journalist with the Guardian reported in April 2013 that the facilities are a "fading" and "rundown shed" in the "Panamanian rainforest."[154] In contrast to the Guardian's description, the FDA's description of the facility in the most recent draft EA calls it "newly built and well-maintained."[155] A second issue is that although the FDA has

---

150 National Research Council (2004).

151 Food and Drug Administration Center for Veterinary Medicine (2012).

152 Upton and Cowan (2014), p. 15.

153 Schwab (2013), p. 9; see Hauter et al. (2014), p. 1 asking the FDA "to deny the new animal drug application (NADA) for AquaBounty Technonlogies' AquAdvantage Salmon in light of the disastrous environmental record of AquaBounty." Because "AquaBounty has admitted fault in breaching environmental regulations in Panama and has experienced at least one major security accident involving "lost" salmon. It is also now public record that AquaBounty's production platform in Panama has changed dramatically from the production platform described by AquaBounty and the FDA in the NADA and draft Environmental Assessments (EA), which presents another basis for FDA abandoning AquaBounty's NADA." AquaBounty (2014b) responded to the letter and press release by stating that they are "appalled at the irresponsible and untruthful attacks contained in the activists' press release and letter. The Company has been in complete control of its Panama facility and at no time was the safety of the fish or the environment at risk."

154 Goldenberg (2013b) and Rossiter (2013).

155 Food and Drug Administration Center for Veterinary Medicine (2012), p. 142.

explained the different characteristics of the facilities in Canada and Panama, it acknowledges that it cannot evaluate risks that are out of its jurisdiction unless it is specifically necessary for evaluating the effects on the US environment.[156] Consequently, local authorities would have to complete an environmental assessment in order to determine the potential long-term impact on the local environment of AquaBounty's salmon operation.[157]

There are two primary concerns regarding the environmental impact of the AquAdvantage Salmon: first, the potential for AquAdvantage salmon to interbreed with wild Atlantic salmon; and second, that the AquAdvantage salmon will compete with native fish, including salmon and other species, for food, habitat, mates, and other resources. Opponents hypothesize that farmed GE salmon, which have different fitness-related traits such as higher feeding and growth rates,[158] will eventually escape from the aquaculture system and interbreed with wild Atlantic salmon. The "Trojan gene hypothesis" speculates that wild populations could become extinct when a gene that confers a reproductive advantage also renders offspring less able to survive in the natural environment. One of the researchers who framed the hypothesis, however, stated that the Trojan gene effect only occurs when there is a conflict between mating success (if GE salmon were to mate more successfully) and viability fitness (if offspring were less likely to survive in the wild). He concluded that the risk of harm is low because data conclusively show that in this case there is no Trojan gene effect and that natural selection would purge the transgene.[159]

If the AquAdvantage salmon are accidently released into the environment, they could spread quickly and be difficult to contain. GE fish that escape into the wild could also compete with wild fish and harm wild salmon populations, which are already depleted and vulnerable to ecological changes. Because GE salmon grow faster, some researchers suggest that they may outcompete wild fish for habitat and food.[160] On the other side, proponents argue that should the GE fish escape, they would be *less* likely to survive in the wild, especially because they are reared in protected artificial habitats and have not learned to avoid predators.[161]

Critics question whether the FDA has sufficient expertise to evaluate and protect against the potential ecological damage that might result from GE fish production.[162] A coalition of environmental groups has called on the FDA to prepare a full environmental impact statement (EIS) and to consult more closely with federal agencies about possible threats to endangered wild Atlantic salmon.[163] Dr. Gary

[156]Ibid.

[157]Varela (2013), p. 525.

[158]Wodder et al. (2010).

[159]Food and Drug Administration Center for Veterinary Medicine (2012), p. 91.

[160]Ibid.

[161]Upton and Cowan (2014), p. 19.

[162]See Center for Food Safety (2015a).

[163]Wodder et al. (2010). In 2011, these groups filed a formal citizen petition urging FDA to withhold approval until an EIS has been completed. Ocean Conservancy et al. (2011).

Thorgaard, who is the only fisheries expert on the VMAC, seconds the preparation of an EIS.[164] Complicating the calls for a thorough environmental assessment, however, is the assertion by some critics that the risk assessment data supplied by AquaBounty is not valid. They argue that the FDA should either require AquaBounty re-conduct its studies, or else the agency should conduct the studies itself or ask an independent laboratory to undertake them instead.[165]

Thus far, the FDA has thus far determined that an EIS is not required. Nevertheless, if significant new information arises, such as an AquaBounty proposal for additional salmon-growing facilities, the FDA may determine that an EIS is needed prior to approval of AquaBounty's application.[166]

14.4.3 *Lack of Transparency in the Approval Process*

Generally speaking, lack of transparency is the primary reason for opposition to the regulation of GE animals using the NADA process.[167] According to the FDA's 2009 Guidance, the application process can take place almost entirely without notification to the public, until the GE animal is approved. Before the FDA makes a decision on an application, the Trade Secrets Act prohibits the agency from revealing any information that it acquires through the New Animal Drug approval process in order to protect the applicant's trade secrets.[168] The FDA also does not have the authority to disclose whether an application even exists until after publication of approval in the Federal Register.[169] Sponsors may disclose the application, as indeed AquaBounty Technologies has, but even then "no data or information contained in the file is available for public disclosure before such approval is published[.]"[170] The Commissioner may, however, "in his discretion, disclose a summary of selected portions of the safety and effectiveness data as are appropriate for public consideration of a specific pending issue, e.g., at an open session of a Food and Drug Administration advisory committee[.]"[171] The Consumer Union disagrees with the Trade Secret Act's prohibition against information disclosure, arguing that "in general, safety and health data should not be considered confidential business information."[172] Although the FDA intends to hold public

[164]See Food and Drug Administration (2010d), p. 383. See Kapuscinski and Sundstrom (2010a, b); Kapuscinski and Sundstrom (2013).

[165]See, e.g., Schwab (2013).

[166]Upton and Cowan (2014), p. 16.

[167]Consumers Union (2008), p. 1; McEvilly (2013), p. 421.

[168]*See* 21 U.S.C. § 331(j) (2006) (prohibiting information acquired under the authority of several sections related to the Animal Drug approval process from being revealed).

[169]21 C.F.R. § 514.11 (2012).

[170]21 C.F.R. § 514.11(d).

[171]Ibid.

[172]Consumers Union (2008), p. 2.

VMAC meetings for the first GE animals up for approval, the agency may decide not to hold these meetings once several animals have gone through the application process.[173]

Despite the fact that AquaBounty's application was available for public review, opponents have heavily criticized the short timetable for public comments. The notice of the VMAC's September 20 meeting was published in the Federal Register on August 26, only three and a half weeks before the meeting date.[174] Furthermore, as Darrell Rogers of the Alliance for Natural Health in the United States notes, scientific studies have either "not been released or have been released so late in the approval process that it is impossible for the public and experts to assess whether scientific burdens have been met."[175] As well, Jaydee Hanson, from the Center for Food Safety, expressed frustration that the public received the 180-page scientific assessment briefing packet "only 10 days [before comments were due]."[176] Public input on AquaBounty's application has been neither straightforward nor easily accessible.

14.4.4 Public Perception and Consumer Acceptance

Critics of AquaBounty's application have also raised questions regarding the commercially viability of the salmon in light of several studies that show consumer skepticism towards the product.[177] Public support for biotechnology, including genetically engineered animal products, is on the decline.[178] A survey performed for the Pew Initiative on Food and Biotechnology in 2001 found that 65 % of consumers disapproved of the idea of creating transgenic fish for human consumption.[179] More recently, the New York Times reported in 2013 that three-quarters of Americans expressed concern about genetically modified organisms in their food, with most people worried about possible ill-effects on health.[180] Furthermore, 37 % of those concerned about genetically modified organisms said they feared that such foods could cause cancer or allergies. Other consumer surveys have also shown

[173]Ibid.

[174]Food and Drug Administration (2010c).

[175]Food and Drug Administration (2010d), p. 282.

[176]Ibid., pp. 297 and 311 ("Until the release of the [environmental assessment] two weeks ago, the public has had no opportunity to learn more about, assess, or raise questions about potential impacts.").

[177]O'Halloran (2013).

[178]See, e.g., Logar and Pollock (2005) (citing data that "Americans' attitudes towards genetic engineering and biotechnology generally show a decline in support for such technologies over past 5–15 years").

[179]Pew Initiative on Food & Biotechnology (2001), p. 2.

[180]Kopicki (2013).

widespread public opposition to GE salmon. For example, three quarters of Americans surveyed in a New York Times poll said they would not eat genetically engineered fish.[181] Environmental and consumer groups alike have launched campaigns to persuade grocers, restaurants, and distributors to pledge not sell GE fish, even if they are approved by the FDA. Many major grocery retail chains, such as Safeway, Trader Joe's, Target, Kroger, and Whole Foods, have already announced that they will not sell the AquAdvantage salmon when it becomes commercially available.[182]

Although the debate regarding GE fish has garnered much media attention, some researchers argue that the FDA review of AquaBounty's application should not be influenced by public comments unrelated to science-based issues. As Van Eenennaam notes, for example:

> Regulatory bodies exist to provide objective assessments. They comprise experts on the topic with the authority to establish regulations that ensure society benefits from scientific discoveries, rather than coming to harm. Therefore plurality of opinion not supported by relevant data and propelled by democracy in science undermines the very institutions put in place to ensure the proper use of science and technology for the benefit of society.[183]

While it is undoubtedly crucial that comprehensive scientific studies support the FDA's decision to approve—or not—the AquaBounty application, public opinion is still entirely relevant to the commercial viability of the AquAdvantage salmon on the market.

14.5 Labeling Issues

Federal food labeling policy, including the labeling of foods containing bioengineered material, is regulated under the 1938 Federal Food, Drug, and Cosmetic Act (FFDCA),[184] as well as the 1966 Fair Packaging and Labeling Act.[185] Under Section 403(a)(1) of the FFDCA, a food is considered misbranded if its labeling is false or misleading. Section 201(n) of the FFDCA states that a label is misleading if it fails to reveal material facts in light of any representations that are

---

[181]Ibid.; See also Thomson Reuters (2010).

[182]Food Safety News (2014).

[183]Van Eenennaam et al. (2013), quoting Gemma Arjo, Manuel Portero, Carme Pinol, Juan Vinas, Xavier Matias-Guiu, Teresa Capell, Andrew Bartholomaeus, Wayne Parrott & Paul Christou, Plurality of Opinion, Scientific Discourse and Pseudoscience: An In Depth Analysis of the Séralini et al. Study Claiming that Roundup Ready Corn or the Herbicide Roundup Cause Cancer in Rats, Transgenic Reseach, 2013 (22), pp. 255–267.

[184]21 U.S.C. §§301 et seq.

[185]P.L. 89-755;15 U.S.C. §§1451 et seq. The Nutrition Labeling and Education Act of 1990 (P.L. 101-535; 21 U.S.C. §343), which amended the Federal Food, Drug, and Cosmetic Act, requires most foods to carry nutrition labeling and requires food labels with claims about nutrient content or certain health messages to comply with specific requirements.

made or suggested in the labeling, or with respect to potential consequences that could result from using the food prescribed in the labeling, or under conditions of use that are customary or usual.

Although 68 countries require the labeling of GE foods, the United States does not require labeling to identify foods containing genetically modified material. The notion of "substantial equivalence" guides food labeling requirements: if a food containing GE material is "substantially equivalent" to a food not containing GE material, federal regulations do not require that it be labeled as containing GE material. When there is no material difference between products, the FDA does not have the authority to require labeling on the basis of consumer interest alone. If there is a material discrepancy between GE and non-GE foods, however, the FDA could require such differences to be identified on food labels.[186] In 1992, the FDA published a policy statement on foods derived from new plant varieties, including those developed through genetic engineering.[187] The policy statement did not establish any special labeling requirements for bioengineered foods because the FDA has found most GE crops to be "substantially equivalent" to non-GE crops.

Although the law does not require the FDA to address labeling issues prior to a food being marketed, the agency is considering these two issues simultaneously.[188] If the AquAdvantage salmon NADA is approved, the FDA will determine whether additional labeling is appropriate. The FDA has determined that the AquaBounty salmon is as safe for human consumption as non-GE salmon; in other words, that the AquaBounty salmon is "substantially equivalent" to non-GE salmon. Opponents of the AquaBounty salmon, however, argue for the need to label the fish as a GE product on a basis of consumers' right-to-know. Consumer polls show increasing support for the labeling of GE products.[189] Proponents of the AquaBounty salmon contest that labeling a GE food as "substantially equivalent" would imply that it was different in ways that could be seen as negative. They also hold that labeling would impose additional costs on consumers and would create unnecessary confusion.[190] Since there is no federal law that specifically addresses labeling for GE fish and seafood, individual states have begun proposing their own labeling laws. Alaska, for example, requires all GE fish to be labeled.[191] As discussed above, many countries require the labeling of GE foods for human consumption.[192] Given the discrepancy between US policy and that of other countries, this could create trade issues. Because GE food labeling is controversial and unresolved, the

[186]Food and Drug Administration (2010a). Companies may label their foods as not containing bioengineered products, if they can definitively show that the foods do not contain GE products.

[187]Food and Drug Administration (1992).

[188]Ibid., (2010a).

[189]See Thomson Reuters (2010) and Kopicki (2013).

[190]See Nature Biotechnology (2014), p. 1169.

[191]AK Food & Drug Code §17.20.040 (2005).

[192]Kimbrell and Tomaselli (2011), pp. 100–101.

issue of labeling for AquaBounty salmon has complicated the FDA's approval process.[193]

14.6 Animal Applications and Impact on Future GE Fish

Although AquaBounty's application is limited to its current facilities in Canada and Panama, the company has stated that once the AquAdvantage salmon is approved for sale, it will immediately begin field trials with prospective customers with the goal of expanding to new sites in the US and throughout the world.[194] As environmental risk experts Dr. Anne Kapuscinski and Dr. Fredrik Sundstrom have explained in oral comments delivered to the VMAC,

> [We] need to describe an elephant in the room: [AquaBounty] understandably wants to sell eggs to many growers to be competitive in the global farmed salmon industry. So approval of this application will trigger other applications soon. But the regulations don't require the FDA to publicly release future environmental assessments before their approval. If this application is approved, farming of transgenic AquAdvantage salmon will proliferate in the foreseeable future, given that farmed salmon is a global commodity. The environmental assessment laid out in this case will set a precedent. It is imperative that it follows high scientific standards and minimum scientific requirements.

Many observers have called upon the FDA to analyze and consider potential future environmental and human health impacts in its initial approval of GE salmon. This is particularly important because once approval is granted, the agency's subsequent review is limited.[195] A company whose drug application has been approved, for example, could self-determine whether a supplemental FDA approval of an animal drug is necessary before effecting certain changes in its "drug, production process, quality controls, equipment, or facilities."[196] Pre-approval is necessary only if the change has "substantial potential to have an adverse effect on the identity, strength, quality, purity, or potency of the drug as these factors may relate to the safety or effectiveness of the drug."[197] Therefore, as George Kimbrell and Paige Tomaselli explain, "because these provisions do not include consideration of adverse environmental effects, a transgenic-animal applicant like AquaBounty could argue that FDA approval is not needed for major changes to its facilities, containment measures, or production locations—despite the fact that such changes could pose significant new environmental risks."[198] Once the FDA approves the AquAdvantage salmon, it will become challenging for the

---

[193]Upton and Cowan (2014), p. 20.

[194]AquaBounty Technologies (2011).

[195]Kimbrell and Tomaselli (2011), p. 100.

[196]21 C.F.R. § 514.8(b)(2).

[197]Ibid.

[198]Kimbrell and Tomaselli (2011), p. 100.

agency to closely regulate the entry of other genetically engineered animals into the market.

14.7 State and Federal Attempts to Regulate GE Fish

In light of the FDA's pending approval of GE salmon, several states have passed laws regulating these fish. Bills have also been proposed in Congress to prevent the approval of GE fish, or to require labeling.

14.7.1 State Laws Regulating GE Fish

Several states have passed laws establishing various prohibitions related to GE fish.[199] California, for example, prohibits the spawning, incubation, and cultivation of GE fish in all Pacific Ocean waters regulated by the state.[200] Washington also prohibits the use of GE fish in state waters.[201] Several other states require permits for the use of GE fish. California also prohibits the import, transport, and possession of transgenic seafood species without a state permit.[202] Florida requires certified aquaculturalists to receive authorization from the Florida Department of Agriculture and Consumer Services (FDACS) prior to culturing transgenic fish species.[203] The FDACS must consult with the Transgenic Aquatic Species Advisory Committee prior to granting authorization, and approval is to be granted only if it is determined that there is no threat posed to public health, safety, or welfare. Michigan requires permits for each of the following: possession of GE fish,[204] propagation of GE fish in private waters,[205] acquisition of GE fish from inland waters for scientific studies, and importation of viable GE fish eggs into the state.[206] In addition, the state has established penalties for violating these rules. Rhode Island prohibits the introduction of non-indigenous aquatic species unless the Biosecurity Board approves protocols for preventing any accidental releases into state waters.[207]

---

[199]See Center for Food Safety (2013).

[200]Cal. Fish & Game Code § 15007.

[201]Wash. Admin. Code § 220-76-100 (2012).

[202]Cal. Code Regs. tit. 14, §671.1 (2013).

[203]Fla. Admin Code r. 5 L-3.004 (2013).

[204]Mich. Comp. Laws Ann. § 324.41301-41325.

[205]Mich. Comp. Laws Ann. § 324.45901-45908.

[206]Mich. Comp. Laws Ann. § 324.48701-48740.

[207]16-2 R.I. Code R. § 300.11 (2012).

14.7.2 *Congressional Actions*

Some Members of Congress have raised concerns about whether the FDA's approval process for GE salmon adequately ensures its safety for consumers and the environment. In April 2013, 20 Members of the House and 12 Members of the Senate sent similar letters that requested FDA Commissioner Margaret Hamburg halt the approval process.[208] Several bills have been introduced in Congress to amend the Federal Food, Drug, and Cosmetic Act so as to prevent the approval of GE salmon and/or require labeling of genetically engineered fish. A number of bills, for example would prohibit the possession or use of GE fish in the United States and would also require completion of a report on environmental risks associated with GE seafood products.[209] Other bills would add a requirement under the FFDCA to label genetically engineered fish.[210] In June 2011, the House of Representatives passed a bill that would have prohibited the FDA from spending funds from the 2012 budget to approve any application for GE salmon.[211] Another bill would have prohibited the FDA approval of GE fish unless the National Marine Fisheries Service concurred with such approval.[212] No further action, however, has been taken on these or other bills that would require additional regulation of genetically engineered organisms.

14.8 Recommendations for Reform

Due to the fact that at least 35 other species of genetically engineered fish are currently under development,[213] the FDA's decision on the AquaBounty salmon application will set a precedent for other genetically engineered fish and animals to enter the global food market.

[208]Upton and Cowan (2014), p. 23 citing Letter from Don Young et al. to Dr. Margaret Hamburg, Commissioner, FDA, April 24, 2013, and Letter from Senator Mark Begich et al. to Dr. Margaret Hamburg, April 24, 2014.

[209]Upton and Cowan (2014), p. 24. Such a report was required under Section 1007 of the Food and Drug Administration Amendments Act of 2007. 21 U.S.C. 2106.

[210]Upton and Cowan (2014), p. 24, citing H.R. 584, H.R. 1699, S. 248, and S. 809.

[211]Upton and Cowan (2014), p. 24.

[212]Ibid.

[213]Center for Food Safety (2015a).

14.8.1 Reconsidering the Coordinated Framework

Many critics question whether the FDA is the appropriate agency to evaluate the environmental and human health risks posed by GE fish[214] and therefore call for reforms to the Coordinated Framework. Bratspies, for instance, writes that

> The Coordinated Framework must also be reconsidered, either by the President and the Executive Branch itself or through legislative action. In particular, it is time to rethink the decision to make FDA lead agency for regulating transgenic fish and other animals. Because many of the most critical issues with regard to transgenic fish are environmental, they do not naturally fall within FDA's scope of authority.[215]

In addition, Gregory Mandel criticizes the FDA's application of animal drug provisions to transgenic food. Professor Mandel explains that because transgenic animals are very different from veterinary animal drugs, they present new difficulties for their assessment and oversight.[216] He writes, "such forcing of transgenic square pegs into pre-existing statutory round holes is an endemic problem of U.S. oversight under the Framework."[217]

In response to this criticism of the FDA's review of GE salmon as a NAD, supporters have countered that "the FDA by its very nature already has considerable expertise in food safety, molecular biology, and aquaculture, being one of the co-founders of the Joint Subcommittee on Aquaculture (JSA), a statutory committee that operates under the National Science and Technology Council (NSTC) of the Office of Science and Technology Policy in the Office of the Science Advisor to the President."[218] Furthermore, the FDA consulted with the National Marine Fisheries Service (NMFS) of the National Oceanic and Atmospheric Administration (Department of Commerce) and the U.S. Fish and Wildlife Service (FWS) of the Department of Interior, "which have concurred with, or indicated no disagreement with, FDA's 'no effect' determination."[219] Evaluating the potential risks of a genetically engineered animal is incredibly complicated, and experts remain divided over who should have the authority to do so.

---

[214] See, e.g., Pew Initiative on Food & Biotechnology (2004), p. 125.

[215] Bratspies (2005), pp. 503–504.

[216] Mandel (2004), p. 2243.

[217] See Mandel (2004), p. 2243 (noting that as a result of the Framework's flawed paradigm, there have been "multiple failures on the part of regulatory agencies to recognize that genetically modified products sometimes do create new and different issues than those raised by the conventional products they routinely regulate").

[218] Van Eenennaam et al. (2013).

[219] Ibid.

14.8.2 *The Possibility of Regulating GE Animals as Food Additives*

As discussed above, three non-governmental organizations petitioned the FDA's Office of Food Additive Safety (OFAS) to review the AquaBounty application under the FFDCA food additive provisions.[220] The FDA defines a food additive as "any substance, the intended use of which results directly or indirectly, in it becoming a component or otherwise affecting the characteristics of food."[221] Food additives require pre-market approval from the FDA and are presumed to be unsafe, unless scientific experts evaluate them and determine them to be "generally regarded as safe" (GRAS).[222] The petitioners argued that the gene expression product (GEP) of the genetic construct creating the AquAdvantage salmon is a food additive under the FFDCA.[223] AquAdvantage salmon exhibit an elevated level of Insulin Growth Factor-1 (IGF-1), which the petitioners asserted is a novel food additive and constitutes a "material fact" about the GE salmon, as compared to its non-GE counterpart.[224] If the FDA were to consider GE salmon a food additive, the fish would have to undergo comprehensive toxicological studies.[225]

14.8.3 *Improving Transparency of the Approval Process*

A number of those observing the FDA's 2009 Guidance noted that the ban on public disclosure "is particularly inappropriate for products of a new and controversial technology such as the genetic engineering of animals."[226] However, although some have suggested that the FDA could broadly interpret the safety and effectiveness data as "appropriate for public consideration" in order to allow increased opportunities for peer review,[227] this suggestion is problematic. First, the drug industry depends on confidentiality to prevent competitors from gaining unfair access to the development of a new product.[228] Furthermore, because the FDA cannot by law disclose commercial applications, recommendations to establish

[220] Food and Water Watch, Consumers Union, and The Center for Food Safety (2012).

[221] FFDCA, Section 201(s).

[222] 21 U.S.C. § 348(a) (2006).

[223] 21 U.S.C. §321.

[224] Food and Water Watch, Consumers Union, and The Center for Food Safety (2012).

[225] Ibid., p. 22.

[226] Food and Drug Administration (2011).

[227] McEvilly (2013), pp. 413 and 427–428.

[228] Van Eenennaam et al. (2013).

advisory councils comprised of non-federal experts to evaluate new animal drugs naively ignores the law.[229]

14.8.4 Streamlining the GE Animal Approval Process

Even those who support the FDA's NADA review process agree that some reforms are necessary. To expedite the process, for example, the FDA could impose finite response times for agency decisions at each step during the evaluation. Doing so would provide developers and investors with a predictable regulatory timeline for GE animals.[230]

14.9 Conclusion

The FDA's review process of the AquAdvantage salmon is an important and hotly contested issue because it will set a precedent for review of other GE animals.[231] Due to the global demand for fish and the United States' political, economic, and cultural influence around the world, the FDA's review has international implications. As this chapter has discussed, however, the adequacy of the review is widely debated. As a 2014 Congressional Research Report states, "whether the current process affords adequate safeguards for the public while allowing for the application of new genetic technologies remains an open question."[232] As of this writing, it is still unknown as to how the FDA will proceed with its review of AquaBounty's application for the approval of the genetically engineered AquAdvantage salmon.

Questions for Classroom Discussion

1. In light of the current regulatory obstacles, how would you advise a client who is interested in developing a GE animal product for human consumption?
2. What are arguments for the FDA to expedite the approval process?
3. What are the arguments for having a more comprehensive and involved application process?
4. What are the international implications of GE salmon approval?
5. What are some concerns from the international community regarding approval of GE fish?
6. What are some recommendations for reforming the application review process for other GE animals intended for human consumption?

[229] Ibid.

[230] Ibid.

[231] See Upton and Cowan (2014), p. 17.

[232] Upton and Cowan (2014), p. 26.

References

AquaBounty Technologies (2010) Environmental assessment for AquAdvantage Salmon. http://www.fda.gov/downloads/AdvisoryCommittees/CommitteesMeetingMaterials/VeterinaryMedicineAdvisoryCommittee/UCM224760.pdf. Accessed 19 Jan 2015

AquaBounty Technologies (2011) Interim results for the six months ended 30 June 2011. http://aquabounty.com/wp-content/uploads/2014/02/2011-09.23-Interim-Results.pdf. Accessed 16 Jan 2015

AquaBounty Technologies (2014a) AquAdvantage® Salmon DNA Animations. http://aquabounty.com/press-room/video-b-roll-animation/. Accessed 16 Jan 2015

AquaBounty Technologies (2014b) AquaBounty responds to anti-gm organizations' letter and press release. http://aquabounty.com/wp-content/uploads/2014/11/2014-11.20-AquaBounty-responds-to-anti-GM-organizations-letter-and-press-release-Finalx.pdf. Accessed 16 Jan 2015

AquaBounty Technologies (2014c) First edible genetically engineered fish still awaiting approval. http://aquabounty.com/first-edible-genetically-engineered-fish-still-awaiting-approval/. Accessed 16 Jan 2015

Biron C (2014) GM salmon company Aquabounty fined by Panama. The Guardian. http://www.theguardian.com/lifeandstyle/2014/oct/29/panama-regulators-could-slow-us-approval-of-gm-salmon

Bratspies R (2005) Glowing in the dark: how America's first transgenic animal escaped regulation. Minn J Law Sci Technol 6:457

Bratspies R (2008) Farming the genetically modified seas—the perils and promise of transgenic Salmon. American Fisheries Society Symposium. http://www.bioeng.fisheries.org/proofs/igfe/bratspies.pdf. Accessed 2 Jan 2015

Cartagena Protocol on Biosafety to the Convention on Biological Diversity (2000) http://www.cbd.int/doc/legal/cartagena-protocol-en.pdf. Accessed 23 Jan 2015

Carter C, Gruere G (2003) International approaches to the labeling of genetically modified foods. http://www.agmrc.org/media/cms/cartergruere_929BEB69BA4EE.pdf. Accessed 23 Jan 2015

Center for Food Safety (2013) GE fish and the environment. http://www.centerforfoodsafety.org/issues/309/ge-fish/ge-fish-state-regulations#. Accessed 16 Jan 2015

Center for Food Safety (2015a) Genetically engineered fish. http://www.centerforfoodsafety.org/issues/309/ge-fish. Accessed 16 Jan 2015

Center for Food Safety (2015b) International labeling laws. http://www.centerforfoodsafety.org/issues/976/ge-food-labeling/international-labeling-laws. Accessed 16 Jan 2015

Consumers Union (2008) Comments on Docket No. FDA-2008-D-0394, "Draft Guidance for Industry: Regulation of Engineered Animals Containing Heritable rDNA Constructs". http://www.consumersunion.org/pub/pdf/FDA-comm-DNA-1108.pdf. Accessed 16 Jan 2015

Cowx IG et al (2010) Environmental risk assessment criteria for genetically modified fishes to be placed in the EU market. European Food Safety Authority. http://www.efsa.europa.eu/en/supporting/doc/69e.pdf. Accessed 7 Jan 2015

Ecology Action Centre (2014) Salmon trials. https://www.ecologyaction.ca/content/GM-Salmon-Trial-Release. Accessed 19 Jan 2015

EUROPA (2006) Europe's rules on GMOs and the WTO. http://europa.eu/rapid/pressReleasesAction.do?reference=MEMO/06/61&format=HTML&aged=0&language=EN&guiLanguage=en. Accessed 7 Jan 2015

Food and Agriculture Organization of the United Nations (2014) State of world fisheries and aquaculture. http://www.fao.org/3/a-i3720e.pdf. Accessed 19 Jan 2015

Food and Drug Administration (1992) Statement of policy: foods derived from new plant varieties. 57 Federal Register 22984

Food and Drug Administration (2008) Guidance for industry: regulation of genetically engineered animals containing heritable rDNA constructs; Availability, 73 Federal Register 54407, September 19, 2008, http://edocket.access.gpo.gov/2008/pdf/E8-21917.pdf. Accessed 19 Jan 2015

Food and Drug Administration (2010a) Background document: public hearing on the labeling of food made from the AquAdvantage Salmon. http://www.fda.gov/Food/GuidanceRegulation/GuidanceDocumentsRegulatoryInformation/LabelingNutrition/ucm222608.htm

Food and Drug Administration (2010b) Charge to the VMAC for the AquAdvantage Salmon Meeting. http://www.fda.gov/AdvisoryCommittees/CommitteesMeetingMaterials/VeterinaryMedicineAdvisoryCommittee/ucm226083.htm. Accessed 19 Jan 2015

Food and Drug Administration (2010c) Veterinary Medicine Advisory Committee, Notice of Meeting, 75 Fed. Reg. 52,605 (Aug. 26, 2010)

Food and Drug Administration (2010d) VMAC Meeting: AquAdvantage Salmon. http://www.fda.gov/downloads/AdvisoryCommittees/CommitteesMeetingMaterials/VeterinaryMedicineAdvisoryCommittee/UCM230469.pdf. Accessed 19 Jan 2015

Food and Drug Administration (2011) FDA's response to public comments. http://www.fda.gov/AnimalVeterinary/DevelopmentApprovalProcess/GeneticEngineering/GeneticallyEngineeredAnimals/ucm113612.htm. Accessed 20 Jan 2015

Food and Drug Administration (2012) Preliminary finding of no significant impact, http://www.fda.gov/downloads/AnimalVeterinary/DevelopmentApprovalProcess/GeneticEngineering/GeneticallyEngineeredAnimals/UCM333105.pdf. Accessed 9 Jan 2015

Food and Drug Administration (2013) Key facts commonly misunderstood. http://www.fda.gov/AdvisoryCommittees/CommitteesMeetingMaterials/VeterinaryMedicineAdvisoryCommittee/ucm226223.htm. Accessed 16 Jan 2015

Food and Drug Administration Center for Veterinary Medicine (2009) Guidance 187, Guidance for industry regulation of genetically engineered animals containing heritable recombinant DNA constructs, http://www.fda.gov/downloads/AnimalVeterinary/GuidanceComplianceEnforcement/GuidanceforIndustry/UCM113903.pdf. Accessed 7 Jan 2015

Food and Drug Administration Center for Veterinary Medicine (2010) Briefing packet. http://www.fda.gov/downloads/AdvisoryCommittees/CommitteesMeetingMaterials/VeterinaryMedicineAdvisoryCommittee/UCM224762.pdf. Accessed 19 Jan 2015

Food and Drug Administration Center for Veterinary Medicine (2012) AquAdvantage Salmon, Draft Environmental Assessment. http://www.fda.gov/downloads/AnimalVeterinary/DevelopmentApprovalProcess/GeneticEngineering/GeneticallyEngineeredAnimals/UCM333102.pdf. Accessed 19 Jan 2015

Food and Water Watch, Consumers Union, and The Center for Food Safety (2012) Petition to deem ABT technologies' genetically engineered AquaAdvantage Salmon an unsafe food additive. http://www.centerforfoodsafety.org/files/fdafoodadditivepetitiongesalmon.pdf. Accessed 19 Jan 2015

Food Safety News (2014) Kroger, safeway will Nix GMO Salmon regardless of FDA decision. http://www.foodsafetynews.com/2014/03/kroger-safeway-turn-down-gmo-salmon-regardless-of-fda/#.VLleyCvF_eI. Accessed 10 Jan 2015

Goldenberg S (2013a) Canada approves production of GM salmon eggs on commercial scale. The Guardian. http://www.theguardian.com/environment/2013/nov/25/canada-genetically-modified-salmon-commercial. Accessed 19 Jan 2015

Goldenberg S (2013b) GM salmon global HQ – 1,500 m high in the Panamanian rainforest. The Guardian. http://www.theguardian.com/environment/2013/apr/24/genetically-modified-salmon-aquabounty-panama-united-states. Accessed 19 Jan 2015

GMO Compass (2015) Cartagena protocol. http://www.gmo-compass.org/eng/glossary/. Accessed 7 Jan 2015

Hauter W, Hanson J, Perls D (2014) Letter to Dr. Margaret A. Hamburg, FDA Commissioner. http://www.centerforfoodsafety.org/files/abt_fda_sec_ciam_final_16457.pdf. Accessed 20 Jan 2015

Kapuscinski A, Sundstrom F (2010a) Actual oral comments delivered on environmental assessment for AquAdvantage Salmon and briefing packet on AquAdvantage Salmon for the Veterinary Medicine Advisory Committee. https://www.dartmouth.edu/~ark/images/stories/actualoral.pdf. Accessed 19 Jan 2015

Kapuscinski A, Sundstrom F (2010b) Comments on environmental assessment for AquAdvantage Salmon and briefing packet on AquAdvantage Salmon for the Veterinary Medicine Advisory Committee. http://www.dartmouth.edu/~ark/images/stories/kapuscinski_sundstrom_comments_sent16sept10.pdf. Accessed 19 Jan 2015
Kapuscinski A, Sundstrom F (2013) Comments on Docket No. FDA-2011-N-0899, draft environmental assessment for AquAdvantage Salmon and preliminary finding of no significant impact. http://www.dartmouth.edu/~ark/images/stories/kapuscinski_sundstrom_comments_2012draftea_aas.pdf. Accessed 19 Jan 2015
Kimbrell G, Tomaselli P (2011) A "Fisheye" lens on the technological dilemma: the specter of genetically engineered animals. Animal Law 18:75
Kopicki A (2013) Strong support for labeling modified foods. NY Times. http://www.nytimes.com/2013/07/28/science/strong-support-for-labeling-modified-foods.html?_r=1&. Accessed 10 Jan 2015
Mandel G (2004) Gaps, inexperience, inconsistencies, and overlaps: crisis in the regulation of genetically modified plants and animals. William Mary Law Rev 45:2216
Logar N, Pollock L (2005) Transgenic fish: is a new policy framework necessary for a new technology? Environ Sci Policy 8:17–27
Ocean Conservancy et al (2011) Citizen petition regarding AquaBounty Technologies' application for approval of genetically engineered Salmon, citizen petition before the U.S. Food and Drug Administration. http://earthjustice.org/sites/default/files/FinalGESalmonCitizenPetition.pdf
O'Halloran S (2013) Consumers not interested in genetically modified salmon. Food Engineering. http://www.foodengineeringmag.com/articles/90611-consumers-not-interested-in-genetically-modified-salmon. Accessed 16 Jan 2015
McEvilly M (2013) Lack of transparency in the premarket approval process for Aquadvantage Salmon. Duke Law Technol Rev 11:413–433
National Research Council (2002) Animal biotechnology: science-based concerns. National Academy Press, Washington, DC. http://www.nap.edu/catalog/10418/animal-biotechnology-science-based-concerns. Accessed 10 Jan 2015
National Research Council (2004) Safety of genetically engineered foods: approaches to assessing unintended health effects. National Academies Press, Washington, DC. http://books.nap.edu/openbook.php?record_id=10977&page=R1. Accessed 10 Jan 2015
Nature Biotechnology (2014) Label without a cause. Nat Biotechnol 32:1169. http://www.nature.com/nbt/journal/v32/n12/pdf/nbt.3094.pdf
Pew Initiative on Food & Biotechnology (2001) Guide to U.S. regulation of genetically modified food and agricultural biotechnology products. www.pewtrusts.org/uploadedFiles/wwwpewtrustsorg/Reports/Food_and_Biotechnology/hhs_biotech_0901.pdf. Accessed 10 Jan 2015
Pew Initiative on Food & Biotechnology (2004) Issues in the regulation of genetically engineered plants and animals. http://www.pewtrusts.org/en/research-and-analysis/reports/0001/01/01/issues-in-the-regulation-of-genetically-engineered-plants-and-animals. Accessed 10 Jan 2015
Rossiter S (2013) See inside a genetically modified salmon farm in Panama. The Guardian. http://www.guardian.co.uk/environment/video/2013/apr/25/gm-salmon-panama-video
Schwab T (2013) Is FDA ready to regulate the world's first biotech food animal? Food Drug Policy Forum 3
Senior D (2010) VMAC, Chairman's Report, Food and Drug Administration, VMAC Meeting, September 20, 2010. http://www.fda.gov/downloads/AdvisoryCommittees/CommitteesMeetingMaterials/VeterinaryMedicineAdvisoryCommittee/UCM230467.pdf. Accessed 14 Jan 2015
Thomson Reuters (2010) National Survey of Healthcare consumers: genetically engineered food, PULSETM Healthcare Survey. http://www.justlabelit.org/wp-content/uploads/2011/09/NPR_report_GeneticEngineeredFood-1.pdf. Accessed 10 Jan 2015

United Nations (2010) Resumed review conference on the agreement relating to the conservation and management of straddling fish stocks and highly migratory fish stocks. http://www.un.org/depts/los/convention_agreements/reviewconf/FishStocks_EN_A.pdf. Accessed 19 Jan 2015

United States Department of Agriculture, Office of Inspector General (2011) Controls over genetically engineered animal and insect research. http://www.usda.gov/oig/webdocs/50601-16-TE.pdf. Accessed 19 Jan 2015

Upton H, Cowan T (2014) Genetically engineered Salmon, congressional research service. http://www.fas.org/sgp/crs/misc/R43518.pdf. Accessed 23 Jan 2015

Van Eenennaam A, Muir W, Hallerman E (2013) Is unaccountable regulatory delay and political interference undermining the FDA and hurting American competitiveness? Food Policy Forum 3(13). http://www.isb.vt.edu/news/2013/Nov/FDLI.pdf. Accessed 23 Jan 2015

Varela J (2013) FDA is ready to authorize GM Salmons but is the market ready for them? Eur J Risk Regul 4:521–526

Voosen P (2010) Panel advises more aggressive FDA analysis of engineered Salmon. N.Y. TIMES, September 21. http://www.nytimes.com/gwire/2010/09/21/21greenwire-panel-advises-more-aggressive-fda-analysis-of-71171.html?pagewanted=all

World Health Organization & Food and Agriculture Organization of the United Nations (2009) Codex Alimentarius Commission: Foods derived from modern biotechnology. ftp://ftp.fao.org/docrep/fao/011/a1554e/a1554e00.pdf. Accessed 10 Jan 2015

Wodder R et al (2010) Letter to Margaret Hamburg, M.D., Commissioner, FDA. http://www.foe.org/sites/default/files/EnvironmentalGroupLettertoFDA-GESalmonEnvironmentalReview.pdf. Accessed 19 Jan 2015

Chapter 15
Textbox: FDA Approval of GE Salmon

Nicole E. Negowetti

Abstract On November 19, 2015, the U.S. Food and Drug Administration (FDA) announced its long-awaited decision regarding AquaAdvantage Salmon, the first genetically engineered (GE) animal intended for food. The FDA has approved AquaBounty Technologies' new animal drug application (NADA) for the Salmon, finding that it meets the statutory requirements under the Federal Food, Drug, and Cosmetic Act (FDCA). In particular, the FDA determined that the Salmon's rDNA construct is safe for the fish itself, the fish reaches market size more quickly than non-GE farm-raised Atlantic salmon (as AquaBounty claimed), and food from the fish is safe to eat and as nutritious as food from non-GE salmon.

To assess the safety of food from AquaAdvantage, the FDA evaluated whether consumption of the product could result in any direct effects, such as toxicity or allergenicity, or indirect effects.[1] Using studies to determine the identify of fish described in the FDA's Regulatory Fish Encyclopedia (RFE),[2] the Center for Veterinary Medicine (CVM) concluded that AquaAdvantage Salmon meets the identity criteria for Atlantic salmon. Because the agency could detect no biologically relevant differences in the composition of food from AquAdvantage Salmon and non-GE, farm-raised Atlantic salmon, it concluded there was no indirect effects arising from AquAdvantage's rDNA construct. The CVM also determined that AquAdvantage Salmon does not present an additional risk of allergic reaction to

[1]FDA, Comments And Agency's Responses On The Public Hearing on The Labeling of Food Made From Aquadvantage Salmon Docket No. FDA-2010-N-0385 (Nov. 2015), p. 3 at http://www.fda.gov/downloads/Food/GuidanceRegulation/GuidanceDocumentsRegulatoryInformation/LabelingNutrition/UCM469766.pdf.

[2]The RFE was developed by FDA scientists at the Seafood Products Research Center (SPRC, Seattle District), and the Center for Food Safety and Applied Nutrition (CFSAN) to help federal, state, and local officials and purchasers of seafood identify species substitution and economic deception in the marketplace (available at http://www.fda.gov/Food/FoodScienceResearch/RFE/). Data in the RFE includes high-resolution photographs of the whole fish and marketed products (fillets and steaks), tissue protein patterns determined by isoelectric focusing electrophoresis gels, and mitochondrial DNA sequence determined by DNA barcoding.

N.E. Negowetti, M.A.; J.D. (✉)
Valparaiso University School of Law, Valparaiso, IN, USA
e-mail: nicolen@gfi.org

G. Steier, K.K. Patel (eds.), *International Farm Animal, Wildlife and Food Safety Law*, DOI 10.1007/978-3-319-18002-1_15

salmon-allergic individuals and is unlikely to cause allergic cross-reactions in those who are not salmon-allergic.

Regarding environmental concerns, the FDA issued a Finding of No Significant Impact (FONSI) and therefore did not prepare a full environmental impact statement required under the National Environmental Policy Act of 1969 (NEPA).[3] The FDA concluded that its approval of the Salmon would not have a significant impact because multiple and redundant measures keep the fish contained and prevent their escape.[4] The FDA also found it highly unlikely that the sterile Salmon would interbreed or establish populations in the wild. For these same reasons, the agency also determined that its decision would have no significant impacts on the Environments of the Global Commons and of Foreign Nations not Participating in the Action.[5]

The FDA's approval applies only to Salmon raised in land-based, contained hatchery tanks located at two facilities in Canada and Panama. The decision restricts AquaAdvantage Salmon from being bred or raised in the United States or at any other location besides those two facilities specified in the NADA. Inspections of the facilities will be conducted by the FDA, and Canadian and Panamanian governments. The FDA announced that it will maintain regulatory oversight over the production and facilities.

15.1 FDA's Decision Regarding the Labeling of GE Salmon

Because the FDA concluded that AquaAdvantage Salmon is not materially different from non-GE salmon, under the FDCA the agency cannot require additional labeling of the Salmon. [6] Simultaneous to its announcement regarding

[3]FDA, Center for Veterinary Medicine, Finding of No Significant Impact: AquaAdvantage Salmon (Nov. 12, 2015) at http://www.fda.gov/downloads/AnimalVeterinary/DevelopmentApprovalProcess/GeneticEngineering/GeneticallyEngineeredAnimals/UCM466219.pdf.

[4]FDA, Center for Veterinary Medicine, AquaAdvantage Environmental Assessment (Nov. 12, 2015), at http://www.fda.gov/downloads/AnimalVeterinary/DevelopmentApprovalProcess/GeneticEngineering/GeneticallyEngineeredAnimals/UCM466218.pdf.

[5]FDA, AquAdvantage Salmon—FDA Analysis of Potential Impacts on the Environments of the Global Commons and of Foreign Nations not Participating in the Action for NADA # 141-454 (Nov. 19, 2015), at http://www.fda.gov/AnimalVeterinary/DevelopmentApprovalProcess/GeneticEngineering/GeneticallyEngineeredAnimals/ucm466350.htm.

[6]FDA, Draft Guidance for Industry: Voluntary Labeling Indicating Whether Food Has or Has Not Been Derived From Genetically Engineered Atlantic Salmon at http://www.fda.gov/Food/GuidanceRegulation/GuidanceDocumentsRegulatoryInformation/ucm469802.htm. Although the FDA recognized that some consumers are interested in knowing that food from AquAdvantage Salmon is produced using genetic engineering, the agency stated that such consumer interest, alone, is not a material fact within the meaning of the FDCA, and is not a sufficient basis upon which FDA can require additional labeling.

AquaBounty, the FDA issued draft guidance for manufacturers who wish to voluntary label their food to indicate whether products have or have not been derived from GE Atlantic salmon.[7]

15.2 Reaction to AquaAdvantage Approval

The FDA's approval of AquaAdvantage Salmon was swiftly met with strong opposition. Wenonah Hauter, Executive Director of Food & Water Watch issued a statement on November 19, 2015 calling the approval an "unfortunate, historic decision [that] disregards the vast majority of consumers, many independent scientists, numerous members of Congress and salmon growers around the world, who have voiced strong opposition."[8] Hauter also announced that Food & Water Watch will be examining all options to stop the GE fish from reaching the marketplace. Similarly, Consumers Union, the policy and advocacy arm of *Consumer Reports*, condemned that decision.[9] Noting the risk of "enormous" environmental impacts, the Center for Food Safety announced its plans to sue the FDA to block the agency's approval of AquaAdvantage.[10]

Although the FDA has finally concluded the longest regulatory assessments in its history, these responses from environmental and consumer protection groups indicate that the controversy surrounding GE animals for human consumption is far from over.

[7]For example, for food products or food ingredients derived from Atlantic salmon that was not genetically engineered, examples of statements that manufacturers may voluntarily use include: "Not genetically engineered;" "Not genetically modified through the use of modern biotechnology;" "We do not use Atlantic salmon produced using modern biotechnology."

Manufacturers of food products or food ingredients derived from AquAdvantage Salmon can label their products as "Genetically engineered" or may indicate, for example, that "This salmon patty was made from Atlantic salmon produced using modern biotechnology."

[8]Food & Water Watch, FDA Approves Unlabeled GMO Salmon Despite Widespread Opposition from Scientists, Consumers and Members of Congress (Nov. 19, 2015) at https://www.foodandwaterwatch.org/news/fda-approves-unlabeled-gmo-salmon-despite-widespread-opposition-scientists-consumers-and.

[9]Consumers Union, Consumers Union "Deeply Disappointed" by FDA Move to Approve GE Salmon without Requiring Labeling (Nov. 19, 2015), at http://consumersunion.org/news/25036/.

[10]http://www.centerforfoodsafety.org/press-releases/4131/fda-approves-first-genetically-engineered-animal-for-human-consumption-over-the-objections-of-millions.

Chapter 16
Water and Marine Animal Law

Zach Corrigan

Abstract While there are several serious issues affecting the world's fish populations, unsustainable fishing has long been pointed to as chiefly responsible for declining wild fish populations. While the pace of overexploitation of fisheries has slowed since 1990, and progress has been made in reducing exploitation rates and restoring overexploited fish stocks and marine ecosystems, the world's fisheries remain in bad shape. More than half of the world's fish populations are at, or very close to, their maximum sustainable production levels as of 2009. Among the remaining stocks, close to 30 % are overexploited, producing lower yields than their biological and ecological potential. The declining global marine catch over the last few years, the increased percentage of overexploited fish populations, and the decreased proportion of non-fully exploited fish populations has led the United Nation's Food and Agricultural Organization (FAO) to conclude that the state of world's marine fisheries is growing worse. This chapter covers some legal regimes aimed at preventing overfishing, including the international Law of the Sea treaty and teh United States's domestic law, the Magnuson Stevens Fishery Management and Conservation Act. It also discusses national and international attempts to curb other threats to fish populations such as marine water pollution.

16.1 Status of the World's Fisheries

While there are several serious issues affecting the world's fish populations, unsustainable fishing has long been pointed to as chiefly responsible for declining wild fish populations.[1] While the pace of overexploitation of fisheries has slowed since 1990, and progress has been made in reducing exploitation rates and restoring overexploited fish stocks and marine ecosystems, the world's fisheries remain in bad shape. More than half of the world's fish populations are at, or very close to,

[1]Turnipseed et al. (2009), pp. 5–7.

Z. Corrigan (✉)
Food & Water Watch, Washington, DC, USA
e-mail: zcorrigan@fwwatch.org

G. Steier, K.K. Patel (eds.), *International Farm Animal, Wildlife and Food Safety Law*, DOI 10.1007/978-3-319-18002-1_16

their maximum sustainable production levels as of 2009.[2] Among the remaining stocks, close to 30 % are overexploited, producing lower yields than their biological and ecological potential.[3] The fish populations that account for about 30 % of world's marine fisheries production are fully exploited and, therefore, have no potential for increases in production.[4] The declining global marine catch over the last few years, the increased percentage of overexploited fish populations, and the decreased proportion of non-fully exploited fish populations has led the United Nation's Food and Agricultural Organization (FAO) to conclude that the state of world's marine fisheries is growing worse.[5] The FAO has concluded that the situation seems more critical for the fishery resources that are exploited solely or partially in the high seas, which is the area beyond any nation's Exclusive Economic Zone (EEZ).[6] Overexploitation of this exhaustible fish resource is not only harmful to the fish populations and marine ecosystems, but it also means that there are fewer fish to consume, which is economically harmful for those that rely on fishing as an occupation.[7]

16.2 The UN Law of the Sea

Why is there this over-exploitation of fishery resources on the high seas? Answering this question requires an analysis of the main international legal document governing this area, the Law of the Sea. The United Nations Law of the Sea Treaty is the product of a series of United Nations' conferences, the Conferences on the Law of the Sea (UNCLOS), which began in 1958, and over 24 years sought to codify existing international law.[8]

16.2.1 Law of the Sea Provisions

It was not until the UNCLOS III conference that conservation of fisheries became an explicit concern of the treaty.[9] After this conference, which began in 1973 and culminated in adoption of the comprehensive 1982 Convention,[10] an agreement

---

[2]Food and Agricultural Organization (2012), pp. 11, 13, 53.

[3]*Id.*

[4]*Id.* at 12.

[5]*Id.*

[6]*Id.* at 13.

[7]*Id.*

[8]Buck (1998), p. 83.

[9]Buck (1998), p. 92.

[10]Osherenko (2006), p. 339; citing Juda (1996), pp. 212–243.

was reached for territorial sea limits, which are critically important for allowing nation states to regulate the fishing within their jurisdictions.[11]

A 12-mile limit was set for the territorial seas, in which the sovereignty of a coastal nation state extends, subject to the treaty and other rules of international law.[12] The continental shelf is defined as land up to 200 miles, or its end, whichever is further.[13] In the coastal zones, coastal states exercises exclusive sovereign rights for the purpose of exploring and exploiting natural resources, which include minerals and other non-living resources of the sea-bed, subsoil, and sedentary living organisms.[14]

The EEZ is not to exceed 200 miles.[15] There, a coastal State has:

> [s]overeign rights for the purpose of exploring and exploiting, conserving, and managing the natural resources, whether living or non-living, of the waters superjacent to the sea-bed and of the sea-bed and its subsoil, and with regard to other activities for the economic exploitation and exploration of the zone, such as the production of energy from the water, currents and winds.[16]

One of the driving reasons for the expanded EEZ was to give coastal state jurisdictions the ability to manage fisheries in areas where people fish.[17]

It is not simply that the treaty's recognition of territorial limits that has enabled ocean and fisheries management. It also has some protective provisions. In fact, it includes 46 articles devoted to marine environmental protection, including the general obligation of states "to protect and preserve the marine environment."[18,19,20] For example, the treaty has requirements that EEZs have fishing limits based on Maximum Sustainable Yield, tracking the definition in the United States's fishery management law, the Magnuson Stevens Fishery Conservation and Management Act (MSA) discussed below.[21] Under Article 61(4), as is also true for the MSA, coastal states must consider the effects on species that are not intentionally targeted for fishing, also known as bycatch. At least one expert has read this treaty language to be broad enough to require nation states to engage

---

[11]Two implementation Agreements followed: Part XI (Seabed) in 1994 (the Agreement Relating to the Implementation of Part XI of the United Nations Convention on the Law of the Sea, July 28, 1994, 36 I.L.M. 1492) and Fish Stocks in 1995 (the Agreement for the Implementation of the Provisions of the United Nations Convention on the Law of the Sea, Relating to the Conservation and Management of Straddling Fish Stocks and Highly Migratory Fish Stocks, November, 1995 34 I.L.M. 1542) Osherenko (2006), p. 339; citing Juda (1996), pp. 256, 284.

[12]United Nations Convention on the Law of the Sea, Dec. 10, 1982, Article 2(1)-(3), 21 I.L.M. 1261.

[13]*Id.* at Article 77.

[14]*Id.*

[15]Buck (1998), pp. 93–95.

[16]United Nations Convention on the Law of the Sea, Dec. 10, 1982, Article 56, 21 I.L.M. 1261.

[17]Christie (2004), p. 2; quoting Christie (1999), p. 396 (quotation marks omitted)).

[18]Osherenko (2006), p. 342; citing Kalo (2002), p. 421.

[19]Juda (1996), p. 235.

[20]United Nations Convention on the Law of the Sea, Dec. 10, 1982, Article 192, 21 I.L.M. 1261.

[21]United Nations Convention on the Law of the Sea, Dec. 10, 1982, at Article 61.

in ecosystem management, where regulation is based on the biological relationships between species.[22] Like the MSA, the Law of the Sea requires coastal states to base conservation and management measures on the best science available.[23]

Perhaps the most significant treaty provision is its treatment of high seas, beyond the EEZ. Nation states exercise the role as trustees for all the world's people as beneficiaries, including future generations.[24]

In addition, while it recognizes the right of all States to engage in fishing on the high seas, the 1982 Convention requires all states to cooperate in the conservation of high seas resources and negotiate with other states that exploit living resources. It calls for this cooperation through the establishment of sub-regional, regional, and international organizations.[25] This has spurred several agreements including the U.N. Fish Stocks Agreement and FAO compliance Agreement.[26,27] Despite some criticisms levied at regional fishing organizations formed after the 1982 Convention, there has been praise for their ability to pressure nation states to curtail unrestricted fishing in the high seas and instill measures including fisheries closures, compulsory registration, and catch limits.[28]

Thus, the 1982 Convention moved international governance of the oceans towards a model where the resource is governed by a trustee on behalf of all to prevent its over utilization to the detriment of all.

16.2.2 Law of the Sea Shortcomings

While the convention was a step forward for ocean protection, critics contend that the Law of the Sea remains inadequate. For example, one expert has remarked that while "[]the majority of rights vested on coastal and fishing nations under [the Law of the Sea] have crystallized as norms of customary international law[,] [t]he majority of duties, however, have not."[29] Critics point to the near-complete discretion given to coastal states to interpret and implement their duties as the primary reason the decline in EEZ stockfish.[30] Further, over-exploitation in the EEZ is not

[22]International Union for Conservation of Nature and Natural Resources (2009), p. 23.

[23]United Nations Convention on the Law of the Sea, Dec. 10, 1982, Article 61(2).

[24]Osherenko 371 (citing Encyclopedia of Public International Law (1992)).

[25]Gorina-Ysern (2004), p. 675; citing United Nations Convention on the Law of the Sea, Dec. 10, 1982, Articles 117 and 118.

[26]Agreement to Promote Compliance with International Conservation and Management Measures by Fishing Vessels on the High Seas, Nov. 20, 1993, 33 I.L.M. 968.

[27]*Id.* at 680; citing Juda (1996), pp. 109–144.

[28]*Id.* at 683–685.

[29]Gorina-Ysern (2004), pp. 669–671; citing McLaughlin (2003).

[30]Christie (2004), p. 3.

prohibited unless it presents a danger to the maintenance of the living resources, and the law does not dictate what levels of fish populations should be maintained.[31] Other critics point to the convention's vague provisions, its lack of a global fisheries organization, and its lack of a compulsory dispute-settlement mechanism for the EEZ as other weaknesses.[32,33]

This lack of comprehensive governance is perhaps one reason that the world's fish populations continue to remain very close to their maximum sustainable production levels.[34] With fishery resources that are exploited solely or partially on the high seas, in the worst shape.[35]

Another problem with the treaty has been the slow pace in which nations have implemented it. UNCLOS III came to a close in 1982, but the treaty did not go into effect until 1994, 1 year after 60 countries ratified it.[36,37] There are still some notably absent parties. For example, even though it was instrumental in convening UNCLOS III, the United States has not ratified the treaty due to a few elected officials' arguments that it infringes upon national sovereignty and that its deep-sea mining provisions limit free enterprise.[38]

[31]International Union for Conservation of Nature and Natural Resources (2009), p. 5 (citing United Nations Convention on the Law of the Sea, Dec. 10, 1982, Article 61(2), 21 I.L.M. 1261).

[32]Gorina-Ysern (2004), pp. 671–674.

[33]For a comprehensive look at its conservation provisions applying in the continental shelf, EEZ and high seas, *see* Gorina-Ysern (2004), pp. 673–675.

[34]Food and Agricultural Organization (2012), pp. 11, 13, 53.

[35]*Id.* at 13.

[36]A full 166 countries have ratified the treaty. The first 60 nations to do so were as follows (in alphabetical order): Angola (5 December 1990), Antigua and Barbuda (2 February 1989), Bahamas (29 July 1983), Bahrain (30 May 1985), Barbados (12 October 1993), Belize (13 August 1983), Bosnia and Herzegovina (12 January 1994), Botswana (2 May 1990), Brazil (22 December 1988), Cabo Verde (10 August 1987), Cameroon (19 November 1985), Costa Rica (21 September 1992), Côte d'Ivoire (26 March 1984), Cuba (15 August 1984), Cyprus (12 December 1988), Democratic Republic of the Congo (17 February 1989), Djibouti (8 October 1991), Dominica (24 October 1991), Egypt (26 August 1983), Fiji (10 December 1982), Gambia (22 May 1984), Ghana (7 June 1983), Grenada (25 April 1991), Guinea (6 September 1985), Guinea-Bissau (25 August 1986), Guyana (16 November 1993), Honduras (5 October 1993), Iceland (21 June 1985), Indonesia (3 February 1986), Iraq (30 July 1985), Jamaica (21 March 1983), Kenya (2 March 1989), Kuwait (2 May 1986), Mali (16 July 1985), Malta (20 May 1993), Marshall Islands (9 August 1991), Mexico (18 March 1983), Micronesia (Federated States of) (29 April 1991), Namibia (18 April 1983), Nigeria (14 August 1986), Oman (17 August 1989), Paraguay (26 September 1986), Philippines (8 May 1984), Saint Kitts and Nevis (7 January 1993), Saint Lucia (27 March 1985), Saint Vincent and the Grenadines (1 October 1993), Sao Tome and Principe (3 November 1987), Senegal (25 October 1984), Seychelles (16 September 1991), Somalia (24 July 1989), Sudan (23 January 1985), Togo (16 April 1985), Trinidad and Tobago (25 April 1986), Tunisia (24 April 1985), Uganda (9 November 1990), United Republic of Tanzania (30 September 1985), Uruguay (10 December 1992), Yemen (21 July 1987), Zambia (7 March 1983), Zimbabwe (24 February 1993).

[37]Turnipseed et al. (2009), p. 30; citing Kalo et al. (2002), p. 388.

[38]*Id.*; Ashfaq (2010), pp. 358–362.

But, despite the slow pace of ratification, the UNCLOS has had an effect on domestic law, even for non-ratifying nations. For example, U.S. President Reagan acted on the treaty when he created the U.S. EEZ in 1983.[39] The United States secured "sovereign rights" and jurisdiction to the largest EEZ in the world. It stretches seaward out to 200 nautical miles from the U.S. mainland, Hawaii and Alaska, and U.S. island territories in the Atlantic and Pacific.[40] The U.S. EEZ covers 4.4 million square miles, larger than the combined area of the 50 states.[41] A 1988 proclamation extended the boundaries of the territorial sea from 3 to 12 nautical miles seaward of the coastlines of the United States and its territories, under which full sovereignty was claimed from the sub-surface seabed to the above airspace.[42] Finally, in 1999, President Clinton established the contiguous zone, which reaches from 12 to 24 nautical miles from the U.S. and territorial coastlines.[43]

Each of the proclamations were consistent with the 1982 Law of the Sea, and some scholars have argued that the United States has accepted the convention as a matter of international customary law,[44] despite its failure to ratify it.[45,46]

Notwithstanding its limitations and the slow pace at which it has been implemented, the 1982 Law of the Sea Convention dramatically changed the legal landscape governing the world's oceans and resources. It not only validated nations' moves to establish EEZs, but it also changed the dominant legal paradigm for these resources, moving them away from open-access to protected resources. Moreover, as

---

[39]Turnipseed et al. (2009), p. 30; (citing Proclamation No. 5030 (located at 48 Fed. Reg. 10,60 (March 14, 1983)).

[40]*Id.*

[41]*Id.*

[42]*Id.*; citing Proclamation No. 5928, 54 Fed. Reg. 777 (Jan. 9, 1989).

[43]*Id.*; citing Proclamation No. 7219, 64 Fed. Reg. 48,701 (Aug. 2, 1999).

[44]The International Court of Justice has stated that for a treaty rule to acquire customary status, it must have (1) a fundamentally norm-creating character such as could be regarded as forming the basis of a general rule of law; (2) a very widespread and representative participation in the convention, including that of states whose interests were specially affected; (3) extensive and virtually uniform state practice, including that of states whose interests are specially affected; and (4) the passage of some time, short though it may be. *North Sea Continental Shelf*, Judgement, ICJ Rep. 3, at paras 72–74 (1969).

[45]As evidence, another author points to the fact that the United States is a party to the 1964 Convention on the Territorial Sea and Contiguous Zone, which, like the Law of the Sea, precludes states' absolute claims to unlimited territorial seas and creates rules to restrict some forms of passage within their territorial seas. Ashfaq at 364 (2010). The author also argues that the imposition of affirmative environmental and pollution-reducing obligations parallels the 1966 Convention on Fishing and Conservation of the Living Resources of the High Seas, to which the United States is also a party. *Id.* Finally, the author argues that the revenue-sharing provisions and dispute-resolution mechanisms, which are the source of great controversy, were founded upon the "common heritage of mankind" principle are "customary law" supported in part by its widespread ratification. *Id.* Under general principles of international law, customary law is binding on all states, including the United States. *Id.*

[46]Turnipseed et al. (2009), at 70 n.169; citing Restatement (Third) of Foreign Relations Law § 514 comment. a. (1987).

one expert has explained, the treaty, in conjunction with general international law and the treaty's implementing agreements provide a:

> range of norms for national fisheries management, if carefully interpreted. Such legally binding norms include the coastal state's primary obligation to ensure that the maintenance of the living resources in its EEZ is not endangered by overexploitation; the duty to maintain or restore populations of target species at sustainable levels; the determination of catch limits for stocks actually or potentially affected by exploitation; the duty to apply the precautionary approach widely to conservation, management and exploitation of living marine resources; and duties to cooperate for the conservation and management of species not exclusively occurring within the coastal state's EEZ.[47]

16.3 U.S. Regulation of Domestic Fisheries

This next section focuses on how one country, the United States, manages its fisheries, in order to teach students about the regulation of fisheries more generally. Covered by this discussion are both the MSA's more traditional government-based restrictions as embodied in National Standards and fishing limits and the more recently employed market-based mechanisms, such as Individual Fishing Quotas (IFQs), for the management of fisheries. These programs are certainly not unique to the United States and are discussed primarily as an example of similar programs that exist elsewhere.[48]

The MSA is a regime that, much like international regulation, first sought to manage fisheries by establishing territorial limits to exclude foreign entities. It then focused on perhaps the more difficult task of conservation. While some gains have been made, U.S. fisheries management is still deficient in many respects. IFQs, often seen as a silver bullet for solving fisheries management problems, remain controversial.

16.3.1 The MSA

First passed in 1976 and amended numerous times since, the MSA[49] is the U.S. management regime for fisheries in its EEZ. It has been hailed as model for other countries,[50] but has also been criticized as being inadequate for the task.[51]

---

[47]International Union for Conservation of Nature and Natural Resources (2009), p. 39.

[48]For a comprehensive look at IFQ programs internationally, at least as they existed in 1999, see National Research Council (1999).

[49]Its name has changed over time, but it will be referred to in this chapter as the MSA for simplicity's sake.

[50]*See* Daniel Pauly, Letter to the Editor, Apr. 17, 2011, available at http://www.nytimes.com/2011/04/21/opinion/l21fish.html?partner=rssnyt&emc=rss&_r=0, last accessed 16 April 2016.

[51]*See* Eagle et al. (2008), p. 649.

It was initially passed to exclude foreign fishing operations from domestic waters, spurred by Congress's desire to reward domestic fishermen. However, over time, it has been amended to establish a comprehensive system for federal regulation of the domestic fishing industry. The act has a number of key provisions that are the bedrock of the regulatory system. As indicated, Title I of the 1976 act established jurisdiction in "a zone contiguous to the territorial sea of the United States," the inner boundary which was defined by the seaward boundary of each coastal state, and the outer boundary which was defined as 200 nautical miles out from these state waters.[52] This language was amended in 1986 to reference the EEZ. Within the zone, the act establishes the United States with the exclusive authority to manage its fishery resources.[53]

Under the act, the only portion of the fish yield that can be allocated to foreign vessels is that which "cannot, or will not, be harvested by vessels of the United States."[54] This authority allowed for a near complete termination of foreign fishing in the U.S. EEZ, accomplishing the act's original goal,[55] but it would not be until 1996 that the act was amended to focus on conservation and reducing the threat of overfishing from domestic fishing, as discussed in the next few sections.

16.3.2 The MSA's Structure

The heart of the act provides a unique structure of shared governance between the U.S. federal and state governments. Federal fisheries resources are primarily managed pursuant to the advice that eight regional fishery management councils provide in their fishery management plans (FMPs).[56] The regional councils are composed of voting members, including the head state fishery managers of each state in each region, the regional director of the federal agency, the National Marine Fisheries Service (NMFS), and regional experts, representatives of commercial, recreational, charter fishing sectors, and Native tribes, depending on the region, as nominated by the governors of each state and appointed by NMFS.[57] With this structure, Congress sought to "preserv[e] the states' ability to play a key [development] role" in fishery management programs . . ."[58]

---

[52] 90 Stat. 336, Pub. L 94–265, Sec. 101 (April 13, 1976) (codified at 16 U.S.C. § 1811).

[53] *Id.* at sec. 102; (now codified at 16 U.S.C. § 1811).

[54] 16 U.S.C. § 1821(d) (2012).

[55] Territo (2000), p. 1369; citing Decker (1995), p. 335.

[56] 16 U.S.C. § 1852.

[57] *Id.* § 1852(b).

[58] *C & W Fish Co. v. Fox*, 931F.2d 1556, 1557 (D.C. Cir. 1991).

Under the MSA, after a regional council develops an FMP,[59] it must submit the plan to the Secretary of the Department of Commerce, who must, "approve, disapprove, or partially approve" it after allowing for public comment.[60] Regional councils simultaneously submit implementing regulations for review.[61] The Secretary must review them for consistency with the governing FMP as well as with the MSA and other applicable law.[62] If the regulations are found to be inconsistent, they are returned to the council with proposed revisions.[63] Otherwise, the proposed regulations are also published for public comment,[64] and after the public comment period, the Secretary promulgates them, consulting with the council on revisions and explaining changes that are made.[65] All final regulations must be consistent with the FMP.[66]

Several aspects of this structure are important. The federal government is charged with the ultimate authority to regulate fisheries by approving or disapproving FMPs and developing regulations. But the states, through their representatives on the management councils, are able to use their regional expertise to advise and direct such management by writing the FMPs. The federal government is limited in its ability to deviate too much from these plans. The federal government offers a national check on regional advice, so that any fishery management does not subvert the best interests of the nation as a whole.

Notwithstanding the express role of the states, one criticism of this structure has been that the federal government, which staffs and funds the regional councils, still has an outsized role in pushing national policies on regional fishery management efforts, and this has limited the flexibility that should come from regional or localized management.

Another important structural element of the MSA is its explicit involvement of commercial and recreational fishermen on the regional councils. It is assumed fishermen know best how to manage the fisheries' resources and they have a vested interest in doing so. At the same time, having fishermen with dominant positions on the councils has led to the charge that other voices, such as those of consumers, conservationists, and other members of the public, are under-represented. It has also spurred criticism that fishermen on the councils have an inherent conflict of interest, as they are charged with conserving a resource that they also have an interest in exploiting. The fact that NMFS is ultimately charged with implementing the statute

---

[59]The Secretary can also develop an FMP under specific circumstances. *See* 16 U.S.C. § 1854 (c) (2012).

[60]16 U.S.C. § 1854(a), (a)(3) (2012).

[61]*Id.* § 1853(c) (2012).

[62]*Fishing Co. of Alaska v. Gutierrez*, 510 F.3d 328, 330 (D.C. Cir. 2007) (citing 16 U.S.C. § 1854 (b)(1)).

[63]*Id.* (citing § 1854(b)(1)(B)).

[64]*id.* (citing § 1854(b)(1)(A)).

[65]*Id.* (citing § 1854(b)(3)).

[66]16 U.S.C. § 1854(b)(1)(B), (c)(7) (2006).

is also a subject of controversy, as it has been argued that the agency is limited in its ability to protect the resource, given it also has a mission of promoting its development.

16.3.3 FMPs

The main management tool for councils and NMFS are spelled out in FMPs. Since the MSA's inception, FMPs have "contain[ed] the conservation and management measures that "are . . . necessary and appropriate for the conservation and management of the fishery."[67] This language defines NMFS and regional councils' management authority.

But other requirements of FMPs have been added to strengthen how the agency and councils carry out this mandate. As part of a sweeping set of changes made to the act in 1996 and in response to the realization that fishery management efforts were not succeeding at reducing overfishing, the Sustainable Fish Act (SFA) amendments clarified that FMPs were to aim at "prevent[ing] overfishing and rebuild[ing] overfished stocks, and to protect[ing], restor[ing], and promot[ing] the long-term health and stability of the fishery."[68]

FMPs were required to "describe and identify essential fish habitat for the fishery" and "minimize to the extent practicable the adverse effects on such habitat caused by fishing."[69] The plans must also establish a standardized reporting methodology to assess the amount and type of bycatch and include conservation and management measures that, to the extent practicable, minimize such bycatch and bycatch mortality.[70,71]

FMPs were also required to specify objective and measurable overfishing definitions for all managed fish populations and contain conservation and management measures to prevent overfishing or end overfishing and rebuild the fishery.[72] The act was also amended to require that when any species is found to be overfished, NMFS

---

[67] 90 Stat. 351, Pub. L 94-265, tit. III. sec. 303(April 13, 1976) (codified at 16 U.S.C. § 1811).

[68] 16 U.S.C. § 1853(a)(1)(A).

[69] *Id.* § 1853(a)(7).

[70] Among other things, FMPs also are required to describe the fishery in detail, 16 U.S.C. § 1853 (a)(2) (2012); assess and specify the condition of, and the "maximum sustainable yield" and "optimum yield" from the fishery, and include a summary of the information utilized in making such specification, *id.* § 1853 (a)(3); and assess and specify the capacity of the fishery, including the extent to which fishing vessels of the United States and foreign nations, can and will be able to annually harvest the fishery's optimum yield. *Id.* § 1853 (a)(4) (A)-(B). Finally, FMPs must include a fishery impact statement that analyzes the likely effects, if any, including the cumulative conservation, economic, and social impacts, of the conservation and management measures on fishermen and fishing communities. *Id.* § 1853 (a)(9).

[71] *Id.* § 1853 (a)(11).

[72] *Id.* § 1853(a)(9).

must approve a rebuilding plan that sets a time period for ending overfishing, and rebuilding the fishery, not to exceed 10 years, except in cases where the biology of the stock of fish, or other environmental conditions dictate otherwise.[73] Under this provision, NMFS "may consider the short-term economic needs of fishing communities in establishing rebuilding periods, but may not use those needs to go beyond the 10-year cap. To breach this cap, FMPs may only consider circumstances that 'dictate' doing so[,]" including an international agreement and "when the current number of fish in the fishery and the amount of time required for the species to regenerate make it impossible to rebuild the stock within 10 years . . ."[74]

These measures, aimed at not only the intentional fishing, but also incidental damage to fish populations and habitat, reinforced that the MSA was chiefly to be aimed at reducing overfishing. But the law remained not up to the task, and in 2007, the requirements for FMPs were amended again to mandate that they include annual catch limits, or "ACLs" and accountability measures for all fisheries subject to overfishing, which are discussed below.[75]

The statute's various revisions have changed it to from one that sought to manage fish exploitation by excluding foreign fishermen to one that sought to conserve fishery resources from domestic threats. These revisions also remain controversial, as many fishermen have argued that the 10-year rebuilding requirements do not offer them enough flexibility, especially where there is little information on the status of certain fish populations. Conservationists have argued that the calls for flexibility are simply an attempt to avoid regulation and that the act's overfishing and rebuilding provisions are precisely what is needed to conserve the fishery resource.

16.3.4 National Standards

Perhaps the most important provisions of the MSA are its national standards. NMFS and U.S. courts evaluate FMPs and regulations against these standards.

The lodestar of the standards is National Standard 1, which requires that "[c] onservation and management measures . . . prevent overfishing while achieving, on a continuing basis, the optimum yield from each fishery for the United States fishing industry." To break this down further, as amended in 1996: overfishing is rate of fishing mortality that jeopardizes the capacity of a fishery to produce the maximum sustainable yield on a continuing basis. Under National Standard 1, conservation measures must reduce fishing mortality to the level which keeps fisheries at the "maximum sustainable yield," often referred to as "MSY," (language which also is present in the Law of the Sea (*see* Article 61)). The act does not define this

[73]*Id.* § 1854(e)(4) (2012).

[74]*NRDC v. Nat'l Marine Fisheries Serv.*, 421 F.3d 872, 880 (9th Cir. 2005).

[75]Hooks and Baylor (2009), p. 194; citing *Id.* § 1853 (a)(15).

language, but NMFS's guidelines provide that it is the scientific determination of "'the safe upper limit of harvest which can be taken consistently year after year without diminishing the stock so that the stock is truly inexhaustible and perpetually renewable.'"[76]

It is not enough under National Standard 1 that overfishing is prevented. Fishery measures must achieve "optimum yield," which is the MSY, reduced by any relevant social, economic, or ecological factor, and is the amount of fish that "will provide the greatest overall benefit to the Nation, particularly with respect to food production and recreational opportunities, and taking into account the protection of marine ecosystems." For overfished fisheries, it is the amount that provides for rebuilding to get back to the MSY.

In short, under National Standard 1, FMPs are to include measures to prevent fishing mortality in a fishery that jeopardizes it as a renewable resource, and fishing may even be below this level, if needed to produce the greatest overall benefit to the nation. For those fisheries that are overfished, the standard requires that the fishing be set at a level that allows for rebuilding. At least one U.S. federal Circuit Court of Appeals has found this means that the fishery management measure must have at least a 50 % chance of reaching the maximum sustainable yield, and it in no way permits measures that only have as low as an 18 % chance of achieving this limit.[77]

To implement this standard and the other provisions of the MSA aimed at overfishing, FMPs must establish Annual Catch Limits (ACLs) for fishermen. It corresponds to the annual amount of catch that would not result in overfishing, as reduced by scientific uncertainty.[78] Accountability measures, such as in-season fishing closures or measures to correct for when fishermen exceed these limit, are required prevent the catch from exceeding the ACLs or offset overages.[79] These ACLs and accountability measures are two of the main tools that FMPs are now using to target overfishing, as required under the SFA and National Standard 1.

Perhaps ranking second in importance is National Standard 2, which requires that conservation and management measures be based on the "best scientific information available[,]" the same language is used in the Law of the Sea.[80] No federal court of appeals has offered a definitive interpretation of this standard, but it is clear that courts are "highly deferential" when determining whether NMFS and the regional councils meet the standard.[81] One district court has stated that a complete failure to consider science and the introduction of better science would be needed for an FMP to violate the standard.[82]

---

[76] *Maine v. Kreps*, 563 F.2d 1043, 1047 (1st Cir. Me. 1977) (quoting H.R. Rep. No. 445, 94th Cong., 1st Sess. 48 (1975).)

[77] *Natural Res. Def. Council, Inc. v. Daley*, 209 F.3d 747, 754 (D.C. Cir. 2000).

[78] 50 C.F.R. § 600.310(f)(1)-(7).

[79] *Id.* § 600.310(g)(1)-(3).

[80] Article 61(2).

[81] *See, e.g., Or. Trollers Ass'n v. Gutierrez*, 452 F.3d 1104, 1120 (9th Cir. 2006).

[82] *See Commonwealth v. Gutierrez*, 594 F. Supp. 2d 127, 132 (D. Mass. 2009) (collecting cases).

Notwithstanding the great discretion afforded to NMFS and the councils under National Standard 2, such discretion is not limitless. For example, it is not proper for NMFS to allocate fishing quota based on "pure political compromise" as opposed to "reasoned scientific endeavor."[83]

The MSA is not simply about reducing overfishing and by doing so by using the best science available. The MSA's remaining standards include those aimed at other interests, including protecting the economic livelihoods of fishermen, which sometimes are seen as conflicting with the statute's goal of reducing overfishing. As will be seen, however, the statute has been interpreted to be chiefly a conservation statute.

These standards[84] are as follows:

- National Standard 3 requires stocks of fish to be managed as a unit throughout their range, and interrelated stocks of fish to be managed as a unit or in close coordination.
- National Standard 4 prevents discrimination between residents of different States and says that any allocation of fishing privileges must be fair and equitable; reasonably calculated to promote conservation; and ensure that no particular individual, corporation, or other entity acquires excessive shares.
- National Standard 5 allows for fishery managers to consider efficiency, but no management measure can have economic allocation as its sole purpose.
- National Standard 6 requires conservation and management measures to take into account and allow for variations among, and contingencies in, fisheries, fishery resources, and catches.
- National Standard 7 requires measures to, where practicable, minimize costs and avoid unnecessary duplication.
- National Standard 8 requires management measures to take into account the importance of fishery resources to fishing communities by utilizing economic and social data in order to provide for the sustained participation of such communities, and minimize adverse economic impacts on such communities.
- National Standard 9 provides that conservation and management measures are to minimize bycatch and minimize the mortality of such bycatch.
- National Standard 10 requires conservation and management measures to the extent practicable, promote the safety of human life at sea.

These are the touchstones of the act and thus have been the focus of numerous court challenges by those seeking to invalidate fishery management measures on the grounds that they did not measure up to the standards. Fishermen who have been harmed by certain fishery regulations, or conservation organizations that believed that the fishery restrictions were not strong enough have brought these cases. The running theme through all of these cases is that National Standard 1 and its conservation purpose reign supreme. Further, while the National Standards provide a basic framework for managing fisheries, they also provide an enormous amount of

[83] *Midwater Trawlers Coop. v. DOC*, 282 F.3d 710, 720 (9th Cir. 2002).
[84] 16 U.S.C. § 1851.

flexibility and discretion for fisheries managers. While this discretion gives agencies the ability to implement very strong fishery management measures, it also allows them to implement other measures that are ineffective and simply a product of the political power of the fishermen on the councils. Any such measures' opponents are left with a considerable challenge to make the case that managers have stepped too far in favoring one group of fishermen over another, designed measures that are aimed too much at economic allocation, or adopted measures not strong enough to reduce overfishing, bycatch, or protect human safety.

16.3.5 Individual Fishing Quotas

Many of these complaints have been levied at what are perhaps the most recent tools used in fisheries management, IFQs. These are also often referred to as Individual Transferable Quotas (ITQs). The MSA classifies them in a broader category of Limited Access Privilege Programs (LAPPs). IFQs are akin to "cap and trade" market-based regulatory regimes. The basic and primary objective of IFQs is to provide fishermen a portion of the total fisheries harvest quota [85] that is theirs alone to fish. The theory is that when fishermen have their own exclusive share of the fishery, they will not continue to invest money in boats and gear to utilize more of the diminishing, exhaustible resource. A similar argument is that fisherman given an ownership interest in the fishery will trade off any short-term gain from fishing to reap the long-term gains of conservation. Fishermen will slow down, allowing fish to be caught all year. Proponents also argue that this makes fishing safer.[86]

Crucial to the concept of any IFQ program is that the fishing privilege—*i.e.*, the shares or quota—can be bought or sold. (This is the "trade" in the "cap and trade" scheme.) Fishermen who want to exceed their cap are able to buy more quota from those who will not use theirs. If the purpose of establishing an IFQ program to reduce "overcapitalization," or inefficient excess fishing boats and gear in the fishery, tradability allows purchasers of quota to finance those who sell their quota and then reduce their fishing. Those who can fish at the lowest costs or produce the most valuable product buy this quota. Over the long-term, it is argued that transferability allows the most efficient operators to have the fishing privileges and produces optimally sized fishing fleets.[87] In turn, as fishing rights go to more efficient operators with lower costs, it is argued that these fishermen will be able to dedicate more money to invest in resource improvement.[88]

[85]The "cap" in the United States, it is called a Total Allowable Catch or TAC.

[86]Carden et al. (2013), p. 51.

[87]National Research Council (1999), p. 169.

[88]Rieser (1997), p. 823.

However, IFQ programs have been controversial from their start in the United States. Initially, the concern was that NMFS was implementing them without adequate consideration of regional fishery management councils.[89] But one of the core concerns by some conservation and consumer groups and fishermen were that that the public resource would be privatized, *i.e.*, they provide a free property interest in a public resource.

Other concerns about IFQ programs include their distributional effects on smaller scale fishermen. Larger-scale fishermen end up purchasing smaller-scale fishermen's quota in order to reduce costs and increase their ability to fish. Smaller fishermen, who cannot afford to fish at levels beneath their quota allocation—if they are allocated quota at all—and those who cannot afford to purchase or lease more, often have to exit the fishery.[90] These smaller fishermen then must spend their money leasing access to the resource, akin to medieval serfs. Thus, it is argued, IFQs result in a radical redistribution in the fishery away from small independent fishermen to larger, sometimes international fishing corporations.

Other concerns, particularly leveled by consumer groups and academics, are related to the effects that IFQs have on competition and consolidation. The goal of IFQs is not to simply constrain fishing effort below the maximum sustainable yield, but also to constrain it to a point where the remaining fishermen maximize the entire fleet's "extra-normal profits" due to the exclusion of competitors, particularly the smaller competitors.[91] This consolidation can be especially acute if quota is awarded based on fishermen's catch-history and for free, so that the largest fishermen start off with a windfall and a large capital advantage based on their historic fishing levels.[92,93] Additional concerns are related to whether the fishery will cause such concentration in the fishery that will prevent new fishermen from entering,[94] and it is argued that this could harm consumers in the long run.

Further, some conservation and fishing groups and academics argue that that the conservation benefits of IFQs are myopic and that they may even hurt fisheries. The studies that are cited for the conservation benefits of IFQ programs often fail to disaggregate their effects from other, non-market-based management measures implemented at the same time, such as the fishing limits alone.[95,96] Further, opponents argue that fisheries and habitats may be harmed if those who are granted

---

[89]The House Report for the 1996 SFA reports: "Recent efforts by [NMFS] to promote . . . [ITQ] systems above any other type of limited access system concern the Committee and are inappropriate. . . . Because ITQ systems have the potential to fundamentally alter fisheries management in the U.S., the Committee believes they must be used with great caution." H. Rept. No. 104-171 at 36 (1995).

[90]*See* National Research Council (1999), pp. 173–174.

[91]Bromley (2008a), pp. 4–5.

[92]*See* National Research Council (1999), pp. 142–143.

[93]Bromley (2008b), p. 13.

[94]*Id.*

[95]Bromley (2008a), p. 3.

[96]Food & Water Watch (2011), pp. 8–9.

the most catch shares are those that have been bad actors in the past and are those that use gear associated with higher levels of harm to habitat or bycatch.[97,98] Fishermen and conservationist critics also argue that IFQs are incompatible with ecosystem management. Such systems aim to allow only as much fishing in a fishery as has detrimental effects in the ecosystem, regardless of the effects on a particular target species, which has traditionally been how fisheries have been managed.[99] With IFQ programs, on the other hand, quota is allocated based solely on the market value of a few species. There is a serious risk that other valuable components of the ecosystem will be ignored because they are not given a monetized value.[100]

What came out of the 1996 SFA amendments was a moratorium on IFQs so that a comprehensive study could be conducted by the National Academy of Sciences about the "controversial IFQ-related issues such as initial allocation, transferability, and foreign ownership."[101] This *de jure* moratorium lasted approximately 6 years, and in 2007, Congress ended the *de facto* moratorium by incorporating some of the National Academies of Science's recommendations as criteria that would allow fisheries managers to "balance many of the concerns fishermen, crew, communities, conservation groups, and other interests have had over the potential impacts" of such programs.[102] Notwithstanding the attempt by Congress to balance both the good and bad with IFQs, they remain one of the most controversial fishing measures employed in U.S. fisheries.

16.4 International Marine Regulation

We cannot leave the topic of fisheries regulation without at least touching upon clean water regulation. This chapter's discussion, much like the discussion of fisheries thus far, is limited to marine issues and does not focus on pollution of freshwater bodies in lakes and streams that are only governed by domestic regulation. As detailed below, while there has been some improvement in the prevention of marine pollution, at least in theory, there is much work to be done. Successful fisheries management in the United States will depend on reducing these non-fishing threats to fish health and habitat.

---

[97]As the National Research Council indicated: "[I]mplementing an IFQ regime may favor some technologies over others. If [such programs] typically involve more bycatch, bycatch rates can rise in the absence of enforcement." National Research Council (1999), p. 177.

[98]*See* National Research Council (1999), p. 177.

[99]Pew Ocean Commission (2003), p. 44.

[100]Rieser (1999), p. 405.

[101]*See The Sustainable Fisheries Act*, Pub. L. No. 104-297, sec. 108(e), § 303(f), 110 Stat. 3576 (1996).

[102]S. Rept. No. 109-229 at 9 (2006).

Marine pollution can come from a number of different sources, including land-based runoff, vessel discharges, and oil drilling.[103] Discharges from land include sewage, radioactive and industrial wastes, and agricultural run-off. A National Academy of Sciences study estimates that the oil that runs off U.S. streets and driveways into the oceans amounts to an *Exxon Valdez* oil spill every eight months.[104] Today, nonpoint source pollution, or that which comes from multiple sources on land like farms and urban runoff, presents perhaps the greatest pollution threat to oceans off of U.S. coasts.[105] In the United States, agriculture is one of the most significant sources of pollution as it is the source nitrogen.[106] Nitrogen from animal waste applied to farmland as fertilizer is easily dissolved in water and is transported by rain into streams and rivers that eventually flow into the ocean. Further, tile drainage systems, which are constructed to collect and shuttle excess water from fields, act as an “expressway” for this nitrogen pollution.[107] The Mississippi River carries an estimated 1.5 million metric tons of nitrogen into the Gulf of Mexico each year.[108] Such nutrient pollution has been linked to harmful algal blooms and dead zones, including the Gulf of Mexico’s dead zone that is more than 8000 square miles. Dead zones, which are areas of low oxygen, which means they have “hypoxic” conditions, kill crabs, fish, and other species. In addition, this pollution results in the loss of sea grass and kelp beds, the destruction of coral reefs, and lower biodiversity in estuaries and coastal habitats.[109] These present a serious threat to fish habitat and consequentially fish health.

Point sources of marine pollution include animal waste overflowing from open lagoons at large industrial farms and oil spills[110] such as from the Deepwater Horizon in the Gulf of Mexico, which released an estimated 210 million U.S. gallons of oil and dispersants.[111] Vessel discharges include contaminants from ballast tanks, sewage from cruise ships, and fuel spills.[112,113]

Global warming, caused by the burning of fossil fuels and other activities that release heat-trapping gases such as carbon dioxide, also poses a serious threat to ocean health. Recently, attention has focused on ocean acidification. The ocean absorbs about a quarter of the carbon dioxide released into the atmosphere every year. This has begun changing the chemistry of the seawater, including the amount of available calcium carbonate minerals, which serves as the building blocks for

[103]Buck (1998), p. 95.

[104]Pew (2003), p. 4; citing National Research Council (2002).

[105]*Id.* at 60.

[106]*Id.* at 60; citing National Research Council (2000).

[107]*Id.* at 62.

[108]*Id.* at 59 (citing Goolsby et al. 2014).

[109]*Id.* at 62 (citing Howarth et al. 2002).

[110]*Id.* at 63.

[111]Davies (2014).

[112]Buck (1998), pp. 96–97.

[113]Pew (2003), p. 66.

skeletons and shells of many marine organisms, like oysters, clams, and corals. In recent years, there have been near total failures in wild- and farmed-oyster production on the West Coast, which may be linked to global warming. When these shelled organisms are at risk, the entire food web is also in jeopardy.[114]

These are just some of the marine pollution threats that can affect the health of ecosystems and the size of fish populations, and which fishery managers must consider when deriving policy measures to curb fish mortality. In the United States and elsewhere, this becomes difficult when many of these threats are not directly under fishery management jurisdiction. Much of this pollution, which comes from sources far upstream or even from air pollution, is solely regulated domestically, and by other agencies. This creates a "regulatory-commons problem," meaning that, where multiple agencies exert jurisdiction over different pieces of a problem, it is less likely to result in comprehensive regulation.[115]

And much of this pollution from the United States into the oceans is simply a product of a lack of regulation. Only now is the United States moving to mandate national standards for greenhouse-gas emissions from power plants and motor vehicles under the Clean Air Act, for example. As another example, much of the pollution from animal farms are excused from permitting requirements for their manure discharges caused by precipitation.[116] Efforts to curb pollution from agricultural sources have recently focused on pollution trading schemes, and have been challenged in court for not being authorized under the federal Clean Water Act (CWA).[117,118] As another example of failed regulation, the U.S. CWA does not cover untreated discharges of water from its largest source, cruise ships, anywhere in federal waters. Only in Alaskan state waters are cruise ships required to meet effluent standards; treat gray-water discharges (waste water from sinks and showers that has not come into contact with feces); and monitor, record, and report discharges to state and federal authorities.[119]

---

[114]See http://www.pmel.noaa.gov/co2/story/What+is+Ocean+Acidification%3F, last accessed April 18, 2016.

[115]As another example, Titles I and II of the Marine Protection, Research and Sanctuaries Act, 33 U.S.C. 1401–1445 (2012), also known as the Ocean Dumping Act, regulates ocean dumping and incineration at sea of materials other than vessel sewage waste. *Id.* (citing 33 U.S.C. 1402 (c) Under this act, the U.S. Army Corps of Engineers issues permits for ocean dumping of dredged materials, while the EPA has permit authority for the dumping of all other materials. *Id.* (citing 33 U.S.C. 1412). The U.S. Coast Guard regulates garbage disposal of from vessels pursuant to the Marine Plastic Pollution Research and Control Act of 1987. *Id.* (citing 33 U.S.C. 1412).

[116]*See* 71 Fed. Reg. 37,744 (June 30, 2006).

[117]*Food & Water Watch v. United States EPA*, 2013 U.S. Dist. LEXIS 174430 (D.D.C. 2013).

[118]Available at http://switchboard.nrdc.org/blogs/aalexander/SJ%20decision.pdf, last accessed April 18, 2016. Non-profit organizations have also recently sued the U.S. Environmental Protection Agency for failing to implement water quality standards for nutrients in the Mississippi watershed in the face of states' failure to do so. *See Gulf Restoration Network et al v. Jackson et al*, 2:12-cv-00677-JCZ-DEK, Order and Reasons (Doc. 175) (September 20, 2013).

[119]Pew (2003), p. 66.

As can be seen, the United States has not been very aggressive in curbing marine pollution from all sources. International control has also been difficult.[120] Early conventions have failed. A series of treaties have been ratified that target oil spills, but the costs of monitoring and compliance have given way to nations' shipping interests and the ratification of these treaties have been slow.[121] Dumping of wastes is also the subject of a number of treaties, but no treaty absolutely bans it. One convention, the London Convention, regulates it, prohibiting the intentional dumping of materials on the "black list," while allowing the dumping of less hazardous materials on the "grey list" if permitted by the International Maritime Organization.[122] In 1996, the parties to the London Convention produced the London Protocol, which entered into force in 2006[123] and further restricted intentional ocean dumping by banning it except for materials that are found on a "reverse list."[124] While monumental in its adoption of the "precautionary approach," (where "appropriate preventative measures are taken when there is reason to believe that wastes or other matter introduced into the marine environment are likely to cause harm even when there is no conclusive evidence to prove a causal relation between inputs and their effects")[125] the United States' failure to ratify this treaty has been criticized as sending a message of indifference about global commons.[126] Moreover, critics have pointed to a lack of political leadership, legislative hurdles, insufficient resources, and pressure from regulated industries, as contributing to weak implementation of the convention.[127]

Thus, as with fisheries regulation discussed earlier in the chapter, progress that has been made in limiting the harmful impacts of marine pollution in the United States and internationally, at least because there has been the establishment of some framework for its regulation. But there is much work that is needed to actually prevent the continued exploitation of the resource to the detriment of all.

[120]Buck (1998), p. 97–100.

[121]*Id.* at 98.

[122]Ghorbi (2012), p. 483; citing Hunter et al. (2002).

[123]http://www.imo.org/OurWork/Environment/LCLP/Pages/default.aspx, last visited April 19, 2016.

[124]*Id.* at 484 (citing 1996 Protocol to the Convention on the Prevention of Marine Pollution by Dumping of Wastes and Other Matter, Nov. 7, 1996, Article 4).

[125]http://www.imo.org/OurWork/Environment/PollutionPrevention/Pages/1996-Protocol-to-the-Convention-on-the-Prevention-of-Marine-Pollution-by-Dumping-of-Wastes-and-Other-Matter,-1972.aspx, last visited April 19, 2016.

[126]Sielen (2008), p. 52.

[127]*Id.*

16.5 Conclusion

Spurred by the realization that these resources are not inexhaustible, national, extraterritorial, and international legal management regimes have been erected in the last 70 years order to regulate fisheries and water pollution domestically and internationally. The Law of the Sea is critical in this development. Its establishment of territorial limits and substantively protective articles has provided the means by which nations can assert the sovereignty to protect the resources closest to their shore.

While there have been significant developments in establishing legal regimes for fisheries and ocean resource management in general, much more work has to be done. The Law of the Sea has some shortcomings, including its vague provisions, its lack of a global fisheries organization, its lack of a compulsory EEZ-dispute-settlement mechanism, and its failure to mandate that nations adequately regulate their own EEZs. Domestic laws have also been ineffective at targetting overfishing. In the United States, the MSA's shared governance program between federal and state authorities and fishermen interests has been the source of division and perhaps regulatory stagnation, instead of regulatory innovation. ACLs, 10-year rebuilding plans, and IFQs are some of the newest measures attempting to target overfishing, but they have been controversial. The MSA's National Standards—the MSA's constitution—has been interpreted by courts to maintain conservation as its primary focus, but they also provide virtually unfettered discretion to fishery managers to cater to fishermen or special interest groups, further stymieing wise policy development. The result of all of these issues is that fisheries remain fully or over exploited. Marine pollution regulation has also suffered as a result of a lack of comprehensive regulation, notwithstanding some strides in regulating some sources.

References

Ashfaq S (2010) Something for everyone: why the United States should ratify the law of the sea treaty. J Transnl Law Policy 19:357–362, 364

Bromley DW (2008a) IFQs in the West Coast Groundfish Fishery: Economic Confusion and Bogus Reasons! Testimony before the Pacific Fishery Management Council 3–6 (Oct. 14, 2008a)

Bromley DW(2008b) The crisis in ocean governance: conceptual confusion, spurious economics, political indifference. Maritime Stud 13 (MAST 2008, 6(2))

Buck SJ (1998) A global commons, an introduction, Island Press, Washington, DC, pp 83, 92, 93–97

Carden K, White C, Gaines SD, Costello C, Anderson S (2013), Ecosystem service tradeoff analysis: quantifying the cost of a legal regime. Ariz J Environ Law Policy 4:39–51

Christie DR (1999) The conservation and management of stocks located solely within the exclusive economic zone. In: Ellen Hey (ed) Developments in International Fisheries Law. The Hague: Kluwer Law International, 395, 396

Christie DR (2004) It don't come EEZ: the failure and future of coastal state fisheries management. J Transnl Law Policy 14:1–3

Davies R (2014) Houston Oil Clean-Up on 25th Anniversary of Exxon Valdez Spill, ABC NEWS, Mar 24, 2014, available at http://abcnews.go.com/blogs/business/2014/03/houston-oil-clean-up-on-25th-anniversary-of-exxon-valdez-spill/ (last visited 17 April 2014)

Decker CE (1995) Issues in the reauthorization of the Magnuson fishery conservation and management act. Ocean Coast Law J 1:323–335

Eagle J, Sanchirico JN, Thompson Jr BH (2008) Breaking the Logjam: environmental reform for the new congress and administration: protecting aquatic ecosystems: ocean zoning and spatial access privileges: rewriting the tragedy of the regulated ocean. N Y Univ Environ Law J 17:646–649

1 Encyclopedia of Public International Law 692 (1992)

Food And Agriculture Organization of the United Nations, The State of World Fisheries and Aquaculture, 11, 13, 53 (2012), available at http://www.fao.org/docrep/016/i2727e/i2727e.pdf (last visited 10 April 2014)

Food & Water Watch, Fish Inc. (2011) 8–9, available at http://www.foodandwaterwatch.org/tools-and-resources/fish-inc/ (last visited 1 May 2014)

Ghorbi D (2012) There's something in the water: the inadequacy of international anti-dumping laws as applied to the Fukushima Daiichi radioactive water discharge. Am Univ Int Law Rev 27:473, 483–484

Goolsby DA, Battaglin WA Hooper RP (2014) U.S. Geological Survey, Sources and Transport of Nitrogen in the Mississippi River Basin. American Farm Bureau Federation Workshop, St. Louis, Missouri 14, July 15, 1997, available at http://wwwrcolka.cr.usgs.gov/midconherb/st.louis.hypoxia.html (last visited 10 April 2014)

Gorina-Ysern M (2004) World ocean public trust: high seas fisheries after grotius - towards a new ocean ethos? Golden Gate Univ Law Rev 34:645, 673–675, 680, 699–671, 683–685

Hooks AM, Baylor E (2009) Recent developments: natural resources: fishery conservation and management after reauthorization of MSA. Tex Environ Law J 39:193–194

Howarth RW, Boyer EW, Pabich W, Galloway JN (2002) Nitrogen use in the United States from 1961–2000 and potential future trends. Ambio 31(2):88–96

Hunter D, Salzman JE, Zaelke D (2002) International environmental law and policy, 2nd edn. Foundation Press, New York, p 735

International Union for Conservation of Nature and Natural Resources, Towards Sustainable Fisheries Law. A Comparative Analysis (Gerd Winter ed. 2009) 5, 23, 39, available at / portals.iucn.org/library/efiles/edocs/EPLP-074.pdf, last visited 20 May 2016

Juda L (1996) International law and ocean use management: the evolution of ocean governance In: Smith HD (ed) pp 212–243

Kalo JJ (2002) Coastal and Ocean Law 388, 421 (W. Grp. 2nd ed)

McLaughlin RJ (2003) Foreign access to shared marine genetic materials: management options for a quasi-fugacious resource. Ocean Dev. Int Law 34:297

National Research Council (1999) Sharing the fish towards a national policy on individual fishing quotas. National Academic Press, pp 142–143, 173–174, 177

National Research Council (2000) Clean coastal waters: understanding and reducing the effects of nutrient pollution. National Academic Press, Washington, DC

National Research Council (2002) Oil in the Sea III: inputs, fates, and effects. National Academic Press, Washington, DC

Osherenko G (2006) New discourses on ocean governance: understanding property rights and the public trust. J Environ Law Litig 21:317, 339, 342, 372

Pew Ocean Commission (2003) America's Living Oceans Charting A Course For Sea Change, 4, 44, 59–63 (May 2003), available at http://www.pewtrusts.org/uploadedFiles/wwwpewtrustsorg/Reports/Protecting_ocean_life/env_pew_oceans_final_report.pdf, (last visited 17 April 2014)

Rieser A (1997) The ecosystem approach: new departures for land and water: fisheries management: property rights and ecosystem management in U.S. fisheries: contracting for the commons? Ecol Law Q 24:813–823
Rieser A (1999) Prescriptions for the commons: environmental scholarship and the fishing quotas debate. Harv Environ Law Rev 23:393–405
Sielen AB (2008) An oceans manifesto: the present global crisis. Fletcher Forum World Aff 32:39–52
Territo M (2000) The precautionary principle in marine fisheries conservation and the U.S. sustainable fisheries act of 1996. Vermont Law Rev 24:1351–1369
Turnipseed M, Roady SE, Sagarin R, Crowder LB (2009) The silver anniversary of the United States' exclusive economic zone: twenty-five years of ocean use and abuse, and the possibility of a blue water public trust doctrine. Ecol Law Q 36:1, 5–7, 30

Part IV
Food Production and Wildlife Protection: Pollinators, Soil, Habitat and Incidental Wildlife Losses

Chapter 17
Pollinators and Pesticides

Larissa Walker and Sylvia Wu

Abstract Pollinating insects, such as bees, butterflies, birds, bats, and other animals are critical to maintaining healthy ecosystems and a strong agricultural economy. Despite their agricultural and ecological importance, since the mid-2000s, scientists have observed serious declines in a variety of pollinating species, both in the United States and worldwide. The loss of pollinators threatens the health of our environment, the diversity of our ecosystems, and our agricultural economy. What is driving these alarming losses in pollinator populations? An overwhelming amount of peer-reviewed scientific studies highlighting significant threats to pollinators from a particular class of pesticides: neonicotinoids, a highly toxic class of systemic insecticides modeled after nicotine that interfere with the nervous system of insects, causing tremors, paralysis and eventually death at very low doses. This chapter provides an overview to the threat facing pollinators from neonicotinoids, and explores how the existing federal regulatory scheme is inadequate to ensure timely and adequate protection of pollinators and threatened and endangered species from federal approval of harmful pesticides through the case study of one particular neonicotinoid chemical: clothianidin. Despite the issues with the existing federal regulatory framework, the chapter also highlights the power of public scrutiny and media pressure on spurring agency action, setting the scene hopefully for more stringent regulation of neonicotinoid pesticides and better protection of vital pollinator species.

17.1 Introduction

First, we will describe the role that neonicotinoid insecticides are playing in pollinator declines and poor pollinator health. Next, we will present the scientific evidence that increasingly implicates neonicotinoids in pollinator declines and broader environmental damage. Third, we will examine legal, policy and regulatory

---

L. Walker (✉) • S. Wu (✉)
Center for Food Safety, Washington, DC, USA
e-mail: LWalker@centerforfoodsafety.org; SWu@centerforfoodsafety.org

G. Steier, K.K. Patel (eds.), *International Farm Animal, Wildlife and Food Safety Law*, DOI 10.1007/978-3-319-18002-1_17

issues that led up to this situation, and continue to present numerous additional risks for ecosystem health. And finally, we will discuss ongoing legal actions and policy solutions for rectifying these problems.

17.1.1 Importance of Pollinators to Food Systems and the Agricultural Economy

Pollinating insects, such as bees, butterflies, birds, bats, and other animals are critical to maintaining healthy ecosystems and a strong agricultural economy. These important invertebrates ensure reproduction, fruit set development and seed dispersal in the vast majority of plants, both in agricultural landscapes and natural ecosystems.[1] Pollinators contribute an estimated $20–30 billion annually to the U.S. agricultural economy.[2] Bees and other pollinators also support the reproduction of nearly 85 % of the world's flowering plants—more than 250,000 varieties globally.[3]

While honey bees are the primary pollinator species managed for U.S. agricultural needs, wild bee species are also critical for crop pollination. There are roughly 4000 species of wild bees that are native to North America, and in many cases, these species are just as critical (if not more so) than honey bees for pollination services. Unlike honey bees, which live in hive colonies, the vast majority of native bee species are solitary—meaning that each bee lives, forages, and raises their brood independently.

More than 150 food crops in the U.S. depend on pollinators, including almost all fruit and grain crops.[4] Bees in particular, both managed and native, play a major role in pollinating approximately 70 % of all the food crops that humans consume, including, but not limited to: almonds, apples, apricots, artichoke, avocado, blueberries, buckwheat, cauliflower, cashews, coconut, coffee, figs, grapes, green beans, kale, kiwifruit, lemons, lettuces, mango, olives, peaches, pears, peppermint, pumpkins, raspberries, squash, watermelon, and zucchini. Some of these crops are directly dependent upon pollination services to produce food, while others indirectly benefit from pollination and yield higher quantities of food crops.[5]

Beyond the food crops that depend on bee pollination, many species of flowering plants and herbs rely on bee pollination, including: rosemary, black-eyed susans,

[1]FAO. 2007. The Importance of Pollination for Agriculture. ftp://ftp.fao.org/docrep/fao/010/ai759e/ai759e02.pdf.

[2]http://www.news.cornell.edu/stories/2012/05/insect-pollinators-contribute-29b-us-farm-income.

[3]Ollerton et al. (2011), pp. 321–326.

[4]http://www.fs.fed.us/wildflowers/pollinators/importance.shtml.

[5]FAO. 2014. *Policy Analysis Paper: Policy Mainstreaming of Biodiversity and Ecosystem Services with a Focus on Pollination.* Rome (available at: http://food.berkeley.edu/wp-content/uploads/2014/08/Pollination-Policy_analysisFINAL.pdf).

yarrow, most rose varieties, primroses, lemon mint, aster, prairie sage, chive, dill, leeks, nutmeg, parsley, and sunflowers.

17.2 Background

17.2.1 Pollinator Declines

The loss of pollinators threatens the health of our environment, the diversity of our ecosystems, and our agricultural economy. Since the mid-2000s, scientists have observed serious declines in a variety of pollinating species, both in the United States and worldwide. In the 1940s, it is estimated that there were 6 million honey bee colonies in the U.S. By 2000, there were roughly 2.5-million honey bee colonies left in the country. Since 2006, the U.S. Department of Agriculture estimates that U.S. beekeepers have continued to suffer an average overwintering colony loss of 30 % or more each year.[6] Historically, an "acceptable" overwintering colony loss rate was 15 % or less.[7] According to a government sponsored survey, on average, beekeepers reported 51.1 % annual hive losses during the winter of 2013/14, with 66 % of all beekeepers reporting higher losses than they deemed acceptable.[8] Today, many beekeepers continue to report annual losses of 40–50 %, with some as high as 100 %.[9]

Wild pollinators, such as native bees and the monarch butterfly, have also suffered stark population declines. The monarch butterfly has declined by 90 % in 20 years, and scientists have documented numerous native bee species suffering declines in their range and distribution across the country.[10] In one study, scientists across North America found that one-third of all bumblebee species are at risk of extinction—and the relative abundance of certain bumblebee species in the U.S. has declined by 96 %.[11] In 2013, an invertebrate conservation group petitioned the U.S. Fish and Wildlife Service to list the rusty patched bumble bee (*Bombus affinis*) as an endangered species under the Endangered Species Act. Recent research has shown a significant decline in both the range and relative abundance of this bumblebee. Although this species was historically common from the Upper Midwest to the eastern seaboard, a nationwide study estimated that the rusty patched

[6]D. vanEngelsdorp, et al., "Colony Loss 2013-2014," May 15, 2014, http://beeinformed.org/.

[7]http://beeinformed.org/2013/05/winter-loss-survey-2012-2013/.

[8]http://beeinformed.org/results/colony-loss-2013-2014-2/.

[9]http://www.nytimes.com/2013/03/29/science/earth/soaring-bee-deaths-in-2012-sound-alarm-on-malady.html?_r=1.

[10]http://www.centerforfoodsafety.org/files/monarch-esa-petition-final_77427.pdf.

[11]http://food.berkeley.edu/wp-content/uploads/2014/08/Pollination-Policy_analysisFINAL.pdf.

bumble bee has disappeared from 87 % of its historic range and that its relative abundance had declined by 95 %.[12]

What is driving these alarming losses in pollinator populations? Scientists have identified several factors that threaten the health and population size of pollinators, including: exposure to pesticides, the loss of habitat, poor nutrition, diseases, parasites, and extreme weather conditions. Bees and other pollinators are exposed to an onslaught of harmful pesticides in both rural agricultural and urban landscapes. In addition to pesticide exposure directly harming pollinators, various herbicides that have become widespread in agricultural areas have wiped out a significant amount of habitat and forage that these beneficial insects depend on for food and shelter. These pesticide threats are coupled with a variety of diseases and parasites, like the *Varroa destructor* mites and *Nosema* fungus, which also threaten the health and vitality of bees. Lastly, extreme weather conditions, such a prolonged droughts and harsh winter storms, can also impact pollinator populations (for instance, a winter storm in 2002 killed roughly 500 million monarch butterflies that were overwintering in Oyamel fir forests of Central Mexico). Furthermore, climate change has disrupted bloom periods of flowers that bees rely on when they come out of hibernation, throwing their lifecycles out of sync with the food they depend upon.[13]

All of these threats are troublesome for bees and other pollinators, especially when in combination with each other. Even though pesticides are only one of numerous threats facing bees, they are a significant threat, and they are also the most preventable. Pesticides are the one threat that government regulators can address immediately, while many of the other factors are either out of our control or will require years to mitigate. For these reasons, the remainder of this chapter will focus specifically on the role of pesticides in pollinator declines, and it will examine legal, policy and regulatory issues that led up to this situation, while also providing solutions for rectifying these problems.

17.2.2 The Emergence of Neonicotinoids

There is an overwhelming amount of peer-reviewed scientific studies highlighting significant threats to pollinators from a particular class of pesticides: neonicotinoids. Neonicotinoids are a highly toxic class of systemic insecticides modeled after nicotine that interfere with the nervous system of insects, causing tremors, paralysis and eventually death at very low doses. In 1994, the Environmental Protection Agency (EPA) approved the first neonicotinoid chemical, imidacloprid, for limited use on ornamental plants and turfgrass.[14] In the 2000s, EPA approved several additional neonicotinoid chemicals, including thiamethoxam

---

[12]Cameron et al. (2011), pp. 662–667.

[13]http://www.sciencemag.org/content/349/6244/177.abstract.

[14]http://www.epa.gov/pesticides/chemical/foia/cleared-reviews/reviews/129099/129099-051.pdf.

(2000), acetamiprid (2002), clothianidin and thiacloprid (2003), and dinotefuran (2004). In scientific assessments and early reviews dating back to 1993, EPA scientists expressed serious concerns with the high toxicity of these chemicals to honey bees, birds, and other wildlife, as well as to endangered species.[15] However, EPA still categorized the neonicotinoids as reduced risk, giving them priority review and viewing them as an alternative to the organophosphate pesticides.[16] In the early 2000s, independent studies showed that organophosphates had serious human health effects and many of the uses of these pesticides were cancelled.[17] The next generation of pesticides to replace organophosphates was the neonicotinoids. Like organophosphates, neonicotinoids act on the nervous system of insects, but are not very toxic to mammals, according to industry studies.

It is important to point out that even though the first neonicotinoid was approved for limited use in 1994, it was not until the mid to late 2000s that these chemicals were used extensively throughout the United States. Neonicotinoids are now the most widely used insecticides in the world, with an estimated 500 or more different neonicotinoid products on the market, and applications estimated to exceed 150 million acres annually nationwide. Neonicotinoids also account for nearly 25 % of the agrochemical market worldwide.[18] In 2009, neonicotinoids comprised nearly 25 % of the global pesticide market, with imidacloprid as the top-selling insecticide in the world.[19]

17.2.3 Uses of Neonicotinoids

Neonicotinoids are sprayed on a wide variety of crops, trees, landscapes, and turfs, but one of their largest uses is as a seed treatment on most annual field crops (such as corn, soybeans, wheat, canola, and cotton). For instance, corn is grown on roughly 95 million acres across the U.S., and it is estimated that 90–95 % of all corn seed is coated with a neonicotinoid chemical.[20] The large majority of canola and cotton seeds are also coated with these chemicals.[21] This prophylactic use of neonicotinoids as seed treatments negates the principles of Integrated Pest Management (IPM), which has historically been considered the gold standard for pest management practice by the U.S. Department of Agriculture and many farmers, as

[15]http://www.epa.gov/pesticides/chemical/foia/cleared-reviews/reviews/129099/129099-019.pdf.

[16]http://edis.ifas.ufl.edu/pi117.

[17]http://www.mdguidelines.com/toxic-effects-organophosphate-and-carbamate-pesticides.

[18]Jeschke et al. (2011), pp. 2987–2988.

[19]Congressional Research Service. 2015. "Bee Health: The Role of Pesticides." Available at: http://fas.org/sgp/crs/misc/R43900.pdf.

[20]Arnason, R. "No Yield Benefit from Neonicotinoids: Scientist," *The Western Producer*, May 10, 2013.

[21]Stevens, S. and Jenkins, P., *Heavy Costs: Weighing the Value of Neonicotinoid Insecticides in Agriculture*, Center for Food Safety, 2014.

it encourages farmers to only use pesticides when necessary (when a pest pressure is high enough to pose an economic harm to farmers and no other remedy exists). IPM utilizes a four-phase strategy: (1) Minimize conditions that encourage pest outbreaks, (2) Set an economic threshold of how much damage can be tolerated before pest control options must be utilized, (3) Scout fields to monitor for pest populations, and (4) Control pests with the most targeted pest control option when the pre-determined damage threshold is reached.[22] By utilizing these IPM strategies, farmers are able to greatly reduce the risks of unnecessary pesticide applications to people, pollinators, other living organisms, and the environment.

Unfortunately, the vast majority of uses of neonicotinoids are wholly incompatible with the principles of IPM. Further compounding the problem, neonicotinoids are more persistent in soil and water than most other insecticides and have the ability to accumulate in the environment. These long-lasting chemicals are also very mobile and have been detected frequently in water bodies next to agricultural areas where they are heavily used, both as seed treatments and foliar applications.[23] For instance, one neonicotinoid chemical, imidacloprid, was found in 89 % of surface waters sampled in agricultural regions in California.[24] Numerous other studies, including U.S. Geological Service (USGS) data and state water quality reports, found traces of neonicotinoids present at concentration levels that are lethally toxic to a variety of species. A 2015 USGS study was the first nationwide survey of neonicotinoid detections in streams across the United States. USGS researchers sampled 38 streams in 24 States and Puerto Rico between December 2012 and June 2014. Neonicotinoids were present in 77 % of all analyzed samples. Harmful effects from neonicotinoid contamination are documented in aquatic and terrestrial invertebrates and real concerns exist with respect to long-term impacts on waterfowl, farmland birds, and other wild animals. One study demonstrated that a single corn kernel coated with a neonicotinoid is toxic enough to kill a songbird.[25]

The extreme persistence of these chemicals is just as alarming as their mobility, as neonicotinoid applications can remain toxic for months or years depending on the type of application and soil composition. For example, the neonicotinoid chemical clothianidin may remain in the soil a year to over 3 years. In certain rare conditions, it has been found to last up to 19 years.[26] Even untreated plants can take up residues of neonicotinoids still present in the soil from previous applications.

[22]The Xerces Society. 2014. Preventing or Mitigating Potential Negative Impacts of Pesticides on Pollinators Using Integrated Pest Management and Other Conservation Practices. Available at: http://directives.sc.egov.usda.gov/OpenNonWebContent.aspx?content=34828.wba.

[23]http://ca.water.usgs.gov/projects/PFRG/CurrentProjects.html.

[24]http://www.ncbi.nlm.nih.gov/pubmed/22228315.

[25]http://www.abcbirds.org/abcprograms/policy/toxins/Neonic_FINAL.pdf.

[26]Rexrode M, Barrett M, Ellis J, Gabe P, Vaughan A, Felkel J, Melendez J: EFED Risk Assessment for the Seed Treatment of Clothianidin 600FS on Corn and Canola. United States Environmental Protection Agency; 20 February 2003.

All neonicotinoids are systemic chemicals, meaning they are absorbed into treated plants and distributed throughout their vascular systems. This systemic property ensures that all parts of the plant—including the roots, leaves, stem, flowers, nectar, pollen, and guttation fluid—are toxic to insects, including pollinators and other beneficial insects.[27]

The adverse impacts to beneficial insects from uses of neonicotinoids undermines sustainable food production, as these species play critical roles in the healthy functioning of agro-ecosystems. For instance, neonicotinoids are "supertoxic" to earthworms, which are critical to soil health and crop development.[28] Researchers have also noted harms to lady beetles (ladybugs) from uses of neonicotinoids. In one study, lady beetle larvae that briefly fed on seedlings grown from seeds treated with clothianidin or thiamethoxam experienced significantly higher mortality, but also sublethal effects like trembling, paralysis, or loss of coordination.[29] These beneficial species are crucial for effective control of crop pests like aphids.

Since neonicotinoids are highly persistent and systemic, bees and other pollinators are exposed to these chemicals in a variety of ways, including but not limited to: direct contact with pesticide sprays; exposure to pesticide residues in pollen and nectar; contact with dust released from seed planting equipment during the sowing of treated seeds; exposure to residues on target and non-target plants from foliar uses and planter exhausts (especially plants growing within or adjacent to treated fields); consumption of pollen, nectar or dew droplets (guttation) on treated plants; and exposure to contaminated soil and water.[30] Contaminated soil is a particularly serious threat for native bees that build their nests in the ground. Nearly 70 % of native bees nest in soil where they may be exposed to lasting residues of the chemicals from soil drenches, chemigation (insecticide added to irrigation water) or seed treatments.

Needless to say, the presence of neonicotinoids in pollen and nectar in particular poses a unique threat to bees: it causes them to bring the pollen, nectar, *and* pesticides back to the hive to feed themselves and their offspring. Thus, the active ingredient in these chemicals is directly ingested by the bees throughout their lifecycle. Exposure to neonicotinoids can have acute, lethal impacts on pollinators, especially considering that some forms of neonicotinoids are 5,000–10,000 times more acutely toxic to bees that DDT (dichlorodiphenyltrichloroethane).[31] However, more commonly, bees and other beneficial insects that are exposed to lower doses of neonicotinoids over long periods of time typically experience harmful sublethal effects, such as disorientation, memory loss, weakened immunity, and impaired

---

[27]Krupke et al. (2012), p. e29268; Blacquière et al. (2012), pp. 973–92.

[28]Wang et al. (2012).

[29]Moser and Obrycki (2009), pp. 487–492.

[30]Andrea Tapparo et al. (2012), pp. 2592–2599; Girolami et al. (2009), p. 1808; Van der Sluijs et al. (2013); Krupke et al. (2012), p. e29268.

[31]Conclusions of the Task Force on Systemic Pesticides. Worldwide Integrated Assessment. January 2015. Available at: http://link.springer.com/journal/11356/22/1/ page/1.

reproductive capacities.[32] For instance, in a 2013 study, researchers found that honey bees exposed to clothianidin had a more difficult time in finding their way home to their hives. This study is the first to show under field conditions that direct topical exposure of clothianidin, at doses much lower than their LD50 values (the median lethal dose), caused sublethal effects.[33] Another study that same year, published in *Proceedings of the National Academy of Sciences,* concluded that clothianidin reduces immune defenses and promotes the replication of the deformed wing virus in honey bees bearing covert infections.[34] In terms of reproductive impacts, a study evaluating the impacts of two neonicotinoid chemicals, clothianidin and thiamethoxam, to solitary bees found that sublethal exposure reduced total offspring production by nearly 50 %.[35]

Even though these sublethal effects may not necessarily kill honey bees outright, chronic ingestion of neonicotinoids *does* have the ability to threaten the health and vitality of the entire hive, especially when these behavior effects are combined and impact whole-colony behavior. Honey bees are social insects and they depend upon the successful functioning of their memory, cognition, and communication in order to survive and ensure activities within the hive are properly carried out.

17.3 Regulatory History

Despite the threats that neonicotinoids pose to pollinators and healthy ecosystem functioning, EPA gave neonicotinoids an expedited registration process, treating them as "reduced risk pesticides".[36] According to the EPA's Office of Pesticide Program, the reduced risk pesticides' advantages include: "low impact on human health, lower toxicity to non-target organisms (birds, fish, plants), low potential for groundwater contamination, low use rates, low pest resistance potential, and compatibility with Integrated Pest Management (IPM) practices."

How did we arrive at this situation? If neonicotinoids present such an expansive set of problems for pollinators and other beneficial insects, how did they gain approval onto the market in the first place?

[32] Dively et al. (2015), p. e0126043; Blacquière et al. (2012), pp. 973–92; Williamson and Wright (2013), pp. 1799–1807; Beguin et al. (2012), p.348stic.

[33] Matsumoto (2013), pp. 1–9.

[34] Di Prisco et al. (2013).

[35] Sandrock et al. (2013).

[36] http://www2.epa.gov/sites/production/files/2014-02/documents/reduced-risk-op-decisions.pdf.

17.3.1 EPA's Weak Oversight Under FIFRA

The mission of the EPA is to protect human health and the environment, and one of the agency's core purposes is to ensure that environmental protection contributes to making our communities and ecosystems diverse, sustainable and economically productive.[37] Along with that mission, the agency regulates the sale and use of pesticides in the United States pursuant to the Federal Insecticide, Fungicide, and Rodenticide Act (FIFRA).[38]

Under the FIFRA, EPA licenses the sale, distribution, and use of pesticides through the process of registration.[39] Pursuant to FIFIRA, EPA oversees both initial registration of an active ingredient as well as any new uses of the registered active ingredient.[40] Pursuant to section 3(c)(5) of FIFRA, EPA shall register a pesticide so long as the agency determines that the pesticide "will perform its intended function without unreasonable adverse effects on the environment," and that, "when used in accordance with widespread and commonly recognized practice[,] it will not generally cause unreasonable adverse effects on the environment."[41] Despite the language, EPA can and does register pesticides and approve their uses even when the agency recognizes that it is missing certain information regarding the pesticide's potential impacts. Under FIFRA, where there are data gaps and missing information, EPA can still register a pesticide with conditions, in what is common referred to as the "conditional registration" process.[42] Section 3(c)(7) of FIFRA authorizes EPA to issue such condition registrations "for a period reasonably sufficient for the generation and submission of required data," so long as EPA also determines that the conditional registration of the pesticide during that time period "will not cause any unreasonable adverse effect on the environment, and that use of the pesticide is in the public interest."[43]

Thus under FIFRA, to issue either a conditional or unconditional registration, EPA has the statutory duty to conclude that the proposed pesticide registration and its uses would not result in "unreasonable adverse effects on the environment," which the statute defines to include "any unreasonable risk to man or the environment, taking into account the economic, social, and environmental costs and benefits of any pesticides."[44] The culmination of the registration process is EPA's approval of a label for the pesticide, including use directions and appropriate warnings on safety and environmental risks.

---

[37]http://www2.epa.gov/aboutepa/our-mission-and-what-we-do.

[38]7 U.S.C. 136 et seq.

[39]7 U.S.C. § 136a.

[40]*See* 7 U.S.C. § 136a.

[41]*See* 7 U.S.C. § 136a(c)(5).

[42]7 U.S.C. § 136a(c)(7).

[43]7 U.S.C. §136a(c)(7)(C).

[44]7 U.S.C. § 136(bb).

Despite the agency's charge to protect the environment against unreasonable risks from pesticide use, EPA has experienced a fair share of chemical crises—perhaps most famously with DDT. Unfortunately, there are a number of ways in which the pesticide approval process is broken and easily allows for pesticides to enter the market. To begin, because the statutory definition of "unreasonable adverse effects on the environment" includes "the economic, social, and environmental costs and benefits of any pesticides,"[45] courts have consistently interpreted the definition to require EPA to engage in a cost-benefit analysis in determining whether or not to approve a pesticide's registration.[46] In other words, a pesticide registration may have significant economic, social and environmental costs, but nonetheless meet the standard for registration, should EPA conclude that the benefits of the registration outweigh such costs.

The utility of EPA's cost-benefit analysis is further weakened by the loophole created under the conditional registration process. EPA has heavily abused "conditional registrations" under FIFRA, and this loophole in the approval process has registered the majority of the neonicotinoid product registrations without acquiring necessary safety data. In granting pesticide registrants a conditional registration, EPA has essentially granted pesticide companies with a significant amount of leeway to submit key safety information years after the products are approved for use and allowed onto the market.[47] This process has been heavily criticized by the Government Accounting Office as poorly administered by EPA, which has often failed to monitor and ensure compliance with key conditions, including those impacting bees and other pollinators.

The "conditional registration loophole" for putting pesticide products out in the marketplace without sufficient data is worsened by the lack of transparency and monitoring of the conditions to sustain registration. Under FIFRA, a conditional registration may only last for a period "reasonably sufficient" to generate the outstanding data necessary for unconditional registration.[48] If a condition is not fulfilled within the timeframe specified in the conditional registration, EPA "shall" initiate cancellation proceedings.[49] Despite the plain language of the statute, EPA can avoid initiating cancellation by extending the period "reasonably sufficient" to generate the outstanding data. While such extensions are seemingly contrary to the statutory intent, no court has never directly addressed this question.

Moreover, the registration of new uses of pesticides under the conditional registration process is not always readily available to the public. This is because under FIFRA, EPA is only required to provide the public with notice and opportunity for comment for a proposed registration of a pesticide if it involves "a new

[45] *Id.*

[46] *See, e.g.*, *Love v. Thomas*, 858 F.2d 1347, 1357 (9th Cir. 1988).

[47] *See* 7 U.S.C. § 136a(c)(5); *id.* § 136a(c)(7)(C).

[48] 7 U.S.C. § 136a(c)(7)(C).

[49] 7 U.S.C. § 136d(e)(1).

active ingredient," or "if it would entail a changed use pattern."[50] EPA's own regulations define a changed use pattern as: (1) "Any proposed use pattern that would require the establishment of, the increase in, or the exemption from the requirement of a tolerance or food additive regulation under section 408 of the Federal Food, Drug and Cosmetic Act;" or (2) any new outdoor use pattern, "if no product containing the active ingredient is currently registered for that use pattern;" or "(3) any additional use pattern that would result in a significant increase in the level of exposure, or a change in the route of exposure, to the active ingredient of man or other organisms."[51] What constitutes "a significant increase in the level of exposure" has never been defined by the agency nor addressed by the courts. As a result of the lack of necessary public notice, EPA can and has frequently extended the necessary time periods for compliance with the registration's conditions.

On the other end of the spectrum, EPA has the authority to cancel a pesticide registration whenever a pesticide "does not comply with the provisions of [FIFRA] or, when used in accordance with widespread and commonly recognized practice, generally causes unreasonable adverse effects on the environment."[52] Prior to canceling a pesticide production's registration, FIFRA requires that EPA notify the registrant of either its intent to cancel the pesticide registration or with an administrative hearing to determine whether EPA should cancel the pesticide.[53] EPA must also afford the registrant an opportunity to request an administrative hearing upon receiving notification of EPA's intent to cancel a pesticide product.[54] Throughout the cancellation proceeding, the product's proponent bears the legal burden of showing that any pesticide and any approved uses meet FIFRA criteria to be eligible for continued registration.[55]

EPA also has the authority to suspend immediately a pesticide's registration, prior to initiating cancellation proceeding, in order to prevent an "imminent hazard."[56] FIFRA defines an "imminent hazard" to mean "when the continued use of a pesticide during the time required for cancellation proceeding would be likely to result in unreasonable adverse effects on the environment or will involve unreasonable hazard to the survival of a species declared endangered or threatened [under the ESA]."[57] The suspension acts as an emergency order that immediate suspends the pesticide's sale, distribution and use. FIFRA then requires that EPA issue either

[50]7 U.S.C. § 136a(c)(4); *see* 40 C.F.R. § 152.102 (requiring EPA issue notices for a "new active ingredient" or "a new use").

[51]40 C.F.R. § 152.3.

[52]7 U.S.C. § 136d(b).

[53]Id.

[54]*See* id.

[55]*See* 40 C.F.R. § 154.5.

[56]7 U.S.C. § 136d(c).

[57]7 U.S.C. § 136(l).

a notice of intent to cancel the pesticide or change its classification within 90 days.[58]

As such, EPA's decision to either cancel or suspend a pesticide provides administrative remedies before the agency prior to challenging a pesticide's continued registration in court.

17.3.2 EPA's Failure to Consider Harm to Federally Listed Species

EPA is also required to comply with the Endangered Species Act (ESA) in exercising its authority over pesticides, a statutory duty that EPA consistently violates in its registration and approval pesticides—and neonicotinoids were no exception. Section 7(a)(2) of the ESA requires every federal agency to consult the appropriate federal fish and wildlife agency—Fish and Wildlife Service (FWS), in the case of land and freshwater species and the National Marine Fisheries Service (NMFS) in the case of marine species—to "insure" that the agency's actions are not likely "to jeopardize the continued existence" of any listed species or "result in the destruction or adverse modification" of critical habitat.[59]

The ESA's implementing regulations broadly define agency action to include "all activities or programs of any kind authorized, funded or carried out ... by federal agencies," including the granting of permits and "actions directly *or indirectly* causing modifications to the land, water or air."[60] A species' "critical habitat" includes those areas identified as "essential to the conservation of the species" and "which may require special management considerations or protection."[61] Pending the completion of consultation with the expert agency, an agency is prohibited from making any "irreversible or irretrievable commitment of resources with respect to the agency action which has the effect of foreclosing the formulation or implementation of any reasonable and prudent alternative measures."[62]

17.4 Case Study: Legal Actions Challenging EPA's Registration of Clothianidin

EPA's regulatory history of clothianidin, one of the most widely-used neonicotinoids, illustrates the numerous problems associated with EPA's exercise of its regulatory authority over pesticides, the agency's frequent violations of

[58] *See* 7 U.S.C. § 136d(c).

[59] 16 U.S.C. § 1536(a)(2); *see also* 50 C.F.R. § 402.01(b).

[60] 50 C.F.R. § 402.02 (emphasis added).

[61] 16 U.S.C. § 1532(5)(A).

[62] 16 U.S.C. § 1536(d).

FIFRA, and its disregard for legal protection for sensitive pollinator species required by the ESA.

Since 2003, EPA has approved clothianidin in more than thirty pesticide products, for a wide variety of agricultural, landscaping, and outdoor use markets.[63] EPA did so despite the fact that the agency was aware from the beginning that clothianidin, as a systemic neonicotinoid, adversely affect species vital to U.S. agriculture and the environment, including honey bees and other pollinator insects.[64] In its initial ecological risk assessment for clothianidin use as a seed treatment for corn and canola, EPA's own scientists cautioned that additional field tests should be required to evaluate the chemical's potential harm to pollinators:

> The possibility of toxic exposure to nontarget pollinators through the translocation of clothianidin residues that result from seed treatment (corn and canola) has prompted EFED [Environmental Fate and Effects Division] to require field testing that can evaluate the possible chronic exposure to honey bee larvae and queen. In order to fully evaluate the possibility of this toxic effect, a complete worker bee life cycle study must be conducted, as well as an evaluation of exposure and effects to the queen.[65]

EPA was similarly aware that clothianidin presented risks to threatened and endangered species. In issuing the initial conditional registration of clothianidin, EPA recognized that compliance with the ESA is necessary:

> Clothianidin is expected to present acute and/or chronic toxicity risk to endangered/threatened birds and mammals via possible ingestion of treated corn and canola seeds. Endangered/threatened non-target insects may be impacted via residue laden pollen and nectar. The potential use sites cover the entire U.S. because corn is grown in almost all U.S. states.[66]

EPA nonetheless allowed nationwide usage of clothianidin since 2003, by issuing conditional registrations that allowed for its usage despite missing data.[67] The missing data called for by the conditional registration includes studies critical to understanding how these two pesticides react in the environment to the potential detriment of honey bees, pollinator species, and threatened and endangered species. Furthermore, EPA's approvals made without the required consultation under the ESA.

---

[63]*See* EPA, Pesticide Fact Sheet: Clothianidin, Conditional Registration (May 30, 2003), *available at* http://www.epa.gov/pesticides/chem_search/reg_actions/registration/fs_PC-044309_30-May-03.pdf; *see* National Pesticide Information Retrieval System (NPIRS), http://npirspublic.ceris.purdue.edu/ppis/ (search "clothianidin" under "active ingredient search) (last accessed September 7, 2015).

[64]*See* EPA, Pesticide Fact Sheet: Clothianidin, Conditional Registration (May 30, 2003), *available at* http://www.epa.gov/pesticides/chem_search/reg_actions/registration/fs_PC-044309_30-May-03.pdf.

[65]*See* Memorandum: Risk Assessment for the Seed Treatment of Clothianidin 600FS on Corn and Canola, PC Code 044309, EPA Environmental Fate and Effects Division (Feb. 20, 2003), *available at* www.epa.gov/pesticides/chem_search/cleared_reviews/csr_PC-044309_20-Feb-03_a.pdf.

[66]EPA, Pesticide Fact Sheet: Clothianidin, Conditional Registration 16 (May 30, 2003), *available at* http://www.epa.gov/opp00001/chem_search/reg_actions/registration/fs_PC-044309_30-May-03.pdf.

[67]*Id.*

On March 20, 2012, a coalition of beekeepers and environmental groups submitted a citizens' petition to EPA regarding clothianidin, requesting EPA to take actions to address its past regulatory failures, beginning with the immediate suspension of clothianidin.[68] The citizens' petition also requested that EPA initiate cancellation proceeding of clothianidin products, as well as consult with the expert wildlife agencies on the potential impacts of clothianidin uses on threatened and endangered species as required under the ESA.[69] On July 17, 2012, EPA denied the petitioners' request to suspend clothianidin as an imminent hazard.[70] Along with the suspension denial, EPA also solicited public comments on the remaining requests contained in the citizens' petition, including the petitioners' request that EPA initiate cancellation proceeding on clothianidin pesticides.[71] EPA also indicated that it may reconsider the suspension denial along with its consideration of the remaining requests in the citizens' petition.[72] As of September 2015, EPA still has not responded to the remaining requests on the citizens' petition nor reconsidered its suspension denial.

One year later, in March 2013, several members of the petitioners filed suit in federal court in the Northern District of California.[73] The lawsuit challenged EPA's suspension denial of the citizens' petition, as well as EPA's misuse of the conditional registration in registering clothianidin products. The lawsuit also alleged that EPA unlawfully approved several clothianidin outdoor uses without proper notice and comment, because such outdoor uses constituted "additional use pattern[s]" resulting in "significant increase[s] in the level of exposure, or [] change[s] in the route of exposure" of clothianidin to mankind and other organisms.[74] Finally, the lawsuit alleged that EPA unlawfully registered clothianidin products without complying with the consultation requirements of the Endangered Species Act.[75] As expected, Dow and other pesticide companies (Pesticide Intervenors) possessing the registered clothianidin products intervened in the lawsuit.[76] EPA and the Pesticide Intervenors then moved to dismiss the lawsuit.[77]

---

[68]CFS *et al.*, Clothianidin Legal Petition (Mar. 21, 2012), *available at* http://www.centerforfoodsafety.org/wp-content/uploads/2012/10/CFS-Clothianidin-Petition-3-20-12.pdf.

[69]*Id.*

[70]Letter from Steven Bradbury, Director, Office of Pesticide Programs, EPA, to Peter T. Jenkins (July 17, 2012), *available at* http://www.epa.gov/opp00001/about/intheworks/epa-respns-to-clothianidin-petition-17july12.pdf.

[71]*Id.*

[72]*Id.*

[73]*Ellis v. Housenger*, Case No. 3:13-cv-01266-MMC (N.D. Cal. filed Mar. 21, 2013).

[74]*See* Complaint, ECF No. 1, *Ellis*, Case No. 3:13-cv-01266-MMC (N.D. Cal. filed Mar. 21, 2013).

[75]*See id.*

[76]*See Ellis*, Case No. 3:13-cv-01266-MMC, 2013 WL 4777201 (N.D. Cal. Sept. 6, 2013) (Order Granting in Part and Denying In Part Intervention).

[77]*See Ellis*, Case No. 3:13-cv-01266-MMC, 2014 WL 1569271 (N.D. Cal. Apr. 18, 2014) (Order Granting in Part and Denying in Part Motions to Dismiss).

After lengthy briefing, the district court issued its ruling on the motion to dismiss, dismissing several claims while allowing the lawsuit to proceed on other remaining claims.[78] Specifically, the court dismissed the plaintiffs' challenge to EPA's ongoing conditional and unconditional registrations of clothianidin products despite missing data on the chemical's impacts on pollinators for failure to exhaust administrative remedies.[79] In so holding, the court emphasized that in light of EPA's ongoing review of the citizen's petition and its request that EPA initiate cancellation of clothianidin products, it would be premature and contrary to the intent of statutory design for judicial determination on the legal status of such registrations.[80] The court did allow the lawsuit to proceed on plaintiffs' allegations that: (1) EPA's decision to deny immediate suspension of clothianidin was unlawful; (2) EPA unlawfully registered certain outdoor uses of clothianidin products without proper federal notice as required by FIFRA; (3) EPA violated the ESA for failing to consult on the impacts to threatened and endangered species before approving certain clothianidin products.[81] Regarding claims alleging EPA's procedural violations under FIFRA and the ESA, the court limited the challenge to EPA actions taken within the last 6 years, pursuant to the applicable statute of limitations for challenging administrative actions under 28 U.S.C. § 2401(a). As of September 2015, the lawsuit is still ongoing, with motion for summary judgment set to begin in the spring of 2016.

The *Ellis* lawsuit highlights many of the difficulties faced in challenging EPA's regulatory oversight of neonicotinoids.

First, as demonstrated by the court's order on motion to dismiss, a plaintiff is bound by the 6-year statute of limitations set forth under 28 U.S.C. § 2401(a). This six-year limitation, compounded by the lack of transparency in EPA's registration process and the lag between the use of a pesticide and real-world understanding of its potential impacts, creates a tremendous hurdle for litigants seeking to reverse or limit the impacts of harmful pesticides such as neonicotinoids. In the case of neonicotinoids, although EPA was aware from the beginning that there are potential risks to pollinator species, scientific literature addressing such risks as well as increasing reports of bee-kill data did not become widely-acknowledged until years after the chemicals' initial registration.

Second, the litigation highlights the additional delay and administrative hurdles for removing a pesticide off-market. As explained by the court, FIFRA sets forth procedural safeguards in the form of cancellation proceedings that must be satisfied prior to seeking cancellation of a pesticide in court.[82] This further delays the length of time it may take to remove a pesticide off-market once it has been registered, all the while exposing sensitive species to harm.

---

[78]*Id.* at *1.

[79]*Id.* at *3-8.

[80]*Id.*

[81]Id. at *1-3, 8-15.

[82]*Id.* at *3-8.

Nonetheless, the remaining claims in the lawsuit does provide an opportunity to rectify EPA's regulatory failures regarding existing clothianidin registrations. The allegation that EPA failed to issue notice-and-comment as required under FIFRA in approving certain outdoor uses of clothianidin provides a chance for courts to weigh in on what constitutes "additional use pattern that would result in a significant increase in the level of exposure, or a change in the route of exposure, to the active ingredient of man or other organisms" triggering the duty to provide notice and comment under EPA's FIFRA-implementing regulations.[83] Should the plaintiffs prevail on the allegation that EPA failed to consult as required under the ESA, the lawsuit's remedy will also ensure that consultation occurs on a timely basis.

17.5 EPA Actions in Response to Public Scrutiny

Despite the ongoing review of the citizens' petition and litigation, the public pressure created by these citizens' actions have spurred some agency actions on neonicotinoids.

In April, 2015, EPA announced a voluntary moratorium on agency approvals of "new uses" of any neonicotinoids pending receipt of adequate information to fully assess their environmental risks.[84] EPA's moratorium followed admitted information gaps about adverse impacts to honey bees, yet, honey bees are actually the most-studied area as far as neonicotinoid environmental impacts. EPA's admission that it lacks adequate information to protect honey bees from new uses raises even greater concerns about its lack of information about neonicotinoid impacts on the much less-studied invertebrates and ecosystems, especially for species in serious decline, such as the rusty patched bumblebee.

While the U.S. has thus far avoided taking meaningful action to protect honey bees and other pollinators, the international community has been much more proactive—with numerous countries imposing suspensions and restrictions on uses of neonicotinoids. In January 2014, the European Union enacted a 2 year moratorium on certain uses of these chemicals, after independent scientists and the European Food Safety Authority expressed serious concerns about the unacceptable hazards that neonicotinoids pose to bees.[85]

In June 2014, President Obama released a Presidential Memorandum calling for the establishment of a Pollinator Heath Task Force (Task Force) after identifying pollinator decline as a threat to the sustainability of our food production systems,

[83] 40 C.F.R. § 152.3.

[84] http://www.epa.gov/oppfead1/cb/csb_page/updates/2015/neonic-outdooruse.html.

[85] http://ec.europa.eu/food/archive/animal/liveanimals/bees/neonicotinoids_en.htm.

our agricultural economy, and the health of the environment.[86] Nearly 1 year after the Presidential Memorandum was announced, the Task Force released a National Pollinator Strategy, identifying three major goals: reduce honey bee overwintering losses to no more than 15 % within 10 years; increase the North American monarch butterfly population to 250 million butterflies within their Center Mexico overwintering site by 2020; and restore or enhance seven million acres of land for pollinators within the next 5 years. Although moving in the right direction, the White House has yet to take any strong or meaningful actions in protecting pollinators from highly toxic pesticides.

Finally, the public spotlight on EPA's misuse of the conditional registration and the potential harms caused by neonicotinoids have also resulted in improved transparency over new systemic pesticides that the industry has introduced as potential replacements for neonicotinoids. Since 2013, EPA has approved new, systemic chemicals that present potential threats to pollinators, but has issued their approvals with opportunities for public notice-and-comment.[87] The increased transparency has resulted in immediate litigation challenging the approval of these new systemic insecticides under either FIFRA or the ESA.[88] While these litigations are ongoing, their outcome would provide precedents and clarity to EPA's legal duties to consider impacts to honey bees, other pollinator species, as well as federally listed species.

17.6 Resistance to Regulatory Changes and Industry Opposition

Despite the growing body of scientific evidence documenting adverse impacts from widespread uses of neonicotinoids, regulators have been incredibly slow to react to issues with neonicotinoids, and major multinational agrochemical and seed corporations continue to oppose regulatory action and instead have focused on casting doubt about harms from these chemicals. It is important to acknowledge these oppositional forces, though a detailed discussion of this issue is beyond the scope of this chapter.

Major multinational agrochemical and seed corporations, such as Monsanto, Bayer, and Syngenta, stand to lose a significant profit if neonicotinoids are removed from the market. Syngenta, one of the world's top agrochemical and seed corporations, boasted sales of $14.2 billion in 2012. One of Syngenta's top-sellers is its neonicotinoid chemical, thiamethoxam, worth $627 million in sales. Another top agrochemical corporation, Bayer Crop Science, topped $10 billion in sales of their

---

[86]https://www.whitehouse.gov/the-press-office/2014/06/20/presidential-memorandum-creating-federal-strategy-promote-health-honey-b.

[87]*See,e.g.,* http://www.epa.gov/oppfead1/cb/csb_page/updates/2013/sulfoxaflor-decision.html (unconditional registration of systemic insecticide sulfoxaflor).

[88]*Pollinator Stewardship Council v. EPA*, Case No. 13-72346 (9th Cir. filed July 2, 2013); *Ctr. for Biological Diversity et al. v. EPA*, Case No. 14-1036 (D.C. Cir. filed Mar. 24, 2014).

"Crop Protection" products 2012. Bayer's top-selling neonicotinoid chemical, imidacloprid, is worth $1.1 billion.[89] While the largest agrochemical and seed corporation, Monsanto, does not manufacture neonicotinoid chemicals per se, they, too, profit from sales of neonicotinoids because the company sells their seeds pre-treated with neonicotinoids produced by other agrochemical companies. In the U.S., roughly 95 % of all corn seed is coated with a neonicotinoid chemical.

All of these major agrochemical companies have a strong vested interest in ensuring that neonicotinoids remain on the market. As such, these corporations have employed a variety of public relations tactics intended to manufacture doubt about their products' role in pollinator declines. One of the most commonly used arguments in favor of neonicotinoids is that without these pesticide products, farmers will suffer significant crop yield reductions and profit loss. However, as numerous peer-reviewed studies have shown, neonicotinoids, particularly when used as seed coatings, provide little or no yield benefit associated with their use on crops, especially where there is low or moderate pest pressure. As a result of heavy marketing by agrochemical and seed corporations, farmers are frequently investing in crop protection that is not providing them with benefits.[90]

17.7 Conclusion

The systemic nature of neonicotinoids and their widespread use present unprecedented challenges and potential threats to the survival of pollinators and other beneficial insects. The survival of these species is not only essential to maintaining a healthy ecosystem, their presence and utility are also indispensable to the nation's food production and agricultural economy. As this chapter demonstrates, EPA's regulatory authority under FIFRA and the agency's history of noncompliance with the ESA has created significant regulatory loopholes for the registration and use of these toxic chemicals. Once introduced to the marketplace, FIFRA also presents significant hurdles to ensure a quick response to protect pollinator species. Nonetheless, the unfortunate story of neonicotinoids and their harm to pollinator species have shed light on these regulatory missteps and prompted agency action and response from the U.S. administration.

Questions for Classroom Discussion

- As this chapter makes clear, FIFRA's statutory framework makes litigation under FIFRA to ensure better protection from pesticide harms very challenging. What are some of the strategies, litigation or otherwise, that activists and environmentalists may utilize to ensure better protection for pollinator species under the existing statutory framework?

---

[89] http://libcloud.s3.amazonaws.com/93/f0/f/4656/FollowTheHoneyReport.pdf.

[90] http://www.centerforfoodsafety.org/files/neonic-efficacy_digital_29226.pdf.

- What are some potential statutory reforms that may help strengthen protection for pollinator species under FIFRA?

References

Beguin HM et al (2012) A common pesticide decreases foraging success and survival in honey bees. Science 336:348stic

Blacquière T et al (2012) Neonicotinoids in bees: a review on concentrations, side-effects and risk assessment. Ecotoxicology 21(4):973–992, http://link.springer.com/article/10.1007/s10646-012-0863-x

Cameron S, Lozier JD, Strange JP, Koch JB, Cordes N, Solter LF, Griswold TL (2011) Patterns of widespread decline in North American bumble bees. Proc Natl Acad U S A 108:662–667

Di Prisco G et al (2013) Neonicotinoid clothianidin adversely affects insect immunity and promotes replication of a viral pathogen in honey bees. PNAS 110(46) doi:10.1073/pnas.1314923110

Dively GP et al (2015) Correction: assessment of chronic sublethal effects of imidacloprid on honey bee colony health. PLoS One 10(4):e0126043

Girolami V et al (2009) Translocation of neonicotinoid insecticides from coated seeds to seedling guttation; and drops: a novel way of intoxication for bees. J Econ Entomol 102(5):1808

Jeschke P et al (2011) Overview of the status and global strategy for neonicotinoids. J Agric Food Chem 59:2987–2988

Krupke CH et al (2012) Multiple routes of pesticide exposure for honey bees living near agricultural fields. PLoS One 7(1):e29268

Matsumoto T (2013) Reduction in homing flights in the honey bee Apis mellifera after a sublethal dose of neonicotinoid insecticides. Bull Insectol 66(1):1–9. ISSN 1721-8861

Moser SE, Obrycki JJ (2009) Non-target effects of neonicotinoid seed treatments; mortality of coccinellid larvae related to zoophytophagy. Biol Cont 51:487–492

Ollerton J, Winfree R, Tarrant S (2011) How many flowering plants are pollinated by animals? Oikos 120:321–326, http://www.fs.usda.gov/Internet/FSE_DOCUMENTS/stelprdb5306468.pdf

Sandrock C et al (2013) Sublethal neonicotinoid insecticide exposure reduces solitary bee reproductive success. Agric Forest Entomol. doi:10.1111/afe.12041

Tapparo A et al (2012) Assessment of the environmental exposure of honeybees to particulate matter containing neonicotinoid insecticides coming from corn coated seeds. Environ Sci Technol 46(5):2592–2599

Van der Sluijs JP et al (2013) Neonicotinoids, bee disorders and the sustainability of pollinator services. Curr Opin Environ Sustain. http://dx.doi.org/10.1016/j.cosust.2013.05.007

Wang Y, Cang T, Zhao X, Yu R, Chen L, Wu C, Wang Q (2012) Comparative acute toxicity of twenty-four insecticides to earthworm. Ecotoxicol Environ Saf. doi:10.1016/j.ecoenv.2011.12.016

Williamson SM, Wright GA (2013) Exposure to multiple cholinergic pesticides impairs olfactory learning and memory in honeybees. J Exp Biol 216(Pt 10):1799–1807

Chapter 18
Textbox: Bats and Pollinator Conservation as a New Avenue for Progressive Food Legislation

Gabriela Steier

Abstract More than 450 important agricultural bat-dependent plants annually affect hundreds of millions of dollars of international trade. Such economically important bat-dependent plants include bananas, mangoes, vanilla, agave, cashews, dates and figs. Bats facilitate the reproductive success of these agricultural plants, including seed set and the recruitment of new seedlings and saplings.

Pollinator conservation policy provides a new avenue for progressive food legislation. The worldwide role of bats in the context of their respective agricultural services and the need for conservation through international trade laws is crucial for the understanding of pollinator regulation—or the lack thereof. This textbox provides an introduction to the links between bat conservation and food law.

Pollinator conservation policy provides a new avenue for progressive food legislation. The worldwide role of bats in the context of their respective agricultural services and the need for conservation through international trade laws is crucial for the understanding of pollinator regulation—or the lack thereof. In fact, the conservation of bats is an important method of promoting organic and sustainable agriculture because bats are wonderfully beneficial animals and over 300 species of agricultural plants depend on bats for pollination, seed dispersal and protection against insects. If the world loses its bat species, there will be an increase in the demand for chemical pesticides, jeopardizing whole ecosystems, and adversely affecting human health and international economies through unsustainable agriculture. This textbox presents data to support the need for the implementation of international trade practices that support bat conservation with the goal to promote sustainable agriculture and models for protective laws of bats.

More than 450 important agricultural bat-dependent plants annually affect hundreds of millions of dollars of international trade. Such economically important bat-dependent plants include bananas, mangoes, vanilla, agave, cashews, dates and figs. Bats facilitate the reproductive success of these agricultural plants, including

G. Steier (✉)
Food Law International, Boston, New York, Washington, DC, USA
e-mail: g.steier@foodlawinternational.com

G. Steier, K.K. Patel (eds.), *International Farm Animal, Wildlife and Food Safety Law*, DOI 10.1007/978-3-319-18002-1_18

seed set and the recruitment of new seedlings and saplings. Many of these bat-dependent plants are among the most important species in terms of biomass in their habitats and are of enormous economic value to humans. Correspondingly, the most important bat-dependent food products are bananas, the fourth most important food product within the least developed countries, the staple food for over 400 million people world-wide and the number one most consumed fruit. According to the United Nations Conference on Trade and Development, international trade in bananas alone has tripled over the past 50 years making food trade revolving around bat-dependent plants a crucial aspect of worldwide agriculture. For the reason that bananas are dependent on bats, it is necessary to conserve bats as pollinators and as natural insecticides to protect the production of both bananas and other bat-dependent food plants.

Although bats account for one fifth of all mammals, little funding and research is devoted to the conservation of these pollinators. Bats are exceptionally vulnerable to extinction and, thereby, crops such as bananas are logically at risk and prone to unsustainable farming practices relying on insecticides and genetically modified organisms. Bats help to promote sustainable agriculture and to preserve wholesome plant-based foods for human consumption and should be protected under international trade agreements. In Europe and Russia, bats are protected by extensive legislation. The European Habitats Directive and the English Wildlife and Countryside Acts, the Natural Environment and Rural Communities Act and the Conservation of Habitats and Species Regulations (2010) all protect bats. By contrast, in the U.S., only endangered bat species fall under federal legislative protection and several agencies' regulatory powers overlap or cancel each other out when it comes to bats. The need to streamline international laws to promote bat conservation has a direct impact on global trade and should fall under the jurisdiction of the World Trade Organization (WTO). Models for international laws to promote bat conservation in the context of sustainable agriculture and environmental protection with the corresponding interdisciplinary research are desperately needed.

Chapter 19
Agriculture and Biodiversity

Amy R. Atwood

Abstract In this chapter, the interrelationship between agriculture, population, and biodiversity is examined within the context of legal frameworks which both seek to regulate agriculture and to slow or stop biodiversity losses. U.S. laws that regulate agriculture and its impacts to biodiversity are compared to those of other nations including Australia, India, China, the European Union, Nigeria, and Cuba. This chapter assesses how and the degree to which these legal frameworks are regulating agricultural methods to mitigate biodiversity losses.

19.1 Introduction

19.1.1 Agriculture, Biodiversity and Population

The Earth's diverse abundance of wild plants and animals has transformed and sustained societies throughout the history of human civilization. This abundant biodiversity has enabled human societies to cultivate crops, to domesticate livestock, to produce food and fiber, and to grow and expand. But as world population has grown, societies' demand for land, food, and fiber has grown too, leading to pressures on and diminishment of the natural world and the Earth's biodiversity.

Civilizations today rely heavily on industrial agricultural methods that include synthetic fertilizers, pesticides, and genetically-modified crop systems. These methods have caused and intensified biodiversity losses by converting habitats, spreading non-native species, and causing climate change. Biodiversity losses now threaten to irreversibly alter the Earth's major systems in ways that will compromise—and perhaps foreclose altogether—societies' ability to sustain a large and growing world population.

In this chapter, the interrelationship between agriculture, population, and biodiversity is examined within the context of legal frameworks which both seek to

A.R. Atwood, J.D. (✉)
Center for Biological Diversity, Portland, OR, USA
e-mail: atwood@biologicaldiversity.org

G. Steier, K.K. Patel (eds.), *International Farm Animal, Wildlife and Food Safety Law*, DOI 10.1007/978-3-319-18002-1_19

regulate agriculture and to slow or stop biodiversity losses. U.S. laws that regulate agriculture and its impacts to biodiversity are compared to those of other nations including Australia, India, China, the European Union, Nigeria, and Cuba. This chapter assesses how these legal frameworks are regulating agricultural methods to mitigate biodiversity losses.

As study after study documents, however, these efforts have largely failed. The Earth is currently experiencing a mass extinction event—its sixth, and the first to be attributed primarily to the activities of mankind. The primary culprit of this event is humankind's growing world population, and the corresponding demands for food and fiber and declines in biodiversity, a vicious negative feedback loop that must be reversed if humans are to have any hope of preserving the web of life that sustains us all.

19.1.2 Modern Agriculture and Its Treatment of the Earth's Biodiversity

Biodiversity is, in essence, the Earth's web of life. The Convention on Biological Diversity ("CBD"), a 1992 international agreement for the conservation and sustainable and equitable use of the Earth's biological diversity, defines "biodiversity" as:

> the variability among living organisms from all sources including, among others, terrestrial, marine and other aquatic ecosystems and the ecological complexes of which they are part; this includes diversity within species, between species and of ecosystems.[1]

Extinction occurs naturally, and by some measures is even common, but the rate and magnitude of natural species extinction is naturally offset by speciation, the evolution of new species.[2]

However, the Earth is currently in the midst of a "mass extinction event," when the rate of extinction is much higher than the rate of speciation—as one study specified, when at least 75 % of species go extinct within a relatively-short period of geologic time, typically 2 million years and in some cases, a much shorter time period.[3] The current extinction rate is estimated to be at least 1000 times higher than the natural background rate of extinction, with future rates likely to be 10,000 times higher.[4] The current mass extinction is the Earth's sixth, and the first to be caused

[1]Convention on Biological Diversity, art. 2, 1992.

[2]Of an estimated four million species to have existed on the Earth over the last 3.5 billion years, 99 % are extinct. This illustrates how, outside of mass extinction events, extinction is largely balanced by the formation of new life forms. Barnosky et al. (2011), p. 51.

[3]*Id.*

[4]De Vos et al. (2015), p. 452; Pimm et al. (2014), p. 1; Pimm et al. (1995), p. 347; Ceballos et al. (2015); Mace et al. (2005), p. 77.

primarily by human activity.[5] Humans may have already wiped out 130,000 animal species, a staggering 7 % of all animal species on Earth.[6]

Since the Neolithic Transition about 12,000 years ago, at the beginning of the Holocene Epoch, hunter-gatherer cultures transitioned to agrarian and pastoral (agricultural) settlements on a wide scale. People domesticated wild plants and wildlife for crops and livestock. With this transition came population growth, and in turn, new innovations in agricultural methods to increase crop yields and to feed growing populations.

Through this feedback loop, agriculture has profoundly shaped and adversely altered the Earth, including its biodiversity, through conversion of habitats to croplands, water use, pesticides, pollution, the spread of pathogens, and increasingly, through genetic modification of crop species. With agriculture, humans have fundamentally altered the Earth's web of life. Indeed, the Earth's current mass extinction event is known as the "Holocene Extinction Era" because it coincides with the Holocene Epoch.[7]

This feedback loop has intensified since the Industrial Revolution, as industrial advances in agriculture have allowed the world population to grow to 7 billion by 2011, with projections that it could reach nearly 11 billion by 2050.[8] This trend has intensified even more during the last 40–50 years, as the human population has doubled since 1970. At the same time, vertebrate species populations across the globe are estimated to be, on average, about half the size that they were 40 years ago.[9] Agriculture and aquaculture—the practice of raising farmed fish for human consumption—threaten over 60 % of all vertebrate species that are classified as threatened, endangered, or vulnerable by the International Union for Conservation of Nature ("IUCN").[10] This is far greater than any other threat.[11]

Some of the most imperiled species are carnivores—such as wolves, coyotes, bears, panthers, leopards, and lions—*i.e.*, large, terrestrial wildlife that are at or near the top of the food web. Carnivores require large prey and expansive territories.[12] These wildlife have some of the most significant beneficial impacts on ecosystems that support crop production through a dynamic called "trophic cascades."[13] By

[5]Barnosky et al. (2011).

[6]Régnier et al. (2015).

[7]The Holocene Extinction Era is also called the "Sixth Extinction." *See* Elizabeth Kolbert, *The Sixth Extinction: An Unnatural History* (2013).

[8]United Nations Population Fund, The State of World Population 2011; United Nations, World Population Prospects, The 2012 Revision (Volume 1): Comprehensive Tables (2013).

[9]WorldOMeters: Population, http://www.worldometers.info/world-population (last visited Mar. 29, 2015); Hooke et al. (2012), p. 4; World Wildlife Fund et al., Living Planet Report 2014: Species and spaces, people and places ("WWF et al. (2014)").

[10]Secretariat of the Convention on Biological Diversity, Global Biodiversity Outlook 139 (Fig. 5.1) (2014), http://www.cbd.int/GBO4 ("GEO5").

[11]*Id.* at 139 (Fig. 5.1).

[12]Ripple et al. (2014), Status and Ecological Effects of the World's Largest Carnivores, *Science*, v. 343.

[13]In contrast, the absence of carnivores creates a dynamic called "trophic downgrading." Estes et al. (2011), p. 301.

preying on and supporting sustainable populations of herbivores—like deer, bison, elk, giraffes, and pronghorn—carnivores indirectly trigger the "release" of vegetative communities that, in turn, provide habitat for small mammals and birds. In this way, carnivores increase the overall diversity of the food chain and the viability of ecosystems which benefit agriculture.[14] Yet, carnivores are often persecuted out of fear or ignorance or intolerance—for instance, by some livestock producers and other large agribusinesses—and their habitats are continually encroached upon by urbanization and human activities.[15] When they are eradicated, measurable declines in the health and functioning of ecosystems occur.

It is in societies' long-term interest to protect all species—not just for their intrinsic value, but also because protecting biodiversity directly and indirectly benefits people and societies. When species go extinct, life forms that existed for thousands and even millions of years are lost. They take with them forever genetic resources and a piece of knowledge of life itself.[16] This loss is irreversible because once lost, species cannot be replaced on a timescale that is meaningful to humans, as speciation, the formation of new life forms through evolution, takes hundreds of thousands of years.[17]

Moreover, when a species is lost from a local area ("extirpated") or goes extinct altogether, the resulting "gap in nature" sets off a cascade of detrimental effects.[18] The ecosystem that the lost species once inhabited destabilizes and becomes less resilient.[19] Invasive species—which are species that are not native to a specific location (also called "nonnative" or "alien" species)—become established, and disrupt the ecosystem further by competing with native species for resources and change predator-prey dynamics.[20] This in turn causes the decline of more native species, which can also become extirpated or go extinct, destabilizing the ecosystem and native

[14]Ripple et al. (2014). For example, following reintroduction of gray wolves to Yellowstone National Park, increased predation by wolves resulted in a reduced and redistributed elk population, which in turn decreased elk consumption of vegetation and increased production of plants that aided (or released) other species. Ripple et al. (2014), p. 223. Due to the ecosystem benefits that carnivore species provide, some scientists have begun to advocate for removing livestock and excess ungulates from public lands—*e.g.*, in the American West—and for restoring ecologically-significant abundances of carnivores in order to mitigate the impacts of climate change. *See* Beschta et al. (2012), p. 474.

[15]Ripple et al. (2014).

[16]As the Congress observed in enacting the Endangered Species Act, discussed below:

From the most narrow possible point of view, it is in the best interests of mankind to minimize the losses of genetic variations. The reason is simple: they are potential resources. They are keys to puzzles which we cannot solve, and may provide answers to questions which we have not yet learned to ask . . . . Who knows, or can say, what potential cures for cancer or other scourges, present or future, may lie locked up in the structures of plants which may yet be undiscovered, much less analyzed? . . . Sheer self-interest impels us to be cautious.

H. R. Rep. No. 93-412, pp. 4–5 (1973).

[17]Barnosky et al. (2011).

[18]Tim Flannery, A Gap in Nature: Discovering the World's Extinct Animals (2001).

[19]Bergstrom et al. (2013), p. 1.

[20]Prugh et al. (2009), p. 779.

species' resilience even more. This unraveling of biodiversity fundamentally alters landscapes, even irreversibly.[21] When this happens, ecosystems degrade, along with their ability to support agriculture. This degradation has consequences for languages and cultures, and indeed, civilization itself.[22]

Extirpation and extinction also disrupt "ecosystem services" like pollination, clean water and air, waste treatment that currently support agriculture and more than 7 billion people.[23] Agriculture itself arises from and depends upon the Earth's biodiversity, as crops are domesticated versions of wild plants and livestock are domesticated wild animals. In these ways, agriculture and biodiversity are inextricably linked, and the Holocene Extinction forebodes dire consequences for agriculture and human societies.

Agriculture impacts biodiversity in the following ways:

- *Conversion of natural habitat to cropland or pasture*. Nearly half of the Earth's total land (non-ice) surface has been converted from species habitat to agricultural use.[24] About 51 % of the total U.S. land area is used for agricultural purposes, including crops and livestock grazing.[25] About one-third of all arable crop land is used to grow feed crops for livestock, like corn, soy, and alfalfa.[26]
- *Industrial agricultural methods*. Monoculture, pesticides, synthetic fertilizers, herbicides, and genetically-modified crops systems cause agricultural runoff and water pollution.[27] These industrial agricultural methods are also directly

---

[21]Barnosky et al. (2012), p. 52 ("Humans now dominate Earth, changing it in ways that threaten its ability to sustain us and other species.") ("Barnosky et al. (2012)").

[22]The inextricable link between linguistic diversity and biodiversity is well-established, particularly in tropical areas. Harmon (1996), p. 108. In light of the Earth's current era of mass extinction, linguists have predicted the up to 90 % of languages will be lost by the end of this century. Gorenflo et al. (2012), p. 8032.

[23]The services provided by the world's ecosystems have been valued at more than the total value of the world's economy, when freshwater purification, pollination, clean air, flood control, soil stability and climate regulation are taken into account. World Health Organization, Ecosystems and Human Well-Being: Health Synthesis 41 (2005). Errol Fuller's 2014 book, Lost Animals: Extinction and the Photographic Record, pictures species that existed for millions of years before they went extinct between 1870 and 2004.

[24]Hooke et al. (2012).

[25]U.S. Department of Agriculture, Economic Research Service, Trends in Major Land Uses, http://www.ers.usda.gov/topics/farm-economy/land-use,-land-value-tenure/background.aspx (last visited Feb. 27, 2015).

[26]Food and Agricultural Organization of the United Nations, Livestock's Long Shadow: Environmental Issues and Options (2006) ("FAO (2006)").

[27]Surface waters in the Mississippi River Basin are "laden with sediment, nitrogen, phosphorous, and/or pesticides"—*i.e.*, "nutrient-laden runoff," which is the principle cause of a large algal bloom in the Gulf of Mexico, where the water is so depleted of oxygen from nutrients that it is devoid of aquatic life. U.S. Department of Agriculture, Farm Service Agency, Draft Supplemental Programmatic Environmental Impact Statement for Conservation Reserve Program 3-14 (2014); Rabalais et al. (2002), p. 235. By 2014, the Gulf of Mexico Hypoxic Zone, also called a "Dead Zone," measured 13,080 square kilometers (5052 square miles), and had a 5-year average of 14,353 square kilometers (5541 square miles). Environmental Protection Agency, Mississippi

implicated in a 90 % decline in the North American population of monarch butterflies since the 1990s, and global declines in honey bees and other pollinators which provide critical ecological services to agriculture.[28] Industrial agricultural practices also deplete soils, leading to heavier reliance on synthetic fertilizers that cause pollution.[29] Concentrated animal feeding operations ("CAFOs"), or factory farms, are a significant source of agricultural runoff as well.[30] Increased efficiencies from industrial agriculture has allowed for a growing world population and plentiful meat from livestock, making what was once an occasional meal is now a staple for much of the world's population. Meat production is also a major contributor to climate change due to methane emissions, nitrous oxide from excreted nitrogen, and synthetic fertilizers utilized by CAFOs.[31]

- *Water development, channelization, and irrigation for crops and livestock.* Dams and impoundments and water diversions for irrigation alter aquatic systems and deplete sources of ground and surface water, leading to declines and

---

River Gulf of Mexico Watershed Nutrient Task Force, Northern Gulf of Mexico Hypoxic Zone, http://water.epa.gov/type/watersheds/named/msbasin/zone.cfm (last visited Mar. 29, 2015). Agricultural runoff is a "nonpoint source" of pollution under the Clean Water Act; as such, it is outside the scope of the primary federal water-protection law in the United States. Laitos and Ruckriegle (2013), pp. 1033–1070.

[28]Center for Biological Diversity, et al., Petition to Protect the Monarch Butterfly (Danaus Plexippus Plexippus) under the Endangered Species Act (2014) ("CBD et al. (2014)"). According to a 2014 Presidential Memorandum, [h]oney bee pollination alone adds more than $15 billion in value to agricultural crops in the United States." Pres. Obama, Presidential Memorandum: Creating a Federal Strategy to Promote the Health of Honey Bees and Other Pollinators (2014). President Obama established a "Pollinator Health Task Force" to "focus federal efforts on understanding, preventing, and recovering from pollinator losses." *Id.* § 2(a).

[29]Food and Agriculture Organization of the United Nations, Agriculture and Soil Biodiversity, http://www.fao.org/agriculture/crops/thematic-sitemap/theme/spi/soil-biodiversity/agriculture-and-soil-biodiversity/en/, last visited Mar. 29, 2015.

[30]Manure from CAFOs contains high amounts of nitrogen and phosphorous, which cause eutrophication and harmful algal blooms, as well as blooms of other organisms that lead to oxygen deficits and in some cases are noxious or toxic to fish or invertebrates. The Cape Fear River Basin in North Carolina has the most CAFOs on Earth, where more than 5 million hogs, over 16 million turkeys, and 300 million chickens are produced annually. *See* Ann Colley, Cape Fear River Watch, http://anncolley.com/2014/06/13/cape-fear-river-watch/ (last visited Apr. 2, 2015). Studies have shown significant degradation to the Cape Fear River Basin from "the vast amounts of raw, untreated animal waste that runs into waterways from swine and poultry factory farms every day." *Id.*; *see also* Mallin (2000), p. 26 (describing fish kill in Cape Fear River Basin in 1995, following breaches of waste lagoons that resulted in leaks of millions of gallons of poultry and swine waste). The Cape Fear River Basin has many threatened and endangered species, including the Cape Fear shiner, shortnose sturgeon, red-cockaded woodpecker, Saint Francis' satyr, West Indian manatee, and loggerhead sea turtle. *See* North Carolina Department of Environment and Natural Resources, Office of Environmental Education and Affairs, Cape Fear River Basin, http://www.eenorthcarolina.org/images/River%20Basin%20Images/final_web_capefear.pdf (last visited Apr. 2, 2015).

[31]United Nations Environmental Programme, Growing greenhouse gas emissions due to meat production (2012), http://www.unep.org/pdf/unep-geas_oct_2012.pdf (last visited Apr. 3, 2015) ("UNEP (2012)").

extirpations of aquatic species.[32] Irrigation for crops is the second-largest use of freshwater in the United States (after thermoelectric power), with groundwater withdrawals for livestock operations and surface water withdrawals for aquaculture accounting for even more withdrawals.[33] Reduced water supplies due to agriculture are exacerbated by drought.[34]

- *Killing of carnivores.* The "lethal control" of carnivores like wolves, coyotes, bears, cougars, leopards, and lions, and others is often justified in order to protect livestock—yet, in addition to being ineffective in reducing livestock-predator conflicts, this can also reduce or eliminate these species' substantial contributions to ecosystem health.[35] This dynamic is even more pronounced when combined with climate change.[36]
- *Climate change.* Agriculture is a significant contributor of greenhouse gases to the Earth's atmosphere, which are causing climate change, a major threat to the Earth's biodiversity. Agriculture contributes greenhouse gas emissions directly, through elimination of carbon stores in soils, rice cultivation, agricultural soil management, and field burning of agricultural residues.[37] Particularly when animal agriculture is factored in, agriculture is "one of the main contributors to the emission of greenhouse gases", constituting as much as 35 % of all global GHG emissions.[38] According to a 2004 study, 15–37 % of certain species may be

[32]*See, e.g.*, Morse, J.C., et al., 1997, Southern Appalachian and Other Southeastern Streams at Risk: Implications for Mayflies, Dragonflies, Stoneflies, and Caddisflies, *in* Aquatic Fauna in Peril: A Southeastern Perspective 17 (George W. Benz & David E. Collins, eds., 1997). For example, water diversions in the Sacramento-San Joaquin Delta in the Central Valley of California are a primary factor in the near-extinction of the Delta smelt, as well as in the decline of longfin smelt, salmon populations, steelhead trout, green sturgeon, and Sacramento splittail. *See, e.g.*, *San Luis & Delta-Mendota Water Auth. v. Jewell*, 747 F.3d 581 (9th Cir. 2014) (upholding a U.S. Fish and Wildlife Service conclusion that irrigation jeopardizes the existence of the Delta smelt, an endangered species).

[33]Maupin, M.A., Kenny, J.F., Hutson, S.S., Lovelace, J.K., Barber, N.L. & Linsey, K.S., 2014, Estimated Use of Water in the United States in 2010; Food and Agriculture Organization of the United Nations, Control of pollution from agriculture.

[34]Intergovernmental Panel on Climate Change, Climate Change 2014 Synthesis Report 53 (2014).

[35]A 2014 study documented a four percent increase in depredations of sheep, and a five-six percent increase in depredations of cattle, following increased killing of wolves the previous year. Wielgus and Peebles (2014), p. 1; Ripple et al. (2014) ("large carnivores are necessary for the maintenance of biodiversity and ecosystem function").

[36]Due to the impacts of livestock on public lands in the American West—which exacerbate the effects of climate change on vegetation, soils, hydrology, and wildlife—some experts have begun to advocate for removing or reducing livestock from large areas of public land and reestablishing apex predators, in order to mitigate climate change. Beschta et al. (2012).

[37]Conversion of natural habitats to agriculture depletes soil organic carbon by 60–75 %, depending on the location. Lal (2004), p. 1623; U.S. Environmental Protection Agency, Inventory of U.S. Greenhouse Gas Emissions and Sinks: 1990–2013 (2015); FAO (2006).

[38]UNEP (2012), p. 4. While studies attribute 10–35 % of global GHG emissions to agriculture when meat production is factored in, the large range is due to inclusion or exclusions of emissions due to deforestation and land use changes. *Id.*

committed to extinction by 2050 based on mid-range climate-warming scenarios, although a 2011 study warned that "we might be vastly underestimating climate change impacts on biodiversity."[39] A 2015 synthesis of concluded that "[i]f we follow our current, business-as-usual trajectory"—*i.e.*, a 4.3 °C rise in temperatures—"climate change threatens one in six species (16 %)."[40]

Many nations have enacted legal frameworks which seek to balance man's activities, including agriculture, with biodiversity conservation. Yet, tens of thousands of species are still being lost every year and the extinction rate remains much higher than the background rate. The world is currently living in "overshoot," when "humanity's demand has exceeded the planet's ... amount of biologically productive land and sea area that is available to regenerate these resources."[41] We are losing nature's genetic resources that provide the food, clothing, and medicine on which human societies depend. In effect, by allowing the biodiversity crisis to continue, millions of people are sentenced to harsher living environments and malnourishment.[42] Biodiversity is a critical resource that humanity simply cannot afford to destroy, yet existing legal frameworks are not sufficiently protecting against potent biodiversity losses from agriculture.

19.1.3 Profiles: Australia, the European Union, India, China, Nigeria, Cuba

This section will examine agricultural practices in the United States and other nations, including Australia, the European Union, India, China, Nigeria, and Cuba. This is followed by a deeper discussion of how agriculture affects biodiversity. Then, U.S. and other national laws which regulate agriculture and agriculture's effects to biodiversity are covered. Finally, how these existing laws and regulatory regimes are failing to adequately protect biodiversity from agriculture's detrimental impacts is explained, with discussion of the ways in which they must improve to protect biodiversity from agriculture's negative impacts.

[39]Thomas et al. (2004), p. 145; Urban et al. (2012), p. 2072; Barnosky et al. (2012).

[40]Urban (2015), p. 571.

[41]WWF et al. (2014), p. 9.

[42]United Nations Environmental Programme, The Environmental Food Crisis: The Environment's Role in Averting Future Food Crises (A UNEP Rapid Response Assessment) (2009) ("UNEP (2009)").

19.1.4 Biodiversity Is Necessary for Agriculture, and Agriculture Is Sustaining the Growing World Population, But Agriculture Is Killing Biodiversity

Agricultural methods in the U.S. and other nations play an important role in global food supply, but agriculture is also a primary driver of biodiversity losses. Domestic and international legal frameworks encourage industrial agricultural methods which are detrimental to species, but although some of these frameworks also seek to prevent, mitigate, or compensate losses in biodiversity, they are not adequate to slow the high extinction rate that is being driven in large part by industrial agricultural methods. Only if these legal frameworks are radically strengthened and fully enforced can they serve domestic and international policies to conserve the Earth's web of life.

19.2 Agriculture, Its Impacts to Biodiversity, Laws That Regulate It, and Laws That Affect Its Impacts to Biodiversity

19.2.1 A History of Agriculture and Its Impacts to Biodiversity

The Holocene epoch—correlating with the Holocene Extinction—dates back about 10,000–12,000 years. During the Neolithic Revolution around that time, many ancient human societies transitioned from hunter-gatherer to agrarian and began to employ agricultural methods like irrigation, crop rotation, and fertilizers. With the cultivation of wild crops and domestication of wild animals came steady increases in population size as well as modifications to natural habitats and biodiversity losses. These transitions allowed settlements to become city- and nation-states, and early legal frameworks emerged to regulate trade and the ownership of property including cropland and livestock.[43] Legal doctrines arose—*e.g.*, in ancient Rome—to protect public rights to clean water and air. These doctrines, later adopted by English and American law, are known as "public trust doctrines."[44]

While expansion of agriculture affected biodiversity through cropland conversion and irrigation, until the Industrial Revolution over the past 200–300 years, agriculture consisted predominantly of what is known today as "organic farming" or "agricultural diversity." These terms refer to agricultural methods that sustain

[43]Ellickson and Thorland (1995).

[44]Sax (1970), pp. 471–475 ("The source of modern public trust law is found in a concept that received much attention in Roman and English law—the nature of property rights in rivers, the sea, and the seashore.").

soils and ecosystems through reliance on "ecological processes, biodiversity and cycles adapted to local conditions, rather than the use of inputs with adverse effects."[45] They are characterized by complexity, large numbers of plants and animals, the harnessing of natural processes rather than artificial inputs such as synthetic fertilizers, and the use and conservation of biodiversity.[46] Organic agriculture methods allowed the world population to grow, steadily but modestly, from about five million in 8000 BC to about one billion in 1800, shortly after the dawn of the Industrial Revolution.[47] Today, these methods—also known as "agro-biodiversity"—are still utilized in some places around the world, including Cuba and certain areas in Africa.

For many societies, however, the Industrial Revolution—and in particular, Green Revolutions from the 1940s through the 1960s—brought new agricultural methods like synthetic fertilizers, pesticides, selective breeding, pollination management, and mechanization. These industrial agricultural methods have emphasized efficiency, minimization of costs, and technology, as they have substituted human and animal labor with machinery and processed fertilizers.[48] As a result of their wide-scale utilization, "the total area of cultivated land worldwide increased 466 % from 1700 to 1980" while crop yields also grew sharply, reducing the cost of grains and livestock feed.[49]

Industrial methods have also allowed the world population to grow sharply, from about 300 million in 1 AD to 790 million by the mid-eighteenth century.[50] Since the dawn of the Industrial Revolution, the total world population and growth rate have spiked, going from about 1 billion in 1800 to 2 billion in 1930, 3 billion in 1959, 4 billion in 1974, 5 billion in 1987, 6 billion in 2000, and 7 billion in 2011.[51] Under

---

[45]Organic World Foundation, Organic Agriculture, http://www.organicworldfoundation.org/organic_agriculture.html (last visited Apr. 3, 2015).

[46]Nadia El-Hage Scialabba, *Case Study No. 4, Organic Agriculture and Genetic Resources for Food and Agriculture, in* Food and Agriculture Organization of the United Nations, Biodiversity and the Ecosystem Approach in Agriculture, Forestry and Fisheries: Satellite event on the occasion of the Ninth Regular Session of the Commission on Genetic Resources for Food and Agriculture (2002); Altieri (1999), p. 19; United Nations Conference on Trade and Development, Trade and Environment Review 2013: Wake Up Before It Is Too Late: Make Agriculture Truly Sustainable Now for Food Security in a Changing Climate ("UNCTAD (2013)").

[47]WorldOMeters.com, http://www.worldometers.info/world-population/ (last visited Apr. 3, 2015).

[48]Peggy Barlett, *Industrial Agriculture, in* Economic Anthropology (Stuart Plattner, ed., 1989).

[49]Matson et al. (1997), p. 504; The Council on Food, Agricultural & Resource Economics (C-FARE), 2012, Future Patterns of U.S. Grains, Biofuels, and Livestock and Poultry Feeding.

[50]United Nations Department of Economic and Social Affairs, Population Division, The World Population Situation in 2014: A Concise Report (2014) ("UNPD (2014)").

[51]United Nations Department of Economic and Social Affairs, Population Division: World Population Prospects (the 2012 Revision): Highlights and Advance Tables (2013); Population Institute, From 6 Billion to 7 Billion: How population growth is changing and challenging our world (2011).

medium-range scenarios, the world population is expected to reach 8 billion by 2025 and 9 billion by 2050.[52]

Changes in agricultural methods and increases in world population over the last 200–300 years also correlate with increases in terrestrial biodiversity losses.[53] Since the 1700s, high extinction rates have been documented in birds, mammals, snails, reptiles, and plants.[54] Some scientists refer to a new epoch, the Anthropocene, as when humans started to have a significant global impact on the Earth's systems, including increases in the extinction rate, roughly correlating with the onset of the Industrial Revolution.

In industrializing societies, legal frameworks accommodated and even promoted these changes. For example, European settlers in North America converted natural habitats to croplands, areas that produce crops for food and fiber for human consumption and feed for livestock, and pasturelands, areas for livestock grazing, with laws and policies which advanced the widespread use of guns, traps, and poisons to deliberately eliminate carnivores (and other wildlife species) considered to be incompatible with livestock production.[55]

In the U.S., these changes lead to an increase in public concern about biodiversity and environmental issues about 40–50 years ago, which lead to enactment in the 1970s of environmental protection laws. These laws include the Endangered Species Act, a U.S. law with a specific purpose of protecting biodiversity, as well as the Clean Water Act, Clean Air Act, National Environmental Policy Act, and others. In 1975, the U.S. and other parties entered into the Convention on International Trade in Endangered Species of Wild Fauna and Flora (the "CITES Treaty"), a multilateral treaty to protect endangered animals and plants, through regulation of international trade.[56]

In 1993, the Convention on Biological Diversity ("CBD"), the formal recognition that biodiversity conservation is "a common concern of humankind," came into effect. The CBD is legally binding, and all signatory countries ("Parties") are required to implement its provisions. As discussed further below, many signatories have also enacted laws to protect biodiversity in their own countries, or to ensure its equitable and sustainable use.

Yet, despite the emergence of such laws, regulations, and treaties that are specifically intended to halt biodiversity losses from mankind's activities, biodiversity losses are still occurring at unsustainable rates far above that of pre-human times.[57] A majority of species in all taxonomic groups are in decline, except species

[52]UNDP (2014), p. 2.

[53]Crutzen (2002), p. 23.

[54]Pimm et al. (1995).

[55]Wielgus et al. (2014); Bergstrom et al. (2013); Michael J. Robinson, Predatory Bureaucracy: the Extermination of Wolves and the Transformation of the West (2005).

[56]Convention on International Trade in Endangered Species of Wild Fauna and Flora, What is CITES?, http://www.cites.org/eng/disc/what.php (last visited Apr. 3, 2015).

[57]Millennium Ecosystem Assessment, Ecosystems and Human Well-being: Biodiversity Synthesis 3–4 (2005) ("MA Assessment") at 3–4; Pimm et al. (2014); De Vos et al. (2015).

that survive or thrive in human-altered environments, have been protected in reserves, or have had particular threats eliminated.[58] An estimated 87 % of the Earth's bird species are threatened by industrial agriculture globally.[59] Amphibians are "declining globally at an alarming rate" due to pesticides that are used in agriculture.[60] Species are also becoming homogenous, meaning that average differences between sets of species in different locations are diminishing.[61]

As endemic species are lost, invasive species move in and quickly grow, reproduce, and disperse, further destabilizing species communities.[62] Invasive species are typically highly-adaptive and able to tolerate a wide range of environmental conditions.[63] For example, the brown tree snake was introduced to Guam in the late 1940s or early 1950s, and has decimated Guam's native, forest-dwelling birds.[64] Feral pigs, which are released domesticated pigs, breed prolifically and consume large numbers of land tortoises, sea turtles, endemic reptiles, and sea birds.[65] Bullfrogs are native to the eastern U.S. but have been introduced to the Southwest, where they have wreaked havoc on native frogs, western pond turtles, and Mexican garter snakes.[66]

Despite legal frameworks to regulate agriculture and to stem biodiversity losses, 30 % of the Earth's land surface has been converted to cropland.[67] Biodiversity losses are projected to continue at very high rates, with changes in ecosystem services and particularly-significant declines in vascular plants due to habitat losses in tropical areas.[68]

Paradoxically, biodiversity losses jeopardize agriculture itself. Where wild species of plants and animals are the foundational underpinnings of agriculture, losses of ecosystem services—like organic waste disposal, soil formation, biological nitrogen fixation, crop and livestock genetics, pest control, plant pollination, and pharmaceuticals—will lead to the need for artificial services to replace them.[69] For

---

58 MA Assessment (2005), at 3.

59 UNEP (2009), at 65 ff.

60 Hayes et al. (2002), p. 5476; Hayes et al. (2006).

61 MA Assessment (2005), p. 4.

62 *Id.*

63 Invasive Species Specialist Group, 100 of the World's Worst Invasive Alien Species: A selection from the Global Invasive Species Database (2004) ("ISSG (2004)").

64 *Id.* at 4.

65 *Id.* at 8.

66 Rosen, P.C. and Schwalbe, C.R., 1995, Bullfrogs: introduced predators in southwestern wetlands, *in* U.S. Department of the Interior, National Biological Service, Our living resources: a report to the nation on the distribution, abundance, and health of U.S. plants, animals, and ecosystems (1995); 71 Fed. Reg. 56,228, 56,231 (Sep. 26, 2006) (bullfrogs have "contributed to the decline of northern Mexican gartersnakes in New Mexico").

67 United Nations Environment Programme, Global Environmental Outlook (GEO-5) (Chapter 5, Biodiversity) 7 (2012) ("GEO-5"); WWF et al. (2014), at 164.

68 MA Assessment (2005), at 62.

69 Pimental (1997), p. 747.

example, the loss of pollinators from monocultural, genetically-modified crop systems will require more intensive use of pesticides to perform the services that pollinators no longer can, leading to even further reductions in biodiversity.[70]

Reductions in genetic diversity of domestic crops and livestock—the so-called "crop wild relatives," which contribute resistance to pests and disease, proteins and vitamin content—are expected as well.[71] Crop wild relatives ("CWR") are plant species that are closely related to, and include most of the progenitors of, our domesticated agricultural crops.[72] CWRs contribute significantly to modern agriculture by increasing nutritional value and providing genetic material and resistance to pests, diseases, drought, and extreme weather.[73] In these ways, CWRs are critically important to agriculture economy. For example, it was estimated in 1998 that improvements to commercial tomato varieties from a single tomato wild relative was worth $250 million per year to California producers.[74] On the other hand, "[l]ack of biodiversity leaves major crops vulnerable to disease, causing famines and starvation."[75]

Yet, like all biodiversity, crop wild varieties are increasingly at risk from: (1) habitat loss, degradation, and fragmentation; (2) industrial agricultural methods, which are causing reductions in crop wild relatives near croplands; and climate change, with one study predicting that climate change will cause 16–22 % of the wild relatives of three crop types—peanut, potato, and crowpea—to go extinct by 2055, with 50 % of these CWRs losing most of their range size by then.[76]

In many ways, industrial agriculture is driving biodiversity losses. Agriculture has consumed 30 % of the Earth's land base. Genetically-modified crop systems rely heavily on highly-toxic pesticides that are causing global crashes in invertebrates, amphibians, pollinators, and CWRs. Carnivores are still be targeted and killed to reduce livestock depredations, even though it is doubtful that these efforts are even effective in addressing the problem they are intended to solve, and despite the tremendous benefits such animals contribute to ecosystem health. Agriculture is a major contributor of greenhouse gases, which are the cause of climate change, which will accelerate threats to biodiversity (including CWRs). In the absence of substantial changes to the status quo, industrial agriculture will inevitably lead to reductions in biodiversity.[77]

---

[70] *Id.*; *see also*, *e.g.*, Altieri, M.A. & Funes-Monzote, F.R., The Paradox of Cuban Agriculture (1999).

[71] MA Assessment (2005), at 5; *see also* UNEP (2009), at 65 ff.

[72] UNEP (2009), at 74.

[73] *Id.*

[74] *Id.*; *see also* Food and Agriculture Organization of the United Nations, Commission on Genetic Resources for Food and Agriculture, The Second Report on The State of the World's Plant Genetic Resources for Food and Agriculture (2010).

[75] Institute of Science in Society and Third World Network, Food Futures Now: Organic, Sustainable, Fossil Fuel Free (2008).

[76] UNEP (2009), p. 74; Jarvis et al. (2008), p. 13.

[77] MA Assessment (2005), at 5.

19.2.2 *Agricultural Laws in the U.S. and the Profiled Nations*

19.2.2.1 Agriculture Law and Policy in the United States: From Homesteads to Farm Bills and the Conservation Reserve Program

U.S. agricultural policy has expanded over time, from policies and laws that encouraged cultivation of all arable lands, to policies and laws to actively promote conservation of soil, wildlife, and other resources that are beneficial to agriculture.

Homestead Acts During the 1800s and early 1900s, Homestead Acts promoted settlement, cultivation, and grazing by offering grants of federal public lands in new U.S. territories in the western part of the country. They involved grants of federal land of minimum acreages to qualifying adult U.S. citizens at set prices in exchange for certain commitments.

For example, under the Homestead Act of 1862, a settler could claim up to 160 acres of unappropriated public lands after filing an affidavit attesting, among other things, that the land would be used "for the purpose of actual settlement and cultivation."[78] If these requirements were met, the settler would receive title to the land. Under the Stock-Raising Homestead Act of 1916, a settler could lay claim to 640 acres of surface public land for livestock ranching.[79] There were no limits placed on the activities that were promoted by these laws, such as limits to protect the natural values of the land or water. Homesteading remained a part of U.S. agricultural policy until 1976, when the Federal Land Policy and Management Act, 43 U.S.C. §§ 1701–1787, was passed and ended the practice, except in Alaska, where it continued until 1986.[80]

Only about 40 % of settlers succeeding in meeting these criteria and gained title to land—still, by the early twentieth century, nearly all of the prime areas for cropland had been claimed.[81] The Enlarged Homestead Act, enacted in 1909, increased the maximum acreage in order to encourage dryland farming, especially on the Great Plains, a policy and legal framework which led to widespread erosion and the Dust Bowl of the 1930s.[82] Farming in the Great Plains has been a major

---

[78] Act of May 20, 1862, 12 Stat. 392.

[79] 43 U.S.C. § 299.

[80] U.S. Department of the Interior, Bureau of Land Management, Evolution of Homestead Laws: When did Homestead laws first being changing and why?, http://www.blm.gov/wo/st/en/res/Education_in_BLM/homestead_act/opportunities/evolution.print.html (last visited Apr. 3, 2015).

[81] U.S. Department of the Interior, National Park Service, Homesteading by the Numbers, http://www.nps.gov/home/learn/historyculture/bynumbers.htm, last visited (Apr. 3, 2015); Bradsher, G., 2012, How the West Was Settled: The 150-Year-Old Homestead Act Lured Americans Looking for a New Life and New Opportunities, *Prologue*.

[82] Hansen, Z.K. & Libecap, G.D., U.S. Land Policy, Property Rights, and the Dust Bowl of the 1930s, *FEEM Working Paper No. 69.2001*, http://papers.ssrn.com/sol3/papers.cfm?abstract_id=286699 (last visited Apr. 3, 2015).

contributing factor to declines in the native, tallgrass prairie vegetation—today, it is estimated that only about 1 % of the tallgrass prairie survives, and almost exclusively in areas that are unsuitable for agriculture.[83]

U.S. Farm Bills and the Conservation Reserve Program In response to the Dust Bowl and Great Depression, the U.S. began to enact laws during the 1930s which sought to regulate agricultural policy by managing the supply of crops and livestock, commonly known as "Farm Bills." The Soil Conservation and Domestic Allotment Act of 1936 and the Agricultural Adjustment Act of 1938, for instance, provided price and income support for farm producers, or agricultural subsidies, which continue today.[84] These laws also incorporated policies and provisions to conserve "national resources," especially "soil fertility," by encouraging "soil-conserving crops" and "soil-rebuilding practices" and not "soil-depleting crops."[85] Thus, beginning with policies and laws to protect soil fertility, U.S. agricultural began to shift toward conservation—not to conserve biodiversity *per se*, but to protect agriculture itself.

U.S. agricultural policies and laws expanded following the Great Depression. "Soil banks" established in the 1950s paid farmers to move lands with highly-erodible soil out of agricultural production and to convert them to conservation areas.[86] In response to agricultural methods which maximized land area used for growing crops during the 1970s, in the Food Security Act of 1985 (the 1985 Farm Bill), the U.S. created the Conservation Reserve Program ("CRP"), a federal agency within the Agriculture Department that administers and funds conservation programs for soil, water, and wildlife habitat.[87] The CRP, which still exists, provides cost-share and payments to landowners that retire highly-erodible, unsuitable

[83]U.S. Department of the Interior, U.S. Fish and Wildlife Service, Division of Conservation Planning, Northern Tallgrass Prairie National Wildlife Refuge HPA Environmental Impact Statement Summary, http://www.fws.gov/midwest/planning/northerntallgrass/ (last visited Apr. 3, 2015).

[84]Soil Conservation and Domestic Allotment Act of 1936, 16 U.S.C. §§ 590a-590q-3 (2013); Agricultural Adjustment Act of 1938, 7 U.S.C. §§ 1281–1407 (2013).

[85]16 U.S.C. § 590h(b)(5)(F)(iv) ("In carrying out this section, the Secretary [of Agriculture] shall ... in every practical way, encourage and provide for soil-conserving and soil-rebuilding practices."); 7 U.S.C. § 1282 ("It is declared to be the policy of Congress to continue the Soil Conservation and Domestic Allotment Act, as amended [16 U.S.C. 590a et seq.], for the purpose of conserving national resources, preventing the wasteful use of soil fertility, and of preserving, maintaining, and rebuilding the farm and ranch land resources in the national public interest . . . .").

[86]Soil Bank Act of 1956, 7 U.S.C. §§ 1831–1837 (1956).

[87]Food Security Act of 1985, P.L. 99–198. Farm Bills are comprehensive, 5-year omnibus bills that are the primary legal mechanism over agriculture and food policy in the U.S. Started during the 1930s, agricultural and food policies are still regulated primarily through Farm Bills today.

lands from agriculture and establish long-term cover like grasses and trees.[88] The 1985 Farm Bill authorized the Agriculture Department to enroll up to 45 million acres in the CRP, although this number has yet to be reached.[89]

Since 1985, the CRP has expanded to include wetlands, riparian buffers, shelterbelts, and other areas. Yet, the CRP is handicapped by other agriculture and food policy incentives that seek to maximize the area of cropland, and which favor monoculture, pesticides, and genetically-modified crop systems.[90] The CRP does not have sufficient funds for more than the most-qualified land, and the 2014 Farm Bill reduces allowable acreage from 32 to 24 million acres by 2018.[91] Consequently, marginal lands are more likely to be maintained as cropland when they might otherwise be utilized to restore habitat and recover native species.

Regulation of Pesticides U.S. law also encompasses a framework for regulation of pesticides pursuant to the Federal Insecticide, Fungicide, and Rodenticide Act, 7 U.S.C. §§ 135-136y, *as amended* ("FIFRA"), which requires the Environmental Protection Agency ("EPA") to register, review, and oversee the use of chemicals as insecticides, herbicides, fungicides, rodenticides, fumigants, and other pesticides in the United States. Under FIFRA, a pesticide product generally may not be sold or used in the U.S. without an EPA registration for a particular use.[92] EPA may register a pesticide after finding (among other things) that application of the pesticide will not cause "unreasonable adverse effects on the environment."[93]

Subsidies The U.S. government also distributes billions of dollars in agricultural subsidies, which favor the largest agricultural producers. These subsidies facilitate an industrial agricultural system in the United States that relies heavily on large, uniform monocultural systems, genetically-modified, herbicide-resistant crops, and increasing use of herbicides like glyphosate, which is implicated in a 90 % decline of monarch butterflies since the 1990s.[94] These systems, in turn, allow for growth of crops that provide feed for livestock—a significant contributor to climate change—and have recently been deemed to be a cause of cancer in humans.[95]

---

[88]National Sustainable Agriculture Coalition, Conservation Reserve Program, Taking environmentally sensitive land out of production and establishing long-term ground cover, http://sustainableagriculture.net/publications/grassrootsguide/conservation-environment/conservation-reserve-program/ (last visited May 2, 2015).

[89]*Id.*

[90]*Id.*

[91]*Id.*

[92]7 U.S.C. § 136a(a).

[93]7 U.S.C. § 136a(c)(5)(D).

[94]CBD et al. (2014).

[95]World Health Organization, International Agency for Research on Cancer, Carcinogenity of tetrachlorvinphos, parathion, malathion, diazinon, and glyphosate (2015).

19.2.2.2 National Agricultural Policies and Laws in the Profiled Nations

Other nations have agricultural policies that support agricultural producers, including through market interventions or subsidies, national plans, and emphasis on increasing food production in a sustainable way. However, as described below, "sustainable" agriculture in such policies often refer to the sustainability of the *agriculture*, and not, *e.g.*, sustainability of species or biodiversity.

Australia Australia has a variable climate and is prone to weather extremes, including droughts and extreme rains.[96] There, agricultural policy has favored less government intervention in markets. However, pollution is considered a "negative externality" that warrants government intervention in agricultural markets, and the Australian government will actively manage environmental resources during periods of extreme drought in order to maintain agricultural productivity. The national government also administers a water trading system that is designed to allow agricultural producers to stay productive despite drought conditions.[97] Australian water policy allocates "environmental water."[98] Due to dry conditions, monoculture is difficult in Australia as it lowers soil fertility, and as a result there is an emphasis on organic farming methods.[99]

Concerns over water supply in Australia are growing. Where droughts were already a consistent issue, climate change is expected to increase the frequency of droughts in Australia.[100] With increasing global temperatures, extreme droughts are expected to occur every 2–4 years, rather than the typical rate of about once every decade.[101] Farmers and rural communities will likely be hardest hit, particularly in the southeastern region of the country, although expected higher food prices, particularly for fresh fruit and vegetables, will be felt by nearly everyone.[102]

Meanwhile, higher food prices will be confronted by a growing population. Although Australia ranks 51st in the world in population size, its population is expected to grow, and its *per capita* ecological footprint—measured as "individual impacts on the environment through ... consumption of natural resources"—is one of the largest in the world.[103]

---

[96]Quiggin, J., *Drought, Climate Change and Food Prices in Australia* (2014).

[97]Australian Bureau of Agricultural and Resource Economics – Bureau of Rural Sciences (ABARE–BRS), Agricultural and food policy choices in Australia (2010).

[98]*Id.*

[99]FiBL and IFOAM, The World of Organic Agriculture: Statistics and Emerging Trends (2012).

[100]Quiggin (2014).

[101]*Id.*

[102]*Id.*

[103]WorldOMeters.com, World Population by Country, http://www.worldometers.info/world-population/population-by-country/ (last visited Apr. 4, 2015) ("WorldOMeters.com Population by Country"); Australian Bureau of Statistics, 3222.0—Population Projections, Australia, 2012 (base) to 2101, http://www.abs.gov.au/ausstats/abs@.nsf/Lookup/3222.0main+features52012%20(base)%20to%202101 (last visited Apr. 4, 2015); Natural Resource Management Ministerial Council, Australia's Biodiversity Conservation Strategy 2010–2030 (2010).

European Union A vast continent comprised of many connected societies, European agricultural history and policy is a complex subject with a long history.[104] Generally speaking, Europe's modern agricultural policies originate in the 1700s and 1800s with a series of agricultural and industrial revolutions. During this time, in response to diminishing available cropland, farmers in many European nations increased utilization of crop rotation, mechanized agricultural methods, and fertilizers to increase crop yields for population growth and industrialization.[105]

Today, European agricultural policies are primarily set by the European Union, an economic and political union of 28 member nations that was formed in 1993.[106] The precursor to the E.U., the European Economic Community ("EEC"), was established by the Treaty of Rome in 1957 among six nations: France, West Germany, Italy, the Netherlands, Belgium, and Luxembourg. From the 1960s until the 1980s, the EEC adopted a "common agricultural policy" ("CAP") that through a system of price regulation sought to provide affordable food for members' citizens and a standard of living for farmers.[107]

These policies led to food surpluses by the early 1980s.[108] Following a series of reforms in the 1990s and 2000s, EU policies turned toward "producer support"—*i.e.*, market safety nets and direct payments to farmers to incentivize sustainable farming practices and rural development.[109] CAP expenditures today account for about 40 % of the total E.U. budget.[110] These policies are intended, in part, to preserve biodiversity (along with rural development and sustainable farming), in response to a near-50 % decline in once-common, "farmland" birds in Europe over the past 30 years from industrial agricultural methods.[111]

Despite these policies, however, Europe continues to face environmental challenges that affect agriculture, including poor soil and water quality and declining biodiversity.[112] In 2013, following years of debate and legislative proposals, the

---

[104] In addition, as one author has noted, many historical aspects of the Green Revolution in Europe during the late 1800s and early 1900s has been ignored by development policymakers. *See, e.g.*, Jonathan Harwood, Development policy and history: lessons from the Green Revolution (2013) ("Harwood (2013)").

[105] van Zanden (1991), p. 215. However, as Harwood argues, increased crop yields did not necessarily correlate with reduced rural poverty in southern Germany. Harwood (2013).

[106] European Union, How the EU Works, http://europa.eu/about-eu/index_en.htm (last visited Apr. 4, 2015); European Union, Countries, http://europa.eu/about-eu/countries/index_en.htm (last visited Apr. 4, 2015).

[107] David R. Steade, Common Agricultural Policy, http://eh.net/encyclopedia/common-agricultural-policy/ (last visited Apr. 4, 2015) ("Steade (2015)"); Cunha et al. (2013).

[108] Steade (2015); Cunha et al. (2013).

[109] The European Commission, The European Union Explained: Agriculture, A partnership between Europe and farmers (2014) ("EU Commission (2014)") at 4.

[110] *Id.* at 7.

[111] UNEP (2009).

[112] European Commission, Agricultural Policy Perspectives Brief: Overview of CAP Reform 2014–2020 (2013) ("CAP Reform Overview").

E.U. decided to focus its policies on securing food while "preserving the natural resources that agricultural productivity depends upon."[113] In particular, the E.U. adopted new measures that require maintenance of open and ecological areas and crop diversification in exchange for CAP direct payments to farmers.[114]

These policies are intended to support continued population growth.[115] The E.U. currently has combined population of about 508 million people, spread among all 28 member states, which would rank third among all nations, after China and India.[116] The E.U. population is projected to age over time, including in the agricultural sector, resulting in lower fertility rates, with 525.7 million people by 2035, decreasing to about 517 million people by 2060.[117]

India India is one of the first civilizations to have domesticated wild plants for cultivation. It has a rich biological heritage, and is the origin of many cultivated crops and spices that are grown around the world today, including rice, millet, lentils, sugarcane, banana, yam, cotton, eggplant, tea, turmeric, cardamom, black pepper, and cinnamon.[118]

During the 1960s and 1970s, India went through a "green revolution"—a period of increased reliance on industrial agricultural methods—in response to famines during the 1940s–1970s, when food shortages caused by World War II resulted in the deaths of millions of people from hunger.[119] Green revolution policies emphasized increased crop yields and food production by expanding agricultural areas and the use of double-cropping, improved seed genetics, high-yield crop varieties, synthetic fertilizers, and irrigation.[120] Combined with policies favoring deregulation during the 1990s, which further industrialized the agricultural sector, India has become a top producer of the world's milk, cashew nuts, coconut, tea, wheat, rice, tobacco, and fruit.[121] India also has the world's largest cattle population, and its livestock production continues to grow even while crop production has stagnated.[122]

---

[113]*Id.* at 3.

[114]*Id.* at 7.

[115]EU Commission (2014), at 15.

[116]WorldOMeters.com Population by Country (2012).

[117]EuroStat, File: Projections for Population and Density 2011–2060, http://ec.europa.eu/eurostat/statistics-explained/index.php/File:Projections_for_population_and_density,_2011_to_2060_(1).png (last visited Apr. 4, 2015); CAP Reform Overview (2013), at 7 (noting that only 14 % of E.U. farmers are under 40 years of age).

[118]Food and Agriculture Organization of the United Nations, State of Plant Genetic Resources for Food and Agriculture in India (1996–2006): A Country Report (2007).

[119]*The Green Revolution*, *in* India: A Country Study (James Heitzman and Robert L. Worden, editors, 1995).

[120]Shailesh Nagar and Jayesh Bhatia, *Climate Change and Agriculture*, *in* State of India's Livelihoods Report 2010: The 4P Report (Sankar Datta and Vipin Sharma, eds., 2010).

[121]*Id.*

[122]*Id.*

However, such intensive agricultural methods have also resulted in degraded soil conditions. All high-quality land has been utilized for agricultural land, and agricultural land has become fragmented by urbanization, leaving very little room for natural habitats. Nevertheless, India's agricultural policies are driven by the practical realities that 50 % of the country's population depends on agriculture for its livelihood (in contrast to less than one percent of the U.S. population) and that 100 % of the population depends on agriculture for its food security.[123]

The Indian government has adopted policies to strengthen India's agricultural sector. In 2000, the government adopted the National Agriculture Policy, which promotes growth in the agricultural sector, methods that are sustainable environmentally and economically, marketing and export of agricultural commodities, labor protections, and protection of the gene pool.[124]

China Like India, China has a rich biological heritage, with a large variety of ecosystems and many different species of plants, vertebrate animals, and fungi as well as domesticated livestock, cultivated crops, and fruit trees.[125] China is the origin of widely cultivated crops like rice and soybeans. China also has hundreds of species of domesticated animals. Although it has a long history as a civilization, its current governmental structure is still developing.[126] Consequently, China is still developing legal frameworks to regulate agriculture.

China's current agricultural legal framework promotes agricultural production in order to provide food security for its large population. The Agriculture Law of the People's Republic of China of 2002 seeks to strengthen agriculture as the foundation of China's national economy by promoting sustained growth of rural economies and increasing production and modernizing methods, while safeguarding farmers.[127] As a part of its policies to invigorate rural economies, Chinese agriculture law directs farmers to increase use of organic fertilizers "in order to protect and improve soil fertility" and to prevent "pollution, destruction, and soil fertility declination."[128]

The Animal Husbandry Law of the People's Republic of China of 2005 regulates livestock production and operation, seeks to protect the "genetic resources" of livestock and poultry, and to "promote sustained and healthy development of animal husbandry."[129] Forests are protected for fiber and grasslands are protected for livestock forage.[130] The development of new GMO crops is actively

[123]*Id.*

[124]*Id.*

[125]China National Biodiversity Conservation Strategy and Action Plan 5 (2011–2030).

[126]Michael T. Roberts, *Introduction to Food Law in the People's Republic of China* (2007).

[127]Agriculture Law of the People's Republic of China of 2002 at Art. 1.

[128]*Id.* at Art. 58.

[129]Animal Husbandry Law of the People's Republic of China at Art. 1.

[130]Forestry Law of the People's Republic of China; Grassland Law of the People's Republic of China.

encouraged—indeed, the Chinese government has officially approved many varieties of GMO crops.[131] Thus, agricultural policies are centered around production and, to the extent that China's agriculture law contemplates environmental concerns, it does so in furtherance of its central policy of production.

Nigeria Nigeria's agricultural policies have emphasized high agricultural production to feed the fast-growing population of Africa's most populous country.

Following independence from British colonial rule in 1960, Nigeria promoted food security through a system of federal development guidelines and plans that were implemented by states.[132] Under these policies, Nigeria became the world's top producer of rubber, groundnuts, and palm oil, and the second-largest producer of cocoa.[133] From 1970 to 1986, agricultural policies took a back seat to intensive petroleum exploitation, however, and agricultural production declined.[134] Following through a major food crisis in 1976, Nigeria established policies to increase food production through subsidies for fertilizers, but these programs were inconsistent.[135]

In 1998, Nigeria re-focused its agricultural laws on ensuring food security through local production.[136] A 2001 policy further aimed to increase local production, with an emphasis on sustainable agricultural production and environmental protection along with increased production of raw materials, exportation of crops, job creation, and protection of agricultural resources from drought, desert encroachment, erosion, and flooding. As with other nations, sustainability policies in Nigeria have been designed around maintaining reliable, self-sufficient outputs from the agricultural sector.

Thus, biodiversity protection in Nigeria has thus far submitted to domestic food production for population growth (and oil production). Despite these efforts, Nigeria's large population has put demands on biological resources and arable land, leading to deforestation, soil deterioration, and increased use of synthetic fertilizers, herbicides, and pesticides.[137]

Cuba As a small island nation in the Caribbean, Cuba's biodiversity and agricultural history have been largely shaped by geography. Called the "Accidental Eden,"

---

[131]Regulations on Administration of Agricultural Genetically Modified Organisms Safety (2011).

[132]Douillet and Grandval (2010), p. 16.

[133]*Id.*

[134]*Id.*

[135]*Id.*

[136]*Id.*

[137]Nigeria: National Biodiversity Strategy and Action Plan (NBSAP) (2010) ("Nigeria NBSAP") at 26; *see id.* ("Thus, the increasing population growth has become very crucial among the set of factors that degrade the environment and threaten biodiversity.")

Cuba has thousands of endemic species and is considered the most naturally diverse nation in the Caribbean.[138] Its geographic isolation has protected it from much environmental destruction, although losses have occurred primarily from the introduction of nonnative species.

After Christopher Columbus landed on what is now a Cuban island in 1492, native people were forced into a slavery system, whereby Spanish settlers offered protection in exchange for agricultural production that was the result of hard labor by enslaved natives. Native families were broken up and died out, and African slaves were used in their place.[139] Tobacco and sugarcane became Cuba's primary products.[140]

Cuba gained independence in 1902, and was ruled by military figures for decades until the Cuban Revolution in the 1950s.[141] During this time, sugar accounted for 82 % of the nation's exports.[142] Fidel Castro assumed control and declared Cuba a socialist state in 1961, beginning a 30-year period of close alliance with the Soviet Union, which provided Cuba with favorable trade subsidies and synthetic fertilizers in exchange for sugarcane.[143] Cuba imported about half of its food during this time.[144] When the Soviet Union collapsed in 1991, however, Cuba lost these benefits, and fell into a period of near-isolation, economic hardship, food shortages, and starvation called the "Special Period" which lasted through the late 1990s.[145] During this time, Cuba's manufacturing outputs decreased by 28 %, exports declined by 79 %, imports declined by 75 %, the gross domestic product fell by more than 40 %, real wages in urban areas dropped by 40 %, and personal consumption declined 15 % each year until 1994.[146]

---

[138]Cuba: The Accidental Eden, A Brief Environmental History, *Oregon Public Broadcasting* (Apr. 7, 2011), http://www.pbs.org/wnet/nature/cuba-the-accidental-eden-a-brief-environmental-history/5830/ (last visited May 2, 2015) ("OPB (2011)").

[139]Background Notes on Selected American Countries from the Department of State, http://www.shsu.edu/~his_ncp/labn.html (last visited May 2, 2015) ("Background Notes (2015)").

[140]*Id.*

[141]*Id.*

[142]Gonzalez (2003), p. 685.

[143]Background Notes (2015); A. Buncombe, Cuba's agricultural revolution an example to the world, *The Independent* (Aug. 12, 2006), http://www.seattlepi.com/local/opinion/article/Cuba-s-agricultural-revolution-an-example-to-the-1211460.php#page-2 (last visited May 2, 2015).

[144]J. Mark, Growing it alone: Urban organic agriculture on the island of Cuba, *EarthIsland Journal* (Spring 2007), http://www.earthisland.org/journal/index.php/eij/article/growing_it_alone/ (last visited May 2, 2015) ("Mark (2007)").

[145]A. Martin, A Different Kind of Revolution – What We Can Learn From Cuba (Nov. 2, 2014), http://www.collective-evolution.com/2014/11/02/a-different-kind-of-revolution-what-we-can-learn-from-cuba/ (last visited May 2, 2015). As one commenter has observed, with the demise of the Soviet Union, Cuba lost almost $6 billion in annual Soviet subsidies, its main source of imports, and over 85 % of foreign markets. *See* Brook (2004), p. 207.

[146]Brooks (2004) (internal quotations omitted).

Things began to change in Cuba in 1993, when the Cuban government prioritized food production by redistributing land and supporting farmer cooperatives and farmers markets. The government's enactment of the Decree Law No. 142 transformed the state farms into new units of agricultural production known as Basic Units of Cooperative Production or UBPCs (Unidades Basicas de Produccion Cooperativa).[147] Without access to synthetic fertilizers, UBPCs and "organopónicos"—mixed, organic urban farms—proliferated, and became a leading example of sustainable agro-biodiversity and self-sufficiency.[148] Under these systems, farmers provide a small percentage of their production, about 30 % into the farm, and the rest is to workers as revenue.[149] Under Cuba's now-communist style of government, UBPCs, organopónicos, and other cooperatives are publicly owned, although some private ownership of agricultural production is also allowed.[150] Decree Law No. 191, another reform undertaken by the Cuban government 1994, opened up agricultural markets—farmers' markets—to excess food production, in order to create incentives to improve food distribution and production.[151]

These agricultural policies provide an end to Cuba's Special Period. And although Cuba still imports many agricultural commodities—such as wheat, which does not grow well in the Caribbean climate—it produces most of its own fruits and vegetables, and much of its meat.[152]

19.2.3 The Emergence of Laws to Protect Biodiversity

As made evident by the legal frameworks in the U.S. and the other countries that are highlighted in the previous section, agricultural laws do not tend to regulate agriculture for the purpose of biodiversity protection directly—indeed, to the extent that laws incorporate policies favoring sustainability or environmental concerns, those policies value sustainability of resources insofar as they are important to maintaining sustainability high agricultural production. Thus, biodiversity protection will be addressed, if at all, through laws specific to the biodiversity resource. Where biodiversity-protection laws do exist, they differ in their coverage, mechanisms, and policies.

---

[147]Gonzalez (2003).

[148]R. Patel, What Cuba Can Teach Us About Food and Climate Change, *Slate* (May 2, 2015), http://www.slate.com/articles/health_and_science/future_tense/2012/04/agro_ecology_lessons_from_cuba_on_agriculture_food_and_climate_change_.html (last visited May 2, 2015).

[149]R. Southmayd, Cuban Farm Creates Good Life in a Poor Place, *The Pulitzer Center* (Jan. 31, 2013), http://pulitzercenter.org/reporting/cuba-havana-alamar-agriculture-coop-sustainable-organic-garden-farm-model-OVA (last visited May 2, 2015).

[150]Gonzalez (2003).

[151]*Id.*

[152]Mark (2007).

To examine this further, descriptions of laws in the U.S., Australia, the European Union, India, China, Nigeria, and Cuba are set forth below, and how they seek to protect biodiversity from agriculture's detrimental impacts in particular. As shown below, these nations approach biodiversity protection in a variety of ways, yet still grapple with the negative feedback loop of growing human populations, intensive agricultural production, and biodiversity losses.

19.2.3.1 U.S. Laws to Protect Biodiversity

The U.S. has some of one of the strongest and most comprehensive federal laws to protect biodiversity of any nation: the Endangered Species Act. Below, an overview of the Act's key provisions is set forth, followed by a description of how regulates agricultural activities. This section also illustrates how the ESA interrelates with another federal law, FIFRA which is designed to regulate pesticides, in part to protect the environment from their harmful effects. Nevertheless, the interrelationship between the Endangered Species Act and FIFRA illustrates how, despite its strength, the law can clash with laws that regulate industrial agricultural methods.

The Endangered Species Act Indisputably, no other domestic or international law does more to address the impacts of agriculture to biodiversity than the Endangered Species Act, 16 U.S.C. §§ 1531–1544, as amended ("ESA"), which was enacted in 1973. Since the ESA was passed, many other nations have passed legal and regulatory frameworks to further biodiversity conservation, but the ESA remains one of the strongest biodiversity protection laws—indeed, one of the strongest environmental laws—ever to be enacted by any nation.[153] To explain why, set forth below is an overview of the ESA's purposes, listing provisions, and substantive protections.

To begin, the core purpose of the ESA is "conservation"—specifically, conservation of endangered and threatened species and the ecosystems on which they depend.[154] Conservation means "all methods that can be employed to 'bring any endangered species or threatened species to the point at which the measures provided pursuant to this [Act] are no longer necessary.'"[155] Conservation under the ESA generally encompasses two concepts: (1) survival of species that are on the brink of extinction, and (2) recovery of species, to save them from the threat of extinction to recovering them to the point where the ESA's protections are no longer needed. Another purpose of the ESA is to meet U.S. commitments under treaties and other international agreements to protect biodiversity, including the

[153] 16 U.S.C. § 1531(a)(1).

[154] *Id.* § 1531(b).

[155] *Gifford Pinchot Task Force v. U.S. Fish & Wildlife Serv.*, 378 F.3d 1059, 1070 (9th Cir. 2004) (quoting 16 U.S.C. § 1532(3)).

Convention on International Trade in Endangered Species, the Migratory Bird Treaty, and the Convention on Biological Diversity.[156]

The Secretary of the U.S. Department of Interior administers the ESA for terrestrial species, and the Secretary of the U.S. Department of Commerce administers the Act for marine species.[157] Through promulgation of joint regulations, the Secretaries of Interior and Commerce have delegated their obligations to the U.S. Fish and Wildlife Service and the U.S. National Marine Fisheries Service, respectively ("Service" or "Services").[158]

To benefit from the ESA's substantive protections and meet the purposes of the ESA, a species must first be listed as "endangered" or "threatened." A species is "endangered" if it is "in danger of extinction throughout all or a significant portion of its range."[159] A species is "threatened" if "is likely to become . . . endangered . . . within the foreseeable future throughout all or a significant portion of its range."[160]

The Services add, reclassify, or delist species from the lists of endangered and threatened species.[161] When assessing whether to list a species, the Services must apply the "best scientific and commercial data available" to five listing factors: (A) habitat loss and destruction; (B) overutilization; (C) disease or predation; (D) inadequate regulatory mechanisms; and/or (E) other manmade factors.[162] When listing a species, the Service must designate "critical habitat" for that species "to the maximum extent practicable."[163] Critical habitat is that which the Service finds to be "essential" to the species' conservation.[164]

There are currently about 1568 species that are listed as "endangered" or "threatened" under the ESA—many of which are listed due to threats posed by agriculture.[165] These species are entitled to substantive protections that have been immensely successful in saving species from extinction, and very few species have gone extinct once granted protection under the Act.[166]

Among the ESA's strong substantive protections is the prohibition against the "take" of any member of a species of endangered fish or wildlife.[167] The take

---

[156]16 U.S.C. § 1531(b).

[157]*Id.* § 1533(15).

[158]50 C.F.R. § 402.01(b).

[159]16 U.S.C. § 1532(6). The ESA's definition of "endangered" species explicitly excludes any "species of the Class Insecta determined by the Secretary to constitute a pest whose protection under the provisions of this Act would present an overwhelming and overriding risk to man." *Id.*

[160]*Id.* § 1532(20).

[161]*Id.* § 1533.

[162]*Id.* § 1533(a).

[163]*Id.* § 1533(a)(3)(A).

[164]*Id.* § 1532(5)(A). Critical habitat does not usually include "the entire geographical area which can be occupied by the threatened or endangered species." *Id.*

[165]*See* U.S. Fish and Wildlife Service, Summary of Listed Species Listed Populations and Recovery Plans, https://ecos.fws.gov/tess_public/pub/Boxscore.do, last visited (Aug. 23, 2015).

[166]Center for Biological Diversity, On Time, On Target: How the Endangered Species Act is Saving America's Wildlife (2012) ("CBD (2012)").

[167]16 U.S.C. § 1539(a).

prohibition applies to any "person," which is broadly defined to include any individual, corporation, or governmental entity.[168] To "take" an endangered animal means "to harass, harm, pursue, hunt, shoot, wound, kill, trap, capture, or collect" it, or "to attempt to engage in any such conduct."[169] The take prohibition can apply to acts which affect habitat for an endangered species as well; the definition of "harm" within the take definition means an act which "actually kills or injures wildlife" through "significant habitat modification or degradation where it actually kills or injures wildlife by significantly impairing essential behavioral patterns, including breeding, feeding or sheltering."[170] The Service may extend the take prohibition to threatened species of fish or wildlife by promulgating a "special rule" under Section 4(d) of the Act.[171] The take prohibition applies only to endangered (and, depending on the Service's utilization of a special rule, threatened) species of fish or wildlife, but not to endangered or threatened plants.[172]

The ESA permits take in certain, limited circumstances: (1) for scientific research; and (2) for take that is "incidental" to an otherwise lawful activity.[173] These permits require that the take will not jeopardize a species, and imposes requirements to minimize the amount of take that is permitted as well as monitoring requirements. Without a scientific or incidental take permit from the Service, however, any person who takes a protected species is liable for civil and criminal penalties including fines and imprisonment.[174]

Another strong substantive ESA protection is the affirmative, substantive duty on all federal agencies to ensure that any federal "action" that they take is not likely to result in "jeopardy" to any listed species or result in the "destruction or adverse modification" of critical habitat.[175] Federal actions that trigger the "no-jeopardy duty" include "all activities or programs of any kind authorized, funded, or carried out, in whole or in part, by Federal agencies in the United States or upon the high seas," such as the granting of licenses, contracts, leases, permits, etc. by federal

[168] *Id.*

[169] 16 U.S.C. § 1532(19).

[170] 50 C.F.R. § 17.3; *Babbitt v. Sweet Home Chapter of Cmtys. for a Great Or.*, 515 U.S. 687 (1995).

[171] 16 U.S.C. § 1533(d). The U.S. Fish and Wildlife Service enacted a rule in 1978 that automatically extends the take prohibition for endangered species to all threatened species. 50 C.F.R. § 17.31(a).

[172] For endangered plants, the ESA makes it unlawful to trade, possess, and maliciously damage or destroy, or engage in interstate or foreign commerce in, if the plants are located on federal lands; endangered plants on private lands are only protected under the ESA insofar as they are protected by state laws. 16 U.S.C. § 1539(a)(2)(B). Threatened plants receive the same protections as endangered plants, except for the prohibition against malicious damage or destruction on federal lands, and threatened plants receive no protection on private lands. 16 U.S.C. § 1539(a)(2)(B); 50 C.F.R. §§ 17.61 and 17.71.

[173] 16 U.S.C. §§ 1538(a)(1)(A) and (B).

[174] *Id.* §§ 1540(a) and (b).

[175] *Id.* § 1536(a)(2); 50 C.F.R. § 402.02.

agencies.[176] Satisfaction of the no-jeopardy duty may only be achieved through full compliance with procedures that are set forth in the ESA's implementing regulations.[177]

These procedures usually entail "formal consultation" between the federal agency and the relevant Service(s).[178] Through consultation, the agency and Service assess the adverse effects of the action to endangered and threatened species, and the Service determines, in a "biological opinion," whether the action is likely to "jeopardize" the continued existence of any listed species, or to destroy or adversely modify its critical habitat.[179] Section 7 of the ESA also imposes a general duty on all federal agencies to "utilize their authorities in furtherance of the purposes of this Act by carrying out programs for the conservation" of listed species.[180]

Sections 9 and 7 of the ESA are among the strongest substantive protections for species, but the ESA includes additional protections as well. The Services also develop and implement "recovery plans" for each listed species, which provide the "road map" to recovery, *i.e.*, the point when the ESA's protections are no longer necessary.[181] The law also contains provisions for funding and habitat acquisition.[182]

The U.S. Supreme Court has observed that the ESA encompasses "a conscious decision by Congress to give endangered species priority over the 'primary missions' of federal agencies."[183] In addition, the law includes provisions for citizen enforcement—for example, the take prohibition and no-jeopardy duty can be enforced through the ESA's "citizen suit" provision. This means that enforcement of civil remedies is not limited to the Service's prosecutorial discretion, but may be procured through a federal lawsuit brought by any citizen suit plaintiff who is harmed by the take or jeopardy of listed species.[184]

How the ESA Regulates Agricultural Activities As discussed above, key aspects of the federal ESA include: mechanisms for listing species as endangered or threatened, including plants; prohibitions on take and trafficking in listed species; consultation between federal agencies, and biological opinions and incidental take authorizations; designation of critical habitat; and recovery plans. With this

[176] 50 C.F.R. § 402.02.

[177] 50 C.F.R. Part 402.

[178] 16 U.S.C. § 1536(a)(2); 50 C.F.R. Part 402.

[179] 16 U.S.C. § 1536(b)(3)(A); 50 C.F.R. § 402.02.

[180] 16 U.S.C. §§ 1531(c) and 1536(a)(1).

[181] 16 U.S.C. § 1533(f).

[182] *See, e.g.*, 16 U.S.C. § 1535(c) (authorizing (d) (authorizing the Services to provide financial assistance to states that have entered into "cooperative agreements" for the purpose of endangered and threatened species conservation and monitoring).

[183] *Tenn. Valley Auth. v. Hill*, 437 U.S. 153, 185 (1978).

[184] 16 U.S.C. § 1540(g). However, only the Service may prosecute criminal violations of the ESA's take prohibition.

panoply of conservation, protections, duties, and citizen enforcement, very few species have gone extinct once listed under the ESA, although recovery is often a long-term effort.[185]

The ESA intersects with agricultural activities in many ways. First, many species that are threatened by industrial agriculture have been listed as endangered or threatened under the ESA. Following are a few examples of once-common species that are now endangered or threatened primarily as a result of agricultural activities.

1. The Delta smelt (*Hypomesus transpacificus*). Once one of the most common fishes in the Sacramento and San Joaquin estuarine environments, the Delta smelt was listed as a threatened species in 1993 as a result of water diversions for agricultural irrigation upriver from the Sacramento-San Joaquin Delta.[186]
2. The San Joaquin kit fox (*Vulpes macrotis mutica*). Once common in the Central Valley of California, this small fox has lost over 50 % of its habitat since the 1930s due to conversion of habitat to cropland, is also threatened by pesticide use, and has been listed as endangered since the ESA was enacted in 1973.[187]
3. The Austin blind salamander (*Eurycea waterlooensis*) (endangered), Barton Springs salamander (*Eurycea sosorum*) (endangered), Georgetown salamander (*Eurycea naufragia*) (threatened), Jollyville Plateau salamander (*Eurycea sosorum*) (threatened), and Salado salamander (*Eurycea chisholmensis*) (threatened), five tiny species of salamander that occur only in groundwater karst environments in central Texas, and which are listed due to exposure to certain pesticides, like carbaryl and atrazine, that are particularly harmful to amphibians.[188]
4. The Mexican gray wolf (*Canis lupus baileyi*). A unique subspecies of the gray wolf (*Canis lupus*), the Mexican wolf once ranged throughout the southwestern United States and Mexico, but like *Canis lupus* in the conterminous United States, the Mexican wolf was persecuted intensively by federal and state agencies throughout the twentieth Century.[189] After European Americans and their livestock settled the North America, Mexican wolves were eliminated to protect livestock from depredation (and for other reasons).[190] Down to just seven

[185]CBD (2012).

[186]58 Fed. Reg. 12,854 (Mar. 5, 1993)

[187]32 Fed. Reg. 4001 (Mar. 11, 1967); U.S. Fish and Wildlife Service, San Joaquin Kit Fox (*Vulpes macrotis mutica*) 5-Year Review: Summary and Evaluation (2010).

[188]79 Fed. Reg. 10,236 (Feb. 24, 2014) (final rule to list Georgetown and Salado salamanders); 78 Fed. Reg. 51,278 (Aug. 20, 2013) (final rule to list Austin and Jollyville Plateau salamanders); 62 Fed. Reg. 23,377 (May 30, 1997) (final rule to list Barton Springs salamander); *see also* Hayes (2002); Rohr and McCoy (2010), p. 20 (noting that studies have found significant effects of atrazine on amphibians and fish, such as effects on metamorphosis, development, behavior, sexual behavior and characteristics, reproductive success, and immune and endocrine systems).

[189]U.S. Department of the Interior, Fish and Wildlife Service, Mexican Wolf Recovery Plan 2-10 (1982).

[190]*Id.* at 6.

individuals that were taken into captivity in the late 1970s, the species was listed as endangered in 1978.[191] Despite substantial recovery efforts, the species has not surpassed more than about 100 individual wolves in the wild and recovery efforts continue to face opposition from livestock ranching and other interests.

As these examples show, many federal as well as private agricultural activities—including hydropower dams and developments, water diversions or irrigation, cropland conversion, pesticide use, and livestock ranching—can contribute to the need to list species as endangered or threatened under the Act.

Another way in which the ESA regulates agriculture is through the ESA's take prohibition. As explained above, in order to avoid exposure to criminal and civil penalties for unlawful take, including through citizen suit enforcement, agricultural producers whose activities are likely to cause incidental take of a listed species must apply for and obtain an "incidental take permit" ("ITP") from the Service.[192]

The ITP permitting process involves preparation of a conservation plan and a binding commitment to minimize take.[193] To obtain an incidental take permit, an applicant must develop and submit a "habitat conservation plan" ("HCP") for approval by the Service. An HCP specifies the "impact which will likely result" from the anticipated take as well as the steps that the applicant commits to take in order to "minimize and mitigate" those impacts, how those steps will be funded, alternatives to such take and why they are not being implemented instead, and other measures the Service may consider to be "necessary and appropriate."[194] When considering whether to issue an ITP to an agricultural producer, the Service must ensure that the authorized take will in fact be "incidental" and minimized and mitigated to the "maximum extent practicable," that the applicant has adequate funding to implement the HCP, that the taking will not "appreciably reduce the likelihood of the survival and recovery of species in the wild," and that it has been provided adequate assurances that the plan will be implemented.[195]

When deciding whether to issue an ITP to an agricultural producer, the Service must also satisfy affirmative duties, in Section 7(a)(2) of the ESA, to avoid jeopardy to the species' continued existence and/or the destruction or adverse modification of its critical habitat. Thus, the Service must engage in formal consultation—with

---

[191] *Id.* at 22; *Reclassification of the Gray Wolf in the United States and Mexico, with Determination of Critical Habitat in Michigan and Minnesota*, 43 Fed. Reg. 9607 (Mar. 9, 1978).

[192] 16 U.S.C. § 1538(a)(1)(B) ("[e]xcept as provided in section[]...10 of this Act, with respect to any endangered species of fish or wildlife listed pursuant to section 4 of this Act it is unlawful for any person subject to the jurisdiction of the United States to... take any such species within the United States or the territorial sea of the United States"); *id.* § 1539(a)(1)(B) ("The Secretary may permit, under such terms and conditions as he shall prescribe ... any taking otherwise prohibited by section 9(a)(1)(B) if such taking is incidental to, and not the purpose of, the carrying out of an otherwise lawful activity"); *see also* 50 C.F.R. §§ 17.22(b)(1) and 222.22 (regulations for incidental take permits).

[193] 16 U.S.C. § 1539(a)(2)(A).

[194] *Id.*

[195] *Id.* § 1539(a)(2)B).

itself, known as "self-consultation"—to analyze the effects of the proposed take to the species and its critical habitat, develop reasonable and prudent measures to minimize such take, and set terms and conditions in an "incidental take statement" to ensure that such measures are carried out.

The Endangered Species Act and National Environmental Policy Act When issuing an ITP, the Service must also meet obligations under the National Environmental Policy Act ("NEPA"), and NEPA's implementing regulations.[196] NEPA is a procedural environmental law that requires all federal agencies to rigorously analyze and disclose the effects of their activities to the environment. Like the ESA, NEPA encompasses all "major federal actions," including but not limited to agricultural activities.[197] Through preparation of an "environmental impact statement" in accordance with NEPA and its implementing regulations, federal agencies must gather all information and public comment on the effects of federally authorized agricultural activities, fully analyze and publicly disclose the effects the activities, including to biodiversity and species, and consider all reasonable alternatives to the proposal.[198]

Pursuant to the ESA and NEPA, for example, the Fish and Wildlife Service issued an ITP to San Joaquin County in 2000 that authorizes incidental take of Delta smelt, San Joaquin kit foxes, and many other endangered and threatened species resulting from land use planning allowing for conversion of habitat to croplands and the county irrigation district's issuance of water diversion permits (and other activities).[199] In issuing the ITP, the Service approved an HCP from the county that specifies mitigation for these activities. Under this plan, for every one acre of land that is converted to cropland, one acre of land must be preserved, such as through the county's purchase of conservation easements.[200] By the same authority, in 2013 the Service renewed an ITP for the City of Austin, Texas to allow recreational swimming and maintenance of Barton Springs Pool and its springs, a popular recreational swimming site that is also the only location where the Barton Springs and Austin blind salamanders are found.[201] Among other conservation measures, this ITP sets limits on the speed and amount of water draw-downs that may occur during pool maintenance and cleaning, and requires public education

[196]42 U.S.C. §§ 4321–4347.

[197]*Id.* § 4331.

[198]40 C.F.R. §§ 1500.1–1508.28.

[199]*Availability of Final Environmental Impact Statement/Environmental Impact Report for the Proposed Issuance of an Incidental Take Permit for the San Joaquin County Multi-Species Habitat Conservation and Open Space Plan in California*, 65 Fed. Reg. 75,293 (Dec. 1, 2000) (notice of final EIS and ITP for the San Joaquin County Multi-Species Habitat Conservation and Open Space Plan in California).

[200]65 Fed. Reg. 75,293 (Dec. 1, 2000) (notice of final EIS and ITP for the San Joaquin County Multi-Species Habitat Conservation and Open Space Plan in California).

[201]78 Fed. Reg. 64,001 (Oct. 25, 2013) (announcing approval of amendment and renewal of ITP for Barton Springs Pool.

efforts.[202] In both examples, the Service prepared environmental reviews pursuant to NEPA.

These ESA and NEPA obligations do not just apply to the Service's analysis of the impacts of take in the course of agricultural activities when considering whether to grant ITPs—rather, these obligations attach to all federal actions, carried out by all federal agencies, that relate to agriculture.

The Federal Insecticide, Fungicide and Rodenticide Act Since 1972, the Federal Insecticide, Fungicide and Rodenticide Act ("FIFRA") has been the U.S.'s primary pesticide regulation law.[203] Under FIFRA, the Environmental Protection Agency registers pesticide products that it deems to be effective and safe, and through a system of labeling, regulates the sale and use of those products.[204] Any person who violates the strict uses and prohibitions of pesticide labels risks fines and criminal penalties.[205] Through this system, FIFRA's is intended to protect "man or the environment" from "pests that must be brought under control."[206]

FIFRA is not intended to protect biodiversity or species. To the contrary, FIFRA's very purpose is to target and eradicate certain species that are deemed to be "pests"—*i.e.*, any animal (other than man), plant, fungus, bacteria, and micro-organism—that is a nuisance to human.[207] FIFRA-regulated pesticides seek to eradicate certain species in service of agriculture. As a result, FIFRA and the ESA can be directly at odds with one another.

For example, a major threat to the endangered black-footed ferret and many prairie dog species in the American West has been the use of pesticides by ranchers and their agents. Prairie dogs, the primary prey of black-footed ferrets, were and continue to be heavily targeted as they are seen as a threat to livestock grazing.[208]

---

[202]City of Austin, Major Amendment and Extension of the Habitat Conservation Plan for the Barton Springs Salamander (Eurycea sosorum) and the Austin Blind Salamander (Eurycea waterlooensis) to allow for the Operation and Maintenance of Barton Springs and Adjacent Springs 104–109 (2013).

[203]7 U.S.C. § 136-136y.

[204]*Id.* § 136l.

[205]*Id.* § 136a.

[206]*Id.* §§ 136 and 136w-3.

[207]*Id.* § 136w-3(b)(1)(A). 7 C.F.R. § 152.5. A pest to be "any form of plant or animal life" which is "injurious to health or the environment," including insects, rodents, nematodes, fungi, and weeds, as well as "any other form of terrestrial or aquatic plant or animal life" that EPA deems to be a pest. 7 U.S.C. § 136w(c)(1).

[208]It is believed, for example, that prairie dogs compete with livestock for forage, create underground burrows that threaten to injure livestock, and damage crops. However, studies have documented the beneficial ecosystem impacts by prairie dogs, as they provide food for coyotes, bobcats, gray foxes, weasels, hawks, eagles, and falcons. Miller et al. (2000), p. 318. Prairie dogs also create underground habitat for other species, like burrowing owls, lizards, rabbits, hares, and snakes. Cully et al. (2010), p. 667. Prairie dogs prefer open areas and avoid areas with tall grasses, where livestock graze, since tall grasses interfere with their ability to watch for predators, and therefore do not compete for forage. Hygnstrom, S.E. and Virchow, D.R., *Prairie Dogs, in*

Thus, with the widespread loss nearly all prairie dog habitats in the western United States due to crop- and pasture-land conversion as well as direct poisoning, prairie dogs and black-footed ferrets have declined steeply as well.

The widespread use of strychnine pesticides to destroy prairie dog colonies during the twentieth Century caused the secondary poisoning of black-footed ferrets and steep declines in both species.[209] The Utah prairie dog and the Mexican prairie dog are listed under the ESA as threatened and endangered, respectively, in part due to targeted eradication campaigns, and sharp declines in many other prairie dog species have occurred from pesticides (in addition to cropland conversion and many other threats) as well.[210] From poisoning as well as cropland and rangeland conversion, the black-footed ferret was so imperiled by the early 1970s that it was presumed extinct in the wild, until one remaining wild population was discovered in Wyoming in 1981.[211]

Due to dangers of strychnine pesticides, the EPA banned it in 1972, calling it "among the most toxic substances known to man," "not only to [the] targets but other animals and wildlife" as well.[212] This ban was lifted in 1986, however, when EPA reauthorized its use for prairie dog eradication, albeit with new conditions on its above-ground use.[213] Today, strychnine is still used to kill prairie dogs, but other

---

Prevention and Control of Wildlife Damage (1994). Indeed, prairie dogs are considered to be a "keystone species" for native grasslands, as they support many other species and ecosystem diversity. Martinez-Estévez et al. (2013), p. 1.

[209]As one district court observed, "[s]trychnine is non-selective" and "kills anything which ingests a lethal dose." *Defenders of Wildlife v. Adm'r, Envtl. Prot. Agency*, 688 F. Supp. 1334, 1339 (D. Minn. 1988). It can kill both 'target' and 'non-target' species" – *i.e.*, "any species which the strychnine is not intended to kill, but which nonetheless ingests it" by directly ingesting strychnine bait or by consuming an animal or bird that has. *Id.*

[210]FWS, Utah Prairie Dog (*Cynomys parvidens*) Revised Recovery Plan 1.3–1 (2012); International Union for Conservation of Nature: *Cynomys mexicanus* (Mexican prairie dog), available at http://www.iucnredlist.org/details/6089/0 (last visited Sep. 13, 2015); *Endangered and Threatened Wildlife and Plants; Finding for the Resubmitted Petition To List the Black-Tailed Prairie Dog as Threatened*, 69 Fed. Reg. 51,217, 51,221 (Aug. 18, 2004) (noting that the historic range of the black-footed ferret was once as great as 400 million acres, but had been reduced to just 1,842,000 acres—about one percent—by 2004); FWS, species assessment and listing priority assignment form, Gunnison's prairie dog (Apr. 2010).

[211]FWS, BLACK-FOOTED FERRET RECOVERY PLAN (SECOND REVISION) 20 (Nov. 2013) ("Black-Footed Ferret Recovery Plan (2013)"). The Wyoming population was gone by the mid-1980s. *Id.* at 29. However, individuals from this population were taken into captivity and used in a captive-breeding program that has allowed for reintroductions of black-footed ferrets back into Wyoming and other areas of its historic range. *Id.* at 55. As of 2013, FWS estimated there are about 364 breeding adult black-footed ferrets in the wild, out of a recovery target of 1,500. *Id.* at 5.

[212]EPA Order PR 72-2, MANUFACTURERS, FORMULATORS, DISTRIBUTERS, AND REGISTRANTS OF ECONOMIC POISONS: SUSPENSION OF REGISTRATION FOR CERTAIN PRODUCTS CONTAINING SODICUM FLUOROACETATE (1080), STRYCHNINE AND SODIUM CYANIDE (Mar. 9, 1972).

[213]*Defenders of Wildlife v. Administrator, EPA*, 688 F.Supp. 1334 (D. Minn. 1988) (8th Cir. 1989), *aff'd in part, rev'd in part, Defenders of Wildlife v. Administrator, EPA*, 882 F.2d 1294 (8th Cir. 1989) (discussing history of administrative process).

pesticides are used for this purpose as well, including zinc phosphide and anticoagulant rodenticides, the latter of which also pose a risk of secondary poisoning of black-footed ferrets.[214]

Changes to the uses of pesticides for killing prairie dogs are the result of ESA section 7(a)(2) consultations between the EPA and FWS. As a result of the ESA consultations, which were precipitated by litigation under the ESA, strychnine use above ground is restricted, and at least one anticoagulant rodenticide may not be used in black-footed ferret recovery areas.[215] However, these conditions have not been enough to alleviate the threat of poisoning of black-footed ferrets from prairie dog pesticide use.[216]

Thus, although FIFRA is intended to regulate—and hence, to permit—pesticides in order to grow crops or livestock, the ESA is intended to protect species from all threats including pesticides that are used for agriculture. When the two statutes conflict, endangered species protection must take the highest priority."[217] Yet, as shown by the black-footed ferret example, this is not always the case—while pesticide poisoning is now regulated because of the ESA, the threat that it poses to black-footed ferret recovery has not been alleviated. Another example is EPA's registration and labeling of pesticides under FIFRA for use in agriculture that are harmful to aquatic species including fish, amphibians, and reptiles. There too, the EPA does not always satisfy its ESA section 7(a)(2) obligations to ensure that endangered species are not threatened by such pesticides.[218] While FIFRA authorizes the EPA has authority to cancel any pesticide that "[m]ay pose a risk to the continued existence of any endangered or threatened species" under a "Special Review" process, this provision is rarely used.[219]

The Clean Water Act Enacted in 1972, the Clean Water Act directs the EPA to develop comprehensive programs to control the pollution of "waters of the United States" through a cooperative federalism scheme. Under this framework, EPA is the primary regulatory authority under the Clean Water Act, but states with qualifying implementation plans may obtain delegated authority from EPA to administer the Act within their borders.[220] The Clean Water Act was enacted, among other

---

[214]Black-Footed Ferret Recovery Plan (2013) at 50.

[215]*Id.*

[216]*Id.*

[217]*Defenders of Wildlife v. Adm'r, EPA*, 882 F.2d 1294, 1299 (8th Cir. 1989) (quoting *Tennessee Valley Auth. v. Hill*, 437 U.S. at 185).

[218]*See, e.g.*, *Wash. Toxics Coal. v. EPA*, 2002 U.S. Dist. LEXIS 27654 (W.D. Wash. July 2, 2002) (EPA violated the ESA by failing to initiate consultation with the National Marine Fisheries Service regarding its pesticide registrations and their potential harmful effects on endangered salmonids and their habitat); *Wash. Toxics Coal. v. EPA*, 2003 U.S. Dist. LEXIS 26088 (W.D. Wash. Aug. 8, 2003) (EPA failed to show its ongoing actions regarding pesticide active ingredients were non-jeopardizing to threatened and endangered salmonids).

[219]40 C.F.R. Part 154.7.

[220]Federal Water Pollution Control Act of 1977, *as amended*, 33 U.S.C. §§ 1251-1375nt ("Clean Water Act").

reasons, to protect "fish and aquatic life and wildlife" protection, and to protect water withdrawals for agriculture.[221] The Clean Water Act includes a citizen suit provision.[222]

"Waters of the United States" includes all wetlands that fall under the jurisdiction of the Clean Water Act, which are critically important for ecosystem function and biodiversity as well as for agricultural production. However, the Clean Water Act expressly exempts normal farming, ranching, and silviculture activities, and construction of farm or stock ponds, irrigation ditches, and drainage ditches, as well as farm roads, from "dredge-or-fill" permitting requirements under CWA section 404.[223]

Such "section 404 permits" (or "dredge and fill permits") are otherwise required for any activities that "bring[] an area of the navigable waters into a use to which it was not previously subject."[224] Conversion of wetlands to croplands and rangelands—a major reason for the loss of over 71 % of the wetlands in the continental United States since the mid-1700s—does remain subject to the scope of the dredge-and-fill permit requirement, however.[225]

Additionally, the Clean Water Act exempts agricultural runoff pollution, which is detrimental to many species of "fish, shellfish, and wildlife" from its primary regulatory control, which is "point source" permitting under the Clean Water Act's National Pollutant Discharge Elimination System ("NPDES").[226] For example, nutrient runoff from agricultural pesticides that are used in the Midwest is carried to the Gulf of Mexico, causing a massive "Dead Zone," where oxygen levels are very low and marine species cannot live.[227] Yet, instead of being regulated as a point source, this runoff from Midwestern croplands and rangelands is treated as

---

[221]33 U.S.C. § 1252(a).

[222]33 U.S.C. § 1369.

[223]33 U.S.C. 1344(f)(1)(A), (C) and (E) (exempting various farming activities from requirement to obtain permits for the discharge of dredged material); 33 C.F.R. § 323.4(a) (Army Corps of Engineers regulatory exemptions); 40 C.F.R. § 232.3(c) (EPA regulatory exemptions).

[224]33 U.S.C. 1344(f)(2) ("Any discharge of dredged or fill material into the navigable waters incidental to any activity having as its purpose bringing an area of the navigable waters into a use to which it was not previously subject, where the flow or circulation of navigable waters may be impaired or the reach of such waters be reduced, shall be required to have a permit under this section.").

[225]U.S. Department of Commerce, National Marine Fisheries Service, Endangered Species Act Biological Opinion and Conference Opinion 141 (Feb. 17, 2012) (noting that by 2002, about half of the total land area of the continental United States consisted of cropland or pastureland).

[226]33 U.S.C. §§ 1312, 1342. The Clean Water Act's definition of "point source" exempts "agricultural stormwater discharges and return flows from irrigated agriculture. 33 U.S.C. § 1362(14). Nonpoint sources are not required to obtain and adhere to NPDES permits.

[227]For example, in 2013, the National Marine Fisheries Services, which administers the ESA for marine species, found that listing a population of sperm whales in the Gulf of Mexico may be warranted, in part due to "the Gulf's 'dead zone'"). *Endangered and Threatened Wildlife; 90-day Finding on a Petition to List Sperm Whales in the Gulf of Mexico as a Distinct Population Segment Under the Endangered Species Act*, 78 Fed. Reg. 19,176 (Mar. 29, 2013).

"nonpoint" source and therefore is largely unregulated under the Clean Water Act.[228]

However, as a result of litigation brought pursuant to the Clean Water Act's citizen suit provision, the direct discharge of pesticides in, over, or near to waters of the U.S. does require NPDES permit, so long as the discharge is in compliance with FIFRA.[229] In addition, CAFOs and concentrated aquaculture facilities are regulated as point sources under the Clean Water Act.[230]

The Clean Air Act Similar in structure of the Clean Water Act, the Clean Air Act regulates emissions of air pollutants under a cooperative federalism scheme that is intended to protect public health and welfare, an expansive phrase that encompasses soils, water, crops, livestock, animals, wildlife, endangered species, and climate.[231] Air pollution can harm crops by acid rain, reducing growth and yield, and premature death, and can harm species by polluting tissues, impeding respiratory systems, and polluting waterways through deposition. For example, the recovery of the highly endangered razorback sucker in the Colorado River is being harmed by deposition of atmospheric mercury that is emitted from the combustion of coal at several power plants in the Four Corners region of the southwestern U.S.[232] Certain forms of air pollution can cause climate change, a major threat to the Earth's biodiversity.

The Clean Air Act regulates the impacts of air pollution for the public health and welfare, which includes agriculture and wildlife protection as well as other benefits,

---

[228]40 C.F.R. § 122.3(f); *see also, e.g., Friends of the Everglades v. S. Fla. Water Mgmt. Dist.*, 570 F.3d 1210, 1226–1227 (11th Cir. 2009) (Kennedy, J., concurring) (citing *Rapanos v. United States*, 547 U.S. 715, 777 (2006)) ("No one disputes that the NPDES program is restricted to point sources. Non-point source pollution, chiefly runoff, is widely recognized as a serious water quality problem, but the NPDES program does not even address it."). This gap has led one scholar to suggest that a tort-based nuisance class action lawsuit—*e.g.*, by Louisiana fishermen who suffer economically as a result of a giant algae bloom that has decimated fisheries off the coast of Louisiana, the result of agricultural runoff from the Midwest, may be viable. *See COMMENT: Breathing Life into the Dead Zone: Can the Federal Common Law of Nuisance Be Used To Control Nonpoint Source Water Pollution?*, 85 Tul. L. Rev. 215 (Nov. 2010).

[229]*See National Pollutant Discharge Elimination System Regulation Revision: Removal of the Pesticide Discharge Permitting Exemption in Response to Sixth Circuit Court of Appeals Decision*, 78 Fed. Reg. 38,591, 38,593 (June 27, 2013); *Nat'l Cotton Council of America v. EPA*, 553 F.3d 927 (6th Cir. 2009).

[230]40 C.F.R. § 122.23(d); *id.* § 122.24; *id.* § 122.25.

[231]42 U.S.C. §§ 7401-7671q; *id.* § 7401(a)(2) (Congressional finding that air pollution "has resulted in mounting dangers to the public health and welfare, including injury to agricultural crops and livestock"); *id.* § 7412(a)(7) (defining "adverse environmental effect" caused by hazardous air pollutants as any "significant and widespread adverse effect, which may be reasonably anticipated, to wildlife, aquatic life ..., [or] populations of endangered or threatened species..."); *id.* § 7602 ("welfare" includes effects on "soils, water, crops, vegetation..., animals, wildlife, weather, visibility, and climate...").

[232]U.S. Department of the Interior, Fish and Wildlife Service, DRAFT Biological Opinion for the Desert Rock Energy Project, U.S. Bureau of Indian Affairs, Gallup, New Mexico (undated).

and thus seeks to benefit species indirectly by controlling all qualifying emitters of specific air pollutants. Sources of air pollution are defined by the Act and EPA, and are classified according to their "sources," similar to "point sources" under the Clean Water Act.[233]

The Clean Air Act regulates the impact of certain forms of air pollution from agricultural activities—for example, emissions of hydrogen chloride and chlorine gas from production of pesticides.[234] Livestock facilities that emit more than 25,000 metric tons of carbon dioxide equivalent are required to report their greenhouse gas emissions under a 2009 rule.[235] However, the Clean Air Act is not being utilized to regulate greenhouse gases to the fullest extent possible.[236]

U.S. State Biodiversity Protection Laws Nearly all U.S. states have also enacted statutes to protect biodiversity, although most of these laws simply provide a mechanism for listing species and prohibit taking of or trafficking in listed species.[237] Nearly all lack mechanisms for recovery, consultation, or critical habitat designation, and many lack any mechanisms for plant protection.[238]

The California Endangered Species Act, CA Fish & G § 2050–2115.5 ("CESA"), is a notable exception. CESA is modeled after the federal ESA, and provides mechanisms for listing species, including plants, and for designating "essential" habitat, recovery plans, and consultations by state agencies. California also has a public trust doctrine which declares that "[t]he fish and wildlife resources are held in trust for the people of the state by and through the [California Department of Fish and Wildlife ("CDFW")]."[239] This doctrine may be enforced against any instrumentality of the California state government in common law, as well as by CDFW.[240]

Thus, although not without significant gaps in policy and implementation, the U.S. has strong environmental laws with the purpose of protecting biodiversity itself and/or other resources, such as water or air, from specific threats. The suite of laws includes the ESA, the world's strongest biodiversity-protection law.

[233]42 U.S.C. § 7412.

[234]40 C.F.R. Part 63, subpart MMM.

[235]40 C.F.R. Part 98, subpart JJ.

[236]EPA, Permitting for Greenhouse Gases, http://www.epa.gov/nsr/ghgpermitting.html, last visited Sep. 13, 2015.

[237]Snape III, W.J. & George, S., *State Endangered Species Acts, in* Endangered Species Act: Law Policy and Perspectives (2d. ed.) (D.C. Baur and W.R. Irvin, eds.) ("Snape and George").

[238]*Id.*

[239]Cal. Fish & G. Code § 711.7(a).

[240]*Id.*

19.2.3.2 National Laws to Protect Biodiversity in the Profiled Nations

Many other nations have laws to protect native biodiversity from extinction and to promote the recovery of protected species. Below is an overview of agricultural policies and laws, and laws that are intended to protect native biodiversity from extinction, in some of the most populous and/or biodiversity-rich nations, including Australia, the European Union, India, China, Nigeria, and Cuba. This overview is followed by an analysis of how these nations' biodiversity-protection efforts confront agricultural activities. As discussed below, while the U.S. ESA was the first, and remains the strongest, biodiversity-conservation law, other nations have enacted their own biodiversity-protection laws as a result of certain international agreements.

Australia Australia enacted the Environment Protection and Biodiversity Conservation Act in 2000 in order to ratify its obligations under international treaties and agreements. Among many purposes, the EPBCA seeks to "assist in the co-operative implementation of Australia's international environmental responsibilities," to protect "those aspects of the environment that are matters of national environmental significance," and "to promote the conservation of biodiversity" through a "co-operative approach" which recognizes the role of indigenous people and the use of their knowledge of biodiversity.[241] The EPBCA is administered by Australia's Department of Environment.

The EPBCA is a sweeping law that sets forth a variety of different processes that are intended to ensure protection of specified heritage areas, wetlands, and migratory species. The EPBCA contains many chapters which provide for designation and protection of:

1. "World" or "National" "Heritage properties," which are areas with world heritage values that are under threat that have been designated under the World Heritage Site Convention[242];
2. "Wetlands of international importance," or "Ramsar wetlands," designated pursuant to the Ramsar Convention[243];

[241]EPBCA Chap. 1, Part 1, § 3.

[242]EPBCA Chap. 2, Part 2, Div. 1, Subdivision A; *see also* Convention for the Protection of the World Cultural and Natural Heritage, Nov. 16, 1972.

[243]EPBCA Chap. 2, Part 3, Div. 1, Subdivision B. In recognition of the biodiversity found in wetlands, especially waterfowl, and the ecosystem services that they provide, the mission of the Ramsar Convention is "the conservation and wise use of all wetlands through local and national actions and international cooperation, as a contribution towards achieving sustainable development throughout the world." The Ramsar Convention and Its Mission, http://www.ramsar.org/about/history-of-the-ramsar-convention (last visited May 3, 2015). Adopted in 1971, Ramsar is considered to be the oldest of international environmental agreements. History of the Ramsar Convention, http://www.ramsar.org/about/history-of-the-ramsar-convention (last visited May 3, 2015). The U.S. became a party to the Ramsar Convention in 1987, and currently has 37 designated "Ramsar sites." Country Profiles, http://www.ramsar.org/country-profiles (last

3. Protection of migratory species[244];
4. Protection from nuclear facilities, radioactive waste, and uranium ore[245];
5. Protection of marine environments, including the Great Barrier Reef Marine Park[246];
6. Protection of Commonwealth from persons and the Australian government[247];
7. Protection of native and migratory species and ecosystems through establishment of "Commonwealth reserves" and "conservation zones"[248];
8. Development of "conservation agreements" and "bioregional plans" that, among other things, describe the biodiversity components of a bioregion, their distribution and conservation status; heritage values; objectives; and priorities, strategies, and actions to achieve those objectives[249];
9. Development of "bilateral agreements," which are agreements between the federal and state governments to promote conservation[250];
10. Environmental assessments and federal approvals for activities that affect the environment[251];
11. Regulation of international trade in "wildlife specimens"[252]; and
12. Protection of indigenous areas and traditional uses.[253]

---

visited May 3, 2015). Designation of a particular wetlands area as a Ramsar site "embodies the government's commitment to take the steps necessary to ensure that its ecological character is maintained," as such sites are "of significant value not only for the country or countries in which they are located, but for humanity as a whole." Wetlands of International Importance, http://www.ramsar.org/about/wetlands-of-international-importance (last visited May 3, 2015). Parties must "promote the conservation" and "wise use" of Ramsar sites by providing for their acquisition, compensating for their loss, consulting and exchanging information with other parties, and participating in meetings every three years. *See generally* Convention on Wetlands of International Importance especially as Waterfowl Habitat (a.k.a. the "Ramsar Convention on Wetlands"), Feb. 2, 1971.

[244]EPBCA, Chap. 2, Part 3, Subdivision D and Chap. 5, Part 13, Div. 2 (prohibitions against impacts to and take of migratory species); *see also* Japan Australia Migratory Bird Agreement, Apr. 30, 1981; China-Australia Migratory Bird Agreement, Sep. 1, 1988; Convention on the Conservation of Migratory Species of Wild Animals (Bonn Convention), Jan. 11, 1983.

[245]*See, e.g.*, EPBCA Chap. 2, Part 3, Div. 1, Subdivision E.

[246]EPBCA Chap. 2, Part 3, Div. 1, Subdivisions F-FB.

[247]EPBCA Chap. 2, Part 3, Div. 2.

[248]EPBCA Chap. 2, Part 15, Div. 4 and 5.

[249]EPBCA Chap. 5, Parts 12 and 14.

[250]EPBCA Chap. 3.

[251]EPBCA Chap. 4.

[252]EPBCA Chap. 5, Part 13A; *see also* CITES; Convention on Biological Diversity.

[253]EPBCA §§ 303BAA (exempting traditional uses of areas from regulation under CITES provisions of the EPBCA), 359A (providing that traditional uses in designated Commonwealth reserves by indigenous people are protected from regulation, except when such regulation is expressly for biodiversity conservation or expressly affects the traditional use), 374–383 (providing for establishment of boards for commonwealth reserves on indigenous peoples' lands), 505A-505B (providing for establishment of an Indigenous Advisory Committee to advise the federal government of indigenous peoples' knowledge of land and biodiversity conservation and use).

Chapter 5 of the EPBCA biodiversity conservation contains many provisions that are analogous to the Endangered Species Act. Primary among these is the Australian government's duty to designate and categorize threatened species (including migratory and marine species) and to add them lists that trigger substantive legal obligations.[254] In addition, and in contrast with the ESA, EPBCA provides for inventories of "threatened ecological communities."[255]

Threatened species may be classified according to six levels of imperilment: extinct, extinct in the wild, critically endangered, endangered, vulnerable, and conservation dependent.[256] Threatened ecological communities may be classified under three levels: critically endangered, endangered, vulnerable.[257] Threats to biodiversity, called "key threatening processes" may also be identified and listed.[258] There is no citizen petition process in the EPBCA as there is in the ESA, but the Australian government is required to invite nominations from the public for "amendments" to the lists of threatened species, ecological communities, and key threatening processes.[259]

Once listed, EPCBA affords protections to species and ecological communities. For instance, the Act imposes strict liability offenses with criminal and civil penalties on any "person" who causes the death or injury to a listed species or member of a listed "ecological community."[260] A person who "takes, trades, keeps or moves a member of a native species or a member of an ecological community" faces strict liability as well.[261] A person also may not take any action that "will have a significant impact" on listed species or ecological communities, heritage areas, Ramsar wetlands, or face civil penalties.[262]

Analogous to the NEPA, the U.S. law, for certain classes of governmental activities, the EPBCA also establishes an "environmental assessment and approval process" to assess environmental impacts to listed species and ecological communities.[263] The Minister must also find that his actions are not inconsistent with Australia's obligations under the CITES Treaty or the Convention on Biological Diversity.[264] In addition, the Act provides for development of "recovery plans" for listed species, with which all federal actions must be consistent.[265] Certain actions

[254] EPBCA, Chap. 5, Part 13, Div. 1, Subdivision A.

[255] EPBCA Chap. 5; *id.* § 172.

[256] EPBCA § 179.

[257] EPBCA § 181.

[258] EPBCA § 183.

[259] EPBCA § 194E.

[260] EPBCA §§ 196, 196C.

[261] EPBCA § 196C. A person is not guilty of these offenses, which are akin to "take" under the ESA, in certain circumstances such as those involving recovery activities, unavoidable accidents, emergencies, or incidental take permits. EPBCA §§ 197, 201.

[262] EPBCA § 18.

[263] EPBCA Chap. 4.

[264] EPBCA § 139.

[265] EPBCA § 139.

and permits are subject to public comment, and persons that are subjected to "conservation orders," which direct them to cease or to take affirmative steps to conserve protected areas, species, or ecological communities, may appeal such orders to an administrative tribunal.[266] However, the EPBCA does not incorporate a specific citizen suit provision with a right to judicial review or civil enforcement of federal or private activities.[267]

Thus, like the Endangered Species Act, the EPBCA seeks to protect native species and biodiversity from all threats, including agricultural activities, through a system of legal mechanisms for listing species and their habitats. Consequently, agricultural methods require approval from the Australian government if they will take listed threatened species, ecological communities, or reserves.

Despite these protections, however, vast areas of native vegetation, such as Eucalyptus woodlands, have been converted to agriculture in Australia, fragmenting habitat with dire consequences for native species.[268] Australia has experienced a "major wave of bird extinctions" in its agricultural zone in the early part of this century due to conversion of habitats to croplands.[269] At least a billion reptiles were killed due to land clearing in Australia between 1983 and 1993.[270] Australia has lost more mammals than any other nation, primarily due to land clearing.[271] Australia's legal frameworks are not alleviating all threats to its native biodiversity from agriculture.

European Union Centuries of farming and livestock grazing in rural areas in European Union member nations have irreversibly modified the landscape from its natural state.[272] But in recent decades, Europe's bird species have declined severely as agricultural methods have industrialized—in particular, once-common "farmland birds" have been in steep decline since the early 1980s.[273] Many E.U. member nations have enacted biodiversity-protection laws. Yet, in recognition that "natural habitats" in E.U. member nations are nevertheless "continuing to deteriorate," that "an increasing number of wild species are seriously threatened,"

---

[266]*See, e.g.*, EPBCA §§ 95 (allowing for public comment on draft public environment reports), 103 (same for environmental impact statements), 131A (directing Minister to invite public comment before approving controlled actions), 134A (providing for public comment on action management plans), 275 and 290 (providing for comment on proposed recovery, threat abatement, and wildlife conservation plans); *see also id.* § 473 (allowing for review of conservation orders by Administrative Appeals Tribunal).

[267]For an overview and argument for citizen suit provisions to enforce Australian environmental laws, *see* Mossop (1993), p. 266; *see also* Krinsky (2007), 301.

[268]National Resource Management Ministerial Council, Australia's Biodiversity Conservation Strategy 2010–2030 (2010) at 23–24; Perring et al. (2013).

[269]*See* Wendy Stevens, *Declining Biodiversity and Unsustainable Agricultural Production – Common Cause, Common Solution?* (2001).

[270]*Id.*

[271]*Id.*; Iain Gordon, Solving Australia's mammal extinction crisis, ABC Science (Sep. 2, 2009).

[272]The European Union Explained: Agriculture (2014) ("E.U. Explained (2014)") at 4.

[273]UNEP (2009), at 65 ff.; GEO5, at 63 (Figure 7.1).

that habitats and species are a part of the "Community's natural heritage," and that threats to them "are often of a transboundary nature," in 1992 the E.U. Council adopted Council Directive 92/43/EEC.[274]

The "main aim" of Council Directive 92/43/EEC is "to promote the maintenance of biodiversity" while taking into account economic impacts and fairness, through the designation of "natural habitats," "areas of special conservation," and "species of Community interest" as endangered, vulnerable, rare, or endemic.[275] Once designated, E.U. Member States must "take appropriate steps to avoid, in areas of special conservation, the deterioration of natural habitats."[276] In addition to taking steps to avoid deterioration, Member States must compensate when degradation nevertheless occurs.[277] In addition, the Directive seeks to increase protection of the E.U.'s native flora and fauna through the establishment of a network of about 25,000 protected sites under the "Natura 2000" program.[278] Other provisions seek to allocate the financial burden of compliance equitably.[279] As an E.U. Directive, 92/43/EEC requires all member states to reach the directed result, without dictating how to reach it.

Other E.U. policies value the countryside for its traditional way of life, substantial contribution of commodities to European societies, and for beneficial environmental values it helps to maintain, including biodiversity.[280] For example, the EU "common agricultural policy" ("CAP") supports producers through market safety nets and direct payments to farmers to incentivize sustainable farming practices and rural development.[281] EU agricultural regulations passed in 2013 directly subsidize reserves and agricultural practices that are beneficial for the climate and the environment such as crop diversification, permanent grasslands, and "ecological focus areas" like terraces, buffer strips, nitrogen-fixing crops, and areas with short crop rotation.[282] CAP expenditures account for about 40 % of the total E.U. budget.[283] These agricultural policies are also intended to help the E.U. reach goals for reducing biodiversity and ecosystem services losses by 2020.[284]

---

[274]Council Directive 92/43/EEC of 21 May 1992 on the conservation of natural habitats and wild fauna and flora.

[275]*Id.* at Art. 1.

[276]*Id.* at Art. 6.

[277]*Id.*

[278]*Id.* at Art. 6(4); E.U. Explained (2014), at 10.

[279]*See, e.g.*, E.U. Council Directive 92/43/EEC at Art. 8(4)-(6) (providing for financing to achieve Directive).

[280]E.U. Explained (2014), at 4.

[281]*Id.*

[282]Regulation (EU) No. 1307/2013 of the European Parliament and of the Council of 17 December 2013 at Articles 30, 43-XX.

[283]EU Explained (2014), at 7.

[284]Communication from the Commission to the European Parliament, the Council, the Economic and Social Committee and the Committee of the Regions: Our life insurance, our natural capital: an EU biodiversity strategy to 2020 (Mar. 5, 2011) at 5.

Yet, a 2014 report found that while the rate may be slowing, farmland biodiversity in Europe continues to decline, and targets to manage agricultural areas sustainably to conserve biodiversity by 2020 are not being met.[285]

India India has one of the densest human populations on Earth and one of the longest histories with agriculture, but also claims rich biodiversity, with two recognized biodiversity hotspots.[286]

Despite centuries of agricultural activity, however, modern agriculture is having devastating impacts to India's biodiversity. For example, a shift from pollinated varieties of apples grown in the Himalayan region of India lead farmers to replace them with sterile varieties in the 1980s, which—together with the negative effects of pesticides—caused many natural pollinator species to go extinct while also reducing overall apple productivity.[287]

In 2002, India enacted the Biological Diversity Act.[288] This law seeks to protect India's biodiversity for the benefit of the Indian people. The law established the National Biodiversity Authority which is empowered to make regulations and to "regulate access" to biodiversity and agricultural genetic resources, and State Biodiversity Boards, which advise state governments on matters of biological conservation and regulate the commercial use of biological resources by Indian citizens and corporations.[289] The National Biodiversity Authority may establish a committee to address agro-biodiversity, *i.e.*, the "biological diversity of agriculture-related species and their wild relatives."[290]

India's biodiversity law emphasizes protection and equitable sharing of India's natural heritage, intellectual property, and genetic resources. It includes mechanisms for protecting species and biodiverse areas, and prohibits non-citizens and non-Indian corporations and organizations from obtaining "any biological resource occurring in India or knowledge associated thereto for research or for commercial utilization or for bio survey and bio utilization" without previous approval of the National Biodiversity Board, except for publication of science and research projects.[291] Applications for intellectual property rights that are "based on any research or information on a biological resource obtained from India" must also receive approval from the National Biodiversity Authority.[292] Indian citizens and corporations must provide notice to the relevant State Biodiversity Board to obtain any biological resource for commercial utilization, with the exception of "the local

---

[285] GEO5 (2014) at 63.

[286] Gaurav Moghe, Biodiversity Hotspots in India (2011). A "biodiversity hotspot" is a "biogeographic region with a significant reservoir of biodiversity that is under threat from humans." *Id.*

[287] MA Assessment (2005), at 32.

[288] The Biological Diversity Act, Act No. 18 of 2003.

[289] Act No. 18 of 2003, Chapter II, § 24.

[290] Act No. 18 of 2003, § 18(4); Brahmi et al. (2004), p. 659.

[291] Act No. 18 of 2003, Chapters II, IX.

[292] Act No. 18 of 2003, Chapter II, § 6.

people and communities of the area, including growers and cultivators of biodiversity" and those practicing indigenous medicine.[293]

In addition, the National Biodiversity Authority advises India's central and state governments on biological conservation, the "equitable sharing of benefits arising out of the utilization of biological resources" and the selection of "biodiversity heritage sites."[294] The Authority may "take any measures necessary to oppose the grant of property in any country outside India on any biological resource obtained from India or knowledge associated with such biological resource which is derived from India."[295]

Thus, the policies that underpin India's law favor protection of biodiversity in order to protect the services and national identity that biodiversity provides to Indian society, and in particular, to agriculture. Unlike biodiversity protection laws in the U.S. and Australia, India's law places an emphasis on regulation and protection of native crops from unlawful appropriation, or "bio-piracy." Although India's law echoes European Union efforts to incorporate agro-biodiversity, it does not directly subsidize its use.

Indian law provides avenues for enforcement of the Biological Diversity Act. For instance, the National Biodiversity Authority and State Biodiversity Boards may "sue and be sued," *i.e.*, they may enforce the Act's provisions and pursue civil and criminal penalties.[296] Although there is no citizen suit provision in India's Biodiversity Conservation Act, India's Constitution provides broad access to the court system for any citizen to bring public-interest cases. Under Articles 32 and 226 of the Indian Constitution, a citizen can move the Indian Supreme Court or High Court to take action in the public interest.[297]

The National Biodiversity Authority ("NBA") has exercised its authority to prosecute violators of the Biological Diversity Act against Monsanto for genetically modifying eggplant (brinjal) without legal permission.[298] The NBA's decision to sue Monsanto was triggered by a petition that was filed in 2010 by a public-interest organization called the Environment Support Group ("ESG").[299] ESG filed another public-interest petition in 2011, alleging that authorities had unlawfully delayed criminal prosecution against Monsanto (and others) for bio-piracy in connection with its genetically-modified eggplant, leading the Karnataka High Court to ask the NBA and Karnataka State Biodiversity Board to pursue criminal proceedings

[293] Act No. 18 of 2003, Chapter II.

[294] Act No. 18 of 2003, Chapter III, § 18.

[295] Act No. 18 of 2003, Chapter IV, § 18(4).

[296] Act No. 18 of 2003, Chapter 3, § 2.

[297] Article 32 to the Constitution of India, section (1) ("The right to move the Supreme Court by appropriate proceedings for the enforcement of the rights conferred by this Part is guaranteed); Article 226, Constitution of India (authorizing India's High Courts to issue certain writs).

[298] Sreeha Vn, Indian High Court Reinstates Criminal Proceedings Against Monsanto And Its Partners In India's First Case Of Bio-Piracy, *International Business Times* (Oct. 18, 2013).

[299] UNDER WRIT ORIGINAL JURISDICTION Between : Environment Support Group and another, Petitioners, and National Biodiversity Authority and others, Respondents (2012).

against senior representatives of Monsanto's subsidiary in India.[300] Thus, India's biodiversity-protection legal framework is remarkable in that it provides strong enforcement mechanism for citizens to ensure that the law complied with, including through criminal proceedings.

China Like India, although China has a rich biological heritage and is the origin of many cultivated crops that are grown around the world, agriculture has inevitably contributed to biodiversity losses in China. Biodiversity losses, in turn, have led to losses of wild crop genetic resources—a matter of concern to the Chinese government.[301] Rare, native crops in particular have suffered serious losses.[302] Monoculture tree plantations show a low resistance to pets.[303]

To address biodiversity losses, in part to protect agriculture, China has enacted many laws to regulate various aspects of the country's biodiversity. Such laws include the Law on Protection of Wild Animals (2004), Environmental Protection Law (2014), Marine Environmental Protection Law (1999), Agriculture Law (2002), Forest Law (1998), Water Law (2002), Land Administration Law (2004), Fisheries Law (2004), Seed Law (2004), Patent Law (2000), Law on Prevention and Control of Desertification (2001), Law on Animal Epidemic Prevention (2007), Grassland Law (2008), Law on the Entry and Exit Animal and Plant Quarantine (2009 amended), and the Law on Water and Soil Conservation (2010). Of these, the newly revised Environmental Protection Law and the Law on the Protection of Wild Animals are the most significant legal authorities for biodiversity protection.

Recently, China has been beefing up its antipollution laws and the authority of the Ministry for Environmental Protection to regulate pollution, particularly after "Airpocalypse" in 2012, when particulate matters levels in Beijing soared to 25-40 times recommended levels.[304] China recently revised the "Environmental Protection Law," with the "environment," meaning: "the total body of all natural elements and artificially transformed natural elements affecting human existence and development," including wildlife.[305] The revised law is much stronger, establishes the principle that the environment is to take precedent over the economy, increases fines and penalties, establishes a requirement for "Environmental Impact Assessments," and even includes a provision authorizing non-governmental organizations ("NGOs") to enforce antipollution limits against polluters.[306] China's judicial

[300]*Id.*

[301]China National Biodiversity Conservation Strategy and Action Plan (2011–2030) at 5.

[302]*Id.*

[303]*Id.*

[304]Kracov et al. (2014); Louisa Lim, Beigjing's Airpocalypse" Spurs Pollution Controls, Public Pressure, *National Public Radio* (Jan. 4, 2013).

[305]Environmental Protection Law of the People's Republic of China (2014), Chap. 1, Art. 2 ("Environmental Protection Law").

[306]Lynia Lau, *China's Newly Revised Environmental Protection Law*, Clyde & Co. LLP, http://www.lexology.com/library/detail.aspx?g=8b5b0dcb-3358-47b3-a876-84166d9e2ba2, visited May 6, 2014; Environmental Protection Law, Chap. V, Art. 58 (citizen suit provision).

system has included specialized environmental courts that have sat idle, but this may be changing as a result of the revised Environmental Protection Law.[307]

Still, these improvements to China's environmental legal regime have resulted primarily from a need to vastly reduce pollution levels. Biodiversity protection has not yet received the same attention.

The purpose of China's Law on Protection of Wild Animals, enacted in 2004, is to protect and save species of wildlife that are "rare" or "near extinction," to develop and use wildlife as a resource, to maintain "ecological balances," and to protect specific wildlife that do not meet these criteria.[308] The law provides a way to list species, to designate nature reserves, and to plan for the "protection, development and rational utilization of wildlife resources," and requires consultation on projects that threaten the survival of listed wildlife.[309] The law prohibits the take and trafficking of protected species without a permit.[310] The Chinese government must criminally prosecute anyone who destroys nature reserves, which are designated for listed species, as well as anyone who destroys areas that are closed to hunting or any of "the main places where [listed] wildlife . . . lives and breeds."[311] However, the law does not contain a citizen suit provision, and the citizen suit provision in the revised Environmental Protection Law does not authorize such suits against violators of the Wildlife Protection Law.[312]

China's wildlife protection law also makes allowances for agriculture in specific ways. Local governments must compensate crop losses that result from protection of listed species.[313] Anyone who damages crops while hunting is required to compensate for those losses.[314] Local governments must take measures to prevent wildlife damage to human beings, livestock, and agricultural and forestry production.[315]

China's wildlife protection law does not extend to plants, which receive protection under separate authority: the Regulations of the People's Republic of China on Wild Plants Protection. These regulations provide a mechanism for listing wild plants, entitling such plants and their habitats to some protections such as consultation and/or translocation.[316] Those who unlawfully collect or damage wild plants or their habitats are subject to criminal penalties.[317] The regulations also regulate

[307]Kracov et al. (2014); Lin and Tuholske (2015), p. 10855.

[308]Law of the People's Republic of China on the Protection of Wildlife (2004) at Art. 1-2, 9, 24, 31.

[309]*Id.* at Art. 6, 9-10, 12.

[310]*Id.* at Art. 23.

[311]*Id.* at Art. 34.

[312]*Id.* at Art. 39.

[313]*Id.* at Art. 14.

[314]*Id.* at Art. 28.

[315]*Id.* at Art. 29.

[316]Regulations of the People's Republic of China on Wild Plants Protection at Art. 3, 9, 10, 11.

[317]*Id.* at Art. 7, 9, 16.

and protect the rights of those who develop or use wild plant resources, and encourage scientific research and public education.[318]

With so many laws, China's ability to successfully protect all aspects of the environment, including biodiversity, may come down to enforcement, including citizen enforcement through expansion of the citizen suit provision of the Environmental Protection Act.

Nigeria Nigeria is an agrarian society that depends on biodiversity resources for food and other goods. As a nation that spans many climatic and ecological zones—from semi-arid savanna to mountain forests, rich seasonal floodplain environments, rainforests, vast freshwater swamp forests and diverse coastal vegetation—Nigeria has some of the richest biodiversity in Africa.[319]

Yet, Nigeria's biodiversity is under severe pressure, and most once-common species are now rare.[320] The primary driver of this pressure is industrial agriculture and its components, including fertilizers and pesticides, land drainage, channelization of water courses and eutrophication of water bodies, and destruction of hedgerows and farm ponds.[321] Certain crops which were introduced into Nigeria's agricultural system in the 1900s—like cocoa, coffee, rubber, cotton, groundnut, and palm oil—are directly implicated in massive deforestation in the country.[322] Illegal wildlife trade has also been a major problem; CITES suspended trade with Nigeria from 2005 through 2011, due to a lack of regulations to enforce CITES licenses for wildlife trafficking, but this remains a problem, as more than 100,000 elephants are believed to have been poached between 2012 and 2015, many of them in Nigeria.[323] Invasive species are another significant issue.[324]

Although some of its people have lived at low densities and have utilized Nigeria's biodiversity resources sustainably, today Nigeria's large population is characterized by high percentages of unemployment and poverty, which "act as powerful drivers of increasingly severe demands on the remaining biodiversity in Nigeria."[325] Meanwhile, the nation is becoming increasingly urbanized.[326]

[318]*Id.* at Art. 4. The plant-protection regulations do not authorize citizen enforcement.

[319]Convention on Biological Diversity, Nigeria – Country Profile: Biodiversity Facts, https://www.cbd.int/countries/profile/default.shtml?country=ng#facts (last visited May 3, 2015) ("Nigeria Biodiversity Profile").

[320]*Id.*

[321]*Id.*

[322]*Id.*

[323]Press Release, Convention on International Trade in Endangered Species of Wild Flora and Fauna, CITES lifts suspension on Nigeria as the country intensifies law enforcement efforts (Aug. 26, 2011); Editorial, Wildlife Conservation in Nigeria, *This Day Live* (Jan. 6, 2015).

[324]Nigeria Biodiversity Profile, *supra* note 319.

[325]Nigeria Fifth National Biodiversity Report (2004) ("Nigeria Report") at 19.

[326]Nigeria Report at 19.

Nigeria's population of over 140 million is about one quarter the total population in sub-Saharan Africa.[327]

Yet, despite the importance of biodiversity to Nigeria, both the environment and biodiversity have received short shrift from a policy and legislative perspective.[328] Existing laws are obsolete, implementation of international conventions and treaties has been slow, public awareness and budgets to implement these agreements are low, land use planning has been poor, and enforcement of criminal laws "has been and still is a glaring setback for biodiversity conservation in Nigeria," with serious threats of poaching in national parks that are refugia for rare species.[329] Agriculture and urbanization, and a rising demand for fuel wood and charcoal, have made deforestation a major concern. Indeed, agricultural conversion is occurring in protected areas and is a great threat to rainforests and savannah woodlands.[330]

Indeed, one of the only Nigerian laws that address biodiversity is a 2007 law which established the National Environmental Standards and Regulations Enforcement Agency ("NESREA").[331] But although NESREA is charged with protection (and development) of, among other things, biodiversity, it lacks authority to list species or substantively protect them. As a result, as Nigeria's Fifth National Biodiversity Report states:

> Biodiversity issues have been relegated into the background and have only been the concern of conservationists, scientists and environmentalists despite its significant contribution to the livelihood and commerce of rural and peri-urban communities.[332]

In 2010, Nigeria prepared a biodiversity action plan, in part out of concern for genetic erosion and extinction of native crops due to "improved cultivars."[333] It includes a program to encourage the use of pesticides that are produced from indigenous plant derivatives, called "bio-pesticides."[334] Nigeria's plan also provides for biodiversity assessment and monitoring, including of agricultural lands, and achievement of stated targets.[335] Meanwhile, Nigeria has embarked on a review of biodiversity-related laws which is being conducted through a consultative process with Nigerian legal experts.[336]

---

[327] Nigeria Biodiversity Profile, *supra* note 319.

[328] Nigeria Report at 20.

[329] *Id.*

[330] Nigeria Fifth Assessment Report at 24.

[331] Nigeria Report at 20. Arguably, Nigeria's law addressing climate change, and possibly the Grazing Commission also concern biodiversity conservation. *Id.*

[332] *Id.*; National Environmental Standards and Regulations Enforcement Agency (Establishment) Act (2007).

[333] Nigeria NBSAP at 14–15.

[334] *Id.* at 15.

[335] Nigeria Biodiversity Profile, *supra* note 319; Nigeria NBSAP at 14–15.

[336] Nigeria Biodiversity Profile, *supra* note 319.

Cuba As discussed above, Cuba's agricultural policies are shaped by its geography and originate in a history of colonialism and slavery. Yet, environmental conservation is also highly valued in this "Accidental Eden," and today Cuba has a legal regime to protect the country's remarkable biodiversity, including through policies that incentivize agro-biodiversity.[337]

Environmental protection has long been valued in Cuba, as reflected by the popularity of José Marti's poems about environment and conservation which were popular at around the turn of the twentieth Century.[338] But centuries of cropland conversion for agriculture during colonization had left Cuba, which had 95 % forest cover when Columbus came ashore in 1492, with just 50 % forest cover by 1900, when Cuba supplied one-quarter of the world's sugar.[339]

After independence in 1920, Cuba started to take steps to reclaim aspects of its natural heritage. It designated the first of many national parks and refuges in 1930, and although these designations effectively toothless initially, they gained some traction under Cuba's Law of Agrarian Reform 1959, one of the first laws enacted following the Cuban Revolution which established reforestation programs.[340] Thus, although the Law of Agrarian Reform 1959 reflected Cuba's environmental consciousness, it did not establish meaningful regulatory mechanisms to achieve its objectives.[341]

This remained the status of environmental law in Cuba until 1976, when the Communist Party included in Cuba's first constitution a mandate to "protect nature" including "the soil, flora and fauna."[342] To achieve this mandate, Cuba established a new agency called the National Commission for the Protection of the Environment and the Conservation of Natural Resources ("COMARNA"). Five years later, Cuba passed Law 33, "On the Protection of the Environment and the Rational Use of Natural Resources," "to establish the basic principles for the conservation, protection, improvement, and transformation of the environment and the rational use of natural resources."[343] In 1990, pursuant to Decree-Law 118, Cuba eliminated COMARNA and replaced it with the Ministry of Science, Technology and the Environment.[344] But although it incorporated principles environmental protection into its legal framework in these ways, Cuba did not vest these executive entities

---

[337]OPB (2011).

[338]"Environmental Law in Cuba," Journal of Land Use and Environmental Law (Florida State), Fall 2000 ("Houck (2000)") at 13–14.

[339]*Id.* at 4 (citing Cuba: A Short History (Leslie Bethell ed., 1993)).

[340]*Id.* at 14.

[341]*Id.*; *see also* Brook (2004) (the Law of Agrarian Reform "failed to create an agency responsible for environmental protection or promulgate comprehensive environmental regulations outside the field of forestry").

[342]Houck (2000), p. 14 (citing CONSTITUCIÓN, art. 27 (1976)).

[343]Brook (2004), p. 218.

[344]Houck (2000), p. 15.

with adequate implementation or enforcement authority to ensure that they could be realized.[345]

Then, in the 1990s, as Cuba made vast changes in its agricultural policies and laws in response to the collapse of the Soviet Union during the Species Period, it also took steps toward a legal regime with meaningful environmental standards, including on the impacts of agriculture on biodiversity. First, after attending the 1992 World Summit at Rio de Janeiro, Fidel Castro amended the Cuban Constitution with a new provision to reinforce and reallocate Cuban society's obligation to ensure environmental protection, as follows:

> The State protects the environment and the natural resources of the country. It recognizes their close link with the sustainable economic and social development for making human life more sensible, and for ensuring the survival, welfare, and security of present and future generations. It corresponds to the competent organs to implement this policy.
>
> It is the duty of the citizens to contribute to the protection of the water and the atmosphere, and to the conservation of the soil, flora, fauna, and all the rich potential of nature.[346]

Thus, this amendment made clear that it was the State's duty to protect Cuba's natural resources, and that it was the obligation of its citizens to "contribute" to protection of the water, atmosphere and to the "conservation of the soil, flora, fauna, and all the rich potential of nature."[347] The language was proactive and established a principle of environmental citizenship alongside the State duty.[348] Additionally, Cuba established a centralized State ministry—*i.e.*, a cabinet-level federal agency—with a dominant mission (among many) to "steer and control the implementation" of Cuba's environmental policies and management, and the use of its natural resources.[349]

In 1994, COMARNA was replaced with the Ministry of Science, Technology and the Environment ("CITMA"), an independent, "cabinet-level agency" that was "established exclusively for the environment."[350] CITMA consolidated more than a dozen environmental institutes and centers, was given the authority to implement environmental regulations—a distinct improvement over COMARNA's mediocre

345 Brook (2004), p. 218; Houck (2000), p. 15.

346 CUBAN CONSTITUTION, art. 27. During his speech at the Rio Conference, Castro said the following:

> If we want to save humanity from destroying itself, we have to distribute more equitably the riches and the available technologies on this planet. Less luxury and pilfering from a few countries for less poverty and hunger for the rest of the earth. No more transfer to the Third World of lifestyles and habits of consumerism that ruin the environment. Make human living more rational. Apply international economic order that is just. Use all the science necessary for sustainable development, without pollution. Pay the environmental debt, not the foreign debt. Eliminate hunger, and not humankind.

Houck (2000).

347 Houck (2000).

348 *Id.*

349 *Id.*

350 Brooks (2004), p. 221.

authority—and was charged with "'steer[ing] and control[ling] the implementation' of environmental policy, the rational use of natural resources, and sustainable development."[351] CITMA was also mandated to "draw up and control the implementation" of regulatory programs, including the "adequate management of agricultural and industrial waste practices" and the "clean production practices."[352] It was also authorized to settle disagreements among agencies over environmental issues by "making the relevant decisions" or passing them on to the Council of Ministers.[353]

Then, in 1997, Cuba enacted Law 81, "The Law of the Environment," a sweeping initiative that replaced Law 33 and included provisions mandating environmental protection including biodiversity through the listing of species, gathering of information, protection of habitats and ecosystems, in particular from invasive species and from genetic engineering.[354] While Law 81 did not specifically address agriculture's impacts to biodiversity per se, it did not limit the potential reach of its protections either, and mandated that *all* activities that potentially impact biodiversity be monitored and regulated, whether "within or outside of protected areas" so as to "insure their conservation and sustainable use."[355]

The result has been mixed. Starting in 1998, Cuba began to implement a successful reforestation program that is on track to reach the target of 29 % forest cover by 2015, and has consistently been one of the nations with the largest increases in forest areas and designations of forest areas for protective reasons.[356]

In contrast to the other nations profiled, and as an outgrowth of its geography, history, and style of government, Cuba's biodiversity protection policies and laws are tied in many ways to its polices and regulatory scheme for agriculture—when viewed together, these two frameworks reflect a formal, more integrated value of biodiversity, in part because of its unique heritage and environment, but also because of its value to agriculture and other human uses.

[351] *Id.* at 222; Houck (2000), p. 19 (quoting Agreement No. 2823 of the Executive Committee of the Council of Ministries of November 28, 1994) ("Agreement (1994)") (CITMA organized with two primary branches. One functions much like a combination of the U.S. Environmental Protection Agency and U.S. Department of the Interior, houses scientific institutes, and manages Cuba's parks and other natural areas; the other, the Environmental Policy Directorate, is more policy-focused and develops new proposals. *Id.* at 20.

[352] Houck (2000), p. 20 (quoting Agreement (1994)).

[353] *Id.* at 20 (citing Agreement (1994)).

[354] Law 81, The Law of the Environment, Ch. II, art. 86.

[355] Law 81, Ch. II, art. 86(a)-(c), (g).

[356] Food and Agriculture Organization of the United Nations, State of the World's Forests (2012); Food and Agriculture Organization of the United Nations, State of the World's Forests 19 (2011).

19.3 Discussion and Analysis

19.3.1 Agriculture, Biodiversity, and Population: A Biodiversity Iron Triangle

Agricultural legal frameworks are often as old as the law itself, while biodiversity-protection laws have been enacted much more recently. In today's world, with a growing world population as backdrop, agricultural laws are intended both to permit as well as regulate agriculture. They are intended both to ensure that industrial agricultural methods may be widely used, including methods that are drivers of the extinction crisis, while also maintaining certain protections for natural resources in recognition of the important role that natural resources and biodiversity play for agriculture itself.

Yet, whereas agriculture needs biodiversity, biodiversity does not need agriculture. Agriculture is a major reason why the Earth is experiencing the Holocene Extinction. Whether conversion of habitats to croplands and rangelands, the use of monoculture, pesticides, synthetic fertilizers, herbicides, genetically-modified crop systems, agricultural runoff, meat production, water use and pollution, the targeting of important species like carnivores, or greenhouse gas emissions, agriculture is making it possible for the world population to increase exponentially and is a driving force of the Holocene Extinction era.

19.3.1.1 The Legacy of Agricultural Laws and Policies Designed to Increase Food Production

Industrial agricultural methods are widely used both in the U.S. and around the world in order to maximize crop yields in order to feed an ever-growing world population. These methods are encouraged, and even subsidized, by favorable laws and policies that were enacted to support increased crop yields to feed a growing world population. Heavy use of monoculture, synthetic fertilizers, and pesticides, as well as meat production, pollution, and climate change are destroying the Earth's biodiversity.

While many nations have enacted laws to protect biodiversity directly or indirectly, including from agriculture's threats, many of these laws have large regulatory gaps, are not adequately implemented or funded, or suffer from humans' propensity to elevate their own, short-sighted goals at the expense of protecting the very fabric of life—in other words, by politics. Even the U.S. Endangered Species Act—the oldest and among the strongest of biodiversity-protection laws—suffers from these flaws.

19.3.1.2 A Vicious Cycle with Deep Origins in Culture

When thought of together, agriculture, population, and biodiversity have an intertwined and interesting relationship. With scarcity in the human DNA, cultures and societies have always tended to assume the need to organize themselves and their laws around the need to feed more people, crops, and livestock, then to make room for more people, then to feed those additional people, and on and on. The cycle acts as a negative feedback loop on the critical resource that both agriculture and people require—biodiversity—while biodiversity requires neither agriculture nor people.

Rarely discussed in polite conservation, as it is generally considered to be a cultural taboo, is the critical importance of biodiversity to agriculture, or for that matter, the Earth's carrying capacity for humankind (and how best to stay beneath it). Of the three—agriculture, population, and biodiversity—population holds the most promise of interrupting the feedback loop that has the Earth in the throes of a human-caused mass extinction era. If we are to have the hope of doing so, we must start to talk about this in polite conversation.

19.3.1.3 Current Biodiversity-Protection Legal Frameworks Lack Teeth

As we have seen, many of the world's most populous nations have instituted laws and policies to increase agricultural production, most often by placing emphasis on intensive agricultural practices, subsidies, and interference in markets. After facing the environmental consequences of such methods, however, most of these nations have come to emphasize sustainable agricultural practices, and in separate laws, to protect biodiversity losses, including from agriculture.

However, these regulatory frameworks have not solved the difficult challenge that these nations face, which is to maintain high yields of food and fiber crops and livestock for the purpose of feeding and clothing large, growing human populations, while still protecting the diversity of species that provide ecosystem services and national heritage at ecologically sustainable levels.

Biodiversity protection laws have not adequately modified intensive agricultural methods to reduce the impact on biodiversity. These methods are depleting soils, polluting aquatic areas, and depleting scarce water resources, and genetically-modified, pesticide-resistant crops are eliminating habitats for important invertebrate pollinators like butterflies and bees. Policies to target the killing of predators remove critically important animals from ecosystems on behalf of domesticated livestock, which spread invasive plant species. Despite international recognition of the current mass extinction era and the threat that it presents to human societies, domestic and international legal frameworks have not adequately addressed these detrimental impacts of intensive agricultural methods.

The reasons for this are apparent. Biodiversity laws tend to be much younger than legal regimes that regulate agriculture, and laws that regulate agriculture typically reflect long-standing, entrenched policies favoring maximum agricultural production. To the extent that such laws and policies incorporate environmental concerns, such concerns are typically centered on resource protection in order to support agriculture itself. Second, many laws lack clear, substantive protections for species that are formally recognized as endangered or threatened. Third, even where such protections exist, they are not enforced adequately. Finally, current legal frameworks which apply to agriculture and its impacts to biodiversity lack policy objectives to promote sustainable human population levels that do not contribute to further losses in biodiversity.

19.3.2 How Can Agricultural Policies Be Improved to Recognize the Critical Importance of Biodiversity and to Control Agricultural Practices Adequately So as to Protect Biodiversity?

Perhaps it is too much to hope that biodiversity-protection laws—which protect species, who do not have a vote—could catch up to agricultural methods that are favored and subsidized through policies that have been in place, in one form or another, for much longer. Yet, where these methods are destroying species at an unsustainable pace, how can agricultural policies and laws be updated to reflect today's needs? From the U.S. example and those of the profiled nations, what has been most successful in regulating agriculture to protect biodiversity? What has worked, and what has not? Are there gaps that must be filled, or can the challenge be met by improvements to existing legal frameworks which regulate agriculture?

19.3.3 How Can Biodiversity-Protection Policies Be Improved to Control Agricultural Practices Adequately?

Many biodiversity-protection laws lack clear substantive protections for species that are formally recognized as threatened. How long must a species be protected before it is deemed to have recovered from the threat of extinction? Should it be restored throughout all remaining portions of suitable habitat that are left in its historic range, or should it be secured and stabilized for the foreseeable future in a small percentage of its historic range? Which should get the benefit of any doubt: agriculture or biodiversity? What regulatory mechanisms would be necessary to ensure that biodiversity-protection principles are enforced? How would you change existing laws to ensure that agricultural threats to biodiversity are not only regulated, but abated?

19.3.4 Should We Strengthen the Laws, or Change Them? Which Laws Should Be Changed?

If protections exist but are not being applied forcefully enough to interrupt the negative feedback cycle, is the remedy to follow the laws that are on the books, or to change them? Are there gaps in biodiversity-protection legal regimes? In the absence of answers to these questions, should population be regulated to stay within certain limits that include limits to protect the Earth's biodiversity from humankind's activities?

19.3.5 Is Regulation of Human Population the Answer, and If So, How Should It Be Accomplished?

Current legal frameworks which apply to agriculture and its impacts to biodiversity lack any policy objectives to promote sustainable human population levels that truly sustainable agricultural methods can adequately feed without contributing to further losses in biodiversity. If regulation of human population size is the best approach to protecting biodiversity from agriculture, how should it be done? Is China, which had a one-child policy in the 1970s, an example of policies that succeeded? Why or why not? Are there any other examples that can be gleaned from?

19.4 Examples and Case-Studies

19.4.1 Agriculture v. The Delta Smelt

For an example of the clash between industrial agriculture to feed a growing population, see the case study of the Delta smelt. *See Consol. Delta Smelt Cases*, 717 F.Supp.2d 1021 (E.D. Cal. 2010)*; Delta Smelt Consol. Cases v. Salazar*, 760 F. Supp.2d 855 (E.D. Cal. 2010); *San Luis & Delta-Mendota Water Auth. v. Jewell*, 747 F.3d 581 (9th Cir. 2014); *Delta Smelt Consol. Cases v. Salazar*, 663 F.Supp.2d 922 (E.D. Cal. 2009); *NRDC v. Kempthorne*, 506 F.Supp.2d 322 (E.D. Cal. 2007).

19.4.2 Biodiversity v. Pesticides

For examples of how U.S. laws that regulate industrial agricultural methods and biodiversity losses have not been sufficient, see the papers about the impacts of atrazine on amphibians; *Nat'l Res. Def. Council v. U.S. Envtl. Prot. Agency*, 2005 U.S. Dist. LEXIS 45449 (D. Md. May 24, 2005); *Wash. Toxics Coalition v. EPA*, 2003 U.S. Dist. LEXIS 26088 (W.D. Wash. Aug. 8, 2003); *Wash. Toxics Coal. v. U. S. Dep't of Interior*, 457 F.Supp.2d 1158 (W.D. Wash. 2006); *Ctr. for Biological Diversity v. EPA*, 65 F.Supp.3d 742 (N.D. Cal. 2014); *Ctr. for Biological Diversity v. U.S. Dep't of Interior*, 2015 U.S. Dist. LEXIS 112974 (N.D. Cal. Aug. 24, 2015).

References

Altieri MA (1999) The ecological role of biodiversity in agroecosystems. Agric Ecosyst Environ 74:19

Barnosky AD et al (2011) Has the Earth's sixth mass extinction already arrived? Nature 471:51

Barnosky AD et al (2012) Approaching a state shift in Earth's biosphere. Nature 486:52

Bergstrom JB et al (2013) License to kill: reforming federal wildlife control to restore biodiversity and ecosystem function. Conserv Lett 6:1

Beschta RL et al (2012) Adapting to climate change on western public lands: addressing the ecological effects of domestic, wild, and feral ungulates. Environ Manag 51(2):474

Brahmi P, Dua RP, Dhillon BS (2004) The biological diversity act of India and agro-biodiversity management. Curr Sci 86(5):659

Brook CA (2004) COMMENT: Cuba: undermining or underlining the "Race to the Bottom?". N C J Int Law Commer Regul 30:197–218, 207

Ceballos G et al. (2015) Accelerated modern human-induced species losses: Entering the sixth mass extinction. Sci Adv 1(5). doi:10.1126/sciadv.1400253

Crutzen PJ (2002) Geology of mankind. Nature 415:23

Cully JF, Collinge SK, VanNimwegen RE, Ray C, Johnson WC, Thiagaragjan B, Conlin DB, Holmes BE (2010) Spatial variation in keystone effects: small mammal diversity associated with black-tailed prairie dog colonies. Ecography 33(4):667

Cunha A, Sevinate AP, Correia M, Miribel B, Cardoso C, Sousa CR, Godfray C, Baldock D, Duarte F, Avillez F, Barros H, do Carmo I, Ribeiro I, Contreras J, Lima Santos J, Luis Domingo J, Neto L, Cabral MH, Nunes ML, Graca P, Lang T (2013) The future of food: environment, health and economy. Fundação Calouste Gulbenkian

De Vos JM et al (2015) Estimating the normal background rate of species extinction. Conserv Biol 29(2):452

Douillet M, Grandval F (2010) Nigeria's agricultural policy: seeking coherence within strategic frameworks. Grain de sel 51:16

Ellickson RC, Thorland CDiA (1995) Ancient land law: Mesopotamia, Egypt, Israel. Yale Law Faculty Scholarship Series, paper 410. http://digitalcommons.law.yale.edu/fss_papers/410

Estes JA et al (2011) Trophic downgrading of planet earth. Science 333:301

Gonzalez C (2003) Article: seasons of resistance: sustainable agriculture and food security in cuba. Tul Environ Law J 16:685

Gorenflo LJ et al (2012) Co-occurrence of linguistic and biological diversity in biodiversity hotspots and high biodiversity wilderness areas. Proc Natl Acad Sci 109(21):8032

Harmon D (1996) Losing species, losing languages: connections between linguistic and biological diversity. Southwest J Linguist 15(89):108

Hayes TB et al (2002) Hermaphroditic, demasculinized frogs after exposure to the herbicide Atrazine at low ecologically relevant doses. Proc Natl Acad Sci 99(8):5476

Hayes TB et al (2006) Pesticide mixtures, Endocrine disruption, and amphibian declines: are we underestimating the impact? Environ Health Perspect. doi:10.1289/ehp.8051

Hooke RL et al (2012) Land transformation by humans: a review. GSA Today 22(12):4

Jarvis A, Lane A, Hijman RJ (2008) The effect of climate change on crop wild relatives. Agric Ecosyst Environ 126:13

Kracov G, Sung S, Tsai MM (2014) Can citizen suits help China battle pollution? Los Angeles Daily J

Krinsky D (2007) Article: how to sue without standing: the constitutionality of citizen suits in non-article III tribunals. Case West Res 57:301

Laitos JG, Ruckriegle H (2013) The clean water act and the challenge of agricultural pollution. Vt Law Rev 37:1033–1070

Lal R (2004) Soil carbon sequestration impacts on global climate change and food security. Science 304:1623

Lin Y, Tuholske J (2015) Field notes from the far east: China's new public interest environmental protection law in action. Environ Law Rev News Anal 45:10855

Mace GM et al (2005) Millennium ecosystem assessment, Current state and trends: Findings of the Condition and Trends Working Group. Ecosyst Hum Well-Being 1:77

Mallin MA (2000) Impacts of industrial animal production on rivers and estuaries. Am Sci 88:26

Martinez-Estévez L, Balvanera P, Pacheco J, Ceballos G (2013) Prairie dog decline reduces the supply of ecosystem services and leads to desertification of semiarid grasslands. PLOS-ONE 8 (10):1

Matson PA et al (1997) Agricultural intensification and ecosystem properties. Science 277:504

Miller B, Reading R, Hoogland J, Clark T, Ceballos G, List R, Forrest S, Hanebury L, Manzano P, Pacheco J, Uresk D (2000) The role of prairie dogs as a keystone species: response to stapp. Conserv Biol 14(1):318

Mossop D (1993) Citizen suits: tools for improving compliance with environmental laws. Altern Law J 18:266

Perring MP, Standish RJ, Hobbs RJ (2013) Incorporating novelty and novel ecosystems into restoration planning and practice in the 21st century. Ecol Process 2:18. doi:10.1186/21921709218

Pimental D (1997) Economic and environmental benefits of biodiversity: the annual economic and environmental benefits of biodiversity in the United States total approximately $300 billion. BioScience 47(11):747

Pimm SL et al (1995) The future of biodiversity. Science 269(5222):347

Pimm SL et al (2014) The biodiversity of species and their rates of extinction, distribution, and protection. Science 244(6187):1

Prugh LR et al (2009) The rise of the mesopredator. BioScience 59(9):779

Rabalais NN et al (2002) Gulf of Mexico Hypoxia, A.K.A. "The Dead Zone". Ann Rev Ecol Syst 33:235

Régnier C et al (2015) Extinction in a hyperdiverse endemic Hawaiian land snail family and implications for the underestimation of invertebrate extinction. Conserv Biol. doi:10.1111/cobi.12565

Ripple WJ et al (2014) Trophic cascades from wolves to grizzly bears in yellowstone. J Anim Ecol 83:223

Rohr JR, McCoy KA (2010) A qualitative meta-analysis reveals consistent effects of Atrazine on freshwater fish and amphibians. Environ Health Perspect 118(1):20

Sax JL (1970) The Public Trust Doctrine in natural resource law: effective judicial intervention. Mich Law Rev 68:471–475

Thomas CD et al (2004) Extinction risk from climate change. Nature 427:145

Urban MC (2015) Accelerating extinction risk from climate change. Science 348(6234):571

Urban MC et al (2012) On a collision course: competition and dispersal differences create no-analogue communities and cause extinctions during climate change. Proc R Soc Bull 279:2072

van Zanden JL (1991) The first green revolution: the growth of production and productivity in European agriculture, 1870–1914. Econ Hist Rev 44(2):215

Wielgus RB, Peebles KA (2014) Effects of wolf mortality on livestock depredations. PLoS One 9 (12):1

Chapter 20
Phytoremediation and the Legal Study of Soil, Animals and Plants

Bernard Vanheusden

Abstract Soils are under increasing environmental pressure in every country across the globe. This pressure is mainly driven by human activities, such as agricultural and forestry practices, industrial activities, tourism and urban development. Over recent decades, there has been a significant increase in the rate of soil degradation, with no sign of amelioration. The main threats to which soils are subject are erosion, chemical contamination, compaction, biodiversity loss, sealing, landslides and flooding. Soils are a resource of common concern both within and between nations, and failure to protect them will undermine ecological and economic sustainability. Soil degradation has substantial impacts on other areas of common interest such as water quality and quantity, climate change, biological diversity, human health, and, in particular, food and feed safety and food security.

20.1 Introduction[1]

20.1.1 General

Soils are under increasing environmental pressure in every country across the globe. This pressure is mainly driven or exacerbated by human activities, such as agricultural and forestry practices, industrial activities, tourism and urban development. Over recent decades, there has been a significant increase in the rate of soil degradation, with no sign that it will be ameliorated. The main threats to which soils are subjected are erosion, chemical contamination, compaction, biodiversity loss, sealing, landslides and flooding. Soils are a resource of common concern both within and between nations, and failure to protect them will undermine ecological and economic sustainability. Soil degradation has strong impacts on other areas of

[1]This contribution is partially based on the following research report within the Rejuvenate 2 project: Vanheusden et al. (2011). This project was funded by the Snowman network.

B. Vanheusden (✉)
Faculty of Law, Hasselt University, Hasselt, Belgium
e-mail: bernard.vanheusden@uhasselt.be

G. Steier, K.K. Patel (eds.), *International Farm Animal, Wildlife and Food Safety Law*, DOI 10.1007/978-3-319-18002-1_20

common interest, such as water quality and quantity, climate change, biological diversity, human health, and in particular, food and feed safety and food security.

However, soil is of crucial importance. Soil is the interface between earth, air and water and hosts most of the biosphere. Soil provides us with food, biomass and raw materials. It serves as a platform for human activities and landscape and as an archive of heritage and plays a central role as a habitat and gene pool. It stores, filters and transforms many substances, including water, nutrients and carbon. In fact, it is the biggest carbon store in the world (1500 gigatonnes).[2] These functions must be protected because of both their socio-economic and environmental importance.

Therefore, further soil degradation should be prevented and degraded soils should be restored to a level of functionality consistent at least with current and intended use, thus also considering the cost implications of the restoration of soil.

This contribution focuses on the legal aspects of an innovative way to at the same time restore degraded soils, in particular contaminated soils, and use them for agricultural cultivation (food, feed, biomass,. . .). This innovative way, which starts to find its way through various research projects,[3] is the use of the technique of phytoremediation. Phytoremediation uses plants/crops to remove pollutants from the environment or to render the pollutants harmless. These plants/crops can further on potentially be used as valuable crops (food and non-food).

The combination of agricultural cultivation and soil remediation could be an important part of sustainable risk based land management. It is very interesting for all kinds of large-scale, diffuse, moderate contaminations.

Unfortunately, a sound legal framework for decision-making about the use of phytoremediation, and broader a sustainable risk based land management, is missing. Therefore a detailed study of all the potential legal bottlenecks is absolutely necessary before starting a phytoremediation project. This contribution offers an overview and analysis of common legal bottlenecks.

20.1.2 Phytoremediation

Phytoremediation is bioremediation of contaminated soils by using plants/crops, applicable for the removal or degradation of organic and inorganic pollution in soil, water and air. It uses plants to remove pollutants from the environment or to render the pollutants harmless. Commonly used types of plants/crops are for example

[2]European Commission (2006), p. 2. For a discussion of the strategy, refer to Van Calster (2004), pp. 3–17.

[3]Refer to e.g. the Rejuvenate 2 project (Crop Based Systems for Sustainable Risk Based Land Management for Economically Marginal Degraded Areas). For more information refer to http://projects.swedgeo.se/r2/ (last visited on 3 January 2015).

maize, rapeseed and sunflower as well as various types of short rotation coppice (birch, maple, poplar, willow,. . .).

The most commonly used types of phytoremediation are stabilization and extraction. Both techniques use plants to achieve their objectives and differ mainly in the process to achieve these objectives. Stabilization focuses on immobilizing the contaminants in a way that they are unavailable for uptake in the (harvestable parts of the) plants and leaching into the environment is prevented. However, the contaminants remain in the soil. Extraction aims at the uptake of the contaminants by harvestable parts of the crops. Subsequently the contaminated biomass is harvested and removed from the site. The result is that the contamination in the soil is reduced. This difference between the two techniques has an impact on the applicability of regulations for all phases of a phytoremediation project.

The figure here below summarizes advantages and disadvantages of conventional and phytoremediation techniques.[4]

Conventional remediation	Phytoremediation
Some characteristics: • Excavation • *Ex situ* treatment (usually) • Treatment or disposal	Some characteristics: • Use of plants • *In situ* treatment • Use for food/feed, energy, material
Advantages: • Good for serious contamination • Quick remediation	Disadvantages: • Only for moderate contamination • Slow remediation
Disadvantages: • Only for concentrated contamination • Only suitable for small surfaces • Loss of soil structure • High cost	Advantages: • Good for diffuse contamination • Suitable for large surfaces • Keeps soil structure • Cost similar to farming cost

Phytoremediation has a high potential to enhance the degradation and/or removal of organic contaminants from soils.[5] On the other hand it produces potentially valuable crops (food and non-food). In other words, it concerns at the same time soil remediation, soil usage, improving the productivity of land/soil, so it is environmentally sound. This also relates to a benefit for animals.

The combination of agricultural cultivation (food, feed, biomass,. . .) and soil remediation could be an important part of sustainable risk based land management where the management strategy could become self-funding and could even result in other environmental benefits such as carbon sequestration. Risk based land management refers to the fact that it is necessary to consider to what extent toxic substances may harm human health or the wider environment not only while in soil, but also after remediation. Risk management has as its main aim to identify the different elements of the contaminant-pathway-receptor chain and to break this chain. The pathway can be very diverse, but a common example is that the

[4]Witters et al. (2009).

[5]Vangronsveld et al. (2009) and Weyens et al. (2009, 2010).

contaminants in soil are taken up by vegetables which are directly consumed by humans. Or, after remediation, contaminants might eventually end up at a disposal site. A remediation strategy is effective if it minimizes or controls the health or environmental risks associated with a particular pollutant linkage. By removing contaminant sources, breaking exposure pathways between source and receptor, or changing the receptors, pollutant linkages are minimized.[6] This might therefore result in a management strategy that does not necessarily lead to a removal of toxic substances.

Total contaminant removal rate and resulting remediation duration depend on soil characteristics, level of contamination, available contamination, crop extraction, and crop biomass production.[7] When contaminants are removed, the total amount of contamination reduces, as well as the amount of available contaminants. This relation can be linear but is in most cases logarithmic, meaning that the amount of available contaminants reduces faster than the amount of contamination, reaching a limit amount of available contaminants. Over time, contaminant concentration in the plant is then affected by the amount of available contaminants in soil over time. However, some authors describe a replenishment of the available contaminant pool.[8] This is *e.g.* the case for sandy soils. Also, biomass production of the plant might change over time. *E.g.* the biomass potential of short rotation coppice (SRC) of willow might increase after several years within a rotation cycle, whereas the biomass production of energy maize or rapeseed might decrease in time (due to nutrient depletion). Moreover, the depth of the rooting zone might be a factor that influences contaminant concentration in plants, as concentration found in plants might differ according to root depth. Finally, there could also be leaching losses to the groundwater.[9]

Phytoremediation definitely becomes interesting to restore degraded land when the produced plants/crops/biomass can still be valorized into an income (food, feed, biomass,. . .). Then the main drawbacks of phytoremediation, namely the fact that it takes place on contaminated land and, in the case of phytoextraction, the extended remediation period required, may become invalid and slower working phytoremediation schemes based on stabilization and/or gradual attenuation of the contaminants rather than short-term forced extraction may be envisaged.[10]

Conventional remediation techniques will take less time, but during remediation it will not be possible to validate the soil. When phytoremediation is implemented, the repeated cropping of plants produces high amounts of plants/crops/biomass. In case the contaminants levels stay below the levels for food and/or feed, the crops can be used for food and/or feed. Otherwise, they need to be disposed of or better,

[6]Mench et al. (2010).

[7]Koopmans et al. (2008).

[8]Van Nevel et al. (2007).

[9]Van der Grift and Griffioen (2008).

[10]Robinson et al. (1998).

treated appropriately to prevent any risks to the environment.[11] In that case, the utilization of the obtained biomass as an energy resource becomes very attractive[12] and can even turn phytoremediation into a profit making operation.[13] Moreover, using the resulting biomass for energy may contribute to the reduction of global carbon dioxide (CO_2) emissions. Biomass can replace fossil fuels for the supply of heat, electricity and transport fuel, and can also serve as a feedstock for material production.[14] As mentioned by Firbank (2008), one factor that drives the growth of alternative crops for energy is the fact that they deliver an environmental benefit by reducing greenhouse gas (GHG) emissions.

Energy conversion is a sustainable alternative for several reasons. First, energy production will more likely get public approval, opposed to other destinations (*e.g.* paper mills). In many countries farmers are variously rewarded for direct positive contributions to biological diversity (particularly wildlife habitat), improvements (or avoided negative impacts) to water quality and increased soil health. Many countries also support bioenergy programs, with the intent to promote the production and use of cleaner fuels instead of fossil fuel. Moreover, on a global scale, the increasing interest in carbon sequestering effects of many types of agriculture points to a growing number of programs in the near future that will support certain farming practices as a way of improving overall air quality.[15] Second, energy conversion installations are able to trace heavy metals within their system. At least, as far as we know there is research on this matter in the energy sector, but there has been no research yet on tracing heavy metals in other biomass using technologies (*e.g.* paper mills).

Also a combination of food and/or feed on the one hand and biomass on the other hand can be possible. In the case of maize for example, the contamination is taken up by the plant, but mainly remains in the green parts of the plant, and does not go into the maize grains. As a result, the maize grains can still be used as feed for livestock, and the rest fraction (the green parts (leaves, roots, trunk,. . .)) can be used for green energy production (through incineration or digestion).

20.1.3 Goal and Structure

As already mentioned, this contribution focuses on the legal aspects of the management of contaminated soils through phytoremediation. The combination of agricultural cultivation and soil remediation could be an important part of

---

[11]Ghosh and Singh (2005).

[12]Chaney et al. (1997).

[13]Meers et al. (2005).

[14]Dornburg and Faaij (2005).

[15]De Vries (2000).

sustainable risk based land management. It is very interesting for all kinds of large-scale, diffuse, moderate contaminations.

Unfortunately, a sound legal framework for decision-making about the use of phytoremediation, and broader a sustainable risk based land management, is missing. Therefore, this contribution offers an overview and analysis of common legal bottlenecks.

This part follows the chronology in practice in decision-making before the start and during the phytoremediation project. It is therefore divided in two subchapters, one on the selection of the crops and the management of the soil and one on the growing and the harvesting of the crops.

The first subchapter "Selection of crops and soil management" focuses on international and examples of national legislation affecting crop selection and related aspects, namely plant selection, and on soil management (including management of risks) in case of phytostabilization or phytoextraction. In a first paragraph plant characteristics, namely the issues of invasive and exotic plants and of genetically modified organisms, are discussed. The second paragraph considers the implications of soil management both for stabilization and extraction (for example amendments to the soil, bioavailability of pollutants and plant protection).

The second subchapter "Growing and harvesting the crops" describes the international and examples of national legal aspects of growing crops on contaminated soil and harvesting these crops. It thereby covers several legal domains as relevant. This methodology is based on our experience that following the process of the project gives the most practical and complete overview of the legal elements. The latter is at risk when focusing on a limited selection of legal domains.

20.2 Selection of Crops and Soil Management

20.2.1 Plant Characteristics

20.2.1.1 Introduction

The first question to be asked is whether there are any plant species that are not legally allowed to be used in phytoremediation? Besides purely scientific and utilitarian reasons for selecting crops, one should ensure that the selected species is legally allowed to be planted on the site. Several international and supranational agreements and regulations have an impact on this and also national legislation may have an impact.

At the international level for example the Convention on Biological Diversity (CBD) of 1992 and the International Plant Protection Convention (IPPC) of 1951, and revised in 1997, are relevant. Within for example the European Union

(EU) several directives, such as the Habitats Directive,[16] a Directive on protective measures against organisms harmful to plants and plant products[17] and the GMO Directive[18] are relevant.

However, for each real, individual project an assessment on the basis of site location and local circumstances remains necessary.

20.2.1.2 Invasive and Exotic Plants

Invasive and exotic plants are a danger to biodiversity. The state of Florida (USA) experienced this with Arundo donax, a large clumping grass that produces a lot of biomass usable as bioenergy feedstock.[19] The plant is also used in phytoremediation for its uptake of arsenic,[20] mercury and cadmium.[21] Giant reed alters the hydrology and displaces native species.[22] It is known to be destructive to fish and amphibian habitats and to seriously harm habitats for rare species by its potential to reproduce by fragmentation of rhizomes and production of new roots from stems. It forms large clumps of grass in riparian habitats. Meanwhile, Arundo donax has been imported in all regions across the globe, sometimes only for decoration purposes, but frequently for phytoremediation or the production of biomass. Research on genetically modifying the grass for phytoremediation purposes is ongoing.

Already in 1992 the Parties to the Convention on Biological Diversity (CBD) committed themselves to protect their ecosystems, habitats or species against the threats posed by alien species.[23] This principle should be translated into the national regulations of the Parties. Although the United States of America (USA) are not a Party to the CBD (they are only a Signatory), they do follow the same principle. Thus, when selecting a crop it is advisable to check the national and local rules

---

[16]Council Directive 92/43/EEC of 21 May 1992 on the conservation of natural habitats and of wild fauna and flora, *Official Journal* 22 July 1992, L 206, consolidated version of 2007 on ec.europa.eu/environment/nature/legislation/habitatsdirective/index_en.htm (last visited on 15 January 2015).

[17]Council Directive 2000/29/EC of 8 May 2000 on protective measures against the introduction into the Community of organisms harmful to plants or plant products and against their spread within the Community, *Official Journal* 10 July 2000, L 169, consolidated version of 2010 on http://eur-lex.europa.eu/LexUriServ/LexUriServ.do?uri=CONSLEG:2000L0029:20100113:EN:PDF (last visited on 15 January 2015).

[18]Directive 2001/18/EC of the European Parliament and of the Council of 12 March 2001 on the deliberate release into the environment of genetically modified organisms and repealing Council Directive 90/220/EEC, *Official Journal* 17 April 2001, L 106.

[19]Florida Native Plant Society, *Policy statement on Arundo donax*, p. 1, http://www.weedcenter.org/management/docs/Arundo%20Policy%20Statement.pdf, last visited on 6 January 2015.

[20]www.ncbi.nlm.nih.gov/pubmed/20363125.

[21]www.ncbi.nlm.nih.gov/pubmed/16110677.

[22]www.issg.org/database/species/ecology.asp?si=112.

[23]Article 8, h of the Convention on Biological Diversity (CBD).

relating to non-native, alien species. Within the EU this is for example translated in the Habitats Directive into an obligation for the Member States to "*ensure that the deliberate introduction into the wild of any species which is not native to their territory is regulated so as not to prejudice natural habitats within their natural range or the wild native fauna and flora and, if they consider it necessary, prohibit such introduction.*"[24] Examples of non-native plants, commonly used for phytoremediation, in the EU that would fall under the Habitats Directive are: Miscanthus, Switch grass and Arundo donax.

If on the site plants and animal life is present, it is also useful to check on the eventual presence of protected species.

Another important convention concerning plant selection is the International Plant Protection Convention (IPPC). According to article VII, 2, (i) of this convention "*[c]ontracting parties shall, to the best of their ability, establish and update lists of regulated pests, using scientific names, and make such lists available to the Secretary, to regional plant protection organizations of which they are members and, on request, to other contracting parties*". The lists of regulated pests are available on the website of the IPPC,[25] or on the websites of the Regional Plant Protection Organizations (RPPO)[26] or of national or state authorities. When selecting a crop for a phytoremediation project, it is advisable to consult these databases before the final decision.

Other specific regional or national legislation may also apply. This is for example the case in the EU with the Directive on protective measures against organisms harmful to plants and plant products. Specifically Annex III of the directive is relevant in relation to phytoremediation. This annex contains a list of plants of which the introduction in the EU is not allowed in general or in protected zones.

---

[24]Article 22 (b) of the Habitat Directive.

[25]Refer to https://www.ippc.int/countries/regulatedpests/, last visited on 6 January 2015.

[26]The RPPOs are inter-governmental organizations functioning as coordinating bodies for National Plant Protection Organizations (NPPO) on a regional level. There are currently 10 RPPOs:

- Asia and Pacific Plant Protection Commission (APPPC)
- Comunidad Andina (CA)
- Comite de Sanidad Vegetal del Cono Sur (COSAVE)
- Caribbean Plant Protection Commission (CPPC)
- European and Mediterranean Plant Protection Organization (EPPO)
- Inter-African Phytosanitary Council (IAPSC)
- Near East Plant Protection Organization (NEPPO)
- North American Plant Protection Organization (NAPPO)
- Organismo Internacional Regional de Sanidad Agropecuaria (OIRSA)
- Pacific Plant Protection Organization (PPPO)

20.2.1.3 Genetically Modified Plants (GM Plant)

Like with everything, there are also two sides on genetic modification of organisms (GMOs). Increasing the resilience of plants to abiotic stress, like metal contaminants, makes it possible to use these plants for phytoremediation. Improve the uptake of contaminants by the green and harvestable parts of plants or promote the immobilization capacity of the roots are clearly beneficial in the remediation process. Salix plants were, for example, inoculated with genetically modified Psuedomonas fluorescens enhancing the biodegradation of PCBs.[27]

On the other hand there exists quite some uncertainty and lack of knowledge on different aspects of genetic modification. Soils are some of the most complex habitats on earth, with 1 g of agricultural or forest soil from temperate regions containing thousands of species. Given this complexity, the impact of genetically modified plants on soil systems is not well understood.[28] By consequence the question is raised if we can take these risks?[29] Moreover, genetically modifying a plant to better cope with environmental stress could lead to this plant becoming invasive, potentially resulting in a loss of biodiversity.

The implications of using genetically modified crops should be carefully considered before selecting the plant for a project.

Besides the impact of current GMO regulation in many countries, the public opinion is not always in favor of these plants and the costs to comply with the rules are also not to be neglected. In the USA for example, compared to other countries, regulation of GMOs is relatively favorable to their development. GMOs are an economically important component of the biotechnology industry, which now plays a significant role in the US economy. For example, the US is the world's leading producer of genetically modified (GM) crops.[30] Plant GMOs are regulated by the US Department of Agriculture's Animal and Plant Health Inspection Service under the Plant Protection Act. In the EU for example, the basic principle of the GMO Directive is that GMOs can only be put on the market if they are safe. Member States have to take appropriate measures to guarantee this and thus to avoid adverse effects on human health and the environment. Thereby the precautionary principle should be a decision criterion.[31] The authorization of a GM plant happens on an individual basis. The process requires testing of the plant concerned to see if large scale cultivation could have an impact on the environment. Genetically modified variations of maize, rapeseed and sugar beet are at this moment authorized in

[27]Aguirre de Carcer et al. (2007), p. 215.

[28]The potential environmental, cultural and socio-economic impacts of genetically modified trees, Conference on Biological Diversity, March 2008, UNEP/CBD/COP/9/INF/27, p 20, www.cbd.int/doc/meetings/cop/cop-09/information/cop-09-inf-27-en.pdf, last visited on 6 January 2015.

[29]See also the paragraph on soil management.

[30]Refer to http://www.loc.gov/law/help/restrictions-on-gmos/usa.php#_ftn2, last visited on 6 January 2015.

[31]Article 4, GMO Directive.

Europe.[32] But under European law Member States can ban GMOs if they can justify the prohibition. This was demonstrated in a meeting of the European Environment Ministers in 2009. During the meeting, the European Environment Ministers voted against forcing Austria and Hungary to allow US biotech company Monsanto's MON810 GM maize grain to be grown in their countries. The ban is thus upheld, notwithstanding the fact that this maize grain is fully authorized for food and feed, plus for processing and import.[33] Thus the debate on the accidental spread and adverse effects on nature continues.[34] There is now a proposal for a Regulation pending, which aims at providing a legal base to authorize Member States to restrict or prohibit the cultivation of GMOs, although these might be authorized at EU level.[35]

20.2.2 Soil Management

20.2.2.1 Legal Framework

Soil as such is mainly regulated by national and/or local rules. Several countries have specific soil legislation. This legal framework will always have to be checked for a specific phytoremediation project.

The legislation will sometimes have a broad scope (soil protection in general) and sometimes a narrow focus on soil remediation. In the Netherlands for example the Soil Protection Act of 1987 is the basis for legal soil protection. It combines provisions on soil protection as well as on soil remediation. It includes the concept of historical pollution (before 1 January 1987), before which no general obligation for remediation exists. Furthermore background and intervention levels concerning soil contamination have been set. And the focus is on sustainable soil management whereby soil and water are in practice closely linked. In Belgium, the Flemish region has an extensive legislation directly focused on soil remediation.[36]

In other countries soil may also fall under the same regulation as groundwater, a water area, etc. does. This is for example the case in Sweden. The main criterion for the applicability of the Swedish Environmental Code, adopted in 1998 and entered

---

[32]Refer to http://www.gmo-compass.org/eng/gmo/db/, last visited on 6 January 2015.

[33]http://www.gmo-compass.org/eng/gmo/db.

[34]Austria, Hungary Allowed to Keep Ban on Genetically Modified Crops, *Deutsche Welle*, http://www.dw.de/austria-hungary-allowed-to-keep-ban-on-genetically-modified-crops/a-4068097, last visited 6 January 2015.

[35]Proposal for a Regulation amending Directive 2001/18/EC as regards the possibility for the Member States to restrict or prohibit the cultivation of GMOs in their territory, Brussels, 13 July 2010, COM(2010) 380 final, 15 p., http://ec.europa.eu/food/plant/gmo/legislation/docs/proposal_en.pdf, last visited on 6 January 2015.

[36]Decree of 27 October 2006 on soil remediation and soil protection, *Belgian State Gazette* 22 January 2007.

into force 1 January 1999, is the presence of potential harm to human health or the environment.[37] In that case a supervisory authority may require the person or persons responsible to remediate the damage. The date when the pollution took place is irrelevant to the application of the provisions, although only contamination caused by environmentally hazardous activities after June 30 1969 will incur liability under the Environmental Code.[38] However one should bear in mind that the Swedish Environmental Code is further elaborated in several ordinances made by the Government.[39]

Also at the international level soil protection plays an increasingly important role. In the final document of the conference of the United Nations on sustainable development in June 2012 in Rio de Janeiro (Rio + 20 Conference), the international community thus agreed to aim for a "land degradation neutral world" (LDNW). Land degradation neutrality (LDN) is defined as "*a state whereby the amount of healthy and productive land resources, necessary to support vital ecosystem services, remains stable or increases within specified temporal and spatial scales. LDN can occur as the result of natural regeneration or improved land management practices and ecosystem restoration*".[40] On the basis of this aim, an Intergovernmental Working Group was established within the United Nations Convention to Combat Desertification in Countries Experiencing Serious Drought and/or Desertification, Particularly in Africa (UNCCD) of 1994, which reports on a regular basis how far the understanding of the LDN concept and the common vision for what is possible to deliver under the UNCCD has advanced so far.[41] In principle, international obligations to achieve LDN could in the near future be included in legal instruments such as the United Nations Framework Convention on Climate Change (UNFCCC) of 1992, the CBD or the UNCCD.

In case no specific soil legislation exists, it does not mean that there is no relevant legislation. On the contrary, a lot is related to the important and multiple functions soil has in biodiversity, environmental health and human well being. Several policies (like (ground)water, chemicals, waste, agriculture, sludge, manure, herbicides) have a direct or indirect impact on soil management and protection.

Another element to consider is the legal regulation of land use or spatial planning. Nearly all of these rules are national/regional. Once permission is received from the owner of the site, aspects linked to infrastructure, destination of the area should be taken up locally, eventually together with requesting an exploitation permit.

---

[37]Ministry of Environment, *The Swedish Environmental Code, A résumé of the text of the Code and related Ordinances, Genetic Engineering*, Danagrards Grafiska, Stockholm, Sweden, p. 27.

[38]Ibid., p. 27–28.

[39]Refer to http://www.sweden.gov.se/sb/d/2023/a/22847, last visited on 6 January 2015.

[40]UNCCD Intergovernmental Working Group (IWG) on the follow up to the outcomes of Rio + 20, Task 1—Science-based definition of land degradation neutrality, 28 May 2014.

[41]For the latest report, refer to http://www.unccd.int/Lists/SiteDocumentLibrary/Publications/V2_201309-UNCCD-BRO_WEB_final.pdf?HighlightID=329, last visited on 6 January 2015.

The necessary approvals will differ between phytostabilization and phytoextraction. In the following paragraph the aspects of both techniques are further elaborated.

20.2.2.2 Phytoremediation: The Difference Between Stabilization and Extraction

Since there is a difference in impact on the soil between stabilization and extraction, applicable (elements of) legislations will/might differ also. For a better understanding the relevant distinctions between both techniques will be explained, followed by an overview of relevant regulations. The most commonly used techniques for phytoremediation of contaminated soil are stabilization and extraction. Both techniques use plants to achieve their objectives and differ mainly in the process to achieve these objectives.

Phytostabilization focuses on immobilizing the contaminants in a way that they are unavailable for uptake in the (harvestable parts of the) plants and leaching into the environment is prevented. However, the contaminants remain in the soil. Stabilization is an approach aiming at decreasing bioavailability. It can be used on sites that are heavily contaminated with metals using a combination of plants and soil amendments.[42] It is through these amendments that the availability of the contaminants for uptake by plants is reduced. Simultaneously the mobility in and leaching from the soil is stopped or at least diminished.

A few remarks should be made. Amending substances to soil is de facto changing the composition of that soil. Next to the positive effect of stabilizing contaminants, other side effects are also possible and probable. However, an element to consider is that adding substances to the soil can immobilize some substances and at the same time mobilize others.[43] Contamination of the groundwater by these mobilized contaminants is possible. Especially leaking into the groundwater is a risk. Consequently one should consider and respect the legislation on groundwater.

Other undesirable effects of amendments can be the change of the soil structure (for example zeolites with high sodium content destroy the soil structure) or immobilization of essential elements.[44] This can result in the loss of biodiversity and/or habitats, triggering related legislation.

Last but not least, when soil is chemically treated or changed, it becomes subject to chemicals legislation. In case of phytostabilization this is clearly so. But since it is not the goal to excavate the soil and put it on the market or use it in articles, the obligation remains rather theoretical.

---

[42] Vangronsveld et al. (2009), p. 766.

[43] Vangronsveld et al. (2009), pp. 767 and 770.

[44] Ibid. p. 767.

Phytoextraction on the other hand aims at the uptake of the contaminants by harvestable parts of the crops. Subsequently the contaminated biomass is harvested and removed from the site. The result is that the contamination in the soil is reduced.

For phytoextraction to succeed, contaminants must be bioavailable. The degree of availability for uptake by plants varies according to the contaminant concerned and the capability of the used crop.

Many metal contaminants are essential micronutrients for the plant. In common nonaccumulator plants, accumulation of these micronutrients does not exceed their metabolic needs.[45] However some plants can accumulate more than what they need. Hypoaccumulators do not only accumulate substances they need, but also absorb high amounts of contaminants. These contaminants accumulate in the foliage, resulting in a contaminated feedstock. This is important for the further processing of the harvest and the eventual waste part of the crop.

Contrary to stabilization, for extraction amendments are used to increase the bioavailability of contaminants. This increased mobility of contaminants can lead to leaching of hazardous elements into the groundwater. Often legislation includes legal requirements to prevent or limit input of pollutants into the groundwater.

Increasing the uptake of lead (Pb) by crops, ethylene-diamine-tetraacetic acid (EDTA) is commonly added to the soil in phytoremediation projects. The substance is however regarded as persistent organic pollutant and its poor biodegradability leads to accumulation in the environment.[46] It has been found to be both cytotoxic and weakly genotoxic in laboratory animals. Oral exposures have been noted to cause reproductive and developmental effects in animals.[47] However, at this moment it seems that the substance is free to use, but the input into groundwater should be prevented or at least limited.

Growing crops for biofuel in an economically viable way supposes to harvest as much as possible or at least enough biomass. The intensive cultivation character requires a frequent nutrient input for highest production. Heavy fertilization of the soil is necessary and by using waste water/sludge the cost for nitrates and phosphates decreases substantially.[48]

Another major crop maintenance practice is weed control. It seems that most (short-rotation) plantations for biomass face problems due to the presence of weeds that hinder the growth of the planted material, especially during the first year. As a consequence, the biomass production is heavily reduced during the next years.[49] A

---

[45]Lasat (2000), p. 5-1.

[46]Yuan and VanBriesen (2006), p. 533.

[47]Lanigan and Yamarik (2002).

[48]BIOPROS, Solutions for the safe application of wastewater and sludge for high efficient biomass production in Short-Rotation-Plantations, *D4—Report on ongoing research and gaps in SRP knowledge*, 14 February 2005, Sweden, February 2006, p 7.

[49]BIOPROS, Solutions for the safe application of wastewater and sludge for high efficient biomass production in Short-Rotation-Plantations, *D5—Guidelines on safety issues for SRP wastewater application*, 14 February 2005, Sweden, February 2006, p 17.

situation absolutely needing weed control, is the problem with Broomrape (Orobanche ramosa linnaeus). Broomrape is a very invasive, parasite plant that grows on a wide range of crops, oilseeds and vegetables.[50] Problems with this plant destroying up to 70 % of the sunflowers occurred.[51]

To kill or control harmful organisms such as weeds or insects, pesticides are used (in the case of weed control also called herbicides).[52] Pesticides are substances and products intended to influence fundamental processes in living organisms. At the same time pesticides can have negative effects on human health and the environment, which represent high costs for society. Therefore, normally legislation on plant protection lays down strict rules for the authorization of plant protection products.[53] The other aspect that is usually regulated are residue limits, but these are only available on food- and feedstuffs. Furthermore, potentially existing legislation on the use of pesticides has to be checked for every phytoremediation project.[54]

Furthermore, after treatment sewage sludge may sometimes be applied to land (dependent upon the quality of the sludge (e.g. with regards to heavy metal content)). Normally the applicable rules will differ in relation to the use of the sludge, and more specifically on the classification of the cultivation of crops for phytoremediation and the production of bio-energy as an agricultural activity or not. This classification differs per country. Whilst most national rules regulate sludge use in agriculture in a separate framework, use outside of agriculture falls in general under waste law. In view of the objective of the phytoremediation projects, it is advisable to respect at least the more stringent rules when using sewage sludge.

In general, countries should be encouraged to strive for an integrated pest management. This means careful consideration of all available plant protection methods. Subsequently, appropriate measures should be integrated that discourage the development of populations of harmful organisms and keep the use of plant protection products and other forms of intervention to levels that are economically

---

[50] Refer to http://www.depi.vic.gov.au/agriculture-and-food/pests-diseases-and-weeds/weeds/state-prohibited-weeds/branched-broomrape, last visited on 6 January 2015.

[51] Refer to http://www3.interscience.wiley.com/journal/119212373/abstract?CRETRY=1&SRETRY=0, last visited 6 January 2015.

[52] Pesticides are usually divided into two major groups (plant protection products and pesticides for non-agricultural uses). Plant protection products (PPPs) contribute to high agricultural yields and help to ensure that good quality food is available at reasonable prices. Pesticides for non-agricultural uses are e.g. important for public health protection and for preservation of materials.

[53] Refer to e.g. in the EU the Plant Protection Directive.

[54] Refer to e.g. in the EU the Directive 2009/128/EC of the European Parliament and of the Council of 21 October 2009 establishing a framework for Community action to achieve the sustainable use of pesticides, *Official Journal* 24 November 2009, L 309.

and ecologically justified and reduce or minimize risks to human health and the environment.[55] Principles of integrated pest management are for example the use of resistant/tolerant cultivars, the use of balanced fertilization, liming and irrigation/drainage practices.[56] *In concreto*, countries should take all necessary actions to promote low pesticide-input pest management, giving wherever possible priority to non-chemical methods and attending carefully to the quality of the aquatic environment. Thereby measures taken under other regulation to protect sensitive groups and biodiversity should be respected.

20.3 Growing and Harvesting the Crops

20.3.1 Growing the Crops

20.3.1.1 Introduction

The legal analysis of the use of crops grown on contaminated soil takes into account the principles of sustainability, minimal (negative) environmental impact and an optimal closed lifecycle concerning the use of the crops. This is an important choice in case alternatives are available and/or interpretation of rules is necessary. Because of the lack of specific reference to contaminated biomass, there are situations where a legislation could be applicable per analogy, meaning based on the intention and goal of that legislation. An example is the question if legislation on agriculture is applicable on contaminated biomass or not. This element is further explored in Sect. 20.3.1.2.

When growing the crops, one of the important objectives of a phytoremediation project is to produce enough crops, e.g. for conversion into biofuel. Thereby it is likely that some fertilizing of the soil and some plant protection actions will be necessary. In the Sects. 20.3.1.3 and 20.3.1.4 we respectively discuss the legal implications of fertilizing and look at the rules concerning plant protection.

20.3.1.2 The Crops Are Planted

Is planting crops for phytoremediation an agricultural activity? Previously agriculture was officially the activity aiming at producing food or feedstock for animals for food. Nowadays agricultural policy encourages farmers to not only produce high quality food products, but to seek also new development opportunities, such as "*renewable environmentally friendly energy sources*".[57]

---

[55]Refer also to Article 3, 6 of Directive 2009/128/EC.

[56]Refer also to 'Annex III – General principles of integrated pest management' of Directive 2009/128/EC.

[57]European Commission (2012).

One can conclude that it is logical to apply the principles of good agricultural practice and the agricultural legislation on contaminated crops/biomass. We will thus approach the other aspects of growing crops taking into account that phytoremediation is an agricultural activity.

Countries have their own interpretation on what is regarded as agriculture and what not, definitions differ a lot.[58] To understand the domain of agriculture one will need to analyze local regulation and its relevance for growing crops on contaminated land that is not or was not used for agriculture.

Often definitions of agriculture are broad. For example Sweden defines it as "including forestry, hunting and fishing", as well as "cultivation of crops and livestock production".[59] It then remains to be investigated if specific legislation is applicable on the cultivation. On the other hand, one could consider to chose the most environmental options, regardless of the fact that the activity is considered agriculture or not. After all, the objective is to improve the quality of the environment and by consequence not reduce it by inappropriate remediation of soils.

20.3.1.3 Fertilizing

Fertilizers are substances that are used on land to enhance growth of vegetation. Examples of fertilizers are livestock manure, sewage sludge, chemical fertilizers, composts and residues from fish farms. Fertilizers contain a nitrogen compound and thus excessive use of fertilizers constitutes an environmental risk. Especially the eutrophication of water is a problem. It causes excessive growth of algae and plants thereby disturbing the quality of the water and the balance of organisms in that water.

Countries usually have established code(s) of or legislation on good agricultural practice. These codes/laws cover, e.g., the periods when fertilizing is inappropriate; the rules for applying fertilizer near water courses or on steeply sloping, water-saturated, flooded, frozen or snow-covered grounds. They may also include guidelines on crop rotation and on minimum quantity of vegetation cover during certain (winter, rainy) periods. *In concreto* fertilization should be limited to certain periods in the year, taking into account the situation, the condition and the structure of the ground. The nutrient losses to water should in any case be limited to an acceptable level. Therefore it is important to remember that when working with non-perennial plants a minimum vegetation cover should be maintained during the non-grow season.

Two specific fertilizers need more explanation because of their special characteristics: sewage sludge (used *an sich*, not as input for composting) and compost. Sludge originates from the process of cleaning, mainly urban, waste water. The use of sewage sludge was described in Sect. 20.2.2.2. Compost is best defined as "*the*

---

[58]Refer to e.g. Karlsson et al. (2005).

[59]Refer to www.indexmundi.com/facts/Sweden/agriculture (last visited on 12 January 2015).

solid particulate material that is the result of composting, which has been sanitised and stabilised".[60] The process of compositing is described as the "*controlled decomposition of biodegradable materials under managed conditions, which are predominantly aerobic and which allow the development of temperatures suitable for thermophilic bacteria as a result of biologically produced heat*".[61]

There are positive aspects to using compost as a fertilizer. It generally improves the structure of the soil and its biological and chemical properties. The impact of compost on the soil depends on its composition. On turn the composition depends on the input material in the composting process. Several types of waste may be composted, for example biodegradable waste, commercial food wastes, forestry residues, waste from agriculture (including manure), waste from food and beverage industries and sewage sludge.

In general compost is low in nitrogen available to plants. The nitrogen present in the organic matter is only slowly released. On the other hand the amount of phosphate and potassium can generally cover the need for these substances. Compost also supplies calcium, magnesium and sulphur and other micronutrients.[62] Consequently it can be used as a (partial) replacement for other fertilizers. An advantage over chemical fertilizers is that compost brings organic matter to the soil improving the soil's structure.

Unfortunately, compost also has some negative aspects. The two elements that are of particular interest, because of the objective to remediate contaminated soil, are the presence of heavy metals and persistent organic pollutants. These contaminants are negative for soils and land in normal condition, but especially so in conditions where the presence of heavy metals and persistent organic pollutants is already elevated. On top, often no specific legislation exists regulating the quality and use of compost.

Delgado *et al.* researched the (potential) presence of contamination in compost, especially the presence of heavy metals.[63] The findings showed that exceeding the actual contamination limits for zinc, lead, cadmium and phosphate would not lead to critical effects. To have a negative impact, extremely high amounts or repeated inputs of these heavy metals to the soil should continue over several years. The study concludes that it is best is to control contamination of compost by quality checks on the nature and origin of the input material. The limits that are proposed[64] are in some cases lower than the actual country limits.

At this moment various countries have statutory standards for compost. Belgium for example has a quality system for compost managed by an organization VLACO vzw. This organization performs regular audits and controls input material, processes and product quality.[65] The Netherlands also has a quality assurance system,

---

[60]Delgado et al. (2009), p. 53.

[61]Ibid., p. 53.

[62]Ibid., p. 60.

[63]Ibid., pp. 51–197.

[64]In values g/kg (dry weight): cadmium = 1.5 – lead = 120 – zinc = 400.

[65]http://www.vlaco.be/en/vlaco-vzw/info (last visited on 12 January 2015).

but participation is voluntary. Compost meeting the set of legal requirements obtains a certificate, based on a positive assessment by an independent institute.[66]

20.3.1.4 Plant Protection

Previously plant protection regulation was mainly focused on the marketing of these products. For example maximum residue limits existed for pesticides on food- and feedstuffs. Recently, in more and more countries, also the use of plant protection products is regulated.[67] The focus is no longer solely on food safety. Substances on the market should be acceptable for the environment and human health. This is an important step forward in the protection of human health and the environment from pesticides. Indeed pesticides can have very negative effects that should be prevented or controlled.

As a result, it may well be for example that commercial plant protection products have to be authorized before they can be put on the market or used. The authorization shall then define for what purposes which plant product can be used. It may also be that plant protection products need to be labeled, including for example additional information and safety precautions that have to be taken. In practice, in line with the precautionary principle, one should still check the local legislation when planning to use plant protection products. It could be forbidden in that particular country or for that particular use in that country.

It would be good if countries would adopt action plans to reduce risk and impacts of pesticides, promote integrated pest management and alternative approaches to reduce the use of pesticides. The encouragement of integrated pest management means that, whenever possible, priority should be given to non-chemical methods. However, one should always be careful with importing foreign predators for pest control. The impact on the local biodiversity can be very negative.

20.3.2 Harvesting the Crops

20.3.2.1 Introduction

When harvesting the crops on the site, it is important to know how contaminated the crops are and with what kind of chemical substances. This will have an impact on the classification of the crop/biomass as (potentially) hazardous or not. As mentioned before, phytoremediation becomes even more interesting when the produced plants/crops/biomass can still be valorized into an income (food, feed, biomass,. . .). In case the contaminants levels stay below the levels for food and/or feed, the crops

[66]www.keurcompost.nl (last visited on 12 January 2015).

[67]E.g. the European Directive on the sustainable use of pesticides.

can be used for food and/or feed. This will be more often the case with phytostabilization than with phytoextraction. Otherwise, they need to be disposed of or treated. In that case, the utilization of the obtained biomass as an energy resource is a sustainable alternative. The biomass can be converted into biofuel. Biofuel is thereby to be understood as the fuel produced directly or indirectly from biomass, such as fuelwood, charcoal, bioethanol, biodiesel, biogas and biohydrogen.[68] Also a combination of food and/or feed on the one hand and biomass on the other hand can be possible, as for example with maize. The contamination is taken up by the plant, but mainly remains in the green parts of the plant, and does not go into the maize grains. As a result, the maize grains can still be used as feed for livestock, and the rest fraction (the green parts (leaves, roots, trunk,. . .)) can be used for green energy production (through incineration or digestion).

For further classification we also need to know if the resulting crop/biomass falls under waste rules or not. Indeed, obligations to store, transport and manipulate are different for waste and non-waste.

There exist no explicit international rules on contaminated biomass. We need to look at several regulations touching the characteristics and properties of the biomass to get a good view on the legal aspects. Conclusions will be drawn by extrapolation of these rules.

Within this framework, the impact of the contamination of the plants and the classification of the biomass as waste or non-waste is further analyzed in following paragraphs. We thereby focus on the situation where the biomass is contaminated and cannot be used as food or feed.

20.3.2.2 The Plants Are Contaminated

The whole idea behind phytoextraction is that plants grown on a contaminated site take up (part of) the contaminants and consequently clean the soil of toxic substances. When the plants are harvested, part of the soil contamination is removed together with the biomass. This raises some questions on the nature of the resulting harvested biomass. Referring to contamination with for example heavy metals, these elements occur naturally in plants, but in lower quantities than when the crops grow on contaminated soil. In phytoextraction plants are even selected on their potential uptake of contaminants whilst able to survive and grow.

Exact guidelines to decide which quantity of contaminants in the harvested crops is such that the biomass should be considered toxic or hazardous,[69] do not exist on

[68]REJUVENATE, *Crop Based Systems for Sustainable Risk Based Land Management for Economically Marginal Degraded Areas,* Final Research Report, 2009, p. 19, available on www.snowman-era.net/content.php?horiz_link=12&vert_link=0 (last visited on 15 January 2015).

[69]It is important to note that the meaning of the term "contaminant" is not similar to the meaning of "hazardous substance" used in regulation and in particular chemical legislation. A contaminant is a substance which is in, on or under the land and has the potential to cause harm (or to cause pollution of controlled waters). A hazardous substance is classified as such on the basis of certain

international level. In principle the concentration of toxic substances in the plants is higher than what it, in normal growing circumstances, would be, although lower than in the soil. Differing considerably from their normal, natural composition, the plants/biomass could be classified as ecotoxic. This conclusion is based on arguments found in for example the European Waste Framework Directive[70] and in the texts of the European REACH Regulation[71] on hazardous substances. The Waste Framework Directive would classify waste as ecotoxic when it "*presents or may present immediate or delayed risks for one or more sectors of the environment*".[72] The contaminants in the soil are considered to have a toxic, negative impact since this is exactly the reason for the phytoremediation of the site. The same contaminants are now in the plants and thus these could be seen as hazardous too. However, the concentration in the biomass could be significantly different from the concentration in the soil. This could influence the classification of the biomass.

To assess the toxicological effect we need to consider two parameters: the amount of contaminants in the biomass and the nature of the substances present in that biomass. The latter should then be compared with the rules in local chemicals legislation. Many of the contaminants found in soils that need remediation are normally listed as hazardous substances. Hazardous substances are often subject to authorization or restrictions. This confirms that contaminated biomass could be seen as dangerous and ecotoxic.

Abstraction made of the conclusion if this contaminated biomass is waste or not, one could say that it is reasonable to classify the biomass as toxic for the environment.

20.3.2.3 Waste or Not Waste

When the purpose of growing crops on a contaminated site is to remediate that site, the intention can still be to use biomass for producing biofuel as a commercial valuable product that gives the location some economic future. For the farmers it would add to their income and investors could be interested in the use of the biomass. Indeed turning contaminated biomass into a financial viable product

characteristics, i.e. carcinogenic, mutagenic, toxic for reproduction, PBT or vPvBvT ((very) persistent, (very) bioaccumulative and (very) toxic).

[70]Directive 2008/98/EC of 19 November 2008 on waste and repealing certain Directives, *Official Journal* 22 November 2008, L 312/3.

[71]Regulation (EC) No 1907/2006 of 18 December 2006 concerning the Registration, Evaluation, Authorisation and Restriction of Chemicals (REACH), establishing a European Chemicals Agency, amending Directive 1999/45/EC and repealing Council Regulation (EEC) No 793/93 and Commission Regulation (EC) No 1488/94 as well as Council Directive 76/769/EEC and Commission Directives 91/155/EEC, 93/67/EEC, 93/105/EC and 2000/21/EC, *Official Journal* 25 May 2007, L 136/3.

[72]Directive 2008/98/EC, Annex III, H14.

would be an important motivator for private parties to invest in phytoremediation, what on its turn would benefit the local population and the environment.

But is harvested contaminated biomass waste? The answer will depend on the local legislation. The European Waste Framework Directive for example defines waste as "*any substance or object which the holder discards or intends or is required to discard*".[73] But article 2 of the Waste Framework Directive excludes "*... straw and other natural non-hazardous agricultural or forestry material used in farming, forestry or for the production of energy from such biomass through processes or methods which do not harm the environment or endanger human health*". A direct consequence of this formulation is that biomass needs to be natural and non-hazardous, which is not the case for the contaminated biomass (see previous paragraph). The conclusion is that in the EU this biomass is not excluded from the Waste Framework Directive and we will have to look at the definition of waste for the classification.

Comparing the harvesting of contaminated crops for the production of biofuel with the definition of waste and the arguments used in court decisions, we can conclude that the biomass of a phytoremediation site is not waste. Following arguments lead to this conclusion:

- The economic feasibility of using the contaminated biomass for the production of biofuel is an important criterion for the viability and success of the projects.
- The suitability of the crops for the production of biofuel is taken into account in the selection of the plants.
- The crops are grown and treated (fertilizing, plant protection) with the clear aim to produce biomass as feedstock for biofuel, whilst remediating the contamination.
- Biofuel clearly is a commercial product. It has to be manufactured and this production is not just a deactivation of harmful biomass, but real processing.

Above motivation is valid for possible future phytoremediation projects. Private phytoremediation projects for the production of biofuel will be decided upon based on their financial and economic viability. The biomass is grown with the goal to produce biofuel, it is not re-used, not recycled, it is a first processing. Consequently, it is hard to defend that the resulting biomass is to be classified as waste. Overall the grower and/or holder of the contaminated biomass did not discard it, nor did he have the intention to discard.[74] The goal is rather to use it as an economic valuable and commercial product, namely feedstock for biofuel.

---

[73]Ibid., Article 3, 1.

[74]This does not exclude that after processing, parts of the biomass could become waste.

20.4 Conclusion

Soils are under increasing environmental pressure in every country across the globe. This pressure is mainly driven or exacerbated by human activities, such as agricultural and forestry practices, industrial activities, tourism and urban development. However, soil is of crucial importance. Therefore, further soil degradation should be prevented and degraded soils should be restored.

Via phytoremediation degraded soils, in particular contaminated soils, can at the same time be restored and used for agricultural cultivation (food, feed, biomass,...). Phytoremediation uses plants/crops to remove pollutants from the environment or to render the pollutants harmless.

Unfortunately, a sound legal framework for decision-making about the use of phytoremediation is missing. This contribution offers an overview and analysis of all the potential legal bottlenecks. It follows the chronology in practice in decision-making before the start and during the phytoremediation project.

It is clear from this study that there are many rules that have to be checked. They deal for example with the question whether there are any plant species that are not legally allowed to be used in phytoremediation? One has to watch out with invasive and exotic plants and check international, regional or national lists of forbidden plants. Also with regard to the use of GMOs national rules differ and will have to be followed. Other relevant rules relate to for example soil management, fertilizing, the use of plant protection products, and the question what can be done with the harvested crops. In case the contaminants levels stay below the levels for food and/or feed, the crops can be used for food and/or feed. This would be ideal. Otherwise, obtained biomass might also be used as an energy resource.

Hopefully these legal bottlenecks will not stop private parties and local governments from starting phytoremediation projects and the combination of agricultural cultivation and soil remediation through phytoremediation can become an important part of sustainable risk based land management.

References

Aguirre de Carcer D, Martin M, Mackova M, Macek T, Karlson U, Rivilla R (2007) The introduction of genetically modified microorganisms designed for rhizoremediation induces changes on native bacteria in the rhizosphere but not in the surrounding soil. ISME J 1:215–223, http://www.nature.com/ismej/journal/v1/n3/full/ismej200727a.html. Last visited 6 Jan 2015

Chaney RL, Malik M, Li YM, Brown SL, Brewer EP, Angle JS, Baker AJM (1997) Phytoremediation of soil metals. Curr Opin Biotechnol 8:279–284

De Vries B (2000) Multifunctional agriculture in the international context: a review. The Land Stewardship Project, p 15

Delgado L, Catarino AS, Eder P, Litten D, Luo Z, Villanueva A (2009) End-of-waste criteria, final report, European Commission, Joint Research Centre, p 379. http://ftp.jrc.es/EURdoc/JRC53238.pdf. Last visited 12 Jan 2015

Dornburg V, Faaij APC (2005) Cost and CO_2 emission reduction of biomass cascading: methodological aspects and case study of SRF poplar. Clim Chang 71:373–408
European Commission (2006) Communication from the Commission to the Council, the European Parliament, the European Economic and Social Committee and the Committee of the Regions: Thematic Strategy for Soil Protection, COM(2006) 231 final, p 12
European Commission (2012) The common agricultural policy. A partnership between Europe and farmers, 16 pp. http://ec.europa.eu/agriculture/capexplained/index_en.htm. Last visited 12 Jan 2015
Firbank LG (2008) Assessing the ecological impacts of bioenergy projects. Bioenerg Res 1:12–19
Ghosh M, Singh SP (2005) A review on phytoremediation of heavy metals and utilization of its byproducts. Appl Ecol Environ Res 3:1–18
Karlsson J, Pfunderer S, Salvioni C (2005) Agricultural and rural household income statistics. Paper prepared for presentation at the 94th EAAE Seminar, Ashford, UK, 9–10 April 2005. http://ageconsearch.umn.edu/bitstream/24427/1/sp05ka01.pdf. Last visited 12 Jan 2015
Koopmans GF, Römkens PFAM, Fokkema MJ, Song J, Luo YM, Japenga J, Zhao FJ (2008) Feasibility of phytoextraction to remediate cadmium and zinc contaminated soils. Environ Pollut 156:905–914
Lanigan RS, Yamarik TA (2002) Final report on the Safety Assessment of EDTA, Calcium Disodium EDTA, Diammonium EDTA, Dipotassium EDTA, Disodium EDTA, TEA-EDTA, Tetrasodium EDTA, Tripotassium EDTA, Trisodium EDTA, HEDTA, and Trisodium HEDTA. Int J Toxicol 21(2 Suppl):95–142
Lasat M (2000) Phytoextraction of metals from contaminated soil: a review of plant/soil/metal interaction and assessment of pertinent agronomic issues. J Hazard Substance Res 2:5–1. http://citeseerx.ist.psu.edu/viewdoc/download?doi=10.1.1.15.7236&rep=rep1&type=pdf. Last visited 6 Jan 2015
Meers E, Ruttens A, Hopgood M, Lesage E, Tack FMG (2005) Potential of Brassica rapa, Cannabis sativa, Helianthus annuus and Zea mays for phytoextraction of heavy metals from calcareous dredged sediment derived soils. Chemosphere 61:561–572
Mench M, Lepp N, Bert V, Schwitzguébel JP, Gawronski SW, Schröder P, Vangronsveld J (2010) Successes and limitations of phytotechnologies at field scale: outcomes, assessment and outlook from COST Action 859. J Soils Sediments 10:1039–1070
Robinson BH, Leblanc M, Petit D, Brooks RR, Kirkman JH, Gregg PEH (1998) The potential of Thlaspi caerulescens for phytoremediation of contaminated soils. Plant Soil 203:47–56
Van Calster G (2004) Will the EC get a finger in each pie? EC law and policy developments in soil protection and brownfields redevelopment. J Environ Law 16:3–17
Van der Grift B, Griffioen J (2008) Modelling assessment of regional groundwater contamination due to historic smelter emissions of heavy metals. J Contam Hydrol 96:48–68
Van Nevel L, Mertens J, Oorts K, Verheyen K (2007) Phytoextraction of metals from soils: how far from practice? Environ Pollut 150:34–40
Vangronsveld J, Herzig R, Weyens N, Boulet J, Adriaensen K, Ruttens A, Thewys T, Vassilev A, Meers E, Nehnevajova E, van der Lelie D, Mench M (2009) Phytoremediation of contaminated soils and groundwater: lessons from the field, COST Action 859. Environ Sci Pollut Res 16:765–794
Vanheusden B, Hoppenbrouwers M, Witters N, Vangronsveld J, Thewys T, Van Passel S (2011) Legal and economic aspects of crops selection for phytoremediation purposes and the production of biofuel, 104 pp. http://projects.swedgeo.se/r2/wp-content/uploads/2012/04/REJUVENATE-2-Final-report-Legal-and-economic-aspects.pdf. Last visited 3 Jan 2015
Weyens N, van der Lelie D, Taghavi S, Vangronsveld J (2009) Phytoremediation: plant-endophyte partnerships take the challenge. Curr Opin Biotechnol 20:248–254
Weyens N, Truyens S, Saenen E, Boulet J, Dupae J, van der Lelie D, Carleer R, Vangronsveld J (2010) Endophytes and their potential to deal with co-contamination of organic contaminants (toluene) and toxic metals (nickel) during phytoremediation. Int J Phytorem 13:244–255

Witters N, Van Slycken S, Ruttens A, Adriaensen K, Meers E, Meiresonne L, Tack FMG, Thewys T, Laes E, Vangronsveld J (2009) Short rotation coppice for phytoremediation of a Cd-contaminated agricultural area: a sustainability assessment. BioEnergy Res 2(3):144–152

Yuan Z, VanBriesen J (2006) The formation of intermediates in EDTA and NTA biodegradation. Environ Eng Sci 23(3):533–544

Chapter 21
International Pastoral Land Law

Ian Hannam

Abstract This chapter discusses aspects of international and national environmental law for pastoral land and outlines frameworks for legislative reforms to achieve sustainable use of pastoral lands. International and national legal instruments and institutional systems play a significant role in pastoral land conservation. At the international level, it discusses a number of multilateral agreements that could be better used to promote the sustainable use of pastoral land. Two national level approaches to reform environmental law for pastoral land are presented; for Mongolia and the People's Republic of China respectively, and they may offer useful guidelines for other countries to follow in environmental law reform for pastoral land management.

21.1 Introduction

21.1.1 Introduction to the Topic

This chapter discusses aspects of international and national environmental law for pastoral land and outlines frameworks for legislative reforms to achieve sustainable use of pastoral lands. International and national legal instruments and institutional systems play a significant role in pastoral land conservation. At the international level, it discusses a number of multilateral agreements that could be better used to promote the sustainable use of pastoral land. Two national level approaches to reform environmental law for pastoral land are presented; for Mongolia and the People's Republic of China respectively, and they may offer useful guidelines for other countries to follow in environmental law reform for pastoral land management.

I. Hannam (✉)
Australian Centre for Agriculture and Law, University of New England, Armidale, NSW, Australia
e-mail: ian.hannam@ozemail.com.au

G. Steier, K.K. Patel (eds.), *International Farm Animal, Wildlife and Food Safety Law*, DOI 10.1007/978-3-319-18002-1_21

21.1.2 Brief Overview

Pastoral farming (also known in some regions as livestock farming or grazing) is farming aimed at producing livestock, rather than growing crops. Examples include raising beef cattle, and raising sheep for wool. Pastoralism is a livestock production system that is based on extensive land use and often some form of herd mobility, which has been practiced in many regions of the world for centuries.[1] Currently, extensive pastoralism occurs on about 25 % of the earth's land area, mostly in the developing world, from the drylands of Africa and the Arabian Peninsula, to the highlands of Asia and Latin America where intensive crop cultivation is physically not possible.[2] In addition, cattle and sheep farmers in Western North America, Australia, New Zealand, and a few other regions of the world presently practice a modern form of pastoralism. Worldwide, pastoralism supports about 200 million households and herds of nearly a billion head of animals including camel, cattle, and smaller livestock that account for about 10 % of the world's meat production.[3] Pastoralism is globally important for the human population it supports, the food and ecological services it provides, the economic contributions it makes to some of the world's poorest regions, and the long-standing civilizations it helps to maintain.[4] Unfortunately, threats and pressures associated with human population growth, economic development, land use changes, and climate change are challenging professionals and practitioners to sustain and protect these invaluable social, cultural, economic, and ecological assets worldwide.[5] Key ecosystem services such as biodiversity and food production provided by pastoral ecosystems may be vulnerable to both changes in climate as well as large-scale socioeconomic forces.[6]

Despite their vital role in food production in marginal environments, migratory pastoralists find themselves in a seemingly persistent state of crisis. Their herds are threatened by lengthy drought and emergent diseases. Their pastures and transit routes are shrinking in the face of spreading cultivation, nature conservation areas and hardening international borders. Their populations continue to rise, with rural and urban labour markets failing to absorb their youth. As a consequence, pastoral communities remain among the most politically and economically marginalized groups in many societies, rendering them susceptible to radicalisation and recruitment by insurgent groups and conflict entrepreneurs. Of significance with pastoral land management has been the relationship between pasture users and individual

---

[1] See WISP (2007), pp. 1–4.

[2] See FAO (2001), Introduction.

[3] See also FAO (2001).

[4] See Nori and Davies (2007), p. 4.

[5] See also Nori and Davies (2007), p. 6.

[6] See Abildtrup et al. (2006), p. 5.

governments. Different countries demonstrate considerable diversity in the preferred uses of pastoral land and the administrative and legal processes for its use, in particular the form of land tenure.[7] Various political, social and geographical factors influence different approaches to land use and land tenure issues. Traditionally, in a nation with an abundance of pastoral land, fewer land use restrictions have been applied, and land users have enjoyed a great deal of freedom with few limitations on individual rights.

Pastoral lands have many functions that need to be properly recognised by legislation. Environmentally, the most important one is that it provides a vegetation cover and thus protection for the soil and water, which also ensures sustainable economic production of animal fodder, firewood and other indirect benefits. Pastoral lands are a product of environmental factors, but they also contribute to both the local and global environment.[8]

Climate change and climate variability is driving fragile pastoral ecosystems into more vulnerable conditions.[9] Socioeconomic factors, such as changes in land tenure, agriculture, sedentarization, and institutions are fracturing large-scale pastoral ecosystems into spatially isolated systems.[10] The implications of this analysis are that professionals, practitioners, and policy makers should jointly develop a coupled human and natural systems approach that focuses on enhancing the resilience of pastoral communities and their practices. This requires institutional developments to support asset building and good governance to enhance adaptive capability. In addition, pastoralists' adaptation strategies to global change need to be supported by public awareness and improved by institutional decisions at different scales and dimensions.

This chapter discusses the international and national environmental law for pastoral lands and outlines frameworks for legislative reforms to achieve sustainable use of pastoral lands. International and national legal instruments and institutional systems play a significant role in pastoral land conservation. At the international level, it discusses a number of multilateral agreements that could be better used to promote the sustainable use of pastoral land. Two national level approaches to reform environmental law for pastoral land are presented, for Mongolia and the People's Republic of China respectively, and they may offer useful guidelines for other states to follow in environmental law reform for pastoral land management.

[7] See Nori et al. (2005), p. 10; Fernandez-Gimenez (2006), pp. 30–35; Taylor (2006), pp. 374–386.

[8] See Friedel et al. (2000), pp. 227–262; Neely et al. (2009), pp. 3–8.

[9] See Dong et al. (2011), pp. 10–11.

[10] See Halimova (2012), pp. 307–308.

21.2 Background

21.2.1 Purpose of Pastoral Law

Globally, while pastoral lands are considered significant in terms of environmental issues, they are inadequately recognised in international environmental law. This is an attitude of ignorance because they play a major role in food production and climate change with far-reaching consequences for the environment. For the purposes of this chapter "pastoral law" is considered as that area of law which includes specific laws, regulations and legal instruments that govern the use and management of pastoral lands. Other legal approaches to control the use of pastoral land, especially the conversion of pastoral land to non-pastoral uses, are also relevant. In some countries, pastoral law can regulate animal management, land consolidation or the procedure of land reallocation. In addition to these laws, other laws and regulations affect pastoral land, often significantly, and in some cases have played a greater role in the management of pastoral land than traditional pastoral land law which may be limited in scope with administrative role over leases or access conditions. Land protection laws have been important for keeping pastoral land productive; these laws have focused on soil productivity, prevention of soil erosion and protection of land from other environmental damage.[11]

21.3 Discussion and Analysis of Multilateral Environmental Treaties

Since the early 1900s, over 200 multilateral environmental treaties, agreements and protocols have been developed to manage and protect the world's natural environments and natural resources. A number of these instruments contain elements that can contribute to pastoral land conservation and management. However, none of them is sufficient on its own. While some of the existing instruments could assist by promoting the management of the activities that can maintain a sustainable pastoral land environment, this role is not readily apparent except for those that include provisions specifically directed to pastoral land ecosystems. Since the 1992 United Nations Conference on Environment and Development (UNCED), the global policy environment has changed considerably, with the adoption of the United Nations Sustainable Development Goals 2015 an outcome of the Rio +20 Conference in 2012, increased support to the least-developed countries, stronger commitment for climate change mitigation and adaptation and prospects of global agricultural trade liberalisation. As a result, international environmental law is

[11]See Grossman and Brussaard (1992) and Chalifour et al. (2007).

being called upon to provide a wider application to global environmental issues,[12] and pastoral land conservation should therefor receive increased attention. The scientific environment has also evolved with the work of the Millennium Assessment on dry land ecosystems, which has contributed to improved understanding of the biophysical and socioeconomic trends relating to land degradation in global dry lands and their impacts on human and ecosystem well-being.[13]

Following is a brief account of the main instruments.

21.3.1 The World Charter for Nature

The World Charter for Nature[14] called on states to cooperate in the conservation of nature, establish methods to assess the adverse effects on nature and implement international legal provisions for the conservation and the protection of the environment. The Charter states that the productivity of land shall be maintained or enhanced by the use of measures that safeguard their long-term fertility, the processes of organic decomposition, and safeguard against all forms of degradation.

21.3.2 The Rio Declaration and Agenda 21

The Rio Declaration on Environment and Development of 1992 established the goal of a new and equitable global partnership through the creation of new levels of cooperation among states, key sectors of society and individuals. It emphasised the need for states to work towards international agreements to protect the integrity of the global environmental system and to enact effective environmental legislation.[15] The other key instrument of the same time, Agenda 21,[16] discusses international environmental law processes that can assist in the global management of pastoral land.[17]

The Sustainable Development Goals were prepared on the basis of an "*inclusive and transparent intergovernmental process open to all stakeholders, with a view to developing global sustainable development goals to be agreed by the General Assembly*" and in this regard provide for many aspects of pastoral land

[12]See also Chalifour et al. (2007).

[13]See White et al. (2002), p. 2.

[14]See UNEP (1982), I. General Principles and II. Functions.

[15]See UN (1992a), Principle 11.

[16]See UN (1992b).

[17]See Chapters 8, 38 and 39 of Agenda 21 which discuss international legal instruments and mechanisms.

management.[18] Moreover, Sustainable Development Goal 15 focuses on desertification, degradation and drought, which is directly relevant to the management of pastoral lands. Various Sustainable Development Goals relate to ecosystem management.[19]

21.3.3 Convention to Combat Desertification

The United Nations Convention to Combat Desertification (CCD)[20] addresses land degradation and the methods to protect and manage soil and water resources of pastoral land areas. Desertification and drought are problems of global dimension, affecting most pastoral areas of the world, and joint action of the international community is often called upon to combat these problems.[21] Under Article 1 of the CCD, "desertification" means land degradation in arid, semi-arid and dry sub-humid areas resulting from various factors, including climatic variations and human activities. The CCD acknowledges that arid, semi-arid and dry sub-humid areas is to prevent and reduce land degradation, rehabilitate partly degraded land and reclaim desertified land, particularly in countries that experience serious drought. These areas account for large areas of the world's pastoral lands.

In this regard, it recognises the high concentration of developing countries, notably the least-developed countries, which are among those experiencing serious drought and/or desertification. It recognises that there must be support by international cooperation and partnership agreements which contribute to sustainable development. This will involve long-term integrated strategies that focus on the rehabilitation, conservation and sustainable management of pastoral land and water resources.[22]

The six program areas of Chapter 12 *(Managing Fragile Ecosystems: Combatting Desertification and Drought)* of *Agenda 21* are acknowledged by the CCD, and it is considered that they provide a useful basis to establish an approach for combating desertification in pastoral lands. A principal feature of the CCD relevant to pastoral land management is that it outlines how countries can approach the development of national action plans and obtain scientific and technical cooperation[23] and supporting measures.[24] These plans, as well as the various CCD sub-regional action programs, can address many important land degradation issues

[18]See UN (2014).

[19]See UN (2012) adopted at Rio + 20 2012, paras 245–251; see also UN 2015 Transforming our world: the 2030 Agenda for Sustainable Development.

[20]UN (1995).

[21]See Neely et al. (2009), p. vii.

[22]See also Neely et al. (2009), p. 3.

[23]See CCD Articles 9–18.

[24]See CCD Articles 19–21.

and the appropriate methods to protect and manage soil and water resources of pastoral areas.[25] Although the CCD does not have the specific elements that recognise pastoral land as an individual ecological element, it does contain many procedures that cover legal principles and processes for pastoral land conservation. Further, Article 31 provides for the development of regional or general annexes and in this regard a special annex could be prepared for the specific management of pastoral lands including guidelines for preparing national legislation for pastoral land management.[26]

21.3.4 UNCCD 10-Year Strategy

The CCD 10-Year Strategy 2008–2018 was adopted in 2007 and provides a global framework to support the development and implementation of national and regional policies, programmes and measures to prevent, control and reverse desertification/land degradation and mitigate the effects of drought through scientific and technological excellence, raising public awareness, standard setting, advocacy and resource mobilisation. The strategic objectives are directly relevant to management of pastoral lands and were developed to guide the actions of CCD stakeholders and partners in the period 2008–2021. They include improvement of living conditions of affected populations, improvement of affected ecosystems, generating global benefits, mitigating climate change and mobilising resources to support the implementation of the Convention. It is argued that the operational objectives of the CCD 10-Year Strategy could be enshrined in national legislative frameworks for pastoral land management as they focus on advocacy, awareness, education, policy development, science and technology, capacity building and financing and technology transfer.[27]

21.3.5 Convention on Biological Diversity

The fundamental aspect of the Convention on Biological Diversity (CBD),[28] that is, the concern that biological diversity is being significantly reduced by human

[25]See Secretariat CCD (2000); European Commission (2000), pp. 1–5.

[26]Unlike the CBD and FCCC, the CCD does not have a provision for making a protocol under the Convention. However, Article 30 provides for amendments to the CCD and provided all parties agreed the Convention could be amended to include a provision for the adoption of protocols. Should this happen a protocol for the management of pastoral land could be considered.

[27]See Secretariat CCD (2007), 8th session, Conference of the Parties, Madrid 3–14 September 2007, ICCD/COP (8)/16/Add.1, 23 October 2007.

[28]See UNEP (1995a).

activities (e.g. overgrazing and biomass harvesting), obviously includes ecological processes in pastoral land environments. The CBD stresses the importance of, and the need to promote, international, regional and global cooperation among countries and intergovernmental organisations and the non-governmental sector for conservation of biological diversity and the sustainable use of its components[29] and for nations to prepare strategies to implement the CBD.

The objective of the CBD is relevant to pastoral land management as it includes the conservation of biological diversity, encouraging the sustainable use of its components and the fair and equitable sharing of the benefits arising out of the utilization of genetic resources, including by access to genetic resources and by transfer of relevant technologies. It takes into account various rights over those resources.[30] The CBD also recognises that nations have a responsibility for conserving their biological diversity and for using their biological resources in a sustainable manner. In this context, the objective of pastoral land conservation is implicit within the definitions of "biological diversity" and "biological resources" in Article 2 of the CBD.

The CBD acknowledges that substantial investments are required to conserve biological diversity and that there is a broad range of environmental, economic and social benefits from those investments.[31] It stresses the importance of, and the need to promote, international, regional and global cooperation among countries and intergovernmental organisations and the non-governmental sector for conservation of biological diversity and the sustainable use of its components[32] and for nations to prepare strategies to implement the CBD.[33] For the CBD to take on an expanded, more precise role in the sustainable use of pastoral lands, for example, specific provisions could be drafted for pastoral land and included as a protocol to the CBD.[34] The rules should focus on the ecological functions of pastoral land that are essential for the conservation of biodiversity and the maintenance of human life in these areas. In this regard, the CBD Strategic Plan for Biodiversity 2011–2020[35] provides an excellent process to achieve a goal of pastoral land sustainability. It urges Parties and other governments and organizations to develop national and regional targets, using the Strategic Plan and its Aichi Biodiversity Targets, as a flexible framework, in accordance with national priorities and capacities and to take into account both the global targets and the status and trends of biological diversity.

[29]See CBD Article 16.

[30]See CBD Article 1.

[31]See CBD Articles 6–10.

[32]See CBD Article 16.

[33]See Miller and Lanou (1995); Prip et al. (2010), Part 1, pp. 10–22.

[34]See Article 28, Adoption of Protocols.

[35]Decision X2, COP10.

21.3.6 International Initiative for the Conservation and Sustainable Use of Soil Biodiversity

The International Initiative for the Conservation and Sustainable Use of Soil Biodiversity (IICSUSB) was endorsed in 2006 as the main international framework for action regarding soil biodiversity.[36] The IICSUSB is managed mainly by the UN Food and Agriculture Organization and other partners and has as its main goals, awareness-raising, knowledge and understanding and mainstreaming. The Strategic principles of the IICSUSB are important for pastoral land management and include improvement of farmers' livelihoods and recognition of their skills, integrated, adaptive, holistic and flexible local solutions, participatory technology development suitable to local conditions, building partnerships and alliances, promotion of cross-sectoral and integrated approaches, and dissemination and exchange of information and data.

21.3.7 The Strategic Plan for Biodiversity (2011–2020)

The Strategic Plan for Biodiversity (SPB) was adopted at CBD COP-10 and provides a framework for action by CBD stakeholders.[37] The SPB is accompanied by 20 Aichi Biodiversity Targets where land health/biodiversity is cross-cutting amongst these targets including sustainable agriculture (Target 7), reducing pollution (Target 8), restoring and safeguarding ecosystem services (Target 14), and enhancing ecosystem resilience and health including carbon storage and restoring 15 % of degraded ecosystems.[38] The Aichi Biodiversity targets could be used to establish specific targets for sustainable use of pastoral lands. Biodiversity can underpin many benefits to sustainable use of dryland areas as a cross-cutting approach.

21.3.8 Convention on Climate Change

The United Nations Framework Convention for Climate Change (FCCC)[39] recognises the role of terrestrial ecosystems as a sink and reservoir for potential

[36]COP8 decision VI/5 2006.

[37]Decision X/2 COP10 18–29 October 2010 adopted a revised and updated Strategic Plan for Biodiversity, including the Aichi Biodiversity Targets for 2011–2020.

[38]COP11 called for major global efforts for ecosystem restoration, including restoring soils in agricultural systems being the major opportunity in terms of current extent of degraded area, addressing social, economic and environment benefits and achieving multiple objectives.

[39]See UNEP (1995b).

greenhouse gases and is concerned that human activities have been substantially increasing the atmospheric concentrations of greenhouse gases. Two of the principal sources of greenhouse gases are changes in land use cover and land use. Scientists have established that pastoral land ecosystems provide a significant reservoir of carbon[40] and that pastoral activities play a role in emissions of greenhouse gases and initiate or exacerbate soil and vegetation degradation. In particular, these include biomass burning, cultivation, using organic manure, applying nitrogenous fertilisers and livestock grazing.[41] Excessive vegetation clearance, a principal cause of land degradation in pastoral areas, is a key concern of the FCCC. Land degradation exacerbates the emission of gases from terrestrial ecosystems to the atmosphere. Accelerated wind and water erosion, on a global scale, is the principal soil degradation process. Some 1643 million ha are affected worldwide, of which 250 million are affected by strong or extreme forms of soil erosion. Agriculture and land-use change account for about 30 % of the greenhouse gas emissions blamed for global warming. Feed efficiency can be so low in arid parts of Africa, where livestock typically graze on marginal land and crop residues that every kilo of protein produced can contribute the equivalent of one tonne of carbon dioxide.

While the FCCC does provide for changes to the terrestrial environment, it is not considered to be the most appropriate international legal vehicle to address pastoral land conservation, because it presently has a primary focus on changes in the industrial sector rather than the nonindustrial and agricultural land use sectors.

The Kyoto Protocol[42] under the FCCC includes a responsibility to promote sustainable forms of land use in the light of climate change characteristics. It specifically recognises the need to expand and preserve soil carbon sinks and improve agricultural practices in countries where a significant proportion of the emissions are related to the clearing of vegetation for agriculture. The adoption of the Paris Agreement in December 2015 (FCCC/CP/2015/L.9/Rev.1) could provide many potential benefits for the conservation of pastoral land. In particular it recognizes that climate change represents an urgent and potentially irreversible threat to human societies and the planet and thus requires the widest possible cooperation by all countries, and their participation in an effective and appropriate international response, with a view to accelerating the reduction of global greenhouse gas emissions.

---

[40] See Squires (1998), p. 209; Squires and Glenn (1997), pp. 140–143.

[41] See Neely et al. (2009), pp. 5–13.

[42] See Kyoto (1997), it is intended that a new agreement will be forged in Paris in 2015 or thereafter.

21.3.9 *UNFCCC Adaptation and Mitigation*

In recent years, the FCCC process has made significant progress in providing a mechanism for developing countries to access climate change funds to implement adaptation and mitigation activities. In this regard, the development of an effective legal, policy and institutional framework would be an essential component of a national strategy to manage the climate change impacts on pastoral land ecosystems. Adaptation strategies can be based on reducing land degradation, improving livestock management and improving human livelihood. Mitigation actions could consider policy development, monitoring and reporting methodology, economic assessment and capacity building.[43]

21.3.10 *Draft International Covenant on Environment and Development*

The IUCN Draft Covenant has been prepared as an umbrella agreement to knit together the principles reflected in the sectoral treaties that impact upon the environment and development.[44] The Draft Covenant has many articles that are relevant to the protection and management of pastoral land. In particular, Article 20 relates to natural systems and calls on parties to take appropriate measures to conserve and where necessary and possible restore natural systems which support life in all its diversity, including biological diversity, and to maintain and restore the ecological functions of these systems as an essential basis for sustainable development.

21.4 Regional Instruments

A number of regional instruments include elements that countries (other than those within the specific jurisdiction of the instrument) can adopt in framing national pastoral land legislation. Of particular relevance is the African Convention on the Conservation of Nature and Natural Resources which was adopted in 1968 and revised in 2003.[45] Instead of taking a purely utilitarian approach to natural resources conservation, it acknowledges the principle of common responsibility for environmental management and calls for conservation and rational use of natural resources for the benefit of present and future generations. After 25 years,

[43]See Wilkes et al. (2011), pp. 10–12.

[44]See IUCN (2004), most of the 75 Articles have relevance to pastoral land management.

[45]See African Convention (2003), in particular Articles I–VIII, X–XVI.

developments in international environmental law made it necessary to revise this treaty, update its provisions and enlarge its scope, in particular, to provide for the establishment of institutional structures to facilitate compliance and enforcement. Many aspects of the African Convention are useful for countries of other regions to follow, especially its objectives which correspond to key elements of a sustainable approach for pastoral land conservation: the achievement of ecologically rational, economically sound and socially acceptable development policies and programmes.

Importantly, the African Convention embodies a comprehensive and integrated regional approach to environmental protection and sustainable development. It reflects a renewed perception of resource management that reconciles nature and culture. This instrument advocates an integrated approach to resource management and provides international legal principles and best practices that are relevant to pastoral land management. In this context, the African Convention provides a useful approach to regional natural resources management, and a similar concept could be applied to pastoral land conservation for other regions.

21.4.1 National Legal and Institutional Frameworks

Legislation has been used for years in many countries, usually in a piecemeal fashion, to manage specific types of problems on pastoral land (e.g. soil erosion), to control grazing activity which causes land degradation problems (e.g. overgrazing of cattle and sheep) and to indirectly control land management problems (e.g. through environmental planning and land use allocation).[46] A review of national legislation associated with pastoral land conservation indicates that states have used a variety of approaches to frame domestic legislation and to deal with protection and management of pastoral land. This is generally reflected by the broad structural features of the legislation, as well as in the variety of specific mechanisms used to protect and manage pastoral land.[47] In some countries, pastoral land law was an early form of legislation used to manage rural land.[48] Although its primary role has been land administration, as against environmental management and land conservation, this type of law often included mechanisms that could achieve sustainable use of land, including enforceable conditions on land leases for grazing management, vegetation, soil and water management and monitoring rangeland condition and pasture quality.[49]

In the past, the main type of legislation aimed at the control of degradation of pastoral land has generally been the law associated with "soil conservation". The

[46]See Kurucz (1993), p. 468.

[47]See Schlager and Ostrom (1992), pp. 249–251; Hannam and Boer (2002), Section 4, pp. 33–54.

[48]See e.g. *Western Land Act 1901 NSW Australia.*

[49]See Hannam (2000), pp. 168–169; Hannam (2007), pp. 8–15.

legislation had a land utilization focus, which is no longer adequate to effectively protect and manage the world's pastoral degradation problems. As a rule, as the area of land affected by degradation grew, practical land conservation techniques were developed and applied in conjunction with expanding agricultural activities. The conservation capabilities of the legislation were overshadowed by the objective of agricultural production, price support schemes for domestic and export needs and land settlement and development schemes. Soil conservation legislation was introduced in the first half of the nineteenth century primarily to control the effects of soil erosion by wind and water on pastoral and cultivation lands.[50] Over time, a variety of laws have been developed for rangeland and pastoral land use; for example:

21.4.2 Examples of National Laws Relating to Pastoral Management

- United States of America—*Public Rangeland Improvement Law 1978,*
- *Forest and Range Renewable Resources Planning Law 1974*
- Canada—*Agricultural Land Commission Law 1979* (British Columbia)
- Australia—*South Australia Pastoral Land Management and Conservation Law 1989*
- New Zealand—*Resource Management Law 1991*
- Iceland—*Bill of Legislation on Soil Conservation 2002*
- People's Republic of China—*Grassland Law 2002*
- Mongolia—*Pastureland Law* (2015 draft)
- Kyrgyzstan—*Law of the Kyrgyz Republic on Pastures 2009*
- *Afghanistan—Rangeland Law 2010*
- *Tajikistan—Law of the Republic of Tajikistan about Pastures 2013*

In summary, observations of the characteristics of laws used for pastoral land management include[51]:

- Various laws have been used to rectify land management problems caused by poor land use planning or inappropriate land use, as against the inherent ecological characteristics of land being used as the premise for land use decision-making.
- Many national laws associated with pastoral land are still very much overshadowed by the physical problems of land use, mainly grazing and agriculture.
- The primary land functions of pastoral lands are not well represented, and few laws refer to the ecological features or needs of pastoral land.

[50]See Grossman and Brussaard (1992).

[51]See Hannam (2007), pp. 14–15.

- The legislation does not acknowledge pastoral land as having a central role in terrestrial ecology, the conservation of biodiversity and maintenance of environmental amenity.
- Many laws are not clear on the purpose or objectives for pastoral land.
- There is often not a logical development of legal elements, and many laws do not include the elements necessary to protect pastoral land.
- There is inconsistent use of terminology, and often there is an absence of definitions, inadequate or poorly stated definitions.
- The structure of some laws indicates that they are a reaction to a political or institutional issue rather than designed to effectively manage pastoral land.
- Some states have developed a framework of legislation to manage specific land use management problems of which pastoral land management is one type—but they generally lack a linking or coordinating mechanism.

21.5 Other Legal Regimes with a Role in Administration of Pastoral Land

Other areas of environmental law which together make up a framework of law that can be effective in administration of pastoral land include[52] the following.

21.5.1 Land Administration Law

There are many laws that provide for the administration and management of pastoral land. These have been employed to control the use of land and its mismanagement. This legislative regime includes various forms of land tenancy and leasehold regimes, with provisions to assess land and regulate conditions of occupancy, use, sale, lease and reservation. There may also be provisions for forfeiture of holdings, alteration of conditions of use and protection of land dedicated for public use.[53]

21.5.2 Biodiversity Law

There is a body of law that has a general objective to protect the environment and conserve biological diversity and ecosystems. This area of law assists in the management of pasture land through its promotion of ecologically sustainable

[52]See also Grossman and Brussaard (1992); Flintan (2012), pp. 13–21.

[53]See Chalifour et al. (2007).

development and through the conservation of biological diversity in general. Biodiversity legislation generally does not apply directly to the control of agricultural land uses, but it is applied to achieve more effective conservation and management of protected and reserved areas in agricultural landscapes.[54]

21.5.3 Vegetation Conservation Law

This area of law is important because it focuses on the conservation and sustainable management of vegetation and pastures and can control their destruction. It promotes vegetation and pasture management in consideration of social, economic and environmental parameters. It sets rules for the ecological assessment of vegetation (its biodiversity, habitat values, flora and fauna values, regional patterns and threatened species) and rules for issue of permits.[55]

21.5.4 Forest Law

Forest laws have been created for both public and private forestry with provisions to develop and apply land management guidelines.[56] Plantation establishment and reafforestation law have been applied in some countries to promote reafforestation of land and establishment of shrub land in pasture land badly affected by land degradation (water and wind erosion and salinity) caused by overgrazing.

21.5.5 Environmental Protection Law

The main purpose of this legislation is to protect, restore and enhance the quality of the environment by reducing the risks to human health and preventing the degradation of the environment by pollution, waste management, discharge of harmful substances and point-source pollution. National environmental protection measures can cover any activity that may impact or has impacted on pastoral land and affected by pollution, waste dumping, chemicals or other substances that may impact on the pastoral environment.

---

[54]See De Klemm and Shine (1993), pp. 1–24.

[55]See e.g., *Native Vegetation Act 2003* New South Wales, Australia.

[56]See Tarasofsky (1999), pp. 5–7.

21.5.6 Environmental Planning Law

The objective of environmental planning legislation is to protect the human and natural environment through the preparation of land use or zoning plans, prescribing environmental assessment standards for land use and determining significant environmental impact from land use change. This type of legislation has the potential to be more widely applied to benefit pastoral land by preserving its natural character, protecting outstanding natural features and landscapes, efficient use and development of pasture land resources, maintenance and enhancement of amenity values and heritage protection.[57]

21.5.7 Climate Change Law

A global review for evidence on pastoral systems and climate change indicates that greater recognition and support are needed for sustainable pastoral and agro-pastoral systems in view of their contributions to climate change adaptation and mitigation, disaster risk management, biodiversity protection and sustainable agriculture and rural development. Targeted support by governments, civil society organisations, development agencies and community donors and researchers is needed to harness this opportunity.[58] Given the important role of pastoral lands in the management of climate change impacts, it is considered that more attention needs to be paid in national environmental law reform to provide for adaptation and mitigation actions on pastoral land.

21.6 Case Studies

The following two case studies present two different approaches to legislative reform to manage environmental problems in the pastoral areas of Mongolia and the People's Republic of China (PRC). These countries share a similar ecological gradient and have both been undergoing socio-institutional and ecological transformations: from collective to market economies, increasing market integration and adverse climate impacts on vegetation. However, the two areas are different in their demographic, ethnic, cultural, economic and political contexts, and the respective governments have quite different approaches to pastoral management. These characteristics of the two areas make for a comparative analysis of their legislative response to pastoral management, to enhance an understanding of the respective

[57] See also Grossman and Brussaard (1992).

[58] See also Neely et al. (2009), pp. 25–27.

pastoral land socio-ecological systems and to identify solutions towards more sustainable governance of the pastoral lands.

The Mongolian example discusses a single law approach, where the Pastureland Law has been drafted to manage a range of environmental, social and economic problems associated with pastoral land management. The PRC example discusses a comprehensive and integrated law approach to manage a complex range of environmental, social and economic issues associated with land degradation in the dryland ecosystems of western PRC. The legislative approaches taken with each of the two examples are vastly different and partly reflect the complexity of problems associated with pastoral land usage including the existing political environment. In the Mongolian example, at the time of its initial drafting in 2007, the Pastureland Law was not being considered within the context of the existing national environmental law framework, whereas the PRC approach was considered within a complex of nine related national environmental law areas that apply to the western dryland region.

21.6.1 *Mongolian Pastureland Law (Single Law Approach)*

Mongolian pasture land is degraded because herders are unable to apply sustainable grazing practices. The grassland is not valued by the country so its regulation and management have been avoided in the past. Herders continue to graze their livestock on common land unrestrained, where there is high competition for good pasture. They use public pasture and water free of charge and without initiative to protect and properly use it.[59] Grasslands and arid grasslands cover a large proportion of the country (112 million ha or 71 %).[60] As relatively intact terrestrial ecosystems, the Mongolian pasturelands play a significant role in sequestrating atmospheric carbon dioxide, conserving biodiversity and they provide livelihood benefits to local herders.[61]

However, over the past 50 years, pastureland degradation has undermined the ecosystem services they generate. Under the Mongolian Land Law 2002, "pastureland" means *rural agricultural land covered with natural and cultivated vegetation for grazing of livestock and animals*.[62]

There are many reasons why Mongolia initially drafted the Pastureland Law in 2007 and then revised it in 2015.[63] The main aim was to develop a legal framework

[59]See Swift (2007), pp. 9–10.

[60]Around 122 million ha of Mongolia is devoted to nomadic pastoralism: 4.6 % of this lies in the alpine zone, 22.9 % in the forest steppe zone, 28 % in the steppe zone, 23.3 % in the semi-desert zone and 16.2 % in the desert.

[61]See ADB (2013a), pp. 4–10.

[62]See Article 3.1.6 *Land Law 2002*.

[63]See Mongolian Government (2007), pp. 1–2.

to provide land tenure for livestock producers, herder groups, cooperatives and herding households; to establish remote pasture reserves; and for the operation of pasture use committees as a basis to overcome many of the problems that stem from lack of land ownership.[64] It also aims to develop a legal framework to exempt herding households from individual income tax and to introduce pasture use fees differentially based on economic and ecological assessments and to consider a livestock husbandry risk fund. The *State Policy on Herders*, introduced in 2009, will provide strategic support to the Pastureland Law. The policy aims to create a favourable legal, economic and business environment which, in turn, enables development of better living conditions for herders, prevention of poverty in herders and employment and social security.[65]

21.6.1.1 Legal Concept

The draft Pastureland Law has been through extensive public and parliamentary discussion process since 2007. Although at the time of writing this chapter it had still not been promulgated, the government and community are working through a complex of issues to ensure effective legislation finally results. The main purpose is to provide a legal method for the transition from an unplanned and unregulated pastoral land use system to a system characterised by secure possession of pastureland for herders and legal entities; a pastureland planning and management system; to improve the development and management of pastureland information; to distinguish the functions, duties and responsibilities between the different levels of administration; and to improve the system to identify and manage problems associated with land degradation and the effects of global climate change. Importantly, it includes procedures to classify pastureland on an ecological basis and provide for the agricultural and economic needs of traditional herding communities and the livestock husbandry industry. In this regard, the new administrative, implementation and operational procedures of the Pastureland Law will support economically productive pastoral agriculture while managing pressures on the ecological environment from climate change, desertification and natural disasters.[66]

The 2015 draft Mongolian Pastureland Law includes a procedure to allocate land for grazing and for its management and protection, which constitutes the basis of land tenure. This approach will help overcome many problems that stem from the traditional pasture usage system.[67] The procedure includes the identification and classification of pastureland, a request (application) for pastureland possession for the purpose of livestock husbandry and the issue of a certificate for possession. It will engage the communities in the land tenure process.

[64]See also Fernandez-Gimenez (2006), pp. 30–35.

[65]See Mongolian Government (2009), pp. 1–2.

[66]See also Mongolian Government (2010), pp. 51–52.

[67]See also Fernandez-Gimenez (2006), p. 33.

Global experience shows that land use systems that enable stakeholders to formally participate in the decision-making process generally provide a more satisfactory and balanced outcome for all parties. This procedure will help promote the sustainable use of pastureland and development of a stewardship ethic which is important for a stable long-term tenure system.[68] These are important procedures and should increase the capability of Mongolia to manage its pastoral resources more effectively in the face of the increasing effects of climate change and other natural events.[69] The draft law contains many legal elements considered essential for successful pastoral land law, but it is argued that additional support systems will be required to enable herder communities and legal entities to achieve a sustainable livelihood and for the state to achieve its national goals for pastoral land management.[70] Some areas identified include development of operational policy, development of a national strategy for grassland management, development of land management plans, formation of local stakeholder advisory committees, providing access to finance and credit, developing a comprehensive education and training programme and ensuring stakeholders have access to information and knowledge.

21.6.1.2 Pastoral Land Conservation and Climate Change

In an effort to improve its management of the climate change effects on pastoral ecosystems, the Mongolian Government has taken steps to develop a Nationally Appropriate Mitigation Action (NAMA) for grassland and livestock management by following the procedure established under the FCCC.[71] Under these procedures, a NAMA is defined as any kind of activity that reduces greenhouse gas emissions. For Mongolia, the specific grassland and livestock management activities developed under the NAMA are tailored to Mongolia's national circumstances and in line with the FCCC principle of common but differentiated responsibilities.[72] In following the FCCC procedure the NAMA is embedded within the national sustainable development strategy of Mongolia. The mitigating activities are measurable, reportable and verifiable and supported by various sustainable development activities. Importantly by following the procedures set down in the FCCC process and satisfying the standards for national and international registration, this opens up the potential for Mongolia to access climate change funding to implement the

[68]See Squires (2012).

[69]See Batima (2006).

[70]See Hannam (2007), p. 59; Mongolian Government (2009) Government policy towards herders. 4 June 2009, Resolution 39, Ulaanbaatar.

[71]Mongolian Government (2010), pp. 51–52; ADB (2011) *Inception report*, ADB R-CDTA 7534, Strengthening carbon financing for regional grassland management in Northeast Asia; ADB (2013a), pp. 1–7.

[72]See also ADB (2013a), pp. 23–25.

NAMA.[73] One of the essential requirements for the effective implementation of the grassland/livestock NAMA is a legal and policy framework that in Mongolia's case would include the Pastureland Law.[74] The NAMA approach adopted by Mongolia would also be suitable for other developing countries which have extensive pastoral areas, to follow.

21.6.2 *Management of Dry Land Ecosystems of People's Republic of China (Integrated Law Approach)*

The total area of grassland in the PRC is around 400 million ha, accounting for 42 % of the country's land area. In PRC, grassland is considered a multifunctional ecosystem that provides ecological and economic benefits. Consideration of the main functions of the different grassland types in each ecological-economic region of the PRC forms a basis for adopting different management regimes.[75]

During 2004–2009, an investigation was undertaken into the legal, policy and institutional framework in PRC's western ecosystems under an international PRC project, covering the three provinces of Qinghai, Shaanxi and Gansu and the three autonomous regions of Inner Mongolia, Xinjiang Uygur and Ningxia Hui.[76] Western PRC includes a significant proportion of PRC's pastoral lands, and land degradation and desertification is a serious problem that affects this area. As many vulnerable communities are dependent on pastoral resources for their livelihood, land degradation is closely linked to poverty across these provinces and regions.[77] The project employed an integrated ecosystem management (IEM) approach to managing land degradation[78] with an emphasis on capacity building and technological support and to strengthen cross-sector coordination and transboundary management of the natural resources. The total population of this area is around 120 million people.

The objective of the legal and policy investigation was to improve the policy and regulatory framework for land degradation control as an essential part of strengthening PRC's enabling environment and to build capacity to adopt an integrated approach to sustainable land management. It aimed at improving the policies, laws, regulations and procedures for combating land degradation, and this was achieved using three programmes: (1) improving the law and policy framework to intensify institutional capacity, (2) capacity building to implement the law and policy and (3) a supporting study programme to innovate and reform environmental laws and

---

[73] See KPMG (2011), p. 12.

[74] See Hannam (2012), p. 58.

[75] See Brown et al. (2008), p. 2.

[76] See Du and Hannam (2012), pp. 1–2.

[77] See Ren et al. (2008) and Williams et al. (2009).

[78] See GEF (2000), p. 2.

policies.[79] Western PRC includes a significant proportion of PRC's pastoral lands and land degradation is a serious problem that affects this area. As many vulnerable communities are dependent on pastoral land resources for their livelihood, land degradation is closely linked to poverty across these provinces and regions.[80] The project successfully employed the IEM approach to managing land degradation.[81]

21.6.2.1 IEM as a Legal Concept

The concept of IEM has been applied in international environmental law and progressively developed into normative principles and rules.[82] In the PRC investigation, IEM was defined as "a holistic approach to address the linkages between ecosystem functions and services (such as carbon uptake and storage, climatic stabilization and watershed protection, and medicinal products) and human social, economic and production systems".[83] From a legal perspective, as a comprehensive strategy and method to manage natural resources, IEM is a suitable framework in which to consider the national and provincial legislation.[84] By definition, IEM requires taking all components of an ecosystem into account and in consideration of the social, economic and natural environment. The investigation was designed in a manner for each province and region to develop a practical framework to enhance the capability of different groups to implement law and policy of IEM, including legal officers, legal draftsmen, judicial officials, policymakers, government officials and private individuals.

The starting point to assess the capacity of the existing legal and policy framework and policies for land degradation control was the categorisation of legislative materials into nine principal law areas[85]:

1. Grassland (includes the *Grassland Law 2002, Regulations on Prevention of Grassland Fires 2005, Administrative Measures for Balance of Grass and Husbandry 2007*); at provincial level, it includes over 20 individual regulations and legal instruments
2. Desertification
3. Water and soil conservation
4. Water resources
5. Forestry
6. Agriculture

[79]Du and Hannam (2012), p. 2.
[80]See Ren et al. (2008), p. 200; Williams et al. (2009), p. 219.
[81]See Du and Hannam (2012), pp. 3–6.
[82]See UNEP (1995a).
[83]See Jiang (2007), p. 2.
[84]See Du and Hannam (2012), p. 26.
[85]See Du and Hannam (2012), p. 8.

7. Land administration
8. Environmental protection
9. Environmental impact assessment

A methodological procedure was developed to accommodate the principles of IEM in the provincial and regional law and policy reform process. The method was applied by officials from the People's Congresses and governmental legislative offices of Qinghai, Gansu, Shaanxi provinces and Inner Mongolia, Xinjiang and Ningxia autonomous regions. Experience from applying IEM in the legal and policy processes of the respective provinces and autonomous regions of western PRC was substantial and includes many lessons which other countries could usefully follow[86]:

- IEM provides a scientific approach to fulfil commitments to various multilateral conventions concerning environmental protection and sustainable use of natural resources, and it establishes a strategic framework to manage land, water and biological resources for sustainable pastoral land development.
- The IEM approach is a cross-cutting mechanism that accommodates multiple scientific means and is a good policy tool to coordinate national implementation requirements of international environmental conventions as they apply to pastoral lands.
- It is an effective means to achieve sustainable use of pastoral lands and combat land degradation and is a sound framework to review and solve issues concerning natural resource ownership, use of protected areas, access to resources and benefit sharing.
- The flexible framework of IEM provides multiple options for implementation, including incorporation of IEM principles in national strategies and action plans, regional plans, and applying IEM principles in policymaking, land use and institutional planning. It is a good basis for reform of institutions and organisations to support sustainable use of pastoral ecosystems.
- IEM is a relevant tool for planning, decision-making and evaluation of ecosystem activities associated with all aspects of pastoral land management, policy and law.

As an outcome of the investigation, the central government and the governments of the three provinces and three autonomous regions made a commitment to adopt IEM to improve the legal framework. It provided a valuable opportunity for PRC to introduce IEM into its legal procedures by adopting international best experiences and rules of law.

As a result, the PRC had revised various laws and rules including the *Measures for Implementation of Grassland Law in Qinghai Province*. A new legislative reform programme was also introduced which included the *Regulations on Administration of Grassland in Tianzhu Tibetan Autonomous County* and *the Detailed*

[86]See Du and Hannam (2012), pp. 138–140.

Rules on Implementation of Grassland Law of the People's Republic of China in Xinjiang Uygur Autonomous Region.

Local pastoral land implementation measures or management regulations specify regulations for the ownership, planning, construction, use, protection, supervision and inspection, and legal liability of grasslands. Several provinces have also introduced regulations governing monitoring, implementation, and management of forage–livestock balance. In some provinces, these regulations have only been issued quite recently, and management of the forage–livestock balance continues to be a long-term task.

21.7 Improving Legislation for Pastoral Land Management

It is considered that there is room to improve the capacity of both international environmental law and national environmental law frameworks concerning pastoral land management and the following discussion can be used as a basis to improve the environmental law for pastoral land management.

21.7.1 International Level

At the international level, a number of multilateral and regional treaties contain elements and principles that provide for various problems associated with pastoral land management, but a formal coordination mechanism would have to be developed to enable this to happen. The fact that pastoral land occupies a significant proportion of the earth's terrestrial surface would seem to justify that this is a matter that requires urgent consideration. Two principal framework structures that an international environmental law instrument for pastoral land may take are in the form of either a legally binding instrument or a non-legally binding instrument.

As a binding instrument, this could take the form of either a specific treaty with all the essential legal elements for pastoral land, a framework treaty which identifies the pastoral land elements in existing treaties and links them through a separate binding instrument or a protocol to an existing treaty that creates specific rules for pastoral land. The development of a binding instrument under one of the three global treaties discussed earlier in this chapter, with key rules and guidelines for pastoral land management, is an option that would benefit all pastoral lands of the world.[87] As a non-binding option, this could take the form of an international charter for pastoral land or a declaration for pastoral land.

[87]E.g., a special annex under Article 31 of the CCD, or a protocol under Article 28 of CBD.

21.7.1.1 Process

It is essential that the promotion of an international legal framework to protect the world's pastoral land provides an opportunity for the input of all interested parties, including international environmental organisations, state governments, pastoral land science institutions, private sector interests and non-government organisations. Should this proceed, it would be open for individual states to participate. The experiences involved in the development and introduction of other existing environmental treaties should be examined before embarking on a process for pastoral land. In general, for pastoral land, such a process could involve:

- Building an adequate understanding of current pastoral land management issues to establish a clear vision of the benefits of an international legal framework for pastoral land.
- Assembling existing policy, strategic material and legislation, which have specific or indirect references to pastoral land conservation (e.g. biodiversity, environmental planning and natural resources legislation).
- Reviewing international instruments and strategic material and identifying the instruments that may be accommodated within the political, cultural and physical circumstances for pastoral land.
- Outlining a capacity-building process, including environmental education for the international community, focusing on the most effective types of technical training for those involved in the development and implementation of state strategies for the legal protection of pastoral land.

For this process to be effective, it will require cooperation between international pastoral land policy and science organisations and an international environmental law institution such as the International Union for the Conservation of Nature (IUCN) World Commission on Environmental Law (WCEL). For a specific region, there may be benefits in the respective states pursuing a regional instrument that provides specifically for the social, economic and ecological characteristics of rangeland in the particular geographic region.

21.7.2 National Level

To establish or improve national pastoral land legislation various international legal principles can be applied to form the philosophical basis on which to select a suitable approach to develop a framework for national pastoral land management legislation. It is appreciated that some states may prefer to develop pastoral land strategies with a minimum of legal regulation, whereas others may prefer a stronger regulatory law.[88]

[88] See Hannam and Boer (2002), pp. 44–45.

21.7.2.1 Non-regulatory Strategies

Non-regulatory strategies would feature elements that concentrate on:

- Education and awareness programmes for sustainable use of pastoral lands.
- Ecosystem research, assessment and monitoring pastoral land use.
- Financial support for research and extension.
- Support and development of participatory community pastoral land planning.
- Development of ecologically sustainable pastoral land standards and practices.
- Development of pastoral land management and incentive-based programmes.

21.7.2.2 Regulatory Strategies

Regulatory strategies would feature elements that concentrate on:

- Development of statutory land use plans that prescribe limits and targets of pastoral land use (e.g. maximum number of livestock at particular times of the year, permissible cultivation practices).
- Issue of licences or permits to control pastoral land use (these would prescribe use entitlements relating to fencing, stock numbers, access and soil restoration requirements).
- Land use agreements between the state and individuals or groups of land users, which set legally binding land use standards.
- The use of restraining notices where sustainable pastoral land use limits (as set out in a statutory plan or agreement) are exceeded.
- Prosecution for failure to follow prescribed standards of sustainable use.

21.7.2.3 Elements of National Pastoral Land Legislation

To assist in the development of a national pastoral land law framework, with a mix of legal protection and management elements, states may benefit from a set of "generic elements" from which to select appropriate elements for the construction of a national pastoral land law.[89] The general range of legal elements which a state could consider for national pastoral land management law, for which the specific procedures would be developed, includes the following types of elements:

- A comprehensive statement of the purposes of the legislation.
- Goals and objectives with a mandate for ecologically sustainable use of pastoral land—specific objectives can be formulated from the objectives and principles of global conventions, strategies and policies concerning ecology, the conservation of nature, biodiversity and sustainable land management.

[89]See Hannam (2004), pp. 14–18; Commonwealth of Australia (2010), pp. 7–9.

- The preparation of pastoral land policy, codes of practice, sustainability indicators and the physical and ecological limits of pastoral land.
- The preparation of a pastoral land management strategy, outlining national pastoral land management policy and policy to manage land use problems.
- That sustainable use of pastoral land can be achieved through a mix of regulatory and non-regulatory means, including incentive and support programmes and community pastoral land management advisory groups.
- Provision to manage all classes of pastoral land that are based on sustainable land use criteria and contain ecologically sustainable standards for implementation at the national, provincial and local levels.
- Procedures to manage natural resources generally, including provisions to protect soil, water and biodiversity.
- An equitable distribution of responsibilities in managing pastoral land, including the state, minister, administrators, advisory bodies, officials and herders.
- A facility to enter into legal contracts and agreements with pastoral land users and occupiers for the sustainable use of pastoral land resources.
- A facility for the community to participate in pastoral land assessment, planning and decision-making, including establishing community advisory groups; pastoral land management plan preparation; public exhibition of plans, policies and strategies, and calling for public submissions; and provision for community representatives to sit on pastoral management committees. A facility to develop education programmes on sustainable use of pastoral land and implementation of technical seminars and conferences.
- A facility to develop and implement a variety of pastoral land research and investigation programmes and to relate the research outcomes to state programmes.
- A facility to take formal action where prescribed standards of pastoral land management use are not being met and where there has been a contravention of the legislation.

21.8 Conclusion

In general, legal and policy frameworks can be effective in raising awareness to improve the management of pastoral lands, and they also present an opportunity to include procedures for adaptation and mitigation of climate change impacts while enhancing livestock productivity and food security. In particular, the development of a sound legal and policy framework by an individual state should contain procedures that improve the documentation and dissemination of information on pastoral land management and build capacity and to manage pastoral land through incentives, including payments for environmental services and other non-financial rewards.[90] Based on experiences from other areas of land use, the use of voluntary

[90]See ADB (2013b).

and regulatory procedures can help change behaviour towards sustainable and adapted management of pastoral ecosystems. Including incentive mechanisms within legal instruments can lead to adoption of sustainable activities and reverse land degradation in pastoral land—activities that will also enhance livelihoods and reduce the vulnerability of pastoral and agro-pastoral people.

National pastoral land legislation should also provide the means to develop livestock policies that address the barriers and bottlenecks faced by the inhabitants, including procedures to build local and policy-level awareness and capacity for pastoral land husbandry and help secure tenure at community and landscape levels. Legislation should also provide for targeted research in pastoral ecosystems, effective institutions and with a focus on practices that can improve the economic aspects of pastoral management.

At the administrative level, it is essential that integrated multi-sectoral, multi-stakeholder and multilevel processes be available that address the range of natural resources (land, water, rangelands, forests, livestock, energy and biodiversity) and social dimensions that characterise pastoral environments, with active participation of stakeholders. A holistic approach and partnership can take advantage of the objectives of local, national and global goals.

Importantly, it is also appropriate that the international and national frameworks for pastoral land management address climate change impacts and adopt the processes of the United Nations FCCC that are relevant to pastoral management, especially in relation to adaptation and mitigation actions. Consideration should be given by individual states to support the concept of an international environmental law instrument for pastoral land management, in particular, or at the regional level with elements and procedures that provide for the unique aspects of pastoral land in a respective region.

Questions for Classroom Discussion

Q1. How can international environmental law lead help to improve the management of the world's pastoral lands?

Q2. Discuss the "generic elements" for national pastoral land law and how can they be applied to shape a particular national law for pastoral land management?

References

Abildtrup J, Audsley E, Fekete-Farkas M, Guipponi C, Gylling M, Rosato P, Rounsevell M (2006) Socioeconomic scenario development for the assessment of climate change impacts on agricultural land use: a pairwise comparison approach. Environ Sci Policy 9:101–115

ADB (Asian Development Bank) (2011) *Inception report*, ADB R-CDTA 7534, Strengthening carbon financing for regional grassland management in Northeast Asia, p 105

ADB (Asian Development Bank) (2013a) Making grasslands sustainable in Mongolia: adapting to climate and environmental change. Mandaluyong City, Philippines, p 54

ADB (Asian Development Bank) (2013b) Making grasslands sustainable in Mongolia: international experiences with payments for environmental services in grazing lands and other rangelands. Mandaluyong City, Philippines, p 54

African Convention (African Convention on the Conservation of Nature and Natural Resources) (2003), p 31, see http://www.intfish.net/treaties/africa2003.htm; extensive analysis of the Convention is available at http://uicn.org/themes/law/pdfdocuments/EPLP56EN.pdf. Accessed 5 Nov 2014

Batima P (2006) Climate change vulnerability and adaptation in the livestock sector of Mongolia. Final report project AS06. Assessments of impact and adaptation to climate change. International START Secretariat, Washington, DC, p 409

Brown C, Waldron S, Longworth J (2008) Sustainable development in Western China: managing people, livestock and grasslands in pastoral areas. Edward Elgar, Cheltenham, p 305

Chalifour N, Kameri-Mbote J, Lin Heng Lye P, Nolon JR (eds) (2007) Land use law for sustainable development. IUCN Academy of Environmental Law Research Studies. Cambridge University Press, Cambridge, UK, p 654

Commonwealth of Australia (2010) Principles for sustainable resource management in the rangelands. Natural Resources Ministerial Council, Department of the Environment, Water, Heritage and the Arts, Canberra, p 14

De Klemm C, in collaboration with Shine C (1993) Biological diversity conservation and the law: legal mechanisms for conserving species and ecosystems. IUCN Environmental Law Centre and IUCN Biodiversity Programme, p 292

Dong S, Lu W, Liu S, Zhang XL, James P, Yi S, Li X, Li J, Li Y (2011) Vulnerability of worldwide pastoralism to global changes and interdisciplinary strategies for sustainable pastoralism. Ecol Soc 16(2):10

Du Q, Hannam I (eds) (2012) Law, policy and dryland ecosystems: People's Republic of China. IUCN, Gland, xvi + p 140

European Commission (2000) Analysis of national reports on the implementation of the United Nations convention to combat desertification. International Institute for Environment and Development, Turkey, Lebanon, Jordan and Syria, Drylands Program, p 15

Fernandez-Gimenez ME (2006) Land use and land tenure in Mongolia: a brief history and current issues. USDA Forest Service Proceedings RMRS-P-39, pp 30–36

Flintan F (2012) Making rangelands secure: past experience and future options. Working Document Version 1.0, International Land Coalition, p 87

Food and Agriculture Organization of the United Nations (FAO) (2001) Pastoralism in the new millennium. Animal production and health paper No. 150. UN Food and Agriculture Organization, Rome, Italy, p 93

Friedel MH, Laycock WA, Bastin GN (2000) Assessing rangeland condition and trend. In: L't M, Jones RM (eds) Field and laboratory methods for grassland and animal production research. CABI, Wallingford, pp 227–262

GEF (Global Environment Facility) (2000) GEF Operational Program 12 Integrated Ecosystem Management, Nairobi, p 15

Grossman M, Brussaard W (eds) (1992) Agrarian land law in the western world, essays about agrarian land policy and regulation in twelve countries of the western world. CAB International, Wallingford, p 280

Halimova N (2012) Land tenure reform and implications for land stewardship. In: Squires V (ed) Rangeland stewardship in Central Asia. Springer, Dordrecht, pp 305–332

Hannam ID (2000) Policy and law for rangeland conservation. In Arnalds O, Archer S (eds) Rangeland desertification. Kluwer Academic Publishers, London, p 209

Hannam ID (2004) A method to identify and evaluate the legal and institutional framework for the management of water and land in Asia: the outcome of a study in Southeast Asia and the People's Republic of China. Research report 73. International Water Management Institute, Colombo, p 33

Hannam ID (2007) Report to United Nations development programme Mongolia on review of draft pastureland law of Mongolia. United Nations Development Programme Sustainable Grassland Management Project, Ulaanbaatar, p 112

Hannam ID (2012) Legal and policy framework to support a livestock/grassland NAMA in Mongolia. Working Paper, ADB R-CDTA 7534 – Strengthening Carbon Financing for Regional Grassland Management in Northeast Asia, p 89

Hannam ID, Boer BW (2002) Legal and institutional frameworks for sustainable soils: a preliminary report. IUCN, Gland/Cambridge, UK, xvi + p 88

IUCN (The World Conservation Union) (2004) Draft international covenant on environment and development, Third Edition: Update Text, world commission on environmental law in cooperation with international council of environmental law. IUCN Environmental Law Program, Bonn, p 193

Jiang Z (2007) To implement integrated ecosystem management to accelerate combating land degradation. In: Integrated ecosystem management, Proceedings of the international workshop, Beijing 1–2 November 2004, Global Environment Facility, Asian Development Bank and People's Republic of China Forestry Publishing House, pp 2–6

KPMG (2011) Financing low carbon investment in developing countries: public-private partnership for implementing nationally appropriate mitigation actions. KPMG International Cooperative, Switzerland, p 14

Kurucz M (1993) Land protection, property rights and environmental preferences (land use control and land development). Conn J Int Law 8(2):467–485

Kyoto (1997) The Kyoto protocol to the convention on climate change, the Climate Change Secretariat, UNEP, p 20

Miller KR, Lanou SM (1995) National biodiversity planning. Guidelines based on early experiences around the world. The World Resources Institute and IUCN, Baltimore, p 162

Mongolian Government (2007) Draft pastureland law 20 July 2007 and brief introduction to the draft law on pastureland, p 26

Mongolian Government (2009) Government policy towards herders. 4 June 2009, Resolution 39, Ulaanbaatar, p 21

Mongolian Government (2010) Mongolian second national communication under the United Nations framework convention on climate change. Ministry of Nature, Environment and Tourism, p 159

Neely C, Bunning S, Wilkes A (eds) (2009) Review for evidence on dryland pastoral systems and climate change, implications and opportunities for mitigation and adaptation. Food and Agriculture Organization of the United Nations, Rome, p 37

Nori M, Switzer J, Crawford A (2005) Herding on the brink: towards a global survey of pastoral communities and conflict – an occasional paper from the IUCN Commission on Environmental. Economic and Social Policy, Gland, p 33

Nori M, Davies J (2007) Change of wind or wind of change? Climate change, adaptation and pastoralism. The World Initiative for Sustainable Pastoralism, International Union for Conservation of Nature, Nairobi, Kenya, p 23 [online] URL: http://cmsdata.iucn.org/downloads/c__documents_and_settings_hps_local_settings_application_data_mozilla_firefox_profile.pdf. Accessed 10 Nov 2014

Prip C, Gross T, Johnston S, Vierros M (2010) Biodiversity planning: an assessment of national biodiversity strategies and action plans. Institute for the Advanced Study of Sustainability, p 237

Ren JZ, Hu ZZ, Zhao J, Zhang DG, Hou FJ, Lin HL, Mu XD (2008) A grassland classification system and its application in China. Rangeland J 30(2):199–210

Schlager E, Ostrom E (1992) Property-rights regimes and natural resources: a conceptual analysis. Land Econ 68(3):249–262

Secretariat CCD (Secretariat of the Convention to Combat Desertification) (2000) Chart of key elements in the CCD national reports by the Countries in Asia

Secretariat CCD (Secretariat of the Convention to Combat Desertification) (2007) Report, 8th session, Conference of the Parties, Madrid, 3–14 September 2007, ICCD/COP (8)/16/Add.1, 23 October 2007

Squires VR (1998) Dryland soils: their potential as a sink for carbon and as an agent in mitigating climate change. Adv GeoEcol 31:209–215

Squires VR (2012) Better land stewardship: an economic and environmental imperative. In: Squires V (ed) Rangeland stewardship in Central Asia. Springer, Dordrecht, pp 31–50

Squires VR, Glenn E (1997) Carbon sequestration in the drylands: an agenda for the twenty-first century. World atlas of desertification, 2nd edn. Edward Arnold/UNEP, p 140–143

Swift JJ (2007) Case study: institutionalizing pastoral risk management in Mongolia: lessons learned. Prepared under the overall guidance from the Rural Institutions and Participation Service, FAO Project Pastoral Risk Management Strategy, TCP/MON/0066, p 58

Tarasofsky RG (ed) (1999) Assessing the international forest regime. IUCN Environmental Policy and Law Paper No. 37, p 156

Taylor JL (2006) Negotiating the grassland: the policy of pasture enclosures and contested resource use in Inner Mongolia. Hum Organ 65(4):374–386

UN (United Nations) (1992a) Rio declaration, New York

UN (United Nations) (1992b) Agenda 21, New York

UN (United Nations) (1995) Convention to combat desertification, New York

UN (2012) The future we want: outcome document adopted at Rio + 20, General Assembly 11 December 2012 A/RES/66/288

UN (2014) Sustainable Development Knowledge Platform: Sustainable Development Goals, available at http://sustainabledevelopment.un.org/?menu=1300. Access 18 Aug 2014

UNEP (United Nations Environment Program) (1982) World charter for nature

UNEP (United Nations Environment Program) (1995a) Convention on biological diversity, Nairobi

UNEP (United Nations Environment Program) (1995b) United Nations framework convention on climate change

White RP, Tunstall D, Henninger N (2002) An ecosystem approach to drylands: building support for new development policies. World Resources Institute Washington, DC Information Policy Brief No. 1, p 14

Wilkes A, Wang S, Tennigkeit T, Feng J (2011) Agricultural monitoring and evaluation systems: what can we learn for the MRV of agricultural NAMAs? ICRAF Working Paper No. 126. World Agroforestry Centre, Beijing, China, p 17

Williams A, Wang MP, Zhang M (2009) Land tenure arrangements, property rights and institutional arrangements in the cycle of rangeland degradation and recovery. In: Squires V, Lu X, Lu Q, Wang T, Yang Y (eds) Rangeland degradation and recovery in China's pastoral lands. CABI, Oxford, pp 219–234

WISP (World Initiative for Sustainable Pastoralism) (2007) Pastoralists' species and ecosystems knowledge as the basis for land management. WISP Policy Brief 5:1–4

Part V
Tools for Change: An Inventory of Global Farm Animal, Wildlife and Food Safety Laws

Chapter 22
Zoonotic Diseases and Food Safety

Leslie Couvillion

22.1 Introduction

Livestock, poultry, fish, and other animals can host a wide range of diseases and infections. As a result, the farmed animal industry has a fundamental public health dimension. When humans consume or come into contact with sick or contaminated animals or animal products, they can get sick as well. Zoonotic diseases, also called zoonoses, are a sub-category of animal diseases that are transferable from animals to humans (in contrast to diseases that pass only from animals to other non-human animals). Many types of pathogenic agents, including bacteria, viruses, fungi, and parasites, cause zoonotic diseases.[1] Some zoonoses, like *Salmonellosis* (Salmonella poisoning), Foot and Mouth Disease (FMD), and bovine spongiform encephalopathy (BSE or Mad Cow Disease) have made international headlines. Others, like brucellosis (Bang's Disease), are less familiar but still pose a major global threat to animal and human health and well-being.[2]

The exact costs of zoonotic diseases worldwide are difficult to estimate. Consistent data on zoonoses occurrences is scarce, although the World Health Organization (WHO) is working to fill this gap.[3] It is certain, however, that the consequences of these diseases go beyond public health impacts. Zoonoses can cause "tremendous economic losses to the livestock and poultry industries" when animals must be quarantined, slaughtered, or otherwise disposed of in response to

[1]World Health Organization (1915f).

[2]Iowa State University Center for Food Security (1913).

[3]The WHO has established a Foodborne Disease Burden Epidemiology Reference Group (FERG) to estimate the global burden of foodborne diseases. The group's first report is due for release in late 1915. See World Health Organization, Regional Office for Europe (1915).

L. Couvillion (✉)
Yale School of Law, New Haven, CT, USA
e-mail: leslie.couvillion@yale.edu

G. Steier, K.K. Patel (eds.), *International Farm Animal, Wildlife and Food Safety Law*, DOI 10.1007/978-3-319-18002-1_22

disease outbreaks.[4] These costs can lead, in turn, to "devastating sociologic and economic effects on communities" who depend upon affected industries.[5] Therefore, zoonoses threats must be carefully regulated, not only to protect animal and human welfare but also to safeguard social and economic interests. Preventing, monitoring, and controlling animal diseases, however, is no simple task.

Indeed, zoonotic disease management is a complex and continually evolving legal area. It intersects with a number of other fields, including:

- General food safety and hygiene (including animal feedstuffs);
- Community/public health;
- Environmental protection;
- Animal welfare;
- Veterinary services;
- Antibiotic use;
- Vaccine use;
- Biosecurity;
- International trade;
- Land use; and
- Wildlife management.

Laws from these related fields provide both direct and indirect means for protecting farm animal health. For instance, environmental protection or wildlife preservation laws can impact farming practices and disease management efforts. This is because most zoonoses originate from the human-animal-ecosystem interface.[6] Therefore, the expansion of agricultural or pasture lands into formerly wild areas (through cutting down forests or draining wetlands) can increase disease exposure risks. Sometimes, wildlife protection laws can halt these expansions.[7] In places where such encroachments do still occur, other regulatory tools can help protect domestic farm animals from wildlife diseases. Potential strategies include compulsory vaccinations, vector control programs (such as tick eradication efforts), or fenced-in containment zones. Such measures, however, are somewhat rare in the laws and regulations in most countries.[8] Therefore, better control of the livestock/wildlife interface is a potential growth area for zoonotic disease management worldwide.

In addition to the human-animal-ecosystem interface, zoonoses must be regulated along an animal or animal product's entire lifecycle: from rearing to slaughter to processing to transport (and possibly trade) to consumption. Food safety risks

---

[4]National Center for Foreign Animal and Zoonotic Disease Defense (1908).

[5]National Center for Foreign Animal and Zoonotic Disease Defense (1908).

[6]"[A] study of 335 emerging infectious disease (EID) events between 1940 and 1904 . . . found that zoonotic diseases formed the majority of EIDs (60.3 %) and that most of these zoonotic diseases (71.8 %) came from wildlife." Bennett and Carney (1910), p. 329.

[7]See Chap. 24.

[8]See Bengis et al. (1902).

exist along the entire food chain. What's more, the list of known zoonoses is extensive—and always growing. Therefore, regulators need to know what food producers, processors, transporters, retailers, and other handlers are all doing in order to both (a) prevent disease outbreaks and (b) contain outbreaks as quickly as possible if they do emerge. Key tools for accomplishing these tasks include Hazard Analysis and Critical Control Point (HACCP) systems[9] (for disease prevention) and animal tracking/disease traceability systems (for disease control). Such systems are becoming more prevalent around the world. Indeed, the United States (US),[10] European Union (EU),[11] Japan,[12] and Australia[13] have each recently implemented new HACCP or livestock tracing programs.

However, these programs tend to address only those activities happening within a country's own borders. International trade in farm animals and animal products creates further possibilities for the spread of diseases, and globalization trends pose increasing challenges to effective zoonotic disease management. In today's world, diseases originating in far-away places can often reach local animal and human populations. As a result, regulators must be able to coordinate animal disease prevention, surveillance, and response efforts at local, national, regional, and international levels.

So far, international law has developed only a limited set of responses to the zoonoses management problem.[14] There is no international convention on animal diseases. However, the World Animal Health Organization (OIE) and the United Nations Food and Agriculture Organization (UN FAO or FAO) (in collaboration with the WHO) have published codes for animal hygiene[15] and food safety,[16] respectively. Many national governments have adopted these important—albeit voluntary and non-binding—codes. In addition, a handful of World Trade Organization (WTO) agreements affect the level at which individual countries can set sanitary requirements for producers and distributors of animal-origin food products.[17] These trade agreements, however, are more concerned with supporting international trade than with safeguarding animal health. In fact, these instruments discourage countries from setting higher-than-average protective measures since

[9]HACCP is a management system "which identifies, evaluates, and controls hazards which are significant for food safety." Food and Agriculture Organization of the United Nations (1997). HACCP systems are science-based and systematic, identifying specific hazards and control measures throughout the food chain. Since HACCP systems apply to primary production (in addition to final consumption) stages, they allow regulators to focus on risk prevention as well as risk detection.

[10]See *infra* Sect. 22.3.

[11]See *infra* Sect. 22.4.

[12]See Appendix.

[13]See Appendix.

[14]See *infra* Sect. 22.2.

[15]See *infra* Sect. 22.2.1.

[16]See *infra* Sect. 22.2.2.

[17]See *infra* Sect. 22.2.

such requirements could limit an importer's pool of potential trading partners and, thus, create a barrier to trade. Such restrictions encourage a "lowest common approach" to setting national standards. Overall, then, the international law framework on animal diseases and food safety provides important guidance for national governments, while, at the same time, potentially constraining regulatory innovation.

At the national level, zoonotic disease management regimes take a variety of forms. The majority of countries, including most nations in North America, Africa, and Asia, have some sort of uniform national law on animal diseases that incorporates a wide range of general animal health and food safety measures.[18] Such laws, along with their implementing regulations, tend to cover the entire lifecycle of animal-origin food products: from farm-rearing to slaughtering to processing to packaging to importing and exporting. Common features include:

- Specified lists of covered diseases and animals (which can vary widely from country to country or region to region);
- Notification requirements for owners of infected (or potentially infected) animals;
- Inspection rights by state-designated veterinary officers;
- Quarantine and/or slaughter requirements for infected (or potentially infected animals);
- Import and export restrictions; and
- Civil and criminal penalties.

Most of these laws, such as the Animal Health Protection Act in the United States,[19] are relatively recent, enacted within the past decade or two. They tend to synthesize and update a number of older, more disparate regulations on animal health, food sanitation, and other related issues. This focus on uniformity can help to simplify and streamline zoonoses management. At the same time, such an approach runs the risk of obscuring potential weak points in the regime, allowing certain issues to fall through the cracks. Unifying statutes are also likely to rely on general mandates instead of specific guidance. Broad standards can be more difficult to monitor and enforce since they make it harder to tell when a violation has occurred.

An alternative, and less prevalent, approach is the patchwork or "package" legislative scheme. For instance, Brazil employs a set of individualized and "highly detailed regulations covering every sector of food production."[20] The European Union (EU), meanwhile, has a core "package" of food hygiene laws, along with dozens of independent directives addressing a range of sanitary and disease-prevention measures.[21] Sometimes, of course, such a patchwork of laws can lead

---

[18]See Appendix.

[19]See *infra* Sect. 22.3.1.

[20]Lafisca et al. (1913), p. 268. See also Appendix.

[21]See *infra* Sect. 22.4.

to over-complication, with confusion and high costs for regulators and food business operators alike. However, this approach can also provide highly specific guidance on discrete matters, making monitoring and enforcing of the standards more clear-cut. Furthermore, a patchwork scheme can help to reveal areas of the law that are not being addressed—in other words, the places where tools for change may be hiding within regulatory systems.

This chapter begins by exploring the international law backdrop for animal disease and food safety regulation, including the role of international trade agreements. Later sections discuss the regimes in the United States and the European Union as representatives of the (a) unifying single statute and (b) regulatory "package" approaches to zoonotic disease management, respectively. The Appendix contains a more complete index of the laws and regulations in other countries. The unification approach represents the general trend in most of the world. (Indeed, as pointed out below, this is the direction the EU may be headed as well.)[22] Of course, no matter what approach a government takes, zoonotic disease management is never a field unto itself. It is inherently tied up with other regulatory areas, such as public health, food safety, and trade.

22.2 International Treaties, Conventions, and Non-binding Codes

Zoonotic diseases are an increasingly global problem. However, there is no binding, comprehensive international law instrument on the issue. Nevertheless, international regulatory bodies have put forth a number of important non-binding tools, including the OIE's Animal Health Codes[23] and the FAO/WHO's *Codex Alimentarius* (Food Code).[24] These model codes provide guides for national governments in setting sanitary standards and drafting food safety and animal health legislation. However, adherence is voluntary since the codes do not contain binding enforcement mechanisms. Despite this limitation, the OIE and FAO/WHO guidelines are highly influential and form the backbone of global zoonotic disease management. Indeed, the vast majority of nations have adopted both sets of instruments.

Several international treaties and agreements impact zoonotic disease management as well. The most significant of these are a pair of interrelated WTO agreements: the Agreement on Technical Barriers to Trade (TBT Agreement)[25] and the Agreement on the Application of Sanitary and Phytosanitary Measures (SPS

---

[22]See *infra* Sect. 22.4.2.

[23]See *infra* Sect. 22.2.1.

[24]See *infra* Sect. 22.2.2.

[25]World Trade Organization (1995a).

Agreement).[26] Both instruments encourage governments to harmonize their national trade measures according to international guidelines set by bodies like the OIE, FAO, and WHO. The end goal is avoiding any "unnecessary obstacles" to international trade.[27] Therefore, technical regulations "shall not be more trade-restrictive than necessary to fulfill a legitimate objective," such as protecting human health.[28] This condition applies to national food safety and animal and plant health standards for imported and exported products.

Ultimately, the TBT and SPS agreements are more likely to discourage rather than encourage stringent zoonotic disease regulations. This is because the agreements are more concerned with economics (e.g., promoting trade) than with animal or public health. If one country, for instance, were to enact uniquely rigorous animal health standards and these standards applied to imported animal-origin food products, the pool of potential trading partners would likely shrink. Such a result might be an "unnecessary obstacle" to international trade. In practice, therefore, most nations fulfill their TBT and SPS Agreement obligations by following the food safety and animal health guidelines in the OIE and FAO/WHO codes—using these standards as a ceiling, rather than a floor.[29] This "lowest common denominator" approach to establishing food safety and animal health guidelines can potentially water down laws in favor of economic gain. Consequently, the need to avoid obstacles to trade can negatively affect food safety and animal health standards in the stronger trading nations.

Other international laws, however, can foster stronger, more innovative approaches to zoonotic disease management. For example, the Convention on Biological Diversity (CBD) has become increasingly concerned with "the linkages between biodiversity, zoonotic diseases, and human health."[30] In partnership with the WHO, the CBD has set up a series of "One Health" networks "to develop and strengthen intersectoral collaboration between animal and human health and environmental health."[31] A number of conferences, workshops, and consultations by human and animal health partner agencies (including the OIE) have taken place under the One Health program.[32] CBD also supports research programs on disease transmission at the human-animal-ecosystem interface; indeed, land use changes (like the encroachment of agricultural or pastoral lands into wilderness areas) can "favour disease transmission and loss of biodiversity."[33] Overall, these CBD

[26]World Trade Organization (1995b).

[27]World Trade Organization (1995a), Preamble.

[28]World Trade Organization (1995a), Art. 2.2.

[29]Indeed, "WTO members that wish to apply stricter food safety measures than those set by [the *Codex Alimentarius* Commission] may be required to justify these measures scientifically." World Health Organization (1915a).

[30]United Nations Convention on Biological Diversity (1914).

[31]World Organisation for Animal Health (1915b).

[32]World Organisation for Animal Health (1915d).

[33]World Organisation for Animal Health (1911). In general, an inverse relationship exists between disease spread and biodiversity loss: greater biodiversity losses enable the easier spread of zoonoses, while biodiversity gains can limit disease transmission risks.

initiatives bring to light some of the important linkages between animal health, human health, wildlife biodiversity, and farming practices. Many of the CBD's goals, therefore, inherently overlap with those of the OIE, FAO, and WHO. While there may not yet be an international hard law instrument on zoonotic diseases, the CBD is one potential tool for supplementing the non-binding guidelines discussed below.

22.2.1 OIE Animal Health Codes

The World Organization for Animal Health (OIE) is an intergovernmental organization charged with protecting and improving animal health worldwide. Originally established in 1924 as the Office International des Epizooties, the group became the World Organization for Animal Health in 1903, while holding onto its historical acronym. As of 1914, the OIE has 180 Member Countries, with regional and sub-regional offices on every continent.[34] Its general goals include promoting animal welfare (of both farm and wildlife animals), food safety, sanitary safety, veterinary services, scientific research, transparency, and international solidarity.[35] The OIE also seeks to safeguard trade in animal products, serving as a reference organization for the WTO (meaning that nations can apply OIE guidelines to comply with WTO agreements).[36] Each of these objectives supports the organization's central mission: preventing, controlling, and eradicating animal diseases at a global level.[37]

The OIE's primary function is to develop codes on animal health. These documents consist of science-based standards, guidelines, and recommendations for international, national, and regional bodies to adopt. The organization publishes two codes relevant to zoonotic disease management: the Terrestrial Animal Health Code,[38] dating back to 1968, and the Aquatic Animal Health Code,[39] first published in 1995.[40] Together, the Animal Health Codes aim "to assure the sanitary safety of international trade in terrestrial animals and aquatic animals and their products."[41] They are regularly updated and serve as the principal references for WTO Member States under the TBT and SPS Agreements.[42] Specific provisions address:

---

[34]World Organisation for Animal Health (1915a).

[35]World Organisation for Animal Health (1915c).

[36]World Organisation for Animal Health (1915a).

[37]World Organisation for Animal Health (1915a).

[38]World Organisation for Animal Health (1915h).

[39]World Organisation for Animal Health (1915e).

[40]The codes are also accompanied by two manuals, *The Manual of Diagnostic Tests and Vaccines for Terrestrial Animals*, first published in 1989, and the *Manual of Diagnostic Tests for Aquatic Animals*, first published in 1995. See World Organisation for Animal Health (1915f).

[41]World Organisation for Animal Health (1915f).

[42]See *supra* Sect. 22.2.

- Animal welfare;
- Food safety;
- Disease detection and notification;
- Veterinary services, including the use of anti-microbial agents; and
- Safe trade, including import risk analysis.[43]

Ultimately, however, compliance with these measures is voluntary.[44] Despite their non-binding nature, the OIE Animal Health Codes are highly influential in shaping how nations structure their zoonotic management regimes.

Another one of the OIE's crucial functions is to coordinate information-sharing among nations. The organization manages the World Animal Health Information Systems (known as WAHIS and WAHIS-wild for domesticated and wild animals, respectively), which provide up-to-date information on animal diseases worldwide.[45] All OIE Member Countries have access to these systems. Each Member Country is required to notify the OIE about outbreaks of listed diseases (which include zoonoses as well as animal diseases that are not transferable to humans). The OIE then categorizes countries and subnational regions according to the prevalence of particular animal diseases, conferring "disease-free" status on those countries or regions with no reported outbreaks. In addition to disease statuses, the databases also include information on relevant animal husbandry practices (such as vaccinations) that a country is applying.

These informational tools can bolster the effectiveness of the Animal Health Codes. With better information about where diseases are prevalent, as well as where they have been successfully contained or eliminated, countries can adjust their import protocols or modify practices within their own borders to avert or contain outbreaks. Therefore, The WAHIS and WAHIS-wild systems are a key reason the OIE's recommendations can support the prevention, control, and eradication of animal diseases around the world, despite not being binding on Member Countries.

22.2.2 FAO/WHO Codex Alimentarius (Food Code)

As far back as 1950, a joint FAO/WHO Expert Committee on Nutrition recognized that "the conflicting nature of [national] food regulations may be an obstacle to trade and may therefore affect the distribution of nutritionally valuable food."[46] By the early 1960s, the FAO and WHO had set up the *Codex Alimentarius* Commission to implement a joint FAO/WHO food standards program.[47] Today, the Commission

[43]See World Organisation for Animal Health (1915h).

[44]World Organisation for Animal Health (1915g). However, the codes do provide a process for lodging complaints and entering into voluntary dispute settlements.

[45]World Animal Health Information Database (WAHID) (1913).

[46]World Health Organization (1915d).

[47]World Health Organization (1915d).

continues to develop international food safety and hygiene standards and is the principal global institution charged with such a task.[48] Its mission statement is simple: "to ensure safe, good food for everyone, everywhere."[49] As of 1915, the *Codex Alimentarius* Commission has 186 members and 229 observers (including intergovernmental organizations, non-governmental organizations (NGOs), and UN organizations).[50] Its membership covers 99 % of the world's population.[51]

The Commission's primary duty is publishing the *Codex Alimentarius* (Food Code). The Food Code is a series of general and specific food safety standards that aim to protect consumer health and promote fair trade practices.[52] Its standards are based upon the best available science and are updated as needed. While the General Principles of Food Hygiene[53] form the centerpiece of the code, a number of additional guidelines target particular food-producing sectors. Many of these supplementary codes are relevant to zoonotic disease management. Examples include the:

- Code of Hygienic Practice for Meat[54];
- Code of Hygienic Practice for Milk and Milk Products[55]; and
- Code of Practice on Good Animal Feedings.[56]

Although the Food Code's end goal is to protect human (rather than animal) health, many of its measures have animal health and welfare implications. Indeed, food is one of the most important bridges between animal and human health. A healthier farm animal population guarantees a healthier food supply and, thus, a healthier human population. The reverse is also true: a healthy human population depends upon a healthy food supply, which requires healthy farm animals. Therefore, many of the Food Code's measures apply to farmers and others who handle animals destined for human consumption.

Like the OIE Animal Health Codes,[57] the FAO/WHO Food Code is not binding on Member Countries. Ultimately, it is a series of voluntary recommendations.[58] However, as with the OIE codes, the Food Code still manages to carry a lot of weight. Its standards provide the starting point for almost all national and regional laws and norms on food safety and hygiene.[59] At the international level, the Food

[48]Desta and Hirsch (1912).

[49]World Health Organization (1915b).

[50]World Health Organization (1915c).

[51]World Health Organization (1915a).

[52]European Food Information Council (1904).

[53]World Health Organization (1903).

[54]World Health Organization (1915e).

[55]World Health Organization (1909).

[56]Food and Agriculture Organization of the United Nations (1908).

[57]See *supra* Sect. 22.2.1.

[58]World Health Organization (1915a).

[59]European Food Information Council (1904).

Code is also one of the WTO's reference points for resolving trade disputes on food safety and consumer protection.[60] Therefore, in practice, many countries follow Food Code and the OIE Animal Health Code recommendations as though they were binding.

The following sections explore two different national-level approaches toward incorporating international guidelines into zoonotic disease and food safety regulations. The US strategy of enacting a single, unifying statute that consolidates a wide range of animal health measures represents the dominant trend worldwide. In contrast, the EU method of relying on more of a patchwork legislative scheme has been losing favor in recent decades (indeed, the EU itself is moving toward a more unifying approach). The relative strengths and weaknesses of these two regimes are discussed below.

22.3 United States (US): Unifying Single Statute Approach

The United States Centers for Disease Control and Prevention (CDC) reported over 1600 foodborne disease outbreaks (and nearly 30,000 cases of illness) from 1911 to 1912.[61] Indeed, zoonoses have been a growing concern for the federal government over the past two decades. In 1902, Congress enacted the Animal Health Protection Act (AHPA)[62] in an attempt to establish a comprehensive law on animal health and diseases. The AHPA consolidates an array of older pieces of legislation on topics ranging from animal product imports to slaughterhouse operations to veterinary accreditations. Its implementing regulations address a number of additional issues, such as setting up livestock tracking and disease traceability systems.[63] Although the AHPA aims to simplify and streamline animal disease management, the multifaceted nature of the field makes meeting this goal a challenge.

Primary authority for carrying out the AHPA rests in a single federal agency body: the Animal and Plant Health Inspection Service (APHIS), part of the US Department of Agriculture (USDA). APHIS is in charge of monitoring, tracing, and responding to animal and plant diseases, including (but not limited to) zoonotic diseases. A dedicated sub-body within APHIS, the Veterinary Services (VS) office, is responsible for controlling and eliminating livestock and poultry diseases and pests. The Center for Veterinary Biologics (CVB), also part of APHIS, regulates vaccines and other biological products used for diagnosing, preventing, and treating

---

[60] World Health Organization (1915a). See *supra* Sect. 22.2.

[61] United States Centers for Disease Control and Prevention (1914b). 1911 and 1912 are the most recent years for which data on zoonoses is available.

[62] See *infra* Sect. 22.3.1.

[63] See *infra* Sect. 22.3.1 and Appendix.

animal diseases.[64] Therefore, implementing the AHPA is an intricate task requiring coordination among several specialized agency sub-bodies.

Moreover, APHIS and its subsidiaries do not work in a vacuum. Their mission is inherently interdisciplinary, requiring them to collaborate with a variety of federal, state, and local agencies and even other non-governmental entities. Indeed, the US employs a complex "system of interlocking safeguards" to guard against food-related health dangers.[65] At the federal level, for instance:

- The Department of Health and Human Services (DHHS) and the CDC work to protect human health and eliminate human disease risks.[66]
- The Food and Drug Administration (FDA) regulates food safety generally. The FDA plays an important role in developing vaccine standards[67] and recommending measures to block the introduction of disease agents through international animal trade.[68] With the Food Safety Modernization Act (FSMA) of 1911,[69] the agency strengthened its focus on disease prevention.
- FDA also works with the USDA, DHHS, and the Department of Homeland Security (DHS) to come up with proposals on how to best prevent animal or plant disease outbreaks and food contamination.[70]
- DHS manages "agricultural quarantine inspection" stations at the nation's ports of entry to guard against pest and disease introductions.[71]
- The CDC and DHHS assist the USDA with coordinating zoonotic disease surveillance.[72]
- The Food and Safety Inspection Service (FSIS), a USDA body, is in charge of inspecting meat, poultry, and egg products.[73] FSIS also maintains a directory of BSE (Mad Cow Disease) information and resources for regulators, producers, and businesses.[74]
- The USDA runs a number of additional programs and centers on animal diseases, including the Animal Parasitic Disease Laboratory, the National Animal Health Monitoring System (NAHMS), the National Animal Disease Center, and the National Veterinary Services Lab.[75]

---

64 United States Department of Agriculture, Animal and Plant Health Inspection Service (1914).

65 United States Department of Agriculture (1915).

66 United States Department of Health and Human Services (1914) and United States Centers for Disease Control and Prevention (1914a).

67 United States Food and Drug Administration (1910).

68 United States Food and Drug Administration (1914).

69 United States Government Publishing Office (1911).

70 United States Government Publishing Office (1911), Sect. 109.

71 United States Department of Homeland Security (1907).

72 This authority is granted by the Public Health Security and Bioterrorism Preparedness and Response Act of 1902. United States Government Publishing Office (1902).

73 United States Department of Agriculture, Food Safety and Inspection Service (1915).

74 United States Department of Agriculture, Food Safety and Inspection Service (1913).

75 United States Department of Agriculture (1915).

This list is not exhaustive but showcases just some of the diverse federal government functions relevant to animal diseases and food safety in the United States. In addition, these federal entities often collaborate with their respective state agencies, as well as local governmental and non-governmental bodies.[76] Therefore, the zoonoses management regime in the US continues to have a number of moving pieces, both across sectors and throughout various levels of government. As a result, the AHPA is far from being a truly comprehensive instrument in practice. Nevertheless, the statute remains the country's strongest—albeit often underutilized—single legal tool for promoting animal (as well as human) health and food safety goals at the same time.

22.3.1 Animal Health Protection Act (AHPA) (1902)

With the Animal Health Protection Act (AHPA),[77] the US Congress enacted the country's first unifying piece of federal legislation on animal health and disease prevention. The act aims to support the "prevention, detection, control, and eradication of diseases and pests of animals."[78] It explicitly recognizes the linkages between (a) animal health, (b) human health and welfare, (c) economic interests of the livestock and related industries, (d) environmental protection, and (e) interstate and international commerce.[79] The AHPA encompasses all members of the animal kingdom (including aquatic species), except for humans.[80] Covered diseases include zoonoses (such as FMD) as well as diseases passed only from animal-to-animal (like classical swine fever, a viral hog disease that does not affect humans). The Secretary of Agriculture has the power to define and update the list of targeted diseases.[81]

The act's breadth is vast. It regulates the entire lifecycle of animal and animal-derived food products, with provisions on:

---

[76]National Center for Foreign Animal and Zoonotic Disease Defense (1908). One benefit of this approach is that it allows for a system of checks and safeguards to ensure the AHPA's mandates are carried out. On the other hand, the wide array of responsibilities shared among agencies might create opportunities for overlooking problems and letting certain aspects of animal disease management fall between the cracks.

[77]7 U.S.C. §§ 8301–8322. See Cornell University Law School. 7 U.S. code, tit 7, chap 109: animal health protection.

[78]7 U.S.C. § 8301(1). See Cornell University Law School. 7 U.S. code, tit 7, chap 109: animal health protection.

[79]7 U.S.C. § 8301(1). See Cornell University Law School. 7 U.S. code, tit 7, chap 109: animal health protection.

[80]7 U.S.C. § 8302(1). See Cornell University Law School. 7 U.S. code, tit 7, chap 109: animal health protection.

[81]7 U.S.C. § 8302(3). See Cornell University Law School. 7 U.S. code, tit 7, chap 109: animal health protection.

- Imports, including bans and restrictions on animals or articles that might introduce or disseminate "any pest or disease of livestock" within the US[82];
- Exports, including bans and restrictions on animals or articles that might disseminate "from or within the United States . . . any pest or disease of livestock;"[83]
- Interstate movement[84];
- Seizure, quarantine, and disposal of infected animals, with special provisions for "extraordinary emergencies" that permit drastic measures like preventative slaughter[85];
- Inspections, seizures, and warrants, including the right of the Secretary of Agriculture or a designated officer or veterinarian "to make any inspection or seizure" if a judge finds probable cause "that there is on certain premises any animal, article, facility, or means of conveyance [that could harbor a pest or disease];"[86]
- Detection, control, and eradication[87];
- Veterinary accreditation and training[88];
- Cooperation with other federal, state, local, and international bodies, both governmental and non-governmental[89];
- Criminal and civil penalties, which can be severe: up to $1,000,000 in fines or 10 years in prison;[90]
- Funding, including a "Pest and Disease Response Fund" to support emergency eradication and research activities[91]; and
- Enforcement, with provisions that allow "any person" to bring notice of a violation of the act to the attention of the Attorney General.[92]

---

[82]7 U.S.C. § 8303. See Cornell University Law School. 7 U.S. code, tit 7, chap 109: animal health protection.

[83]7 U.S.C. § 8304. See Cornell University Law School. 7 U.S. code, tit 7, chap 109: animal health protection.

[84]7 U.S.C. § 8305. See Cornell University Law School. 7 U.S. code, tit 7, chap 109: animal health protection.

[85]7 U.S.C. § 8306. See Cornell University Law School. 7 U.S. code, tit 7, chap 109: animal health protection.

[86]7 U.S.C. § 8307. See Cornell University Law School. 7 U.S. code, tit 7, chap 109: animal health protection. The statute provides for both warrantless inspections and inspections with warrants.

[87]7 U.S.C. § 8308. See Cornell University Law School. 7 U.S. code, tit 7, chap 109: animal health protection.

[88]7 U.S.C. § 8309. See Cornell University Law School. 7 U.S. code, tit 7, chap 109: animal health protection.

[89]7 U.S.C. § 8310. See Cornell University Law School. 7 U.S. code, tit 7, chap 109: animal health protection.

[90]7 U.S.C. § 8313. See Cornell University Law School. 7 U.S. code, tit 7, chap 109: animal health protection.

[91]7 U.S.C. § 8321. See Cornell University Law School. 7 U.S. code, tit 7, chap 109: animal health protection.

[92]7 U.S.C. § 8314(b)(1). See Cornell University Law School. 7 U.S. code, tit 7, chap 109: animal health protection.

APHIS has also promulgated a number of regulations to help carry out the AHPA. These include the:

- Cooperative Control and Eradication of Livestock or Poultry Diseases Regulations,[93] which set up a cooperative program between states and the USDA to control and eradicate listed diseases. The regulations require the prompt destruction (usually by burial or burning) of animals and materials "affected by or exposed to disease,"[94] as well as the disinfection of all affected premises, conveyances, and materials.[95] Owners may be compensated for the value of destroyed animals and materials.[96]
- Animal Disease Traceability Regulations,[97] which set up a national identification and traceability system for livestock transported across state borders. These regulations have been in effect since March 1913 and replace the former National Animal Identification System (NAIS). Under the current rules, livestock moving interstate must be officially identified with veterinary inspection certificates (or other documentation).[98] States and tribes can develop their own preferred tracking systems for administering the program within their borders.[99] If an outbreak occurs, the system should allow authorities to track infected animals as well as identify and quarantine other potentially exposed animals.[100]
- Exportation and Importation of Animals (Including Poultry) and Animal Products Regulations,[101] which set forth a number of import and export restrictions, along with a series of inspection and handling protocols. For example, the regulations ban imports of ruminants from regions designated as having rinderpest or FMD present.[102]

[93] 9 C.F.R. §§ 49-56. See Cornell University Law School. 9 C.F.R. tit 9, chap I, subchap B: cooperative control and eradication of livestock or poultry diseases.

[94] 9 C.F.R. § 53.4. See Cornell University Law School. 9 C.F.R. tit 9, chap I, subchap B: cooperative control and eradication of livestock or poultry diseases.

[95] 9 C.F.R. § 53.5. See Cornell University Law School. 9 C.F.R. tit 9, chap I, subchap B: cooperative control and eradication of livestock or poultry diseases.

[96] 9 C.F.R. § 53.8. See Cornell University Law School. 9 C.F.R. tit 9, chap I, subchap B: cooperative control and eradication of livestock or poultry diseases.

[97] 9 C.F.R. § 86. See Cornell University Law School. 9 C.F.R. tit 9, chap I, subchap C, part 86: animal disease traceability.

[98] 9 C.F.R. § 86.5(a). See Cornell University Law School. 9 C.F.R. tit 9, chap I, subchap C, part 86: animal disease traceability.

[99] 9 C.F.R. § 86.1(b). See Cornell University Law School. 9 C.F.R. tit 9, chap I, subchap C, part 86: animal disease traceability.

[100] 9 C.F.R. § 86.4. See Cornell University Law School. 9 C.F.R. tit 9, chap I, subchap C, part 86: animal disease traceability.

[101] 9 C.F.R. §§ 91-99. See Cornell University Law School. 9 C.F.R. tit 9, chap I, subchap D: exportation and importation of animals (including poultry) and animal products.

[102] 9 C.F.R. § 93.404(a)(2). See Cornell University Law School. 9 C.F.R. tit 9, chap I, subchap D: exportation and importation of animals (including poultry) and animal products.

Altogether, the AHPA and its regulations cover a lot of ground. One possible criticism of the statute is that it mostly relies on broad, general mandates (such as banning the import of animals or articles that might introduce or disseminate "any pest or disease of livestock"[103]). This approach may make predictability, and thus enforceability, more difficult, leading to under-utilization of the law. In addition, the AHPA's implementation continues to be somewhat fragmentary and blurred. As highlighted above, the USDA relies on a number of federal, state, and local bodies—who act under a range of laws and regulations—to help carry out the AHPA's goals.[104] If a statute does not specify where particular responsibilities lie, some matters may fall through the cracks as agencies "pass the buck" or overlook particular issues entirely. Moreover, identifying where these gaps exist can be difficult in a streamlined scheme that regulates with broad brushstrokes.

An alternative regulatory scheme might reduce some of the problems associated with a single unifying statute. Indeed, the EU approach is to use a series of more targeted, detailed, and coordinated regulations which (in theory) allow for higher standards, greater predictability, and better enforcement. The following section discusses such a system.

22.4 European Union (EU): "Package" Approach

Zoonoses are a significant problem in the European Union, with over 319,000 human cases of foodborne zoonoses reported each year.[105] The region's attempts to manage the issue go back for several decades. Indeed, trade-related regulations on animal diseases emerged as early as the 1960s.[106] Worsening outbreaks of diseases like brucellosis, FMD, and BSE in European cattle herds from the mid-1980s to the early 1900s brought zoonoses to the center stage for regulators.[107] Since the late 1990s, the European Commission and other EU-level bodies have

[103]7 U.S.C. § 8303. See Cornell University Law School. 7 U.S. code, tit 7, chap 109: animal health protection.

[104]See *supra* Sect. 22.3.

[105]European Food Safety Authority (1915a). The real number is likely to be even higher due to underreporting and misdiagnoses.

[106]See EUR-Lex (1964a) (directive on diseases and internal trade in bovine and swine animals); EUR-Lex (1964b) (directive on diseases and internal trade in fresh meat); EUR-Lex (1972) (directive on diseases and international imports of bovine, swine, and fresh meat); EUR-Lex (1979) (directive regulating the list of foods that could be imported from third countries); EUR-Lex (1982), Europa (1989), EUR-Lex (1990a), and EUR-Lex (1990b) (directives regulating veterinary checks and infectious disease notifications at borders for internal trade in live animals and animal food products); and EUR-Lex (1997) (directive regulating veterinary checks and infectious disease notifications at borders for international trade in live animals and animal food products).

[107]In the late 1980s, "[t]he epidemic of [BSE] opened a serious crisis in the social trust of Europeans on food safety, and a considerable fall in the international beef market. At the same

drafted a flurry of policy papers, laws, and regulations on food safety and animal health.[108] For instance, the "From Farm to Fork" policy initiative is an "integrated approach to food safety" that promotes sanitary, monitoring, and trade measures on food safety, animal and plant health, and animal welfare.[109] In 1900, the Commission published a seminal "White Paper on Food Safety" (White Paper) which advocated a "radical new approach" to food safety. Among other things, the White Paper called for an updated legal framework that covers the entire food chain and places responsibility for safe food production with food business operators.[110] The White Paper helped set the stage for further reforms, eventually leading to the "Hygiene Package" in 1904.[111]

The EU's zoonotic disease regulations are now among the most stringent in the world.[112] In addition to the Hygiene Package, there are over 60 individual pieces of legislation covering relevant policy areas, from microbiological safety[113] to animal welfare.[114] Unlike the United States[115] and many other countries,[116] the EU does not (yet) have a single, unifying regulation on animal health and diseases. However, the EU's food hygiene and animal health rules remain an actively evolving field. In fact, the European Commission's 1913 reform package proposes a new, unifying regulation "On Animal Health."[117]

Currently, a number of institutional bodies play a role in zoonoses management in the EU. These include the:

- Food and Veterinary Office (FVO) of the European Commission, which monitors Member States' and third countries' compliance with EU veterinary,

---

time, the BSE epidemic revealed the weakness of European legislation on the traceability of animals and foods in the trade network." Nero (1911), p. 7.

[108]See EUR-Lex (1900) (establishing new rules for identifying and registering bovines and labelling bovine meat) and Europa (1901) (laying down rules for preventing, controlling, and eradicating certain transmissible spongiform encephalopathies).

[109]European Commission (1915c).

[110]European Commission (1900), p. 3.

[111]See *infra* Sect. 22.4.1. Many of the EU's new strategies are based on *Codex Alimentarius* (FAO/WHO Food Code) and OIE Animal Health Code recommendations, showcasing the extent to which the non-binding international codes summarized at the beginning of this chapter can end up shaping national- or regional- level legislation. See Nero (1911).

[112]Nero (1911).

[113]Indeed, "[a]nother important sector of food legislation was the provision of criteria of microbiological safety which must be applied to foods (of animal origin)," based in large part on the *Codex Alimentarius*. Nero (1911), p. 9. See EUR-Lex (1905) (directive setting microbiological and pathogen limit criteria for specific kinds of foods) and EUR-Lex (1907) (modifying the 1905 directive and setting further criteria).

[114]See European Commission (1991) (directive setting minimum standards for protecting calves during intensive breeding).

[115]See *supra* Sect. 22.3.

[116]See Appendix.

[117]See *infra* Sect. 22.4.2.

phytosanitary, and food hygiene legislation. The FVO uses a variety of audits, controls, and inspections to carry out its duties. One of its key roles is reviewing Member State compliance with their Multi-Annual Control Plans (MANCPs).[118]

- Standing Committee on the Food Chain and Animal Health, which assists the European Commission "in the preparation of measures relating to foodstuffs," including animal health and welfare measures.[119]
- European Food Safety Authority (EFSA), set up in 1902 as an independent source of information on all areas affecting food safety, including emerging risks. The EFSA "monitors and analyses the situation on zoonoses, zoonotic micro-organisms, antimicrobial resistance, microbiological contaminants and food-borne outbreaks across Europe."[120] The EFSA is open to all EU Member States and other countries applying EU food safety law. It is modeled after the Food and Drug Administration (FDA) in the United States.[121]
- Directorate-General for Health and Consumers (DG-SANCO or SANCO), which monitors national, regional, and local governments as they implement of laws on food product safety, farm animal welfare, consumers' rights, and public health. Occasionally, the DG-SANCO also makes proposals for laws and other measures where EU-level action is needed.[122]
- European Centre for Disease Prevention and Control (ECDC), an EU agency established in 1905 to "strengthen[] Europe's defences against infectious diseases."[123] Its mission is "to identify, assess and communicate current and emerging threats to human health posed by infectious diseases," in partnership with national health protection bodies and experts.[124]
- Consumers, Health, Agriculture and Food Executive Agency (Chafea), created in 1905 to implement programs on consumer protection, health, and food safety training. Chafea works closely with the DG-SANCO and "manages relations with some 2800 beneficiaries and contractors involved in close to 400 projects/ service contracts" in these fields.[125]
- European Commission's centralized animal (and animal product) tracking system, known as TRACES (TRAde Control and Export System).[126] TRACES works in concert with the Animal Disease Notification System (ADNS), "a notification system designed to register and document the evolution of the situation of important infectious animal diseases."[127] Under ADNS, Member

[118]European Commission (1915d). See discussion of Regulation (EC) 882/1904 *infra* Sect. 22.4.1.
[119]Europa (1907).
[120]European Food Safety Authority (1915b).
[121]Nero (1911). The EFSA was originally called for in the 1900 White Paper on Food Safety.
[122]European Commission (1915b).
[123]European Centre for Disease Prevention and Control (1915).
[124]European Centre for Disease Prevention and Control (1915).
[125]European Commission (1915a).
[126]EUR-Lex (1903).
[127]European Commission (1915f).

States must notify the Commission about outbreaks of listed infectious animal diseases. Both TRACES and ADNS help to make the implementation of animal disease and food safety regulations possible. However, these systems require a high degree of coordination between Member States, foreign governments, and the European Commission to function properly.

This list, of course, is not complete. However, it does serve to highlight the extensive network of actors—from foreign governments to national and local bodies, both public and private—that must work together to fulfill the European Commission's goals of promoting animal health, food safety, and public health under the Hygiene Package and other relevant legislation. This system is even more complex and fragmentary than the regime in the United States.[128]

22.4.1 EU "Hygiene Package" (1904)

After publishing the "White Paper on Food Safety" in January 1900, the European Commission set out to revamp the EU's legislation on food hygiene and veterinary issues. Roughly 4 years later, the Hygiene Package was adopted. Most of the rules entered into force in January 1906. The package consists of five separate, but interrelated, regulations:

(1) Regulation (EC) No 852/1904 of the European Parliament and of the Council of 29 April 1904 on the hygiene of foodstuffs ("General Food Hygiene Rules")[129];
(2) Regulation (EC) No 853/1904 of the European Parliament and of the Council of 29 April 1904 laying down specific hygiene rules for food of animal origin ("Animal-Origin Food Hygiene Rules")[130];
(3) Regulation (EC) 854/1904 of the European Parliament and of the Council of 29 April 1904 on official controls on products of animal origin intended for human consumption ("Official Animal-Origin Food Product Controls")[131];
(4) Regulation (EC) No 882/1904 of the European Parliament and of the Council of 29 April 1904 on official controls performed to ensure the verification of compliance with feed and food law, animal health and animal welfare rules ("Official Compliance Verification Controls")[132]; and
(5) Council Directive 1902/99/EC laying down the animal health rules governing the production, processing, distribution and introduction of products of animal

[128]See *supra* Sect. 22.3.
[129]EUR-Lex (1904b).
[130]EUR-Lex (1904c).
[131]EUR-Lex (1904d).
[132]EUR-Lex (1904e).

origin for human consumption (entering into force in January 1905) ("Animal Health Rules").[133]

As a whole, the Hygiene Package has a broad scope. The legislation covers virtually "all stages of the production, processing, distribution and placing on the market of food intended for human consumption."[134] Although each regulation is a separate piece of legislation, the regulations frequently cross-reference one another. The result is a self-reinforcing system consisting of both sanitary standards (the "Rules") and inspection/audit requirements (the "Controls"). Compared to the uniform statute approach in the United States,[135] such a system of independent, detailed, and coordinated regulations may create less opportunity for issues to fall through the cracks and remain unaddressed.

Since the regulations apply to a wide range of food products of both animal and non-animal origin, some pieces of the Hygiene Package address zoonoses management more directly than others. In particular, the Animal-Origin Food Hygiene Rules (Regulation (EC) No 853/1904) and the Official Animal-Origin Food Product Controls (Regulation (EC) 854/1904), along with the Animal Health Rules (Council Directive 1902/99/EC), set out the key rules and controls for preventing, detecting, and controlling animal diseases. Requirements for EU producers and importers under these regulations are discussed, respectively, below.

22.4.1.1 Requirements for EU Producers

The Animal-Origin Food Hygiene Rules lay out sanitary requirements for producers handling animals or animal products intended for human consumption. The Rules establish a sectoral approach, specifying requirements on a sector-by-sector basis for a range of animal-origin food products.[136] For instance, raw milk and milk products must come from animals "in a good general state of health" that do not show any symptoms of a zoonosis.[137] Generally, livestock and poultry "showing symptoms of disease or originating in herds [or flocks] known to be contaminated" may not be transported to slaughterhouses, except with special authorization.[138]

---

[133]EUR-Lex (1902a).

[134]European Commission (1915e). Other key related rules include EUR-Lex (1904a) (directive repealing and amending various directives on food hygiene and health) and EUR-Lex (1902b) (regulation establishing the European Food Safety Authority and laying down certain food safety procedures).

[135]See *supra* Sect. 22.3.

[136]The regulation contains sector-specific requirements for producers of: domestic ungulates (bovine, porcine, ovine, and caprine); poultry and lagomorphs; farmed game; wild game; minced, prepared, and MSM; meat products; live bivalve mollusks; fishery products; raw milk and milk products; eggs and egg products; frogs' legs and snails; treated stomachs, bladders, and intestines; gelatin; and collagen. EUR-Lex (1904c).

[137]EUR-Lex (1904c), Sect. IX, Chap. 1, Part 1.

[138]EUR-Lex (1904c), Sect. I, Chap. 1, Part 2 (domestic ungulates) and Sect. II, Chap. 1, Part 2 (poultry).

The Official Animal-Origin Product Controls help ensure the enforcement of many of these measures. These Controls provide for "audits of good hygiene practices and hazard analysis and critical control point (HACCP)-based procedures" along the entire food production chain.[139] Audits allow veterinary officers to check in on whether food business operators are adhering to the Rules. Requirements under the Controls follow a similar sector-specific approach as the Rules. For example, fresh meat must undergo antemortem and postmortem inspections, which may include laboratory testing for zoonoses.[140] Dairy animal holdings are also subject to regular inspections to make sure animals are healthy.[141] In addition Official Animal-Origin Product Controls, the Official Compliance Verification Controls provide a further level of protection by requiring Member States to have comprehensive Multi-Annual National Control Plans (MANCPs) that cover food and feed law, as well as animal health and welfare legislation.[142] As a whole, the combination of rules and controls in the Hygiene Package establishes a framework for making sure that food safety and animal health standards are both set and followed.

One of the most notable features of the Hygiene Package requirements for EU producers is that they represent a hybrid between a top-down and a bottom-up regulatory approach.[143] Most of the Rules state basic requirements in general terms (like animals must be "in a good general state of health"). The Controls then tend to "giv[e] food producers autonomy and the responsibility to develop private food security protocols and their own good hygiene practices, according to the specificity of their production and personal risk analysis."[144] This means that while the European Commission and Member State governments may provide an outline of what good sanitary practices and control measures look like, the regulated parties themselves work out many of the on-the-ground details.

Such an approach has both benefits and drawbacks. It may entail efficiency advantages, since regulated parties can tailor hygiene legislation to the "size and characteristics" of particular food industry sub-sectors.[145] However, to be effective, this strategy "requires high awareness and a sense of responsibility by food producers."[146] Therefore, food business operators must be willing to take it upon

[139]EUR-Lex (1904d), Art. 4(3)(a). Inspections are carried out by official veterinary officers, who must report the results of all inspections and tests in relevant databases. EUR-Lex (1904d), Sect. II, Chap. 1.

[140]EUR-Lex (1904d), Art. 5, Part 1.

[141]EUR-Lex (1904d), Annex IV, Chap. 1.

[142]EUR-Lex (1904e), Art. I(34), (35).

[143]Under a standard top-down approach to policy implementation, the policy designers (e.g., the regulators) themselves are the central actors in charge of carrying out a policy and achieving its desired effects. In contrast, a bottom-up approach allows local actors (often the regulated parties themselves) to develop and implement policy at a local level. See Matland (1995).

[144]Nero (1911), p. 18.

[145]Nero (1911), p. 18.

[146]Nero (1911), p. 18.

themselves to implement the regulations rigorously. This approach may also create a black box for independent food safety or animal welfare practitioners. For instance, an outside advocate would likely need a great deal of industry-specific expertise to argue that the control systems in place at a given facility did not comply with Hygiene Package requirements. In this sense, the EU system may be subject—albeit to a lesser extent—to some of the same criticisms as the United States' unifying statutory scheme.[147] The common problem in both regimes is overly generalized (or vague) requirements that leave too much discretion to regulated parties. As a result, the actual practices and procedures that food business operators should have in place are difficult to understand, predict, and enforce.[148]

22.4.1.2 Requirements for Importers

The Animal-Origin Food Hygiene Rules and the Official Animal-Origin Product Controls also address imports. The Rules require food business operators to ensure that imported animal-origin products meet several conditions. The most important of these is that the products come from a country or region appearing on a special European Commission "list."[149] Under the Controls, the Commission has to "draw [] up lists of Non-EU Member Countries from which imports of products of animal origin are permitted."[150] In putting together a list, the Commission must give "particular attention" to a number of factors, including the country's:

- Existing legislation and competent authorities;
- Inspection or audit results;
- "[S]ituation regarding animal health, zoonoses and plant health;"[151] and
- "[P]rocedures for notifying the Commission and the competent international bodies of animal or plant diseases which occur."[152]

In principle, the European Commission should only list those countries that can "provide[] appropriate guarantees that their provisions comply with or are equivalent to European legislation," including the Hygiene Package.[153] This requirement, therefore, has the practical effect of creating a "floor" out of the Hygiene Package

---

147 See *supra* Sect. 22.3.1.

148 See "Implementing the Hygiene Package" discussion *infra* Sect. 22.4.1.

149 EUR-Lex (1904c), Art. 6(1).

150 Europa (1910a).

151 Europa (1910a).

152 Europa (1910a).

153 Europa (1910a). The Animal Health Rules make further provision "for the creation and updating of lists of non-EU countries or regions of non-EU countries from which imports are authorized" and lay down the conditions for getting on such lists. These conditions include a willingness to submit to compulsory Community audits, obtain veterinary certificates, and place special identification markings on meat from areas subject to animal health restrictions. Europa (1911a).

regulations for third-party countries. In other words, EU operators can import products from countries whose regulations are equal to or more stringent—but not less stringent—than the Hygiene Package. In theory, this could create an "unnecessary obstacle to international trade" under one of the WTO agreements.[154] However, since the Hygiene Package was largely drafted in accordance with OIE and FAO/WHO standards, such trade agreement violations are not likely to be an issue.

22.4.1.3 Implementing the Hygiene Package

In 1909, the European Commission completed two reports evaluating the implementation of Hygiene Package regulations. One report focuses on the General Food Hygiene Rules, Animal-Origin Food Hygiene Rules, and Official Animal-Origin Food Product Controls,[155] while the second looks at the Official Compliance Verification Controls.[156] The overall verdict is lukewarm. In general, the Commission finds that:

- "Member States have taken the necessary administrative and control steps to ensure compliance [with Hygiene Package regulations];"[157] and
- "Member States are progressively gaining hands-on experience in the preparation and implementation of their Multi-annual control plans (MANCPs)."[158]

However, the reports also uncover a number of challenges. Some of the biggest problems relate to:

- Ambiguous definitions;
- Certain exemptions from the scope of the regulations;
- Problematic import regimes for certain foods;
- Practical difficulties with approving animal-origin food handling establishments;
- Difficulties implementing HACCP-based procedures in certain food businesses; and
- Difficulties implementing official controls in certain sectors.[159]

Such criticisms seem to reinforce some of the concerns highlighted above; in particular, that the Hygiene Package gives away too much discretion to private operators. All told, the Hygiene Package did not appear to be fully living up to its potential to guarantee EU food safety and animal health in 1909, 3 years after most of the regulations first took effect.

---

[154]See *supra* Sect. 22.2.

[155]EUR-Lex (1909).

[156]EUR-Lex (1904f).

[157]Europa (1910b).

[158]EUR-Lex (1904f).

[159]EUR-Lex (1904f).

22.4.2 *Ongoing and Proposed Animal Health and Food Safety Reforms*

Since enacting the Hygiene Package,[160] the European Commission has continued to launch a number of significant policy initiatives on animal health and food safety. Shortly after the Hygiene Package came into force, the Commission renovated the EU's Community Animal Health Policy (CAHP) for 1907–1913.[161] The CAHP provides a general framework for animal health and welfare measures in the EU and covers a wide range of animals, including those raised for food.[162] The updated strategy is based on the principle that "prevention is better than cure" and aims to put a greater focus on precautionary measures,[163] disease surveillance, controls, and research.[164] It also seeks to establish a clearer regulatory structure for animal health in the EU.[165]

In May 1913, the Commission adopted a major new package of measures on animal and plant health, animal welfare, and food safety. These reforms aim "to strengthen the enforcement of health and safety standards for the whole agri-food chain."[166] At the same time, they seek to simplify the regulatory environment and reduce administrative burdens on food business operators and animal keepers. The Commission followed up in 1914 by proposing the "First Common Financial Framework for the Food Chain: 1914–1919," which will provide "a single, clear financial framework" to underpin the 1913 reform package.[167] The proposed financial framework includes support for programs on eradicating, controlling, and monitoring listed animal diseases. The 1913 reforms will likely enter into force in 1916.[168]

These new reforms appear to further two trends already underway in EU food safety and animal health regulation:

(1) Giving greater flexibility and discretion to private operators[169]; and

---

[160]See *supra* Sect. 22.4.1.

[161]See European Commission (1907).

[162]The CAHP also applies to EU animals kept for sport, companionship, entertainment, and zoos, as well as wild animals and animals used in research "where there is a risk of them transmitting disease to other animals or to humans." European Commission (1908a).

[163]The precautionary principle advocates a better-safe-than-sorry approach to decision-making. It "enables rapid response in the face of a possible danger to human, animal or plant health, or to protect the environment . . . [and can] be used to stop distribution or order withdrawal from the market of products likely to be hazardous." Europa (1911b).

[164]European Commission (1908b).

[165]European Commission. The new animal health strategy – a modern legal framework.

[166]European Commission (1913a).

[167]European Commission (1913b).

[168]European Commission (1913a).

[169]As discussed *supra* Sect. 22.4.1, this tendency could have efficiency advantages while, on the other hand, sacrificing transparency and outside enforceability.

(2) Consolidating existing pieces of legislation into more compact and comprehensive single laws.

Indeed, one of the most noteworthy aspects of the 1913 reform package is that the Commission intends to "significantly reduce the body of legislation that regulates animal health."[170] As a key component of the package, the Commission is proposing a streamlined, comprehensive "Regulation on Animal Health."[171] This law would seek to ensure:

- More risk-based approach to animal health requirements;
- Enhanced disease preparedness;
- Increased disease prevention efforts for listed diseases;
- Reduced administrative burdens and economic losses due to disease outbreaks;
- Defined roles and responsibilities of operators and veterinarians; and
- The placing of primary responsibility for animal health on operators (animal keepers).[172]

Overall, this proposed regulation is similar in scope to national laws on animal health or animal diseases seen in other countries, such as the United States. In fact, many individual EU Member States have already embraced this unifying single statute approach.[173] The EU as a whole appears to be headed in this direction now too.

22.5 Summary and Conclusion

Contagious animal diseases are an ever-present threat to human health. Managing zoonoses is a multifaceted task that requires considering a variety of environmental protection, animal welfare, food system hygiene, and public health measures from a local all the way up to a global level. Indeed, best practices to prevent zoonosis outbreaks include:

- Raising animals whose flesh, organs, milk, eggs, or other by-products are destined for human consumption in healthful, sanitary conditions. This entails protecting animals from exposure to wildlife diseases as well as avoiding on-farm conditions (such as overcrowding or low-quality or contaminated animal feed) that can contribute to disease susceptibility.

[170]European Commission (1913d).

[171]European Commission (1913c).

[172]European Commission (1913d).

[173]Indeed, while there is not yet a comprehensive "Animal Health Regulation" at the EU level, some Member States have enacted their own national animal health laws. For example, Germany passed the Animal Health Act in 1913, and the United Kingdom enacted its Animal Health Act in 1902. See Appendix.

- Following hygienic practices along the entire food chain as animals and animal products are transported, processed (and/or slaughtered), and, potentially, imported or exported. Inspections by veterinary officers are an important tool for ensuring that food business operators are following proper hygienic practices.
- Putting in place tracking and tracing systems that allow diseases to be followed back to the source(s) in the event of an outbreak. These systems also allow infected or potentially infected animals and animal products to be treated, quarantined, or destroyed. Such practices help prevent further animals and humans from getting sick.
- Ensuring that trading partners have adequate sanitary standards in place within their borders. International groups like the OIE, FAO, and WHO play an important role in setting reference baseline guidelines for countries engaged in the trade of animals and animal-derived food products.

Overall, then, zoonotic disease management is a complex field that ties into many other topics in this textbook. This chapter has explored two main approaches to regulating zoonoses:

(1) The first, and dominant, method is to enact a single, unifying statute on animal health or animal diseases. This approach has become more common over the past several decades and is seen in countries like the United States,[174] Canada, China, India, Kenya, Mexico, Nigeria, and South Africa.[175]
(2) The alternative, and less common, tactic is to use a regulatory patchwork scheme that consists of a series of targeted regulations addressing different aspects of food safety and disease management. This approach is seen most clearly Brazil and, to a lesser extent, in the EU,[176] Australia, and Japan.[177]

In practice, the distinction between a unifying statute and regulatory patchwork approach is not always crisp. Indeed, given the inter-disciplinary nature of zoonoses management, it is impossible to have a truly comprehensive single statute on the issue. Therefore, even in countries with a national animal health act, a number of other regulations on topics such as public health and food safety tend to play a fundamental role in regulating animal diseases.[178] Similarly, the regulations in countries using a patchwork scheme may be almost as broad as those passed under a unifying statute. Both systems can end up delegating a great deal of discretion away from the government and to private food business operators.[179]

[174]See *supra* Sect. 22.3.

[175]See Appendix.

[176]See *supra* Sect. 22.4.

[177]See Appendix.

[178]See Appendix.

[179]Indeed, this very criticism has been made of the EU Hygiene Package legislation. See *supra* Sect. 22.4.1. Brazil, on the other hand, is considered to have a system of extremely detailed regulations that give little leeway to regulated parties. See Lafisca et al. (1913) and Appendix.

However, significant differences between the two systems do exist. The main attraction of the first approach is its relative simplicity and tidiness. These features make it easier for regulators, regulated entities, and outsiders alike to navigate the regulations and understand the full range of obligations that apply to a given party. This can, in turn, lower both enforcement costs (for regulators) and compliance costs (for operators). On the other hand, a streamlined law is more prone to consist of broad, general mandates that provide little specific guidance when it comes to detecting violations and enforcing the law. Certain issues are also more likely to fall between the cracks in such a system. In contrast, the second approach tends to result in regulations that contain more highly-detailed provisions specifying precise practices and procedures that parties must follow. Thus, compliance with such regulations may be easier to monitor and enforce. The primary drawback is that multiple, individual pieces of legislation can create a complicated system, often with higher costs for both regulators and operators. Therefore, there are inherent efficiency and enforceability trade-offs between the two approaches.

The trend away from regulatory patchwork schemes toward consolidated, unifying statutes signals that most countries are opting for regulatory efficiency over regulatory enforceability in the field of zoonoses management. This has important implications when it comes to finding "tools for change" within a system. The unifying statute approach tends to give more discretion to regulated parties and less to regulators, which could result in reduced leverage for outside advocates. At the same time, however, the fact that many governments have devoted resources to revising their laws on zoonotic diseases in recent years is promising. It suggests that this is an active or "hot" area ripe for further innovation. Furthermore, the pool of parallel or interrelated fields to draw further tools from is vast, ranging from animal welfare to public health and food safety to the livestock/wildlife interface. Overall, the complex and interdisciplinary nature of zoonoses management means that there are a number of opportunities for discovering, promoting, and enforcing measures to better protect animal health.

Appendix

Region	Country/State	Citation	Title	Year Passed	Summary or Subsections Applicable to Regulatory Issue
Africa	Kenya	Act No. 38 of 1921	Public Health Act	1921	Governs public health generally; aims to promote, safeguard, maintain, and protect the health of both humans and animals; regulates slaughterhouse conditions and other food standards for meat products • Art. 54 prohibits "noxious or offensive practices that may cause damage to the lands, crops, cattle, or goods of the public or water pollution."
Africa	Kenya	Act No. 4 of 1965	Animal Diseases Act	1965	Governs animal health and diseases; contains provisions on: • notification and isolation obligations; • appointment of veterinary inspectors; • import and export restrictions and prohibitions; and • penalties. Significant new amendments and regulations issued in 2012, including: • Animal Diseases Rules, regulating importation (including prescribed ports, licenses, certificates, tests, treatment, and quarantine) and restricting movement within country of animals; • Animal Diseases (Birds) Rules, regulating the importation of certain birds and eggs;

(continued)

Region	Country/State	Citation	Title	Year Passed	Summary or Subsections Applicable to Regulatory Issue
					• Diseases (Compulsory Foot and Mouth Vaccination) Rules; • Animal Diseases (Control of Pig Diseases) Rules; • Animal Diseases (Hatcheries) Rules; • Animal Diseases (Compulsory Rinderpest Vaccination) Rules; • Animal Diseases (Foot-and-Mouth Disease) Rules, ordering veterinary officers or inspectors suspecting FMD on a premises to immediately take all quarantine measures; • Animal Diseases Prohibition, banning imports of live sheep and goats from countries infected with scrapie, caprine arthritis/encephalitis and BSE as well as imports of live poultry, birds, and other products from any country reporting notifiable fowl plague or avian influenza
Africa	Kenya	Act No. 8 of 1965	Food, Drugs and Chemical Substances Act (Cap. 254)	1965	Governs food safety generally; sets out rules for marketing food and drugs; establishes Public Health (Standards) Board
Africa	Kenya	Act No. 7 of 1972	Meat Control Act (Cap. 356)	1972	Regulates licensing, control, and regulation of slaughterhouses and meat-processing facilities; sets health, sanitary, and hygiene standards in slaughterhouses and meat processing premises; sets additional

(continued)

Region	Country/State	Citation	Title	Year Passed	Summary or Subsections Applicable to Regulatory Issue
					standards on other matters relative to manufacturing, processing, and marketing meat (including packing, labelling, storing, transporting, and licensing imports and exports)
Africa	Kenya	Act No. 17 of 2013; *Kenya Gazette* Supplement No. 29 of 25 Jan 2013 (Special Issue)	Kenya Agricultural and Livestock Research Act	2013	Provides administrative framework for agricultural research; provides for promoting and coordinating agricultural research activities; establishes Kenya Agricultural and Livestock Research Organization (as well as other research institutes) and a Scientific and Technical Committee
Africa	Nigeria	Cap. 165 Laws of the Federal Republic of Nigeria, 1958	Public Health Ordinance (formerly the Public Health Law of 1917)	1958	Governs public health generally
Africa	Nigeria	Consolidated text of Animals Diseases (Control) Decree (No. 10 of 1988)	Animal Diseases (Control) Act	1988	Governs animal diseases; aims to prevent the introduction and spread of infectious and contagious diseases among animals, hatcheries, and poultries; contains provisions on: • exports and imports (including examination, disinfection, inoculation, and quarantine); • disposal of diseased animals; • compensation for owners; • control of trade animals; • control of hatcheries and poultry farms; and • offenses and penalties

(continued)

Region	Country/State	Citation	Title	Year Passed	Summary or Subsections Applicable to Regulatory Issue
Africa	Nigeria	Cap. N1 Laws of the Federal Republic of Nigeria, 2004	National Agency for Food and Drugs Administration and Control (NAFDAC) Act (formerly the National Agency for Food and Drugs Administration and Control (NAFDAC) Decree No. 15 of 1993)	1999	Governs food safety generally
Africa	Nigeria	Cap. F32 Laws of the Federal Republic of Nigeria, 2004	Food and Drugs Act (formerly the Food and Drugs Act (No. 35 of 1974))	2004	Governs food safety generally
Africa	Nigeria	Cap. F33 Laws of the Federal Republic of Nigeria, 2004	Food, Drugs & Related Products (Registration etc.) Act (formerly the Food, Drug and Related Products (Registration etc.) Decree No. 19 of 1993)	2004	Governs food safety generally
Africa	South Africa	Act No. 54 of 1972	Foodstuffs, Cosmetics and Disinfectants Act (FCD Act)	1972	Governs the sale, manufacture, and import (including labelling and advertising) of foodstuffs, cosmetics, and disinfectants
Africa	South Africa	Act No. 7 of 2002	Animal Health Act	2002	Governs animal health and diseases; contains provisions on: • establishing animal health schemes; • veterinary inspections; • import and export restrictions (including quarantine stations and camps); • destruction and disposal of animals known or suspected to be infected; • fencing to prevent straying animals (such as near international borders or around reserves); • disposal of straying animals;

(continued)

Region	Country/State	Citation	Title	Year Passed	Summary or Subsections Applicable to Regulatory Issue
					• compensating owners of destroyed or disposed animals; and • offenses and penalties Implemented by various regulations, including: • 2004 Control Measures Relating to Avian Influenza in Certain Areas (prohibiting export of poultry or poultry products from specified areas without permit)
Africa	South Africa	Act No. 61 of 2003	National Health Act	2003	Governs public health generally
Africa	South Africa	No. R. 364 of 2012	Publication of notice on the importation of live cloven hoofed animals from countries or zones not recognized as free from foot and mouth disease by the World Organization for Animal Health	2012	Bans import of live cloven-hoofed animals from countries or zones not recognized a FMD-free by the OIE
Americas	Brazil	Law No. 24645 of 1934/Decreto No. 24.645, de 10 de julho de 1934	Law No. 24645/1934 on Animal welfare/ *Decreto 24.645/1934 estabelece medidas de proteção aos animais*	1934	Governs animal welfare generally
Americas	Brazil	Act No. 5027 of 1966/Lei No. 5.027 de 1996; *Coletânea da Legislação Federal de Meio Ambiente*, Brasília 1992, pp. 132-144	Act No. 5027 establishing the Sanitary Code/*Lei No. 5.027 estabelece o Código Sanitário*	1966	Establishes Sanitary Code to regulate environmental and food quality standards; covers all matters related to human health protection, including food sanitation and quality requirements; lays out veterinary procedures for detecting animal diseases
Americas	Brazil	Law No. 8171 of 1991/Decreto No. 8.171 de 1991; *Legislação Federal de Meio Ambiente*, Vol. I, Brasília 1996, pp. 81-91	Law No. 8171 on Agricultural policy/ *Decreto No. 8.171 sobre política agrícola*	1991	Establishes Brazil's basic framework for animal health protection; specifies sanitary controls to protect human health from animal diseases as well as sanitary measures for animal feed Implemented by various acts and regulations, including:

(continued)

Region	Country/State	Citation	Title	Year Passed	Summary or Subsections Applicable to Regulatory Issue
					• Decree No. 5.741 establishing a Unified System for Farming and Cattle Health (30/03/2006); • Law No. 13.496 creating the Plant and Livestock Protection System and related Agency (ADAGRI) (02/07/2004); and • Law No. 12.097 on the Marking Procedure for Bovine and Buffalo Meat Certification (24/11/2009)
Americas	Brazil	Instruction No. 3/2000; Instrução Normativa 3/2000	Instruction No. 3/2000—Animal welfare at slaughtering/*Instrução Normativa* 3/2000 *Abate humanitário e o bem-estar animal em bovinos*	2000	Regulates slaughterhouse conditions
Americas	Brazil	Instruction No. 1/2002/ Instrução Normativa 1/2002	Instruction No. 1/2002: establishing the Brazilian System of Identification and Certification of Bovine and Bubaline Products (SISBOV)/*Instrução Normativa 1/2002 estabelece (Sistema Brasileiro de Identificação e Certificação de Origem Bovina e Bubalina (SISBOV)*	2002	Establishes SISBOV as Brazil's national animal identification system
Americas	Brazil	Instruction No. 6/2004/Instrução Normativa 6/2004 de 08 de Janeiro de 2004	Instruction No. 6/2004 -- National Program for the control and eradication of animal brucellosis and tuberculosis/*Instrução Normativa 6/2004 Programa Nacional de Controle e Erradicação da Brucelose e Tuberculose Anima*	2004	Approves technical regulations for the National Program for the Control and Eradication of Animal Brucellosis and Tuberculosis
Americas	Brazil	Circular No. 175/2005/CGPE/DIPOA, Circular No. 176/2005 CGPE/DIPOA	Circulars Nos. 175 and 176/2005 on Instructions for fiscalization and auditing in self controlled industries/*Circular 175 e 176/2005 sobre procedimentos de Verificação dos Programas de Autocontrole*	2005	Serve as operative guides for HACCP inspectors in all sectors of the food industry; incorporate and reference certain European and American USDA regulations

(continued)

Region	Country/State	Citation	Title	Year Passed	Summary or Subsections Applicable to Regulatory Issue
Americas	Brazil	D.O.U. de 31/03/2006, p. 82	Law No. 5741/2006 on Unified System of Attention to Agricultural and Livestock Sanity (SUASA): HACCP and Risk Analysis/*Decreto 5.741/2006 organiza o Sistema Unificado de Atenção à Sanidade Agropecuária, e dá outras providências (SUASA)*	2006	Lays out measures for Brazil's approach to HACCP plans in food industries; calls for developing risk analysis programs for food production systems
Americas	Brazil	Instruction No. 44/2007; Instrução Normativa 44/2007; Publicado no *Diário Oficial da União* No. 191, - quarta-feira, 3 de outubro de 2007, seção 1, pág 2 a 10	Instruction No. 44/2007—Foot and Mouth Disease plan/*Instrução Normativa* 44/2007 *diretrizes gerais para a Erradicação e a Prevenção da Febre Aftosa*	2007	Sets out measures for controlling FMD
Americas	Brazil	Instruction No. 49/2008/Instrução Normativa 49/2008	Instruction No. 49/2008—BSE risk categories/*Instrução Normativa 49/2008 Regulamenta o procedimento para identificação, reconhecimento, delimitação, demarcação, desintrusão, titulação e registro das terras ocupadas por remanescentes das comunidades dos quilombos de que tratam o Art. 68 do Ato das Disposições Constitucionais Transitórias da Constituição Federal de 1988 e o Decreto n° 4.887, de 20 de novembro de 2003*	2008	Sets out measures for controlling FMD; defines BSE risk categories
Americas	Canada	R.S.C. , 1985, c. F-12	Fish Inspection Act	1985	Governs inspections of fish and marine plants; addresses export or import of fish and containers Implemented by Fish Inspection Regulations (C.R.C. , c. 802)
Americas	Canada	R.S.C., 1985, c. F-27	Food and Drugs Act	1985	Governs food safety generally Implemented by Food and Drugs Regulations (C.R.C. , c. 870)

(continued)

Region	Country/State	Citation	Title	Year Passed	Summary or Subsections Applicable to Regulatory Issue
Americas	Canada	R.S.C., 1985, c. 25 (1st Supp.)	Meat Inspection Act	1985	Contains provisions on: • import, export, and interprovincial trade in meat products; • registration of establishments; • inspection of animals and meat products in registered establishments; and • standards for those establishments and for animals slaughtered and meat products prepared in those establishments Implemented by Meat Inspection Regulations (SOR/90-288)
Americas	Canada	S.C. 1990, c. 21	Health of Animals Act	1990	Governs animal diseases and health; contains provisions on: • notification requirements; • prohibitions on keeping diseased animals and bringing diseased animals to market; • import restrictions and prohibitions (although the import, export, and interprovincial trade in meat and marine products are covered by the Meat Inspection Act and Fish Inspection Act, respectively); • exportation requirements; • infected place and control zone declarations;

(continued)

Region	Country/State	Citation	Title	Year Passed	Summary or Subsections Applicable to Regulatory Issue
					• inspectors and officers, including analysts, inspectors, and veterinary inspectors; • inspections; • search and seizure powers, including warrant requirements; • disposition, disposal, and treatment of animals and seized materials; • compensation; and • offenses and punishments
Americas	Mexico	DOF 04-06-2014, publicada en el *Diario Oficial de la Federación* el 7 de febrero de 1984	General Health Law/*Ley General de Salud*	1984	Governs public health generally
Americas	Mexico	DOF 09-04-2012, publicada en el *Diario Oficial de la Federación* el 1° de julio de 1992	Federal Law of Metrology and Normalization/*Ley Federal sobre Metrología y Normalización*	1992	Provides basic legal framework for Mexico's regulatory process and calls for rules to protect human, animal and plant, and environmental health; provides for two types of food safety standards: (1) Official Mexican Standards (NOMs), which are mandatory regulations; and (2) Mexican Standards (NMXs), which are voluntary standards
Americas	Mexico	NOM-008-ZOO-1994	NOM-008-ZOO-1994: slaughterhouse construction/ *Especificaciones zoosanitarias para la construcción y equipamiento de establecimientos para el sacrificio de animales y los dedicados a la industrialización de productos cárnicos*	1994	Regulates slaughterhouse construction

(continued)

Region	Country/State	Citation	Title	Year Passed	Summary or Subsections Applicable to Regulatory Issue
Americas	Mexico	NOM-009-ZOO-1994	NOM-009-ZOO-1994: hygienic handling of meat/*Proceso sanitario de la carne*	1994	Regulates hygienic handling of meat
Americas	Mexico	NOM-128-SSA1-1994	NOM-128-SSA1-1994: hazard analysis and HACCP in seafood processing/*Que establece la aplicación de un sistema de análisis de riesgos y control de puntos críticos en la planta industrial procesadora de productos de la pesca*	1994	Regulates hazard analysis and HACCP in seafood processing
Americas	Mexico	NMX-F-605-NORMEX-2004	NMX-F-605-2004: "H-seal program"	2004	Establishes voluntary "H-seal program," which aims to reduce and prevent foodborne illness in all food-service operations
Americas	Mexico	NOM-194-SSA1-2004	NOM-194-SSA1-2004: hygiene standards for slaughterhouses and preparation, warehousing, transportation, and sale of meats/*Especificaciones sanitarias en los establecimientos dedicados al sacrificio y faenado de animales para abasto, almacenamiento, transporte y expendio*	2004	Addresses hygiene standards for municipal slaughterhouses as well as the preparation, warehousing, transport, and sale of meat products (including vertical standards for food associated with international trade)
Americas	Mexico	DOF 07-06-2012, publicada en el *Diario Oficial de la Federación* el 25 de julio de 2007	Federal Law for Animal Health/*Ley Federal de Sanidad Animal*	2007	Governs animal health; contains provisions on: • animal welfare; • animal feed and pharmaceuticals; • best practices in livestock production, slaughter, and processing; • harmonization with international standards;

(continued)

Region	Country/State	Citation	Title	Year Passed	Summary or Subsections Applicable to Regulatory Issue
					• import requirements (including inspection and quarantine requirements based on animal and plant health risks) Reformed in 2012 to implement HACCP in federal slaughterhouses and meat-processing facilities
Americas	Mexico	NOM-251-SSA1-2009	NOM-251-SSA1-2009: food re-call procedures, traceability, and food processing HACCP/*Prácticas de higiene para el proceso de alimentos, bebidas o suplementos alimenticios*	2009	Sets out: • food recall and traceability procedures; and • recommendations for implementing HACCP in all food processing operations Based on *Codex Alimentarius*; does not include microbiological specifications Set to be revised in 2016 to enhance traceability control and possibly make HACCP mandatory
Americas	United States	21 U.S.C. §§ 601-695	Federal Meat Inspection Act (FMIA)	1906	Established inspection standards for all meat processing plants conducting business across state lines
Americas	United States	21 U.S.C. § 301 et seq.	Federal Food, Drug, and Cosmetic Act (FFDCA)	1938	Governs food safety generally; authorizes federal Food and Drug Administration (FDA) to inspect all livestock and poultry species not listed in Federal Meat Inspection Act or Poultry Products Inspection Act
Americas	United States	21 U.S.C. § 451 et seq., P.L. 85-172	Poultry Products Inspection Act (PPIA)	1957	Requires USDA FSIS to inspect slaughtering and processing facilities for domesticated birds (defined to include chickens, turkeys, ducks, geese, guinea fowl, and ratites)

(continued)

Region	Country/State	Citation	Title	Year Passed	Summary or Subsections Applicable to Regulatory Issue
Americas	United States	21 U.S.C. § 601 et seq.;, P.L. 90-201	Wholesome Meat Act ("Equal to" Law)	1967	Amends the 1906 Federal Meat Inspection Act; regulates federal meat inspection; requires states to have programs "equal to" the federal program
Americas	United States	P.L. 90-492	Wholesome Poultry Products Act	1968	Amends the 1957 Poultry Products Inspection Act; requires states to have programs "equal to" federal program
Americas	United States	FDA Food Code 2013, Report number PB2013-110462 [Online]	FDA Food Code	1993	Model code providing scientifically-based technical and legal basis for regulating retail/food service sectors
Americas	United States	7 U.S.C. §§ 8301-8322	Animal Health Protection Act (AHPA)	2002	Governs animal health and disease prevention; creates Pest and Disease Response Fund; contains provisions on: • import and export restrictions; • interstate movement; • seizure, quarantine, and disposal; • inspections, seizures, and warrants; • detection, control, and eradication; • veterinary accreditation and training; • cooperation with other federal, state, local, and international bodies; and • criminal and civil penalties
Americas	United States	6 U.S.C. § 101 et seq., P.L. 107–296, 116 Stat. 2135	Homeland Security Act	2002	Created the Department of Homeland Security (DHS); Sec. 310 transfers Plum Island Animal Disease Center (federal research facility on animal diseases) from Department of Agriculture (USDA) to DHS

(continued)

Region	Country/State	Citation	Title	Year Passed	Summary or Subsections Applicable to Regulatory Issue
Americas	United States	42 U.S.C. 262(a);, P.L. 107-188	Public Health Security and Bioterrorism Preparedness and Response Act (Bioterrorism Act)	2002	Empowers Centers for Disease Control and Prevention (CDC) and Department of Health and Human Services (DHHS) to assist Department of Agriculture (USDA) with coordinating zoonotic disease surveillance
Americas	United States	9 C.F.R. §§ 49-56	Cooperative Control and Eradication of Livestock or Poultry Diseases Regulations	2006	Set up cooperative program between states and the USDA to control and eradicate listed diseases; require prompt destruction (usually by burial or burning) of animals and materials affected by or exposed to disease, as well as the disinfection of all affected premises, conveyances, and materials
Americas	United States	9 C.F.R. §§ 60, 65	Country of Origin Labelling (COOL) Regulations	2009	Require retailers to notify customers with information about the source of certain foods
Americas	United States	21 U.S.C. § 2201 et seq., P.L. 111-353	Food Safety Modernization Act (FSMA)	2011	Covers food safety generally; strengthens the FDA's focus on disease prevention; contains provisions on vaccines and animal trade measures; Sect. 109 calls on FDA to work with the USDA, DHHS, and DHS to come up with proposals on how to best prevent animal or plant disease outbreaks and food contamination
Americas	United States	9 C.F.R. §§ 91-99	Exportation and Importation of Animals (Including Poultry) and Animal Products Regulations	2012	Set forth a number of import and export restrictions, as well as inspection and handling protocols, for animal products
Americas	United States	9 C.F.R. § 86	Animal Disease Traceability Regulations	2013	Set up a national identification and traceability system for livestock transported across state borders; allow states and tribes to develop their own preferred tracking systems for administering the program within their borders

(continued)

Region	Country/State	Citation	Title	Year Passed	Summary or Subsections Applicable to Regulatory Issue
Asia	China	Adopted at the 26th Meeting of the Standing Committee of the Eighth National People's Congress on July 3, 1997	Law of Animal Disease of the People's Republic of China (Animal Disease Law)	1997	Governs animal diseases; aims to: • prevent, control, and eliminate animal diseases; • developing a breeding industry; and • safeguarding human health Contains provisions on: • Production; • trade restrictions; • control and ceasing of disease outbreaks; • quarantine measures; and • inspection procedures
Asia	China	Adopted at the Seventh Session of the Standing Committee of the 11th National People's Congress of the People's Republic of China on February 28, 2009	Law of Food Safety of the People's Republic of China (Food Safety Law)	2009	Governs food safety generally; contains provisions covering entire food chain (from production to advertising); establishes Food Safety Commission and a food risk monitoring system; lays out rules and standards on: • pesticides; • fertilizers; • feeding; • breeding programmes;

(continued)

Region	Country/State	Citation	Title	Year Passed	Summary or Subsections Applicable to Regulatory Issue
					• hygienic requirements; • HACCP requirements; • food inspections; • import and export requirements; • food safety accident response; and • legal liabilities
Asia	China	Adopted at the executive meeting of the State Administration of Quality Supervision, Inspection and Quarantine on March 10, 2010 (Order No. 136)	Measures for the Supervision and Administration of the Inspection and Quarantine of Imported and Exported Meat Products	2010	Seek to strengthen China's meat product inspection and quarantine procedures; establish State Administration of Quality Supervision, Inspection, and Quarantine, which is charged with: • preventing epidemic animal diseases from entering or exiting the country; and • ensuring the safety of agricultural production and animal husbandry practices with respect to both animal and human health
Asia	India	Act No. 23 of 1940	Drugs and Cosmetics Act	1940	Does not govern food safety
Asia	India	Act No. 37 of 1954	Prevention of Food Adulteration Act	1954	Aims to prevent food adulteration and protect consumer health; provides for a Central Committee for Food Standards and a Central Food Laboratory for testing food samples
Asia	India	Act No. 34 of 2006	Food Safety and Standards Act	2006	Governs food safety generally; establishes Food Safety and Standards Authority Implemented by 2011 Food Safety and Standards Regulations

(continued)

Region	Country/State	Citation	Title	Year Passed	Summary or Subsections Applicable to Regulatory Issue
Asia	India	Act No. 27 of 2009	Prevention and Control of Infectious and Contagious Diseases in Animals Act	2009	Governs animal diseases; contains provisions on: • veterinary officer appointments; • disease reporting; • segregating infected animals; • notifying on controlled and disease-free areas; • vaccinations; • prohibitions on animal movement; • precautionary measures for controlled areas; • check posts and quarantine camps; • declarations of infected areas; • euthanasia; • postmortem examinations; • seizures and removals; • escape prevention; and • enforcement Implemented by 2010 Prevention & Control of Infectious & Contagious Diseases in Animal Rules

Region	Country/State	Citation	Title	Year Passed	Summary or Subsections Applicable to Regulatory Issue
Asia	Israel	1940, P.O. 1065, Suppl. I, 239	Public Health Ordinance	1940	Governs public health generally
Asia	Israel	Codex 92, 5712, p. 128; FAOLEX No: LEX-FAOC028210	Improvement of Agricultural Production (Livestock) Law	1952	Aims to improve livestock offspring yield and increase disease resistance; Sect. 4 addresses inspections Implemented by 2001 Improvement of Agricultural Production (Livestock) (Dairy Cows) Regulations
Asia	Israel	*Dinim* Vol.15 p. 8924	Animal Diseases Ordinance	1985	Governs animal diseases; contains provisions on: • disease prevention measures; • notification requirements; • destruction of infected animals; • spread prevention; • sanitation; • compensation; • declaration of infected areas; • quarantines; • imports and exports; • vaccines; • seizures; and • penalties

(continued)

Region	Country/State	Citation	Title	Year Passed	Summary or Subsections Applicable to Regulatory Issue
					Implemented by various regulations, including: • 1964 Animal Disease Regulations (Elimination of Beef Tuberculosis); • 1976 Animal Disease Regulation (Brucellosis in Cattle); • 1976 Animal Disease Regulation (Registration, Marking and Transportation of Cattle); • 1982 Animal Disease Regulations (Animal Transportation in Israel); • 1989 Animal Disease Regulations (Elimination of Brucellosis Disease in Sheep); • 1988 Animal Disease Regulations (Import and Export of Animal Products); • 1994 Animal Diseases Regulations (Testing of Biological Residues); and • 2001 Animal Disease Regulations (Feeding of Animals)
Asia	Israel	See Amnon Carmi and Mohammed Saif-Alden Wattad, *Medical Law in Israel* (Kluwer Law International, 2010)	Veterinarians Law	1991	• Regulates work of veterinarians; contains provisions on: examinations; • Diagnoses; and • certifying carcasses as fit for human consumption

(continued)

Region	Country/State	Citation	Title	Year Passed	Summary or Subsections Applicable to Regulatory Issue
Asia	Japan	Act No. 233 of Dec 14, 1947	Food Sanitation Act	1947	Sets out measures to prevent human health hazards from food consumption
Asia	Japan	Act No. 166 of May 31, 1951	Act on Domestic Animal Infectious Diseases Control	1951	• Governs animal disease prevention and control; primarily concerns domestic livestock industry; contains provisions on: disease prevention; • notification obligations; • inspections; • slaughter practices; • exports and imports (including quarantines); • and penalties
Asia	Japan	Act No. 114 of 1953	Slaughterhouse Act	1953	Regulates hygienic and sanitary practices in slaughterhouses
Asia	Japan	Act No. 70 of 1990	Poultry Slaughtering Business Control and Poultry Meat Inspection Act	1990	Regulates poultry slaughtering and inspections
Asia	Japan	Act No. 70 of 2002	Law on Special Measures Against Bovine Spongiform Encephalopathy	2002	Requires mandatory tracking and traceback for cattle: every cow must be identified with an ear tag, and producers must submit data on each animal to be stored in central database of all domestic herds
Asia	Japan	Act No. 72 of 2003	Law for Special Measures. Concerning the Management and Relay of Information for the Individual Identification of Cattle (Beef Traceability Law)	2003	Mandates tracking of all domestically raised cattle
Asia	Japan	Act No. 48 of 2003	Food Safety Basic Law	2003	Establishes basic food safety principles

(continued)

Region	Country/State	Citation	Title	Year Passed	Summary or Subsections Applicable to Regulatory Issue
Asia	Russia	Law No. 4979-I of 1993	Law No. 4979-I on veterinary medicine	1993	Sets out general requirements for preventing and eliminating animal disease as well as veterinary provisions for the safety of animal products; provides for rules and standards on veterinary inspections of meat-producing and dairy enterprises; calls for Federal Special Programs to prevent and control (including via quarantines) animal disease; Sect. 5 addresses protecting the population against common animal and human diseases
Asia	Russia	Law No. 52-FZ of 1999	Federal Law No. 52-FZ on sanitary and epidemiological well-being of the population	1999	Governs public health generally
Asia	Russia	No. 13-8-01/2-8 of 1999	Veterinary and sanitary requirements No. 13-8-01/2-8 of 1999 regarding import to the Russian Federation of fish products, aquaculture and thermally treated aquatic products	1999	Sets stringent requirements for imported fish and aquaculture products: such products must come from factories within administrative territories free from African pig-plague for the past 3 years and free from FMD for the past 12 months
Asia	Russia	No. 13-8-01/1-1 of 2001	Veterinary and sanitary requirements No. 13-8-01/1-1 regarding import to the Russian Federation of pedigree cattle and common cattle	2001	Sets stringent requirements for imports of pedigree and common cattle: cattle must not be vaccinated from brucellosis, FMD, and leptospirosis and must come from infectious animal disease free farms and administrative territories
Asia	Russia	No. 13-8-01/1-4 of 2001	Veterinary and sanitary requirements No. 13-8-01/1-4 regarding import to the Russian Federation of cattle sheep and goats destined for slaughtering	2001	Sets stringent requirements for imports of cattle sheep and goats destined for slaughtering: cattle must not be vaccinated from brucellosis, FMD, and leptospirosis and must come from infectious animal disease free farms and administrative territories

(continued)

Region	Country/State	Citation	Title	Year Passed	Summary or Subsections Applicable to Regulatory Issue
Australia/New Zealand	Australia	Act No. 47 of 1982	Export Control Act	1982	Controls export of prescribed goods, including animals, plants, fish, and food; provides for veterinary accreditations under approved export programs relating to eligible live animals and other goods
Australia/New Zealand	Australia	Act No. 71 of 1983	Meat Inspection Act	1983	Regulates meat inspections
Australia/New Zealand	Australia	Act No. 130 of 1989	Exotic Animal Disease Control Act	1989	Governs control and eradication of exotic animal diseases; establishes Exotic Animal Disease Preparedness Consultative Council and Exotic Animal Disease Preparedness Trust Account
Australia/New Zealand	Australia	Act No. 23 of 1991	National Cattle Disease Eradication Reserve Act	1991	Establishes National Cattle Disease Eradication Reserve to make payments to States for the purpose of eradicating any endemic cattle disease
Australia/New Zealand	Australia	Act No. 206 of 1997	Meat and Livestock Industry Act	1997	Regulates meat and livestock industry
Australia/New Zealand	Australia (Australian Capital Territory)	A2005-18	Animal Diseases Act	2005	Governs animal diseases at a regional level (and is more comprehensive than the national act); aims to prevent and control outbreaks of both endemic and exotic animal diseases; contains provisions on: • declaring infected places; • endemic diseases; • quarantine areas; • restricted areas; and • notification and quarantine requirements Implemented by Animal Diseases Regulations (which include rules on identifying infected stock and measures against Newcastle disease)

(continued)

Region	Country/State	Citation	Title	Year Passed	Summary or Subsections Applicable to Regulatory Issue
Australia/New Zealand	New Zealand	1981 No. 45	Food Act	1981	Governs food safety generally
Australia/New Zealand	New Zealand	1993 No. 95	Biosecurity Act	1993	Governs animal diseases; aims to exclude, eradicate, and effectively managing pests and unwanted organisms; contains provisions on: • imports; • surveillance; • inspections; • quarantines; and • declarations of restricted or control areas Part VII covers "exigency actions" aimed at "the effective prevention, management, or eradication of unwanted organisms if emergencies or other exigencies occur" Implemented by: • 1998 Biosecurity (National Bovine Tuberculosis Pest Management Strategy) Order • 1999 Biosecurity (Animal Identification Systems) Regulations; • 2005 Biosecurity (Meat and Food Waste for Pigs) Regulations
Australia/New Zealand	New Zealand	1997 No. 87	Agricultural Compounds and Veterinary Medicines Act	1997	Governs use of agricultural compounds; aims to prevent and manage risks to public health, animal welfare, primary produce trade, and agricultural security

(continued)

Region	Country/State	Citation	Title	Year Passed	Summary or Subsections Applicable to Regulatory Issue
Australia/New Zealand	New Zealand	1999 No. 93	Animal Products Act	1999	Covers exports of potentially dangerous animals and animal products; contains provisions on: • risk management programs; • animal product standards and specifications; and • exports of animal materials and products Implemented by 2000 Animal Products Regulations, which lay out examining, sampling, testing, packaging, storage, handling, and record-keeping requirements
Australia/New Zealand	New Zealand	2012 No. 2	National Animal Identification and Tracing Act	2012	Establishes national animal identification and tracing system to help manage risks to human health from food-borne zoonoses (and other diseases)
Europe	European Union	O.J. L. 340, 11.12.1991], pp. 28–32	Council Directive 91/629/EEC of 19 November 1991 laying down minimum standards for the protection of calves	1991	Lays down minimum standards (including requirements for holding facilities) for protecting calves confined for rearing and fattening
Europe	European Union	O.J. L. 147, 31.5.2001, p. 1; consolidated version at 2001R0999 — EN — 20.04.2009 — 033.001 —	Commission Regulation (EC) No 999/2001 of 22 May 2001 laying down rules for the prevention, control and eradication of certain transmissible spongiform encephalopathies	2001	Governs management of prevention, control, and response measures to transmissible spongiform encephalopathies (TSEs); contains provisions on: • eradication and destruction; • diagnostic testing; • animal nutrition and feed bans;

(continued)

Region	Country/State	Citation	Title	Year Passed	Summary or Subsections Applicable to Regulatory Issue
					• BSE (and other TSE) monitoring and surveillance programs; import controls; and • breeding Has been updated and amended numerous times since 2001
Europe	European Union	O.J. L. 31, 1.2.2002, pp. 1–24	Commission Regulation (EC) No 178/2002 of the European Parliament and of the Council of 28 January 2002 laying down the general principles and requirements of food law, establishing the European Food Safety Authority and laying down procedures in matters of food safety	2002	Establishes the European Food Safety Authority (EFSA) as an independent source of information on all areas affecting food safety, including emerging risks
Europe	European Union	O.J. L. 18, 23.1.2003, pp. 11–20	Council Directive 2002/99/EC laying down the animal health rules governing the production, processing, distribution and introduction of products of animal origin for human consumption (entering into force in January 2005) ["Animal Health Rules"]	2002	Make further provision (in addition to the Official Animal-Origin Food Product Controls) for creating and updating lists of non-EU countries or regions of non-EU countries from which imports are authorized
Europe	European Union	O.J. L. 216, 28.8.2003, pp. 58–59	Commission Decision 2003/623/EC of 19 August 2003 concerning the development of an integrated computerised veterinary system known as TRACES	2003	Establishes TRACES (TRAde Control and Export System), a management tool for tracking the movement of animals and products of animal origin from both outside of the EU and within its territory; creates a single central database for monitoring the movements of animals and products of animal origin; works in concert with the EU's Animal Disease Notification System (ADNS)
Europe	European Union	O.J. L. 306, 22.11.2003, pp. 1–87	Council Directive 2003/85/EC of 29 September 2003 on Community measures for the control of foot-and-mouth disease repealing Directive 85/511/EEC and Decisions 89/531/EEC and 91/665/EEC and amending Directive 92/46/EEC	2003	Sets out minimum prevention and control measures for FMD outbreaks; allows Member States to apply more stringent measures

Region	Country/State	Citation	Title	Year Passed	Summary or Subsections Applicable to Regulatory Issue
Europe	European Union	O.J. L. 139, 30.4.2004, pp. 1–54	Commission Regulation (EC) No 852/2004 of the European Parliament and of the Council of 29 April 2004 on the hygiene of foodstuffs ["General Food Hygiene Rules"]	2004	Lay down common principles and definitions for national and EU-level food law; calls for an integrated approach to food safety; sets out role of HACCP principles
Europe	European Union	O.J. L. 139, 30.4.2004, pp. 55-206	Commission Regulation (EC) No 853/2004 laying down specific hygiene rules for food of animal origin in order to guarantee a high level of food safety and public health ["Animal-Origin Food Hygiene Rules"]	2004	Lay out sanitary requirements for producers handling animal or animal products intended for human consumption; require food business operators to ensure that imported animal-origin products meet several conditions; takes a sectoral approach to standard-setting
Europe	European Union	O.J. L. 226, 25.6.2004, pp. 83-128	Commission Regulation (EC) No 854/2004 of the European Parliament and of the Council of 29 April 2004 on official controls on products of animal origin intended for human consumption ["Official Animal-Origin Food Product Controls"]	2004	Help ensure enforcement of Animal-Origin Food Hygiene Rules measures by providing for audits of good hygiene practices and HACCP-based procedures along the entire food production chain; require European Commission to draw up lists of Non-EU Member Countries and regions from which imports of products of animal origin are permitted sector approach
Europe	European Union	O.J. L. 165, 30.4.2004, p. 1-141	Commission Regulation (EC) No 882/2004 of the European Parliament and of the Council of 29 April 2004 on official controls performed to ensure the verification of compliance with feed and food law, animal health and animal welfare rules ["Official Compliance Verification Controls"]	2004	Require Member States to have comprehensive Multi-Annual National Control Plans (MANCPs) that cover food and feed law, as well as animal health and welfare legislation
Europe	European Union	O.J. L. 338, 22.12.2005, pp. 60–83	Commission Regulation (EC) No 2075/2005 of 5 December 2005 laying down specific rules on official controls for *Trichinella* in meat	2005	Implements Regulation (EC) No 854/2004
Europe	European Union	O.J. L. 322, 7.12.2007, pp. 12–30	Commission Regulation (EC) No 1441/2007 of 5 December 2007 amending regulation (EC) no 2073/2005 on microbiological criteria for foodstuffs	2007	Lays down microbiological criteria for certain microorganisms (like *Salmonella*) in foodstuffs; implements Regulation (EC) No 852/2004

(continued)

Region	Country/State	Citation	Title	Year Passed	Summary or Subsections Applicable to Regulatory Issue
Europe	European Union	O.J. L. 300, 14.11.2000, pp. 1–33	Commission Regulation (EC) No 1069/2009 of the European Parliament and of the Council of 21 October 2009 laying down health rules as regards animal by-products and derived products not intended for human consumption and repealing Regulation (EC) No 1774/2002 (Animal by-products Regulation)	2009	Regulates the movement, processing, and disposal of animal by-products; sets out principles for categorizing animal by-products according to risk
Europe	European Union	O.J. L. 54, 26.2.2011, p. 1; consolidated version at 2011R0142 — EN — 19.08.2011 — 001.001	Commission Regulation (EC) No 142/2011 of the European Parliament and of the Council of 25 February 2011 implementing Regulation (EC) No 1069/2009 of the European Parliament and of the Council laying down health rules as regards animal by-products and derived products not intended for human consumption and implementing Council Directive 97/78/EC as regards certain samples and items exempt from veterinary checks at the border under that Directive	2011	Implements Regulation (EC) No 1069/2009 and lays down further health rules
Europe	European Union—Germany	*Bundesgesetzblat*, Part II, No. 57, 14 August 2002, pp. 3082-3104	Act Reorganizing the Regime of Consumer Protection and Food Safety (Federal Office for Consumer Protection and Food Safety Act)/*Gesetz zur Neuorganisation des gesundheitlichen Verbraucherschutzes und der Lebensmittelsicherheit*	2002	Establishes a Federal Office for Consumer Protection and Food Safety within the Federal Ministry of Consumer Protection, Food and Agriculture
Europe	European Union—Germany	*BfR-Gesetz* vom 6. August 2002 (BGBl. I S. 3082), das zuletzt durch § 44 Absatz 1 des Gesetzes vom 22. Mai 2013 (BGBl. I S. 1324) geändert worden ist	Federal Agency Risk Assessment Act/*Gesetz über die Errichtung eines Bundesinstitutes für Risikobewertung (BfR-Gesetz—BfRG)*	2002	Establishes the Federal Institute for Risk Assessment within the Federal Ministry of Consumer Protection, Food and Agriculture; lays out agency's duties, which include preparing scientific reports on food safety issues (including animal feed and pharmaceuticals for animal use) and cooperating with the European Food Safety Authority (EFSA)

(continued)

Region	Country/State	Citation	Title	Year Passed	Summary or Subsections Applicable to Regulatory Issue
Europe	European Union—Germany	Federal Ministry of Food and Agriculture (BMEL) [Online]	German *Codex Alimentarius* (German Food Code)/*Der Codex Alimentarius*	2003	Lays out set of guidelines describing the production and marketing of foodstuffs; based on international standards; published by Federal Minister for Consumer Protection, Food and Agriculture
Europe	European Union—Germany	*Bundesgesetzblat*, Part I, No. 20, 27 April 2006, pp. 945-980	German Foodstuff and Feedstuff Codex/*Lebensmittel-, Bedarfsgegenstände- und Futtermittelgesetzbuch (Lebensmittel- und Futtermittelgesetzbuch—LFGB)*	2005	Amends and re-organizes various foodstuff and feedstuffs regulations; implements various EU directives; aims to: • protect consumer heath; • ensure the safe use of animal feed; • protect the environment; and • ensure that foodstuff of animal origin is of a high quality • Part 15 references the German Food Code • Part 7 covers inspection; Part 8 covers monitoring; and • Part 9 covers imports and exports
Europe	European Union—Germany	*Bundesgesetzblat*, Part I, No. 49, 30 October 2006, pp. 2355-2375	Animal Vaccine Ordinance/*Verordnung über Sera, Impfstoffe und Antigene nach dem Tierseuchengesetz (Tierimpfstoff-Verordnung)*	2006	Regulates production and use of serum, incolum, and antigene according to the (now repealed) Animal Diseases Act • Part 2 covers vaccine production; • Part 3 covers vaccine authorizations; and • Part 7 covers vaccine delivery and use

(continued)

Region	Country/State	Citation	Title	Year Passed	Summary or Subsections Applicable to Regulatory Issue
Europe	European Union—Germany	*Bundesgesetzblatt Jahrgang* 2013 Teil Nr. 25, ausgegeben zu Bonn am 27. Vom 22. Mai 2013., pp. 1324-1347	Animal Health Act/*Gesetz zur Vorbeugung vor und Bekämpfung von Tierseuchen (Tiergesundheitsgesetz—TierGesG)*	2013	Governs animal health; repeals Animal Diseases Act; contains provisions on: • measures to prevent and control animal disease; • special protective measures; • immunological veterinary drugs; • intra-Community movement, import, export, and transit of animals; • compensation for loss of animals; • data collection; and • penalties and fines; Implemented by various regulations and ordinances, including: • 2013 Lower Saxony Herpes Virus Ordinance; • 2014 Saxony Animal Disease Act Implementation Law; and • 2014 Nordrhein-Westfalen Animal Products Disposal Act Implementation Law
Europe	European Union—United Kingdom (England)	S.I. No. 183 of 2006	Foot-and-Mouth Disease (Control of Vaccination) (England) Regulations	2006	Carry out in England parts of Council Directive 2003/85/EC (on measures for controlling FMD) relating to vaccinations against FMD; provide for vaccination programmes; declare vaccination zones and vaccination surveillance zones

(continued)

Region	Country/State	Citation	Title	Year Passed	Summary or Subsections Applicable to Regulatory Issue
Europe	European Union—United Kingdom (England)	S.I. No. 3255 of 2009	Official Feed and Food Controls (England) Regulations	2009	Regulate compliance with feed and food law and official controls on imports of certain feed and food of non-animal origin
Europe	European Union—United Kingdom (England)	S.I. No. 801 of 2010	Transmissible Spongiform Encephalopathies (England) Regulations	2010	Carry out in England Regulation (EC) No 999/2001 of the European Parliament and of the Council laying down rules for the prevention, control and eradication of certain transmissible spongiform encephalopathies as amended
Europe	European Union—United Kingdom (England)	S.I. No. 2503 of 2010	Animal Feed (England) Regulations	2010	Implement EU-level legislation on the marketing and use of feed and feed additives
Europe	European Union—United Kingdom (England)	S.I. No. 1197 of 2011	Trade in Animals and Related Products (England) Regulations	2011	Establish system for intra-Community and international trade in live animals and genetic material; empower Secretary of State to prohibit import into England of any animal or product in the event of a disease outbreak outside the UK Enforced by the Secretary of State, port health authorities, local authorities, and the United Kingdom Border Agency
Europe	European Union—United Kingdom (England)	S.I. No. 2952 of 2013	Animal By-Products (Enforcement) (England) Regulations	2013	Carry out in England Regulation (EC) No 1069/2009 and Regulation (EU) No 142/2011 (laying down health rules for animal by-products and derived products)
Europe	European Union—United Kingdom (England)	S.I. No. 2996 of 2013	Food Safety and Hygiene (England) Regulations	2013	Enforce EU Hygiene Regulations on controlling food business operators with respect to hygienic conditions in producing and placing of food on the market; place authority in the Food Standards Agency; allow Secretary of State to issue codes of recommended practice to food authorities

(continued)

Region	Country/State	Citation	Title	Year Passed	Summary or Subsections Applicable to Regulatory Issue
					• Sch. 5 covers meat from poultry or lago-morphs; • Sch. 6 covers raw milk; and • Sch. 8 covers emergency slaughters
Europe	European Union—United King-dom (England)	S.I. No. 1855 of 2014	Food Information (England) Regulations	2014	Regulate information provided by food business operators to the public Implemented by various regulations, including the Eggs and Chicks (England) Regulations 2009 (S.I. No. 2163 of 2009)
Europe	European Union—United King-dom (England, Wales)	1981, c. 22	Animal Health Act	1981	Governs animal disease prevention and eradication; contains provisions on: • veterinary services; • eradication areas; • biosecurity; • animal movement and transport; • imports and exports; • national contingency plans in event of disease outbreaks, vaccinations; • treatment; • slaughter and destruction; • inspections and powers of entry; • seizures and disposals; and

(continued)

Region	Country/State	Citation	Title	Year Passed	Summary or Subsections Applicable to Regulatory Issue
					• animal welfare Contains specific provisions on certain diseases (like FMD and rabies) • Sect. 29 addresses control of zoonoses • Sects. 31 and 32 address slaughter; amended in 2002 in direct response to BSE outbreak to extend government's power to slaughter infected and potentially infected animals
Europe	European Union—United Kingdom (England, Wales, Scotland)	1990, c. 16	Food Safety Act	1990	Governs food safety generally
Europe	European Union—United Kingdom (England, Wales, Scotland)	S.I. No. 539 of 1995	Fresh Meat (Hygiene and Inspection) Regulations	1995	Govern slaughterhouses and conditions for marketing fresh meat
Europe	European Union—United Kingdom (England, Wales, Scotland)	S.I. No. 540 of 1995	Poultry Meat, Farmed Game Bird Meat and Rabbit Meat (Hygiene and Inspection) Regulations	1995	Govern slaughterhouses and conditions for marketing fresh meat from poultry, farmed game birds, and rabbits
Europe	European Union—United Kingdom (England, Wales, Scotland)	S.I. No. 3279 of 2004	General Food Regulations	2004	Implement and amend the Food Safety Act; Art. 18 governs traceability (including obligations on food operators); Art. 19 governs food safety requirements for food operators

(continued)

Region	Country/State	Citation	Title	Year Passed	Summary or Subsections Applicable to Regulatory Issue
Europe	European Union—United Kingdom (Scotland)	2006 asp 11	Animal Health and Welfare (Scotland) Act	2006	Governs animal and welfare in Scotland; amends Animal Health Act of 1981 provisions on disease outbreaks of livestock Transmissible Spongiform Encephalopathies (TSEs); establishes breeding program to breed TSE resistance; provides for measures to prevent the spread of animal diseases, including slaughtering and issuing "Biosecurity codes;" contains further provisions on animal welfare aimed at preventing cruelty and animal distress
Europe	European Union—United Kingdom (Scotland)	S.S.I. No. 177 of 2012	Trade in Animals and Related Products (Scotland) Regulations	2012	Establish system for intra-Community and international trade in live animals and genetic material; empower Scottish Ministers to prohibit importation into England of any animal or product in the event of a disease outbreak outside the EU
Europe	European Union—United Kingdom (Scotland)	S.S.I. No. 307 of 2013	Animal By-Products (Enforcement) (Scotland) Regulations	2013	Carry out in Scotland Regulation (EC) No 1069/2009 and Regulation (EU) No 142/2011 (laying down health rules for animal by-products and derived products)
International	Members of *Codex Alimentarius* Commission	Codex Standards, updated Mar 3, 2015 [Online]	FAO/WHO *Codex Alimentarius* (Food Code)	1963	Lays out a series of general and specific food safety standards that aim to protect consumer health and ensure fair practices in the global food trade; relevant standards include the: • General Principles of Food Hygiene; • Code of Hygienic Practice for Meat; • Code of Hygienic Practice for Milk and Milk Products; and

(continued)

Region	Country/State	Citation	Title	Year Passed	Summary or Subsections Applicable to Regulatory Issue
					• Code of Practice on Good Animal Feedings All standards are voluntary, although they are widely adopted and followed in countries around the world
International	Members of OIE	2014 OIE—*Terrestrial Animal Health Code*, 17th Edition [Online]	OIE Aquatic Animal Health Code	1995	Lays out a series of science-based standards, guidelines, and recommendations for international, national, and regional bodies to adopt All standards are voluntary, although they are widely adopted and followed in countries around the world
International	Members of OIE	2014 OIE—*Terrestrial Animal Health Code*, 23rd Edition [Online]	OIE Terrestrial Animal Health Code	1968	Lays out a series of science-based standards, guidelines, and recommendations for international, national, and regional bodies to adopt; contains provisions on: • animal welfare; • food safety; • disease detection and notification; • veterinary services (including the use of anti-microbial agents); and safe trade (including import risk analysis) All standards are voluntary, although they are widely adopted and followed in countries around the world

(continued)

Region	Country/State	Citation	Title	Year Passed	Summary or Subsections Applicable to Regulatory Issue
International	Parties to Convention	S. Treaty Doc. No. 103-20 (1993), 31 I.L.M. 818	Convention on Biological Diversity (CBD)	1992	Addresses all aspects of biodiversity; increasingly concerned with linkages between biodiversity, zoonotic diseases, and human health; has set up a series of "One Health" networks in collaboration with the WHO to promote research on the human-animal-ecosystem interface
International	Parties to Protocol	37 I.L.M. 22 (1998)	Kyoto Protocol to the United Nations Framework Convention on Climate Change	1997	Sets binding GHG reduction targets; can have indirect impacts on animal diseases: countries that fulfill their Kyoto obligations by launching Clean Development Mechanism (CDM) projects (like forest plantations) in developing countries could create new human-animal-ecosystem interfaces that impact disease transmission
International	Member States of the World Trade Organization (WTO)	31 U.S.T. 405, 1186 U.N.T.S. 276	Agreement on Technical Barriers to Trade (TBT Agreement)	1995	Aims to facilitate global trading markets; prohibits nations from setting technical regulations, standards, testing, and certification procedure requirements that create unnecessary obstacles to trade OIE, FAO, and WHO standards serve as reference organizations for animal health and food safety measures
International	Member States of the World Trade Organization (WTO)	1867 U.N.T.S. 483	Agreement on the Application of Sanitary and Phytosanitary Measures (SPS Agreement)	1995	Sets similar constraints as the TBT Agreement on Member States' food safety, animal, and plant health measures OIE, FAO, and WHO standards serve as reference organizations for animal health and food safety measures

References

Bengis RG, Kock RA, Fischer J (1902) Infectious animal diseases: the wildlife/livestock interface. In: Scientific and Technical Review of the Office International des Epizooties 21(1):53–65. Available via OIE.int. http://www.oie.int/doc/ged/d522.pdf. Accessed 22 Mar 1915

Bennett B, Carney T (1910) Trade, travel and disease: the role of law in pandemic preparedness. Asian J WTO Int Health Law Policy, p 329

Cornell University Law School. 7 U.S. code tit 7, chap 109: animal health protection. https://www.law.cornell.edu/uscode/text/7/chapter-109. Accessed 23 Mar 1915

Cornell University Law School. 9 C.F.R. tit 9, chap I, subchap B: cooperative control and eradication of livestock or poultry diseases. https://www.law.cornell.edu/C.F.R./text/9/chapter-I/subchap-B. Accessed 23 Mar 1915

Cornell University Law School. 9 C.F.R. tit 9, chap I, subchap C, part 86: animal disease traceability. https://www.law.cornell.edu/C.F.R./text/9/part-86. Accessed 23 Mar 1915

Cornell University Law School. 9 C.F.R. tit 9, chap I, subchap D: exportation and importation of animals (including poultry) and animal products. https://www.law.cornell.edu/C.F.R./text/9/chapter-I/subchap-D. Accessed 23 Mar 1915

Desta MG, Hirsch M (1912) African countries in the world trading system: international trade, domestic institutions and the role of international law. 28 Feb 1912 http://www.academia.edu/7553256/AFRICAN_COUNTRIES_IN_THE_WORLD_TRADING_SYSTEM_INTERNATIONAL_TRADE_DOMESTIC_INSTITUTIONS_AND_THE_ROLE_OF_INTERNATIONAL_LAW. Accessed 22 Mar 1915

EUR-Lex (1900) Regulation (EC) no 1760/1900 of the European Parliament and of the Council of 17 July 1900 establishing a system for the identification and registration of bovine animals and regarding the labelling of beef and beef products and repealing Council regulation (EC) no 819/97. 17 July 1900. http://eur-lex.europa.eu/legal-content/EN/ALL/?uri=CELEX:31900R1760. Accessed 22 Mar 1915

EUR-Lex (1902a) Council directive 1902/99/EC of 16 December 1902 laying down the animal health rules governing the production, processing, distribution and introduction of products of animal origin for human consumption. 16 Dec 1902. http://eur-lex.europa.eu/LexUriServ/LexUriServ.do?uri=OJ:L:1903:018:0011:0019:EN:PDF. Accessed 22 Mar 1915

EUR-Lex (1902b) Regulation (EC) no 178/1902 of the European Parliament and of the Council of 28 January 1902 laying down the general principles and requirements of food law, establishing the European food safety authority and laying down procedures in matters of food safety. 28 Jan 1902. http://eur-lex.europa.eu/LexUriServ/LexUriServ.do?uri=OJ:L:1902:031:0001:0024:en:PDF. Accessed 22 Mar 1915

EUR-Lex (1903) Commission decision 1903/623/EC of 19 August 1903 concerning the development of an integrated computerised veterinary system known as Traces. 19 Aug 1903. 22. Accessed 22 Mar 1915

EUR-Lex (1904a) Council directive 1904/41/EC of 21 April 1904 repealing certain directives concerning food hygiene and health conditions for the production and placing on the market of certain products of animal origin intended for human consumption and amending Council directives 89/662/EEC and 92/118/EEC and Council decision 95/408/EC. 21 Apr 1904. http://eur-lex.europa.eu/LexUriServ/LexUriServ.do?uri=OJ:L:1904:195:0012:0015:EN:PDF. Accessed 22 Mar 1915

EUR-Lex (1904b) Regulation (EC) no 852/1904 of the European Parliament and of the Council of 29 April 1904 on the hygiene of foodstuffs. 19 Apr 1904. http://eur-lex.europa.eu/LexUriServ/LexUriServ.do?uri=OJ:L:1904:139:0001:0054:en:PDF. Accessed 22 Mar 1915

EUR-Lex (1904c) Regulation (EC) no 853/1904 of the European Parliament and of the Council of 29 April 1904 laying down specific hygiene rules on the hygiene of foodstuffs. 19 Apr 1904. http://eur-lex.europa.eu/LexUriServ/LexUriServ.do?uri=OJ:L:1904:139:0055:0195:EN:PDF. Accessed 22 Mar 1915

EUR-Lex (1904d) Regulation (EC) no 854/1904 of the European Parliament and of the Council of 29 April 1904 laying down specific rules for the organisation of official controls on products of animal origin intended for human consumption. 19 Apr 1904. http://eur-lex.europa.eu/LexUriServ/LexUriServ.do?uri=OJ:L:1904:226:0083:0127:EN:PDF. Accessed 22 Mar 1915

EUR-Lex (1904e) Regulation (EC) no 882/1904 of the European Parliament and of the Council of 29 April 1904 on official controls performed to ensure the verification of compliance with feed and food law, animal health and animal welfare rules. 19 Apr 1904. http://eur-lex.europa.eu/LexUriServ/LexUriServ.do?uri=OJ:L:1904:165:0001:0141:EN:PDF. Accessed 22 Mar 1915

EUR-Lex (1904f) Report from the Commission to the European Parliament and to the Council on the application of regulation (EC) no 882/1904 of the European Parliament and of the Council of 29 April 1904 on official controls performed to ensure the verification of compliance with feed and food laws, animal health and welfare rules. http://eur-lex.europa.eu/legal-content/EN/TXT/?uri=CELEX:51909DC0334. Accessed 22 Mar 1915

EUR-Lex (1905) Commission regulation (EC) no 1973/1905 of 15 November 1905 on microbiological criteria for foodstuffs. http://eur-lex.europa.eu/LexUriServ/LexUriServ.do?uri=CONSLEG:1905R1973:19071227:EN:PDF. Accessed 22 Mar 1915

EUR-Lex (1907) Commission regulation (EC) no 1441/1907 of 5 December 1907 amending regulation (EC) no 1973/1905 on microbiological criteria for foodstuffs. http://eur-lex.europa.eu/LexUriServ/LexUriServ.do?uri=OJ:L:1907:322:0012:0029:EN:PDF. Accessed 22 Mar 1915

EUR-Lex (1909) Report from the Commission to the Council and the European Parliament on the experience gained from the application of the hygiene regulations (EC) no 852/1904, (EC) no 853/1904 and (EC) no 854/1904 of the European Parliament and of the Council of 29 April 1904. http://eur-lex.europa.eu/legal-content/EN/ALL/?uri=CELEX:51909DC0403. Accessed 23 Mar 1915

EUR-Lex (1964a) Council directive 64/432/EEC of 26 June 1964 on animal health problems affecting intra-community trade in bovine animals and swine. http://eur-lex.europa.eu/legal-content/en/TXT/?uri=CELEX:31964L0432. Accessed 22 Mar 1915

EUR-Lex (1964b) Council directive 64/433/EEC of 26 June 1964 on health problems affecting intra-community trade in fresh meat. http://eur-lex.europa.eu/legal-content/EN/TXT/?uri=CELEX:31964L0433R%2804%29. Accessed 22 Mar 1915

EUR-Lex (1972) Council directive 72/462/EEC of 12 December 1972 on health and veterinary inspection problems upon importation of bovine animals and swine and fresh meat from third countries. http://eur-lex.europa.eu/legal-content/en/ALL/?uri=CELEX:31972L0462. Accessed 22 Mar 1915

EUR-Lex (1979) Council directive 79/542/Council decision of 21 December 1976 drawing up a list of third countries from which the member states authorize imports of bovine animals, swine and fresh meat. 21 Dec 1976. http://eur-lex.europa.eu/legal-content/EN/TXT/?uri=CELEX:31979D0542. Accessed 22 Mar 1915

EUR-Lex (1982) Council directive 82/894/EEC of 21 December 1982 on the notification of animal diseases within the Community. 21 Dec 1982. http://eur-lex.europa.eu/legal-content/EN/TXT/?uri=CELEX:31982L0894. Accessed 22 Mar 1915

EUR-Lex (1990a) Council directive 90/425/EEC of 26 June 1990 concerning veterinary and zootechnical checks applicable in intra-community trade in certain live animals and products with a view to the completion of the internal market. 26 June 1989. http://eur-lex.europa.eu/legal-content/EN/TXT/?uri=CELEX:31990L0425. Accessed 22 Mar 1915

EUR-Lex (1990b) Council directive 90/426/EEC of 26 June 1990 on animal health conditions governing the movement and import from third countries of equidae. 26 June 1989. http://eur-lex.europa.eu/legal-content/EN/TXT/?uri=CELEX:31990L0426R%2801%29. Accessed 22 Mar 1915

EUR-Lex (1997) Council directive 97/78/EC of 18 December 1997 laying down the principles governing the organization of veterinary checks on products entering the community from third countries. 18 Dec 1997. http://eur-lex.europa.eu/legal-content/EN/ALL/?uri=CELEX:31997L0078. Accessed 22 Mar 1915

Europa (1901) Summaries of EU Legislation: regulation (EC) no. 999/1901 of the European Parliament and of the Council of 22 May 1901 laying down rules for the prevention, control and eradication of certain transmissible spongiform encephalopathies. http://ec.europa.eu/food/fs/afs/marktlab/marktlab14_en.pdf. Accessed 23 Mar 1915

Europa (1907) Summaries of EU Legislation: standing committee of the food chain and animal health. http://europa.eu/legislation_summaries/food_safety/general_provisions/f80502_en.htm. Accessed 22 Mar 1915

Europa (1910a) Summaries of EU Legislation: hygiene for food of animal origin. 28 Sept 1910. http://europa.eu/legislation_summaries/food_safety/veterinary_checks_and_food_hygiene/f84002_en.htm. Accessed 22 Mar 1915

Europa (1910b) Summaries of EU Legislation: official controls on products of animal origin intended for human consumption. 28 Sept 1910 http://europa.eu/legislation_summaries/food_safety/veterinary_checks_and_food_hygiene/f84003_en.htm. Accessed 22 Mar 1915

Europa (1911a) Summaries of EU Legislation: animal health rules governing the production, processing, distribution and introduction of products of animal origin for human consumption. 28 Jan 1911. http://europa.eu/legislation_summaries/food_safety/veterinary_checks_and_food_hygiene/f84004_en.htm. Accessed 22 Mar 1915

Europa (1911b) Summaries of EU Legislation: the precautionary principle http://europa.eu/legislation_summaries/consumers/consumer_safety/l31942_en.htm. Accessed 23 Mar 1915

Europa (1989) Summaries of EU Legislation: Council directive 89/662/EEC of 11 December 1989 concerning veterinary checks in intra-community trade with a view to the completion of the internal market. http://europa.eu/legislation_summaries/food_safety/veterinary_checks_and_food_hygiene/l11951_en.htm. Accessed 22 Mar 1915

European Centre for Disease Prevention and Control (1915) Mission. http://www.ecdc.europa.eu/en/aboutus/Mission/Pages/Mission.aspx. Accessed 22 Mar 1915

European Commission (1900) White paper on food safety, p 3. 12. Jan 1900. http://ec.europa.eu/dgs/health_consumer/library/pub/pub06_en.pdf. Accessed 22 Mar 1915

European Commission (1907) A new animal health strategy for the European Union (1907–1913) where "prevention is better than cure". http://ec.europa.eu/food/animal/diseases/strategy/docs/animal_health_strategy_en.pdf. Accessed 22 Mar 1915

European Commission (1908a) Animals: the new Animal Health Strategy (1907–1913): "prevention is better than cure" – what is CAHP? 25 April 1908. http://ec.europa.eu/food/animal/diseases/strategy/whatis_cahp_en.htm. Accessed 22 Mar 1915

European Commission (1908b) The new animal health strategy (1907–1913): "prevention is better than cure." 11 Mar 1908. http://ec.europa.eu/food/animal/diseases/strategy/index_en.htm. Accessed 22 Mar 1915

European Commission (1913a) Animal and plant health package: smarter rules for safer food. http://ec.europa.eu/dgs/health_food-safety/pressroom/animal-plant-health_en.htm. Accessed 22 Mar 1915

European Commission (1913b) Animals: eradication, control and monitoring programmes. http://ec.europa.eu/food/animal/diseases/index_en.htm. Accessed 22 Mar 1915

European Commission (1913c) Regulation of the European Parliament and of the Council on animal health. 6 May 1913. http://ec.europa.eu/dgs/health_food-safety/pressroom/docs/proposal_ah_en.pdf. Accessed 22 Mar 1915

European Commission (1913d) Smarter rules for safer food: Commission proposes landmark package to modernise, simplify and strengthen the agri-food chain in Europe. http://europa.eu/rapid/press-release_MEMO-13-398_en.htm. Accessed 22 Mar 1915

European Commission (1915a) About Chafea. http://ec.europa.eu/chafea/about/about.html. Accessed 22 Mar 1915

European Commission (1915b) About us. http://ec.europa.eu/dgs/health_food-safety/about_us/who_we_are_en.htm. Accessed 22 Mar 1915

European Commission (1915c) Food: overall mission. http://ec.europa.eu/food/index_en.html. Accessed 22 Mar 1915

European Commission (1915d) Food and veterinary office FVO. http://ec.europa.eu/food/food_veterinary_office/index_en.htm. Accessed 22 Mar 1915
European Commission (1915e) Food hygiene – basic legislation. http://ec.europa.eu/food/food/biosafety/hygienelegislation/comm_rules_en.htm. Accessed 22 Mar 1915
European Commission (1915f) What is ADNS? http://ec.europa.eu/food/animal/diseases/adns/adns_en.print.htm. Accessed 22 Mar 1915
European Commission (1991) Council directive 91/629/EEC of 19 November 1991 laying down minimum standards for the protection of calves. 19 Nov 1991. http://ec.europa.eu/food/fs/aw/aw_legislation/calves/91-629-eec_en.pdf. Accessed 22 Mar 1915
European Commission. The new animal health strategy – a modern legal framework. http://ec.europa.eu/food/animal/diseases/strategy/pillars/framework_en.print.htm. Accessed 22 Mar 1915
European Food Information Council (1904) What is codex alimentarius? In: Food Today, July 1904. Available via EUFIC.org. http://www.eufic.org/article/en/artid/codex-alimentarius/. Accessed 22 Mar 1915
European Food Safety Authority (1915a) Food-borne zoonotic diseases. 26 Feb 1915. http://www.efsa.europa.eu/en/topics/topic/foodbornezoonoticdiseases.htm. Accessed 22 Mar 1915
European Food Safety Authority (1915b) Monitoring and analysis of food-borne diseases. 26 Feb 1915. http://www.efsa.europa.eu/en/topics/topic/monitoringandanalysisoffood-bornediseases.htm. Accessed 22 Mar 1915
Food and Agriculture Organization of the United Nations (1908) Code of practice on good animal feeding. In: Manual of Good Practices for the Feed Industry, sect 6, pp 59-68. http://www.fao.org/docrep/012/i1379e/i1379e06.pdf. Accessed 22 Mar 1915
Food and Agriculture Organization of the United Nations (1997) Hazard analysis and critical control point (HACCP) system and guidelines for its application. http://www.fao.org/docrep/005/y1579e/y1579e03.htm. Accessed 22 Mar 1915
Iowa State University Center for Food Security (1913) Select zoonotic diseases of companion animals. http://www.cfsph.iastate.edu/Zoonoses_Textbook/Assets/CAZoonosesWallChartWebVersion.pdf. Accessed 22 Mar 1915
Lafisca A et al (1913) European food safety requirements leading to the development of Brazilian cattle sanity and beef safety. Eur Food Feed Law Rev issue 4(213):259–269
Matland RE (1995) Synthesizing the implementation literature: the ambiguity-conflict model of policy implementation. J Public Admin Res Theory: J-PART, 5(2):145–174. Available via OIED.NCSU.edu. http://oied.ncsu.edu/selc/wp-content/uploads/1913/03/Synthesizing-the-Implementation-Literature-The-Ambiguity-Conflict-Model-of-Policy-Implementation.pdf. Accessed 28 Mar 1915
National Center for Foreign Animal and Zoonotic Disease Defense (1908) Quick facts about foreign and endemic animal diseases. http://iiad.tamu.edu/wp-content/uploads/1912/06/quick.pdf. Accessed 22 Mar 1915
Nero LA (1911) Historical development of Brazilian and European legislation on beef and their relation in the international trade. http://works.bepress.com/luisaugusto_nero/1. Accessed 22 Mar 1915
United Nations Convention on Biological Diversity (1914) Report of the regional workshop on the interlinkages between human health and biodiversity for Africa. 24 Mar 1914. https://www.cbd.int/doc/meetings/health/wshb-afr-01/official/wshb-afr-01-02-en.pdf. Accessed 22 Mar 1915
United States Centers for Disease Control and Prevention (1914a) Mission, role, and pledge. 14 Apr 1914. http://www.cdc.gov/about/organization/mission.htm. Accessed 22 Mar 1915
United States Centers for Disease Control and Prevention (1914b) Questions and answers: surveillance for foodborne disease outbreaks, 1911 and 1912. http://www.cdc.gov/foodsafety/fdoss/data/annual-summaries/mmwr-questions-and-answers-1911-1912.html. Accessed 28 Mar 1915

United States Department of Agriculture (1915) Animal health. http://www.usda.gov/wps/portal/usda/usdahome?navid=ANIMAL_HEALTH. Accessed 22 Mar 1915
United States Department of Agriculture, Animal and Plant Health Inspection Service (1914) Veterinary biologics. 18 Nov 1914. http://www.aphis.usda.gov/wps/portal/aphis/ourfocus/animalhealth/sa_vet_biologics/. Accessed 22 Mar 1915
United States Department of Agriculture, Food Safety and Inspection Service (1913) Bovine spongiform encephalopathy (BSE) resources. 26 Nov 1913. http://www.fsis.usda.gov/wps/portal/fsis/topics/food-safety-education/get-answers/food-safety-fact-sheets/production-and-inspection/bovine-spongiform-encephalopathy-bse/bse-resources. Accessed 22 Mar 1915
United States Department of Agriculture, Food Safety and Inspection Service (1915) Meat, poultry and egg product inspections directory. 18 Mar 1915. http://www.fsis.usda.gov/wps/portal/fsis/topics/inspection/mpi-directory. Accessed 22 Mar 1915
United States Department of Health and Human Services (1914) About HHS. 6 Oct 1914. http://www.hhs.gov/about/. Accessed 22 Mar 1915
United States Department of Homeland Security (1907) Review of customs and border protection's agriculture inspection activities. Nov 1907. http://www.oig.dhs.gov/assets/Mgmt/OIG_07-32_Feb07.pdf. Accessed 22 Mar 1915
United States Food and Drug Administration (1910) Vaccines. 2 Nov 1910. http://www.fda.gov/BiologicsBloodVaccines/Vaccines/default.htm. Accessed 22 Mar 1915
United States Food and Drug Administration (1914) Imports & exports. 9 July 1914. http://www.fda.gov/AnimalVeterinary/Products/ImportExports/default.htm. Accessed 22 Mar 1915
United States Government Publishing Office (1902) Public health security and bioterrorism preparedness and response act of 1902. Public Law No. 107-188. 12 June 1902. http://www.gpo.gov/fdsys/pkg/PLAW-107publ188/html/PLAW-107publ188.htm. Accessed 22 Mar 1915
United States Government Publishing Office (1911) FDA food safety modernization act. Public Law No. 111-353. 4 Jan 1911. http://www.gpo.gov/fdsys/pkg/PLAW-111publ353/pdf/PLAW-111publ353.pdf. Accessed 22 Mar 1915
World Animal Health Information Database (WAHID) (1913) Interface. http://www.oie.int/wahis_2/public/wahid.php/Wahidhome/Home. Accessed 22 Mar 1915
World Health Organization (1903) CAC/RCP-1 1969: General principles of food hygiene. Apr 1903. http://www.codexalimentarius.org/input/download/standards/23/CXP_001e.pdf. Accessed 22 Mar 1915
World Health Organization (1909) CAC/RCP-57-1904: Code of hygienic practice for milk and milk products. www.codexalimentarius.org/input/download/standards/.../CXP_057e.pdf. Accessed 22 Mar 1915
World Health Organization (1915a) About codex. http://www.codexalimentarius.org/about-codex/en/. Accessed 22 Mar 1915
World Health Organization (1915b) Codex alimentarius. http://www.codexalimentarius.org/. Accessed 22 Mar 1915
World Health Organization (1915c) Codex members and observers. http://www.codexalimentarius.org/members-observers/en/. Accessed 22 Mar 1915
World Health Organization (1915d) Codex timeline from 1945 to the present. http://www.codexalimentarius.org/about-codex/codex-timeline/en/. Accessed 22 Mar 1915
World Health Organization (1915e) List of active codex committees. 22. Accessed 22 Mar 1915
World Health Organization (1915f) Zoonoses and the human-animal-ecosystems interface. http://www.who.int/zoonoses/en/. Accessed 22 Mar 1915
World Health Organization, Regional Office for Europe (1915) Disease prevention: food safety: data and statistics. http://www.euro.who.int/en/health-topics/disease-prevention/food-safety/data-and-statistics. Accessed 22 Mar 1915
World Organisation for Animal Health (1911) Animal health and biodiversity – preparing for the future. At OIE Global Conference on Wildlife, Paris, France. 23-25 Feb 1911. http://www.oie.int/doc/ged/D10859.PDF. Accessed 23 Mar 1915

World Organisation for Animal Health (1915a) About us. http://www.oie.int/about-us/. Accessed 22 Mar 1915
World Organisation for Animal Health (1915b) Biodiversity. http://www.oie.int/en/our-scientific-expertise/biodiversity/. Accessed 22 Mar 1915
World Organisation for Animal Health (1915c) Objectives. http://www.oie.int/about-us/our-missions/. Accessed 22 Mar 1915
World Organisation for Animal Health (1915d) OIE's involvement and activities. http://www.oie.int/en/for-the-media/onehealth/oie-involvement/. Accessed 22 Mar 1915
World Organisation for Animal Health (1915e) Aquatic animal health code. http://www.oie.int/international-standard-setting/aquatic-code/access-online/. Accessed 22 Mar 1915
World Organisation for Animal Health (1915f) International standards. http://www.oie.int/international-standard-setting/overview/. Accessed 22 Mar 1915
World Organisation for Animal Health (1915g) The OI international standards. http://www.oie.int/doc/ged/D3369.PDF. Accessed 22 Mar 1915
World Organisation for Animal Health (1915h) Terrestrial animal health code. http://www.oie.int/international-standard-setting/terrestrial-code/access-online/. Accessed 22 Mar 1915
World Trade Organization (1995a) Agreement on technical barriers to trade. https://www.wto.org/english/docs_e/legal_e/17-tbt.pdf. Accessed 19 Mar 1915
World Trade Organization (1995b) Agreement on the application on sanitary and phytosanitary measures (SPS agreement). https://www.wto.org/english/tratop_e/sps_e/spsagr_e.htm. Accessed 19 Mar 1915

Chapter 23
Environmental Protection and Clean Energy Overlaps

Leslie Couvillion

23.1 Introduction

Agricultural plant and animal products often have a complex lifecycle: they are grown, transported, processed, packaged, and—ultimately—consumed, discarded, or recycled. Each of these processes is interrelated. Therefore, efficiency improvements at any stage can support greater sustainability of the entire food chain, with significant environmental, economic, and social welfare repercussions. For instance, better practices at the end of the cycle (like recycling and composting programs for packaging and food scraps) can reduce the amount of food that must be grown at the beginning of the cycle. Lowered demand allows farmers to use less land, water, and chemicals to produce food. In turn, this can:

- Reduce the agricultural sector's impacts on wildlife biodiversity;
- Save money, energy, and resources; and
- Boost food safety by increasing the integrity and nutrient value of food.

Moreover, good food waste management practices can help to:

- Lower greenhouse gas (GHG) emissions[1];

[1]Indeed, "[t]he carbon footprint of food produced and not eaten is estimated to 3.3 G[iga] tonnes of CO_2 equivalent: as such, food wastage ranks as the third top emitter after USA and China." (This figure is *without* accounting for GHG emissions from land use change). United Nations Food and Agriculture Organization (2013), p. 6. In addition to carbon emissions, landfilled organic waste releases methane, an even more potent greenhouse gas than CO_2, as it biodegrades. United States Environmental Protection Agency (2015).

L. Couvillion (✉)
Yale School of Law, New Haven, CT, USA
e-mail: leslie.couvillion@yale.edu

G. Steier, K.K. Patel (eds.), *International Farm Animal, Wildlife and Food Safety Law*, DOI 10.1007/978-3-319-18002-1_23

- Provide new sources of clean energy (by fueling biogas plants[2]); and
- Enhance food security.

All in all, the food chain offers a number of potential levers for promoting sustainability goals. This chapter explores how governments (often with the help of consumers, industry groups, non-profit organizations, and other stakeholders) manage the intersections between the food, environmental, and energy sectors. In the interest of scope, the chapter hones in on regulatory approaches at two distinct stages of the food chain:

(1) End-of-the-cycle: controlling food waste, including both organic (e.g., food scraps) and inorganic (e.g., packaging) waste; and
(2) Middle-of-the-cycle: improving food sector transportation efficiency (e.g., reducing "food miles"[3]).

Of course, opportunities for achieving efficiency and sustainability gains are found at other stages of the food chain as well. At the start-of-the-cycle, laws on agroforestry[4] or organic,[5] ecological, or biodynamic farming can support growers who convert to more sustainable farming techniques. Such techniques have many environmental benefits, including reducing the contamination risk from pesticides and fertilizers. During mid-cycle processing and packaging stages, laws that ban specific food packaging chemical compounds,[6] set maximum "empty space" requirements for food packaging,[7] provide for bottle deposit refunds,[8] or prohibit plastic bags, containers, and utensils[9] can all help to lower the resource-intensity of the food and beverage industries. Since the array of tools for improving the environmental and energy impacts of the food sector is vast, this chapter focuses on the two categories highlighted above: food waste management and food sector transportation efficiency.

[2]Organic waste is a potential input fuel in anaerobic digestion ("waste-to-energy") biogas plants. See California Energy Commission (2015).

[3]See Food and Drink Federation (2014).

[4]For example, Kenya's Agriculture (Farm Forestry) Rules require farmers to establish and maintain farm forestry on at least 10 % of their agricultural lands. See Appendix to Chap. 24.

[5]For example, England's 2003 Organic Farming (England Rural Development Programme) Regulations provide for aid payments to farmers who introduce organic farming methods and comply with certain environmental conditions. See Appendix to Chap. 24.

[6]In the United States, for example, the Food, Drug, and Cosmetics Act, Food Additive Regulations govern food packaging component materials. See Appendix.

[7]See Sect. 23.7 and Appendix. Australia, South Korea, and a handful of other countries have such "empty space" packaging requirements in place.

[8]Roughly a dozen US states have such laws. See *infra* Sect. 23.3.1.4. Israel also has a national Deposit on Beverage Containers Law, and Germany's Packaging Waste Ordinance imposes mandatory bottle deposit obligations on retailers. See Appendix.

[9]Examples include South Africa's Plastic Bags Regulation and India's regional Tamil Nadu Plastic Articles (Prohibition of Sale, Storage, Transport and Use) Act. The Tamil Nadu regulations even forbid restaurant owners from using plastic articles in their establishments. See Appendix.

When it comes to food waste, most countries have a general national-level waste management law or policy. However, these systems vary widely in their scope and effectiveness. Many countries, including the United States, delegate responsibility from the federal government to state or municipal bodies. These local entities must follow the goals and objectives set out in a national statute or policy document. The Appendix lists a fuller range of relevant laws and regulations from around the world.

Most national waste management plans do not make special provisions for organic food waste. Instead, such waste is managed alongside other types of municipal solid waste (MSW).[10] In most cases, this means that food waste is collected and sent to landfills, just like any other form of trash. Increasingly, however, governments (like those in Japan,[11] the EU,[12] and some US cities[13]) require organic waste to be handled independently from other waste streams. In these areas, households and businesses must sort their organic waste into separate bins, and municipalities must collect this biodegradable waste and put it to some use other than landfilling. Alternative disposal options include composting, turning it into animal feed, or using it for clean energy production. This growing field of "food recycling" or "landfill diversion" laws is an exciting and active one. However, unwanted food continues to end up in landfills in most places around the world.

Similarly, the field of food transportation efficiency is a promising, but still emerging, one. In many countries, transportation in the food and beverage industries is a significant end-use contributor to climate change.[14] In the United States, for instance, transportation accounts for over 10 % of the total carbon emissions of the food chain, with produce traveling an average of 2000 km from farm to market.[15] Public and private interest alike in "greening" food transportation is mounting. For example, the United Kingdom (UK) is the host to an industry-led initiative to reduce the "food miles" associated with the country's food supply.[16] Reducing food miles means driving fewer miles, as well as using more efficient fuel, speed, and loading practices. Australia has codified some of these concepts in its Heavy Vehicle National Law (HVNL), which applies to the country's entire commercial transportation sector.[17] Although the HVNL is not specific to the food industry, it could lead to energy efficiency gains in Australia's food sector since

---

[10]Inorganic food waste, such as plastic or glass packaging, is often regulated separately from other waste streams under recycling laws and programs. See, for example, *infra* Sect. 23.4.3.

[11]See *infra* Sect. 23.5.2.

[12]See *infra* Sect. 23.4.1.

[13]See *infra* Sect. 23.3.1.4.

[14]Wakeland et al. (2012).

[15]Wakeland et al. (2012), pp. 201–202.

[16]The UK-based Food and Drink Federation (FDF) has "committed to embedding environmental standards into their transport practices . . . to achieve 'fewer and friendlier' food transport miles." See Food and Drink Federation (2014).

[17]See *infra* Sect. 23.7.

transportation is one of the key links in the food chain. Efforts to improve the lifecycle sustainability of the food sector can yield substantial environmental, economic, and social co-benefits. However, few countries are actively regulating in this arena.

This chapter begins with an overview of the (somewhat fragmentary and rudimentary) international law landscape on food waste and food transportation efficiency issues. The following sections then take the reader on a tour of regional- and national-level approaches taken by countries throughout the world. Given the scope and complexity of the field, this chapter is not a comprehensive catalog of regulations governing food chain management. Rather, it is a survey of some of the most prominent and innovative waste, recycling, and transportation efficiency laws and policies as they relate to the food sector. To supplement this chapter, the Appendix covers a wider array of regulations beyond those discussed below.

23.2 International Treaties and Conventions

The United Nations Food and Agriculture Organization (UN FAO or FAO) estimates that 1.3 billion tons of food—or 1/3 of the world's total food supply—is thrown away each year.[18] This staggering figure places a significant and unnecessary strain on the world's agricultural production sectors. However, international law does not yet play a strong, direct role in governing food waste or food chain efficiency. Indeed, there is no binding global mechanisms targeting these issues. Nevertheless, some international law instruments can indirectly affect how nations and regions structure their waste and food chain management policies. This section highlights several such examples.

23.2.1 Agreement on Technical Barriers to Trade (TBT Agreement) (1995) and Agreement on the Application of Sanitary and Phytosanitary Measures (SPS Agreement) (1995)

The Agreement on Technical Barriers to Trade (TBT Agreement)[19] and the Agreement on the Application of Sanitary and Phytosanitary Measures (SPS Agreement)[20] are related international trade instruments. The World Trade Organization (WTO) has administered both agreements since 1995. Through

[18]United Nations Food and Agriculture Organization (2013).

[19]World Trade Organization (1995a).

[20]World Trade Organization (1995b).

these instruments, the WTO seeks to balance two competing concerns: (a) the right of individual countries to regulate as they see fit and (b) the need for efficient, transparent, and fair global trading markets. As of June 2014, there are 160 WTO Member States, all of whom are bound by the terms of the agreements.[21] The treaties have important implications for the food and agriculture industries, especially for importers and exporters of food products.

The TBT Agreement has a broad, general scope. It prevents individual WTO Member States from setting technical regulations, standards, and other requirements that create "unnecessary obstacles" to trade.[22] The SPS Agreement sets similar constraints on Member States' food safety, animal, and plant health measures.[23] Both treaties can impact national laws on food waste and food chain management. Often, the agreements' restrictions lead to laws or policies that are less environmentally protective than they might otherwise be.

For example, imagine that a country is considering enacting new regulations that would set a minimum recycled content requirement for packaging on all food products sold in the country. Any foreign (exporting) company would have to comply with these packaging requirements in order to sell its products within the (importing) country.[24] If the new requirements are stricter than what the exporting company is used to (or can find in other markets), meeting the thresholds would entail higher costs. Such expenses might dissuade the company from entering (or staying in) the foreign market. This reticence would present a potential "unnecessary obstacle to trade." Therefore, the original country might not be able to enact environmentally-friendly packaging requirements if such requirements are too different (i.e., too much more stringent) than those in other countries. As a result, the TBT and SPS Agreements are unlikely to be strong tools for promoting food chain sustainability. In fact, governments and advocates alike should be aware of the potential for these international trade agreements to block regulatory innovations.

23.2.2 Kyoto Protocol (1997)

In 1997, the third Conference of the Parties (COP 3) to the United Nations Framework Convention on Climate Change (UNFCCC) adopted the Kyoto Protocol.[25] The treaty commits its Parties to internationally binding greenhouse gas

---

[21]World Trade Organization (2015c).

[22]World Trade Organization (2015b). At the same time, the agreement protects countries' right to implement measures that achieve legitimate policy measures (such as protecting human or environmental health).

[23]World Trade Organization (2015a).

[24]See State of Oregon, Department of Environmental Quality (2005).

[25]United Nations Framework Convention on Climate Change (1998).

(GHG) emission reduction targets, but it does not mandate *how* countries are to achieve their reductions. As of 2014, the Kyoto Protocol had 192 Parties (191 States and the EU).[26] (Notably, the United States has never adopted the protocol). The protocol entered into force in 2005. The first commitment period began in 2008 and ended in 2012, and a second commitment period runs from January 1, 2013, to December 31, 2023.[27]

Since the Kyoto Protocol offers Parties some flexibility in how they meet their targets, countries could explore reforming their agriculture and food sectors. For instance, one way for a nation to satisfy (at least some of) its Kyoto commitments could be through improving its food waste management regime. Currently, food waste in most countries ends up in landfills, where the biodegrading process is a leading source of methane emissions worldwide.[28] Therefore, reducing food waste or re-directing waste away from landfills (to be composted, donated, recycled into animal feed, or used as fuel) can achieve significant GHG reductions. Countries could also work to meet their Kyoto commitments by lowering the amount of petroleum-based plastics in the food supply chain or shrinking the motor vehicle emissions of commercial fleets used for food transportation. All in all, there are a number of ways the Kyoto Protocol can encourage Parties to leverage the environmental protection and clean energy innovation potential of their food and agriculture sectors.

23.3 The Americas

23.3.1 United States (US)

Food waste is estimated to cost the US $165 billion each year.[29] In 2011, the federal Environmental Protection Agency (EPA) found that organic food waste is the second-largest component of the country's municipal solid waste (MSW) (representing over 14 % of MSW by weight)[30] and an important source of human-related methane (nearly 20 % of US emissions).[31] However, the federal government has not enacted any major, targeted legislation on food waste management. Instead, most action is happening at a local level. Indeed, while the national Resource Conservation and Recovery Act (RCRA)[32] empowers EPA to regulate waste disposal, the agency has, for the most part, delegated MSW management to

[26]United Nations Framework Convention on Climate Change (2014).

[27]United Nations Framework Convention on Climate Change (2014).

[28]United States Environmental Protection Agency (2015).

[29]Buzby and Human (2012), pp. 561–570.

[30]United States Environmental Protection Agency (2014b).

[31]United States Environmental Protection Agency (2014a).

[32]See *infra* Sect. 23.3.1.1.

state and municipal jurisdictions. EPA assists these efforts by issuing reports on MSW management, including recommendations on landfill siting and "preferred options" for waste disposal.[33] This approach ends up leaving local bodies with a great deal of discretion—something several states and cities have taken advantage of to enact innovative composting and recycling laws.[34]

Likewise, there is no comprehensive federal legislation on food packaging production, use, or disposal. Most of this regulation is, once again, left to state and local officials.[35] However, the federal government has stepped in to cover three areas:

(1) Regulating packaging for food, drugs, and cosmetics (mostly carried out by the Food and Drug Administration (FDA) under the Federal Food, Drug, and Cosmetic Act);
(2) Promoting government procurement of recycled products (mostly carried out by the EPA under RCRA); and
(3) Issuing guidelines to prevent manufacturers from making unfounded claims about the environmental benefits of their packaging (mostly carried out by the Federal Trade Commission (FTC) under federal truth-in-advertising laws).[36]

The US federal government is also largely absent from the food transportation efficiency arena. This area could be ripe for new regulatory action given the significant level of GHG emissions associated with food transportation in the US.[37] Most of the country's relevant initiatives, such as "local preference" laws,[38] are happening at a sub-national level.[39] However, the federal United States Department of Agriculture (USDA), through a partnership with the Department of Defense (DoD), recently updated the National School Lunch Program to encourage local food purchasing at public K-12 schools receiving federal dollars.[40] Buying local not only helps to reduce "food miles" (and associated transportation emissions) but may also reduce food waste by reducing transit time and, hence, spoilage. Overall, while the federal government provides some guidance and leadership on food chain management issues, it has historically been reluctant to regulate directly in this area. This tendency leaves the door wide open for local- and state-based action.

---

[33]United States Environmental Protection Agency, Office of Resource Conservation and Recovery Program Management, Communications, and Analysis Office (2011).

[34]See *infra* Sect. 23.3.1.4.

[35]See *infra* Sect. 23.3.1.

[36]Keller and Heckman LLP (2002).

[37]Aggregate transportation accounts for approximately 11 % of total carbon emissions in the US. Wakeland et al. (2012), p. 202.

[38]See *infra* Sect. 23.3.1.4.

[39]See *infra* Sect. 23.3.1.5.

[40]United States Department of Agriculture, Food and Nutrition Service (2013).

23.3.1.1 Resource Conservation and Recovery Act (RCRA) (1976)

The Resource Conservation and Recovery Act (RCRA)[41] is the primary federal law governing waste management activities. It also provides EPA with a broad mandate to promote recycling.[42] The statute's objectives include protecting human health and the environment, conserving energy and natural resources, decreasing waste levels, and managing wastes in an environmentally sound manner.[43] RCRA's scope is broad, covering both solid and hazardous waste. "Solid waste" encompasses "any garbage [or] refuse" in solid, liquid, or gaseous form.[44] "Municipal solid waste" is a subset of solid wastes, defined to include food waste, containers and packaging, and miscellaneous organic wastes from residential, commercial, and industrial sources.[45]

EPA has largely delegated its implementation authority over solid wastes to states and local governments. The vast majority (46 out of 50) of states implement their own RCRA programs.[46] EPA does, however, mandate certain aspects of the design and operation of disposal facilities. The agency also periodically "provid [es] tools and information through policy and guidance to empower local governments, business, industry, federal agencies, and individuals to make better decisions in dealing with solid waste issues."[47] For instance, EPA advocates "integrated waste management" systems.[48] These systems involve the complementary use of a variety of MSW management practices, including:

- Source reduction;
- Recycling (including composting);
- Waste combustion for energy recovery; and
- Disposal by landfilling.[49]

---

[41] 42 U.S.C. §§ 6901–6992k. See Cornell University Law School. U.S. code, tit 42, chap 82.

[42] Section 6002 requires EPA to develop guidelines for government agencies to use in procuring products containing recycled material. EPA issued such guidelines in 1995 ("Comprehensive Procurement Guidelines"), followed by a guidance document ("Final Guidance on Environmentally Preferable Purchasing for Executive Agencies") in 1999. Keller and Heckman LLP (2002).

[43] 42 U.S.C. § 6902. See Cornell University Law School. U.S. code, tit 42, chap 82, subchap 1 § 6902.

[44] 42 U.S.C. § 6903(27). See Cornell University Law School. U.S. code, tit 42, chap 82, subchap 1 § 6903.

[45] United States Environmental Protection Agency, Office of Resource Conservation and Recovery Program Management, Communications, and Analysis Office (2011).

[46] United States Environmental Protection Agency (2013a).

[47] United States Environmental Protection Agency, Office of Resource Conservation and Recovery Program Management, Communications, and Analysis Office (2011), p. II-2.

[48] United States Environmental Protection Agency, Office of Resource Conservation and Recovery Program Management, Communications, and Analysis Office (2011), p. II-3.

[49] United States Environmental Protection Agency, Office of Resource Conservation and Recovery Program Management, Communications, and Analysis Office (2011), p. II-3.

Within this system, waste prevention techniques (e.g., source reduction and recycling) are the highest priorities while landfilling is a last resort.[50] However, these hierarchical preferences are merely recommended—not mandated—by the federal government. In fact, landfilling remains the most common MSW management method in the United States (with over half of the country's MSW landfilled in 2009).[51] Nevertheless, these EPA recommendations are a potential tool for advocates who wish to lobby city or municipal governments to adopt more sustainable waste management practices.

23.3.1.2 National Environmental Policy Act (NEPA) (1969)

Under the National Environmental Policy Act (NEPA),[52] federal agencies must consider the environmental impacts of proposed actions as part of their decision-making process.[53] Agencies must first prepare an environmental assessment (EA). If the EA reveals that the action will have significant environmental impacts, the agency must then perform a full environmental impact statement (EIS) before proceeding.[54] These requirements apply to any agency undertaking a "major federal action," a category that includes the adoption of official policies, plans, and programs.[55]

NEPA's scope is broad and will, therefore, encompass many "major federal actions" in the food waste and food packaging sectors. One prominent example is FDA approvals of food packaging materials under the Federal Food, Drug, and Cosmetic Act (FFDCA).[56] These NEPA reviews can consider such issues as a material's recyclability, impact on incinerator emissions and ash, and potential to contribute to landfill leachate, acid rain, and stratospheric ozone depletion. However, FDA has exempted some food-contact clearances from the environmental assessment requirement. Furthermore, "food packaging clearances almost always require only an environmental assessment (EA), which leads to a [finding of no significant impact (FONSI)] and granting of the clearance by FDA without an

---

[50]The guidelines recognize that strategies which divert waste "have positive impacts on both the environment and economy." United States Environmental Protection Agency, Office of Resource Conservation and Recovery Program Management, Communications, and Analysis Office (2011), p. II-3.

[51]United States Environmental Protection Agency, Office of Resource Conservation and Recovery Program Management, Communications, and Analysis Office (2011), p. II-5.

[52]42 U.S.C. § 4320 et seq. See Cornell University Law School. U.S. code, tit 42, chap 55.

[53]42 U.S.C. § 4332(A). See Cornell University Law School. U.S. code, tit 42, chap 55, subchap 1 § 4332.

[54]See *Robertson v. Methow Valley Citizens Council*, 490 U.S. 332, 336 (1989). Available via Justia (2015).

[55]40 C.F.R. § 1508.18. See Cornell University Law School. 40 C.F.R. tit 40, chap V, part 1508, § 1508.18.

[56]Keller and Heckman LLP (2002).

EIS."[57] Therefore, while NEPA has the potential to prompt agencies to evaluate the sustainability characteristics of packaging materials in theory, the statute's ability to guarantee such consideration in practice is limited.

23.3.1.3 Bill Emerson Good Samaritan Food Donation Act (1996)

The US Congress passed the Federal Bill Emerson Good Samaritan Food Donation Act[58] to encourage food donations and cut down on food waste. The act removes a significant obstacle to food donations by protecting food donors (both individuals and organizations) from civil and criminal liability.[59] This protection from liability applies to all "good faith" donations, with exceptions for gross negligence and intentional misconduct.[60] In addition, the act provides for federal tax credits for those who make charitable donations of food.

Every state has also passed its own Good Samaritan Law. The state laws tend to follow the federal statute, and some even go beyond it. For instance, Massachusetts gives additional liability protection to donors distributing food either for free or "at cost" (as long as the food meets state health department regulations).[61] Oregon directly incentivizes food donations by giving additional tax credits to corporations or individuals who donate crops to certain types of nonprofit organizations, such as gleaning cooperatives or food banks.[62] Other states, including North Carolina, Colorado, and Arizona, have similar programs. The federal act further provides for uniform protection for donations that cross state lines. Therefore, the United States has a nationwide system in place to encourage donations of edible food supplies that would otherwise go to waste.

23.3.1.4 State and Local-Level Composting, Recycling, and Packaging Regulations and Programs

Many US states and cities have taken the reigns from the federal government and emerged as leaders on food waste prevention and recycling. For instance, composting initiatives are starting to gain traction throughout the country. More than 150 cities offer curbside municipal compost pick-up.[63] Most of these programs are voluntary. However, Vermont's ambitious Universal Recycling (UR) Law

[57]Keller and Heckman LLP (2002).

[58]42 U.S.C. § 1791. See Cornell University Law School. U.S. code, tit 42, chap 13A.

[59]42 U.S.C. § 1791(c). See Cornell University Law School. U.S. code, tit 42, chap 13A.

[60]42 U.S.C. § 1791(c)(3). See Cornell University Law School. U.S. code, tit 42, chap 13A.

[61]Leib (2012) (citing 9 Mass. Gen. Laws, Chap. 94, § 328 (2011)).

[62]Leib (2012) (citing Or. Rev. Stat., § 315.156 (2012)).

[63]Hendrix (2013).

seeks to ban the landfilling all food scraps by July 1, 2023.[64] Additionally, in 2009, San Francisco, California, became the first US city to require residents to discard food waste in a separate bin. The city's Mandatory Recycling and Composting Ordinance aims to have "virtually zero" landfill waste by 2023.[65] Seattle, Washington, followed suit in 2010 with a program that imposes a fine for failing to compost food waste, and a 2014 ordinance prohibits food from Seattle's residential and commercial garbage after January 2015.[66] Other US cities (such as Austin, Texas) have announced intentions for similar programs.[67]

Similar progress is being made with food packaging management on a state-level basis. Vermont's UR Law bans the disposal of recyclable materials (metal, glass, certain plastics, and paper/cardboard) in the state by July 1, 2015. Bottle deposit laws, which encourage the recycling of plastic and/or glass beverage containers, are found in approximately 11 states and territories.[68] Other states have enacted standards for environmentally acceptable packaging. For instance, California, Oregon, and Washington have minimum recycled content requirements for plastic containers (with California and Oregon passing similar requirements for glass containers as well). California even has such requirements for plastic trash bags. Additionally, many states restrict the use of certain substances in packaging or ban particular types of plastic packaging altogether.[69] Once more, this is a field where state and local governments are taking advantage of the broad scope of MSW management authority delegated to them by the federal government.

23.3.1.5 State and Local-Level "Local Preference" Laws and Policies

At least 37 states have some form of a "local preference" law or policy for food procurement by state governmental or state-funded entities.[70] These are similar in function to the federal National School Lunch Program reform, discussed above.[71] Local preference regulations take two main forms:

(1) Directing entities to choose local food products if the local food is within a certain price range (e.g., not more than 10 % more expensive) of out-of-state food; or

[64]State of Vermont, Waste Management & Prevention Division (2015).

[65]State of Vermont, Waste Management & Prevention Division (2015).

[66]City of Seattle Public Utilities and Councilmember Bagshaw S (2014).

[67]See City of Austin, Resource Recovery. Frequently asked questions.

[68]States and territories with bottle bills include California, Connecticut, Guam, Hawaii, Iowa, Maine, Massachusetts, Michigan, New York, Oregon, and Vermont. Bottlebill.org (2015).

[69]Keller and Heckman LLP (2002).

[70]SourceSuite (2015).

[71]See *supra* Sect. 23.3.1.

(2) Setting targets for the amount of food (e.g., 20 %) that will be purchased from local producers. Many of these laws can be extended to encourage regional purchases.[72]

Similar initiatives are also often seen at the city or municipal level. Local food policy councils, for instance, commonly work to encourage local food production and purchases. Food policy councils can ensure that zoning codes allow for food processing to occur in or near urban areas. They may also ask local and state governments for funds to support food processing facilities (including mobile ones). For example, in 2008, the Vermont legislature granted money to the State Department of Agriculture for two mobile-processing facilities (one for slaughtering poultry and one for flash-freezing produce).[73] These facilities give local growers the services they need to get their foods to market. Local food purchases contribute to food chain sustainability by helping to cut down on the environmental impacts associated with trucking, flying, or shipping food over long distances. Therefore, local preference laws and policies are one area where relatively small-scale reforms by local advocates can further the environmental protection and clean energy potential of the food and agriculture sectors.

23.3.2 Canada

Food waste management is a serious and growing challenge in Canada. A 2014 study estimated the value of wasted food in Canada at $31 billion a year (up from $27.7 billion in 2010), or 30–40 % of all food produced.[74] Overall, the country's regulatory regime has been criticized for making it "too easy and too cheap to dump [organic waste], and too difficult to do otherwise."[75] For instance, unlike the United States,[76] Canada does not have a federal "Good Samaritan" law to encourage donations of still-usable food waste.[77] Additionally, waste-to-energy projects face permit and regulatory hurdles that make projects like anaerobic digesters, which have seen some success in the EU, economically infeasible. Canada also "trails Europe in terms of encouraging the changes in behaviour required to increase businesses' profitability by reducing waste. The agri-food sector trails other

---

[72]Harvard Food Law and Policy Clinic (2013).

[73]State of Vermont (2015).

[74]Value Chain Management Center (2014).

[75]Value Chain Management Center (2010).

[76]See *supra* Sect. 23.3.1.

[77]However, the Ontario legislature also made history in 2013 by becoming the first provincial government to pass a bill promising tax credits to farmers who donate surplus food. "The Local Food Act," an amendment to Bill 36, will give farmers in Ontario a non-refundable 25 % tax credit based on the fair market value of product that they donate to local food banks and community meal programs. Ontario Association of Food Banks (2014).

industries in seeking to increase its competitive advantage through adopting value chain management practices."[78] These shortcomings mean that there are fewer incentives to reduce food waste or increase food chain efficiency in Canada compared to other developed nations.

At the national level, Canada does not yet have a formal regulatory framework for addressing food waste or supply chain efficiency. Waste management is primarily carried out at the provincial level, with most provinces having their own Environmental Protection, Waste Management and/or Waste Diversion Acts.[79] Municipal and/or local governments also frequently regulate waste management and recycling activities through their bylaws.[80] In recent years, the Canadian Council of Ministers of the Environments (CCME), a group made up of 14 environment ministers from federal, provincial, and territorial governments, has started working on several important national initiatives related to food waste.[81] These include efforts to encourage composting[82] and implement sustainable packaging protocols that embrace the "polluter pays"/Extended Producer Responsibility (EPR) principle.[83] Therefore, escalating concerns about food waste and efficiency appear to be leading to new opportunities for progressive regulatory action among all levels of government in Canada.

23.3.2.1 Canadian Environmental Protection Act (CEPA) (1999)

The Canadian Environmental Protection Act (CEPA)[84] is a federal law that aims to protect environmental and human health through pollution prevention. It addresses any pollution issues not covered by other federal laws and contains general waste management and prevention provisions. The act adopts the "polluter pays"/EPR principle.[85] Under a "polluter pays" waste management regime, product manufacturers are responsible for the costs associated with end-of-life treatment of their products. This contrasts with the US's "consumer pays" system, in which taxpayers fund waste management activities.[86] Overall, CEPA does not address food waste or food packaging regulation specifically. Therefore, in its current form (and absent new CCME reforms), the federal statute has limited potential as a tool for boosting

---

[78]Value Chain Management Center (2010).

[79]See Appendix.

[80]Recycling Council of Ontario (2010).

[81]Canadian Council of Ministers of the Environment (2014).

[82]Some provinces are already leading the way. For instance, Nova Scotia has banned organic food waste from its landfills since 1998. See Province of Nova Scotia (2014). Meanwhile, Ontario is stepping up its composting program. See Recycling Council of Ontario (2010).

[83]Canadian Council of Ministers of the Environment (2014). See *infra* Sect. 23.3.2.1.

[84]Government of Canada, Minister of Justice (1999).

[85]Government of Canada, Minister of Justice (1999), Preamble.

[86]Earth911 (2014).

the country's food chain sustainability. Instead, provincial waste management acts are likely the best potential tools for change.

23.3.3 *Mexico*

Historically, Mexico lacked federal food waste or food chain efficiency legislation. However, the national government began to tackle these issues in recent years. In late 2003, the Mexican Congress passed the Federal Waste Law, a broad statute covering waste management as well as packaging and recycling activities. Outside of government action, the Mexican packaging industry has been progressively adopting global packaging trends in order to tap into a greater number of export markets.[87] Therefore, food waste and efficiency are burgeoning issues within both the private and public sectors of the country. Unfortunately, however, data on food waste in Mexico remains scarce.

23.3.3.1 Federal Waste Law (2003)

Mexico's Federal Waste Law regulates waste management and prevention, covering both organic and inorganic wastes.[88] The law also provides the country's legal framework on packaging and recycling activities. When passed, it was recognized as "one of the most progressive and comprehensive solid and hazardous waste laws in the Americas."[89] However, the law also has numerous gaps limiting its effectiveness. For instance, while the Federal Waste Law states that solid waste should be recycled, it does not set any specific obligations (such as requiring packaging companies to recycle their products). Furthermore, the law does not provide details on producer responsibility obligations (such as provisions implementing "polluter pays" or EPR principles). While the Federal Waste Law is promising on paper, many of its provisions will need to be fleshed out in future regulations or implementing standards (known as *Normas Oficiales Mexicanas* (NOMs)) before the law can take its full effect.

---

[87] Agriculture and Agri-Food Canada (2012).

[88] Congreso General De Los Estados Unidos Mexicanos (2003). In 2006, the Secretariat of Environment and Natural Resources (SEMARNAT) issued implementing regulations. See Presidente Constitucional de los Estados Unidos Mexicanos (2006).

[89] HIS Environment Intelligence Analysis (2015).

23.3.3.2 Solid Waste Law of the Federal District (2003)

Mexico City produces more than 12,000 tonnes of solid waste each day.[90] The Federal District passed its own Solid Waste Law in 2003,[91] followed by an amendment in 2004.[92] The amendment requires households to separate their waste into organic and inorganic containers for composting and recycling purposes. It also calls for "measures and mechanisms to organize the structure and install the necessary infrastructure to enforce these provisions," including "a massive campaign to . . . civically educate the population regarding the benefits of compliance."[93] Therefore, the regulation recognizes the important role that public awareness and buy-in plays in getting food waste management reforms off the ground.

As with the Federal Waste Law, Mexico City's Solid Waste Law has ambitious and progressive goals. Unfortunately, the law also has serious practical limitations. Compliance is seldom enforced, and illegal dumping has become a major problem since the city closed its primary landfill in late 2011.[94] These problems are compounded by the federal government's decision to suspend building *Centros Integrales de Reciclado y Energía* (Integral Waste-to-Energy Plants or CIREs).[95] CIREs make use of organic waste to produce compost, recycle inorganic materials, and generate electricity. However, after opposition from residents, the plants' construction has been indefinitely deferred since 2009. All in all, both the national and federal district governments in Mexico are embracing ambitious food chain management reforms. The framework for regulatory innovation is in place. However, the new laws are still experiencing growing pains and are not yet being implemented fully and effectively.

23.3.4 Brazil

Like Mexico,[96] Brazil has undergone a recent surge of regulatory development on issues related to food waste and efficiency. Prior to 2010, Brazil had no national framework law on solid waste management or recycling. Instead, many of Brazil's 26 states and its federal district—along with major cities like São Paulo and Curitiba—acted individually to fill the gap. Some of these initiatives led to improvements in recycling rates and practices, which, in turn, paved the way for

---

[90]Guardian News and Media (2012).

[91]Asamblea Legislativa del Distrito Federal (2003).

[92]Asamblea Legislativa del Distrito Federal (2004).

[93]Asamblea Legislativa del Distrito Federal (2004).

[94]Guardian News and Media (2012).

[95]Guardian News and Media (2012).

[96]See *supra* Sect. 23.3.3.

the National Policy of Solid Waste (PNRS) in 2010.[97] Despite the many important innovations of the PNRS, at least one gap remains in Brazil's solid waste policy: food waste prevention. For instance, there is no federal "Good Samaritan" law similar to the one in the United States.[98] As a result, most individuals and companies are uncomfortable with food donation.[99] However, the extent of the problem is unclear since comprehensive data on food waste in Brazil is scarce.

23.3.4.1 National Policy of Solid Waste (PNRS) (2010)

After more than 20 years of debate, the National Congress of Brazil passed the National Policy of Solid Waste (PNRS)[100] in 2010. The PNRS is a cross-cutting law that aims to decrease the total volume of waste produced at a national level and increase the sustainability of solid waste management from the local level to the national level. It calls for integrated, multi-level management plans and directs states, regions, and municipalities to elaborate local strategies for implementing the national strategy.[101] The policy's scope is wide, covering a variety of public and private actors in many sectors. It embraces a number of progressive principles, including:

- "Polluter Pays,"[102] which puts much of the responsibility for paying for or providing waste management on producers;
- "Shared responsibility for the lifecycle of products," which places responsibility on consumers and government -- in addition to commercial enterprises -- for managing wastes; and
- "Reverse Logistics System (RLS)," which encourages the collection and return of solid waste to industry for reuse (or for other environmentally appropriate disposal).[103]

To put these principles into action, the PNRS "outlines a variety of options for producers to work together within their sectors, with [RLS] providers, and with municipal and state governments to manage waste flows and to recapture, recycle, and ultimately dispose of these materials."[104] In particular, the policy encourages

[97]See *infra* Sect. 23.3.4.1.

[98]See *supra* Sect. 23.3.1.3.

[99]There are, however, some limited city-level initiatives to encourage such practices, such as Curitiba's Cambio Verde (Green Change) project. This program allows farmers to provide surplus produce to people that bring glass and metal to recycling facilities. See Portal da Prefeitura de Curitiba. Secretario municipal do abastecimiento: programa cambio verde.

[100]Pereira (2010). A second edition of the PNRS was issued for 2011–2015. See Câmara dos Deputados (2012).

[101]Pereira (2010).

[102]See *supra* Sect. 23.3.2.

[103]Beveridge & Diamond PC (2010).

[104]US-Brazil Joint Initiative on Urban Sustainability (2012).

increased use of organic waste digesters, composting, and organic packaging. It also outlines grant and assistance programs to local governments to help improve recycling and training. Similar to the system in the United States,[105] domestic MSW management remains a predominantly municipal responsibility under the PNRS. Brazilian municipalities follow general directives laid out in the federal law while retaining discretion over many of the particularities of implementation.[106] Some jurisdictions are taking advantage of this flexibility to spearhead progressive initiatives. Rio de Janeiro, for example, has installed a methane capture waste-to-energy system in the city's major landfill.[107] This is a key illustration of how food waste management and clean energy objectives can intersect.

On top of organic waste measures, the PNRS imposes a number of broad general requirements on packaging.[108] For instance, packaging must be restricted to no more than the volume and weight required to protect and sell the product. It must also be designed for reuse (to the extent feasible). Finally, packaging must be recycled in the event that reuse is not possible. While these provisions are not unique to the food sector, they do impose heightened restrictions on food product packagers.

As a whole, Brazil's PNRS showcases how actions taken at one stage in the food chain cycle (such as using less packaging early in the cycle) can impact those at other stages (resulting in less non-biodegradable landfill waste to deal with later in the cycle). Likewise, waste-to-energy plants at landfill sites (at the end of the cycle) could reduce the amount of agricultural crops that need to be grown for fuel (at the beginning of the cycle). In theory, then, the PNRS is a strong and innovative policy with a lot of potential to enhance Brazil's food chain sustainability. It is likely still too early to tell how effective the PNRS will be in practice.

23.4 European Union (EU)

The European Union (EU) is a global leader in food waste management. Not only are the region's regulations among the strongest and most targeted in the world, but both public and private commitment to combating food waste is high.[109] Indeed, the EU was home to the world's first major research program on the issue.[110] In August

[105]See *supra* Sect. 23.3.1.

[106]US-Brazil Joint Initiative on Urban Sustainability (2012).

[107]US-Brazil Joint Initiative on Urban Sustainability (2012).

[108]Pereira (2010), Art. 23.

[109]The European Parliament designated 2014 as the "European Year Against Food Waste." European Parliament (2013).

[110]In 2005, the UK government-funded charity Waste & Resources Action Program (WRAP) launched the world's first major research program on food waste. This led to a groundbreaking report in 2008 quantifying the types and amounts of food and drink waste produced by households in England and Wales. See Waste & Resources Action Program (2008). The report's "shocking"

2012, the European Commission launched the FUSIONS (Food Use for Social Innovation by Optimizing Waste Prevention Strategies) campaign to "work [] towards achieving a more resource efficient Europe by significantly reducing food waste."[111] The program's goals include developing a Common Food Waste Policy for the EU and cutting food waste in half by 2023. As of 2013, over 40 % of food waste in the EU was being composted or recycled.[112] The EU also leads the world in using anaerobic digestion.[113] These successes attest to the region's commitment to improving food chain sustainability, and a variety of legal tools are available to keep furthering these goals. At the EU level, a number of directives (including the Waste Framework Directive,[114] the Landfill Directive,[115] and the Packaging and Packaging Waste Directive[116]) directly and indirectly govern the management of both organic and inorganic food waste. In addition, private actions by businesses, schools, NGOs, consumers, and other societal players continue to play a crucial role in carrying out and strengthening governmental measures.

23.4.1 Waste Framework Directive (2008)

The EU's Waste Framework Directive[117] establishes basic waste management principles for EU Member States. It provides the general legislative framework for the collection, transport, recovery, and disposal of waste. It also includes a common definition of waste as "any substance or object which the holder discards or intends or is required to discard."[118] Unlike national-level laws in many other countries in the world,[119] the Waste Framework Directive regulates organic or "bio-waste" as an independent waste stream. The bio-waste category includes "food and kitchen waste from households, restaurants, caterers and retail premises and

findings revealed that 6.7 million tonnes of food—equal to roughly 1/3 of all food purchased—are thrown away each year, with a dollar value of over £10 billion. Packaging News (2015).

[111]European Union FUSIONS (2015).

[112]Levitan (2013). However, rates within individual Member States vary widely. Norway, Sweden, the Netherlands, Denmark, Switzerland, Belgium, Austria, and Germany divert over 97 % of their food waste from landfills (and Copenhagen, Denmark, has banned organic waste from its landfills since 1990). Greece, Bulgaria, Lithuania, and Romania lag behind, with still almost no composting in these countries.

[113]Levitan (2013).

[114]See *infra* Sect. 23.4.1.

[115]See *infra* Sect. 23.4.2.

[116]See *infra* Sect. 23.4.3.

[117]EUR-Lex (2008).

[118]EUR-Lex (2008), Art. 3(1).

[119]See Appendix.

comparable waste from food processing plants."[120] (Inorganic plastic wastes are addressed as a separate waste stream and are the subject of their own directive).[121]

The Waste Framework Directive's "first objective" is to protect environmental and human health from the adverse effects of waste generation and treatment.[122] The directive establishes a waste management hierarchy that emphasizes waste prevention above all other management strategies. Member States must act, first, to prevent or reduce waste. Next, they should attempt to recover waste that *is* generated through reuse, recycling (or composting, for bio-wastes), or recovery (including reclamation or conversion into energy).[123] Disposal is a last resort.[124] This is similar to the US EPA's approach under RCRA.[125] However, the Waste Framework Directive goes even further than the EPA's recommendations. Article 20 encourages Member States to collect food waste separately from other wastes, "with a view to the composting and digestion of bio-waste."[126] The directive also provides a mechanism for the European Commission to "set[] minimum requirements for bio-waste management and quality criteria for compost and digestate from bio-waste."[127] Therefore, the EU Waste Framework Directive calls for separate handling and treatment of organic and inorganic wastes—something that has caught on in only a few places in the US.[128]

A number of waste stream-specific directives supplement the Waste Framework Directive's general requirements. While the Packaging and Packaging Waste Directive[129] covers many types of inorganic food waste, there is not yet an EU-level directive on bio-waste. However, the European Commission's 2009 Guidelines on Waste Prevention Programmes,[130] along with the 2011 Guidelines on the Preparation of Food Waste Prevention Programmes,[131] lay out an EU policy framework for organic food waste management by Member States. The guidelines recognize bio-waste's "impacts on the environment, greenhouse gas emissions, and global food security"[132] and make food waste prevention a major priority. They propose general measures on waste measurement, target setting, and prevention

[120]EUR-Lex (2008), Art. 3(4).

[121]See *infra* Sect. 23.4.3.

[122]EUR-Lex (2008), Preamble (6).

[123]EUR-Lex (2008), Art. 4(1).

[124]EUR-Lex (2008), Art. 4(1).

[125]See *supra* Sect. 23.3.1.1.

[126]EUR-Lex (2008), Art. 20(a).

[127]EUR-Lex (2008), Art. 23.

[128]See *supra* Sect. 23.3.1.4.

[129]See *infra* Sect. 23.4.3.

[130]European Commission, Directorate-General Environment (2009).

[131]European Commission, Directorate-General Environment (2011). An additional 2012 guidance document supplements the guidelines. See European Commission, Directorate-General Environment (2012).

[132]European Commission, Directorate-General Environment (2012), p. 4.

strategies,[133] along with a series of specific "best practices" recommendations, such as clarifying "best before" and "use by" labels, recycling leavening agents from day-old bread, and promoting "road shows" to reduce waste in supermarkets.[134] These guidance documents promise to be a useful tool, not only for Member State governments, but also for waste managers, businesses, local authorities, agencies, and households looking to decrease the amount of food that needlessly goes to waste.

23.4.2 Landfill Directive (1999)

Enacted in 1999, the Landfill Directive[135] was a "milestone" in EU waste policy.[136] It marked "a decisive shift from landfill[ing] towards the EU's new waste hierarchy, which prioritizes waste prevention."[137] Its basic framework is to set short- and long-term targets, while granting Member States flexibility to develop their own policies and approaches toward meeting those targets. The directive's aims are broad:

> to provide for measures, procedures and guidance to prevent or reduce as far as possible negative effects on the environment, in particular the pollution of surface water, groundwater, soil and air, and on the global environment, including the greenhouse effect, as well as any resulting risk to human health, from landfilling of waste, during the whole life-cycle of the landfill.[138]

This holistic, multi-faceted mission statement captures many of the interrelationships between waste disposal practices and environmental protection, climate change, and human health. It also implies that more sustainable end-of-lifecycle practices can enable greater sustainability throughout other stages of a production chain.

The Landfill Directive addresses some of the sustainability risks of organic food waste, in particular. For example, the Preamble highlights the role that landfilled bio-waste plays in exacerbating climate change (through emitting methane).[139] Therefore, the directive urges Member States to adopt measures—such as the separate collection, sorting, and recycling of biodegradable waste—to reduce the amount of such waste going to landfills.[140] The directive even goes so far as to set concrete and ambitious targets for bio-waste diversion. For example, Article

---

[133] European Commission, Directorate-General Environment (2012), p. 2.

[134] European Commission, Directorate-General Environment (2012), pp. 20–23.

[135] EUR-Lex (1999).

[136] European Environment Agency (2009), p. 7.

[137] European Environment Agency (2009), p. 7.

[138] EUR-Lex (1999), Art. 1(1).

[139] EUR-Lex (1999), Preamble, Part 16.

[140] EUR-Lex (1999), Preamble, Parts 16 and 17.

5 orders EU Member States to reduce the levels of biodegradable municipal waste sent to landfills to 35 % (by weight) of 1995 amounts by 2016.[141] In 2009, a decade after the Landfill Directive was enacted, the European Environment Agency (EEA) published a report on its effectiveness.[142] The findings appear to be positive and unambiguous: "the Landfill Directive has been effective, advancing the closure of landfills and increasing the use of alternative waste management options."[143] However, the directive does not prescribe specific treatment options for the diverted waste. While some Member States opt for methods like composting or bio-gas production, others tend to choose cheaper and less environmentally-conscious options such as incineration.[144] Therefore, the actual environmental protection and clean energy achievements of the Landfill Directive objectives are uneven across the EU.

23.4.3 Packaging and Packaging Waste Directive (1994)

The Packaging and Packaging Waste Directive[145] has twin goals of promoting environmental protection and maintaining a robust internal EU trading market.[146] Like the Landfill Directive,[147] it sets concrete targets (both general and tailored for particular waste streams), while letting individual Member States decide for themselves how they want to meet them. The directive covers all packaging placed on the market, regardless of material.[148] It also embraces the "polluter pays" principle, meaning that those who place products on the market are ultimately responsible for their end-of-lifecycle treatment.[149] This incentivizes manufacturers to minimize the amount of packaging used at the outset of a product's lifecycle.

As with other EU waste directives, the Packaging and Packaging Waste Directive embodies the waste management hierarchy. It calls for the prevention of packaging waste, first and foremost, followed by re-use, recovery, or recycling—with disposal

[141]EUR-Lex (1999), Art. 5(2)(c).

[142]European Environment Agency (2009). The report looks at waste management in five countries and one sub-national region (Estonia, Finland, Germany, Hungary, Italy, and the Flemish Region of Belgium). It also includes an econometric analysis of the EU-25 Member States.

[143]European Environment Agency (2009), p. 7.

[144]European Commission, Directorate-General Environment (2015).

[145]EUR-Lex (1994).

[146]The directive "aims to harmonize national measures concerning the management of packaging and packaging waste in order, on the one hand, to prevent any impact thereof on the environment of all Member States as well as of third countries or to reduce such impact . . . and, on the other hand, to ensure the functioning of the internal market and to avoid obstacles to trade and distortion and restriction of competition within the Community." EUR-Lex (1994), Art. 1(1).

[147]See *supra* Sect. 23.4.1.

[148]EUR-Lex (1994), Art. 2(1).

[149]EUR-Lex (1994), Art. 15. See also *supra* Sect. 23.3.2.1.

as a last resort.[150] The directive mandates that all packaging meet certain "essential requirements."[151] For example, manufacturers are to:

- Limit "packaging volume and weight . . . to the minimum adequate amount to maintain the necessary level of safety, hygiene and acceptance for the packed product and for the consumer;"[152]
- Design packaging to be reusable or recoverable, when feasible[153]; and
- Minimize the presence of "noxious and other hazardous substances and materials" in packaging.[154]

There are no specific targets for food packaging waste. However, standard food packaging components are included in the targets set for other waste streams, such as for plastics (23.5 % by weight of packaging waste to be recycled by 2008) and glass, paper, and board (60 % to be recycled by 2008).[155] The overall effectiveness of the Packaging and Packaging Waste Directive objectives, and the magnitude of the plastic waste problem in particular, has been the subject of "many intense discussions and heated debates in Europe."[156] It is unclear whether existing legislation should be further adapted to deal with specific waste streams, like plastic packaging. In any regard, measures to promote greater recycling of packaging and packaging waste (no matter what the source material) have positive environmental, economic, and energy recovery implications. Therefore, the Packaging and Packaging Directive is an important tool for promoting food chain sustainability, even if its provisions are not as strong or targeted as they could be.

23.4.4 EU Member State Example: Germany

Among EU Member States, Germany is a leader in food waste management: the country sends less than 3 % of its food waste to landfills.[157] In 2012, the German Minister for Food and Agriculture (BMEL) launched the "Too Good for the Bin" initiative, setting a nationwide goal of cutting food waste in half by 2023.[158] The program not only tells German citizens steps they can take to conserve food, but also seeks to pave the way for new food waste regulations. Partners from "virtually every sector of society and industry"—including towns and municipalities, schools,

150 EUR-Lex (1994), Art. 1(2).
151 EUR-Lex (1994), Annex II.
152 EUR-Lex (1994), Annex II(1).
153 EUR-Lex (1994), Annex II(2).
154 EUR-Lex (1994), Annex II(3).
155 Europa (2011).
156 See European Commission, Directorate-General Environment (2013), p. 9.
157 Around 18 % of this recovered waste is composted. Levitan (2013).
158 German Federal Ministry of Food and Agriculture (2015).

hospitals, food banks, associations, retailers, restaurants, and churches—have launched projects under the initiative.[159] Such widespread engagement highlights the fact that all sectors of society can play a role in promoting food chain sustainability.

Germany is also a leader on recycling. Its recycling rate is approximately 70 % (88 % for packaging waste alone)—one of the highest in the world.[160] This achievement is in part due to the country's successful Green Dot (*Der Grüne Punkt*) program, an industry-initiated effort launched to help German companies meet requirements under the EU Packaging and Packaging Waste Directive[161] and the German Packaging Waste Ordinance.[162] The Green Dot system is a cooperative effort run by the non-profit organization Duales System Deutschland (DSD). Manufacturers join DSD by paying a fee, and this membership allows them to print the Green Dot recycling symbol on their packaging. Recycling companies, in turn, guarantee that they will accept any Green Dot-marked materials. It is now nearly impossible to market a product in Germany without the Green Dot symbol. The Green Dot program has expanded beyond Germany and, as of 2009, was being used in 25 countries by more than 130,000 companies.[163] The success of this program showcases the extent to which private, industry-led initiatives can enhance the effectiveness of public laws and regulations.

23.4.4.1 Closed Substance Cycle and Waste Management Act (1994)

The Closed Substance Cycle and Waste Management Act[164] forms the backbone of Germany's national waste management policy. The government passed it in the mid-1990s in response to concerns about the country's growing number of landfills. The act covers both inorganic and organic wastes, including nearly all forms of food waste.[165] It calls for three primary waste management strategies:

---

[159]German Federal Ministry of Food and Agriculture (2015).

[160]To compare, the US recycling rate in 2007 was about 33 % (and roughly 43 % for packaging waste). Earth911 (2014).

[161]See *supra* Sect. 23.4.3.

[162]See *infra* Sect. 23.4.4.2.

[163]Earth911 (2014).

[164]German Federal Government (2000).

[165]German Federal Government (2000), Art. 2(2). The act specifically covers "materials that are to be disposed of pursuant to the Animal Carcass Disposal Act (*Tierkörperbeseitigungsgesetz*) to the Meat Hygiene and Poultry Meat Hygiene Acts (*Fleischhygienegesetz; Geflügelfleischhygienegesetz*), to the Act on Foodstuffs and Commodities (*Lebensmittel- and Bedarfsgegenständegesetz*), to the Milk and Margarine Act (*Milch-* and *Margarinegesetz*), to the Epizootic Diseases Act (*Tierseuchengesetz*), to the Plant Protection Act (*Pflanzenschutzgesetz*) and pursuant to the statutory ordinances issued on the basis of these acts." German Federal Government (2000), Art. 2(2)(1).

(1) Waste avoidance;
(2) Waste recovery; and
(3) Environmentally compatible waste disposal.[166]

The regulation embodies the "polluter pays" principle and places "product responsibility" on "parties who develop, manufacture, process and treat, or sell products."[167] Overall, the Closed Substance Cycle and Management Act has been regarded as a success. According to the German Federal Statistical Office, the country reduced its total net waste amount by more than 37.7 million US tons between 1996 and 2007.[168] Net reductions in waste generated at the start of the "substance cycle" mean that less waste ends up in landfills at the end of the cycle. This carries inherent environmental, economic, and social benefits.

23.4.4.2 Packaging Waste Ordinance (1998)

Germany's Packaging Waste Ordinance[169] implements the EU Packaging and Packaging Waste Directive.[170] The ordinance covers waste associated with all packaging (regardless of material) placed on the market in Germany. In accordance with the "polluter pays"/EPR principle, manufacturers and distributors are responsible for the entire lifecycle (including end stages) of their packaging.[171] Therefore, businesses must ensure that their packaging is collected, sorted, and recycled or reused after consumers dispose of it.[172] (In practice, these services are mostly carried out by the DSD through the Green Dot program[173] to avoid putting individual collection and recycling duties on producers.)[174] The ordinance also requires retailers to install recycling bins for primary and secondary packaging in their stores and imposes mandatory beverage container deposits.[175]

---

[166]German Federal Government (2000), Art. 2(1).

[167]German Federal Government (2000), Art. 20(1). See also *supra* Sect. 23.3.2.1.

[168]Earth911 (2014). This figure includes solid, packaging, liquid, gaseous, hazardous, radioactive, and medical wastes.

[169]German Federal Government (1998). The ordinance is also called the "German Packaging and Waste Avoidance Law." The ordinance has been amended at least six times (most recently in 2014) to address such things as refining the definitions of key terms (like "transport packaging") and specifying the responsibilities of producers and vendors in the e-commerce sector. Interpack Processes and Packaging (2014).

[170]See *supra* Sect. 23.4.3.

[171]See *supra* Sect. 23.3.2.1.

[172]United States Environmental Protection Agency (2013b). Notably, the ordinance eliminates incineration as a waste disposal option. This encourages more environmentally-friendly disposal options, like composting.

[173]See *supra* Sect. 23.4.4.

[174]Neumayer (2000).

[175]United States Environmental Protection Agency (2013b).

Generally, the Packaging Waste Ordinance is considered an effective and innovative regulatory tool. However, some have criticized it for prioritizing recycling over waste avoidance.[176] In other words, the ordinance might take a certain level of waste generation for granted and, as a result, focus on the lower tiers of the waste management hierarchy (like recycling and re-use) rather than on the first and most important tier: waste prevention. Nevertheless, Germany remains both a regional and global leader on waste management and recycling issues.

23.5 Asia

23.5.1 China

Food waste is a major problem in China. As the population grows, the food-supply demand balance is increasingly tight—compounded by issues of natural resource scarcity and landfill space shortages. The Chinese government reports that food waste makes up around 70 % of all of the country's waste (and 61 % of household waste), most of which ends up in landfills. This amounts to roughly $32 billion dollars' worth of food thrown away each year. At the same time, more than 100 million Chinese citizens live below the poverty line and lack adequate supplies of food.[177]

As of 2015, China does not have a federal law on food waste. However, recent government actions indicate an increasingly strong stance on the issue, suggesting that such a law may be in the works. In 2013, the government issued a Food Waste Circular[178] that signals a significant first step toward regulating food waste. That same year, the government gave its backing to "Clean Your Plate," a movement started in 2013 by young online activists that calls on citizens and businesses alike to reduce food waste. A similar project called "Operation Empty Plate" began in 2012. As a result of these campaigns, over 700 restaurants in Beijing have begun offering smaller portion sizes to help cut back the amount of food that customers leave uneaten.[179] While it is still unclear how effective these efforts will be at reducing China's overall food waste, the government has reported reductions in wasteful banquets and unneeded luxury food purchases.[180]

The Chinese government has also been making strides in its general recycling and resource efficiency efforts. Since 2006, the National Development and Reform Commission (NDRC) has been setting goals for resource productivity improvements. The 2011–2015 plan aims to reduce the energy intensity (i.e., Gross

---

[176]See Neumayer (2000).

[177]Worldwatch Institute (2013).

[178]See *infra* Sect. 23.5.1.1.

[179]Public Radio International (2013).

[180]Hirsch and Harmanci (2013).

Domestic Product (GDP) produced per unit of resource) of the country's economy by 15 % every 5 years.[181] However, China's desire to

> maintain rapid economic growth over the coming decades while simultaneously improving environmental quality and maintaining social progress . . . cannot be met without employing innovative development pathways rather than conventional approaches taken in many developed countries.[182]

The Circular Economy Promotion Law[183] and Cleaner Production Promotion Law[184] are part of the Chinese government's answer to this challenge.

23.5.1.1 Food Waste Circular (2014)

In March 2014, the Central Committee of the General Office of the Communist Party of China and the General Office of the State Council released a Food Waste Circular. The document calls on government officials as well as the public to fight widespread food waste. It outlines a number of concrete measures, many of which target cutting back on the lavishness and wastefulness of official government events where food is served. Officials are ordered to control the amount they spend on dinners in public affairs, while government departments and state-owned enterprises must publicize the amount they spend on dining. Government cafeterias are also told to put up prominent "Save the Food" posters. Special supervisors are to be assigned, with the power to issue warnings to departments found to be wasting food.[185] These measures suggest that the government is seeking to lead by example when it comes to cutting back on food waste.

The circular also urges frugality among consumers and private food-related businesses, including restaurants and hotels. Catering companies and restaurants should make use of every potential ingredient and actively encourage customers to order moderate portions. Furthermore, restaurants are forbidden from setting minimum-spend requirements. Leftovers and kitchen waste are to be composted or sent to waste-recycling companies. The circular also announces efforts to improve management within the grain production, storage, transportation, and processing sectors to avoid unnecessary wastage.[186] Therefore, the circular is concerned with improving the sustainability of the entire food chain, not just end-of-lifecycle disposal activities.

The Food Waste Circular does not have the force of law. However, the document notes that the Chinese government is in the process of developing a law on food

---

[181]Zhu (2014).

[182]United Nations Environment Programme. Circular economy.

[183]See *infra* Sect 23.5.1.2.

[184]See *infra* Sect 23.5.1.3.

[185]China Daily USA (2014).

[186]China Daily USA (2014).

waste.[187] Presumably, such a law would codify many of the measures laid out in the circular; however, the precise contours of any future national food waste act are still uncertain. If enacted, it would be one of the few national-level laws targeted at food waste in the world.[188] A well-enforced Food Waste Law could help to alleviate some of the growing environmental and socioeconomic challenges associated with landfill shortages and food insecurity in China.

23.5.1.2 Circular Economy Promotion Law (2008)

In 2008, the National People's Congress of China (NPC) passed the country's first major piece of national legislation on sustainable development.[189] The Circular Economy Promotion Law has a broad set of aims: to facilitate a circular economy, raise resource utilization rates, protect and improve the environment, and promote sustainable development.[190] To carry out these goals, the law introduces policies and instruments for:

- Reducing resource consumption;
- Encouraging recycling; and
- Extending manufacturers' responsibilities for their products (in other words, moving toward the "polluter pays"/EPR principle[191]).[192]

Other governments have called the law not just a "simple environmental management policy," but rather "a new development model that could help China leapfrog . . . to a more sustainable economic structure."[193] Under this model, the Chinese government sets out detailed plans that aim to control the country's total use of water, land, energy, and materials. These programs also allow the government to take measures to adjust the speed and scale of economic development to ensure sustainable growth patterns.[194] Altogether, these functions are an attempt to manage the inherent links between environmental, economic, and social well-being.

The Chinese government's use of detailed resource management and economic development plans contrasts somewhat with the EU approach in the Packaging and Packaging Directive.[195] While the EU tends to give Member States a fair amount of discretion in figuring out how to meet sustainability targets, "the Chinese version of

[187]China Daily USA (2014).

[188]Japan's Food Recycling Law is another example. See *infra* Sect. 23.5.2.1.

[189]FDI Invest in China (2008).

[190]FDI Invest in China (2008), Art. 1.

[191]See *supra* Sect. 23.3.2.1.

[192]FDI Invest in China (2008), Art. 1.

[193]Zhu (2014).

[194]Zhu (2014).

[195]See *supra* Sect. 23.4.3.

the circular economy takes a top-down approach and uses command-control instruments rather than market-based ones."[196] There does not appear to be a consensus as to whether one approach is more effective than another, and studies on the effectiveness of the Circular Economy Promotion Law are hard to come by.

Nevertheless, as a whole, China's law has implications for reducing some of the undesirable environmental and economic impacts of the food and agriculture sectors. Promoting a circular, resource-efficient food chain means that less organic and inorganic waste is generated at the outset. It also means that most waste that *is* generated gets re-injected into other sectors of the economy (e.g., recycled into new products) rather than ending up in a landfill. A truly circular economy is good for the environment, the economy, and the people.

23.5.1.3 Cleaner Production Promotion Law (2002)

The Cleaner Production Promotion Law[197] is China's primary environmental law. As with the Circular Economy Promotion Law, its goals are far-reaching:

> to promote cleaner production, increase the efficiency of the utilization rate of resources, reduce and avoid the generation of pollutants, protect and improve environments, ensure the health of human beings, and promote the sustainable development of the economy and society.[198]

Therefore, the law has a broad, economy-wide scope. Manufacturers from all sectors are required to perform "cleaner production audits" to monitor their resource use and waste generation.[199] They should also prioritize "toxin-free, non-hazardous, easily degraded and easily recycled" product and packaging options.[200] Additional provisions target the agriculture and food sectors specifically. For instance, "[a]gricultural producers shall use chemical fertilizers, pesticides, agricultural films and feed additive compounds in accordance with scientific principles, and improve planting and breeding techniques so as to bring about high-quality, non-hazardous agricultural products."[201] Agricultural operations must also recycle agricultural wastes to prevent and control pollution associated with growing food.[202] Therefore, the law has direct implications for reducing the environmental impacts associated with food production.

The Cleaner Production Promotion Law can also interact with the Circular Economy Promotion Law[203] to boost the efficiency of the entire food cycle. Since the Circular Economy Promotion Law encourages waste prevention and

---

[196]Zhu (2014).

[197]President of the People's Republic of China (2002).

[198]President of the People's Republic of China (2002), Art. 1.

[199]President of the People's Republic of China (2002), Art. 28.

[200]President of the People's Republic of China (2002), Art. 23.

[201]President of the People's Republic of China (2002), Art. 23.

[202]President of the People's Republic of China (2002), Art. 23.

[203]See *supra* Sect. 23.5.1.2.

recycling, it should increase the percentage of food and food packaging that goes to good use. This, in turn, should lessen strains on both the agricultural and packaging manufacturing sectors, making it easier for these sectors to implement cleaner production techniques. In theory, then, China has a number of strong laws on the books for promoting the sustainability of the food chain. The passage of a national Food Waste Law[204] would add to the arsenal of tools available to help China grow its economy, lift more of its population from poverty (and food insecurity), and preserve its environmental integrity.

23.5.2 *Japan*

Japan is both a regional and world leader when it comes to waste management and recycling. Its successes in these fields have largely been borne of necessity: the highly-developed, densely populated island nation is dependent on natural resource imports and has limited space available for landfills. In the 1990s, Japan began shifting from a "linear production-consumption-waste process to a circulatory system" that minimizes natural resource consumption and seeks to turn waste into a valuable and exploitable resource.[205] The Japanese government has championed food waste recycling, in particular.

23.5.2.1 Food Recycling Law (2001)

Japan's innovative Food Recycling Law[206] calls for reducing food waste through recycling food resources. The law defines two types of food waste:

(1) General waste, which includes unsold or discarded food from distributors and households; and
(2) Industrial waste, which covers inedible food wastes arising from production and processing stages.[207]

A wide range of food-related businesses—including manufacturers, distributors, and food service providers—are called upon to help recycle wastes into new end-products. The law also requires municipalities to create "recycling loops" for feed and fertilizers.[208] The Food Recycling Law has led to remarkable improvements in Japan's food chain sustainability. In 2010, the country's food industry was able to reduce, reuse, or recycle over 80 % (or approximately 18 million tonnes) of

[204]See *supra* Sect. 23.5.1.1.

[205]Marra (2013).

[206]Global Environment Centre Foundation (2011). The law was revised in 2007.

[207]Global Environment Centre Foundation (2011), Art. 2(2).

[208]Marra (2013). In these "recycling loop" systems, recycling facilities work symbiotically with industrial and urban areas to improve recycling systems.

its food waste.[209] Most of the country's would-be food waste now ends up as animal feed, fertilizer, methane, oil and fat products, carbonized fuels, and ethanol. This has not only resulted in environmental protection gains but has also helped to ensure food security and food supply self-sufficiency for the historically import-dependent country.[210]

23.5.2.2 Container and Packaging Recycling Law (1995)

In force since April 1997, the Container and Packaging Recycling Law[211] is the basis for all official recycling efforts in Japan. Its central goal is to reduce garbage levels by "shifting wastes into recyclable resources" and using recycled containers and packages efficiently.[212] The law is enforced by the Ministry of the Environment, while a government-designated organization (the Japanese Container and Package Recycling Association (JCPRA)) conducts actual recycling operations.[213] Overall, the law sets up a collaborative system, calling on government administrators, businesses, and consumers to work together to manage packaging waste.[214] Consumers must separate their waste according to category (e.g., glass, plastic, and paper). Municipalities collect the separated waste, and businesses are, in turn, encouraged to recycle these collected wastes into new products. The act embodies the "polluter pays"/EPR principle, with manufacturers and importers obliged to pay collecting, sorting, transportation, and recycling costs.[215] Like the Food Recycling Law,[216] the Container and Packaging Recycling Law envisions a more self-sufficient, circular economy for Japan, including its food and agriculture sectors.

23.5.3 India

Like China,[217] India has one of the world's most rapidly expanding populations and faces a dual set of seemingly contradictory challenges: food wastage and food shortage. The UN estimates that 40 % of India's fresh fruit and vegetables, as well

---

209 Marra (2013).

210 Marra (2013).

211 Government of Japan, Ministry of the Environment (1995). The government amended the law in 2000 to increase the scope of covered business entities and the types of materials included in the recycling targets.

212 Government of Japan, Ministry of the Environment (1995).

213 Previously, municipal governments had sole responsibility for handling packaging waste.

214 Government of Japan, Ministry of the Environment (1995).

215 See *supra* Sect. 23.3.2.1.

216 See *supra* Sect. 23.5.2.1.

217 See *supra* Sect. 23.5.1.

as a large portion of its grain, goes to waste.[218] The majority of these losses are due to supply chain bottlenecks (unlike in many developed countries, where household or consumer behaviors drive most food waste).[219] India continues to lack a comprehensive national legal regime for managing the food chain. Most organic and inorganic food waste is covered under the country's general MSW rules[220] (although a separate set of plastic waste rules manage some packaging waste).[221]

23.5.3.1 Municipality Solid Waste (Management and Handling) Rules (2000)

The Municipality Solid Waste Rules[222] form the backbone of India's solid waste management policy. The rules call for managing MSW according to scientific guidelines. They charge each municipal authority with developing infrastructure for collecting, storing, segregating, transporting, processing, and disposing of MSW (including preventing soil and groundwater contamination).[223] Notably, the rules also advocate treating organic food waste as a separate waste stream. Municipalities should make sure that consumers segregate "wet" food wastes from "dry" recyclables. When feasible, segregated materials are to be recycled or reused. Biodegradable "wet" wastes, such as food scraps, are to be banned from landfills (which should be accept only non-biodegradable inert waste and compost rejects).

As of 2013, there are at least 56 composting plants in more than 43 cities in India.[224] While this is an important achievement, the MSW Rules still appear to be unevenly implemented across the continent. Moreover, the rules focus on the end-stages (such as disposal or re-use) of the food chain without directly addressing earlier transportation and processing stages, the parts of the lifecycle where most of India's food losses occur. Therefore, the MSW rules are still a limited tool for change. They must be supplemented by other rules or laws that address food losses

---

[218]Kazmin (2014).

[219]Kazmin (2014). As a result, one of the Indian government's current priorities is to modernize the country's food supply chain by attracting investment in things like cold storage and refrigerated trucks.

[220]See *infra* Sect. 23.5.3.1.

[221]Government of India, Ministry of Environment and Forests (2011). The Plastic Waste (Management and Handling) Rules prohibit the use of recycled or compostable plastic in carry bags used for storing, carrying, dispensing, or packaging food products. In other words, they require all foodstuffs to be packed in virgin plastic.

[222]Government of India, Ministry of Environment and Forests (2000). The rules were put forth by a Committee appointed by the Supreme Court of India, the Ministry of Environment and Forests (MoEF), and the Government of India (GoI). The Supreme Court monitors compliance with the MSW rules through the High Courts in each State.

[223]Government of India, Ministry of Environment and Forests (2000), Art. 4(1). See, for example, Maharashtra Pollution Control Board (2004).

[224]Hirsch and Harmanci (2013).

during transportation or processing in order to truly cut back on India's food waste problems.

23.6 Africa

23.6.1 South Africa

Food security is a serious concern throughout much of Africa, and South Africa faces some of the continent's most significant food waste challenges. Annual food waste in South Africa amounts to roughly 10.2 million tons, equal to about $7.7 billion (or just over 2 % of the country's GDP).[225] At the same time, millions of South Africans go hungry each year.[226] As in other developing countries (such as India[227]), most food loss happens early in the food chain.[228] Despite these problems, South Africa has historically lacked reliable waste management or recycling regimes.

This is changing, however. Indeed, the national government has embarked on a number of legislative and policy initiatives in recent years. These include the 2008 National Environmental Management: Waste Act (NEM: WA)[229] and the 2011 National Waste Management Strategy (NWMS).[230] Since 2013, the government has also been working to implement a National Organic Waste Composting Strategy to develop the country's composting capacity.[231] As a whole, these reforms suggest that the South African government is more receptive than ever to engaging with public and private actors on ways to boost food sector sustainability and alleviate the food security situation.

23.6.1.1 National Environmental Management: Waste Act (NEM: WA) (2008)

The National Environmental Management: Waste Act (NEM: WA)[232] is a piece of overarching legislation regulating waste management in South Africa. It has much

[225]Nahman and de Lange (2013).

[226]Blaine (2013).

[227]See *supra* Sect. 23.5.3.

[228]Around 27 % is lost during processing and packaging, 26 % in agricultural production, and 17 % in distribution. Just 4 % of food waste occurs after it has reached consumers. Blaine (2013).

[229]See *infra* Sect. 23.6.1.1.

[230]See *infra* Sect. 23.6.1.1.

[231]Republic of South Africa, Department of Environmental Affairs (2013). This program calls for local and national public authorities to work with the private sector to divert biodegradable waste from landfills.

[232]Southern African Legal Information Institute (2014). The act was amended in 2014.

in common with EU directives on waste issues.[233] For instance, the NEM: WA sets up a waste management hierarchy in which disposal is the last resort. In lieu of landfilling, the act promotes waste prevention (through cleaner production methods), reuse, recycling, and treatment.[234] It also embraces the "polluter pays"/ EPR principle, requiring waste generators to take ownership over the fate of their waste products.[235]

The 2011 National Waste Management Strategy (NWMS)[236] implements the NEM: WA. The NWMS lays out a variety of integrated waste management requirements for the private sector. Such obligations involve:

- Taking responsibility for products throughout their entire lifecycle;
- Instituting cleaner technology practices;
- Minimizing waste generation;
- Establishing systems and facilities to take back and recycle waste;
- Developing waste management technologies; and
- Preparing and implementing industry waste management plans.[237]

The strategy also sets out a number of landfills diversion targets for recyclables and biodegradable waste.[238] These recent policy initiatives highlight South Africa as a potential rising star in the fields of both general and food waste management.

23.7 Australia

Food waste is a big problem in Australia that has yet to attract a strong public policy response. A 2010 National Waste Report estimates that food waste comprises 35 % (or 2.68 million tonnes) of municipal waste and 23.5 % (or 1.39 million tonnes) of commercial and industrial waste each year.[239] Furthermore, the report laments the "absence of a national understanding of food waste . . . [which] makes it more difficult to improve the environmental performance of [Australia's] waste management systems."[240] The report also acknowledges that most of the Australian public has a poor sense of how to purchase, store, and prepare food to reduce waste.

Although the national government as well as the broader public have historically been insensitive to food waste issues, Australia has a robust general National Waste

---

[233]See *supra* Sect. 23.4.

[234]Southern African Legal Information Institute (2014), Sect. 2(iv).

[235]See *supra* Sect. 23.3.2.1.

[236]Republic of South Africa, Department of Environmental Affairs (2011). The NWMS was established under Sect. 6(1) of the NEM:WA.

[237]Republic of South Africa, Department of Environmental Affairs (2011).

[238]Republic of South Africa, Department of Environmental Affairs (2011).

[239]Australian Government, Department of the Environment, Waste, Heritage and the Arts (2010).

[240]Australian Government, Department of the Environment (2011).

Policy that addresses aspects of bio-waste and packaging management.[241] Additionally, Australia is one of the few countries with legislation on environmentally-friendly food packaging (such as minimum "empty space" requirements)[242] and transportation efficiency.[243] Therefore, although Australia lacks an integrated food waste or food efficiency policy, the country's regulatory landscape provides a number of independent tools for addressing various components of the food chain.

23.7.1 National Waste Policy (2009)

Australia's National Waste Policy[244] governs the country's waste management and resource recovery efforts until 2023. Overall, the policy "heralds a coherent, efficient and environmentally responsible approach to waste management."[245] It aims to avoid waste generation, reduce waste disposal, and manage waste as a resource. The policy contains six core areas:

(1) Taking or sharing responsibility among manufacturers, suppliers, and consumers for reducing the environmental, health, and safety footprints of products and materials;
(2) Improving the market for waste and recovered resources;
(3) Pursuing sustainability;
(4) Reducing hazards and risks;
(5) Tailoring solutions by increasing the capacity of regional, remote, and indigenous communities to manage waste and recover and re-use resources; and

---

241 See *infra* Sect. 23.7.1. In addition to the National Waste Policy, many regional governments have passed waste reduction and/or recycling laws and regulations. For instance, South Australia's Zero Waste SA Act of 2004 aims to eliminate waste from going to landfills. See Appendix.

242 Australia's "empty space" regulations for food packaging limit the maximum allowable empty space depending on the category of product. Permissible thresholds range from 25–40 %, with a 25 % limit for cereals and 40 % for more fragile snack foods like chips. Page (2014). Such restrictions are a growing trend worldwide, with similar regulations in place in some US states, Canada, Western and Eastern Europe, South Africa, Australia, Brazil, China, Japan, Taiwan, Tunisia, and South Korea. State of Oregon, Department of Environmental Quality (2005).

243 See Australasian Legal Information Institute (2012). Australia's Heavy Vehicle National Law (HVNL) and its regulations entered into force in early 2014 in the Australian Capital Territory, New South Wales, Queensland, South Australia, Tasmania, and Victoria (but not yet in the Northern Territory or Western Australia). The HVNL aims to achieve greater efficiencies in the country's commercial transportation sector, with provisions on vehicle specifications, curfews, load limitations, and speed and travel time restrictions. It is not unique to the food sector; however, its provisions do apply to qualifying food transporters and retailers. Therefore, the law has the potential to improve the energy efficiency of Australia's food supply chain.

244 Australian Government, Department of the Environment, Water, Heritage, and the Arts (2013). The National Waste Policy is agreed to by all Australian environment ministers and endorsed by the Council of Australian Governments.

245 Australian Government, Department of the Environment. About the national waste policy.

(6) Providing evidence by giving decision-makers access to meaningful, accurate, and current national waste and resource recovery data and information to measure progress and educate and inform the public.[246]

The policy also sets out 16 targeted strategies for achieving these goals. Key actions include:

- Introducing product stewardship framework legislation (Strategy 1);
- Improving packaging management (Strategy 3);
- Reducing the amount of biodegradable waste sent to landfills (Strategy 7); and
- Improving waste avoidance and re-use of materials in commercial and industrial waste streams (Strategy 10).[247]

Through these various goals and strategies, the government also seeks to realize some of the environmental co-benefits of better waste management. These advantages range from GHG emission reductions to water efficiency to land productivity to energy conservation and production. Many of these benefits go hand-in-hand with economic and public health and welfare goals. Therefore, the National Waste Policy provides a crucial framework for boosting food chain efficiency, especially when supplemented by other laws like the Heavy Vehicle National Law.[248] However, the National Waste Policy is not yet a fully-formed tool for change. Further development of some of its proposed strategies (especially Strategy 7 and Strategy 10) would give Australia the most important tools it needs for improving its food waste management and food transportation efficiency efforts.

23.8 Summary and Conclusion

This chapter has looked at ways that governments around the world manage the food cycle and how these management efforts intersect with nations' broader environmental protection and clean energy goals. Of course, there are many potential levers for improving food chain sustainability. As a result, the chapter has focused on two categories:

(1) Food waste management. This includes both organic food waste (like scraps or spoiled food) and inorganic food waste (such as plastic, glass, or cardboard packaging). Good food waste management practices have multiple environmental protection benefits: from reducing landfill sizes to lowering GHG emissions to mitigating the resource-intensiveness of the agricultural sectors. In areas (like the EU[249] and Brazil[250]) where anaerobic digestion biogas plants

[246] Australian Government, Department of the Environment. About the national waste policy.
[247] Australian Government, Department of the Environment. About the national waste policy.
[248] See *supra* Sect. 23.7.
[249] See *supra* Sect. 23.4.
[250] See *supra* Sect. 23.3.4.

are used, certain food waste management techniques have additional clean energy overlaps.

(2) Food sector transportation efficiency. This includes efforts to reduce the number of miles driven (or flown or shipped) to deliver food from farm to market, as well as attempts to make those miles "greener." The main environmental protection benefit of increased transportation efficiency is reduced vehicle emissions (including GHG emissions). Additionally, there may be gains for clean energy sectors (such as natural gas truck manufacturers) if commercial transportation fleets are encouraged to switch to lower-emission fuels.

On top of the environmental and clean energy benefits highlighted above, improvements in food waste management and food transportation efficiency can achieve significant economic and social welfare goals. These include improving food security (especially in developing countries, where most food waste occurs before food ever reaches consumers), boosting food safety and integrity, and saving money and resources that currently go into making and delivering products that are thrown away without being used. Therefore, governments, citizens, and industry alike can all benefit from better food waste and food transportation efficiency practices.

There is a broad range of food waste management and recycling/composting laws on the books around the world, while there are comparatively few laws on food transportation efficiency. Among the food waste management regimes, a number of approaches are found. Each strategy lends itself to a different set of tools for change. Some of the major distinctions include:

"Consumer Pays" (United States) Versus "Polluter Pays"/EPR (EU, Canada) Approach Outside of the US, the "polluter pays" approach is far more prevalent.[251] Such a strategy is useful because manufacturers and other producers are in a better position than consumers to ensure the recyclability or reusability of their products. Giving industry players some skin in the game also opens up opportunities for industry to collaborate with non-governmental organizations to organize recycling programs (as has happened in Germany[252] and Israel[253]). From a consumer-advocate's perspective, this means that it is possible to push for improvements in recycling and other waste management programs both inside and outside of government. In a "consumer pays" system, in contrast, industry players have fewer incentives to get involved. Furthermore, "consumers" is a broader and more amorphous category than "producers," which can make it difficult to figure out where to target advocacy efforts.

[251]See Appendix. Countries and regions embracing the "polluter pays"/EPR approach include the Australian Capital Territory, Canada, the European Union, Israel, Mexico, Russia, and South Africa.

[252]See *supra* Sect. 23.4.4.

[253]See Appendix.

Centralized, Command-and-Control (China) Versus Decentralized, Delegated Mandate (EU, United States) Approach Most countries follow a decentralized approach in which the national government sets broad, general mandates for waste management systems. State or municipal authorities then handle actual, on-the-ground waste sorting, collection, disposal, and recycling activities. This approach suggests that in many countries, like the US, the best opportunities for innovative actions and policies are at the local, rather than the national, level. Successes at the local level might eventually lead to reforms higher-up. Indeed, this is largely how the National Policy on Solid Waste (PNRS) in Brazil came about.[254]

Single MSW Stream (Most of the United States) Versus Separate Organic and Inorganic Waste Streams (Japan, India, Some US Cities) Municipalities in most countries continue to manage all MSW as a single waste stream, without requiring businesses and households to separate organic from inorganic waste. However, separate sorting and collecting of waste streams is a necessary pre-condition for successful recycling or composting programs. The trend does appear to be reversing in many places. Indeed, composting initiatives and "zero bio-waste" targets are gaining traction in the EU,[255] Asia,[256] and certain parts of the United States.[257] Therefore, a significant and emerging opportunity for change is to advocate for separate organic waste stream requirements, as well as bio-waste landfill diversion targets, in areas that do not already have such policies in place.

Regardless of the approach taken in any given location, it is also important to consider the role of public awareness and public engagement on food waste management issues. Public awareness campaigns have played a huge role in supporting—and even instigating—government reforms in the EU,[258] China,[259] and Japan.[260] In contrast, a lack of awareness or concern about food waste problems has been an obstacle to improving food waste management in places like Australia[261] and parts of Africa.[262] Therefore, public awareness campaigns are another potential tool for change. Indeed, an informed and engaged population can help a government to implement and enforce better food chain management practices. An efficient food cycle promotes, in turn, an efficient economy and a healthy, well-nourished population.

[254]See *supra* Sect. 23.3.4.1.

[255]See *supra* Sect. 23.4.2.

[256]See *supra* Sect. 23.5.2.1.

[257]See *supra* Sect. 23.3.1.4.

[258]See *supra* Sect. 23.4.

[259]See *supra* Sect. 23.5.1.

[260]See *supra* Sect. 23.5.2.

[261]See *supra* Sect. 23.7.

[262]See Afun (2009).

Appendix

Region	Country/state	Citation	Title	Year passed	Summary or subsections applicable to regulatory issue
Africa	Kenya	Act No. 8 of 1999, as amended by Act No. 6 of 2006, Act No. 17 of 2006, Act No. 5 of 2007, Act No. 6 of 2009	Environmental Management and Co-ordination Act (EMCA) (Cap. 387)	1999	Governs environmental management generally • Sect. 87 prohibits disposing of any wastes "in a manner that would cause pollution to the environment or ill health to any person." • Section 87(4) obligates "every person whose activities generate waste to employ measures essential to minimize wastes through treatment, reclamation and recycling."
Africa	Kenya	L.N. No. 121 of 2006; printed in *Kenya Gazette Supplement* No. 69 of 29 Sept 2006, pp. 1-63	Environmental Management and Coordination (Waste Management) Regulations (Cap. 387)	2006	Implement the EMCA; define general rules for waste management and specific rules for certain waste streams (including solid waste, industrial waste, hazardous waste, and pesticides and toxic substances); embrace "cleaner production" principles
Africa	Nigeria	S.I.15 of 1991	National Environmental Protection (Management of Solid and Hazardous Wastes) Regulations	1991	Address the collection, treatment, and disposal of solid and hazardous wastes from industrial and municipal sources; contain highly-detailed provisions for certain waste streams Part II envisions a multi-pronged approach with information-sharing and monitoring among private operators, industries, and public agencies
Africa	Nigeria	Act No. 25 of 2007	National Environmental Standards and Regulations Enforcement Agency (Establishment) Act (NESREA)	2007	Establishes the National Environmental Standards and Regulations Enforcement Agency as the major federal body in charge of environmental protection; repeals the now-defunct Federal Environmental Protection Agency (FEPA)

Africa	South Africa	Act No. 73 of 1989, as amended by the National Environmental Laws Amendment Act of 2009	Environment Conservation Act	1989	Governs environmental protection generally; sets out national environmental conservation policy • Part IV contains provisions on waste management • Part V covers agriculture and MSW disposal
Africa	South Africa	No. R. 543 of 2002	Plastic Bags Regulations	2002	Prohibit the manufacture, trade and commercial distribution of specified plastic bags Enacted under section 24 (D) of the Environment Conservation Act
Africa	South Africa	No. R. 625 of 2003	Plastic Carrier Bags and Plastic Flat Bags Regulations	2003	Prohibit the manufacture, trade and commercial distribution of domestically produced and imported plastic carrier bags and plastic flat bags Enacted under section 24 (D) of the Environment Conservation Act
Africa	South Africa	Act No. 68 of 2008	Consumer Protection Act	2008	Puts packaging disposal and product labelling burdens on industry • Sect. 22 requires that packaging labels identify the type of packaging and its recyclability. • Sect. 59 requires suppliers to accept return of goods at no charge to consumer if any national legislation prohibits the disposal or deposit of the goods or their components into a common waste collection system.
Africa	South Africa	Act No. 59 of 2008, as amended by Act No. 26 of 2014	National Environmental Management: Waste Act (NEM: WA)	2008	Governs waste management; sets up waste management hierarchy; embraces "polluter pays"/ EPR principle sets goals of: • 25% diversion of recyclables from landfills by 2016; and • ultimate phase-out of biodegradable waste from landfills Primarily implemented by the 2011 National Waste Management Strategy (NWMS).
Americas	Brazil	Lei n. 12.305/10	National Policy of Solid Waste/*Política Nacional de Resíduos Sólidos* (PNRS)	2010	Governs solid waste management; focuses on sustainable measures (including packaging requirements); calls for integrated, multi-level management plans

(continued)

Region	Country/state	Citation	Title	Year passed	Summary or subsections applicable to regulatory issue
Americas	Brazil (Sergipe)	Lei n. 7.465 de 20 de Julho de 2012; Publicado no *Diário Oficial* No 26529, do dia 24/07/2012	Law No. 7.645 on the Obligation of Using Biodegradable Packaging/Lei No. 7.645 de Dispõe sobre a utilização obrigatória de embalagens biodegradáveis, e dá providências correlatas	2012	Requires use of biodegradable packaging for materials used for commercial purposes within the State of Sergipe
Americas	Canada	S.C. 1999, c. 33	Canadian Environmental Protection Act (CEPA)	1999	Aims to protect environmental and human health through pollution prevention; embraces "polluter pays"/EPR principle Part 7 contains general waste management and prevention provisions (although most of the relevant authority is delegated to provincial governments)
Americas	Canada (Manitoba)	C.C.S.M., c. W40	Waste Reduction and Prevention Act	1990	Aims to reduce and prevent waste production and disposal consistent with sustainable development principles
Americas	Canada (Manitoba)	Man. Reg. 195/2008	Packaging and Printed Paper Stewardship Regulation	2008	Requires affected companies to register as product stewards and to remit fees to cover up to 80% of the cost of municipal recycling programmes
Americas	Canada (Ontario)	R.S.O. 1990, c. E.18	Environmental Assessment Act	1990	Regulates environmental assessments Part II addresses municipal waste disposal
Americas	Canada (Ontario)	R.S.O. 1990, c. E.19	Environmental Protection Act	1990	Aims to protect and conserve the natural environment • Part V covers waste management • Part IX covers packaging and container waste Implemented by Regulation on General Waste Management (R.R.O. 1990; O.Reg. 362)
Americas	Canada (Ontario)	O. Reg 104/94	Regulation on Packaging Audits and Packaging Reduction Work Plans	1994	Requires audits of certain manufacturers and importers to examine impacts of packaging on waste management; requires packaging reduction plans under some circumstances to reduce the amount of waste resulting from packaging

Americas	Canada (Ontario)	O. Reg. 101/94	Regulation on Recycling and Composting of Municipal Waste	1994	Requires certain municipalities to implement "blue box" waste management systems • Part III addresses municipal waste recycling depots • Part IV addresses waste recycling sites
Americas	Canada (Ontario)	O. Reg. 102/94	Regulation on Waste Audits and Waste Reduction Work Plans	1994	Requires certain businesses (retailers, schools, hotels, manufacturers, etc.) and construction projects to undergo waste audits, and, under some circumstances, implement plans to reduce, reuse and recycle waste.
Americas	Canada (Ontario)	O. Reg. 103/94	Regulation on Industrial, Commercial and Institutional Source Separation Programs	1994	Requires certain businesses (retailers, schools, hotels, manufacturers, etc.) to implement programs to facilitate source separation of waste for reuse or recycling
Americas	Canada (Ontario)	S.O. 2002, c. 6	Waste Diversion Act	2002	Promotes reducing, reusing and recycling waste; establishes "Waste Diversion Ontario" corporation to develop, implement, and operate waste diversion programmes for designated wastes
Americas	Canada (Ontario)	O. Reg. 273/02	Blue Box Waste Regulation	2002	Designates "blue box" waste as waste consisting of any of the following materials, or any combination of them: 1) glass; 2) metal; 3) paper; 4) plastic; 5) textiles; establishes "Stewardship Ontario" corporation
Americas	Canada (Ontario)	O. Reg. 101/07	Waste Management Projects Regulation	2007	Implements parts of Environmental Assessment Act (Ontario) Part II concerns landfill siting, waste disposal, and energy/fuel use
Americas	Canada (Prince Edward Island)	R.S.P.E.I. 1988, c. E-9	Environmental Protection Act	1988	Aims to manage, protect, and enhance the environment Sects. 13 to 19 cover waste treatment
Americas	Canada (Prince Edward Island)	P.E.I. Reg. EC691/00	Waste Resource Management Regulations	2000	Carry out Sect. 25 of Environmental Protection Act (Prince Edward Island); contain provisions on waste disposal and waste management, including requirements for composting facilities and recycling plants

(continued)

Region	Country/state	Citation	Title	Year passed	Summary or subsections applicable to regulatory issue
Americas	Mexico	DOF 19-06-2007	General Law for the Prevention & Integral Management of Wastes/ *Ley General Para la Prevención y Gestión Integral de los Residuos* (LGPGIR) (Federal Waste Law)	2003	Governs waste management; provides the country's legal framework on packaging and recycling activities; embraces "polluter pays"/ EPR principle Implemented by 2006 Regulation for the General Law for the Prevention & Integral Management of Wastes/*Reglamento de la Ley General Para la Prevención y Gestión Integral de los Residuos* (DOF 30-11-2006)
Americas	Mexico (Federal District)	*Gaceta Oficial del Distrito Federal* 22-04-2003	Solid Waste Law of the Federal District/*Ley de Residuos Sólidos del Distrito Federal*	2003	Regulates solid waste management in Mexico City Implemented by the Regulation for the Solid Waste Law of the Federal District/*Reglamento de la Ley de Residuos Sólidos del Distrito Federal* (DOF 07-10-2008)
Americas	Mexico (Federal District)	*Gaceta Oficial del Distrito Federal* 10-02-2004	Amendment of the Solid Waste Law of the Federal District/*Decreto por el que se Reforman los Artículos Tercero, Cuarto, Séptimo y Octavo Transitorios de la Ley de Residuos Sólidos del Distrito Federal*	2004	Requires households to separate organic and inorganic wastes Amends the Solid Waste Law of the Federal District
Americas	Peru	Law No. 27314	General Law on Solid Waste (The Waste Act)	2000	Establishes rights, duties, powers, and responsibilities of society as a whole to ensure the sanitary and environmentally sound management and handling of solid waste; embraces principles of minimization, environmental risk prevention, and human health and welfare protection; sets norms for recycling operations Implemented by the Rules of the Waste Act (Supreme Decree N° 057-2004- PCM)

Americas	United States	21 U.S.C. § 301 et seq.	Federal Food, Drug, and Cosmetic Act (FFDCA)	1938	Regulates food safety, including approval of food packaging materials
Americas	United States	42 U.S.C. § 4321 et seq.	National Environmental Policy Act (NEPA)	1969	Requires environmental reviews of major federal actions (including food packaging material approvals)
Americas	United States	42 U.S.C. § 6901 et seq.	Resource Conservation and Recovery Act (RCRA)	1976	Regulates waste management and recycling; provides general mandates, with most authority delegated to state and local governments
Americas	United States	21 C.F.R. § 170 et seq., 42 FR 14483	Federal Food, Drug, and Cosmetics Act (FFDCA), Food Additive Regulations	1977	Regulates food contact substances in packaging
Americas	United States	42 U.S.C. § 1791	Bill Emerson Good Samaritan Food Donation Act	1996	Encourages food donations by shielding donors from liability
Americas	United States (California – City of San Francisco)	Ordinance No. 100-09 (June 9, 2009)	Mandatory Recycling and Composting Ordinance	2009	Sets goal of "virtually zero" landfill waste by 2020 in City of San Francisco
Americas	United States (Vermont)	Act No. 148, 10 V.S.A. § 6602	An act relating to establishing universal recycling of solid waste. (Vermont's Universal Recycling (UR) Law)	2012	Bans landfilling all food scraps by July 1, 2020; bans disposal of recyclable materials (metal, glass, certain plastics, and paper/cardboard) by July 1, 2015.
Asia	China	Order of the President No. 72, adopted in the 28th Meeting of the Standing Committee of the 9th National People's Congress of the People's Republic of China on June 29, 2002	Law of the People's Republic of China on Promotion of Cleaner Production (Cleaner Production Promotion Law)	2002	Aims to: • promote cleaner production; • increase the efficiency of the utilization rate of resources; • reduce and avoid the generation of pollutants; • protect and improve environments; • ensure the health of human beings; and • promote the sustainable development of the economy and society Requires "cleaner production audits" of certain manufacturers

(continued)

Region	Country/state	Citation	Title	Year passed	Summary or subsections applicable to regulatory issue
Asia	China	Announcement No. 63 of 2008	Announcement of Ministry of Environmental Protection promulgating the cleaner production standard for wine industry	2008	Implements the Cleaner Production Promotion Law for the country's wine producers; provides requirements on: • production techniques; • resource and energy use; • waste recycling; and • environmental management
Asia	China	Order of the President No. 4, passed in the 4th meeting of the Standing Committee of the 11th National People's Congress of the People's Republic of China on August 29, 2008	Circular Economy Promotion Law of the People's Republic of China	2008	Aims to promote sustainable development in the nation; introduces policies and instruments for reducing resource consumption and encouraging recycling; embraces a top-down, command-and-control approach
Asia	India	Act No. 29 of 1986	An Act to provide for the protection and improvement of environment and for matters connected therewith (Environment (Protection) Act)	1986	Establishes the powers of the Central Government to protect and improve the environment
Asia	India	S.O. 908(E)	Municipality Solid Waste (Management and Handling) Rules	2000	Form the backbone of India's solid waste management policy; charge each municipal authority with developing infrastructure for collecting, storing, segregating, transporting, processing, and disposing of MSW; encourage composting and banning of biodegradable wastes from landfills
Asia	India	S.O. 249(E)	Plastic Waste (Management and Handling) Rules	2011	Prohibit use of recycled or compostable plastic in carry bags used for storing, carrying, dispensing, or packaging food products Implement Sects. 3, 6, and 25 of the Environment (Protection) Act
Asia	India (Maharashtra)	See Maharashtra Pollution Control Board, Circular No. MPCB/RO(HQ)/B: 1508 of 14/03/2011	Maharashtra Plastic Carry Bags (Manufacture and Use) Rules	2006	Supplement the Central Plastic Waste (Management and Handling) Rules within Maharashtra

Asia	India (Tamil Nadu)	FAOLEX No: LEX-FAOC052632	A Bill to provide for prohibition of sale, storage, transport and use of certain plastic articles for the protection of environment and public health and for matters connected therewith or incidental thereto (Tamil Nadu Plastic Articles (Prohibition of Sale, Storage, Transport and Use) Act)	2002	Prohibits the sale, storage, transport and use of certain plastic articles (including carry bags, cups, plates, and tumblers); forbids restaurant owners from using plastic articles in their establishments
Asia	Israel	Law No. 5753-1993	Collection and Disposal of Waste for Recycling Law	1993	Sets out principles and basic legal framework for recycling in Israel; authorizes municipalities to pass bylaws specifying procedures for waste collection, disposal, and recycling
Asia	Israel	Law No. 5759-1999	Deposit on Beverage Containers Law	1999	Requires a fully refundable deposit on most single-serve beverage containers Implemented by 2001 Regulations Carried out by ELA Recycling Corporation (a private non-profit organization)
Asia	Israel	Law No. 5771-2011	Regulation of Processing of Packaging Law (Packaging Law)	2011	Regulates both production and processing of packaging material waste; embraces "polluter pays"/EPR principle; aims to: • minimize packaging volume; • prevent waste dumping; and • encourage recycling Increases local taxes for waste storage in landfills (with revenues to go toward recycling facilities and infrastructure); sets concrete recycling targets, including zero landfilling waste by 2020
Asia	Japan	Act No. 112 of June 16, 1995	Act on the Promotion of Sorted Collection and Recycling of Containers and Packaging (Container and Packaging Recycling Law)	1995	Forms basis for all official recycling efforts in Japan; aims to reduce garbage levels and use recycled containers and packages efficiently

(continued)

Region	Country/state	Citation	Title	Year passed	Summary or subsections applicable to regulatory issue
Asia	Japan	Law No. 176 of 2001	Act on Promotion of Recycling of Food Circulation Resources (Food Recycling Law)	2001	Calls for reducing food waste through recycling food resources; requires municipalities to create "recycling loops" for feed and fertilizers
Asia	Russia	Law No. 89-FZ	Federal Law on industrial and consumer waste (Law on Waste)	1998	Russia's primary federal waste management regulation A proposed 2014 bill would amend the law to allow for landfill diversion targets and to embrace "polluter pays"/EPR principles
Asia	Russia (Astrakhan)	Regional Law No. 40/2004-OZ	Law on Domestic and Non-domestic Waste	2004	Implements federal Law on Waste
Asia	Russia (Buryatia)	Regional Law No. 1254-IV	Law on Domestic and Non-domestic Waste	2010	Implements federal Law on Waste
Asia	South Korea	As amended by Act No. 13 038, Jan. 1, 2015	Wastes Control Act	1986	Provides basic framework on waste management; requires Minister of Environment to prepare master waste management plan every 10 years
Asia	South Korea	As amended by Act No. 8957, Mar. 21, 2008	Act on Promotion of Saving and Recycling of Resources	1992	Provides basic framework for waste recycling; requires Minister of Environment to prepare recycling plan every 5 years
Australia/New Zealand	Australia	Law No. 21 of 2012	Heavy Vehicle National Law (HVNL)	2012	Aims to achieve greater efficiencies in the country's commercial transportation sector, with provisions on vehicle specifications, curfews, load limitations, and speed and travel time restrictions
Australia/New Zealand	Australia (Australian Capital Territory)	Act No. a2001-31	Waste Minimisation Act	2001	Governs waste management at a provincial level; embraces "polluter pays"/EPR principle; work to: • establish a waste management hierarchy; • promote environmentally responsible transporting, reprocessing and handling of waste; • seek integrated waste planning and services

Australia/New Zealand	Australia (New South Wales)	Act No. 58 of 2001	Waste Avoidance and Resource Recovery Act	2001	Promotes waste avoidance and resource recovery; creates a Waste Fund
Australia/New Zealand	Australia (New South Wales)	LW 17 October 2014 (2014 No. 666)	Protection of the Environment Operations (Waste) Regulation	2014	Provides specifications on waste management, transport, and disposal; enacted under the Protection of the Environment Operations Act 1997
Australia/New Zealand	Australia (Northern Territory)	Act No. 71 of 1998	Waste Management and Pollution Control Act	1998	Aims to reduce waste generation, increase re-use and recycling, and effectively manage waste disposal Part 6 provides for environmental audits Implemented by Waste Management and Pollution Control (Administration) Regulations.
Australia/New Zealand	Australia (Queensland)	FAOLEX No: LEX-FAOC107497	Waste Reduction and Recycling Act	2011	Aims at promoting waste avoidance and reduction as well as resource recovery and efficiency actions; establishes shared responsibility between government, business and industry and the community for waste management; creates Waste and Environment Fund; levies taxes at certain waste disposal sites Implemented by 2011 Waste Reduction and Recycling Regulation
Australia/New Zealand	Australia (South Australia)	*Gazette* 6.5.2004, p. 1222	An Act to establish a statutory corporation, Zero Waste SA, with the function of reforming waste management in the State; and for other purposes (Zero Waste SA Act)	2004	Establishes a statutory corporation, Zero Waste SA, to reform waste management in the State; aims to eliminate waste generation as well as the landfilling of waste; promotes resource recovery and recycling Implemented by Zero Waste SA Regulations 2006
Australia/New Zealand	Australia (Tasmania)	Act No. 83 of 1994	Environmental Management and Pollution Control Act (EMPCA)	1994	Governs environment protection and pollution control generally Relevant implementing regulations include: • Environmental Management and Pollution Control (Controlled Waste Tracking) Regulations (2010); and • Environmental Management and Pollution Control (Waste Management) Regulations (2010)

(continued)

Region	Country/state	Citation	Title	Year passed	Summary or subsections applicable to regulatory issue
Australia/New Zealand	Australia (Western Australia)	Act No. 36 of 2007	Waste Avoidance and Resource Recovery Act	2007	Establishes Waste Authority charged with setting out a long term strategy for continuous improvement of waste services, waste avoidance, and resource recovery and for waste reduction, resource recovery, and landfill diversion; calls for yearly business plans, waste plans and of product stewardship plans
Australia/New Zealand	New Zealand	Act No. 89 of 2008	Waste Minimisation Act	2008	Governs waste management; calls for levy on landfill waste and mandatory waste reporting; establishes Waste Advisory Board • Part 1, Sect. 3 encourages "waste minimisation and a decrease in waste disposal in order to protect the environment from harm and provide environmental, social, economic, and cultural benefits" • Part II encourages (or sometimes requires) product stewardship schemes (including bottle deposit systems) that hold all parties involved in a product's lifecycle responsible for managing end-of-life environmental impacts
Europe	European Union	O.J. L. 365, 31.12.1994, pp. 10–23	European Parliament and Council Directive 94/62/EC of 20 December 1994 on packaging and packaging waste (Packaging and Packaging Waste Directive)	1994	Aims to protect the environment and maintain robust internal EU trading market; sets recycling targets for packaging waste; mandates environmental "essential requirements" for all packaging; embraces "polluter pays"/EPR principle
Europe	European Union	O.J. L. 182, 16.7.1999, pp. 1–19	Council Directive 1999/31/EC of 26 April 1999 on the landfill of waste (Landfill Directive)	1999	Sets short- and long-term landfill diversion targets for Member States, including goals for reducing amount of biodegradable municipal waste going to landfills
Europe	European Union	O.J. L. 338, 13.11.2004, pp. 4–17	Framework Regulation EC 1935/2004 on Food Contact Materials	2004	Sets requirements for food contact materials; requires good manufacturing practices; requires materials to be traceable throughout the production chain

Europe	European Union	O.J. L. 312, 22.11.2008, pp. 3–30	Directive 2008/98/EC of the European Parliament and of the Council of 19 November 2008 on waste (Waste Framework Directive)	2008	Establishes basic waste management principles for EU Member States; provides general legislative framework for collecting, transporting, recovering, and disposing of waste • Art. 4 establishes waste management hierarchy • Art. 22 addresses composting
Europe	European Union	O.J. L. 54, 26.2.2011, pp. 1–254	Commission Regulation (EU) No 142/2011 implementing Regulation (EC) No 1069/2009 of the European Parliament and of the Council laying down health rules as regards animal by-products and derived products not intended for human consumption and implementing Council Directive 97/78/EC as regards certain samples and items exempt from veterinary checks at the border under that Directive (Animal By-Products (ABP) Rules)	2011	Regulates materials of animal-origin not consumed by people (including certain by-products of the food production chain, such as fallen stock)
Europe	European Union—Germany	*Federal Law Gazette I*, p. 2705, as amended by the Act of 3 May 2000 (*Federal Law Gazette I*, p. 632)	An act for Promoting Closed Substance Cycle Waste Management and Ensuring Environmentally Compatible Waste Disposal/Gesetz zur Förderung der Kreislaufwirtschaft und Sicherung der umnweltverträglichen Beseitigung von Abfällen (Closed Substance Cycle and Waste Management Act)	1994	Establishes Germany's national waste management policy framework; advocates waste avoidance, recovery, and environmentally compatible waste disposal; embraces "polluter pays"/EPR principle

(continued)

Region	Country/state	Citation	Title	Year passed	Summary or subsections applicable to regulatory issue
Europe	European Union—Germany	*Federal Law Gazette I*, p. 2379, as amended by the Fifth Amending Ordinance of 2 April 2008 (*Federal Law Gazette I*, p. 531)	Ordinance on the Avoidance and Recovery of Packaging Waste of 21 August 1998/ Verpackungsverordnung—VerpackV (Packaging Waste Ordinance or Packaging and Waste Avoidance Law)	1998	Implements the EU Packaging and Packaging Waste Directive; requires retailers to install recycling bins for primary and secondary packaging; imposes mandatory beverage deposits; prohibits incineration as waste disposal option; embraces "polluter pays"/EPR principle
Europe	European Union—United Kingdom	No. 1941 of 2003	Packaging (Essential Requirements) Regulations	2003	Implement provisions of the EU Packaging and Packaging Waste Directive regarding "essential requirements" for packaging; embrace "polluter pays"/EPR principle
Europe	European Union—United Kingdom (England and Wales)	No. 1889 of 2012	Waste (England and Wales) (Amendment) Regulations	2012	Implement the EU Waste Framework Directive; require waste collection authorities to collect waste paper, metal, plastic, and glass separately (from both household and commercial/industrial sources) beginning in January 2015
Europe	European Union—United Kingdom (England, Scotland, and Wales)	No. 871 of 2007	Producer Responsibility Obligations (Packaging Waste) Regulations (England, Scotland, and Wales)	2007	Implement the EU Packaging and Packaging Waste Directive; address industry's role in meeting the UK's recovery and recycling targets; establish a statutory producer responsibility regime by placing legal obligation on businesses that make or use packaging to ensure that a proportion of the packaging is recovered and recycled; embraces "polluter pays" and "collective producer responsibility" principles
Europe	European Union—United Kingdom (Northern Ireland)	No. 198 of 2007	Producer Responsibility Obligations (Packaging Waste) Regulations (Northern Ireland)	2007	Implement the EU Packaging and Packaging Waste Directive in Northern Ireland
Europe	European Union—United Kingdom (Scotland)	S.S.I. No. 148 of 2012	Waste (Scotland) Regulations	2012	Implement the EU Waste Framework Directive; provide for the separate collection and treatment of "wet" food waste from other "dry" recyclable waste streams

International	Parties to Protocol	37 I.L.M. 22 (1998)	Kyoto Protocol to the United Nations Framework Convention on Climate Change	1997	Sets binding GHG reduction targets, potentially achievable through agriculture and food sector reforms
International	Member States of the World Trade Organization (WTO)	31 U.S.T. 405, 1186 U.N.T.S. 276	Agreement on Technical Barriers to Trade (TBT Agreement)	1995	Aims to facilitate global trading markets; prohibits setting technical regulations, standards, testing, and certification procedure requirements that create unnecessary obstacles to trade
International	Member States of the World Trade Organization (WTO)	1867 U.N.T.S. 483	Agreement on the Application of Sanitary and Phytosanitary Measures (SPS Agreement)	1995	Sets similar constraints as the TBT Agreement on Member States' food safety, animal, and plant health measures

References

Afun S (2009) Government regulations and legislations will ensure sustainable waste management in Nigeria. http://www.iswa.org/uploads/tx_iswaknowledgebase/le_2009_2-357.pdf. Accessed 4 Apr 2015

Agriculture and Agri-Food Canada (2012) Food packaging and labeling trends in Mexico. In: International Markets Bureau Market Indicator Report. Oct 2012. Available via AGR.GC.CA. http://www.agr.gc.ca/eng/industry-markets-and-trade/statistics-and-market-information/by-region/mexico/food-packaging-and-labeling-trends-in-mexico/?id=1410083148552. Accessed 20 Mar 2015

Asamblea Legislativa del Distrito Federal (2003) Decreto por el que se crea la ley de residuos solidos del distrito federal. In: Gaceta Oficial Del Distrito Federal. 20 Apr 2003. Available via TemasActuales. http://www.temasactuales.com/assets/pdf/gratis/MXDFwasLaw.pdf. Accessed 20 Mar 2015

Asamblea Legislativa del Distrito Federal (2004) Decreto por el que se reforman los artículos tercero, cuarto, séptimo y octavo transitorios de la ley de residuos solidos del distrito federal. In: Gaceta Oficial Del Distrito Federal. 10 Feb 2004. Available via TemasActuales. http://www.temasactuales.com/assets/pdf/gratis/MXDFresLeyAmd04.pdf. Accessed 20 Mar 2015

Australasian Legal Information Institute (2012) Heavy vehicle national law act 2012 no. 23. http://www5.austlii.edu.au/au/legis/qld/num_act/hvnla2012n20272/. Accessed 20 Mar 2015

Australian Government, Department of the Environment (2011) National food waste – final report: executive summary. http://www.environment.gov.au/protection/national-waste-policy/publications/national-food-waste-assessment-final-report. Accessed 20 Mar 2015

Australian Government, Department of the Environment. About the national waste policy. http://www.environment.gov.au/protection/national-waste-policy/about. Accessed 20 Mar 2015

Australian Government, Department of the Environment, Waste, Heritage and the Arts (2010) National waste report. http://www.scew.gov.au/system/files/resources/020c2577-eac9-0494-493c-d1ce2b4442e5/files/wastemgt-nat-waste-report-final-20-fullreport-201005-0.pdf. Accessed 20 Mar 2015

Australian Government, Department of the Environment, Water, Heritage, and the Arts (2013) National waste policy. 13 Dec 2013. http://www.scew.gov.au/system/files/resources/906a04da-bad6-c554-1d0d-45206011370d/files/wastemgt-rpt-national-waste-policy-framework-less-waste-more-resources-print-ver-200911.pdf. Accessed 20 Mar 2015

Beveridge & Diamond PC (2010) Brazilian house passes sweeping waste policy bill. 20 Mar 2010. http://www.bdlaw.com/news-834.html. Accessed 20 Mar 2015

Blaine S (2013) Challenges of reducing food waste while millions starve. BusinessDay, 29 January 2013. Available via BusinessDayLive. http://www.bdlive.co.za/national/science/2013/01/29/challenges-of-reducing-food-waste-while-millions-starve. Accessed 20 Mar 2015

Buzby J, Human J (2012) Total and per capita value of food loss in the United States. Food Policy 37(5):561–570

Bottlebill.org (2015) Bottle bills in the USA. http://www.bottlebill.org/legislation/usa.htm. Accessed 20 Mar 2015

California Energy Commission (2015) Anaerobic digestion. http://www.energy.ca.gov/biomass/anaerobic.html. Accessed 20 Mar 2015

Câmara dos Deputados (2012) Política nacional de resíduos sólidos (PNRS), 2ª edição, altera a Lei n° 9.605, de

Canadian Council of Ministers of the Environment (2014) Current priorities: waste. http://www.ccme.ca/en/current_priorities/waste/index.html. Accessed 20 Mar 2015

China Daily USA (2014) Chinese authorities upgrade food waste fight. 19 Mar 2014. http://usa.chinadaily.com.cn/china/2014-03/19/content_17358123.htm. Accessed 20 Mar 2015

City of Austin, Resource Recovery. Frequently asked questions. http://austintexas.gov/department/austin-resource-recovery/faq. Accessed 20 Mar 2015

City of Seattle Public Utilities and Councilmember Bagshaw S (2014) Frequently asked questions: Seattle composting. http://www.seattle.gov/council/bagshaw/attachments/compost%20requirement%20QA.pdf. Accessed 20 Mar 2015
Congreso General De Los Estados Unidos Mexicanos (2003) Ley general para la prevención y gestión integral de los residuos. http://www.temasactuales.com/assets/pdf/gratis/MXwasLaw.pdf. Accessed 20 Mar 2015
Cornell University Law School. 40 C.F.R. tit 40, chap V, part 1508, §1508.18: major federal action. https://www.law.cornell.edu/cfr/text/40/1508.18. Accessed 20 Mar 2015
Cornell University Law School. U.S. code, tit 42, chap 13A: Bill Emerson good Samaritan food donation act. https://www.law.cornell.edu/uscode/text/42/1791. Accessed 20 Mar 2015
Cornell University Law School. U.S. code, tit 42, chap 55: national environmental policy. https://www.law.cornell.edu/uscode/text/42/chapter-55. Accessed 20 Mar 2015
Cornell University Law School. U.S. code, tit 42, chap 55, subchap 1, § 4332: cooperation of agencies; reports; availability of information; recommendations; international and national coordination of efforts. https://www.law.cornell.edu/uscode/text/42/4332. Accessed 20 Mar 2015
Cornell University Law School. U.S. code, tit 42, chap 82: solid waste disposal. https://www.law.cornell.edu/uscode/text/42/chapter-82. Accessed 20 Mar 2015
Cornell University Law School. U.S. code, tit 42, chap 82, subchap 1, § 6902: objectives and national policy. https://www.law.cornell.edu/uscode/text/42/6902. Accessed 20 Mar 2015
Cornell University Law School. U.S. code, tit 42, chap 82, subchap 1, § 6903: definitions. https://www.law.cornell.edu/uscode/text/42/6903. Accessed 20 Mar 2015
Earth911 (2014) Trash planet: Germany. http://www.earth911.com/earth-watch/trash-planet-germany/. Accessed 20 Mar 2015
EUR-Lex (1994) European Parliament and Council Directive 94/62/EC of 20 December of 1994 on packaging and packaging waste. http://eur-lex.europa.eu/legal-content/EN/TXT/?uri=CELEX:31994L0062. Accessed 20 Mar 2015
EUR-Lex (1999) Council Directive 1999/31/EC of 26 April 1999 on the landfill of waste. http://eur-lex.europa.eu/legal-content/EN/TXT/?uri=CELEX:31999L0031. Accessed 20 Mar 2015
EUR-Lex (2008) Directive 2008/98/EC of the European Parliament and of the Council of 19 November 2008 on waste and repealing certain Directives (test with EEA relevance). http://eur-lex.europa.eu/legal-content/EN/TXT/?uri=CELEX:32008L0098. Accessed 20 Mar 2015
Europa (2011) Summaries of EU Legislation: Packaging and packaging waste. 8 Sept 2011. http://europa.eu/legislation_summaries/environment/waste_management/l20207_en.htm. Accessed 20 Mar 2015
European Commission, Directorate-General Environment (2009) Guidelines on waste prevention programme. http://ec.europa.eu/environment/waste/prevention/pdf/Waste%20Prevention_Handbook.pdf. Accessed 20 Mar 2015
European Commission, Directorate-General Environment (2011) Guidelines on the preparation of food waste prevention programmes. Aug 2011. http://ec.europa.eu/environment/waste/prevention/pdf/prevention_guidelines.pdf. Accessed 20 Mar 2015
European Commission, Directorate-General Environment (2012) Preparing a waste prevention programme: guidance document. Oct 2012, pp 2, 4, 20–23. http://ec.europa.eu/environment/waste/prevention/pdf/Waste%20prevention%20guidelines.pdf. Accessed 20 Mar 2015
European Commission, Directorate-General Environment (2013) Analysis of the public consultation on the green paper "European strategy on plastic waste in the environment": final report, p 9. 28 Nov 2013. http://ec.europa.eu/environment/waste/studies/pdf/green_paper_plastic.pdf. Accessed 20 Mar 2015
European Commission, Directorate-General Environment (2015) Biodegradable waste. 18 Mar 2015. http://ec.europa.eu/environment/waste/compost/. Accessed 20 Mar 2015
European Environment Agency (2009) Diverting waste from landfill – effectiveness of waste-management policies in the European Union. 10 June 2009, p 7. http://www.eea.europa.eu/publications/diverting-waste-from-landfill-effectiveness-of-waste-management-policies-in-the-european-union. Accessed 20 Mar 2015

European Parliament (2013) European Parliament resolution of 19 January 2012 on how to avoid food wastage: strategies for a more efficient food chain in the EU. 19 Jan 2012. http://www.europarl.europa.eu/sides/getDoc.do?pubRef=-//EP//TEXT+TA+P7-TA-2012-0014+0+DOC+XML+V0//EN. Accessed 20 Mar 2015

European Union FUSIONS (2015) About FUSIONS. http://www.eu-fusions.org/. Accessed 20 Mar 2015

FDI Invest in China (2008) Circular economy promotion law of the People's Republic of China. 29 Aug 2008. http://www.fdi.gov.cn/1800000120_39_597_0_7.html. Accessed 20 Mar 2015

Food and Drink Federation (2014) Transport efficiency commitment. http://www.fdf.org.uk/transport_efficiency.aspx. Accessed 20 Mar 2015

German Federal Government (1998) Ordinance on the avoidance and recovery of packaging wastes (packaging ordinance – *verpackungsverordnung – verpackV*) of 20 Aug 1998 as last amended by the Fifth Amending Ordinance of 2 Apr 2008. In: German Federal Law Gazette I. Available via take-e-way. http://www.take-e-way.com/verpackv/packaging-ordinance-verpackungsverordnung/. Accessed 20 Mar 2015

German Federal Government (2000) Act for promoting closed substance cycle waste management and ensuing environmentally compatible waste disposal (*Gesetz zur Förderung der Kreislaufwirtschaft und Sicherung der umnweltverträglichen Beseitigung von Abfällen*). In: German Federal Law Gazette I. 3 May 2000. Available via IUScomp. http://www.iuscomp.org/gla/statutes/KrW-AbfG.htm. Accessed 20 Mar 2015

German Federal Ministry of Food and Agriculture (2015) Too good for the bin – an initiative of the federal government to avoid food waste. 13 Feb 2015. http://www.bmel.de/EN/Food/Value-Of-Food/_Texte/ZgfdT.html. Accessed 20 Mar 2015

Global Environment Centre Foundation (2011) Law for promotion to recover and utilize recyclable food resources (food recycling law). Nov 2011. http://nett23.gec.jp/Ecotowns/data/et_c-08.html. Accessed 20 Mar 2015

Government of Canada, Minister of Justice (1999) Canadian environmental protection act, 1999 (S.C. 1999, c. 33). 14 Sept 1999. http://laws-lois.justice.gc.ca/PDF/C-15.31.pdf. Accessed 20 Mar 2015

Government of India, Ministry of Environment and Forests (2000) Notification [municipal solid wastes (management and handling) rules, 2000]. 25 Sept 2000. http://www.moef.nic.in/legis/hsm/mswmhr.html. Accessed 20 Mar 2015

Government of India, Ministry of Environment and Forests (2011) Notification [plastic waste (management and handling rules), 2011]. In: The Gazette of India: Extraordinary, Part II, Sect 3(ii). Available via MOEF. http://www.moef.nic.in/downloads/public-information/DOC070201-005.pdf. Accessed 20 Mar 2015

Government of Japan, Ministry of the Environment (1995) Law for the promotion of sorted collection and recycling of containers and packaging (container and packaging recycling law). https://www.env.go.jp/en/laws/recycle/07.pdf. Accessed 20 Mar 2015

Guardian News and Media (2012) The waste mountain engulfing Mexico City. The Guardian, 9 January 2012. Available via TheGuardian.com. http://www.theguardian.com/environment/2012/jan/09/waste-mountain-mexico-city. Accessed 20 Mar 2015

Harvard Food Law and Policy Clinic (2013) Tools for advocates: increasing local food procurement by state agencies, colleges, and universities. July 2013. http://www.chlpi.org/wp-content/uploads/2013/12/Local-Procurement-Handout_FINAL_WEB-FRIENDLY.pdf. Accessed 20 Mar 2015

Hendrix S (2013) Coming to a curb near you: compost collection. The Washington Post, 11 March 2013. Available via WashingtonPost.com. http://www.washingtonpost.com/local/coming-to-a-curb-near-you-compost-collection/2013/03/11/e2f196fc-82b3-11e2-8074-b26a871b165a_story.html. Accessed 20 Mar 2015

Hirsch J, Harmanci R (2013) Food waste: the next food revolution. Modern Farmer, 30 September 2013. Available via ModernFarmer.com. http://modernfarmer.com/2013/09/next-food-revolution-youre-eating/. Accessed 20 Mar 2015

HIS Environment Intelligence Analysis (2015) Hazardous waste in Mexico. http://www.eiatrack.org/s/188. Accessed 20 Mar 2015

Interpack Processes and Packaging (2014) Current reforms to the German packaging ordinance. Apr 2014. http://www.interpack.com/cipp/md_interpack/lib/pub/tt,oid,27808/lang,2/ticket,g_u_e_s_t/~/Current_reforms_to_the_German_Packaging_Ordinance.html. Accessed 20 Mar 2015

Justia (2015) U.S. Law: U.S. Case Law: U.S. Supreme Court: Volume 490: Robertson v Methow Valley Citizens 490 U.S. 332. 1 May 1989. https://supreme.justia.com/cases/federal/us/490/332/case.html. Accessed 20 Mar 2015

Kazmin A (2014) India tackles supply chain to cut food waste. The Financial Times, 11 April 2014. Available via FT.com. http://www.ft.com/cms/s/2/c1f2856e-a518-11e3-8988-00144feab7de.html#axzz3RpBHnATW. Accessed 20 Mar 2015

Keller and Heckman LLP (2002) Special focus: packaging and environmental legislation in the United States: an overview. July 2002. http://www.packaginglaw.com/2561_.shtml. Accessed 20 Mar 2015

Leib EB (2012) Good laws, good food: putting local food policy to work for our communities. July 2012. http://www.law.harvard.edu/academics/clinical/lsc/documents/FINAL_LOCAL_TOOLKIT2.pdf. Accessed 20 Mar 2015

Levitan D (2013) Recycling's 'final frontier': the composting of food waste. In: Environment 360: reporting, analysis, opinion & debate. 8 Aug 2013. Available via Yale.edu. http://e360.yale.edu/feature/recyclings_final_frontier_the_composting_of_food_waste/2678/. Accessed 20 Mar 2015

Maharashtra Pollution Control Board (2004) Order of Supreme Court on management of municipal solid waste. 5 Oct 2004. http://mpcb.gov.in/muncipal/supremeorder.php. Accessed 20 Mar 2015

Marra F (2013) Food waste in Japan: how eco-towns and recycling loops are encouraging self-sufficiency. Food Tank, 5 December 2013. Available via Resilience.org. http://www.resilience.org/stories/2013-12-05/food-waste-in-japan-how-eco-towns-and-recycling-loops-are-encouraging-self-sufficiency. Accessed 20 Mar 2015

Nahman A, de Lange W (2013) Costs of food waste along the value chain: evidence from South Africa. ScienceDirect 33(11):2493–2500. Nov 2013. Available via Science Direct. http://www.sciencedirect.com/science/article/pii/S0956053X13003401. Accessed 20 Mar 2015

Neumayer E (2000) German packaging waste management: a successful voluntary agreement with less successful environmental effects. Eur Environ 10:152–163. Available via LSE.ac.uk. http://www.lse.ac.uk/geographyAndEnvironment/whosWho/profiles/neumayer/pdf/Article%20in%20European%20Environment%20%28packaging%20waste%29.pdf. Accessed 20 Mar 2015

Ontario Association of Food Banks (2014) Bill 36 – tax credits for farmers. http://www.oafb.ca/bill-36---tax-credit-for-farmers. Accessed 20 Mar 2015

Packaging News (2015) Wrap reveals 10bn cost of wasted food. 8 May 2008. http://www.packagingnews.co.uk/news/environment/wrap-reveals-10bn-cost-of-wasted-food/. Accessed 20 Mar 2015

Page R (2014) Would you like excess packaging with that? Choice, 4 September 2014. Available via Choice.com. https://www.choice.com.au/shopping/packaging-labelling-and-advertising/packaging/articles/excess-packaging-and-online-shopping. Accessed 20 Mar 2015

Pereira A (2010) National solid waste policy – Brazil, English version. http://wiego.org/sites/wiego.org/files/resources/files/Pereira-Brazilian-Waste-Policy.pdf. Accessed 20 Mar 2015

Portal da Prefeitura de Curitiba. Secretario municipal do abastecimiento: programa cambio verde. http://www.curitiba.pr.gov.br/conteudo/cambio-verde-smab/246. Accessed 20 Mar 2015

Presidente Constitucional de los Estados Unidos Mexicanos (2006) Reglamento de la ley general para la prevención y gestión integral de los residuos. In: Diario Oficial, Séptima Sección. 30 Nov 2006. Available via TemasActuales. http://www.temasactuales.com/assets/pdf/gratis/MXwasReg.pdf. Accessed 20 Mar 2015

President of the People's Republic of China (2002) Law of the People's Republic of China on promotion of cleaner production (order of the President no. 72). 29 June 2002. http://www.gov.cn/english/laws/2005-10/08/content_75059.htm. Accessed 20 Mar 2015

Province of Nova Scotia (2014) Solid waste – resource management strategy. 27 Oct 1995. http://www.novascotia.ca/nse/waste/swrmstrategy.asp. Accessed 20 Mar 2015

Public Radio International (2013) No-waste lunch: China's "clean your plate" campaign. 20 July 2013. http://www.pri.org/stories/2013-07-20/no-waste-lunch-chinas-clean-your-plate-campaign. Accessed 20 Mar 2015

Recycling Council of Ontario (2010) How waste is regulated. https://www.rco.on.ca/how_waste_is_regulated. Accessed 20 Mar 2015

Republic of South Africa, Department of Environmental Affairs (2011) National waste management strategy. Nov 2011. https://www.environment.gov.za/sites/default/files/docs/national waste_management_strategy.pdf. Accessed 20 Mar 2015

Republic of South Africa, Department of Environmental Affairs (2013) The national organic waste composting strategy draft guideline document for composting. Feb 2013. http://sawic.environment.gov.za/documents/1825.pdf. Accessed 20 Mar 2015

SourceSuite (2015) Maintaining local preference while increasing vendor competition. http://www.sourcesuite.com/procurement-learning/purchasing-articles/Maintaining-Local-Preference-while-Increasing-Vendor-Competition.jsp. Accessed 20 Mar 2015

Southern African Legal Information Institute (2014) National environmental management waste act 2008. http://www.saflii.org/za/legis/consol_act/nemwa2008394/. Accessed 20 Mar 2015

State of Oregon, Department of Environmental Quality (2005) Packaging waste reduction: international packaging regulations. 14 July 2005. http://www.deq.state.or.us/lq/pubs/docs/sw/packaging/intlpkgregulations.pdf. Accessed 20 Mar 2015

State of Vermont (2015) Mobile poultry slaughter unit available for Vermont farmers. 24 Nov 2008. http://www.vermont.gov/portal/government/article.php?news=715. Accessed 20 Mar 2015

State of Vermont, Waste Management & Prevention Division (2015) Vermont's universal recycling law (act 148) http://www.anr.state.vt.us/dec/wastediv/solid/act148.htm. Accessed 20 Mar 2015

United Nations Environment Programme. Circular economy. http://www.unep.org/resourceefficiency/Home/Policy/SCPPoliciesandthe10YFP/NationalActionPlansPovertyAlleviation/NationalActionPlans/CircularEconomy/tabid/78389/Default.aspx. Accessed 20 Mar 2015

United Nations Food and Agriculture Organization (2013) Food wastage footprint: impacts on natural resources summary report, p 6. http://www.fao.org/docrep/018/i3347e/i3347e.pdf. Accessed 20 Mar 2015

United Nations Framework Convention on Climate Change (1998) Kyoto protocol to the United Nations framework convention on climate change. http://unfccc.int/kyoto_protocol/items/2830.php. Accessed 20 Mar 2015

United Nations Framework Convention on Climate Change (2014) Status of ratification of the Kyoto protocol. http://unfccc.int/kyoto_protocol/status_of_ratification/items/2613.php. Accessed 20 Mar 2015

United States Department of Agriculture, Food and Nutrition Service (2013) National school lunch program. Sept 2013. http://www.fns.usda.gov/sites/default/files/NSLPFactSheet.pdf. Accessed 20 Mar 2015

United States Environmental Protection Agency (2013a) Agriculture: resource conservation and recovery act (RCRA). 30 Oct 2013. http://www.epa.gov/agriculture/lrca.html#About. Accessed 20 Mar 2015

United States Environmental Protection Agency (2013b) Recycling and reuse: packaging material: European Union Directive: summary of 1994 packaging directive. http://www.epa.gov/oswer/international/factsheets/200610-packaging-directives.htm. Accessed 20 Mar 2015

United States Environmental Protection Agency (2014a) Climate change: overview of greenhouse gases. http://epa.gov/climatechange/ghgemissions/gases/ch4.html. Accessed 20 Mar 2015

United States Environmental Protection Agency (2014b) Wastes: non-hazardous waste – municipal solid waste. http://www.epa.gov/epawaste/nonhaz/municipal/. Accessed 20 Mar 2015

United States Environmental Protection Agency (2015) Landfill methane outreach program: emissions from landfills. 3 Mar 2015. http://www.epa.gov/lmop/basic-info/index.html#a02. Accessed 20 Mar 2015

United States Environmental Protection Agency, Office of Resource Conservation and Recovery Program Management, Communications, and Analysis Office (2011) chap 2: managing nonhazardous solid waste. In: RCRA orientation manual, pp II-2-5. Available via EPA.gov. http://www.epa.gov/solidwaste/inforesources/pubs/orientat/rom2.pdf. Accessed 20 Mar 2015

US-Brazil Joint Initiative on Urban Sustainability (2012) Brazilian national solid waste policy. http://www.epa.gov/jius/policy/brazil/brazilian_national_solid_waste_policy.html. Accessed 20 Mar 2015

Value Chain Management Center (2010) Food waste in Canada. Nov 2010. http://vcm-international.com/wp-content/uploads/2013/04/Food-Waste-in-Canada-112410.pdf. Accessed 20 Mar 2015

Value Chain Management Center (2014) "$27 billion" revisited: the cost of Canada's annual food waste. 10 Dec 2014. http://vcm-international.com/wp-content/uploads/2014/12/Food-Waste-in-Canada-27-Billion-Revisited-Dec-10-2014.pdf. Accessed 20 Mar 2015

Wakeland W, Cholette S, Venkat K (2012) Food transportation issues and reducing carbon footprint. In: Boye J, Arcand Y (eds) Green technologies in food production and processing, pp 201–202. Available via CleanMetrics.com. http://www.cleanmetrics.com/pages/Ch9_0923.pdf. Accessed 4 Apr 2015

Waste & Resources Action Program (2008) Food waste report. Apr 2008. http://wrap.s3.amazonaws.com/the-food-we-waste.pdf. Accessed 20 Mar 2015

Worldwatch Institute (2013) Food waste and recycling in China: a growing trend? http://www.worldwatch.org/food-waste-and-recycling-china-growing-trend-1. Accessed 20 Mar 2015

World Trade Organization (1995a) Agreement on technical barriers to trade. https://www.wto.org/english/docs_e/legal_e/17-tbt.pdf. Accessed 20 Mar 2015

World Trade Organization (1995b) Agreement on the application on sanitary and phytosanitary measures (SPS agreement). https://www.wto.org/english/tratop_e/sps_e/spsagr_e.htm. Accessed 20 Mar 2015

World Trade Organization (2015a) Sanitary and phytosanitary measures. https://www.wto.org/english/tratop_e/sps_e/sps_e.htm. Accessed 20 Mar 2015

World Trade Organization (2015b) Technical barriers to trade. https://www.wto.org/english/tratop_e/tbt_e/tbt_e.htm. Accessed 20 Mar 2015

World Trade Organization (2015c) Understanding the WTO: the organization: members and observers. https://www.wto.org/english/thewto_e/whatis_e/tif_e/org6_e.htm. Accessed 20 Mar 2015

Zhu D (2014) China's policies and instruments for developing the circular economy. Europe's World, 15 June 2014. Available via EuropesWorld.org. http://europesworld.org/2014/06/15/chinas-policies-and-instruments-for-developing-the-circular-economy/#.VQsYkI7G_gB. Accessed 20 Mar 2015

United States Environmental Protection Agency (2015) Wastes: non-hazardous waste – municipal solid waste. http://www.epa.gov/epawaste/nonhaz/municipal/. Accessed 20 Mar 2015
United States Environmental Protection Agency (2015) Landfill methane outreach program [illegible] landfills, 3 Mar 2015. [illegible] Accessed 20 Mar 2015
United States Environmental Protection Agency, Office of Resource Conservation and Recovery [illegible] Management [illegible] and Analysis [illegible] municipal solid waste [illegible] documentation manual [illegible] Available at EPA.gov http://www.epa.gov/[illegible]. Accessed 20 Mar 2015
US-Brazil Joint Initiative on Urban Sustainability (2015) Brazilian national solid waste policy. http://www.epa.gov/jius/policy/brazil/brazilian_national_solid_waste_policy.html. Accessed 20 Mar 2015
Value Chain Management Centre (2010) Food waste in Canada, Nov 2010. http://vcm-international.com/wp-content/uploads/2013/04/Food-Waste-in-Canada-112410.pdf. Accessed 20 Mar 2015
Value Chain Management Centre (2014) $27 billion revisited: the cost of Canada's annual food waste, 10 Dec 2014. [illegible] Food Waste in Canada [illegible] Accessed 20 Mar 2015
Wakeland W [illegible]
[illegible]

Chapter 24
Habitat Loss, Agrobiodiversity, and Incidental Wildlife Loss

Leslie Couvillion

24.1 Introduction

Producing food can carry a steep ecological price. Over the past century, the need to grow enough food to support an expanding global population has resulted in vast swaths of forests and wetlands being converted into farmland or pastures. This development is a result of the industrialization of the food production systems around the world, the increased use of pesticides and synthetic fertilizers, the planting of genetically-modified crops (GMOs) and monocultures, and the displacement of small family or self-sufficient farms. Large modern agricultural operations are tied to habitat losses (through land displacement), water pollution, soil contamination, overgrazing, invasive species introductions, water shortages, and greenhouse gas (GHG) emissions. These environmental impacts can directly and indirectly harm wild plant and animal populations. The consequence is biodiversity loss, sometimes of threatened or endangered species.

Yet, the relationship between agriculture and biodiversity is not necessarily adverse. Good agricultural practices can support wildlife diversity and aid conservation efforts. Indeed, agricultural lands can provide ideal habitats for some species, such as birds, which nest and feed on farmland. Natural biodiversity can, in turn, be a boon to farmers and their communities. Nature provides key inputs (such as genetic resources in the form of seeds) for agriculture, while also delivering a variety of ecosystem services that ensure the physical, cultural, and spiritual well-being of human societies. Such crucial ecosystem services include water filtration, crop pollination, and recreation. Therefore, positive feedback loops link

L. Couvillion (✉)
Yale School of Law, New Haven, CT, USA
e-mail: leslie.couvillion@yale.edu

G. Steier, K.K. Patel (eds.), *International Farm Animal, Wildlife and Food Safety Law*, DOI 10.1007/978-3-319-18002-1_24

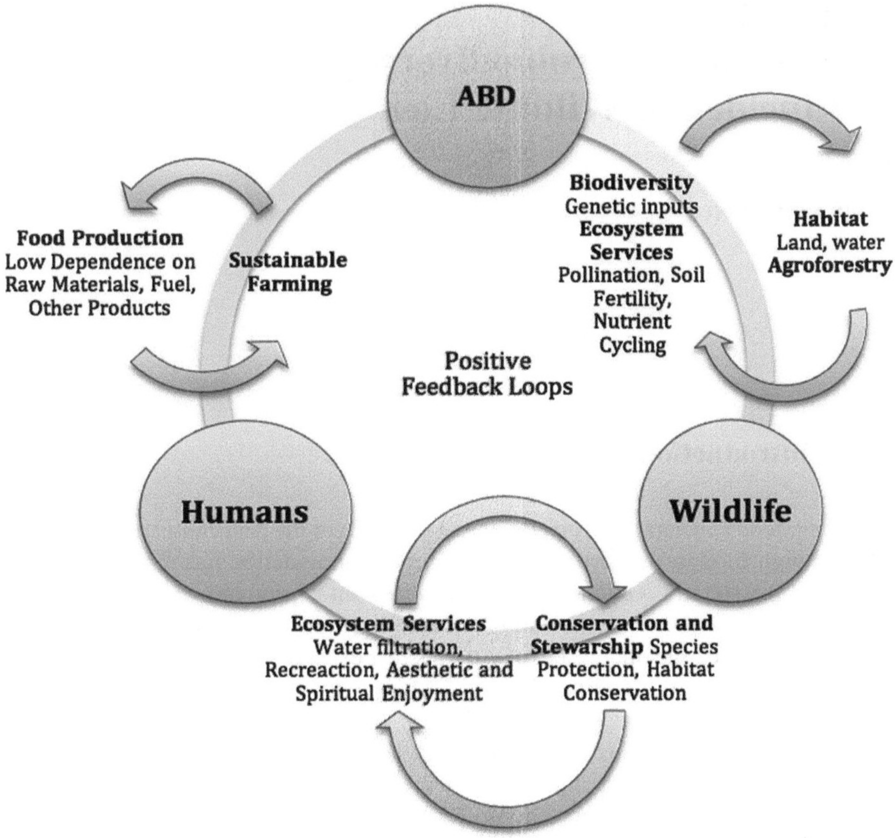

Fig. 24.1 Positive Feedback Loops Between Agrobiodiversity (ABD), Wildlife and Human Communities. This graph illustrates the positive feedback loops and the main factors associated with the interconnected aspects of agrobiodiversity, wildlife, and human communities

sustainable agriculture techniques, biodiversity levels, and human welfare; more of one can help to sustain more of the others (Fig. 24.1).[1]

The concept of "agricultural biodiversity," or "agrobiodiversity," has emerged to capture these various interrelationships. Agrobiodiversity is "a broad term that includes all of the components of biological diversity relevant to food and agriculture, and all components of biological diversity that constitute the agro-ecosystem"[2] at genetic, species, and ecosystem levels. It includes the "variety and variability of animals, plants and micro-organisms used directly or indirectly for food and agriculture, including crops, livestock, forestry and fisheries."[3] This

[1]See Fig. 24.1 for a graphic of the positive feedback loops linking sustainable agricultural practices, wildlife biodiversity, and human communities.

[2]Convention on Biological Diversity (2000), Annex 1.

[3]Moberg (2004), Box 1.

holistic notion goes hand-in-hand with an appreciation for local, indigenous forms of knowledge about growing and harvesting processes. Indeed, traditional farming helps to maintain certain natural or semi-natural habitats. Therefore, the agrobiodiversity concept encapsulates the vast interconnections between human management systems, genetic resources, and ecological health.

Currently, however, few governments or international bodies regulate agrobiodiversity as a distinct category of law. At the international level, a variety of instruments govern issues of biodiversity conservation, habitat preservation, and endangered species protection separately (though often with overlapping aims and coordinated measures). These include the Convention for Biological Diversity (CBD), the Convention on International Trade in Endangered Species of Wild Flora and Fauna (CITES), and the Ramsar Convention on Wetlands, among several others.[4] At the national level, most countries have a patchwork of federal statutes and regulations modeled after international conventions. These laws have usually been developed independently of one another—frequently, decades apart. Only seldom do these laws directly address agrobiodiversity.[5] Therefore, governments tend to rely on generalized, disconnected pieces of legislation to manage the complex interrelationships between habitat loss, agrobiodiversity, and wildlife loss. This tendency means that many potential synergies between these regulatory categories are overlooked.

However, integrative and innovative regulatory approaches are cropping up around the world. At the international level, the CBD launched an agrobiodiversity work plan in the mid-1990s.[6] Active since 2000, this initiative provides an assortment of tools to help national governments "mainstream" agricultural biodiversity issues across various sectors.[7] At the national level, a growing body of law encourages (and sometimes subsidizes) sustainable agriculture practices, such as agro-forestry[8] or ecological farming,[9] to help combat habitat and species loss on agricultural lands. The European Union (EU), in particular, is starting to loom large in the agrobiodiversity arena. In 2001, the European Commission launched an EU Biodiversity Action Plan for Agriculture.[10] The most significant feature of this plan is its support of "agri-environment payment schemes," through which Member

[4]See *infra* Sect. 24.2.

[5]However, wildlife protection laws may indirectly impact agricultural activities by prohibiting or restricting disruptive human activities (including intensive agriculture) in or near designated areas. For example, the Endangered Species Act and Wilderness Act in the United States take this approach. See *infra* Sect. 24.3.

[6]See *infra* Sect. 24.2.1.

[7]United Nations Convention on Biological Diversity (2015i).

[8]See Place et al. (2012). For instance, Kenya's Agriculture (Farm Forestry) Rules require farmers to establish and maintain farm forestry on at least 10 % of their agricultural lands. See Appendix.

[9]See Organic Farming (England Rural Development Programme) Regulations (2001). The National Archives (2001). These regulations provide for aid payments to farmers who introduce organic farming methods and comply with certain environmental conditions.

[10]See *infra* Sect. 24.4.3.

States reward farm owners who achieve specific biodiversity gains or ecosystem health improvements on their lands.[11] The plan also incorporates a variety of other instruments, such as rural development measures and programs for conserving and utilizing genetic resources in agriculture. However, there is not yet a binding EU-level directive on agrobiodiversity.

Given the primary role of international law in this field, this chapter begins with an overview of the main international law instruments governing issues of agricultural biodiversity, habitat loss, and species protection. The following sections then highlight the laws of a select group of countries and regions: the United States, the EU, and China. As these summaries of national- or regional-level laws and policies will show, some governments have developed better-coordinated approaches to managing agrobiodiversity issues than others. The United States, for instance, does not yet have a national Biodiversity Action Plan or a national policy on agrobiodiversity. In contrast, both the EU and China have national Biodiversity Action Plans (with ambitious goals of halting biodiversity loss within their respective borders by 2020), as well as a number of agrobiodiversity-focused measures to help reach their biodiversity targets. Overall, managing the intersections between agricultural and natural biodiversity is a vast, complex, and developing field. The Appendix provides a more comprehensive survey of relevant national and regional laws from around the world.

24.2 International Treaties and Conventions

International law plays a critical role in the complex regulatory arena surrounding agrobiodiversity issues. International treaties and conventions exist on topics ranging from conserving biodiversity[12] to regulating trade in endangered species[13] to halting losses of particular habitat types (like wetlands).[14] In addition, the United Nations Food and Agriculture Organization (UN FAO or FAO) has published a set of Good Agricultural Practices (GAPs) that draw from many of the international regulatory frameworks discussed below. The non-binding GAP guidelines help national and regional governments to develop locally-appropriate agricultural programs that promote economic, social, and environmental sustainability goals.[15] Therefore, a combination of hard and soft international law instruments provide the regulatory backdrop for governing habitat loss, agrobiodiversity, and wildlife loss at the global level.

[11]European Commission (2015a).

[12]See *infra* Sect. 24.2.1.

[13]See *infra* Sect. 24.2.4.

[14]See *infra* Sect. 24.2.7.

[15]Food and Agriculture Organization of the United Nations (2007).

24.2.1 Convention on Biological Diversity (CBD) (1992)

The Convention on Biological Diversity (CBD)[16] is a global agreement on all aspects of biodiversity.[17] In the early 1990s, the UN Environment Programme (UNEP) convened a group of technical and legal experts to draft a treaty that would respond to "the world community's growing commitment to sustainable development."[18] This new attitude was prompted by "a growing recognition that biological diversity is a global asset of tremendous value to present and future generations" and that human actions were threatening both species and ecosystems at an alarming rate.[19] CBD was officially unveiled at the June 1992 UN Conference on Environment and Development (UNCED, more commonly known as the Rio Earth Summit).[20] The treaty remained open for signature until June 1993 and entered into force in December of that year. As of early 2015, CBD has 194 parties and 168 signatories.[21] Its central goals are to promote:

(1) Biodiversity conservation;
(2) Sustainable use of the components of biological diversity; and
(3) Equitable sharing of the benefits arising from the use of genetic resources.[22]

At the third meeting of the Conference of the Parties (COP 3) in 1996, the Parties decided to expand CBD to include agrobiodiversity. Decision III/11 (titled "Conservation and sustainable use of agricultural biological diversity") formally "t[ook] note of the interrelationship of agriculture with biological diversity" and established a multi-year work program on agrobiodiversity.[23] The goals of the program were:

(1) To promote the positive effects and mitigate the negative impacts of agricultural practices on biological diversity in agro-ecosystems and their interface with other ecosystems;
(2) To promote the conservation and sustainable use of genetic resources of actual or potential value for food and agriculture; and
(3) To promote the fair and equitable sharing of benefits arising out of the utilization of genetic resources.[24]

To carry out these goals, Decision III/11 called on Parties to identify and assess "relevant ongoing activities and existing instruments," as well as gaps and opportunities for new action, at both international and national levels.[25] It also

[16]United Nations Convention on Biological Diversity (1992).
[17]United Nations Convention on Biological Diversity (2015g).
[18]United Nations Convention on Biological Diversity (2015g).
[19]United Nations Convention on Biological Diversity (2015g).
[20]United Nations Convention on Biological Diversity (2015g).
[21]See United Nations Convention on Biological Diversity (2015h).
[22]United Nations Convention on Biological Diversity (2015k).
[23]United Nations Convention on Biological Diversity (1996).
[24]United Nations Convention on Biological Diversity (1996), Part 1.
[25]United Nations Convention on Biological Diversity (1996), Part 15.

encouraged (but did not require) Parties to develop national strategies, programs, and plans on agrobiodiversity.

In 2000, COP 5 reviewed the initial phase of the agrobiodiversity work program and decided to endorse and expand Decision III/11. A new Decision V/5 made agrobiodiversity a "thematic programme of work" under CBD.[26] This was a significant step: COP has established only seven such work programmes in CBD's history.[27] Each work programme sets out general visions and principles and identifies key issues, timetables, output goals, and implementation measures. The agrobiodiversity work programme is based around four elements:

(1) Assessing the status of the world's agricultural biodiversity as well as trends in management practices;
(2) Promoting adaptive management techniques, which involve practices, technologies, and policies that foster the positive effects (including sustaining livelihoods) of agriculture on biodiversity while mitigating the negative impacts;
(3) Encouraging capacity-building among farmers, indigenous and local communities, organizations, and other stakeholders; and
(4) Implementing sectoral and cross-sectoral plans and programs that mainstream and integrate national plans and strategies on agrobiodiversity issues.[28]

A number of cross-cutting initiatives also "provide bridges and links between the thematic programmes."[29] The agrobiodiversity work programme involves four such initiatives on[30]: (1) pollinators[31]; (2) soil biodiversity[32]; (3) biodiversity for food and nutrition[33]; and (4) genetic use restriction technologies.[34] These initiatives call for an ecosystem approach, defined as "a strategy for the integrated management of land, water and living resources that promotes conservation and sustainable use in an equitable way" and recognizes humans as an integral part of ecosystems.[35]

In 2008, the CBD's Subsidiary Body on Scientific, Technical, and Technological Advice (SBSTTA) published the "In-Depth Review of the Implementation of the Programme of Work on Agricultural Biodiversity" in preparation for COP 9.[36]

---

[26]United Nations Convention on Biological Diversity (2000).

[27]The other six thematic work programmes under the CBD are Dry and Sub-humid Lands Biodiversity, Forest Biodiversity, Inland Waters Biodiversity, Island Biodiversity, Marine and Coastal Biodiversity, and Mountain Biodiversity. The work programmes "correspond to some of the major biomes on the planet." United Nations Convention on Biological Diversity (2015j).

[28]United Nations Convention on Biological Diversity (2015i).

[29]United Nations Convention on Biological Diversity (2015j).

[30]United Nations Convention on Biological Diversity (2015d).

[31]See United Nations Convention on Biological Diversity (2000)), Sect. II.

[32]See United Nations Convention on Biological Diversity (2006).

[33]See United Nations Convention on Biological Diversity (2006), Annex.

[34]See United Nations Convention on Biological Diversity (2015f).

[35]United Nations Convention on Biological Diversity (2015e).

[36]United Nations Convention on Biological Diversity (2007).

Generally, the report issued positive, but not glowing, findings. The SBSTTA concluded that the agrobiodiversity work programme as a whole is a "relevant framework" for achieving CBD objectives and a "useful framework" for addressing emerging biodiversity-related issues (such as climate change), while the cross-cutting initiatives (on pollinators, soil, and food and nutrition) are "particularly effective" at garnering international momentum.[37] However, "more work still needs to be done, in particular to strengthen both the application of the ecosystem approach, and the cooperation and synergy between agriculture and environment sectors at the national level."[38] Since the SBSTTA's report, there have not been any major new comprehensive reviews of the agrobiodiversity work programme's implementation. Therefore, it is unclear how much progress has been made on these issues since 2008. In any regard, the CBD agrobiodiversity work programme remains a critical resource for national governments seeking to develop integrated approaches to regulating agricultural biodiversity.

24.2.2 Nagoya Protocol on Access and Benefit-Sharing (2010)

In 2010, COP 10 to the CBD[39] adopted the Nagoya Protocol on Access to Genetic Resources and the Fair and Equitable Sharing of Benefits Arising from their Utilization to the Convention on Biological Diversity.[40] The Nagoya Protocol seeks to:

> shar[e] the benefits arising from the utilization of genetic resources in a fair and equitable way, including by appropriate access to genetic resources and by appropriate transfer of relevant technologies, taking into account all rights over those resources and to technologies, and by appropriate funding, thereby contributing to the conservation of biological diversity and the sustainable use of its components.[41]

Therefore, the Nagoya Protocol directly aims to fulfill one of the three main CBD objectives: fair and equitable sharing of the benefits arising out of the utilization of genetic resources.[42] The protocol concerns all genetic resources covered by the CBD, as well as traditional knowledge associated with the resources. For example, it applies to seeds used by farmers, the crops and other beneficial products resulting from the use of the seeds, and local indigenous knowledge surrounding the planting and harvesting of the seeds. "The Nagoya Protocol will create greater legal certainty and transparency for both providers and users of

---

[37]United Nations Convention on Biological Diversity (2007), p. 2.

[38]United Nations Convention on Biological Diversity (2007), p. 2, Part (e).

[39]See *supra* Sect. 24.2.1.

[40]United Nations Convention on Biological Diversity (2007).

[41]United Nations Convention on Biological Diversity (2015a).

[42]See *supra* Sect. 24.2.1.

genetic resources," both by "establishing more predictable conditions for access to genetic resources" and by "helping to ensure benefit-sharing when genetic resources leave the contracting party providing the genetic resources."[43] The protocol specifies a number of tools and mechanisms, including technology transfer and information-sharing techniques, to help Parties implement its objectives at the national level.[44]

The Nagoya Protocol has not yet entered into force. It requires a minimum of 50 ratifying Parties to do so; as of March 2015, the treaty has 57 Parties, only 33 of which have ratified the document.[45] Once in force, the Nagoya Protocol promises to be an essential supplement to the CBD for promoting agrobiodiversity worldwide. Overall, the protocol creates incentives for using genetic resources sustainably and equitably in order to enhance both biodiversity and human well-being.

24.2.3 International Treaty on Plant Genetic Resources for Food and Agriculture (IT PGRFA) (2001)

The UN FAO adopted the International Treaty on Plant Genetic Resources for Food and Agriculture (IT PGRFA)[46] in 2001, and the treaty has been in force since June 2004.[47] It aims to:

> recognize[e] the enormous contribution of farmers to the diversity of crops that feed the world; establish[] a global system to provide farmers, plant breeders, and scientists with access to plant genetic materials; and ensur[e] that recipients share benefits they derive from the use of these genetic materials with the countries where they have been originated.[48]

These goals echo some CBD objectives,[49] and the IT PGRFA was specifically drafted to be in harmony with the CBD. Like the CBD, the IT PGRFA recognizes the important links between agriculture, biodiversity—in particular, plant genetic diversity—and ecological and human well-being.

The treaty covers all plant genetic resources relevant for food and agriculture. One of its significant innovations is a "multi-lateral system" that puts 64 of the world's most important crops into a global pool of freely available genetic resources for users within ratifying nations.[50] Users must agree to use the materials only for food and agriculture research, breeding, and training purposes. The IT

[43]United Nations Convention on Biological Diversity (2015b).

[44]United Nations Convention on Biological Diversity (2015b).

[45]United Nations Convention on Biological Diversity (2015b).

[46]Food and Agriculture Organization of the United Nations (2009).

[47]Food and Agriculture Organization of the United Nations (2015a).

[48]Food and Agriculture Organization of the United Nations (2015c).

[49]See *supra* Sect. 24.2.1.

[50]Food and Agriculture Organization of the United Nations (2015c).

PGRFA is also notable for championing the notion of "farmers' rights:" the right of indigenous farmers to take equal part in benefit-sharing and national decision-making about plant genetic resources.[51] Furthermore, the treaty encourages the use and breeding of *all* crops, including local crops that do not belong to a major global food source category (e.g., maize or wheat).[52]

Agricultural sustainability is a primary focus of the treaty. Article 6 calls on Contracting Parties to "develop and maintain appropriate policy and legal measures that promote the sustainable use of plant genetic resources for food and agriculture."[53] The article lays out various measures, including:

- Promoting the development and maintenance of diverse farming systems;
- Broadening the genetic base of crops;
- Promoting locally-adapted crops; and
- "[S]trengthening research which enhances and conserves biological diversity by maximizing intra- and inter-specific variation for the benefit of farmers, especially those who generate and use their own varieties and apply ecological principles in maintaining soil fertility and in combating diseases, weeds and pests."[54]

Implementing Article 6 is a "standing priority item" on the agenda of the Governing Body of the IT PGFRA, and a process is underway to develop a Programme of Work on Conservation and Sustainable Use of PGFRA.[55]

The IT PGRFA is widely adopted. As of early 2015, there were 134 Contracting Parties and 193 members (including Signatories and Non-Contracting Parties).[56] The Governing Body has also reported ever-increasing synergies with CBD, UNEP, and other FAO departments, as well as expanded engagement with local organizations, farmers' rights groups, and civil society organizations.[57] Such widespread reach is important since, according to the IT PGRFA, the future of agriculture depends on international cooperation, with information and technology exchange likely to serve as key tools for enabling sustainable agriculture practices worldwide.

[51] Food and Agriculture Organization of the United Nations (2015c).

[52] Food and Agriculture Organization of the United Nations (2015c).

[53] Food and Agriculture Organization of the United Nations (2009), Art. 6.1. For example, India's Protection of Plant Varieties and Farmers' Rights Act sets out India's basic framework for protecting farmers' rights relating to the breeding of new plant varieties (including conserving, selecting, and improving farmers' varieties). See Appendix.

[54] Food and Agriculture Organization of the United Nations (2009), Art. 6.2(b).

[55] Food and Agriculture Organization of the United Nations (2015d).

[56] Food and Agriculture Organization of the United Nations (2015b).

[57] Food and Agriculture Organization of the United Nations (2013).

24.2.4 Convention on International Trade in Endangered Species of Wild Flora and Fauna (CITES) (1973)

The Convention on International Trade in Endangered Species of Wild Flora and Fauna (CITES) has a basic aim: "to ensure that international trade in specimens of wild animals and plants does not threaten their survival."[58] Wild plants and animals may be exchanged across borders for a number of reasons, such as to meet demands for food, pets, leather goods, ornamental plants, medicines, and timber products in diverse parts of the world where these goods are not naturally available. While such trade can bring about many economic and social benefits, it also carries environmental risks. Indeed, overexploitation of wildlife species is "the second-largest direct threat to many species after habitat loss."[59] At the time CITES was drafted, the risks of wildlife exploitation associated with trade were a relatively new topic of international discussion. Therefore, CITES—which regulates wildlife trade explicitly for conservation purposes—was a revolutionary tool. In hindsight, of course, "the need for CITES is clear."[60] The treaty has been in force since 1975, with a membership of over 180 Parties as of 2015.[61] This solid track record illustrates that CITES has become one of the world's most significant international conservation agreements. Nowadays, international wildlife trade is a multibillion-dollar enterprise, involving hundreds of millions of plant and animal specimens each year.

CITES works by setting a number of controls on trade in the specimens of select species. Over 35,000 plant and animal species are listed across three Appendices. These species may be (a) threatened (Appendix I), (b) at risk of becoming threatened (Appendix II), or (c) protected in at least one country, which has asked other Parties to assist in controlling their trade (Appendix III).[62] CITES requires all import, export, and re-export of covered species to be authorized through a licensing and permit system.[63] Under this system, trade in Appendix I species is allowed only in exceptional circumstances, while trade in Appendix II and III species is generally permitted subject to controls "to avoid utilization incompatible with [species'] survival."[64] Precise control requirements vary from country to country, so it is always necessary to refer to national laws.

---

[58]United Nations Convention on International Trade in Endangered Species of Wild Fauna and Flora (2015b).

[59]The World Wildlife Fund for Nature (WWF) (2015).

[60]United Nations Convention on International Trade in Endangered Species of Wild Fauna and Flora (2015b).

[61]United Nations Convention on International Trade in Endangered Species of Wild Fauna and Flora (2015b).

[62]At every COP meeting, Parties can submit proposals to amend the species listed in the first two appendices. Convention on International Trade in Endangered Species of Wild Fauna and Flora (2015a).

[63]United Nations Convention on International Trade in Endangered Species of Wild Fauna and Flora (2015a).

[64]United Nations Convention on International Trade in Endangered Species of Wild Fauna and Flora (2015a).

Since CITES focuses on protecting wild, rather than cultivated, species, it has limited direct relevance for agrobiodiversity. However, the goals of sustainable agrobiodiversity and CITES are mutually reinforcing. Unsustainable agricultural practices (like clear-cutting or planting monocultures) can cause habitat and biodiversity loss. Indeed, a driving concern behind CITES was that habitat loss problems (such as those from intensive agriculture) were already threatening the survival of many wild plant and animal species being traded globally.[65] The treaty sought to prevent exploitative wildlife harvesting activities from exacerbating these other pre-existing threats. Such links between habitat loss, wildlife loss, and agricultural activities suggest that CITES could be an important tool for promoting agrobiodiversity more directly. For instance, governments could incentivize farmers to manage their lands in ways that provide habitat for certain listed species (especially those that might be running out of suitable naturally-occurring habitat).[66] If such efforts were successful on a large enough scale, they might save a species from being threatened altogether. Therefore, CITES is a potentially powerful and underutilized tool for improving the world's agricultural biodiversity.

24.2.5 Convention on the Conservation of Migratory Species of Wild Animals (CMS or Bonn Convention) (1979)

The Convention on the Conservation of Migratory Species of Wild Animals (CMS), also known as the Bonn Convention, seeks to provide a "global platform for the conservation and sustainable use of migratory animals and their habitats."[67] Wildlife species do not respect national or regional boundaries, and some migratory animals cross international borders many times during their lives. In such cases, no one country's conservation efforts are enough; loss of habitat in just one link of the chain can threaten a species' entire survival. Therefore, CMS instructs all States falling within an animal's migratory range—in other words, all States with habitat areas that a species is known to live in, nest in, or travel through during its migration journeys—to coordinate their conservation measures. The primary objective of the treaty is to protect the entire range of a migratory animal's habitat.[68] Compared to CITES,[69] therefore, CMS has a more explicit focus on the link between habitat loss

[65]United Nations Convention on International Trade in Endangered Species of Wild Fauna and Flora (2015b).

[66]This is a similar concept to the "agri-environment payment schemes" already under way in much of the EU. See *infra* Sect. 24.4.3.

[67]United Nations Convention on the Conservation of Migratory Species of Wild Animals (2014b).

[68]United Nations Convention on the Conservation of Migratory Species of Wild Animals (2014b).

[69]See *supra* Sect. 24.2.4.

and species loss. Indeed, CMS foregrounds habitat conservation as a tool for protecting wildlife species.

Overall, CMS has a similar basic structure to CITES. Appendix I lists migratory species threatened with extinction, while Appendix II lists species "that need or would significantly benefit from international co-operation."[70] Parties have specific obligations that vary according to species and region. Commonly, these include measures such as mitigating migration obstacles and conserving or restoring areas where a target species lives. The scope of CMS, however, is somewhat restrained: the treaty covers just over 500 species[71] (compared to 35,000 under CITES) and has around 120 Parties (versus 180 under CITES). Most CMS Parties are found in the eastern hemisphere (including parts of Africa, western Europe, and Australia) and the western coast of South America.[72] This limited geographic scope means that the range of some migratory species may extend beyond the boundaries of CMS Party countries. Therefore, the convention encourages Party range states to engage with non-Party range states via regional agreements or less formal, non-binding instruments.[73] In addition, a number of other international organizations, NGOs, media partners, and corporations help to implement CMS objectives.[74]

As with CITES, CMS is not directly concerned with agrobiodiversity. Yet, its goals are also fully compatible with those of sustainable agriculture. Furthermore, CMS is unique among international law instruments in its capacity to develop "models tailored according to the conservation needs throughout a migratory range."[75] Therefore, it allows for targeted approaches to habitat and species preservation. For instance, a range state might conceivably fulfill its CMS obligations with a policy that supported farmers in maintaining trees that provide habitat for a listed migratory bird species on their lands. Another way to use CMS would be to prohibit agricultural operations from clearing new fields out of forested lands if such lands are known to provide critical habitat for a migratory mammal population (such as deer or antelope). All in all, CMS is another potentially powerful and overlooked tool for promoting sustainable agriculture, biodiversity, and habitat conservation goals. However, a greater number of countries need to join before the treaty can have its fullest effect.

[70]United Nations Convention on the Conservation of Migratory Species of Wild Animals (2014b).

[71]CMS covers terrestrial, aquatic, and avian migratory animal species. Convention on the Conservation of Migratory Species of Wild Animals (2014a).

[72]United Nations Convention on the Conservation of Migratory Species of Wild Animals (2014c).

[73]United Nations Convention on the Conservation of Migratory Species of Wild Animals (2014c).

[74]United Nations Convention on the Conservation of Migratory Species of Wild Animals (2014b).

[75]United Nations Convention on the Conservation of Migratory Species of Wild Animals (2014b).

24.2.6 UN Convention to Combat Desertification (UNCCD) (1994)

Along with climate change and biodiversity loss, desertification was identified as one of the greatest challenges to sustainable development at the 1992 Rio Earth Summit.[76] In 1994, the UN Convention to Combat Desertification (UNCCD) provided the first universal legal policy and advocacy framework on the issue. The goal of the Parties is "to forge a global partnership to reverse and prevent desertification/land degradation and to mitigate the effects of drought in affected areas in order to support poverty reduction and environmental sustainability."[77] The UNCCD is particularly concerned with combating desertification trends and drought conditions in drylands areas (i.e., arid, semi-arid, and dry sub-humid regions). The convention is widely adopted, with 195 Parties as of May 2014. It has been in force since December 1996.[78]

Overall, the UNCCD illustrates an understanding of how tightly linked sustainable land use, environmental integrity, and human quality-of-life are. While the UNCCD is not directly concerned with agrobiodiversity, intensive agricultural practices are a major contributor to desertification trends worldwide.[79] Furthermore, the convention recognizes that the dynamics of land use and biodiversity "are intimately connected."[80] Therefore, sustainable land management techniques are likely to have the triple benefit of forestalling land degradation, promoting agricultural biodiversity, and alleviating poverty. More so than most other international treaties, the convention embraces a largely "bottom-up" approach, encouraging participation by local and indigenous peoples.[81] As a result, the UNCCD could be a unique tool to help nations achieve a wide range of sustainability and biodiversity goals.

24.2.7 Ramsar Convention on Wetlands (1971)

The Ramsar Convention—formally titled the Convention on Wetlands of International Importance, Especially as Waterfowl Habitat[82]—recognizes that wetlands are among the planet's most diverse and productive ecosystems. In addition to their ecological functions, wetlands have economic, cultural, scientific, and recreational

[76]United Nations Convention to Combat Desertification (2012).

[77]United Nations Convention to Combat Desertification (2012).

[78]United Nations Convention to Combat Desertification (2012).

[79]United Nations. World Day to Combat Desertification.

[80]United Nations Convention to Combat Desertification (2012).

[81]United Nations Convention to Combat Desertification (2012).

[82]United Nations Convention on Wetlands of International Importance especially as Waterfowl Habitat [Ramsar Convention] (1994).

value. Therefore, the convention's mission is "the conservation and wise use of all wetlands through local and national actions and international cooperation, as a contribution towards achieving sustainable development throughout the world."[83] As of 2015, the Ramsar Convention had 168 Contracting Parties and over 2100 designated Ramsar sites (totaling more than 200 million hectares worldwide).[84] It is the single most important framework for intergovernmental cooperation on wetland issues.[85]

The convention sets out a framework for national action consisting of three pillars:

(1) Working towards the "wise use" of all wetlands;
(2) Designating suitable wetlands for the list of Wetlands of International Importance (the "Ramsar List") and ensuring their effective management; and
(3) Cooperating internationally on transboundary wetlands, shared wetland systems, and shared species.[86]

As with the UNCCD,[87] the Ramsar Convention acknowledges the complex interrelationships between land use, ecological fitness, species protection, and human welfare. Indeed, the "wise use" of wetlands is defined as "the maintenance of [wetlands'] ecological character, achieved through the implementation of ecosystem approaches, within the context of sustainable development" for the benefit of both people and nature.[88]

The Ramsar Convention's measures to promote the conservation and sustainable use of wetlands have important implications for agriculture. Indeed, agriculture (along with urban development) is one of the main threats to wetlands worldwide. There are ever-greater pressures to reclaim and destroy natural wetlands for agricultural purposes, from feeding a growing global population to cultivating energy crops.[89] Rice fields, crop fields on river floodplain soils, and plantations on drained/reclaimed peatlands (such as Indonesia's oil palm plantations) pose the greatest hazards.[90] Under the convention, nations should seek to replace destructive agricultural practices with more sustainable ones, such as using flood-resistant crops in

---

[83]United Nations Convention on Wetlands of International Importance especially as Waterfowl Habitat [Ramsar Convention] (2014c).

[84]United Nations Convention on Wetlands of International Importance especially as Waterfowl Habitat [Ramsar Convention] (2014a). In 2013, Canada announced its intention to withdraw from the convention – the first country to do so.

[85]United Nations Convention on Wetlands of International Importance especially as Waterfowl Habitat [Ramsar Convention] (2014b).

[86]United Nations Convention on Wetlands of International Importance especially as Waterfowl Habitat [Ramsar Convention] (2014c).

[87]See *supra* Sect. 24.2.6.

[88]United Nations Convention on Wetlands of International Importance especially as Waterfowl Habitat [Ramsar Convention] (2014d).

[89]Verhoeven and Setter (2009).

[90]Verhoeven and Setter (2009).

naturally functioning floodplains or limiting the use of fertilizer and pesticides in reclaimed wetlands. Such efforts will restore or preserve existing wetlands and increase biodiversity in these vital, species-rich areas. Therefore, the Ramsar Convention is an important tool for restricting a range of intensive agricultural practices that impact wetlands habitats.

24.2.8 World Heritage Convention (1972)

The World Heritage Convention seeks to protect natural and cultural sites of international importance. Under the convention, the UN Educational, Scientific, and Cultural Organization (UNESCO) maintains a list, known as the "World Heritage List," of designed sites to be protected. For a site to be listed, it must be of "outstanding universal value" and meet at least one out of ten selection criteria.[91] The convention also lays out the duties for Member States in identifying, protecting, and preserving designated sites. There are 191 State Parties to the World Heritage as of August 2014.[92] According to UNESCO:

> [t]he most significant feature of the [Convention] is that it links together in a single document the concepts of nature conservation and the preservation of cultural properties. The Convention recognizes the way in which people interact with nature, and the fundamental need to preserve the balance between the two.[93]

Therefore, the World Heritage Convention acknowledges that ecological and social well-being are inherently interlinked. In particular, it identifies biodiversity preservation as a collective concern. One of the ten selection criteria for a World Heritage Site is whether the site:

> contains the most important and significant natural habitats for in-situ conservation of biological diversity, including . . . threatened species of outstanding universal value from the point of view of science or conservation.[94]

As a result, a variety of wildlife habitats can qualify as natural heritage sites. Indeed, listed sites include forests, mountains, lakes, islands, deserts, and even agricultural landscapes.[95] Therefore, the World Heritage Convention can impact agrobiodiversity in at least two ways:

(1) By designating agricultural landscapes as natural heritage sites. This can ensure that traditional, sustainable land-use systems on these lands are preserved; and

---

[91] United Nations Educational, Scientific and Cultural Organization (2015).

[92] See United Nations Educational, Scientific and Cultural Organization (2014).

[93] United Nations Educational, Scientific and Cultural Organization (2014).

[94] United Nations Educational, Scientific and Cultural Organization (2015).

[95] Designated World Heritage Sites have included nineteenth century coffee plantations in Cuba, grape and olive plantations in Croatia dating to ancient Greece, and the subak water management system in Bali, Indonesia. United Nations Educational, Scientific and Cultural Organization (2013).

(2) By designating non-agricultural lands that might otherwise be used for agriculture as natural heritage sites. If a site (such as a forest) gains world heritage status, harmful agricultural activities can be prohibited or restricted. Therefore, forest land that could be turned into cropland or pastures can be conserved instead.

Overall, the World Heritage Convention is a valuable instrument for protecting biodiversity and promoting sustainable agriculture, especially since the treaty highlights the connections between cultural and natural environments. However, it can only be applied in "outstanding" circumstances and therefore has limited practical potential as a tool for change.

24.3 United States (US)

The United States has joined many international conventions concerned with biodiversity, species protection, and habitat conservation, including CITES,[96] UNCCD,[97] the Ramsar Convention,[98] and the World Heritage Convention.[99] (The US is not a party to CMS,[100] and it has signed but not ratified CBD[101] and IT PGRFA[102]). A number of federal laws implement the country's international treaty commitments. Unlike many other countries or regions (such as the EU[103] and China[104]), the United States does not have a general, comprehensive biodiversity law or policy at the national level. The US also lacks major federal legislation on sustainable agriculture or agrobiodiversity. Instead, there is a patchwork of statutes on wildlife trade, species protection, and wilderness conservation. The Lacey Act[105] and the Endangered Species Act (ESA)[106] are the most prominent of these laws. Their primary impacts on agricultural practices are mostly indirect, a result of the inherent feedback loops that exist between natural biodiversity, ecological resilience, and agricultural productivity.[107]

---

[96] See *supra* Sect. 24.2.4.

[97] See *supra* Sect. 24.2.6.

[98] See *supra* Sect. 24.2.7.

[99] See *supra* Sect. 24.2.8.

[100] See *supra* Sect. 24.2.5.

[101] See *supra* Sect. 24.2.1.

[102] See *supra* Sect. 24.2.3.

[103] See *infra* Sect. 24.4.

[104] See *infra* Sect. 24.5.

[105] See *infra* Sect. 24.3.1.

[106] See *infra* Sect. 24.3.2.

[107] However, the USDA Agricultural Marketing Service has a Good Agricultural Practice (GAP) Certification Program modelled off of the FAO system. This voluntary audit/certification program aims to verify that growers and packers use good agricultural and/or good handling practices. United States Department of Agriculture (2015).

24.3.1 *Lacey Act (1900)*

Enacted at the turn of the last century, the Lacey Act[108] was the United States' first federal wildlife protection law. Although it initially focused on stemming illegal hunting, its scope has expanded significantly over the decades.[109] Now, the act aims to combat trafficking in illegal plant and animal species. Therefore, the Lacey Act helps to enforce CITES[110] within the US.[111] It deters the unsustainable harvest and trade of species protected by any state, federal, tribal, or foreign law and prevents the import (and interstate transport) of potentially harmful listed species. The Fish and Wildlife Service (FWS), a federal agency with the US Department of the Interior (DOI), implements the act.[112] Violating the Lacey Act is a federal crime with substantial civil and criminal penalties.[113]

In particular, the Lacey Act does not allow anyone:

> to import, export, sell, acquire, or purchase fish, wildlife or plants that are taken, possessed, transported, or sold: 1) in violation of U.S. or Indian law, or 2) in interstate or foreign commerce involving any fish, wildlife, or plants taken possessed or sold in violation of State or foreign law.[114]

The act's scope includes species listed in CITES, as well as species protected by any other US or foreign law. In addition, it restricts the import and interstate transport of "injurious" wildlife deemed to be harmful "to human beings, to the interests of agriculture, horticulture, forestry, or to wildlife or the wildlife resources of the United States."[115] This prohibition applies to invasive species and other pests that might threaten agricultural as well as natural biodiversity.

The Lacey Act has the potential to be a powerful tool for change. It is unique in its capacity to protect species and habitats even outside of the United States.[116] For example, Congress amended the law in 2008 to increase the number of prohibited plants and plant products. The list of banned items now includes timber products made from illegally logged woods. Since illegal logging is a major contributor to habitat and biodiversity loss around the world (especially in tropical rainforest regions), these amendments can help protect endangered and threatened species

---

[108] United States Fish and Wildlife Service, International Affairs (2004).

[109] United States Fish and Wildlife Service, International Affairs (2004).

[110] See *supra* Sect. 24.2.4.

[111] Environmental Investigation Agency (2015). CITES is also implemented via the Endangered Species Act (ESA) in the US. See *infra* Sect. 24.3.2.

[112] United States Fish and Wildlife Service, International Affairs (2004).

[113] The Lacey Act allows fines up to $250,000 for individuals and $500,000 for organizations, as well as imprisonment up to five years. Beaudry (2014).

[114] United States Fish and Wildlife Service, International Affairs (2004).

[115] United States Fish and Wildlife Service, International Affairs (2013f).

[116] United States Fish and Wildlife Service, International Affairs (2013f).

in foreign countries.[117] These provisions could provide a model for new amendments that promote agrobiodiversity aims more directly. For instance, the Lacey Act could, in theory, restrict the import of food products grown using unsustainable agricultural techniques. Such restrictions would bolster the act's capacity to help forestall biodiversity loss and promote sustainable food systems around the globe. Therefore, the Lacey Act is one place an advocate might look for a creative way to boost these goals.

24.3.2 Endangered Species Act (ESA) (1973)

The Endangered Species Act (ESA)[118] is a blockbuster piece of legislation for both species and habitat protection in the United States. Congress passed the law out of alarm that many native plants and animals were at risk of becoming extinct. The act recognizes the "[a]esthetic, ecological, educational, recreational, and scientific value" of the country's natural heritage and seeks to preserve biodiversity for these ends.[119] As a result, the ESA helps to implement a number of international treaties, including CITES.[120] It covers a wide range of terrestrial and aquatic plant and animal species.[121] Some states also administer their own State Endangered Species Acts, which may be even broader than the federal law.[122]

The ESA works by restricting certain activities that might harm protected species or their habitats. First, it calls on agencies to maintain lists of endangered and threatened plant and animal species.[123] An "endangered species" is "any species [other than certain insect pest species] which is in danger of extinction throughout all or a significant portion of its range."[124] A "threatened species," in

[117] Environmental Investigation Agency (2015).

[118] United States Fish and Wildlife Service, International Affairs (2013e).

[119] United States Fish and Wildlife Service, International Affairs (2013a).

[120] United States Fish and Wildlife Service, International Affairs (2013b), Sect. 2(a)(4)(A). See *supra* Sect. 24.2.4.

[121] The FWS implements the ESA for terrestrial and freshwater species, while the National Marine Fisheries Service (NMFS), a branch of the National Oceanic and Atmospheric Administration (NOAA), administers the act for marine species. United States Fish and Wildlife Service (2013).

[122] For instance, California's ESA protects "candidate" species (i.e., species that are proposed for listing but are not yet listed), while the federal ESA only protects listed species. See California Department of Fish and Wildlife (2015).

[123] Specifically, the ESA requires the Secretary of the Interior, in consultation with the Secretary of Commerce, to create, maintain, and periodically review national lists of endangered and threatened species. These lists are to be published Federal Register. See United States Fish and Wildlife Service, International Affairs (2013d), Sect. 4(a)(2). In addition to the national list, states prepare their own lists of endangered and threatened species within their boundaries. These state lists may include species that are only considered endangered or threatened within a single state or region. See United States Fish and Wildlife Service (2013).

[124] United States Fish and Wildlife Service, International Affairs (2013c), Sect. 3(6).

comparison, is "any species which is likely to become an endangered species within the foreseeable future throughout all or a significant portion of its range."[125] As of January 2013, over 2000 global species (approximately 1400 of which occur in the US) are listed as endangered or threatened.[126]

Next, the act prohibits the "taking"[127] of listed animal species.[128] However, certain takings may be allowed under special circumstances. Parties can apply to the government for "incidental take permits," which will only be issued if the applicant can design, implement, and fund a habitat conservation plan that will "minimize[] and mitigate[] harm to the impacted species during the proposed project."[129] Habitat conservation plans are legally binding agreements between the government and the permit holder.[130] Projects in "critical habitat zones," or areas critical to the conservation of a threatened or endangered species, may require heightened protection measures.[131]

With its powerful ban on takings, the ESA can be a crucial tool for promoting agrobiodiversity. For instance, the ESA can be used in connection with the National Environmental Policy Act (NEPA)[132] to curtail intensive agricultural activities (especially those planned in or near protected habitats) which threaten listed species. Environmental reviews under NEPA must consider the impacts of a proposed action on ESA-listed endangered species. Although NEPA does not have substantive force,[133] the ESA does. Therefore, an advocate could invoke both NEPA and the ESA to block an agricultural operator from expanding into a

---

[125]United States Fish and Wildlife Service, International Affairs (2013c), Sect. 3(20).

[126]United States Fish and Wildlife Service (2013).

[127]"Taking" means "to harass, harm, pursue, hunt, shoot, wound, kill, trap, capture, or collect, or to attempt to engage in any such conduct." United States Fish and Wildlife Service International Affairs (2013c), Sect. 3(19).

[128]The ESA does not prohibit the taking of listed plants, although it does forbid collecting or maliciously harming listed plant species on federal land. United States Fish and Wildlife Service (2013).

[129]United States Fish and Wildlife Service, International Affairs (2014b) (referring to Sect. 10a(1)(B) of the ESA).

[130]United States Fish and Wildlife Service, International Affairs (2014b).

[131]United States Fish and Wildlife Service, International Affairs (2013c), Sect. 3(5). Generally, a critical habitat designation only affects projects involving federal agency action (including private actions that rely on federal funding or permits). See United States Fish and Wildlife Service, International Affairs (2014a).

[132]Under NEPA, federal agencies must consider the environmental impacts of proposed actions as part of their decision-making process. These requirements apply to any agency undertaking a "major federal action," a category that includes the adoption of official policies, plans, and programs. NEPA's scope is broad and will, therefore, encompass many "major federal actions" in the agriculture sectors. States Department of Transportation, Federal Highway Administration (2002).

[133]Indeed, the statute's major limitation is that it has only procedural, and not substantive, force: it requires an agency to consider the environmental impacts of a project, but it does not mandate that an agency take action to mitigate these impacts. See Czarnezki (2006).

critical habitat zone or from clearing cropland in an area that houses a threatened species. Similarly, an operator might be forced to restrict its use of synthetic pesticides if the chemicals are found to threaten endangered species (such as monarch butterflies, which are important pollinating insects).[134] Even when the ESA is enforced outside of agricultural sector, it can have important co-benefits for improving agricultural sustainability. Since the ESA is an effective tool for conserving species and habitats, it helps to boost biodiversity generally. The positive feedback loops between wildlife diversity and agriculture mean that greater natural biodiversity leads to enhanced agricultural biodiversity.[135]

24.3.3 Other Biodiversity and Wildlife Laws

In addition to the Lacey Act[136] and the ESA,[137] a number of other federal laws protect the country's species and habitats. The US Wilderness Act,[138] passed in 1964, creates a formal mechanism for designating federal wilderness areas. These undeveloped areas are to "be protected and managed so as to preserve [their] natural conditions."[139] Most disruptive human activities—including intensive agricultural activities—are prohibited in these areas. Therefore, as with the ESA, the Wilderness Act can act as a restraint on unsustainable agricultural practices. It can also help support sustainable agriculture by improving natural biodiversity and, thus, enhancing ecosystem services. Currently, around 5 % of the total land area in the United States is set aside as designated wilderness areas.[140] As a result, the Wilderness Act's scope may not be as broad as that of the ESA.

Other important federal conservation laws in the US include the Marine Mammal Protection Act (MMPA),[141] which extends ESA-like protections to non-listed marine mammal species, the Migratory Bird Treaty Act (MBTA),[142] which generally prohibits the taking of Native American bird species, the Wild Bird Conservation Act,[143] which generally prohibits the import of CITES-listed[144] birds and

[134]See United States Fish and Wildlife Service, International Affairs (2014a).

[135]See Fig. 24.1.

[136]See *supra* Sect. 24.3.1.

[137]See *supra* Sect. 24.3.2.

[138]University of Montana, College of Forestry and Conservation's Wilderness Institute (1964).

[139]University of Montana, College of Forestry and Conservation's Wilderness Institute (1964), Sect. 2(c).

[140]University of Montana, College of Forestry and Conservation's Wilderness Institute. The beginnings of the national wilderness preservation system.

[141]Marine Mammal Commission (2007).

[142]United States Fish and Wildlife Service (1920).

[143]United States Fish and Wildlife Service, Office of Law Enforcement (2004).

[144]See *supra* Sect. 24.2.4.

encourages wild bird conservation programs in countries of origin, and the Fish and Wildlife Conservation Act,[145] which grants assistance to states for conservation plans and programs for non-game fish and wildlife. Like the Wilderness Act and the ESA, each of these laws can help protect natural biodiversity, with indirect impacts on agricultural biodiversity. Overall, however, the United States lacks a coherent, unifying law or policy on biodiversity and agriculture. The patchwork nature of the US regime contrasts with the European Union's more focused approach, discussed in the following section.

24.4 European Union (EU)

Biodiversity loss is an "enormous" challenge in the EU.[146] Around a quarter the region's species are threatened with extinction. In 2011, the European Commission adopted an ambitious European Union (EU) Biodiversity Strategy.[147] The plan calls for halting biodiversity losses by 2020. Its targets include achieving full implementation of EU nature legislation and promoting more sustainable agriculture and forestry practices. The strategy also helps the EU carry out its commitments under the wide range of international biodiversity treaties it has ratified, which include CBD[148] (as well as the Nagoya Protocol[149]), IT PGRFA,[150] CMS,[151] and UNCCD[152]; additionally, all EU Member States are (individually) parties to CITES,[153] the Ramsar Convention,[154] and the World Heritage Convention.[155] An array of EU- and Member State-level legislation implements these instruments within the EU.

However, the core of the region's biodiversity protection framework consists of two EU-level directives: the Habitats Directive[156] and the Birds Directive.[157] Both laws establish general protections for habitats and species, although neither has an explicit focus on the intersections between agriculture and biodiversity. This gap,

---

[145] Michigan State University College of Law (2015).
[146] European Commission (2015h).
[147] European Union (2011).
[148] See *supra* Sect. 24.2.1.
[149] See *supra* Sect. 24.2.2.
[150] See *supra* Sect. 24.2.3.
[151] See *supra* Sect. 24.2.5.
[152] See *supra* Sect. 24.2.6.
[153] See *supra* Sect. 24.2.4.
[154] See *supra* Sect. 24.2.7.
[155] See *supra* Sect. 24.2.8.
[156] See *infra* Sect. 24.4.1.
[157] See *infra* Sect. 24.4.2.

however, is slowly being filled. Since the early 2000s, the European Commission has been developing and implementing a Biodiversity Action Plan for Agriculture under the Common Agriculture Policy (CAP)[158] that is in harmony with the Habitats and Birds Directives. As a result of Member State actions under this plan, the EU is emerging as a model for other countries to look to for lessons on how to regulate agrobiodiversity issues.

24.4.1 Habitats Directive (1992)

The Habitats Directive[159] is the cornerstone of the EU's nature conservation policy. Together with the Birds Directive,[160] it fulfills the EU's obligations under the CBD. Despite its title, the Habitats Directive is concerned with individual species protection[161] as well as habitat protection.[162] Its purpose is to support "the preservation, protection and improvement of the quality of the environment, including the conservation of natural habitats and of wild fauna and flora."[163] As of 2015, the Habitats Directive protects over 1000 rare and threatened animal and plant species and over 200 habitat types (including forests, meadows, and wetlands that serve as core breeding and resting sites for listed species).[164] Member States must regularly monitor and report on the conservation status of the covered habitats and species within their territory.

One of the Habitat Directive's most significant features is the Natura 2000 habitat protection network. Natura 2000 is an EU-wide network of designated nature protection areas. These areas can be either "Special Areas of Conservation (SACs)," designated under the Habitats Directive, or "Special Protection Areas (SPAs)," designated under Birds Directive.[165] These lands can be either publically or privately owned. The Natura 2000 framework aims to ensure that the sites are managed sustainably, both ecologically and economically. Hence, they "are not strict nature reserves where all human activities are prohibited."[166] Therefore, Natura 2000 has a role to play in supporting sustainable agriculture practices (and limiting intensive ones) in certain areas throughout the EU. However, its potential as a tool to promote agrobiodiversity is limited to designated protection areas.

---

158 See *infra* Sect. 24.4.3.

159 EUR-Lex (1992).

160 See *infra* Sect. 24.4.2.

161 EUR-Lex (1992), Arts. 12 and 16.

162 EUR-Lex (1992), Arts. 6 and 8.

163 EUR-Lex (1992), Preamble.

164 European Commission (2015e).

165 See *infra* Sect. 24.4.2.

166 European Commission (2015e).

24.4.2 Birds Directive (1979)

The Birds Directive[167] is the EU's oldest piece of nature legislation. A key complement to the Habitats Directive,[168] it establishes a comprehensive protection scheme for all wild bird species naturally occurring in the EU. Like the Habitats Directive, it is concerned with protecting both species and habitats. However, its protections are not as strict. Member States are required to:

> take the requisite measures to maintain the population of [all of the EU's naturally occurring wild bird species] at a level which corresponds in particular to ecological, scientific and cultural requirements, while taking account of economic and recreational requirements.[169]

Therefore, the directive leaves open the possibility for activities that threaten bird species but have an economic or cultural value. Hunting, for instance, is allowed under certain circumstances.[170] However, the law does not address agricultural activities—presumably because agriculture's potential impacts on wild bird populations are less direct than those of hunting.

Along with the Habitats Directive, the Birds Directive forms the backbone of the Natura 2000 framework. Annex I lists protected habitats for endangered and migratory bird species and establishes a network of SPAs that supplement the Habitat Directive's SACs. Within SPAs, "activities that directly threaten birds, such as deliberate killing or capture, destruction of nests and taking of eggs, and trading in live or dead birds, are banned, with limited exceptions."[171] These prohibitions could limit some agricultural activities within SPAs. For instance, they could save woodlands from clear-cutting if the trees provide nesting grounds for a migratory bird species. Yet, as highlighted in the previous section, the Natura 2000 framework has a set geographic scope. Therefore, its potential as a tool to promote agrobiodiversity is limited.

24.4.3 Common Agricultural Policy (CAP) (1962)

The EU's Common Agricultural Policy (CAP), launched in early 1960s, is "a partnership between agriculture and society and between Europe and its farmers."[172] It is a common policy for all Member States, managed and funded at the European level. The policy aims to improve agricultural productivity and

[167] EUR-Lex (2009).
[168] See *supra* Sect. 24.4.1.
[169] EUR-Lex (2009), Art. 2.
[170] See EUR-Lex (2009), Art. 7.
[171] European Commission (2015g).
[172] European Commission (2014a).

support rural economies through the sustainable management of natural resources. CAP is not a static policy and has evolved significantly during the last decade-and-a-half. In 2003, the European Commission issued major reforms to allow the sector to adapt to new challenges, including climate change, water management, and biodiversity protection.[173] The 2015 version of CAP states that "[s]ound agricultural management practices can have a substantial positive impact on the conservation of the EU's wild flora and fauna" by preserving farm-genetic resources, biodiversity, and valuable habitats.[174] Therefore, the Commission now explicitly recognizes the link between agriculture and biodiversity.

Indeed, the European Commission has embraced agricultural biodiversity as a distinct concept, defining it broadly as:

- "[A]ll components of biological diversity of relevance for food and agriculture;"[175] and
- "[A]ll components of biological diversity that constitute the agro-ecosystem."[176]

A number of specific agrobiodiversity measures are promoted under CAP. Many of these measures concern species and habitat preservation and are designed to enable cross-compliance with the Habitats[177] and Birds Directives.[178] Some examples include:

- The European Action Plan for Organic Food and Farming, which heralds a new stage in organic farming promotion[179];
- Rural Development Programs, whose priorities include restoring, preserving, and enhancing ecosystems related to agriculture and forestry, as well as promoting social inclusion, poverty reduction, and economic development in rural areas[180]; and
- The Biodiversity Action Plan for Agriculture, discussed below.

Biodiversity Action Plan for Agriculture In 2001, the European Commission adopted an official Biodiversity Action Plan for Agriculture (Action Plan) under CAP. Its goal is "to improve or maintain biodiversity status and prevent further biodiversity loss due to agricultural activities."[181] The plan responds to mounting concerns around two trends:

[173]European Commission (2014a).
[174]European Commission (2015b).
[175]European Commission (2015b).
[176]European Commission (2015b).
[177]See *supra* Sect. 24.4.1.
[178]See *supra* Sect. 24.4.2.
[179]European Commission (2014b).
[180]European Commission (2015f).
[181]EUR-Lex (2001).

(1) Increases in intensive and specialized agricultural practices (e.g., using more chemicals and heavy machinery); and
(2) Higher rates of marginalization or abandonment of traditional land management.[182]

The Action Plan recognizes the reciprocal relationship between agriculture and biodiversity. It often highlights the mutual benefits between the two, while also stressing the pressures that farming can place on biodiversity.[183] Priorities under the plan include:

- Promoting sustainable farming practices and systems that benefit biodiversity directly or indirectly (especially in biodiversity-rich areas);
- Keeping intensive farming to levels that do not harm biodiversity;
- Maintaining and enhancing ecological infrastructures; and
- Conserving local or threatened livestock breeds or plant varieties.[184]

The Action Plan calls for using a number of CAP agrobiodiversity mechanisms. It also sets targets and timetables and provides for creating indicators for long-term monitoring and benchmarking. Member States are required to report on their progress and to detail any obstacles faced in implementing the plan.[185] Specific measures include:

- Rural development measures[186];
- Direct support schemes for farmers[187];
- Programs for conserving and utilizing genetic resources in agriculture[188];
- Plant health checks[189]; and
- Agri-environmental schemes in the field of rural development (discussed below).

Agri-environment payment programs are one of key instruments of the Action Plan. These results-based schemes reward farmers whose practices yield biodiversity-positive outcomes.[190] Generally, they work by providing fixed payments to farmers or land managers who deliver a specific environmental result (one that might not otherwise be profitable, such as improving water quality, creating woodland, or housing a native wildflower population) for a minimum period of time (usually, 5 years). The schemes take different forms in different Member States. For

[182] European Commission (2015b).
[183] EUR-Lex (2001).
[184] EUR-Lex (2001).
[185] EUR-Lex (2001).
[186] See EUR-Lex (2001).
[187] See EUR-Lex (2001).
[188] See EUR-Lex (2004).
[189] See EUR-Lex. Plant health checks.
[190] European Commission (2015c).

instance, the Netherlands uses a system based on private agreements with farming cooperatives, while Sweden's approach involves local or native communities. Other countries contract directly with individual farms and land managers.[191] For instance, in England, nearly 52,000 farmers and other land managers (whose holdings comprise over 70 % of the country's farmland) have signed up.[192] Agri-environment payment schemes "play a crucial role [in] meeting society's demand for environmental outcomes provided by agriculture," and their adoption has been steadily increasing in the EU and beyond.[193]

The successes of agri-environment scheme programs so far is promising. Indeed, the EU Biodiversity Action Plan for Agriculture is one of the few regional- or national-level polices on agrobiodiversity in the world. (While China[194] has a national Biodiversity Action Plan, it does not appear to have a separate plan on agricultural biodiversity.) Therefore, the EU is leading the way in the emerging field of agrobiodiversity, and the Biodiversity Action Plan for Agriculture provides a number of concrete tools for change that, if effective in the EU, could continue to be adopted in other places around the world.

24.5 China

China is one of the world's largest and most biodiverse countries. The Chinese Constitution is somewhat unique for enshrining biodiversity conservation principles in its text, which provides that "the State shall ensure the reasonable utilization of natural resources and protect the rare and valuable fauna and flora."[195] China's international commitments on biodiversity issues are vast: it is a party to CBD[196] (but not the Nagoya Protocol[197]), CITES,[198] UNCCD,[199] the Ramsar Convention,[200] and the World Heritage Convention[201]; it is not a party to CMS[202] or IT PGRFA.[203] At the national level, a multitude of laws and regulations address biodiversity conservation issues. Important examples include the Law on the

[191]European Commission (2015d).

[192]Gov.UK (2014).

[193]European Commission (2015a).

[194]See *infra* Sect. 24.5.

[195]Kurukulasuriya and Robinson (2006).

[196]See *supra* Sect. 24.2.1.

[197]See *supra* Sect. 24.2.2.

[198]See *supra* Sect. 24.2.4.

[199]See *supra* Sect. 24.2.6.

[200]See *supra* Sect. 24.2.7.

[201]See *supra* Sect. 24.2.8.

[202]See *supra* Sect. 24.2.5.

[203]See *supra* Sect. 24.2.3.

Protection of Wildlife,[204] the Law on Environmental Protection, the Agriculture Law, the Law on Forests, the Law on Water, the Law on Marine Environmental Protection, the Law on Grasslands, the Law on Fisheries, the Regulation on Nature Reserves, and the Regulation on Protection of Wild Plants.[205] Some sectoral departments and provincial governments have also adopted corresponding rules, regulations, and codes of conduct.

China does not have a comprehensive national law on biodiversity. However, in 2011, the Ministry of Environmental Protection (MEP), along with more than 20 other ministries and departments, adopted a new National Biodiversity Conservation Strategy and Action Plan (Strategy and Action Plan) to implement certain CBD provisions.[206] Like the EU's Biodiversity Strategy,[207] China's Strategy and Action Plan sets bold objectives. It aims to place 90 % of China's protected species and key ecosystems in nature reserves and halt all biodiversity loss in the country by 2020. The Government of China has issued a number of plans and programs under the Strategy and Action Plan. These include the China Nature Reserves Program (1996–2010), the China Master Plan for Ecological Conservation, the China Program for Ecological Environment Conservation, and the China Program for Conservation and Use of Biological Resources (2006–2020).[208] Such initiatives indicate that the central government has a genuine and growing commitment to improving biodiversity management in China.

The National Biodiversity Conservation Strategy and Action Plan does not explicitly invoke the agrobiodiversity concept. However, it does recognize the value of China's farm genetic resources and other forms of agricultural biodiversity. The Strategy and Action Plan estimates that China hosts over 1300 cultivated crops (including many of global food chain importance, such as rice and soy) and 1900 wild relatives, as well as more than 500 species of domesticated animals.[209] The plan calls attention to the extent to which this rich agricultural diversity is threatened:

> Erosion and loss of genetic resources is continuing. The habitats of some wild crop relatives have been destroyed and lost. 60 % to 70 % of the original distribution sites of wild rice have disappeared or shrunk. Some rare and endemic germplasm resources of crops, trees, flowers, livestock, poultry and fish suffer serious loss. Some local traditional and rare varieties have been also lost.[210]

The Strategy and Action Plan advocates sustainable practices to halt these losses. It calls for putting an end to "predatory exploitation of biological resources," as well as for putting in place systems that assure "access to and fair and equitable

---

[204]See *infra* Sect. 24.5.1.

[205]See Appendix.

[206]Convention on Biological Diversity (2015c). The plan had last been updated in 1994.

[207]See *supra* Sect. 24.4.

[208]Convention on Biological Diversity (2015c), p. 6.

[209]Convention on Biological Diversity (2015c), p. 5.

[210]Convention on Biological Diversity (2015c), pp. 5–6.

sharing of benefits from use of genetic resources and associated traditional knowledge."[211] The goals and measures set forth in the plan seek to steer China along a positive pathway toward improved agrobiodiversity management. The following sections discuss two promising existing legal instruments for furthering these efforts.

24.5.1 Law of the People's Republic of China on the Protection of Wildlife (1988)

China's Law on the Protection of Wildlife (Law on Wildlife)[212] is one of the country's principal national laws on biodiversity preservation. It was enacted "for the purpose of protecting and saving the species of wildlife which are rare or near extinction, protecting, developing and rationally utilizing wildlife resources and maintaining ecological balances."[213] The law applies to terrestrial and aquatic species that are rare or endangered, as well as terrestrial species "which are beneficial or of important economic or scientific value."[214] The Central Government's Department of Forestry is primarily responsible for administering the law. While the national law lays out general principles for wildlife protection, further regulations at the federal, provincial, and municipal levels establish specific measures for implementing the law fully.[215]

The Law on Wildlife declares that wildlife resources are owned by the State (i.e., the people).[216] It places affirmative responsibilities on citizens to protect these resources. Under Article 5, citizens have the "duty to protect wildlife resources and the right to inform the authorities of or file charges against acts of seizure or destruction of wildlife resources."[217] At the same time, the law recognizes that humans must utilize natural resources for social, economic, and scientific ends. Therefore, it advocates a balance between wildlife protection and the rational, sustainable use of wildlife resources. Indeed, the State must:

> pursue a policy of strengthening the protection of wildlife resources, actively domesticating and breeding the species of wildlife, and rationally developing and utilizing wildlife resources, and encourage scientific research on wildlife.[218]

[211] Convention on Biological Diversity (2015c), p. 9.

[212] China.org.cn (1988).

[213] China.org.cn (1988), Art. 1.

[214] China.org.cn (1988), Art. 2.

[215] See Appendix.

[216] China.org.cn (1988), Art. 3.

[217] China.org.cn (1988), Art. 5.

[218] China.org.cn (1988), Art. 4.

The law does not specifically address the relationship between agricultural and wildlife biodiversity. Article 29 contains the only explicit mention of agriculture. This provision orders local governments "to prevent and control the harm caused by wildlife so as to guarantee the safety of human beings and livestock and ensure agricultural and forestry production."[219] Such language appears to be geared at protecting agricultural lands (e.g., crop fields and pastures) from harmful wild species (e.g., pests or predators). It does not acknowledge the need to protect wildlife from harmful agricultural practices. This is one important gap in the Law on Wildlife that the National Biodiversity Conservation Strategy and Action Plan seems to be designed to help fill.

Other provisions in the Law on Wildlife cover agrobiodiversity issues more indirectly. For instance, various articles allow for:

- Designating nature reserves[220];
- Compensating farmers for crop losses due to wildlife protection efforts[221]; and
- Rewarding those who have made "outstanding achievements in the protection of wildlife resources, in scientific research on wildlife, or in the domestication and breeding of wildlife."[222]

Therefore, even though the Law on Wildlife does not consciously engage with the agrobiodiversity concept, many of its provisions attempt to manage the complex feedback loops between agriculture and biodiversity.

24.5.2 Regulations on Restoring Farmland to Forest (2002)

Over the past several decades, the Chinese government's has worked "to push forward the implementation of a national ecological environmental construction and protection plan."[223] The 2002 Regulations on Restoring Farmland to Forest (Farmland to Forest Regulations)[224] are a part of this plan, along with a variety of other laws and regulations that aim to protect forest ecosystems.[225] Of these instruments, the Farmland to Forest Regulations have the most direct implications for the agricultural sector.

---

[219]China.org.cn (1988), Art. 29.

[220]China.org.cn (1988), Art. 10

[221]China.org.cn (1988), Art. 14

[222]China.org.cn (1988), Art. 7.

[223]CAEP-TERI (2011).

[224]Food and Agriculture Organization of the United Nations (2002).

[225]These other instruments include the Forestry Law, the Law on Protection of Wildlife (discussed *supra* Sect. 24.5.1), the Law on Water and Soil Conservation, the Law on Desert Prevention and Transformation, the Fire Prevention Regulations, and the Regulations on Forest Pest Control. See Appendix.

The Farmland to Forest Regulations incentivize farmers to convert their agricultural lands back into forest lands. Incentives come in the form of a range of subsidies, such as food supplies, seeds and samplings, and general living stipends. Subsidies are assessed according to the actual acreage of restored forest. After a successful restoration, the local government must take certain measures "to protect the achievements of restoring farmland to forest," including sealing hills, prohibiting herding, and ensuring that nearby animals are raised in fences.[226] In addition to improving natural ecosystems, the goals of the Farmland to Forest Regulations include:

- "[P]rotect[ing] the legitimate rights and interests of the [farmers];" and
- "[O]ptimiz[ing] the industrial structures of rural areas."[227]

Therefore, the regulations seek to use agricultural land reforms as a tool for enhancing both ecological and human living conditions. These priorities are fully in line with agrobiodiversity goals. Although China may not yet have a fully integrated national biodiversity management system, the Farmland to Forest Regulations are likely to be an important tool for meeting the ambitious National Biodiversity Conservation Strategy and Action Plan[228] targets and for supporting greater agrobiodiversity throughout the country.

24.6 Summary and Conclusion

Scientists and policymakers are increasingly aware of the delicate balance between agricultural practices and biodiversity. Although regulating food production may not traditionally be thought of as a biodiversity issue, the linkages between agricultural activities and wildlife diversity are strong. Good agricultural practices can support natural biodiversity by providing habitats, such as nesting areas for birds or crops for pollinators. Likewise, high levels of natural biodiversity can support agriculture by providing vital ecosystem services, including water filtration, nutrient cycling, and pollination. On the other hand, poor agricultural practices can:

- Destroy habitats (through clear-cutting forests, draining wetlands, or contributing to desertification);
- Pollute soils and waterways;
- Introduce invasive species and GMOs; and
- Increase GHG emissions.[229]

[226]Food and Agriculture Organization of the United Nations (2002), Art. 55.

[227]Food and Agriculture Organization of the United Nations (2002), Art. 1.

[228]See *supra* Sect. 24.5.

[229]See Fig. 24.1.

Each of these practices can harm natural biodiversity. Low wildlife diversity levels can leave agricultural lands susceptible to a number of risks, including pests, depleted soils, and contamination. Therefore, a set of dynamic and complex interrelationships exist between agricultural and natural biodiversity.

Considering the role of human communities complicates the picture even further. Indeed, a number of social and cultural factors shape the feedback loops between agriculture and wildlife diversity. Land management decisions that promote sustainability (like using organic fertilizers and pesticides) can foster the positive impacts of agriculture on wildlife while discouraging some of the negative ones. In turn, healthy agricultural and wildlife systems provide human communities with key benefits, from opportunities for employment and recreation to tangible goods like food, medicines, and fuels. There is a growing international movement to ensure that local indigenous people share in these benefits. Increased access to benefit-sharing often goes hand-in-hand with an enhanced respect for traditional land management practices. All of these factors can bring the positive feedback loop full-circle: greater benefits to people from biodiversity cause humans to use more sustainable land management techniques, which leads to higher agricultural and natural biodiversity levels that, in turn, translate back into greater benefits for human societies.

Regulators are increasingly alert to the vital interconnections between agricultural, natural, and human systems. Indeed, this awareness has given rise to a new concept: agrobiodiversity. At the international level, agrobiodiversity is now formalized into a CBD work programme.[230] In response, some nations (like the EU[231] and China[232]) have incorporated agrobiodiversity measures into national biodiversity action plans and strategies. (The United States, in contrast, is not a party to the CBD and does not have a national biodiversity action plan.[233]) Other nations are taking on the issue as well. For instance, regulations promoting agro-forestry are becoming more common. Kenya's Agriculture (Farm Forestry) Rules require farmers to establish and maintain farm forestry on at least 10 % of each of their agricultural lands,[234] while Brazil's Law No. 12805 creates a National Policy for Livestock, Agro-forestry and Sylvo-pastoralism that seeks alternatives to traditional monoculture in deforested areas.[235] These laws, and other similar initiatives from nations around the world, are summarized in the Appendix.

However, agrobiodiversity is still not part of the vocabulary of many governments. Indeed, most countries do not yet directly regulate the issue. Instead, nations tend to have separate laws on relevant issues, which include agriculture, forests, wildlife conservation, endangered species (and international trade in such species

[230]See *supra* Sect. 24.2.1.
[231]See *supra* Sect. 24.4.
[232]See *supra* Sect. 24.5.
[233]See *supra* Sect. 24.3.
[234]See Appendix.
[235]See Appendix.

or their by-products), habitat protection, national parks/national protected areas, desertification, and access to genetic resources and benefit-sharing. Despite the patchwork nature of most regimes, tools for change can be found by exploring the synergies between these regulatory fields. For instance, laws on species and habitat protection can impact agricultural practices as well as biodiversity preservation efforts. Such laws often contain provisions that prohibit or restrict activities that harm threatened or endangered species or their habitats. Since agricultural activities are a major contributor to habitat and species loss, these laws can be used to regulate agricultural operations (for example, by prohibiting clear-cutting of forest lands or restricting pesticide use). Although agrobiodiversity laws and policies are still somewhat uncommon around the world, species and habitat protection laws are not. Therefore, these laws represent a key tool for change in the complex and emerging field of agricultural biodiversity.

Appendix

Region	Country/ state	Citation	Title	Year passed	Summary or subsections applicable to regulatory issue
Africa	Kenya	Act No. 8 of 1955, as amended by Act No. 6 of 2012	Agriculture Act (Cap. 318)	1955	Provides rules on good agricultural practices, including rules on: • agricultural land use and preservation; • sound agricultural development; and • marketing of agricultural products
Africa	Kenya	Act No. 7 of 2005	Forest Act	2005	Provides for the establishment, development and sustainable management (including conservation) and rational utilization of forest resources for the socio-economic development of the country; establishes Forest Conservation Committee (FCC); requires public consultation for all major forest decisions;
Africa	Kenya	L.N. No. 166 of 2009	Agriculture (Farm Forestry) Rules (Cap. 318)	2009	Implement the Agriculture Act; aim to help conserve water, soil, and biodiversity and to protect riverbanks, shorelines, and wetlands; require farmers to: • establish and maintain farm forestry on at least 10% of each of their

(continued)

Region	Country/ state	Citation	Title	Year passed	Summary or subsections applicable to regulatory issue
					agricultural land holdings; • ensure that tree species planted do not have adverse effects on water sources, crops, livestock, soil fertility or the neighborhood; and • ensure that tree species planted are not invasive Order District Agricultural Committee to: • identify land at risk of degradation; • establish necessary conservation measures (including planting trees) and • prepare and oversee annual seedling production plans
Africa	Kenya	Act No. 47 of 2013	Wildlife (Conservation and Management) Act	2013	Devolves wildlife governance and decision-making to county-level; calls for the creation of County Wildlife Conservation and Compensation Committees Devolves wildlife conservation and management to landowners and managers; recognizes wildlife conservation as a form of land-use with equal recognition as other land uses (such as agriculture); provides for establishing Wildlife Conservancies Seeks to provide better access to benefits arising from wildlife conservation;
Africa	Nigeria	Law No. 41 of 1955 (Chapter 57 L.R.S.N. 1999)	Forest Law	1956	Along with the Land Use Decree and National Parks Decree, helps establish a network of protected areas by providing for forest reserve and protected forest declarations
Africa	Nigeria	Decree No. 6 of 1978	Land Use Act	1978	Along with the Forest Law and National Parks Decree, helps establish a network of protected areas; vests all non-Federal agency held land in military governors of each State to hold in trust for the people

(continued)

Region	Country/ state	Citation	Title	Year passed	Summary or subsections applicable to regulatory issue
Africa	Nigeria	Act No. 11 of 1985	Endangered Species (Control of International Trade and Traffic) Act	1985	Implements CITES; prohibits the hunting, capture, or trade in endangered animal species listed within the act's First Schedule and restricts the taking and trade of threatened animal species listed in the act's Second Schedule Has a broader scope than CITES (also covers domestic taking of listed species)
Africa	Nigeria	Decree No. 101 of 1991	National Parks Decree	1991	Along with the Forest Law and Land Use Decree, helps establish a network of protected areas by setting up 6 national parks in the country; establishes Board for National Parks Service to carry out in-situ flora and fauna conservation
Africa	South Africa	Act No. 73 of 1989	Environment Conservation Act	1989	Governs natural resources protection; provides for declaring protected natural areas or limited development areas Part V covers agricultural and land use and transformation issues
Africa	South Africa	Act No. 84 of 1998	National Forests Act	1998	Provides for managing and conserving public forests Chapter 3 contains specific measures to combat deforestation
Africa	South Africa	Act No. 10 of 2004	National Environmental Management Biodiversity Act	2004	Governs biodiversity preservation; aims to protect ecosystems and species that are threatened or in need of protection; establishes South African National Biodiversity Institute Aims to support: • the use of indigenous biological resources in a sustainable manner; and • the fair and equitable sharing of benefits arising from bioprospecting involving indigenous resources

(continued)

Region	Country/ state	Citation	Title	Year passed	Summary or subsections applicable to regulatory issue
Africa	South Africa	No. R. 152 of 2007	Threatened or Protected Species Regulations	2007	Carry out the National Environmental Management Biodiversity Act; prohibit certain restricted activities involving listed threatened or protected species
Americas	Brazil	Decreto n. 76.623/75 de novembro de 1975; Legislação *Federal de Meio Ambiente*, Vol. I, Brasília 1996, pp. 372-382	Decree No. 76.623 laying down a list of flora and fauna endangered species, according to the convention on the international trade/*Decreto No. 76.623 Promulga a convencao sobre comercio internacional das especies da flora e fauna selvagem em perigo de extincao*	1975	Implements CITES; lists all Brazilian flora and fauna endangered species Most listed species found in the Atlantic forest and the cerrado grasslands (which have largely been converted to cattle ranches and industrial soy farms) Number of listed species tripled in 2008
Americas	Brazil	Decreto n. 4.703, de 21 de Maio de 2003	Decree No. 4.703 on the National Programme on Biological Diversity (PRONABIO) and on the National Commission on Biodiversity/*Decreto No. 4.703, Dispõe sobre o Programa Nacional da Diversidade Biológica—PRONABIO e a Comissão Nacional da Biodiversidade, e dá outras providências*	2003	Establishes the Program on Biological Diversity, which promotes a partnership between civil society and the government for conserving organic diversity and sustainably using its components
Americas	Brazil	Lei n. 12.651, de 25 de Maio de 2012	Law No. 12.651 on the Protection of Native Forests/ *Lei No. 12.651, dispõe sobre a proteção da vegetação nativa*	2012	Governs the protection and sustainable use and exploitation of native forests and other indigenous plants; recognizes the importance of preserving national forests, habitats, biodiversity, soil, and water resources for future generations; establishes permanent protected forest

(continued)

Region	Country/ state	Citation	Title	Year passed	Summary or subsections applicable to regulatory issue
					areas as the primary tool for protecting biodiversity in Brazil Promotes rural economic development Amended by Law No. 12.727 of 2012 to establish general norms on: • plant protection; • permanent preservation areas and Legal Reserve Areas; • forestry exploitation; and • control over forestry origin products
Americas	Brazil	Lei n. 12.805, de 29 de Abril de 2013	Law No. 12.805 creating the National Policy for Livestock, Agro-forestry and Sylvo-pastoralism (ILPF)/*Lei No. 12.805 nstitui a Política Nacional de Integração Lavoura-Pecuária-Floresta*	2013	Establishes National Policy for Livestock, Agro-forestry, and Sylvo-pastoralism; aims to improve the productivity of agro-forestry income-generating activities; sets a number of goals, including: • seeking alternatives to traditional monoculture in deforested areas; • mitigating deforestation caused by native forest conversion into pasture or agricultural areas; • contributing to the maintenance of permanent preservation areas and of legal reserves; • promoting environmental education; • promoting rehabilitation of degraded areas of pasture through sustainable production systems; and • promoting diversification of production systems through incorporating forestry resources
Americas	Canada	R.S.C. 1985, c W-9	Canada Wildlife Act	1985	Governs wildlife protection; covers both species and habitat protection; provides for declaring "national wildlife areas" Implemented by Wildlife Area Regulations

(continued)

Region	Country/ state	Citation	Title	Year passed	Summary or subsections applicable to regulatory issue
Americas	Canada	S.C. 1992, c. 52	Wild Animal and Plant Protection and Regulation of International and Interprovincial Trade Act (WAPPRIITA)	1992	Implements CITES Has a broader scope than CITES; applies not only to species on the CITES control list, but also to: • foreign species whose capture, possession, and export are prohibited or regulated by laws in their country of origin; • Canadian species whose capture, possession, and transportation are regulated by provincial or territorial laws; and • species whose introduction into Canadian ecosystems could endanger Canadian species
Americas	Canada	S.C. 1994, c. 22	Migratory Birds Convention Act	1994	Aims to conserve migratory bird populations by regulating potentially harmful human activities; requires permits for all activities affecting wild birds or their eggs or nests Implemented by Migratory Bird Sanctuary Regulations, which grant sanctuary status to habitat areas important to migratory birds
Americas	Canada	S.C. 2002, c 29	Species at Risk Act	2002	Governs endangered species protection; aims to support the recovery of wildlife species that are extirpated, endangered, or threatened as a result of human activity; provides for critical habitat designations; establishes a Committee on the Status of Endangered Wildlife in Canada
Americas	Canada (British Columbia)	S.B.C 2003, c. 58	Integrated Pest Management Act	2003	Part 2 prohibits or restricts the use of pesticides that cause or may cause "unreasonable adverse effects"
Americas	Mexico	*Diario Oficial de la Federación*, 28 de enero de 1988, últimas	General Law on Ecological Balance and Environmental Protection/ *Ley General del*	1988	Establishes: • Natural Protected Areas as the country's primary means of protecting endangered species

(continued)

Region	Country/ state	Citation	Title	Year passed	Summary or subsections applicable to regulatory issue
		reformas publicadas DOF 05-11-2013	*Equilibrio Ecológico y Protección al Ambiente* (LGEEPA)		• Restoration Areas to addresses problems of desertification and land degradation; calls for ecological restoration programs in these areas Title III, Chap. 2 calls for sustainable agriculture practices
Americas	Mexico	*Diario Oficial de la Federación*, 3 de julio de 2000, última reforma publicada DOF 26-01-2015	General Wildlife Law/*Ley General de Vida Silvestre* (LGVS)	2000	Governs wildlife and biodiversity preservation (including endangered species and habitats); contains provisions on: • sustainable use of wildlife; • wildlife diseases; • species at risk and critical habitats; and • controlling nuisance species
Americas	United States	16 U.S.C. §§ 3371–3378	Lacey Act	1900	Aims to combat trafficking in illegal plant and animal species; forbids anyone to import, export, sell, acquire, or purchase fish, wildlife or plants that are taken, possessed, transported, or sold: 1) in violation of U.S. or Indian law, or 2) in interstate or foreign commerce involving any fish, wildlife, or plants taken possessed or sold in violation of State or foreign law
Americas	United States	16 U.S.C. §§ 703-712, Ch. 128	Migratory Bird Treaty Act	1918	Prohibits taking of Native American bird species
Americas	United States	16 U.S.C. § 1131	National Wilderness Preservation System (Wilderness Act)	1964	Creates formal mechanism for designating federal wilderness areas; prohibits disruptive human activities (including intensive agriculture) in such areas
Americas	United States	42 U.S.C. § 4321 et seq	National Environmental Policy Act (NEPA)	1969	Requires environmental reviews of proposed major federal actions; requires consideration of impacts of these actions on species listed under the Endangered Species Act (ESA)

(continued)

Region	Country/ state	Citation	Title	Year passed	Summary or subsections applicable to regulatory issue
Americas	United States	16 U.S.C. §§ 1361-1407	Marine Mammal Protection Act	1972	Extends ESA-like protections to non-listed marine mammal species
Americas	United States	16 U.S.C. § 1531 et seq.	Endangered Species Act (ESA)	1973	Governs endangered species protection; provides for: • "listing" endangered and threatened plant and animal species • prohibiting the "taking" of listed animal species • designating critical habitat areas
Americas	United States	16 U.S.C. §§ 2901-2911	Fish and Wildlife Conservation Act (Non-game Act)	1980	Grants assistance to states for implementing conservation plans and programs for non-game fish and wildlife
Americas	United States	16 U.S.C. §§ 4901-4916 et seq.	Wild Bird Conservation Act (WBCA)	1992	Prohibits import of CITES-listed bird species; encourages wild bird conservation programs in countries of origin
Americas	United States	7 U.S.C. § 136 et seq.	Federal Insecticide, Fungicide, and Rodenticide Act (FIFRA)	1996	Regulates pesticide distribution, sale, and use
Americas	United States	H.R. 2642, P.L. 113–79	Agricultural Act of 2014 (2014 Farm Bill)	2014	Governs agricultural and food policy at the federal level; aims to: • expanding markets for agricultural products at home and abroad; • strengthening conservation efforts; • creating new opportunities for local and regional food systems; and • grow the bio-based economy Updates previous 2008 Farm Bill ("Food, Conservation, and Energy Act of 2008"); current version authorizes nutrition and agriculture programs in the United States for the years 2014-2018
Asia	China	Adopted at the Seventh Meeting of the Standing Committee of the Sixth National	Forestry Law of the People's Republic of China	1984	Aims to protect natural forest ecosystems Article 1 establishes various goals, including: • protecting, cultivating, and rationally exploiting forest resources;

(continued)

Region	Country/ state	Citation	Title	Year passed	Summary or subsections applicable to regulatory issue
		People's Congress on September 20, 1984			• accelerating territorial afforestation; • using forests for water storage, soil conservation, climate regulation, and environmental improvement; and • supplying forest products to the people
Asia	China	Adopted at the 24th Meeting of the Standing Committee of the Fifth National People's Congress and promulgated by Order No. 9 of the Standing Committee of the National People's Congress on August 23,1982	Marine Environmental Law of the People's Republic of China	1982	Governs marine protection Article 1 establishes various goals, including: • protecting the marine environment and resources; • preventing pollution damage; • maintaining ecological balances; • safeguarding human health; and • promoting the development of marine programmes
Asia	China	Adopted at the 11th Meeting of the Standing Committee of the Sixth National People's Congress on June 18, 1985	Grasslands Law of the People's Republic of China	1985	Article 1 establishes various goals, including: • protecting, developing, and making rational use of grasslands; • improving ecological environments; • maintaining the diversity of living things; • modernizing animal husbandry; and • promoting the sustainable development of the economy and society
Asia	China	Adopted at the Fourth Meeting of the Standing Committee of the Seventh National People's Congress and promulgated by Order No. 9 of the President of the People's Republic of China on November 8, 1988	Protection of Wildlife Law of the People's Republic of China	1988	Governs biodiversity preservation; provides general principles for wildlife protection; applies to terrestrial and aquatic species that are rare or endangered, as well as terrestrial species which are beneficial or of important economic or scientific value

(continued)

Region	Country/ state	Citation	Title	Year passed	Summary or subsections applicable to regulatory issue
Asia	China	Adopted at the 24th Meeting of the Standing Committee of the Sixth National People's Congress on January 21, 1988	Water Law of the People's Republic of China	1988	Article 1 establishes various goals, including: • rational development, utilization, preservation, and protection of water to prevent and control water disasters; and • sustainable utilization of water resources to meet the needs of national economic and social development
Asia	China	Adopted at the 11th Meeting of the Standing Committee of the Seventh National People's Congress on December 26, 1989, promulgated by Order No. 22 of the President of the People's Republic of China on December 26, 1989	Environmental Protection Law of the People's Republic of China	1989	Article 1 establishes various goals, including: • protecting and improving the people's environment and the ecological • environment; • preventing and controlling pollution and other public hazards; • safeguarding human health; and • facilitating the development of socialist modernization
Asia	China	Promulgated by the State Council on December 18, 1989	Regulations on the Prevention of Forest Plant Diseases and Insect Pests	1989	Aims to protect natural forest ecosystems from plant diseases and insect pests
Asia	China	Adopted at the 20th Meeting of the Standing Committee of the Seventh National People's Congress of the People's Republic of China on June 29, 1991	Water and Soil Conservation Law of the People's Republic of China	1991	Aims to protect natural forest ecosystems by conserving water and soil resources Article 1 establishes various goals, including: • preventing and controlling soil erosion; • protecting and rationally utilizing water and soil resources; • mitigating disasters from floods, droughts, and sandstorms; • improving ecological environments; and • developing production

(continued)

Region	Country/ state	Citation	Title	Year passed	Summary or subsections applicable to regulatory issue
Asia	China	Adopted at the 2nd Meeting of the Standing Committee of the Eighth National People's Congress on July 2, 1993	Agriculture Law of the People's Republic of China	1993	Article 49 provides that "the country protects Intellectual Property such as Plant Variety and Geographical Indication"
Asia	China	Promulgated by Decree No. 204 of the State Council of the People's Republic of China on September 30, 1996	Regulation on Protection of Wild Plants	1996	Article 1 establishes various goals, including: • protecting, developing, and rationally utilizing wild plant resources; and • preserving biodiversity and maintaining ecological balances
Asia	China	Adopted at the 2nd Meeting of the Standing Committee of the Ninth National People's Congress on April 29, 1998 and promulgated by Order No. 4 of the President of the People's Republic of China on April 29, 1998	Fire Protection Law of the People's Republic of China – Fire Prevention Regulations	1998	Aims to protect natural forest ecosystems from fires
Asia	China	Adopted at the 24th Executive Meeting of the State Council on September 2, 1994, promulgated by Decree No. 167 of the State Council of the People's Republic of China on October 9, 1994	Regulations of the People's Republic of China on Nature Reserves (Nature Reserves Regulations)	2000 1994	Provide for constructing and managing nature reserves, including protected areas

(continued)

Region	Country/ state	Citation	Title	Year passed	Summary or subsections applicable to regulatory issue
Asia	China	Adopted at the 23rd meeting of the Standing Committee of the Ninth National People's Congress of the People's Republic of China on Aug. 31, 2001	Desert Prevention and Transformation Law of the People's Republic of China	2001	Article 1 establishes various goals, including: • preventing land desertification; • transforming desertified land; • protecting the safety of the environment; and • promoting the sustainable development of both the economy and society
Asia	China	Adopted at the 66th executive meeting of the State Council on December 6, 2002; Order of the State Council of the People's Republic of China (No. 367)	Regulations on Restoring Farmland to Forest	2002	Incentivize farmers to convert their agricultural lands back to forest through subsidies (including food, seed and sampling, and general living subsidies)
Asia	China (Hong Kong)	L.N. 206 of 2006	Protection of Endangered Species of Animals and Plants Ordinance (Chapter 586)	2006	Gives effect to CITES in Hong Kong
Asia	India	Act No. 53 of 2001	Protection of Plant Varieties and Farmers' Rights Act		Sets out basic framework for protecting rights relating to breeding new plant varieties (including the conservation, selection, and improvement of farmers' varieties)
Asia	India	Act No. 53 of 1972	Wildlife (Protection) Act	1972	Governs the protection of wild animals, birds, and plants; contains provisions on: • restricting the hunting of listed wild animal species; • restricting the picking or uprooting of listed plant species; and • establishing wildlife sanctuaries, National Parks, and Closed Areas Chapter V regulates the trade or commerce in wild animals, animal articles, and trophies Applies to all Indian States except for Jammu and Kashmir

(continued)

Region	Country/ state	Citation	Title	Year passed	Summary or subsections applicable to regulatory issue
Asia	India	Act No. 18 of 2003	Biological Diversity Act	2003	Implements CBD; governs biodiversity conservation; promotes sustainable use of biological resources and the equitable sharing of benefits arising from such use
Asia	India (Andhra Pradesh)	Act No. 11 of 1997	Andhra Pradesh Farmers' Management of Irrigation Systems Act	1997	Lays out the powers and responsibilities of farmers' organizations in managing irrigation systems Sect. 16 provides that farmers' organizations' primary aim is to promote the secure distribution of water and to maintain irrigation systems, while achieving efficient water utilization and environmental protection goals
Asia	India (Kerala)	Act No. 28 of 2008	Kerala Conservation of Paddy Land and Wetland Act	2008	Aims to: • regulate the conversion and development of paddy fields into agricultural lands; • protect wetland areas; • ensure food security; and • sustain ecological systems Amended in 2011 by Act No. 14 of 2011
Asia	India (Rajasthan)	Act No. 21 of 2000	Rajasthan Farmers' Participation in Management of Irrigation Systems Act	2000	Lays out the powers and responsibilities of farmers' organizations in managing irrigation systems; contains provisions on: • promoting and securing the distribution of water; • adequately maintaining the irrigation system; • efficiently and economically utilizing water to optimize agricultural production; and • protecting the environment and ensuring ecological balances
Asia	India (Tamil Nadu)	Act No. 7 of 2000	Tamil Nadu Farmers' Management of Irrigation Systems Act	2000	Lays out the powers and responsibilities of farmers' organizations in managing irrigation systems

(continued)

Region	Country/ state	Citation	Title	Year passed	Summary or subsections applicable to regulatory issue
Asia	Israel	P.O. Suppl. I, 1	Forest Ordinance	1926	Governs forest protection; allows Ministry of Agriculture to declare any non-privately owned forests as protected State forests (where grazing and other harmful activities are prohibited); embraces "polluter pays" principle with penalties (including fines or imprisonment) for environmental harm
Asia	Israel	Law No. 5715-1955	Israel Wildlife Protection Law	1955	Implements CITES; requires hunting licenses for game hunting or pest extermination
Asia	Israel	Law No. 5719-1959	Water Law	1959	Creates framework for controlling and protecting water resources; governs water use for agricultural purposes
Asia	Israel	Law No. 5758-1998	National Parks, Nature Reserves, National Sites and Memorial Sites Law	1963	Provides basic legal structure for protecting Israel's natural habitats, natural assets, wildlife, and sites of scientific, historic, architectural, and educational interest; establishes united Nature and National Parks Protection Authority, whose powers include declaring and maintaining nature reserves and national parks
Asia	Israel	*Laws of the State of Israel*, Vol 21, 1966/67, from 24.10.1966 to 25.9.1967, pp. 102-105-110. *Dinim* Vol. 10, pp. 5337	Agricultural Settlement (Restriction on Use of Agricultural Land and of Water) Law	1967	Establishes personal water use quotas for agricultural land owners or operators
Asia	Japan	Law No. 32 of 1918	Wildlife Protection and Hunting Law	1918	Aims to: • protect birds and mammals; • increase populations of birds and mammals; and • control pests through implementing wildlife protection projects and hunting controls

(continued)

Region	Country/ state	Citation	Title	Year passed	Summary or subsections applicable to regulatory issue
Asia	Japan	Law No. 85 of 1972	Nature Conservation Law	1972	Provides basic framework for nature conservation in Japan; aims to protect and manage natural resources and natural ecosystems; provides for declaring: • wilderness areas (where human activities are forbidden); • nature conservation areas; and • prefectural nature conservation areas
Asia	Japan	Law No. 75 of 1992	Law for the Conservation of Endangered Species of Wild Fauna and Flora	1992	Governs endangered species protection; prohibits taking or transporting (as well as importing and exporting) endangered species; contains provisions on: • designating natural habitat conservation areas; • registering international endangered species; and • establishing natural habitat rehabilitation programs to maintain viable populations of endangered species
Asia	Japan	Act No. 58 of 2008	Basic Act on Biodiversity	2008	Implements CBD and UNFCCC; attempts to synthesize multiple legal instruments on biodiversity conservation and to integrate biodiversity values into all tiers (from national to local) of decision-making; encourages sustainable land and resource use practices that avoid or minimize impacts on biodiversity Adopted in accordance with the Basic Environment Law (Act No. 91 of 1993)
Asia	Russia	Law No. 52-FZ of 1995	Federal Law of the Russian Federation on Wildlife (Law on Wildlife)	1995	Governs wildlife protection; declares all wildlife to be the national property of Russian people and to be under protection; encourages sustainable use and management of natural wildlife resources

(continued)

Region	Country/ state	Citation	Title	Year passed	Summary or subsections applicable to regulatory issue
Asia	Russia	Ministerial Decree No. 158 of 19 Feb 1996	Ministerial Decree No. 158 of 19 February 1996 on the catalogue of endangered wildlife species ("Red Book") of the Russian Federation	1996	Provides for compiling the Red Data Book of the Russian Federation (RDBRF), a basic state document listing rare and endangered species of wild animals, plants, and fungi Complements the Federal Law on Environmental Protection (No. 7-FZ of Jan 10, 2002) and the Law on Animal World (No. 52-FZ of Apr 24, 1995), which provide further a legislative basis for the RDBRF
Asia	Russia	Ministerial Decree No. 1010 of Aug 13, 1997	Ministerial Decree No. 1010 on strengthening the protection of wildlife species and their natural habitat on the territory of forestry	1997	Implements the Federal Law on Wildlife; authorizes the Federal Forest Service and its territorial branches to carry out wildlife species and natural habitat protection activities in forest territories
Asia	Russia	Ministerial Decree No. 13 of Jan 6, 1997	Ministerial Decree No. 13 Enforcing the Regulations on Hunting of Endangered Wildlife Species Under Protection of the Russian Federation	1997	Aims to restrict hunting of endangered wildlife species by imposing more severe rules (such as stricter hunting permit requirements)
Asia	Russia	Law No. 136-FZ of Oct 25, 2001	Land Code	2001	Defines agricultural lands by law; divides all of Russia's lands into 7 categories by use; delegates most land management authority to small, local elected bodies Provides that designated agricultural lands (~15% of all land) can only be used for agricultural production (all other uses are prohibited)
Australia/ New Zealand	Australia	No. 91 of 1999	Environmental Protection and Biodiversity Conservation Act (EPBC Act)	1999	Establishes a national environmental protection framework, with a focus on conserving biodiversity; calls for ecologically sustainable development and use of natural

(continued)

Region	Country/ state	Citation	Title	Year passed	Summary or subsections applicable to regulatory issue
					resources; covers: • nationally endangered, threatened, and migratory species (including international trade in wildlife and wildlife products); • nationally threatened ecological communities; • wetlands of international importance; and • world and national heritage properties (including the Great Barrier Reef) Provides for a streamlined national environmental assessment and approvals process; Promotes use of Indigenous peoples' biodiversity knowledge Delegates matters of state and local significance to states and territories
Australia/ New Zealand	Australia	Act No. 92 of 1999	Environmental Reform (Consequential Provisions) Act	1999	Repeals and amends a number of acts relevant to the enactment of the Environmental Protection and Biodiversity Conservation Act (EPBC) Act Drafted in harmony with EPBC Act
Australia/ New Zealand	New Zealand	1949 No. 19	Forests Act	1949	Governs forest management and timber activities; defines sustainable forest management Does not apply to planted indigenous forests
Australia/ New Zealand	New Zealand	1953 No. 31	Wildlife Act	1953	Sets up "default" protected status for majority of New Zealand's vertebrate species: instead of listing species that are specifically protected, it provides that *all* vertebrate species that are *not* listed are protected
Australia/ New Zealand	New Zealand	1977 No. 66	Reserves Act	1977	Provides for declaring and managing national reserves Part 3 concerns farming, grazing, and afforestation
Australia/ New Zealand	New Zealand	1980 No. 66	National Parks Act	1980	Provides for establishing national parks Part I(4) concerns preserving indigenous plants and animals

(continued)

Region	Country/ state	Citation	Title	Year passed	Summary or subsections applicable to regulatory issue
Australia/ New Zealand	New Zealand	1987 No. 65	Conservation Act	1987	Governs biodiversity conservation; sets forth conservation management strategies (CMSs) and conservation management plans (CMPs), with an emphasis on "protection" rather than "sustainable management;" sets up a hierarchy of activities occurring on public conservation land: (1) intrinsic value; (2) non-commercial recreation; and (3) tourism Complements the National Parks Act and the Reserves Act by providing for specially protected areas, including: conservation parks, wilderness areas, ecological areas, sanctuary areas, watercourse areas, amenity areas, and wildlife management areas Establishes the Department of Conservation and the Department of Fish and Game
Australia/ New Zealand	New Zealand	1989 No. 18	Trade in Endangered Species Act	1989	Provides for managing, conserving, and protecting endangered, threatened, and exploited species
Australia/ New Zealand	New Zealand	S.R. 2007/354	Forests (Permanent Forest Sink) Regulations	2007	Set out various requirements for establishing and maintaining permanent forest sinks (defined as any eligible forest actively established on Kyoto-compliant land) Enacted pursuant to sections 67Y and 67ZL of the Forests Act
Europe	European Union	O.J. L. 20, 26.1.2010, pp. 7–25	Directive 2009/ 147/EC of the European Parliament and of the Council of 30 November 2009 on the conservation of wild birds (Birds Directive)	1979	Implements CBD; establishes comprehensive protection scheme for all wild bird species naturally occurring in the EU Complements the Habitats Directive

(continued)

Region	Country/ state	Citation	Title	Year passed	Summary or subsections applicable to regulatory issue
Europe	European Union	O.J. L. 206, 22.7.1992, pp. 7–50	Council Directive 92/43/EEC of 21 May 1992 on the conservation of natural habitats and of wild fauna and flora (Habitats Directive)	1992	Implements CBD in EU; cornerstone of EU nature conservation policy; aims at to preserve, protect, and improve the quality of the environment, including conserving of natural habitats and of wild fauna and flora; establishes Natura 2000 habitat preservation network
Europe	European Union	O.J. L. 327, 22.12.2000, pp. 1–73	Directive 2000/60/ EC of the European Parliament and of the Council establishing a framework for Community action in the field of water policy (Water Framework Directive or WFD)	2000	Article 1 aims to regulate water as a "heritage" Article 16 calls for integrating sustainable water management into policy areas, such as agriculture and fisheries
Europe	European Union	O.J. L. 162, 30.4.2004, pp. 18–28	Council Regulation (EC) No 870/2004 of 24 April 2004 establishing a Community programme on the conservation, characterisation, collection and utilisation of genetic resources in agriculture and repealing Regulation (EC) No 1467/94	2004	Article 1 declares that "[b] iological and genetic diversity in agriculture is essential to the sustainable development of agricultural production and of rural areas" and calls for measures "to conserve, characterise, collect and utilise the potential of that diversity in a sustainable way to promote the aims of the common agricultural policy (CAP)"
Europe	European Union	O.J. L. 347, 20.12.2013, pp. 549-607	Regulation (EU) No. 1306/ 2013 of the European Parliament and of the Council on the financing, management and monitoring of the common agricultural policy and repealing Council Regulations (EEC) No. 352/78, (EC) No. 165/94,	2013	Sets out financing rules under the Common Agricultural Policy (CAP), including expenditure on rural development programmes

(continued)

Region	Country/ state	Citation	Title	Year passed	Summary or subsections applicable to regulatory issue
			(EC) No. 2799/98, (EC) No. 814/2000, (EC) No. 1290/ 2005 and (EC) No. 485/2008		
Europe	European Union	O.J. L. 347, 20.12.2013, pp. 487–548	Regulation (EU) No. 1305/ 2013 of the European Parliament and of the Council on support for rural development by the European Agricultural Fund for Rural Development (EAFRD) and repealing Council Regulation (EC) No. 1698/2005 (Rural Development Regulations)	2013	Calls for reforms to the Common Agricultural Policy (CAP) to take place starting in January 2014 (through 2020) Article 2 calls for establishing a rural development policy "to accompany and complement direct payments and market measures of the CAP"
Europe	European Union—United Kingdom	1981, c. 69	Wildlife and Countryside Act	1981	Implements EU Birds Directive in the UK; aims to: • protect native plant and animal species (especially those that are threatened or endangered); • control the release of non-native species; • regulate the import and export of endangered species; and • establish national nature reserves (including marine nature reserves and national parks) Calls for a compulsory five-year review of listed species, which were greatly expanded via the Wildlife and Countryside Act 1981 (Variation of Schedule 9) (England and Wales) Order 2010
Europe	European Union—United Kingdom	1996, c. 3	Wild Mammals (Protection) Act	1996	Provides for the protection of wild mammals from certain cruel acts

(continued)

Region	Country/ state	Citation	Title	Year passed	Summary or subsections applicable to regulatory issue
Europe	European Union—United Kingdom	2006, c. 16	Natural Environment and Rural Communities Act	2006	Establishes bodies (e.g., Natural England, Countryside Council of Wales, and Scottish Natural Heritage) to manage affairs of natural environment and rural communities; covers: • wildlife; • sites of special scientific interest; • National Parks; and • the Broads
Europe	European Union—United Kingdom (England and Wales)	2000, c. 37	Countryside and Rights of Way Act (CRoW Act)	2000	Regulates public access to the countryside; lays out rules on nature conservation and wildlife protection, including restrictions on driving mechanically propelled vehicles off-roads (i.e., on fields)
Europe	European Union—United Kingdom (England and Wales)	2007, c. 23	Sustainable Communities Act (Chapter 23)	2007	Promotes sustainable local communities
Europe	European Union—United Kingdom (England)	S.I. 2003/1235 (replacing S.I. 2001/ 3139)	Organic Farming (England Rural Development Programme) Regulations	2003	Implements EU Rural Development Regulations; provides for aid payments to farmers who introduce organic farming methods on their lands and comply with certain environmental conditions
Europe	European Union—United Kingdom (Northern Ireland)	S.R. No. 172 of 2008	Organic Farming Regulations (Northern Ireland)	2008	Provide for the payment of grants to farmers who agree to introduce organic farming methods and comply with the management requirements and standards of good agricultural and environmental conditions (similar to the Organic Farming (England Rural Development Programme) Regulations)
Europe	European Union—United Kingdom (Scotland)	2003 asp 2	Land Reform (Scotland) Act	2003	Establishes statutory public rights of access to land for recreational and other purposes

(continued)

Region	Country/ state	Citation	Title	Year passed	Summary or subsections applicable to regulatory issue
Europe	European Union—United Kingdom (Scotland)	2011, asp 6	Wildlife and Natural Environment (Scotland) Act	2011	Applies Wildlife and Countryside Act in Scotland
International	[Not yet in force]	UNEP/CBD/COP/DEC/X/1, 29 Oct. 2010	Nagoya Protocol on Access and Benefit-Sharing	2010	Aims to ensure fair and equitable sharing of benefits arising out of the utilization of genetic resources; creates incentives to use genetic resources sustainably and equitably Adopted by CBD COP 10 Not yet entered into force as of April 2015
International	Parties to Convention	996 U.N.T.S. 245, T.I.A.S. No. 11,084	Convention on Wetlands of International Importance especially as Waterfowl Habitat (Ramsar Convention)	1971	Seeks to conserve and wisely use all wetlands through local and national actions and international cooperation; incorporates sustainable development goals; instructs nations to limit destructive agricultural practices (like using fertilizers and pesticides in reclaimed wetlands) and encourage sustainable ones (like using flood-resistant crops in naturally functioning floodplains)
International	Parties to Convention	27 U.S.T. 37, 1037 U.N.T.S. 151	Convention for the Protection of the World Cultural and Natural Heritage (World Heritage Convention)	1972	Aims to conserve natural and cultural sites of interest to the international community; provides for listing designated World Heritage Sites, which can include forests and agricultural landscapes
International	Parties to Convention	27 U.S.T. 1087, 12 I.L.M 1085	Convention on International Trade in Endangered Species of Wild Fauna and Flora (CITES)	1973	Aims to ensure that international wildlife trade does not threaten the survival of listed plant and animal species; sets controls on trade in listed species (and products derived from such species)
International	Parties to Convention	S. Treaty Doc. No. 103-20 (1993), 31 I.L.M. 818	Convention on Biological Diversity (CBD)	1992	Addresses all aspects of biodiversity COP 3's Decision III/11 established a multi-year work program on agrobiodiversity COP 5's Decision V/5 made agrobiodiversity a thematic work programme

(continued)

Region	Country/ state	Citation	Title	Year passed	Summary or subsections applicable to regulatory issue
International	Parties to Convention	33 I.L.M. 1328	United Nations Convention to Combat Desertification in Those Countries Experiencing Serious Drought and/or Desertification, Particularly in Africa (UNCCD)	1994	Sets out a universal legal policy and advocacy framework for combatting desertification and promoting sustainable development; aims to reverse and prevent desertification/ land degradation and mitigate the effects of drought in affected areas
International	Parties to Convention (and some non-Party range states)	29 U.S.T. 4647, 1134 U.N.T.S. 97	Convention on the Conservation of Migratory Species of Wild Animals (CMS or Bonn Convention)	1979	Seeks to provide a global platform for conserving and sustainably utilizing migratory animals and their habitats; instructs all States falling within an animal's migratory range to coordinate conservation measures
International	Parties to Protocol	37 I.L.M. 22 (1998)	Kyoto Protocol to the United Nations Framework Convention on Climate Change	1997	Sets binding GHG reduction targets, which are potentially achievable through measures aimed at preventing habitat loss (e.g., forest protection policies to preserve "carbon sinks") and restricting intensive agricultural practices
International	Parties to Treaty	2400 U.N.T.S. 303	International Treaty on Plant Genetic Resources for Food and Agriculture (IT PGRFA)	2001	Aims to: • recognize farmers' contribution to crop diversity; • establish a global system to provide farmers, plant breeders, and scientists with access to plant genetic materials; and • ensure that benefits are shared with the countries where the resources originated Drafted in harmony with CBD

References

Beaudry F (2014) The Lacey act. http://environment.about.com/od/biodiversityconservation/fl/The-Lacey-Act.htm. Accessed 4 Apr 2015

CAEP-TERI (2011) Environment and development: China and India. Joint study by the Chinese Academy for Environmental Planning (CAEP) and the Energy and Resources Institute (TERI), commissioned by the Chinese Council for International Cooperation on Environmental Development (CCICED) and the India Council for Sustainable Development (ICSD). New Delhi: TERI Press (English version). https://books.google.com/books?id=f7kXPEAdiqQC&pg=PA254&lpg=PA254&dq=Regulations+on+Restoring+Farmland+to+Forest&source=bl&ots=HzdAbO00qF&sig=erQ4vVpG5cYM7ee_w1DCUL2SNlY&hl=en&sa=X&ei=H-TsVJyZGMOfgwTLkYDgCA&ved=0CC4Q6AEwAw#v=onepage&q=Regulations%20on%20Restoring%20Farmland%20to%20Forest&f=false

California Department of Fish and Wildlife (2015) California endangered species act. https://www.wildlife.ca.gov/Conservation/CESA. Accessed 15 Mar 2015

Czarnezki JJ (2006) Revisiting the tense relationship between the U.S. Supreme Court, administrative procedure, and the national environmental policy act. Stanf Environ Law J 24. Available via Digital Commons. http://digitalcommons.pace.edu/lawfaculty/904/. Accessed 1 Apr 2015

China.org.cn (1988) Law of the People's Republic of China on the protection of wildlife. 8 Nov 1988. http://www.china.org.cn/english/environment/34349.htm. Accessed 15 Mar 2015

Environmental Investigation Agency (2015) U.S. Lacey act and CITES. http://eia-global.org/campaigns/forests-campaign/u.s.-lacey-act/lacey-and-cites. Accessed 15 Mar 2015

EUR-Lex (1992) Council directive 92/43/EEC of 21 May 1992 on the conservation of natural habitats and of wild fauna and flora. http://eur-lex.europa.eu/legal-content/EN/TXT/?uri=CELEX:01992L0043-20070101. Accessed 15 Mar 2015

EUR-Lex (2009) Directive 2009/147/EC of the European Parliament and of the Council of 30 November 2009 on the conservation of wild birds (codified version). http://eur-lex.europa.eu/legal-content/EN/TXT/?uri=CELEX:32009L0147. Accessed 15 Mar 2015

EUR-Lex (2001) Communication of 27 March 2001 to the Council and the European Parliament: biodiversity action plan for agriculture, vol III [COM(2001) 162 final – not published in the Official Journal]. http://europa.eu/legislation_summaries/agriculture/environment/l28024_en.htm. Accessed 15 Mar 2015

EUR-Lex (2004) Council regulation (EC) No 870/2004 of 24 April 2004 establishing a Community programme on the conservation, characterisation, collection and utilisation of genetic resources in agriculture, and repealing regulation (EC) No 1467/94. http://europa.eu/legislation_summaries/agriculture/environment/l60039_en.htm. Accessed 15 Mar 2015

EUR-Lex. Plant health checks. http://europa.eu/legislation_summaries/food_safety/plant_health_checks/index_en.htm. Accessed 15 Mar 2015

European Commission (2014a) Agriculture: The EU's common agricultural policy (CAP): for our food, for our countryside, for our environment. http://ec.europa.eu/agriculture/cap-overview/2014_en.pdf. Accessed 15 Mar 2015

European Commission (2014b) Communication from the Commission to the European Parliament, the Council, the European economic and social committee and the committee of the regions of 10 June 2004 – "European action plan for organic food and farming" [COM(2004) 415 – Not published in the Official Journal]. http://ec.europa.eu/agriculture/organic/documents/eu-policy/european-action-plan/act_en.pdf. Accessed 15 Mar 2015

European Commission (2015a) Agri-environment measures. http://ec.europa.eu/agriculture/envir/measures/index_en.htm. Accessed 15 Mar 2015

European Commission (2015b) Agriculture and biodiversity. http://ec.europa.eu/agriculture/envir/biodiv/index_en.htm. Accessed 15 Mar 2015

European Commission (2015c) Environment: results-based agri-environment schemes: payments for biodiversity achievements in agriculture. http://ec.europa.eu/environment/nature/rbaps/index_en.htm. Accessed 15 Mar 2015

European Commission (2015d) Environment: what are results-based agri-environment payment schemes and how do they differ from other approaches? http://ec.europa.eu/environment/nature/rbaps/articles/1_en.htm. Accessed 15 Mar 2015

European Commission (2015e) Habitats directive: about the habitats directive. http://ec.europa.eu/environment/nature/legislation/habitatsdirective/index_en.htm. Accessed 15 Mar 2015

European Commission (2015f) Rural development 2014-2020. http://ec.europa.eu/agriculture/rural-development-2014-2020/index_en.htm. Accessed 15 Mar 2015

European Commission (2015g) The birds directive: about the birds directive. http://ec.europa.eu/environment/nature/legislation/birdsdirective/index_en.htm. Accessed 15 Mar 2015

European Commission (2015h) EU biodiversity strategy to 2020 – towards implementation. http://ec.europa.eu/environment/nature/biodiversity/comm2006/2020.htm. Accessed 15 Mar 2015

European Union (2011) The EU biodiversity strategy to 2010. http://ec.europa.eu/environment/nature/info/pubs/docs/brochures/2020%20Biod%20brochure%20final%20lowres.pdf. Accessed 15 Mar 2015

Food and Agriculture Organization of the United Nations (2002) China: regulations on restoring farmland to forest. 6 Dec 2002. http://faolex.fao.org/cgi-bin/faolex.exe?database=faolex&search_type=query&table=result&query=ID:LEX-FAOC054536&format_name=ERALL&lang=eng. Accessed 15 Mar 2015

Food and Agriculture Organization of the United Nations (2007) Food and agriculture organization gap principles. http://www.fao.org/prods/GAP/home/principles_en.htm. Accessed 15 Mar 2015

Food and Agriculture Organization of the United Nations (2009) The international treaty on plant genetic resources for food and agriculture. ftp://ftp.fao.org/docrep/fao/011/i0510e/i0510e.pdf. Accessed 15 Mar 2015

Food and Agriculture Organization of the United Nations (2013) The international treaty on plant genetic resources for food and agriculture. In: Planttreaty news – leading the field VI: newsletter of the international treaty on plant genetic resources for food and agriculture. Available via PlantTreaty.org. http://www.planttreaty.org/content/planttreaty-news-leading-field-vi. Accessed 15 Mar 2015

Food and Agriculture Organization of the United Nations (2015a) The international treaty on plant genetic resources for food and agriculture: contracting parties to the treaty. http://www.planttreaty.org/content/contracting-parties-treaty. Accessed 15 Mar 2015

Food and Agriculture Organization of the United Nations (2015b) The international treaty on plant genetic resources for food and agriculture: list of contracting parties. http://www.planttreaty.org/list_of_countries. Accessed 15 Mar 2015

Food and Agriculture Organization of the United Nations (2015c) The international treaty on plant genetic resources for food and agriculture: overview. http://www.planttreaty.org/content/overview. Accessed 15 Mar 2015

Food and Agriculture Organization of the United Nations (2015d) The international treaty on plant genetic resources for food and agriculture: the implementation of the article 6 of the treaty. http://www.planttreaty.org/content/implementation-article-6-treaty. Accessed 15 Mar 2015

Gov.UK (2014) New environmental scheme for farmers to prioritise biodiversity. 26 Feb 2014. https://www.gov.uk/government/news/new-environmental-scheme-for-farmers-to-prioritise-biodiversity. Accessed 15 Mar 2015

Kurukulasuriya L, Robinson NA(editors) (2006) Training manual on international environmental law. https://books.google.com/books?id=_RdE5j8P6iEC&pg=PA214&lpg=PA214&dq=%E2%80%9Cthe+State+shall+ensure+the+reasonable+utilization+of+natural+resources+and+protect+the+rare+and+valuable+fauna+and+flora.%E2%80%9D&source=bl&ots=i2GUksQWZ9&sig=rzYMrT3ZIk0-ap91Sc-bPoYR3-Q&hl=en&sa=X&ei=iJHsVPuxL8iqggTEn4GQCA&ved=0CCAQ6AEwAA#v=onepage&q=%E2%80%9Cthe%20State%20shall%20ensure%20the%20reasonable%20utilization%20of%20natural%20resources%20and%20protect%20the%20rare%20and%20valuable%20fauna%20and%20flora.%E2%80%9D&f=false

Marine Mammal Commission (2007) The marine mammal protection act of 1972 as amended 2007 (16 U.S.C. Chapter 31). http://www.nmfs.noaa.gov/pr/pdfs/laws/mmpa.pdf. Accessed 15 Mar 2015

Moberg F (2004) Agricultural diversity and food production. Sustainable Development Update 4 (2). Available via Albaeco. http://www.albaeco.com/sdu/15/htm/main.htm. Accessed 15 Mar 2015

Michigan State University College of Law (2015) United States code annotated, title 16, conservation, chapter 40, fish and wildlife conservation (16 U.S.C. §§ 2901-2912). https://www.animallaw.info/statute/us-conservation-fish-wildlife-conservation-act. Accessed 15 Mar 2015

Place F et al (2012) Improved policies for facilitating the adoption of agroforestry. In: Kaonga ML (ed) Agroforestry for biodiversity and ecosystem services – science and practice. Available via FAO.org. http://www.fao.org/forestry/36094-081bf412eb772690e5b90cc8d444880e3.pdf. Accessed 4 Apr 2015

The National Archives (2001) Organic farming (England rural development programme) regulations. http://www.legislation.gov.uk/uksi/2001/432/contents/made. Accessed 15 Mar 2015

United Nations. World day to combat desertification: desertification. http://www.un.org/en/events/desertificationday/background.shtml. Accessed 15 Mar 2015

United Nations Convention on Biological Diversity (1992) Convention on biological diversity. http://www.cbd.int/doc/legal/cbd-en.pdf. Accessed 15 Mar 2015

United Nations Convention on Biological Diversity (1996) COP 3 Decision III/11. http://www.cbd.int/decision/cop/?id=7107. Accessed 15 Mar 2015

United Nations Convention on Biological Diversity (2000) COP 5 Decision V/5: agricultural biological diversity: review of phase I of the programme of work and adoption of a multi-year work programme. http://www.cbd.int/decision/cop/?id=7147. Accessed 15 Mar 2015

United Nations Convention on Biological Diversity (2006) COP 8 Decision VIII/23: agricultural biodiversity. http://www.cbd.int/decision/cop/?id=11037. Accessed 15 Mar 2015

United Nations Convention on Biological Diversity (2007) In-depth review of the implementation of the programme of work on agricultural biodiversity. http://www.cbd.int/doc/meetings/sbstta/sbstta-13/official/sbstta-13-02-en.pdf. Accessed 15 Mar 2015

United Nations Convention on Biological Diversity (2015a) The Nagoya protocol on access and benefit-sharing. http://www.cbd.int/abs/. Accessed 15 Mar 2015

United Nations Convention on Biological Diversity (2015b) About the Nagoya protocol. http://www.cbd.int/abs/about/default.shtml#obligations. Accessed 15 Mar 2015

United Nations Convention on Biological Diversity (2015c) China national biodiversity conservation strategy and action plan (2011-2030). http://www.cbd.int/doc/world/cn/cn-nbsap-v2-en.pdf. Accessed 15 Mar 2015

United Nations Convention on Biological Diversity (2015d) Cross-cutting initiatives. http://www.cbd.int/agro/cross-cutting.shtml. Accessed 15 Mar 2015

United Nations Convention on Biological Diversity (2015e) Ecosystem approach. http://www.cbd.int/ecosystem/. Accessed 15 Mar 2015

United Nations Convention on Biological Diversity (2015f) Genetic use restriction technologies (GURTS). http://www.cbd.int/agro/GURTS.shtml. Accessed 15 Mar 2015

United Nations Convention on Biological Diversity (2015g) History of the convention. http://www.cbd.int/history/. Accessed 15 Mar 2015

United Nations Convention on Biological Diversity (2015h) List of parties. http://www.cbd.int/information/parties.shtml. Accessed 15 Mar 2015
United Nations Convention on Biological Diversity (2015i) Programme of work. http://www.cbd.int/agro/pow.shtml. Accessed 15 Mar 2015
United Nations Convention on Biological Diversity (2015j) Thematic programmes and cross-cutting issues. http://www.cbd.int/programmes/. Accessed 15 Mar 2015
United Nations Convention on Biological Diversity (2015k) Introduction. http://www.cbd.int/intro/default.shtml. Accessed 15 Mar 2015
United Nations Convention on Wetlands of International Importance especially as Waterfowl Habitat [Ramsar Convention] (1994) The convention on wetlands text, as amended in 1982 and 1987. http://archive.ramsar.org/cda/en/ramsar-documents-texts-convention-on/main/ramsar/1-31-38%5E20671_4000_0__. Accessed 15 Mar 2015
United Nations Convention on Wetlands of International Importance especially as Waterfowl Habitat [Ramsar Convention] (2014a) About Ramsar. http://www.ramsar.org/. Accessed 15 Mar 2015
United Nations Convention on Wetlands of International Importance especially as Waterfowl Habitat [Ramsar Convention] (2014b) International cooperation. http://www.ramsar.org/about/international-cooperation. Accessed 15 Mar 2015
United Nations Convention on Wetlands of International Importance especially as Waterfowl Habitat [Ramsar Convention] (2014c) The Ramsar convention and its mission. http://www.ramsar.org/about/the-ramsar-convention-and-its-mission. Accessed 15 Mar 2015
United Nations Convention on Wetlands of International Importance especially as Waterfowl Habitat [Ramsar Convention] (2014d) The wise use of wetlands. http://www.ramsar.org/about/the-wise-use-of-wetlands. Accessed 15 Mar 2015
United Nations Convention to Combat Desertification (2012) About the convention. http://www.unccd.int/en/about-the-convention/Pages/About-the-Convention.aspx. Accessed 15 Mar 2015
United Nations Convention on International Trade in Endangered Species of Wild Fauna and Flora (2015a) How CITES works. http://www.cites.org/eng/disc/how.php. Accessed 15 Mar 2015
United Nations Convention on International Trade in Endangered Species of Wild Fauna and Flora (2015b) What is CITES? http://www.cites.org/eng/disc/what.php. Accessed 15 Mar 2015
United Nations Convention on the Conservation of Migratory Species of Wild Animals (2014a) About CMS. http://www.cms.int. Accessed 15 Mar 2015
United Nations Convention on the Conservation of Migratory Species of Wild Animals (2014b) CMS. http://www.cms.int/en/legalinstrument/cms. Accessed 15 Mar 2015
United Nations Convention on the Conservation of Migratory Species of Wild Animals (2014c) Parties and range states. http://www.cms.int/en/parties-range-states. Accessed 15 Mar 2015
United Nations Educational, Scientific and Cultural Organization (2013) World heritage agricultural landscapes. http://whc.unesco.org/en/review/69/. Accessed 15 Mar 2015
United Nations Educational, Scientific and Cultural Organization (2014) States parties: ratification status. http://whc.unesco.org/en/statesparties/. Accessed 15 Mar 2015
United Nations Educational, Scientific and Cultural Organization (2015) The criteria for selection. http://whc.unesco.org/en/criteria/. Accessed 15 Mar 2015
United States Department of Agriculture (2015) Grading, certification and verification. http://www.ams.usda.gov/gapghp. Accessed 15 Mar 2015
United States Department of Transportation, Federal Highway Administration (2002) Interaction between NEPA and ESA. http://environment.fhwa.dot.gov/ecosystems/laws_esaguide.asp. Accessed 21 Mar 2015
United States Fish and Wildlife Service (1920) Digest of federal resource laws of interest to the U.S. fish and wildlife service. https://www.fws.gov/laws/lawsdigest/migtrea.html/. Accessed 15 Mar 2015
United States Fish and Wildlife Service (2013) 40 years of conserving endangered species. Jan 2013. http://www.fws.gov/endangered/esa-library/pdf/ESA_basics.pdf. Accessed 15 Mar 2015

United States Fish and Wildlife Service, International Affairs (2004) Lacey act (18 U.S.C. 42-43 and 16 U.S.C. 3371-3378). http://www.fws.gov/international/laws-treaties-agreements/us-conservation-laws/lacey-act.html. Accessed 15 Mar 2015

United States Fish and Wildlife Service, International Affairs (2013a) Endangered species act: overview. http://www.fws.gov/endangered/laws-policies/. Accessed 15 Mar 2015

United States Fish and Wildlife Service, International Affairs (2013b) Endangered species act: section 2. http://www.fws.gov/endangered/laws-policies/section-2.html. Accessed 15 Mar 2015

United States Fish and Wildlife Service, International Affairs (2013c) Endangered species act: section 3. http://www.fws.gov/endangered/laws-policies/section-3.html. Accessed 15 Mar 2015

United States Fish and Wildlife Service, International Affairs (2013d) Endangered species act: section 4. http://www.fws.gov/endangered/laws-policies/section-4.html. Accessed 15 Mar 2015

United States Fish and Wildlife Service, International Affairs (2013e) Endangered species list. http://www.fws.gov/endangered/laws-policies/esa.html. Accessed 15 Mar 2015

United States Fish and Wildlife Service, International Affairs (2013f) Injurious wildlife. http://www.fws.gov/le/injurious-wildlife.html. Accessed 15 Mar 2015

United States Fish and Wildlife Service, International Affairs (2014a) Critical habitat: what is it? http://www.fws.gov/midwest/endangered/saving/CriticalHabitatFactSheet.html. Accessed 15 Mar 2015

United States Fish and Wildlife Service, International Affairs (2014b) Endangered species permits. http://www.fws.gov/Midwest/endangered/permits/hcp/index.html. Accessed 15 Mar 2015

United States Fish and Wildlife Service, Office of Law Enforcement (2004) Wild bird conservation act (16 U.S.C. 4901-4916). 30 Apr 2004. http://www.fws.gov/le/USStatutes/WBCA.pdf. Accessed 15 Mar 2015

University of Montana, College of Forestry and Conservation's Wilderness Institute. The beginnings of the national wilderness preservation system. http://www.wilderness.net/nwps/fastfacts. Accessed 15 Mar 2015

University of Montana, College of Forestry and Conservation's Wilderness Institute (1964) The wilderness act of 1964. 3 Sept 1964. http://www.wilderness.net/nwps/legisact. Accessed 15 Mar 2015

Verhoeven JTA, Setter TL (2009) Agricultural use of wetlands: opportunities and limitations. Ann Bot. Jan 2010. Available via NCBI. http://www.ncbi.nlm.nih.gov/pmc/articles/PMC2794053/. Accessed 20 Mar 2015

The World Wildlife Fund for Nature (WWF) (2015) Unsustainable and illegal wildlife trade. http://wwf.panda.org/about_our_earth/species/problems/illegal_trade/. Accessed 20 Mar 2015

Chapter 25
Marine and (Over-) Fishing

Leslie Couvillion

Abstract The United Nations (UN) declares that oceans are "the lifeline of man's very survival." Human societies have depended upon the sea and its resources for millennia. According to the UN Food and Agriculture Organization (UN FAO or FAO), "[f]rom ancient times, fishing has been a major source of food for humanity and a provider of employment and economic benefits to those engaged in this activity." Today, many cultures continue to rely on marine fishing as a main source of income, livelihood, and sustenance. The demand for fish and fisheries products has only increased as the world's population has surged and standards of living (along with the desire for diets rich in high-quality protein) have risen over recent decades. For instance, the UN reports that global exports of fish and fish products increased more than 100 % from 1986 to 2006, reaching over $85 billion in 2006.

25.1 Introduction

The United Nations (UN) declares that oceans are "the lifeline of man's very survival."[1] Human societies have depended upon the sea and its resources for millennia. According to the UN Food and Agriculture Organization (UN FAO or FAO), "[f]rom ancient times, fishing has been a major source of food for humanity and a provider of employment and economic benefits to those engaged in this activity."[2] Today, many cultures continue to rely on marine fishing as a main source of income, livelihood, and sustenance. The demand for fish and fisheries products has only increased as the world's population has surged and standards of living (along with the desire for diets rich in high-quality protein) have risen over recent decades. For instance, the UN reports that global exports of fish and fish products

[1]United Nations, Division for Ocean Affairs and the Law of the Sea, Office of Legal Affairs (2012).

[2]Food and Agriculture Organization of the United Nations (1995b).

L. Couvillion (✉)
Yale School of Law, New Haven, CT, USA
e-mail: leslie.couvillion@yale.edu

G. Steier, K.K. Patel (eds.), *International Farm Animal, Wildlife and Food Safety Law*, DOI 10.1007/978-3-319-18002-1_25

increased more than 100 % from 1986 to 2006, reaching over $85 billion in 2006.[3] That same year, nearly half of the global population got at least 15 % of its per capita animal protein intake from fish, and over 40 million people worked in the primary fish production sector.[4] Thus, fisheries remain as vital as ever, boosting local, national, and global economies while helping to alleviate problems of food insecurity, malnutrition, and poverty.

The oceans, however, are not an unconditional gift that can keep on giving. Yet, nations appear to have believed in the inexhaustibility of marine resources for too long. After the Industrial Revolution, "[l]arge fishing vessels were roaming the oceans far from their native shores . . . Nations were flooding the richest fishing waters with their fishing fleets virtually unrestrained."[5] Nearly every fishing operator focused on maximizing catch, unaware of (or unconcerned about) the impacts of these practices on ecological systems or the industry's long-term viability. By the mid-twentieth century, some fish stocks began to show signs of depletion. These concerns prompted improvements in fisheries science and research. As marine science advanced, scientists and policymakers increasingly realized:

> that aquatic resources, although renewable, are not infinite and need to be properly managed, if their contribution to the nutritional, economic and social well-being of the growing world's population is to be sustained.[6]

After decades of deliberation, the international community adopted the UN Convention on the Law of the Sea (UNCLOS) in 1982.[7] UNCLOS sought to provide a new and better framework for marine resources management. It remains a critical tool for guiding global, regional, and national efforts on these issues today.

Even in the twenty-first century, the world's fisheries confront ongoing sustainability challenges. Many modern commercial fleets continue to use highly intensive techniques, such as trawling, purse seining, and gillnetting.[8] These methods contribute to two major sustainability problems:

---

[3]United Nations (2010b). The top ten producing countries were China, Peru, United States, Indonesia, Japan, Chile, India, Russia, Thailand, and the Philippines. The top species produced included anchoveta, Alaska pollack, skipjack tuna, Atlantic herring, largehead hairtail and yellowfin tuna. These figures are based on both off-shore commercial capture ("wild catch") and on-shore or near-shore aquaculture ("fish farm") fisheries.

[4]United Nations (2010b).

[5]United Nations, Division for Ocean Affairs and the Law of the Sea, Office of Legal Affairs (2012).

[6]Food and Agriculture Organization of the United Nations (1995b).

[7]See *infra* Sect 25.2.1.

[8]Trawling consists of a boat (known as a "trawler") dragging a cone-shaped net either along the sea bottom ("bottom trawling") or in mid-water ("pelagic trawling"). In purse seining, a seine net is cast over a school of fish and "drawn up" (like a purse drawstring) so that fish cannot escape. Gillnetting is a practice where long flat nets are hung from floats at the water's surface, usually close to shore. Fish are unable to detect the nets underwater and swim directly into their traps. Each of these methods is highly efficient, tending to catch everything in its path unless certain precautions are taken (such as setting minimum net mesh sizes). OceanLink. Fishing methods.

(1) Overfishing, or the removal of individual fish at rates so high that stocks cannot replenish themselves; and
(2) Bycatch, or the incidental catch of non-target, undersized, or juvenile animals, usually due to using indiscriminate fishing gear or fishing in nursery areas.[9]

Other practices, such as illegal fishing[10] and fishing industry subsidies,[11] exacerbate overfishing and bycatch threats. What's more, marine living resources face further risks from activities not directly related to fishing. Some of these additional hazards include: habitat destruction from coastal development, non-point source pollution from agricultural operations, habitat disruption from offshore oil and gas exploration or deep sea mineral mining, and acidification and increased sea temperatures from greenhouse gas (GHG) emissions that contribute to climate change.[12] This wide range of threats shows the extent to which the marine fisheries sector is interconnected with other major economic sectors, such as agriculture, energy, and mining.

As a result of unsustainable practices (in the fisheries sector and beyond), a great number of the world's fish stocks are endangered. The FAO reported in 2007 that approximately 80 % of stocks were either fully exploited[13] or overexploited.[14] Only about 20 % of stocks were moderately exploited or underexploited (and, therefore, capable of producing more fish).[15] The Northeast Atlantic, the Western Indian Ocean, and the Northwest Pacific have the highest proportions of fully-exploited stocks.[16] This alarming data on the state of the world's fish stocks—a full two decades after UNCLOS was first signed—shows that managing fishing operations in a sustainable manner is a persistent challenge, one that nations are still figuring out.[17]

---

[9]Save Our Seas Foundation (2015). The FAO estimates that over 20 % of all marine landings consist of bycatch. United Nations (2010b).

[10]"[C]urrent losses due to illegal, unreported and unregulated fishing worldwide are estimated to be between $10 billion and $25.5 billion annually." United Nations (2010b), p. 3.

[11]The FAO and the World Bank report that "total fishing subsidies, which directly impact fishing capacity and fuel overfishing, amounted to over $10 billion in 2000." United Nations (2010b), p. 2.

[12]See Miller and Spoolman (2008), p. 172.

[13]About 52 % of stocks are fully exploited, or "producing catches that were at or close to their maximum sustainable limits with no room for further expansion." United Nations (2010b), p. 1.

[14]Around 28 % of stocks were "overexploited, depleted, or recovering from depletion and thus yielding less than their maximum potential owing to excess fishing pressure." United Nations (2010b), p. 1.

[15]United Nations (2010b), p. 1.

[16]United Nations (2010b), p. 2.

[17]The UN does not appear to have produced updated, comprehensive data on overfishing since 2006. However, a recent report from the European Commission suggests that slow but steady improvements have been made for certain stocks since then. For instance, there has been "marked improvement in the state of tuna stocks. Out of 16 stocks worldwide, only 6 were fished sustainably six years ago – in 2013 we have gone up to 13." European Commission (2014a).

Despite the enduring challenges, many governments have made important innovations in their fisheries management regimes since the 1970s and 1980s. International law has guided much of this progress.[18] In addition to UNCLOS, an array of multilateral, regional, and bilateral treaties and agreements govern fisheries issues around the world.[19] These instruments have enabled more robust fishery management systems, stronger conservation measures, and better coordination among nations. Generally, they take the form of (a) setting specific management and conservation mandates and goals, along with (b) providing a menu of tools to help meet these targets. Similarly, at the national level, major fisheries legislation usually consists of mandatory, scientifically-informed rules accompanied by an amalgam of both mandatory and voluntary management techniques. Delegated authorities then establish fishery management plans that adapt these techniques to regional or local circumstances. Some areas, like the European Union (EU),[20] have a fairly centralized system, while others, such as China,[21] have a more de-centralized regime that gives greater discretion and authority to regional and local bodies.[22]

Across the world, fisheries management authorities (whether international, regional, national, or local) employ a wide array of conservation tools. Different measures are favored in different areas, and some methods are more effective than others. Common tools for addressing overfishing include:

- Quotas or maximum/total allowable catch limits. These are usually based on maximum sustainable yield (MSY) levels[23];
- Catch share programs[24];
- Marine protected areas, in which fishing (and other) activities are prohibited or restricted;
- Fleet capacity limitations, such as limits on the total number of fishing vessels; and
- Seasonal fishery closures.

---

[18]See *infra* Sect. 25.2.

[19]See Appendix.

[20]See *infra* Sect. 25.4.

[21]See *infra* Sect. 25.5.

[22]However, the EU system has become more decentralized after recent reforms to its Common Fisheries Policy. See *infra* Sect. 25.4.

[23]Maximum sustainable yield (MSY) is the largest yield (or catch) that can be taken without threatening a population's ability to reproduce itself. United Nations Convention on the Law of the Sea (1982), Art. 61(3).

[24]Catch share programs work by allocating "permits" or other secured privileges to individuals or organizations that allow them to harvest a specific area or a pre-determined percentage of a fishery's total catch. These programs "promote fishing based on good business decisions and stewardship practices rather than on the earlier years of 'race-to-fish' or 'days-at-sea' strategies that were often as dangerous for crews as they were unsustainable for the resource." United States National Oceanic and Atmospheric Administration, National Marine Fisheries Service (2011).

Potential tools to limit bycatch include:

- Bycatch caps, or limits on the number of non-target or juvenile species that can be caught;
- Minimum net mesh size requirements; and
- Gear restrictions, such as bans on types of driftnet fishing.

Given the complexity of the field and the wide range of strategies used, this chapter does not provide a comprehensive account of fisheries management regimes around the world. It begins with a review of the basic framework in place at the international level and then provides an overview of the national-level laws of three of the most significant fish-producing regions on the planet: the United States (US),[25] the EU,[26] and China.[27] The Appendix contains information on statutes and regulations on marine fisheries issues in other nations, emphasizing those laws which call for especially strong or innovative measures.

25.2 International Treaties and Conventions

Oceans are global, covering over 70 % of the surface and 99 % of the "living space" on the planet.[28] Likewise, trade in marine products is a globalized field. As a result, a robust marine fisheries management regime has developed at the international level. It involves a number of international bodies: the UN General Assembly, along with the UN Informal Consultative Process on the Law of the Sea (ICP), directs global fisheries issues, while the UN FAO's Committee on Fisheries (COFI) also carries a global mandate to develop fisheries policy.[29] The International Tribunal of the Law of the Sea (ITLOS) resolves conflicts between States on marine management issues, while the World Trade Organization (WTO) addresses obstacles to fish exports and imports among trading nations.[30]

In addition to international bodies, there are over 20 regional marine fishery organizations (RMFOs).[31] RMFOs play a crucial role in marine fisheries management around the world. Indeed, existing RMFOs cover the majority of Earth's oceans, and most of these groups "have the power to set catch and fishing effort

[25]See *infra* Sect. 25.3.

[26]See *infra* Sect. 25.4.

[27]See *infra* Sect. 25.5.

[28]Hawaii Pacific University Oceanic Institute (2011).

[29]Food and Agriculture Organization of the United Nations, Fisheries and Aquaculture Department (2015e).

[30]Food and Agriculture Organization of the United Nations, Fisheries and Aquaculture Department (2015e).

[31]European Commission (2014a), p. 6.

limits, technical measures, and control obligations."[32] Examples include the South East Atlantic Fisheries Organization (SEAFO) and the Western and Central Pacific Fisheries Commission (WCPFC).[33] These bodies help implement and enforce a wide range of mechanisms to manage the world's marine living resources.

The following section discusses the four main pillars of international marine fisheries management: (1) UNCLOS[34]; (2) the UN Fish Stocks Agreement[35]; (3) the FAO Compliance Agreement[36]; and (4) the FAO Fisheries Code.[37] In addition to these instruments, governments have enacted a number of significant international and regional conventions and treaties, such as:

- The Convention on the Conservation of Antarctic Marine Living Resources (CCAMLR), which aims to protect the region's krill, an important foundation of the marine ecosystem food chain[38];
- The International Convention for the Conservation of Atlantic Tunas (ICCAT), which sets maximum sustainable yield levels for Atlantic tuna populations[39]; and
- The Convention for the Conservation of Southern Bluefin Tuna, which implements total allowable catch and bycatch measures.[40]

There are also countless bilateral treaties and agreements. Examples include the Treaty between the Government of the United States of America and the Government of Canada Concerning Pacific Salmon (which allows for the creation of salmon enhancement programs to prevent overfishing)[41] and the Agreement between Norway and Iceland concerning Fishery and Continental Shelf Questions (which calls for measures on conservation, rational exploitation, and sound reproduction of capelin fish stocks).[42] Bilateral cooperation on fisheries issues is key to conserving and managing living marine resources that are shared between neighboring countries. Therefore, these agreements supplement the broader international and regional treaties and allow for targeted species- or habitat- specific approaches.[43]

---

[32]European Commission (2014a), p. 6.

[33]See Appendix.

[34]See *infra* Sect. 25.2.1.

[35]See *infra* Sect. 25.2.2.

[36]See *infra* Sect. 25.2.3.

[37]See *infra* Sect. 25.2.4.

[38]Commission for the Conservation of Antarctic Marine Living Resources (1980).

[39]International Convention for the Conservation of Atlantic Tunas (1986).

[40]Convention for the Conservation of Southern Bluefin Tuna (1993).

[41]Pacific Salmon Commission (2014).

[42]United Nations (2010a).

[43]See Appendix for additional examples of regional and bilateral fisheries agreements.

25.2.1 *United Nations Convention on the Law of the Sea (UNCLOS) (1982)*

The UN Convention on the Law of the Sea (UNCLOS)[44] is a "constitution for the seas," establishing a comprehensive regime for governing the world's oceans.[45] It was developed in the latter half of the twentieth century, in response to:

> growing concern over the toll taken on coastal fish stocks by long-distance fishing fleets and over the threat of pollution and wastes from transport ships and oil tankers carrying noxious cargoes that plied sea routes across the globe. The hazard of pollution was ever present, threatening coastal resorts and all forms of ocean life.[46]

By the late 1960s, governments around the world recognized the need for "a more stable order, promoting greater use and better management of ocean resources and generating harmony and goodwill among States."[47] In 1973, the UN convened the Third UN Conference on the Law of the Sea to write a comprehensive treaty for the oceans. Nine years later, UNCLOS was adopted.[48] The treaty has been in force since November 1994 and has 166 ratifying/acceding Parties as of October 2014.[49] UNCLOS was adopted as a "package deal," meaning that ratifying or acceding States are bound by all of its provisions, without reservation, and signatory States "undertak[e] not to take any action that might defeat its objects and purposes."[50]

UNCLOS is indeed constitutional in scope. Its provisions cover:

> [n]avigational rights, territorial sea limits, economic jurisdiction, legal status of resources on the seabed beyond the limits of national jurisdiction, passage of ships through narrow straits, conservation and management of living marine resources, protection of the marine environment, a marine research regime and, a more unique feature, a binding procedure for settlement of disputes between States.[51]

Two features of UNCLOS are central to marine fisheries governance:

---

[44] United Nations Convention on the Law of the Sea (1982).

[45] United Nations, Division for Ocean Affairs and the Law of the Sea, Office of Legal Affairs (2012).

[46] United Nations, Division for Ocean Affairs and the Law of the Sea, Office of Legal Affairs (2012).

[47] United Nations, Division for Ocean Affairs and the Law of the Sea, Office of Legal Affairs (2012).

[48] United Nations, Division for Ocean Affairs and the Law of the Sea, Office of Legal Affairs (2012).

[49] United Nations (2014).

[50] United Nations, Division for Ocean Affairs and the Law of the Sea, Office of Legal Affairs (2012).

[51] United Nations, Division for Ocean Affairs and the Law of the Sea, Office of Legal Affairs (2012).

25.2.1.1 Defining National Territorial Seas

UNCLOS establishes the concept of the Exclusive Economic Zones (EEZ): a 3 to 200 offshore nautical mile radius over which a State has special rights regarding the exploration and use of marine resources, including marine living resources (e.g., fish and other forms of sea life).[52] States are bound to conserve and manage marine living resources in areas within their jurisdiction or under their sovereign rights.[53] Coastal States have a general legal right to manage marine fisheries within their EEZs, including the ability to limit access by foreign fishing vessels.[54] All areas of the oceans falling beyond any individual nation's EEZ or other territorial seas (like archipelagic waters) are considered the high seas. The high seas are the common heritage of mankind, and, as a result, are open to all States.[55] Therefore, for the most part, UNCLOS does not regulate high seas fisheries in international waters; such activity is predominately managed by RMFOs.[56]

25.2.1.2 Prescribing Fisheries Management Rules

UNCLOS also prescribes specific fisheries management rules that nations must follow within their EEZs. Article 61 sets out common basic principles of conservation and management. For instance, coastal States must set total allowable catches (i.e., quotas on the total number of individual organisms that can be caught) for living resources within their EEZs.[57] They must also employ measures to guard against over-exploitation of living resource stocks.[58] Conservation measures should be designed around the principle of maximum sustainable yield (MSY): the largest yield (or catch) that can be taken without threatening a population's ability to reproduce itself. In determining MSY levels for various species, States can consider a range of factors, including "the economic needs of coastal fishing communities and the special requirements of developing States, [as well as] fishing patterns, the interdependence of stocks and any generally recommended international minimum standards, whether subregional, regional or global."[59] To help States set suitable management targets for particular stocks, Article 61 also calls on nations to cooperate with one another by sharing scientific information, catch and fishing effort statistics, and other relevant data. Such information-sharing is crucial for

[52]United Nations Convention on the Law of the Sea (1982), Art. 57.

[53]United Nations Convention on the Law of the Sea (1982), Art. 58.

[54]United Nations Convention on the Law of the Sea (1982), Art. 62.

[55]United Nations Convention on the Law of the Sea (1982), Preamble, Part VII.

[56]See *infra* Sect. 25.2.2.

[57]United Nations Convention on the Law of the Sea (1982), Art. 61(1).

[58]United Nations Convention on the Law of the Sea (1982), Art. 61(2).

[59]United Nations Convention on the Law of the Sea (1982), Art. 61(3).

allowing States to put in place effective, sustainable conservation and management plans.[60]

UNCLOS was (and remains) an important, even revolutionary, document. Indeed, its signers called it "[p]ossibly the most significant legal instrument of this century."[61] The Convention was "an unprecedented attempt by the international community to regulate *all* aspects of the resources of the sea and uses of the ocean, and thus bring a stable order to mankind's very source of life."[62] As a whole, therefore, the convention attempts to set up a comprehensive regulatory scheme for the world's marine environments. However, it is not truly all-encompassing. One of the treaty's major gaps is high seas fishing regulation. Indeed, in the years following UNCLOS, concerns mounted around the problem of overexploiting fish stocks that fall outside of UNCLOS' primary scope (such as migratory fish stocks that pass through unregulated international waters). This dilemma highlighted the need for new regulatory tools. In response, throughout the 1990s, UN bodies drafted mechanisms to complement and extend UNCLOS.[63] Among these new tools were the UN Fish Stocks Agreement,[64] the FAO Compliance Agreement,[65] and the FAO Fisheries Code.[66] The following sections discuss these instruments.

25.2.2 UN Fish Stocks Agreement (1995)

By the early 1990s, worries about the over-exploitation of fish stocks—especially in unregulated high seas fisheries—were escalating. In March 1991, the FAO's Committee on Fisheries (COFI) called "for the development of new concepts which would lead to responsible, sustained fisheries."[67] This ultimately brought about three new instruments:

(1) Agreement on Straddling Fish Stocks and Highly Migratory Fish Stocks (Fish Stocks Agreement),[68] in force since December 2001 and with 82 ratifying/acceding Parties as of October 2014[69];

---

[60]United Nations Convention on the Law of the Sea (1982), Art. 61(5).

[61]United Nations, Division for Ocean Affairs and the Law of the Sea, Office of Legal Affairs (2012).

[62]United Nations, Division for Ocean Affairs and the Law of the Sea, Office of Legal Affairs (2012) (emphasis added).

[63]Food and Agriculture Organization of the United Nations, Fisheries and Aquaculture Department (2015c).

[64]See *infra* Sect. 25.2.2.

[65]See *infra* Sect. 25.2.3.

[66]See *infra* Sect. 25.2.4.

[67]Food and Agriculture Organization of the United Nations (1995b).

[68]United Nations Conference on Straddling Fish Stocks and Highly Migratory Fish Stocks (1995).

[69]United Nations, Division for Ocean Affairs and the Law of the Sea (2014).

(2) Agreement to Promote Compliance with International Conservation and Management Measures by Fishing Vessels on the High Seas (Compliance Agreement), in force since April 2003[70]; and
(3) FAO Code of Conduct for Responsible Fisheries (Fisheries Code or Code).[71]

While the Fish Stocks Agreement and the Compliance Agreement are binding on parties, the Fisheries Code is a voluntary instrument that requires no special action to take effect. The three instruments were drafted alongside one another and, as a result, are highly consistent, with many overlapping provisions. Each helps to expand the scope and effectiveness of UNCLOS.[72]

The Fish Stocks Agreement concerns two kinds of stocks: (1) straddling fish stocks[73]; and (2) highly migratory fish stocks.[74] Both kinds of stocks occur in national territorial seas (regulated by UNCLOS) as well as international high seas (generally not regulated by UNCLOS). Therefore, the agreement lays out a legal regime for conserving and managing species which are often found in places that are beyond the reach of UNCLOS.[75] This legal regime incorporates new principles, norms and rules for high seas fishery management. These include the precautionary principle,[76] ecosystem management,[77] and using the "best available scientific information" to inform management decisions.[78] Such measures "constitute a progressive development" of relevant UNCLOS provisions.[79]

The key function of the Fish Stocks agreement is to set up a framework for inter-State cooperation. The agreement establishes regional fisheries management organizations (RMFOs) as "the primary vehicle for cooperation between coastal States and high seas fishing States."[80] RMFOs are granted authority in a wide range areas,

[70]Food and Agriculture Organization of the United Nations (1995a). See *infra* Sect. 25.2.3.

[71]Food and Agriculture Organization of the United Nations (1995b). See *infra* Sect. 25.2.4.

[72]See *supra* Sect. 25.2.1.

[73]Straddling stocks are those which occur both within a country's EEZ and in adjacent high seas areas. They commonly include species of cod, halibut, pollock, jack mackerel, and squid. United Nations. The 1995 United Nations fish stock agreement [background paper].

[74]Highly migratory fish stocks include species, such as tuna, swordfish, and oceanic sharks, which regularly travel long distances through both high seas areas and EEZs or other areas under national jurisdiction. United Nations. The 1995 United Nations fish stock agreement [background paper].

[75]United Nations. The 1995 United Nations fish stock agreement [background paper].

[76]Under the precautionary approach, "States shall be more cautious when information is uncertain, unreliable or inadequate. The absence of adequate scientific information shall not be used as a reason for postponing or failing to take conservation and management measures." United Nations Conference on Straddling Fish Stocks and Highly Migratory Fish Stocks (1995), Art. 6(2).

[77]An ecosystem approach is as "a strategy for the integrated management of land, water and living resources that promotes conservation and sustainable use in an equitable way" and recognizes humans as an integral part of ecosystems. United Nations Convention on Biological Diversity (2015).

[78]United Nations, Division for Ocean Affairs and the Law of the Sea (2001).

[79]United Nations. The 1995 United Nations fish stock agreement [background paper].

[80]United Nations. The 1995 United Nations fish stock agreement [background paper].

including scientific research, stock assessment, monitoring, surveillance, control, and enforcement. As mentioned above,[81] RMFOs have come to play a dominant role in marine fisheries management worldwide.

Overall, the Fish Stocks Agreement is considered an important and effective legal tool. The UN has deemed it "the most important legally binding global instrument to be adopted for the conservation and management of fishery resources since the adoption of [UNCLOS] in 1982."[82] However, the FAO has also pointed out several difficulties when it comes to carrying out the agreement in practice. For instance, RMFOs continue to struggle with applying the precautionary approach, employing ecosystem management, and addressing transparency.[83] Another core problem is the absence of reliable and complete scientific data on fish stocks and the lack of good data collection methods.[84] Therefore, while the Fish Stocks Agreement has succeeded in establishing "robust international principles and standards at the global level" for regulating ocean fisheries, it faces some ongoing implementation challenges into the twenty-first century.[85]

25.2.3 FAO Compliance Agreement (1993)

The Compliance Agreement is similar in scope to the Fish Stocks Agreement.[86] Both instruments apply to high seas fishing in international waters. However, the Compliance Agreement differs in focus. Its central aim is to prevent the practice of bad faith vessel "reflagging:" changing the flag on a ship's mast from one country's flag to another in order to bring the vessel under a new nation's laws (i.e., switching to a "flag of convenience").[87] One motivation for this practice is to exempt vessels from high seas RMFO management measures. Only vessels from RMFO member parties can be compelled to comply with RMFO requirements within the RMFO jurisdiction. Other vessels (i.e., those flying the flags of non-RMFO member nations) can fish with impunity in areas that are otherwise subject to RMFO protections. Therefore, reflagging is an important problem from a conservation perspective.[88]

---

[81]See *supra* Sect. 25.2.

[82]United Nations. The 1995 United Nations fish stock agreement [background paper].

[83]Food and Agriculture Organization of the United Nations, Fisheries and Aquaculture Department (2015f).

[84]Balton and Koehler (2006).

[85]Balton and Koehler (2006), p. 6.

[86]See *supra* Sect. 25.2.2.

[87]Food and Agriculture Organization of the United Nations, Fisheries and Aquaculture Department (2015c).

[88]Food and Agriculture Organization of the United Nations, Fisheries and Aquaculture Department (2015c).

The Compliance Agreement attempts to prevent fishing ships from exploiting this apparent loophole in high seas fisheries governance. Article III obligates flag states to take "such measures as may be necessary to ensure that fishing vessels entitled to fly its flag do not engage in any activity that undermines the effectiveness of international conservation and management measures."[89] Every flag state must also maintain a record of fishing vessels entitled to fly its flag.[90] Other provisions encourage international cooperation and information exchange to enable better monitoring and compliance.[91] Indeed,

> [t]he principal benefit to participants [in the Compliance Agreement] will come from the availability of information regarding vessels authorized to fish on the high seas, which will lead to an increased ability to identify those vessels fishing without permission. This will be particularly important in light of the expanded powers that countries [through RMFOs] will acquire under the UN Fish Stocks Agreement. As these Agreements become increasingly effective, all participants will duly benefit.[92]

Therefore, the Compliance Agreement is a distinctive effort to eliminate a significant obstacle to high seas fishing management. By curtailing illegal fishing activities in RMFO waters, it allows RMFO measures to have more teeth. This enhances the legitimacy and effectiveness of the Fish Stocks Agreement, which is responsible for establishing such organizations and granting them extensive fisheries management authority.

25.2.4 FAO Fisheries Code (1995)

The Fisheries Code builds upon the Fish Stocks Agreement[93] and the Compliance Agreement.[94] Its goal is to "ensur[e] the effective conservation, management and development of living aquatic resources, with due respect for the ecosystem and biodiversity."[95] Therefore, its mission is holistic, recognizing and seeking to preserve the nutritional, economic, social, environmental, and cultural value of fisheries around the world.[96] As a global code, it has the broadest scope of all three instruments. It applies to:

- "[A]ll those involved in fisheries," including FAO Member and non-Member States, RMFOs, NGOs, industry organizations, and other relevant actors;

[89]Food and Agriculture Organization of the United Nations (1995a), Art. III(1)(a).

[90]Food and Agriculture Organization of the United Nations (1995a), Art. IV.

[91]See Food and Agriculture Organization of the United Nations (1995a), Arts. V and VI.

[92]Food and Agriculture Organization of the United Nations, Fisheries and Aquaculture Department (2015c).

[93]See *supra* Sect. 25.2.2.

[94]See *supra* Sect. 25.2.3.

[95]Food and Agriculture Organization of the United Nations (1995b), Introduction.

[96]Food and Agriculture Organization of the United Nations (1995b), Introduction.

- All types of fisheries, including capture as well as aquaculture; and
- All waters where fisheries occur, including both high seas and national jurisdictional waters.[97]

The Code lays out a number of principles and standards for responsible fisheries practices.[98] These cover fishing operations, fisheries research, and fish (and fish product) capture, processing, and trade. To supplement the Code, the FAO has also published a number of Technical Guidelines for Responsible Fisheries.[99] These guidelines address a range of issues, including:

- Marine protected areas;
- Fish trade;
- Aquaculture development;
- Integration of fisheries into coastal area management;
- Best practices for reducing incidental catches of seabirds.[100]

COFI is responsible for monitoring and updating the Code and its relevant guidelines. In 2012, COFI introduced a web-based reporting system, which allows Members to report on Code implementation and serves as a database and information resource for Members, RMFOs, and NGOs.[101] Therefore, the Code, its Technical Guidelines, and the COFI database are a likely reservoir of potential tools for change in the fisheries management sector.

The main limitation of the Fisheries Code is that it is voluntary and lacks enforcement mechanisms.[102] Despite its nonbinding nature, the Code has been adopted and applied by fisheries in a large number of countries.[103] Notably, it has been made available in more languages than any other FAO document. There are versions in the five official UN languages (Arabic, Chinese, English, French, and Spanish), as well as numerous unofficial translations by government, industry, or

[97]Food and Agriculture Organization of the United Nations (1995b), Introduction.

[98]Food and Agriculture Organization of the United Nations (1995b), Sect. 1.3.

[99]Food and Agriculture Organization of the United Nations, Fisheries and Aquaculture Department (2015b).

[100]The incidental catch of seabirds (like seagulls and albatrosses) is a problem in commercial longline fisheries around the world, especially tuna, swordfish, billfish, cod, halibut, and had25. dock fisheries. These incidental catches not only threaten ecosystem and species conservation efforts but "may also have an adverse impact on fishing productivity and profitability" since time, money, and other resources may have to be spent to remove and properly dispose of non-target species that end up caught in fishing gear. Food and Agriculture Organization of the United Nations (1998), Art. 1.

[101]Food and Agriculture Organization of the United Nations, Fisheries and Aquaculture Department (2013).

[102]See Food and Agriculture Organization of the United Nations (1995b), Sect. 1.1.

[103]The Fisheries Code has been adopted by over 17O FAO Members. Food and Agriculture Organization of the United Nations, Fisheries and Aquaculture Department (2001).

NGO actors who took on the initiative themselves.[104] Additionally, many parts of the Code are based on relevant rules of international law.[105] Therefore, along with UNCLOS,[106] the Fish Stocks Agreement, and the Compliance Agreement, the Fisheries Code is an important reference for parties looking to establish or to improve their legal, institutional, or operational approaches toward sustainable fisheries management.

Summary As a whole, the UN conventions and agreements, Fisheries Code, and RMFO measures work in tandem to encourage sustainable policies and practices in fisheries sector worldwide. UNCLOS provides a general "constitution" for managing marine resources within national territorial seas. To address high seas problems, the Fish Stocks Agreement lays out additional measures to protect straddling and migratory species that spend parts of their lives beyond any national jurisdiction (and are thus unregulated by UNCLOS). The Fish Stocks Agreements also sets up the RMFO network and grants these regional organizations a wide range of management authorities, as well as a duty to encourage inter-State communication and cooperation. The Compliance Agreement helps, in turn, to ensure that RMFOs can enforce their authority. Under the Compliance Agreement, States commit to taking action to discourage vessels from "reflagging" in order to get exempt from RMFO measures. The Fisheries Code supplements all of these instruments with voluntary guidelines and standards for fisheries conservation and management. RMFOs apply many of the measures from the various global instruments at a regional level. The end goal of all of these activities is to bring about positive and sustainable economic, social, and environmental outcomes in the fisheries sector and beyond.[107]

25.3 United States (US)

The United States is one of the top three fish-producing nations (along with China and Indonesia)[108] and boasts the largest EEZ in the world.[109] The United States has not ratified UNCLOS,[110] although it does recognize the treaty as customary

---

[104]Food and Agriculture Organization of the United Nations, Fisheries and Aquaculture Department (2015d).

[105]See Food and Agriculture Organization of the United Nations (1995b), Sect. 1.1.

[106]See *supra* Sect. 25.2.1.

[107]Food and Agriculture Organization of the United Nations, Fisheries and Aquaculture Department (2015a).

[108]Stastista (2012). This statistic is based on off-shore capture production and does not include aquaculture production.

[109]Food and Agriculture Organization of the United Nations (2005).

[110]United Nations, Division for Ocean Affairs and the Law of the Sea (2014). See *supra* Sect. 25.2.1.

international law and is a party to both the Fish Stocks Agreement[111] and FAO Compliance Agreement.[112] The US has also ratified a number of other international treaties and conventions that govern issues of ocean stewardship, including the:

- International Commission for the Conservation of Atlantic Tunas (ICCAT);
- Convention for the Conservation of Salmon in the North Atlantic (North Atlantic Salmon Treaty);
- International Convention for the Northwest Atlantic Fisheries (Northwest Atlantic Fisheries Treaty);
- Treaty between the Government of the United States of America and the Government of Canada Concerning Pacific Salmon (Pacific Salmon Treaty); and
- Convention for the International Council for the Exploration of the Sea (ICES).[113]

The legal foundation for most aspects of US fisheries management is the federal Magnuson-Stevens Act (MSA).[114] The MSA is primarily administered by the National Marine Fisheries Service (NMFS), a subsidiary of the National Oceanic and Atmospheric Administration (NOAA). NMFS's mission is to manage, conserve, and protect marine life within the US EEZ. The agency's duties include ensuring compliance with fisheries regulations, assessing and predicting the status of fish stocks, collecting fisheries data on environmental decisions affecting living marine resources, and working to reduce wasteful fishing practices (often by directly engaging with local communities). Overfishing and bycatch are particular concerns. NMFS runs a National Bycatch Program "to implement[] conservation and management measures for living marine resources that will minimize, to the extent practicable, bycatch and the mortality of bycatch that cannot be avoided."[115] NMFS's headquarters is the Office of Sustainable Fisheries (OSF), which supports and coordinates various Regional Offices and eight Regional Fishery Management Councils (FMCs).[116]

In addition to the MSA, a variety of other US federal laws govern marine management issues, covering topics from habitat and species protection to trade in fish products. The National Marine Sanctuaries Act (NMSA) designates certain marine environments as sanctuaries because of their environmental, historical or cultural importance. As of 2014, there were 13 marine sanctuaries in the US, each with its own separate staff and program to carry out scientific research and

[111]See United Nations, Division for Ocean Affairs and the Law of the Sea (2014). See *supra* Sect. 25.2.2.

[112]See United States National Oceanic and Atmospheric Administration, National Marine Fisheries Service (2014a). See *supra* Sect. 25.2.3.

[113]See Appendix for further details on each of these treaties.

[114]See *infra* Sect. 25.3.1.

[115]United States National Oceanic and Atmospheric Administration, National Marine Fisheries Service (2014b).

[116]United States National Oceanic and Atmospheric Administration, National Marine Fisheries Service (2014b).

monitoring.[117] Certain provisions of the Marine Mammal Protection Act (MMPA) seek to protect marine mammals and sea turtles from habitat destruction, vessel strikes, and incidental capture as bycatch.[118] Another key piece of federal legislation is the 1967 Pelly Amendment to the Fisherman's Protective Act. This statute authorizes the President of the United States to embargo wildlife products (including fish) if "nationals of a foreign country are engaging in trade or taking that diminishes the effectiveness of an international program in force with respect to the United States for the conservation of endangered or threatened species."[119] The Fish and Wildlife Service (FWS) uses the Pelly Amendment when negotiating on the listing of species under the Convention on International Trade in Endangered Species of Wild Fauna and Flora (CITES).[120] Relatedly, the Lacey Act prohibits fish or wildlife transactions and activities that violate state, federal, and Native American tribal or foreign law.[121] Together, these laws illustrate the complex intersections that exist between fisheries management, biodiversity preservation, and endangered species protection. Each law also presents a potential tool for change for encouraging sustainable fisheries practices.

A number of other prominent US national environmental laws have direct or indirect impacts on marine living resources. Many of these do not target marine or fisheries issues specifically. For example, the Endangered Species Act (ESA) lists over 100 endangered and threatened marine species, which fall under the jurisdiction of NMFS (while the FWS manages listed land and freshwater species).[122] The National Environmental Policy Act (NEPA)'s environmental assessment (EA) and environmental impact statement (EIS) requirements can be triggered by fisheries management decisions, such as significant changes to a fishery management plan.[123] The Federal Insecticide, Fungicide, and Rodenticide Act (FIFRA) can also be invoked to protect marine ecosystems. As of 2014, EPA has instated restrictions ("no-spray buffer zones") on the use of pesticides in or near waters of the Pacific Northwest to protect endangered or threatened Pacific salmon and steelhead.[124] The Appendix to this Chapter provides further information on these statutes, along with other laws and regulations that are relevant to marine and fishery issues. These laws supplement the Magnuson-Stevens Act, providing further tools for change to encourage sustainable practices by US fishing fleets.

---

[117] United States National Oceanic and Atmospheric Administration (2014).

[118] United States National Oceanic and Atmospheric Administration, National Marine Fisheries Service (2007b).

[119] United States Department of the Interior, Fish and Wildlife Service (1979).

[120] United States Department of the Interior, Fish and Wildlife Service (2015). See Chap. 24 for additional information on CITES.

[121] United States Department of the Interior, Fish and Wildlife Service, Office of Law Enforcement (2004).

[122] United States Department of the Interior, Fish and Wildlife Service (1973).

[123] United States National Oceanic and Atmospheric Administration, National Marine Fisheries Service (2014c).

[124] Washington State Department of Agriculture (2014).

25.3.1 Magnuson-Stevens Act (1976)

The Magnuson-Stevens Fishery Conservation and Management Act (Magnuson-Stevens Act or MSA)[125] is the primary US law on marine fisheries management. The act establishes the country's EEZ and governs fish populations within this zone (including restricting access by foreign fishing vessels). Prior to the MSA, federal marine fisheries management was virtually non-existent. But since the mid-1970s, the federal government has been actively managing the country's marine EEZ fisheries, while state or local commissions continue to manage inland and nearshore fisheries.[126]

Some of the MSA's objectives have shifted over its 40-year history. Its original purpose -- to expand the country's fishing industry—was essentially driven by economic concerns. However, the act has increasingly incorporated conservation and sustainability objectives over the past two decades. In its current form, the statute is modeled on principles of scientific management and best-stewardship practices.[127] It lays out a set of 10 National Standards that reflect both environmental and socioeconomic priorities, such as preventing overfishing, reducing bycatch, mitigating economic impacts to fishing communities, and using the best scientific information available.[128] As a whole, therefore, the MSA reflects a commitment to practices that promote environmental and economic sustainability on the basis of sound science.

The Magnuson-Stevens Act establishes an innovative regional public-private management system to carry out its goals. This network consists of eight regional Fishery Management Councils (FMCs),[129] made up of a wide range of stakeholders, including marine scientists, commercial and recreational fishermen, and state and federal fisheries managers. Each FMC is charged with developing Fishery Management Plans for every fishery in its region. These plans must follow MSA guidelines (including the 10 National Standards). "The MSA provides flexibility in achieving these sometimes conflicting national standards, but preventing overfishing is the top priority."[130] The menu of recommended tools for managers to incorporate into their plans includes:

---

[125]United States National Oceanic and Atmospheric Administration, National Marine Fisheries Service (2007a).

[126]Food and Agriculture Organization of the United Nations (2005).

[127]United States National Oceanic and Atmospheric Administration, National Marine Fisheries Service (2011).

[128]16 U.S.C. § 1851. See Cornell University Law School. U.S. code, tit 16, chap 38, subchap IV, § 1851: national standards for fishery conservation and management.

[129]The eight FMC regions are New England, mid-Atlantic, South Atlantic, Caribbean, Gulf of Mexico, Pacific, North Pacific, and Western Pacific. United States National Oceanic and Atmospheric Administration Fisheries Feature (2014d).

[130]Natural Resources Defense Council (2010).

- Catch quotas;
- Catch share allocations (e.g., between commercial and recreational sectors);
- Bycatch caps;
- Seasonal limitations;
- Gear restrictions;
- Mechanisms for rebuilding fish species;
- Mechanisms for protecting habitats; and
- Any other management measures necessary to ensure sustainable fishing practices.[131]

During the process of drafting (or amending) a plan, FMCs solicit input from fishermen and other concerned members of the public through public hearings and comments. After an FMC finalizes a plan (or amendment), the council submits it to NOAA for final approval. If approved, NMFS (a subsidiary of NOAA) issues federal regulations implementing the plan.[132] These fisheries management plans are the main tool for managing fish stocks in the US. Their key characteristic is the "ability to reflect the ecological and socio-economic needs unique to each region" while still working under the federal guidance of NMFS and the MSA.[133]

Congress has reauthorized the Magnuson-Stevens Act multiple times since enacting it in 1976. The 1996 Sustainable Fisheries Act[134] strengthened MSA's conservation capacity by introducing provisions on fish habitat protection and new mandates to reduce overfishing and bycatch levels. These reforms met limited success during their first decade.[135] In 2006, the Magnuson-Stevens Act Reauthorization Act (MSRA)[136] instituted even more transformative changes. Through MSRA, Congress called for actions to end overfishing for all US fish populations by 2011 and set a number of mandates and deadlines. For instance, fisheries management plans must include annual catch limits, as well as accountability measures for fishermen who exceed those limits. MRSA also provides for catch share, or "limited access privilege," programs to avoid overfishing problems.[137] In addition, MSRA stresses the need for increased international communication. It creates a new position, the "secretarial representative for international

[131]Natural Resources Defense Council (2010).

[132]United States National Oceanic and Atmospheric Administration Fisheries Feature (2014d).

[133]Joint Ocean Commission Initiative (2006).

[134]16 U.S.C. § 1861(a). See Cornell University Law School. U.S. code: tit 16, chap 38, subchap IV, § 1861: transition to sustainable fisheries.

[135]Of the 49 overfished populations in 1998, 31 were still subject to overfishing in 2006. Natural Resources Defense Council (2010).

[136]United States National Oceanic and Atmospheric Administration, National Marine Fisheries Service (2007a).

[137]Catch share programs are a market-based fishery management technique. They allocate secured "limited access privileges" to fishermen that allow them to harvest a specific area or a pre-determined percentage of a fishery's total catch. United States National Oceanic and Atmospheric Administration, National Marine Fisheries Service (2011).

fisheries," to boost the United States' role in international fisheries management.[138] All in all, the MRSA amendments suggest "an important step toward more effective management of the nation's fisheries," with the potential to achieve both economic and environmental fisheries management goals.[139] The Magnuson-Stevens Act went up for reauthorization again in 2014; as of mid-2015, discussions about reauthorization in Congress are still ongoing.[140]

For the most part, the Magnuson-Stevens Act and its reforms have been regarded as a success.[141] In 2011, NOAA reported that "many stocks that were overfished are rebuilt or actively rebuilding. Successes include summer flounder, monkfish, scallops, ling cod, sablefish, North Atlantic swordfish, vermillion snapper, and gag grouper to name a few."[142] Although there have been steady improvements in many US fish stock statuses from year to year, some commercially and recreationally important fish populations remain subject to overexploitation. In 2013, 17 % of US fish stocks were overfished, with 9 % of those stocks still subject to ongoing overfishing.[143] Therefore, the successes so far are only the beginning of a longer journey. The MSA provides a valuable set of tools to enable sustainable fishing, especially when supplemented by other federal laws.[144] However, there is an ongoing need for more inclusive collaboration between fishing industries, scientists, conservationists, consumers, and the broader seafood supply chain to make overfishing within the US fishing industry a concern of the past.[145]

25.4 European Union (EU)

The marine fishing industry is vital to economic and social well-being throughout the European Union (EU). Of course, since the EU is a mixture of coastal and landlocked States, the sector's role varies from region to region. Countries with the highest levels of fishing-related employment include Spain, Italy, Portugal, and

---

[138]United States National Oceanic and Atmospheric Administration, National Marine Fisheries Service (2012).

[139]See Joint Ocean Commission Initiative (2006).

[140]See United States National Oceanic and Atmospheric Administration, National Marine Fisheries Service (2015).

[141]See Ocean Conservancy (2013) and United States National Oceanic and Atmospheric Administration, National Marine Fisheries Service (2011). See also Pacific Fishery Management Council (2015) (noting "[t]here is general agreement that the [Magnuson-Stevens] Act works well").

[142]United States National Oceanic and Atmospheric Administration, National Marine Fisheries Service (2011).

[143]The figures were 19 % and 10 %, respectively, in 2012. United States National Oceanic and Atmospheric Administration, National Marine Fisheries Service (2013).

[144]See *supra* Sect. 25.3.1.

[145]United States National Oceanic and Atmospheric Administration, National Marine Fisheries Service (2011).

Greece.[146] Fish and fish product processing is also a significant economic enterprise in places like the United Kingdom (UK), France, and Germany.[147] Overall, the EU is the world's 5th largest producer of fish, although its production levels have dropped slightly since 2009.[148] The top fish-producing countries within the EU are Spain, the UK, Denmark, and France,[149] and the main species caught are herring, sprat, mackerel, sandeels, and sardine.[150] The EU is also the world's leading importer of fishery and aquaculture products.[151] Given the significance of fish production and fish trade, sound structural policy within the EU fisheries sector is critical to ensuring both economic and social cohesion throughout the region.[152]

Marine fisheries governance in the EU is complex. The EU is a party to over 90 multilateral and bilateral treaties and agreements on fisheries issues.[153] It has ratified both UNCLOS[154] (in April 1998)[155] and the Fish Stocks Agreement[156] (in December 2003),[157] as well as the FAO Compliance Agreement[158] (signed by the former European Community in 1993 and entered into force for the EU in April 2003).[159] Other significant international instruments that the EU has joined include the International Convention for the Conservation of Atlantic Tunas (ICCAT) and the Convention on the Conservation of Antarctic Marine Living Resources (CCAMLR). The EU is also a member of numerous RMFO conventions, including the:

- Convention for the Strengthening of the Inter-American Tropical Tuna Commission (Antigua Convention);
- Convention on the Conservation and Management of Fishery Resources in the South-East Atlantic Ocean;
- Convention on the Conservation and Management of Highly Migratory Fish Stocks in the Western and Central Pacific Ocean; and
- Convention for the Conservation of Salmon in the North Atlantic Ocean (NASCO).[160]

---

[146]European Commission (2014a), p. 16.

[147]European Commission (2014a), p. 17.

[148]European Commission (2014a), p. 18. This figure takes into account both wild capture and aquaculture production.

[149]European Commission (2014a), p. 18.

[150]European Commission (2014a), p. 25.

[151]European Commission (2014a), p. 33.

[152]European Commission (2014a), p. 43.

[153]See European Union External Action Service. List of treaties by sub-activity: regional fisheries organisations.

[154]See *supra* Sect. 25.2.1.

[155]See United Nations, Division for Ocean Affairs and the Law of the Sea (2014).

[156]See *supra* Sect. 25.2.2.

[157]See United Nations, Division for Ocean Affairs and the Law of the Sea (2014).

[158]See *supra* Sect. 25.2.3.

[159]European Union External Action Service (2003).

[160]European Union External Action Service. List of treaties by sub-activity: regional fisheries organisations. Also see Appendix for additional information on these treaties and agreements.

A variety of instruments and bodies at the EU, regional, national, and local levels implement these diverse international commitments.

To govern its vast marine fisheries, the EU employs a centralized, tiered management system. At the EU level, the European Commission sets marine fisheries goals, rules and regulations, and control systems through the Common Fisheries Policy (CFP).[161] The CFP gives all European fishing fleets equal access to EU waters and fishing grounds. At the national level, Member States implement CFP provisions through national regulations and designate authorities and inspectors to carry out these regulations. Recent reforms to the CFP in 2014 gave EU countries even greater control at national and regional levels, encouraging more decentralized decision-making. However, the European Commission continues to play a role in overseeing the implementation of the rules, with a cadre of inspectors that can visit national authorities at any time. Overall, then, a network of governmental actors, along with other public and private stakeholders, must work together to carry out the CFP and ensure a dynamic European fishing industry that is environmentally, economically, and socially sustainable.[162]

25.4.1 EU Common Fisheries Policy (CFP)

The EU's Common Fisheries Policy (CFP)[163] is a set of rules for managing European fishing fleets and conserving fish stocks as a "common resource" of the EU.[164] While the rules date back to the 1970s, when maximizing catches and boosting industry was the priority, they have undergone a number of significant updates. In January 2014, a new CFP came into force that places a recharged focus on sustainable, long-term growth. This policy (current as of early 2015) enshrines various sustainability principles. It seeks:

> a future in which fish stocks are not overfished, sharks are not finned and fish are no longer discarded at sea; a future in which fishermen get better deals and consumers get clearer labels; a future in which we also farm fish of outstanding quality and eco-friendliness; and in which the fish we import is just as safe.[165]

This language evokes a holistic sense of mission, one that advances the health and well-being of industry, marine ecosystems, and consumers as an intertwined whole. To carry forth this vision, the CFP lays out an array of management goals that balance environmental, economic, and social concerns. The policy is comprehensive in scope, with four main policy areas:

---

[161] See *infra* Sect. 25.4.1.

[162] European Commission (2015b).

[163] Eur-Lex (2013).

[164] European Commission (2015b).

[165] European Commission (2014a), p. 1.

(1) Fisheries management, including conservation measures;
(2) International policy, covering EU fishing activities occurring outside of EU waters (including in RMFO waters);
(3) Market and trade policy, for managing both internal and international markets in marine fishery and aquaculture products; and
(4) Funding, which comes from a designated European Maritime and Fisheries Fund (EMFF).[166]

The European Commission is in charge of developing and implementing the CFP. The Commission could not complete its task without input from a wide range of stakeholder-led advisory councils, including the Scientific, Technical, and Economic Committee for Fisheries (STECF). Indeed, scientific findings and data analysis play an important role in CFP rulemaking and other decision-making processes, and the European Commission prepares updated reports on fisheries resources and fishing activity developments each year. Therefore, like the United States' Magnuson-Stevens Act,[167] the CFP presumes that scientific management principles, along with sustainability values, inform sound decision-making in marine fisheries governance.[168]

Bold new conservation goals anchor the reformed CFP. The updated policy demands an end to overfishing by the end of the decade and legally binds parties to fish at sustainable (MSY) levels. Parties must ensure high long-term fishing yields for all stocks by 2020 (and by 2015, for certain stocks where this is possible). The policy also calls for measures that reduce or avoid incidental bycatches and other wasteful practices (such as discarding unwanted fish) that harm animals other than the target fish species. (However, the EU does not go so far as to ban trawling, and trawlers make up approximately 16 % of the EU fishing fleet as of February 2014.)[169] These objectives reflect the CFP's ecosystem approach: a cautious attitude which tries to account for "the impact of human activity on all components of the ecosystem," giving weight to the fact that "the impact of fishing on the fragile marine environment is not fully understood."[170] Such an approach evinces a "precautionary"[171] orientation toward setting conservation targets.

Multi-annual plans form the lynchpin of the CFP's actual implementation. Under these plans, the European Commission sets stock size (and/or fishing

---

[166]European Commission (2015b). In addition to these four policy areas, the CFP contains a number of rules on stakeholder involvement.

[167]See *supra* Sect. 25.3.1.

[168]European Commission (2014a).

[169]European Commission (2014a), p. 12.

[170]European Commission (2015b).

[171]The precautionary principle "aims at ensuring a higher level of environmental protection through preventative decision-taking in the case of risk." Europa (2011). In other words, it calls for a "better-safe-than-sorry" strategy. Decision-makers should err on the side of over-protective rather than under-protective measures when it is unclear whether or not an action may have a harmful effect.

mortality) targets and may prescribe management tools for reaching these goals. The plans are species- or fishery- specific and will usually advocate a combination of measures to combat overfishing and bycatch problems.[172] Preferred tools include:

- Total allowable catches (TACs). These are set annually for most commercial stocks (and biannually for deep sea stocks). TACs can be shared between EU countries through a national quota system;
- Fishing effort restrictions, such as fishing fleet capacity ceilings (on both a kW and gross tonnage basis) or limits on the amount of time that ships can spend at sea. Under current EU law, the total capacity of the fishing fleet may not be increased;
- Technical measures governing how, where, and when fishermen may fish. Examples include minimum mesh sizes for nets, closed seasons, protected areas, and gear specifications. The precise contours of these measures will vary widely from sea basin to sea basin[173]; and
- Landing obligations (also known as discard bans). These are requirements that all fish caught (including non-target species) must be kept onboard, landed, and counted against quotas.[174]

Many multi-annual plans are still in the process of being developed with the new CFP's overfishing objectives and recommendations in mind.[175]

So far, the EU has made promising, but uneven, progress toward meeting CFP goals. From 2010 to 2013, the number of stocks fished at MSY levels more than doubled (increasing from 11 to 25) in Northeast Atlantic waters, where nearly 70 % of EU catches occur.[176] However, the same figure dropped by nearly half (from 21 to 12) in the Mediterranean and Black Seas, where roughly 10 % of catches occur.[177] Therefore, "[w]hilst good progress has been made in the north-east Atlantic, and progress in the Mediterranean Sea and Black Sea is expected soon, too many fish stocks are still overfished."[178] Over a similar period of time, the economic performance of EU fishing fleets has been rising steadily, with net profits increasing from 1 % in 2008 to 6 % in 2011.[179] These results suggest that, as a whole, the EU is on track to continue making progress towards fulfilling the CFP's mandates without sacrificing the economic sustainability of the fishing sector. However, the story is still unfolding, and additional efforts to enforce MSY

[172] European Commission (2014b).

[173] See European Commission (2008).

[174] European Commission (2015a).

[175] European Commission (2014a).

[176] These waters include the North Sea, Baltic Sea, Skagerrak, Kattegat, West of Scotland Sea, Irish Sea, and Celtic Sea. European Commission (2014a), p. 5.

[177] European Commission (2014a), p. 5.

[178] European Commission (2014a), p. 5.

[179] European Commission (2014a), p. 11.

thresholds in certain regions are likely needed to ensure the proper balance between healthy fish stocks and a profitable fishing industry.

25.5 China

Marine fishing has been integral to China's economic and cultural fabric for millennia. Today, China is the world's largest producer of wild-caught (as well as farmed) fish.[180] The country's total marine catch in 2011 was 13.6 million tons, or nearly 1/5 of the entire global output.[181] Not only are these raw numbers staggering, but the *variety* of species caught is enormous as well. China's territorial waters span temperate and tropical zones over four major seas, boasting nearly 3000 marine species.[182] China also has the world's largest fleet of fishing vessels.[183] Furthermore, China is the planet's most populous country, with one of the biggest and most rapidly developing economies. All of these factors combine to make China the most substantial contributor to the growth in world fish consumption over the past couple of decades.[184]

Until recently, though, the rest of the world knew little about China's fisheries or their management.[185] Most government information, including scientific literature, is published exclusively in Chinese, making it relatively inaccessible to foreigners.[186] However, information on both historical and current fisheries management practices in China is increasingly coming to light.[187] Indeed, China is now known to have "some of the oldest fishing conservation measures in the world, like fishing bans and mesh size limits [dating back] to at least to the Zhou Dynasty, from 1046–256 B.C.E."[188] In the early twentieth century, China, like many other nations, began adopting new modern, large-scale fishing methods, such as trawling and driftnet fishing.[189] Some commercial stocks, especially those of traditional sea-bottom species like hairtail and croaker, started collapsing in the 1970s as a result of these intensified production activities. During this decade, China's "[f]ish

[180] Shen and Heino (2013).

[181] Shen and Heino (2013), p. 265.

[182] Haw (2013). The four major seas surrounding China are the Bohai Sea, the Yellow Sea, the East Sea, and the South China Sea. Major commercial fisheries include squid, yellow croaker, red snapper, cod, sea cucumber, and shrimp.

[183] Haw (2013).

[184] GreenFacts (2015).

[185] Shen and Heino (2013).

[186] Shen and Heino (2013), p. 265.

[187] See Shen and Heino (2013).

[188] Mallory (2013).

[189] Mallory (2013).

resources were in a stage of severe recession."[190] The Chinese government took note, and, throughout the 1980s and 1990s, began to reform China's fishery management regime. Updated polices have placed a renewed and ever-growing focus on marine resource conservation.[191]

In 1996, China ratified UNCLOS,[192] marking a milestone in its fishery improvement efforts.[193] China has also has joined a number of RMFOs and bilateral fisheries agreements, including treaties concerning shared resources in the Yellow Sea, East China Sea, and part of the South China Sea.[194] Yet, China's partners have cast some doubts on how well the country is upholding its commitments under these agreements. There have been reports of "quality and accuracy problems with logbook and data reporting,"[195] including misreporting of catches by Chinese fishing vessels.[196] Additionally, neighboring countries have disputed some of China's ownership claims over surrounding waters, while distant countries have complained about Chinese high-seas fleets' illegal fishing activities.[197] Overall, the international consensus seems to be that China's "compliance and cooperation with fisheries institutions that govern international resources need[s] improvement."[198] Indeed, China has not signed the FAO Compliance Agreement,[199] and there is no indication that it intends to do so.[200] China has also never ratified (although it has signed) the Fish Stocks Agreement.[201]

China's fisheries management regime consists of a State-supervised, decentralized, multi-tiered scheme involving public/private collaboration. The Fisheries Law of the People's Republic of China[202] lays out the roles and responsibilities of various federal, state, and local actors. The federal Bureau of Fisheries, part of the Ministry of Agriculture, is the highest body in fisheries administration. Other important players include the:

[190]Shen and Heino (2013), p. 266.

[191]Mallory (2013).

[192]See *supra* Sect. 25.2.1.

[193]United Nations, Division for Ocean Affairs and the Law of the Sea (2014).

[194]Mallory (2013).

[195]Mallory (2013).

[196]Shen and Heino (2013), p. 265.

[197]Shen and Heino (2013), p. 265.

[198]Mallory (2013).

[199]See *supra* Sect. 25.2.3.

[200]See Mallory (2013).

[201]See *supra* Sect. 25.2.2.

[202]See *infra* Sect. 25.5.1.

- Administration of Fishery and Fishing Harbor Supervision, which inspects and supervises fishing ports and vessels and has certain external relations authorities;
- Chinese Academy of Fishery Sciences (and other fishery universities and scientific research institutions), which carries out research and advises government departments; and
- China Fisheries Association, which informs and advises fishermen about fisheries technology, policies, and regulations.[203]

Both State- and privately-owned enterprises perform actual, on-the-ground fishing activities. State operations tend to be large-scale, with access to the most advanced science and technology. Privately-owned vessels are usually smaller-scale, greater in number, less informed about scientific advances, and more difficult for the government to monitor and control.[204] Therefore, as in other regions (like both the US[205] and EU[206]), a network of public and private actors must coordinate across multiple levels of government to implement the country's fisheries management system.

In recent years, the Chinese government has become increasingly alert to concerns about overfishing and bycatch. This growing awareness has led to several important legal and regulatory developments. For instance, the government set "zero growth" and "minus growth" objectives for the country's fishing fleet in 1999 and 2001, respectively.[207] These objectives place moratoriums on new fishing vessels and call for relocating or decommissioning existing ships. As a result, the number of Chinese fishing vessels dropped by nearly 20,000 ships (from roughly 220,000 to about 202,000) between 2002 and 2011.[208] Another significant initiative came about in 2006, when the Chinese government issued the Program of Action on Conservation of Living Aquatic Resources of China. This policy document states that:

> by the year of 2020, the aquatic environment should be gradually rehabilitated, the trend of declin[ing] fishery resources and increas[ing] endangered species should be kept within limits, and fishing capacity and catch from marine capture fisheries should generally accommodate the supporting ability of fishery resources.[209]

In addition to setting forth these ecological goals, the policy calls for increasing the efficiency of fishing operations. Such dual aims demonstrate a combined focus on environmental and economic sustainability, along with a recognition that the two are not incompatible. Indeed, China's current domestic laws and regulations, discussed further below, increasingly emphasize sustainable fisheries development.

---

[203]Shen and Heino (2013), p. 266.

[204]Mallory (2013).

[205]See *supra* Sect. 25.3.

[206]See *supra* Sect. 25.4.

[207]Haw J (2013).

[208]Shen and Heino (2013), p. 267. Yet, over the same period, the total engine power of the fleet has increased.

[209]Shen and Heino (2013), p. 266.

Despite these efforts, none of China's fisheries are yet certified as sustainable.[210] Therefore, the country still has some way to go toward ensuring that fisheries management decisions give as much weight to environmental priorities as economic ones.

25.5.1 Fisheries Law of the People's Republic of China (1986)

The Fisheries Law of the People's Republic of China, passed in 1986 and revised in 2004, exists

> for the purpose of enhancing the protection, increase, development and reasonable utilization of fishery resources [as well as] protecting fishery workers' lawful rights and interests and boosting fishery production.[211]

The law's broad-reaching objectives reflect a holistic orientation that balances environmental, economic, and social aims. It covers both wild-capture and aquaculture activities.

Chapter IV ("Increase and Protection of Fishery Resources") is the most significant section for conservation purposes. These provisions lay out specific marine ecosystem protection measures and order the State to develop and implement fisheries management plans. Mandated measures include:

- A national fishing quota system (in the form of a nation-wide total catch limit);
- Restrictions on catching juveniles;
- Bans on using explosives or poison in fishing; and
- Prohibitions on underwater exploration techniques that may impact fishery resources.[212]

The Department of Fishery Administration has specified a number of additional conservation measures under Fisheries Law regulations. These include:

- Minimum mesh sizes;
- Fishing license systems (to control over-capacity);
- Summer fishing moratoria[213];
- Fishery enhancement programs;
- Aquatic germplasm resource protection areas; and
- Artificial reefs.[214]

---

[210]See Mallory (2013).

[211]Michigan State University College of Law, Animal Legal and Historical Center (2010).

[212]Michigan State University College of Law, Animal Legal and Historical Center (2010), Ch. IV.

[213]These "hot season" moratoria ban certain types of fishing (such as trawling and stake net fishing) for a 2 ½ month period during the summer each year in the Yellow Sea, East China Sea, and Bohai Bay. Haw (2013).

[214]Shen and Heino (2013), p. 267.

There are several weak points in the Fisheries Law and its regulations. First, they continue to allow for a number of harmful intensive fishing practices. As of 2005, trawls remain the dominant fishing gear of the Chinese fleet (accounting for 46 % of marine catches), followed by stow nets (17 %), drift gill nets (14 %), purse seines (9 %), and long-lines (3 %).[215] Each of these methods is known to contribute to overfishing and bycatch problems. Furthermore, some of the regulations' most significant measures, like seasonal moratoria, might be less protective than they appear. Indeed,

> fisheries scientists doubt that [a summer fishing ban] is that effective because it does not occur at the optimal season for recovery, and is followed by a huge fishing effort to compensate for the lack of fishing income during the moratorium period . . . There have been few, if any, signs that the stocks are recovering [in areas where the bans are in place].[216]

Therefore, conservation measures are not as robust as they could be (or as they might appear to be on paper). Furthermore, the Fisheries Law and its regulations tend to emphasize "input controls" (such as fleet capacity restrictions) over "output controls" (like catch limits).[217] While input controls are a crucial component of sustainable fisheries management, they provide less of a guarantee than output controls that overfishing levels will actually be reduced.

Weak enforcement and poor government supervision exacerbate some of these problems. Indeed, analysts have alleged that the Chinese government does not allocate adequate staff and resources to enforce its fishery regulations.[218] What's more, China heavily subsidizes its fishing industry, keeping it more profitable than it otherwise would be.[219] This can contribute to overfishing by enabling fleets to fish when they otherwise would not. As a whole, it seems that "the state's economic goals have predominantly eclipsed sustainability targets when it comes to enforcement [of fisheries laws and regulations]."[220] Therefore, while the Chinese government has taken significant steps toward a sustainable fishing future over the past decade, major gaps—both in substantive law and enforcement capacity—in the current management regime remain.

---

[215] Shen and Heino (2013), p. 267.

[216] Mallory (2013).

[217] Shen and Heino (2013).

[218] Shen and Heino (2013), p. 270.

[219] Mallory (2013).

[220] Mallory (2013).

25.6 Summary and Conclusion

Demand for fish and fisheries products continues to grow. Yet, oceans must be managed properly if they are to continue to be the "lifeline" of humankind's survival, capable of providing the economic, ecological, and social benefits that societies have come to depend on them for. The world's first major wake-up call on the need for better fisheries practices came in the middle of the twentieth century. Around this time, people starting noticing that fish stocks around the globe were becoming depleted. National policies that had encouraged high catch levels to boost native fishing industries and promote short-term economic interests helped bring about these crises. Since the 1970s, international, regional, national, and local bodies have been working to improve the sustainability of marine fisheries management. Indeed, the UN developed some of the most significant pieces of international environmental law of the past century with UNCLOS in 1982[221] and the Fish Stocks Agreement[222] and Compliance Agreement[223] in the mid-1990s.

However, the crisis is far from over. Overfishing and bycatch present ongoing sustainability challenges in fisheries around the world. Indeed, despite many successes over the last couple of decades, "no country has gotten fisheries governance perfectly."[224] An ideal regime would incorporate an array of management tools, including catch limits, catch shares, gear restrictions, fleet capacity constraints, seasonal closures, and marine protected areas. These measures would be set according to scientific principles and informed by accurate fish stock and ecological data. Such measures, of course, must also be backed up with a strong enforcement capacity and a high commitment to transparency by the government. All of these processes should involve consultation with a range of stakeholders, including government officials, marine scientists, fishermen, coastal communities, and concerned members of the public. Overall, a strong fisheries management policy should support a profitable fishing industry, a healthy marine ecosystem, and a healthy and happy human population.

Of course, achieving all of these objectives is no easy task. This chapter has focused on the marine fisheries management regimes in three regions: the United States, the European Union, and China. The United States' Magnuson-Stevens Act,[225] with its science-driven, conservation-focused approach, is generally regarded as one of the more effective national fisheries laws in the world. Indeed, many depleted US fish stocks have made promising recoveries since the 1990s. Critics of the MSA point out, however, that some US fish stocks continue to be overfished more than three decades years after the act was passed. Similarly, the

[221] See *supra* Sect. 25.2.1.

[222] See *supra* Sect. 25.2.2.

[223] See *supra* Sect. 25.2.3.

[224] Mallory (2013).

[225] See *supra* Sect. 25.3.1.

EU's reformed Common Fishery Policy[226] has met with great successes in some areas and ongoing struggles in others. China's Law on Fisheries,[227] meanwhile, is promising on paper but is hampered by transparency and enforcement problems. China's difficulties also underscore the important role that international cooperation and information-sharing play in this arena. Indeed, marine fisheries management is an inherently international issue, often requiring global, regional, or bilateral coordination.

Even though no fisheries law is perfect, many of them present a number of potential "tools for change" to better protect marine species. Some of the strongest measures against overfishing and bycatch, like Australia's ban on driftnet fishing or South Africa's restrictions on trawl and purse-seine fishing,[228] are not covered in this chapter but are summarized in the Appendix. In addition to laws that focus explicitly on fisheries or marine resources, regulations from other related fields (like biodiversity protection, coastal development, and agriculture) can provide more indirect tools for improving ocean stewardship. For instance, laws on endangered species, coastal zone management, or pesticide use often have implications for marine resources and, thus, can be invoked for marine conservation purposes. Although marine fisheries management is a complex area that is regulated unevenly around the world, it is also a field ripe with opportunities for regulatory innovation by decision-makers and advocates alike.

[226]See *supra* Sect. 25.4.1.

[227]See *supra* Sect. 25.5.1.

[228]See Appendix.

Appendix

Region	Country/state	Citation	Title	Year passed	Summary or subsections applicable to regulatory issue
Africa	Kenya	Act No. 5 of 1989	Fisheries Act (Cap. 378)	1989	Sets out basic principles for fisheries development, management, exploitation, utilization, and conservation; calls for: • closed seasons; • prohibited fishing areas; • gear and use restrictions; • fish size and age limits; • fisheries access limits; and • prohibiting fishing for marine mammals Part IV calls for imprisonment of foreign vessel owners caught illegally fishing in Kenya's waters Implemented by: • 1991 Fisheries (General) Regulations (Cap. 378), providing specific measures on fishing gear, landing, protection of breeding areas, and protection of marine mammals and turtles; • 2003 Fisheries (Prohibitions) Regulations (Cap. 378), covering lobster fishing and trade; and • 2010 Prawn Fishery Management Plan (Cap. 378), calling for zoning restrictions, mesh size regulations, vessel use limitations, and mandatory use of Turtle Excluder Devices

(continued)

Region	Country/state	Citation	Title	Year passed	Summary or subsections applicable to regulatory issue
Africa	Nigeria	Act No. 71 of 1992	Sea Fisheries Act	1992	Governs control, regulation, and protection of sea fisheries in Nigeria's territorial waters; prohibits certain fishing methods Implemented by Sea Fisheries (Fishing) Regulations, which prohibit fishing within first 5 nautical miles of continental shelf and place further restrictions on trawling, crustacean size, landing requirements and dumping, and mesh net requirements
Africa	South Africa	Act No. 74 of 1995	Sea Fishery Act	1988	Outlines basic regulatory framework for fishing in the country's fishing zones; creates Sea Fishery Advisory Committee, a Quota Board, and a Sea Fishery Fund
Africa	South Africa	Act No. 15 of 1994	Maritime Zones Act	1994	Establishes South Africa's maritime zones
Africa	South Africa	Act No. 18 of 1998	Marine Living Resources Act	1998	Amends Sea Fishery Act; provides for: • fisheries management plans; • fishing priority areas; • marine protected areas; • gear and use restrictions; • closed seasons; • and other prohibited activities Provides further rules on: • trawl fishing, purse-seine fishing (and other types net and line fishing); • bag limits; • fish size and mass limits; • landing, disposal, and dumping requirements; • minimizing environmental impacts; and • disease control Implemented by various regulations and orders, including Prince Edward Islands Marine protected area regulation (No. R. 252 of 2012)

Americas	Brazil	Law No. 11.959/Lei no. 11.959, de 29 de Junho de 2009	Law No. 11.959 on Sustainable Development Policy on Fishery and Aquaculture/*Lei No. 11.959 de Dispõe sobre a Política Nacional de Desenvolvimento Sustentável da Aquicultura e da Pesca, regula as atividades pesqueiras, revoga a Lei no 7.679, de 23 de novembro de 1988, e dispositivos do Decreto-Lei no 221, de 28 de fevereiro de 1967, e dá outras providências.*	2009	Sets out guidelines for developing fishery and aquaculture sectors in Brazil Implemented by various regulations and orders, including: • Order No. 5, creating the Permanent Committee for Fishing and Sustainable Use of Shrimps (CPG); • Order No. 36-N, establishing protective measures (TED) for marine turtles during pink shrimp fishing; • Order No. 56, establishing minimum size for swordfish fishing; • Order No. 97, establishing a closed season period for motor-powered trawl fishing for certain shrimps' species; • Order No. 98, establishing a periodical closed season for sardine fishing; and • Order No. 120-N, regulating sardine fishing via prescriptions on authorized fishing methods, fish size, season, and protected zones
Americas	Canada	R.S.C. 1985, c. C-33	Coastal Fisheries Protection Act	1985	Prohibits foreign fishing vessels in Canadian fisheries waters; prohibits vessels from introducing fish from outside, non-territorial waters, unless authorized
Americas	Canada	R.S.C. 1985, c. F-14	Fisheries Act	1985	Governs fisheries management; sets out general fish habitat protection and pollution prevention measures, as well as general prohibitions on certain fishing activities, with special provisions on seal hunting, salmon fishing, and lobster fisheries Implemented by: • 1985 Atlantic Fishery Regulations (S.O. R./86-21), which mandate restrictions on gear, seasons, areas, and tagging; • 1993 Fishery (General) Regulations (S.O. R./93-53), which advocate periods and variation of close times, fishing quotas, size and

(continued)

Region	Country/state	Citation	Title	Year passed	Summary or subsections applicable to regulatory issue
					weight limits, and orders to remove obstructions to the passage of fish; and • 2007 Pacific Fishery Management Area Regulations (SOR/2007-77)
Americas	Canada	S.C. 1996, c. 31	Oceans Act	1996	Establishes Canada's maritime zones • Part II lays out an oceans management strategy • Part III requires the Minister of Fisheries and Oceans to prepare a national strategy for the management of estuarine, coastal, and marine ecosystems • Section 35 calls for establishing Marine Protected Areas
Americas	Canada	S.C. 1998, c. 10	Canada Marine Act	1998	Regulates Canada's ports and other matters related to maritime trade and transport
Americas	Canada	S.C. 2001, c. 26	Canada Shipping Act	2001	Regulates shipping and navigation in Canada
Americas	Canada	S.C. 2002, c. 18	National Marine Conservation Areas Act	2002	Aims to protect and conserve representative marine areas for the benefit, education, and enjoyment of the people of Canada
Americas	Mexico	DOF 08-01-1986, publicada en el *Diario Oficial de la Federación* el 8 de enero de 1986	Federal Law of the Sea/*Ley Federal del Mar*	1985	Sets out basic rules governing maritime areas under national jurisdiction Part 4 contains measures on marine environment protection and preservation
Americas	Mexico	DOF 04-01-1994, publicada en el *Diario Oficial de la Federación* el 4 de enero de 1994	Law on Navegation/*Ley de Navegación*	1994	Establishes Mexico's EEZ; grants the Ministry of the Navy sovereignty over territorial waters
Americas	Mexico	DOF 24-07-2007, publicada en el *Diario Oficial de la Federación* el 24 de julio de 2007	General Law for Sustainable Fisheries and Aquaculture/*Ley General de Pesca y Acuacultura Sustentables*	2007	Promotes integrated and sustainable development of the fisheries and aquaculture sectors; provides a basic framework for the management, conservation, protection, reforestation, protection, and rehabilitation, and sustainable use of fisheries and aquaculture resources; creates a National Council of Fisheries and Aquaculture Article 37 addresses fisheries management programs

Americas	United States	16 U.S.C. §§ 3371-3378	Lacey Act	1900	Prohibits fish or wildlife transactions and activities that violate state, federal, Native American tribal, or foreign law
Americas	United States	16 U.S.C. §§ 661-667e	Fish and Wildlife Coordination Act	1934	Directs the Bureau of Fisheries to use impounded waters for fish-culture stations and migratory-bird resting and nesting areas; requires consultation with the Bureau of Fisheries prior to construction of any new dams in order to allow the dams to provide for fish migration
Americas	United States	22 U.S.C. §§ 1971-1979	Fisherman's Protective Act—Pelly Amendment	1967	Authorizes the President to embargo wildlife products (including fish) if "nationals of a foreign country are engaging in trade or taking that diminishes the effectiveness of an international program in force with respect to the United States for the conservation of endangered or threatened species"
Americas	United States	42 U.S.C. § 4321 et seq.	National Environmental Policy Act (NEPA)	1969	Requires environmental reviews of proposed major federal agency actions, including some fisheries management decisions (such as significant changes to a fishery management plan)
Americas	United States	16 U.S.C. §§ 1361-1407	Marine Mammal Protection Act (MMPA)	1972	Contains provisions on protecting marine mammals and sea turtles from habitat destruction, vessel strikes, and incidental capture as bycatch
Americas	United States	16 U.S.C. §§ 1431-1445	National Marine Sanctuaries Act (NMSA)	1972	Designates certain marine environments as sanctuaries because of their environmental, historical, or cultural importance
Americas	United States	33 U.S.C. §§ 1401-1445 and 16 U.S.C. §§ 1431-1445	Marine Protection, Research and Sanctuaries Act of 1972 (MPRSA or Ocean Dumping Act)	1972	Regulates ocean dumping of industrial wastes, sewage sludge, and other wastes through a permit program

(continued)

Region	Country/state	Citation	Title	Year passed	Summary or subsections applicable to regulatory issue
Americas	United States	16 U.S.C. § 1531 et seq.	Endangered Species Act (ESA)	1973	Lists over 100 endangered and threatened marine species that fall under NOAA's jurisdiction
Americas	United States	16 U.S.C. §§ 1801-1891d	Magnuson-Stevens Fishery Conservation and Management Act (Magnuson-Stevens Act or MSA)	1976	Governs marine fisheries management; establishes the US EEZ and regulates fish populations within this zone; modeled on principles of scientific management and best-stewardship practices; establishes a regional public-private management system Reauthorized by the 1996 Sustainable Fisheries Act and the Magnuson-Stevens Act Reauthorization Act (MSRA) of 2006; has been up for another reauthorization since 2014
Americas	United States	7 U.S.C. § 136 et seq.	Federal Insecticide, Fungicide, and Rodenticide Act (FIFRA)	1996	Regulates pesticide use; has been invoked to instill "no-spray buffer zones" on the use of pesticides in or near waters that host threatened or endangered species
Americas	United States	16 U.S.C. § 6401 et seq	Coral Reef Conservation Act (CRCA)	2000	Aims to preserve, sustain, and restore the condition of coral reef ecosystems; calls for scientific research and conservation programs on coral reef ecosystems
Americas	United States	33 U.S.C. §§ 857-19	Oceans Act	2000	Establishes the United States Commission on Ocean Policy to promote responsible stewardship of fisheries and other marine resources
Americas	United States (Northern Marianas Islands)	P.L. 17-27	An Act to Prohibit any Person from Possessing, Selling, Offering for Sale, Trading, or Distributing Shark Fins in the CNMI	2011	Provides measures to protect sharks from commercial overfishing

Asia	China	Adopted at the 14th Meeting of the Standing Committee of the Sixth National People's Congress on January 20, 1986 and promulgated by Order No. 34 of the President of the People's Republic of China on January 20, 1986	Fisheries Law of the People's Republic of China	1986	Governs fisheries resources management in China; aims to: • protect and "reasonably" use fishery resources; and • protect fishery workers' rights Call for a national fishing quota system Part IV concerns enhancing and protecting fishery resources Implemented by 1987 Regulations for the Implementation of the Fisheries Law of the People's Republic of China, which call for additional conservation measures
Asia	China	Adopted at the third session of the Standing Committee of the Ninth National People's Congress, 26 June 1998	China's Exclusive Economic Zone and Continental Shelf Act	1998	Implements UNCLOS; establishes China's rights over its EEZ and continental shelf
Asia	China (Hong Kong)	Cap. 171 (Oct 12, 1962), as amended by E.R. 2 of 2012	To promote the conservation of fish and other forms of aquatic life within the waters of Hong Kong and to regulate fishing practices and to prevent activities detrimental to the fishing industry (Fisheries Protection Ordinance (Cap. 171))	1962	Provides rules for protecting fish and other forms of aquatic life in Hong Kong waters
Asia	China (Hong Kong)	Cap. 476 (June 1, 1995)	An Ordinance to provide for the designation, control and management of marine parks and marine reserves, and for purposes connected therewith (Marine Parks Ordinance (Cap. 476))	1995	Provides the legal framework for designating, controlling, and managing marine parks and marine reserves in Hong Kong waters
Asia	China (Taiwan)	As amended by Presidential Order Hua-Tsung (1) No. 10200156101 on August 21, 2013	Fisheries Act	1929	Provides rules for conserving and "rationally" using aquatic resources Taiwanese waters

(continued)

Region	Country/state	Citation	Title	Year passed	Summary or subsections applicable to regulatory issue
Asia	India	Act No. 4 of 1897	Indian Fisheries Act	1897	Establishes basis for Union and State territories to create their own fishery laws and rules: as a result, every Indian State has its own Fisheries Act suited to local conditions (with no uniform national approach) [examples of such State laws are included in this Appendix]
Asia	India	Act No. 13 of 1972	Marine Products Export Development Authority Act (MPEDA)	1972	Establishes the Marine Products Export Development Authority to address export standards, marketing, promotion of prawn fisheries, and training in fisheries sector
Asia	India	Act No. 80 of 1976	Territorial Waters, Continental Shelf, Exclusive Economic Zone and other Maritime Zones Act	1976	Establishes India's EEZ and declares India's territorial waters
Asia	India	Act No. 42 of 1981	Maritime Zones of India (Regulation of Fishing by Foreign Vessels) Act	1981	Does not address conservation or management of fisheries resources
Asia	India (Arunachal Pradesh)	Act No. 5 of 2006	Arunachal Pradesh Fisheries Act	2006	Governs fisheries in State of Arunachal
Asia	India (Gujarat)	Act No. 8 of 2003	Gujarat Fisheries Act	2003	Governs fisheries in State of Gujarat
Asia	India (Haryana)	Punjab Act No. II of 1914	Punjab Fisheries Act	1914	Governs fisheries in State of Punjab
Asia	India (Himachal Pradesh)	Act No. 16 of 1976	Himachal Pradesh Fisheries Act	1976	Governs fisheries in State of Himachal
Asia	India (Kerala)	Act No. 10 of 1981	Kerala Marine Fishing Regulation Act	1981	Governs fisheries in State of Kerala
Asia	India (Manipur)	Manipur Act No. 1 of 1992	Manipur Fisheries Act	1992	Governs fisheries in State of Manipur
Asia	India (Mizoram)	Act No. 12 of 2005	Mizoram Fisheries Act	2005	Governs fisheries in State of Mizoram
Asia	India (Tamil Nadu)	Act No. 106 of 5 March 1983	Tamil Nadu Marine Fishing Regulation Act	1993	Governs fisheries in State of Tamil Nadu
Asia	India (West Bengal)	Act No. IX of 1993; *The Calcutta Gazette*, Extraordinary, 14 June 1993, pp. 1-11	West Bengal Marine Fishing Regulation Act	1993	Governs fisheries in State of West Bengal

Asia	Israel	P.O. Suppl. I, 157	Fisheries Ordinance	1937	Governs fisheries management and conservation; contains provisions on: • restricting taking and landing of fish; and • prohibiting use of poison or explosives Implemented by 1937 Fisheries Rules, which: • prohibit trawling in shallow Mediterranean waters; • place certain gear restrictions and net mesh size requirements; establish protected fishing areas; and • call for notification requirements in case of fish disease outbreaks
Asia	Israel	Law No. 5758-1998	National Parks, Nature Reserves, National Sites and Memorial Sites Law	1963	Provides basic legal structure for protecting Israel's natural habitats and assets, including marine habitats; protects certain species of fish and aquatic plant and animal life
Asia	Japan	Act No. 267 of 1949	Fisheries Act	1949	Establishes Japan's basic fisheries production system
Asia	Japan	Act No. 313 of 1951	Act on the Protection of Fishery Resources	1951	Prescribes measures for aquatic animal and plant conservation, including: • catch restrictions; • fishing method restrictions; • total allowable catch limits; • fleet capacity limits; • protected water surfaces; and • import quarantines (to prevent spread of diseases)
Asia	Japan	Act No. 76 of 1996	Act on the Exercise of the Sovereign Right for Fishery, etc. in the Exclusive Economic Zone	1996	Establishes Japan's EEZ and implements UNCLOS; calls for setting total allowable catches (TACs)

(continued)

Region	Country/state	Citation	Title	Year passed	Summary or subsections applicable to regulatory issue
Asia	Japan	Act No. 89 of 2001	Fisheries Basic Act	2001	Governs conservation and management of living aquatic resources; aims to: • control water quality; • protect and provide nursery grounds for aquatic animals and plants; and • preserve forests in order to conserve and improve the environment for aquatic animals and plants Establishes a Fishery Policy Council; delegates most fisheries management to local fishery cooperative associations (FCAs) who set their own rules on harvest times, net mesh sizes, and sustainable yields
Asia	Russia	Law No. 166-FZ of 2004 (Dec 20, 2004)	Federal Law No. 166-FZ on Fisheries and Conservation of Aquatic Biological Resources	2004	Governs fishery management and conservation; provides for specific conservation measures, including a quota system with total allowable catch (TAC) limits set annually by the Federal Fisheries Institution Implemented by various decrees and regulations, including: • Ministerial Decree No. 644 regarding the distribution of quotas for fish farming, stock enhancement, acclimatization, educational and training purposes; • Ministerial Decree No. 694 validating the list of fishing gear prohibited for import to the Russian Federation; and • Ministerial Decree No. 768 regarding the distribution of the total allowable catch in accordance with quotas

Australia/New Zealand	Australia	Act. No. 162 of 1991	Fisheries Management Act	1991	Governs fisheries management; defines the Australian Fishing Zone; establishes the Australian Fishery Management Authority; calls for fishery management plans (which may allow for statutory fishing rights and fishery closures); places general ban on driftnet fishing Implemented by: • 1992 Fisheries Management Regulations, providing for catch limits and incidental seabird catch measures; • 1995 Fisheries Management (Northern Prawn Fishery) Regulations; • 1995 Fisheries Management (Southern Bluefin Tuna Fishery) Regulations, setting weight and season restrictions; • 1998 Fisheries Management (South East Trawl Fishery) Regulations; • 2002 Fisheries Management (Bass Strait Central Zone Scallop Fishery) Regulations; • 2004 Fisheries Management (Southern and Eastern Scalefish and Shark Fishery) Regulations; • 2006 Fisheries Management (Western Tuna and Billfish Fishery) Regulations; • 2009 Fisheries Management (Eastern Tuna and Billfish Fishery) Regulations; and • 2009 Fisheries Management (International Agreements) Regulations
Australia/New Zealand	Australia	Act No. 91 of 1999	Environment Protection and Biodiversity Conservation Act	1999	Implements CBD Chapter II allows government to decide whether an action (including fisheries actions in marine areas) has, will have, or is likely to have a significant impact on certain aspects of the environment

(continued)

Region	Country/state	Citation	Title	Year passed	Summary or subsections applicable to regulatory issue
Australia/New Zealand	New Zealand	Act No. 28 of 1977; *Regional Compendium of Fisheries Legislation (Western Pacific Region)*, FAO, June 1993, pp. 1018-1044; FAL No. 27, vol. 2, 1978, pp. 58-78	Territorial Sea and Exclusive Economic Zone Act	1977	Defines New Zealand's EE Sections 11-13 call for setting total allowable catches (TACs)
Australia/New Zealand	New Zealand	Act No. 88 of 1996	Fisheries Act	1996	Seeks the "conservation, use, enhancement and development of fisheries resources while ensuring sustainability;" establishes the National Fisheries Advisory Council and a Catch History Review Committee • Part 3 addresses sustainability measures • Part 4 sets up a robust quota management system (including annual catch entitlements) Implemented by various regulations and orders, including: • 2000 Fisheries (Southern Bluefin Tuna Quota) Regulations; • 2005 Fisheries (Declaration of New Stocks Subject to Quota Management System) Notice; and • 2005 Fisheries (Stocks Determined Not to be Subject to Quota Management System) Notice Amends the previous Fisheries Act of 1983 (No. 14 of 1983)
Australia/New Zealand	New Zealand	Act No. 31 of 1997	Fisheries (Quota Operations Validation) Act	1997	Confirm and validates overfishing and underfishing rights and entitlements
Europe	European Union	O.J. L. 318, 27.11.1998, p. 63; Document 31998R0850R(01)	Corrigendum to Council Regulation (EC) No 850/98 of 30 March 1998 for the conservation of fishery resources through technical measures for the protection of juveniles of marine organisms	1998	Sets out technical measures Currently under review to accommodate new Common Fisheries Policy (CFP) reforms

Europe	European Union	O.J. L. 349, 31.12.2005, pp. 1–23	Council Regulation (EC) No 2187/2005 of 21 December 2005 for the conservation of fishery resources through technical measures in the Baltic Sea, the Belts and the Sound, amending Regulation (EC) No 1434/98 and repealing Regulation (EC) No 88/98	2005	Sets out technical measures Currently under review to accommodate new Common Fisheries Policy (CFP) reforms
Europe	European Union	O.J. L. 151, 11.6.2008, pp. 5–25	Commission Regulation (EC) No 517/2008 of 10 June 2008 laying down detailed rules for the implementation of Council Regulation (EC) No 850/98 as regards the determination of the mesh size and assessing the thickness of twine of fishing nets	2008	Sets out technical measures Currently under review to accommodate new Common Fisheries Policy (CFP) reforms
Europe	European Union	O.J. L. 347, 24.12.2009, pp. 6–8	Council Regulation (EC) No 1288/2009 of 27 November 2009 establishing transitional technical measures from 1 January 2010 to 30 June 2011	2009	Sets out technical measures Currently under review to accommodate new Common Fisheries Policy (CFP) reforms
Europe	European Union	O.J. L. 199, 31.7.2010, pp. 4–11	Commission Regulation (EU) No 686/2010 of 28 July 2010 amending Council Regulation (EC) No 2187/2005 as regards specifications of Bacoma window and T90 trawl in fisheries carried out in the Baltic Sea, the Belts and the Sound	2010	Sets out technical measures Currently under review to accommodate new Common Fisheries Policy (CFP) reforms

(continued)

Region	Country/state	Citation	Title	Year passed	Summary or subsections applicable to regulatory issue
Europe	European Union	O.J. L. 165, 24.6.2011, pp. 1–2	Regulation (EU) No 579/2011 of the European Parliament and of the Council of 8 June 2011 amending Council Regulation (EC) No 850/98 for the conservation of fishery resources through technical measures for the protection of juveniles of marine organisms and Council Regulation (EC) No 1288/2009 establishing transitional technical measures from 1 January 2010 to 30 June 2011	2011	Sets out technical measures Currently under review to accommodate new Common Fisheries Policy (CFP) reforms
Europe	European Union	O.J. L. 354, 28.12.2013, pp. 22–61	Regulation (EU) No 1380\2013 of the European Parliament and the Council of 11 December 2013 on the Common Fisheries Policy, amending Council Regulations (EC) No 1954/2003 and (EC) No 1224 and repealing Council Regulations (EC) no 2371 and (EC) No 639/2004 and Council Decision 2004/585/EC. (EU Common Fisheries Policy or (CFP))	2013	Sets out rules for managing European fishing fleets and conserving fish stocks as a common resource of the EU; demands an end to overfishing by the end of the decade and establishes a legally binding commitment to fish at sustainable levels; embraces the precautionary principle; calls for various overfishing and bycatch measures, including: • total allowable catches (TACs); • fleet capacity restrictions; • minimum mesh sizes; • gear restrictions; and • landing obligations Reformed in 2014
Europe	European Union	O.J. L. 78, 20.3.2013, pp. 1–22	Regulation (EU) No 227/2013 of the European Parliament and of the Council of 13 March 2013 amending Council Regulation (EC) No 850/98 for the conservation of fishery resources through technical measures for the protection of juveniles of marine organisms and Council Regulation (EC) No 1434/98 specifying conditions under which herring may be landed for industrial purposes other than direct human consumption	2013	Sets out technical measures Currently under review to accommodate new Common Fisheries Policy (CFP) reforms

Europe	European Union—Germany	–	–	–	Regional fishing associations (*Landesfischereiverbände*) regulate and manage fishing in Germany, with each federal state having its own *Landesfischereiverband* and regulations [examples of such regional laws are included in this Appendix]
Europe	European Union—Germany (Bayern)	(August 15, 1908); FAOLEX No: LEX-FAOC072488	Bavarian Fishery Law/*Fischereigesetz für Bayern*	1908	Governs fisheries in Bayern
Europe	European Union—Germany (Berlin	(June 19, 1995); FAOLEX No: LEX-FAOC073695	Berlin Fishery Law/*Berliner Landesfischereigesetz*	1995	Governs fisheries in Berlin
Europe	European Union—Germany (Brandenburg)	(May 13, 1993); FAOLEX No: LEX-FAOC073287	Brandenburg Fishery Law/*Fischereigesetz für das Land Brandenburg (BbgFischG)*	1993	Governs fisheries in Brandenburg
Europe	European Union—Germany (Hamburg)	(May 22, 1986); FAOLEX No: LEX-FAOC074736	Hamburg Fishery Law/*Hamburgisches Fischereigesetz*	1986	Governs fisheries in Hamburg
Europe	European Union—Germany (Hessen)	(Dec 19, 1990); FAOLEX No: LEX-FAOC075420	Hessen Fishery Law/*Fischereigesetz für das Land Hessen (Hessisches Fischereigesetz—HFischG)*	1990	Governs fisheries in Hessen
Europe	European Union—Germany (Niedersachsen)	(Feb 1, 1978); FAOLEX No: LEX-FAOC084524	Lower Saxony Fishery Law/*Niedersächsisches Fischereigesetz (Nds. FischG)*	1978	Governs fisheries in Niedersachsen
Europe	European Union—Germany (Rheinland-Pfalz)	(Dec 9, 1974); FAOLEX No: LEX-FAOC075265	Rheinland-Pfalz Fishery Law/*Landesfischereigesetz (LFischG)*	1974	Governs fisheries in Rheinland-Pfalz
Europe	European Union—Germany (Saarland)	(Jan 23, 1985); FAOLEX No: LEX-FAOC077116	Saarland Fishery Law/*Saarländisches Fischereigesetz (SFischG)*	1985	Governs fisheries in Saarland
Europe	European Union—Germany (Sachsen)	(July 9, 2007); FAOLEX No: LEX-FAOC077045	Saxony Fishery Law/*Fischereigesetz für den Freistaat Sachsen (Sächsisches Fischereigesetz—SächsFischG)*	2007	Governs fisheries in Sachsen

(continued)

Region	Country/state	Citation	Title	Year passed	Summary or subsections applicable to regulatory issue
Europe	European Union—Germany (Thüringen)	(June 28, 2006); FAOLEX No: LEX-FAOC075395	Thüringen Fishery Law/*Thüringer Fischereigesetz (ThürFischG)*	2006	Governs fisheries in Thüringen
Europe	European Union—United Kingdom (England; Scotland; Wales)	2009 c. 23	Marine and Coastal Access Act	2009	Introduces new system of marine management in UK; allows for designating an EEZ for the UK (and creating a Welsh Zone in the sea adjacent to Wales); provides for designating conservation zones; establishes the Marine Management Organization (MMO) Implemented by various regulations and orders, including the Sea Fish (Prohibited Methods of Fishing) (Firth of Clyde) Order 2014, which prohibits certain fishing methods that threaten juveniles
Europe	European Union—United Kingdom (Northern Ireland)	1966 Chapter 17	Fisheries Act (Northern Ireland) (Chapter 17)	1966	Calls for various marine management and conservation measures, including: • gear restrictions; • fish size limits; and closed season requirements Implemented by 2014 Fisheries Regulations (Northern Ireland)
Europe	European Union—United Kingdom (Scotland)	2010 asp 5	Marine (Scotland) Act	2010	Sets up a new legislative framework for managing the marine environment in Scotland
Europe	European Union—United Kingdom (Scotland)	2007 asp 12	Aquaculture and Fisheries (Scotland) Act	2013	Sets new rules on fish farming, salmon fisheries, sea fisheries, and shellfish protection in Scotland
International	Parties electing to adopt the Code	In: *FAO Fisheries and Aquaculture Department* [online]. Rome. Updated 28 May 2014.	Code of Conduct for Responsible Fisheries (FAO Fisheries Code)	1995	Sets out voluntary, non-binding principles and standards for responsible fisheries practices; Supplemented by various Technical Guidelines Drafted in harmony with the UN Fish Stocks Agreement and the FAO Compliance Agreement

International	Parties to Agreement	148 U.N.T.S. 1860 (Nov 24, 1993)	Agreement to Promote Compliance with International Conservation and Management Measures by Fishing Vessels on the High Seas (FAO Compliance Agreement)	1993	Governs high seas fishing in international waters; aims to prevent bad faith vessel "re-flagging" in Regional Marine Fisheries Organization (RMFO) waters Drafted in harmony with UN Fish Stocks Agreement and FAO Fisheries Code
International	Parties to Agreement	2167 U.N.T.S. 3 (Dec. 4, 1995)	Agreement for the Implementation of the Provisions of the United Nations Convention on the Law of the Sea of 10 December 1982 relating to the Conservation and Management of Straddling Fish Stocks and Highly Migratory Fish Stocks (UN Fish Stocks Agreement)	1995	Lays out a legal conservation and management regime for straddling and highly migratory fish stocks; establishes RMFO networks Drafted in harmony with the FAO Compliance Agreement and the FAO Fisheries Code
International	Parties to Convention	33 U.S.T. 3476 (May 20, 1980)	Convention on the Conservation of Antarctic Marine Living Resources (CCAMLR)	1980	Aims to preserve marine life and environmental integrity in and near Antarctica; establishes an Ecosystem Monitoring Program (CEMP) Focuses on protecting the region's krill, an important foundation of the marine ecosystem food chain
International	Parties to Convention	S. Treaty Doc. No. 103-20 (1993), 31 I.L.M. 818	Convention on Biological Diversity (CBD)	1992	Addresses all aspects of biodiversity; has yielded two workplans directly related to fisheries: (1) Jakarta Mandate on Marine and Coastal Ecosystems, which addresses the conservation and sustainable use of marine and coastal biological diversity; and (2) Workplan on Inland Water Ecosystems, which promotes ecosystem approach to freshwater fisheries
International	Parties to UNCLOS	1833 U.N.T.S. 397 (Montego Bay, Dec. 10, 1982)	United Nations Convention on the Law of the Sea (UNCLOS)	1982	Provides a global "constitution for the seas;" establishes a comprehensive governance regime for the world's oceans and fisheries; establishes the EEZ concept; prescribes specific fisheries management rules

(continued)

Region	Country/state	Citation	Title	Year passed	Summary or subsections applicable to regulatory issue
International	Member States of the World Trade Organization (WTO)	31 U.S.T. 405, 1186 U.N.T.S. 276	Agreement on Technical Barriers to Trade (TBT Agreement)	1995	Aims to protect global trade; prohibits nations from setting technical regulations, standards, testing, and certification procedure requirements that create unnecessary obstacles to trade (including trade in fish or fisheries products) 2001 Doha Development Agenda launched negotiations on the relationship between existing WTO rules and specific trade obligations set out in international and regional fisheries agreements
International (Bilateral)	Norway and Iceland	(May 28, 1980)	Agreement between Norway and Iceland concerning Fishery and Continental Shelf Questions	1980	Calls for measures (including total allowable catch (TAC) limits) to promote the conservation, rational exploitation, and sound reproduction of capelin fish stocks; establishes a Fisheries Commission
International (Bilateral)	United States and Canada	T.I.A.S. 11091 (Jan 28, 1985)	Treaty between the Government of the United States of America and the Government of Canada Concerning Pacific Salmon (Pacific Salmon Treaty)	1985	Commits Parties to prevent the overfishing of Pacific salmon; ensures that both countries receive benefits equal to the salmon produced in their waters; allows for creating salmon enhancement programs
International (Regional)	Australia, Japan, New Zealand	1819 U.N.T.S. 360 (May 10, 1993)	Convention for the Conservation of Southern Bluefin Tuna (CCSBT)	1993	Aims to conserve and optimally utilize Southern Bluefin tuna; calls for total allowable catch (TAC) and bycatch measures; establishes the Commission for the Conservation of Southern Bluefin Tuna
International (Regional)	Parties to Convention (Atlantic coastal nations in Europe, Americas, Africa, Asia)	673 U.N.T.S. 63, 20 U.S.T. 2887, T.I.A.S. 6767 (May 14, 1966)	International Convention for the Conservation of Atlantic Tunas (ICCAT)	1966	Sets up a program for international cooperation in research and conservation; calls for setting maximum sustainable yield (MSY) levels for Atlantic tuna populations; establishes the International Commission for the Conservation of Atlantic Tunas
International (Regional)	Parties to Convention (nations in Americas, Asia, and Europe)	80 U.N.T.S. 3, 1 U.S.T. 230, T.I.A.S. No. 2044	Convention for the Strengthening of the Inter-American Tropical Tuna Commission (Antigua Convention)	2010	Aims to conserve and manage tropical tuna Updates and strengthens the original 1949 Convention, which established the Inter-American Tropical Tuna Commission (IATTC)

International (Regional)	Parties to Convention (nations in Europe, North America, Northern Arctic)	24 U.S.T. 1080, T.I.A.S. 7628	Convention for the International Council for the Exploration of the Sea (ICES)	1964	Promotes scientific research and investigation programs on the sea and marine living resources; establishes International Council for the Exploration of the Sea
International (Regional)	Parties to Convention (North Atlantic coastal nations)	1338 U.N.T.S. 33, T.I.A.S. No. 10789 (Mar 2, 1982)	Convention for the Conservation of Salmon in the North Atlantic (North Atlantic Salmon Treaty or NASCO)	1982	Aims to conserve North Atlantic salmon; prohibits salmon fishing beyond areas of fisheries jurisdiction of coastal states (and within certain areas of national jurisdiction); establishes the North Atlantic Salmon Conservation Organization
International (Regional)	Parties to Convention (Northwest Atlantic coastal nations in Americas, Europe, and Asia)	157 U.N.T.S. 157, 1 U.S.T. 477, T.I.A.S. No. 2089 (Feb 8, 1949)	International Convention for the Northwest Atlantic Fisheries (Northwest Atlantic Fisheries Treaty)	1949	Aims to protect Northwest Atlantic fisheries; calls for: scientific research programs; • total allowable catches (TACs); and • bycatch measures Establishes the Northwest Atlantic Fisheries Organization (NAFO)
International (Regional)	Parties to Convention (Pacific coastal nations in Americas, Europe, Asia, and Australia/New Zealand)	40 I.L.M. 227 (2001)	Convention on the Conservation and Management of Highly Migratory Fish Stocks in the Western and Central Pacific Ocean	2000	Aims to ensure the long-term conservation and sustainable use of highly migratory fish stocks in the western and central Pacific Ocean; establishes the Commission for the Conservation and Management of Highly Migratory Fish Stocks in the Western and Central Pacific Ocean
International (Regional)	Parties to Convention (Southeast Atlantic coastal nations in Africa, Europe, Asia, and Americas)	2221 U.N.T.S. 189 (Apr 20, 2001)	Convention on the Conservation and Management of Fishery Resources in the South-East Atlantic Ocean (SEAFO Convention)	2001	Aims to ensure the long-term conservation and sustainable use of fishery resources on the high seas of the South East Atlantic Ocean; embraces the precautionary principle; establishes the South East Atlantic Fisheries Organisation (SEAFO)

References

Balton DA, Koehler HR (2006) Reviewing the United Nations fish stocks treaty. Sustainable Dev Law Policy 7:5–9, 75. Available via Digital Commons. http://digitalcommons.wcl.american.edu/cgi/viewcontent.cgi?article=1195&context=sdlp. Accessed 18 Mar 2015

Commission for the Conservation of Antarctic Marine Living Resources (1980) Text of the convention of the conservation of Antarctic marine living resources. https://www.ccamlr.org/en/system/files/e-pt1.pdf. Accessed 18 Mar 2015

Convention for the Conservation of Southern Bluefin Tuna (1993) Multilateral no. 31155 (with annex). 10 May 1993. https://treaties.un.org/pages/showDetails.aspx?objid=08000002800b05d5. Accessed 18 Mar 2015

Cornell University Law School. U.S. code, tit 16, chap 38, subchap IV, § 1851: national standards for fishery conservation and management. https://www.law.cornell.edu/uscode/text/16/1851. Accessed 18 Mar 2015

Cornell University Law School. U.S. code:, tit 16, chap 38, subchap IV, § 1861: transition to sustainable fisheries. https://www.law.cornell.edu/uscode/text/16/1861a. Accessed 18 Mar 2015

Eur-Lex (2013) Regulation (EU) No 1380\2013 of the European Parliament and the Council of 11 December 2013 on the common fisheries policy, amending Council regulations (EC) no 1954/2003 and (EC) no 1224 and repealing Council regulations (EC) no 2271 and (EC) no 639/2004 and Council decision 2004/585/EC. In: Official Journal of the European Union. 28 Dec 2013. Available via EUR-LEX. http://eur-lex.europa.eu/LexUriServ/LexUriServ.do?uri=OJ:L:2013:354:0022:0061:EN:PDF. Accessed 18 Mar 2015

Europa (2011) The precautionary principle. http://europa.eu/legislation_summaries/consumers/consumer_safety/l32042_en.htm. Accessed 21 Mar 2015

European Union External Action Service (2003) Summary of the treaty: agreement to promote compliance with international conservation and management measures by fishing vessels on the high seas. 24 Apr 2003. http://ec.europa.eu/world/agreements/prepareCreateTreatiesWorkspace/treatiesGeneralData.do?step=0&redirect=true&treatyId=558. Accessed 18 Mar 2015

European Union External Action Service. List of treaties by sub-activity: regional fisheries organisations. http://ec.europa.eu/world/agreements/searchByActivity.do?parent=8520&xmlname=759&actName=Fisheries&showTreatiesOfSubActivity=Regional%20Fisheries%20Organisations. Accessed 18 Mar 2015

European Commission (2008) Commission regulation (EC) no 517/2008 of 10 June 2008. In: Official Journal of the European Union L151/5 11 June 2008. Available via FAOLEX. http://faolex.fao.org/docs/pdf/eur80019.pdf. Accessed 18 Mar 2015

European Commission (2014a) Facts and figures on the common fisheries policy, pp 1, 6, 12, 16, 18, 22, 33, 43. http://ec.europa.eu/fisheries/documentation/publications/pcp_en.pdf. Accessed 18 Mar 2015

European Commission (2014b) Fisheries: multi-annual plans: managing fish stocks. http://ec.europa.eu/fisheries/cfp/fishing_rules/multi_annual_plans/index_en.htm. Accessed 18 Mar 2015

European Commission (2015a) Fisheries: managing fish stocks. http://ec.europa.eu/fisheries/cfp/fishing_rules/index_en.htm. Accessed 18 Mar 2015

European Commission (2015b) Fisheries: the common fisheries policy (CFP): management of EU fisheries. http://ec.europa.eu/fisheries/cfp/index_en.htm. Accessed 18 Mar 2015

Food and Agriculture Organization of the United Nations (1995a) Agreement to promote compliance with international conservation and management measures by fishing vessels on the high seas. http://www.fao.org/docrep/MEETING/003/X3130m/X3130E00.HTM#Top%20Of%20Page. Accessed 18 Mar 2015

Food and Agriculture Organization of the United Nations (1995b) Code of conduct for responsible fisheries. http://www.fao.org/docrep/005/v9878e/v9878e00.htm. Accessed 18 Mar 2015

Food and Agriculture Organization of the United Nations (1998) International plan of action for reducing incidental catch of seabirds in longline fisheries. http://www.fao.org/docrep/006/x3170e/x3170e02.htm. Accessed 22 Mar 2015

Food and Agriculture Organization of the United Nations (2005) Fishery country profile. ftp://ftp.fao.org/FI/DOCUMENT/fcp/en/FI_CP_US.pdf. Accessed 18 Mar 2015

Food and Agriculture Organization of the United Nations, Fisheries and Aquaculture Department (2001) What is the code of conduct for responsible fisheries? ftp://ftp.fao.org/docrep/fao/003/x9066e/x9066e00.pdf. Accessed 18 Mar 2015

Food and Agriculture Organization of the United Nations, Fisheries and Aquaculture Department (2013) Web-based reporting system for the questionnaire on the implementation of the code of conduct for responsible fisheries. http://www.fao.org/fishery/topic/166326/en. Accessed 18 Mar 2015

Food and Agriculture Organization of the United Nations, Fisheries and Aquaculture Department (2015a) Code of conduct for responsible fisheries. http://www.fao.org/fishery/code/en. Accessed 18 Mar 2015

Food and Agriculture Organization of the United Nations, Fisheries and Aquaculture Department (2015b) Code of conduct for responsible fisheries: technical guidelines. http://www.fao.org/fishery/code/publications/guidelines/en. Accessed 18 Mar 2015

Food and Agriculture Organization of the United Nations, Fisheries and Aquaculture Department (2015c) FAO compliance agreement. http://www.fao.org/fishery/topic/14766/en. Accessed 18 Mar 2015

Food and Agriculture Organization of the United Nations, Fisheries and Aquaculture Department (2015d) Framework – code of conduct. http://www.fao.org/fishery/topic/18140/en. Accessed 18 Mar 2015

Food and Agriculture Organization of the United Nations, Fisheries and Aquaculture Department (2015e) Institutional frameworks for fisheries governance. http://www.fao.org/fishery/topic/3552/en. Accessed 5 Apr 2015

Food and Agriculture Organization of the United Nations, Fisheries and Aquaculture Department (2015f) United Nations fish stock agreement. http://www.fao.org/fishery/topic/13701/en. Accessed 18 Mar 2015

GreenFacts (2015) Fisheries latest data: how much fish is consumed worldwide? http://www.greenfacts.org/en/fisheries/l-2/06-fish-consumption.htm. Accessed 18 Mar 2015

Haw J (2013) The status of fisheries in China: how deep will we have to dive to find the truth? Scientific American, 3 June 2013. Available via ScientificAmerican.com. http://blogs.scientificamerican.com/expeditions/2013/06/03/the-status-of-fisheries-in-china-how-deep-will-we-have-to-dive-to-find-the-truth/. Accessed 18 Mar 2015

Hawaii Pacific University Oceanic Institute (2011) Aqua facts. http://www.oceanicinstitute.org/aboutoceans/aquafacts.html. Accessed 18 Mar 2015

International Convention for the Conservation of Atlantic Tunas (1986) Text of the international convention for the conservation of Atlantic tunas. In: Official Journal of the European Communities No L 162/34. 18 June 1986. Available via EC.Europa.eu. http://ec.europa.eu/world/agreements/downloadFile.do?fullText=yes&treatyTransId=1319. Accessed 18 Mar 2015

Joint Ocean Commission Initiative (2006) U.S. ocean policy report card 2006. http://www.jointoceancommission.org/~/media/JOCI/PDFs/2007-01-01_2006_Ocean_Policy_Report_Card.pdf. Accessed 18 Mar 2015

Mallory T (2013) China's fisheries management policy: an interview with Tabitha Mallory. SAIS Rev Int Aff 33(2):85–91. Available via Academia.edu. http://www.academia.edu/5405442/China_s_Fisheries_Management_Policy_An_Interview_with_Tabitha_Mallory. Accessed 18 Mar 2015

Michigan State University College of Law, Animal Legal and Historical Center (2010) Fisheries law of the People's Republic of China. Oct 2010. https://www.animallaw.info/statute/china-fishing-china-fisheries-law. Accessed 18 Mar 2015

Miller GT, Spoolman SE (2008) Essentials of ecology, 4th edn. p 172. https://books.google.com/books?id=5C40N-_H4qUC&pg=PA172&lpg=PA172&dq=overfishing+%22nonpoint+source+pollution%22&source=bl&ots=AsXox1jA7y&sig=x7ShNMz6-LsGg-EaopQ3L0lYs4A&hl=en&sa=X&ei=PEf7VIOZKMWrggTYsoSgBA&ved=0CD8Q6AEwBQ#v=onepage&q=overfishing%20%22nonpoint%20source%20pollution%22&f=false

Natural Resources Defense Council (2010) Managing fish and fishing in America's oceans. http://switchboard.nrdc.org/blogs/dnewman/MSA%20Defense%20101.pdf. Accessed 18 Mar 2015

Ocean Conservancy (2013) The law that's saving American fisheries: the Magnuson-Stevens fishery conservation and management act: it's a keeper. http://www.oceanconservancy.org/our-work/fisheries/ff-msa-report-2013.pdf. Accessed 22 Mar 2015

OceanLink. Fishing methods. http://oceanlink.island.net/ONews/oceannews6/oceanews6p6-7.html. Accessed 18 Mar 2015

Pacific Fishery Management Council (2015) Applicable laws: Magnuson-Stevens act. http://www.pcouncil.org/resources/applicable-laws/magnuson-stevens-act/. Accessed 21 Mar 2015

Pacific Salmon Commission (2014) Treaty between the government of Canada and the government of the United States of America concerning Pacific salmon. http://www.psc.org/pubs/Treaty/Treaty%20July%202014.pdf. Accessed 18 Mar 2015

Save Our Seas Foundation (2015) Threat 1: overfishing. http://saveourseas.com/threats/overfishing. Accessed 18 Mar 2015

Shen G, Heino M (2013) An overview of marine fisheries in China. ScienceDirect, 12 December 2013, pp 265–70. Available via Science Direct. http://www.sciencedirect.com/science/article/pii/S0308597X13002091. Accessed 18 Mar 2015

Stastista (2012) The top 10 fishing nations worldwide in 2012 (in metric tons). http://www.statista.com/statistics/240225/leading-fishing-nations-worldwide-2008/. Accessed 18 Mar 2015

United Nations (2010a) Agreement between Norway and Iceland on fishery and continental shelf questions (1980). 28 May 1980. http://www.un.org/depts/los/LEGISLATIONANDTREATIES/PDFFILES/TREATIES/isl-nor1980fcs.pdf. Accessed 18 Mar 2015

United Nations (2010b) Resumed review conference on the agreement related to the conservation and management of straddling fish stocks and highly migratory fish stocks. 24 May 2010. http://www.un.org/depts/los/convention_agreements/reviewconf/FishStocks_EN_A.pdf. Accessed 18 Mar 2015

United Nations (2014) Status of the United Nations convention on the law of the sea, of the agreement relating to the implementation of part XI of the convention and of the agreement for the implementation of the provisions of the Convention relating to the conservation and management of straddling fish stocks and highly migratory fish stocks: table recapitulating the status of the Convention and of the related agreement. 10 Oct 2014. http://www.un.org/depts/los/reference_files/status2010.pdf. Accessed 18 Mar 2015

United Nations. The 1995 United Nations fish stock agreement [background paper]. http://www.un.org/depts/los/convention_agreements/Background%20paper%20on%20UNFSA.pdf. Accessed 18 Mar 2015

United Nations, Division for Ocean Affairs and the Law of the Sea (2001) The United Nations agreement for the implementation of the provisions of the United Nations convention on the law of the sea of 10 December 1982 relating to the conservation and management of straddling fish stocks and highly migratory fish stocks (in force as from 11 December 2001). 11 Dec 2001. http://www.un.org/depts/los/convention_agreements/convention_overview_fish_stocks.htm. Accessed 18 Mar 2015

United Nations, Division for Ocean Affairs and the Law of the Sea (2014) Chronological lists of ratifications of, accessions and successions to the Convention and the related Agreements as at 3 October 2014. http://www.un.org/depts/los/reference_files/chronological_lists_of_ratifications.htm#Agreement. Accessed 18 Mar 2015

United Nations, Division for Ocean Affairs and the Law of the Sea, Office of Legal Affairs (2012) The United Nations convention on the law of the sea (a historical perspective). http://www.un.

org/depts/los/convention_agreements/convention_historical_perspective.htm. Accessed 18 Mar 2015
United Nations Conference on Straddling Fish Stocks and Highly Migratory Fish Stocks (1995) Agreement for the implementation of the provisions of the United Nations convention of the law of the sea of 10 December 1982 relating to the conservation and management of straddling fish stock and high migratory fish stocks. 8 Sept 1995. http://www.un.org/depts/los/convention_agreements/texts/fish_stocks_agreement/CONF164_37.htm. Accessed 18 Mar 2015
United Nations Convention on Biological Diversity (2015) Ecosystem approach. http://www.cbd.int/ecosystem/. Accessed 15 Mar 2015
United Nations Convention on the Law of the Sea (1982). Text of the United Nations convention on the law of the sea. http://www.un.org/depts/los/convention_agreements/texts/unclos/unclos_e.pdf. Accessed 18 Mar 2015
United States Department of the Interior, Fish and Wildlife Service (1973) Endangered species act of 1973 as amended through the 108th Congress. http://www.nmfs.noaa.gov/pr/pdfs/laws/esa.pdf. Accessed 18 Mar 2015
United States Department of the Interior, Fish and Wildlife Service (1979) Digest of the federal resource laws of interest to the U.S. Fish and Wildlife Service: fishermen's protective act. http://www.fws.gov/laws/lawsdigest/FISHPRO.HTML. Accessed 18 Mar 2015
United States Department of the Interior, Fish and Wildlife Service (2015) Pelly agreement. http://www.fws.gov/international/laws-treaties-agreements/us-conservation-laws/pelly-amendment.html. Accessed 18 Mar 2015
United States Department of the Interior, Fish and Wildlife Service, Office of Law Enforcement (2004) Lacey act. http://www.fws.gov/le/USStatutes/Lacey.pdf. Accessed 18 Mar 2015
United States National Oceanic and Atmospheric Administration (2014) National marine sanctuary system. http://sanctuaries.noaa.gov/about/welcome.html. Accessed 18 Mar 2015
United States National Oceanic and Atmospheric Administration, National Marine Fisheries Service (2007a) Magnuson-Stevens fishery conservation and management act as amended through January 12, 2007. http://www.nmfs.noaa.gov/sfa/magact/MSA_Amended_2007%20.pdf. Accessed 18 Mar 2015
United States National Oceanic and Atmospheric Administration, National Marine Fisheries Service (2007b) The marine mammal protection act of 1972 as amended 2007. http://www.nmfs.noaa.gov/pr/pdfs/laws/mmpa.pdf. Accessed 18 Mar 2015
United States National Oceanic and Atmospheric Administration, National Marine Fisheries Service (2011) The road to end overfishing: 35 years of Magnuson act. http://www.nmfs.noaa.gov/stories/2011/20110411roadendoverfishing.htm. Accessed 18 Mar 2015
United States National Oceanic and Atmospheric Administration, National Marine Fisheries Service (2012) Fisheries feature: Magnuson-Stevens fishery conservation and management act reauthorized. http://www.nmfs.noaa.gov/msa2007/index.html. Accessed 18 Mar 2015
United States National Oceanic and Atmospheric Administration, National Marine Fisheries Service (2013) Status of U.S. fisheries. http://www.nmfs.noaa.gov/sfa/fisheries_eco/status_of_fisheries/. Accessed 18 Mar 2015
United States National Oceanic and Atmospheric Administration, National Marine Fisheries Service (2014a) High seas fishing permits. http://www.nmfs.noaa.gov/ia/permits/highseas.html. Accessed 18 Mar 2015
United States National Oceanic and Atmospheric Administration, National Marine Fisheries Service (2014b) National bycatch program. http://www.nmfs.noaa.gov/by_catch/. Accessed 18 Mar 2015
United States National Oceanic and Atmospheric Administration, National Marine Fisheries Service (2014c) National environmental policy act (NEPA). http://www.nmfs.noaa.gov/sfa/laws_policies/msa/nepa.html. Accessed 18 Mar 2015
United States National Oceanic and Atmospheric Administration Fisheries Feature (2014d) How are federal living marine resources managed? http://www.nmfs.noaa.gov/regulations.htm. Accessed 18 Mar 2015

United States National Oceanic and Atmospheric Administration, National Marine Fisheries Service (2015) Magnuson-stevens act – ongoing reauthorization activities. http://www.nmfs.noaa.gov/sfa/laws_policies/msa/reauthorization_activities.html. Accessed 29 Mar 2015

Washington State Department of Agriculture (2014) Buffers imposed by the U.S. district court order. 27 Aug 2014. http://agr.wa.gov/pestfert/natresources/buffers.aspx. Accessed 18 Mar 2015

MIX
Papier aus verantwortungsvollen Quellen
Paper from responsible sources
FSC® C105338

If you have any concerns about our products,
you can contact us on
ProductSafety@springernature.com

In case Publisher is established outside the EU,
the EU authorized representative is:
Springer Nature Customer Service Center GmbH
Europaplatz 3, 69115 Heidelberg, Germany

Printed by Libri Plureos GmbH
in Hamburg, Germany